Biology and Biotechnology

Science, Applications, and Issues

Biology and Biotechnology

Science, Applications, and Issues

Helen Kreuzer

Department of Biology, University of Utah,
Salt Lake City, Utah

Adrianne Massey

A. Massey & Associates,
Chapel Hill, North Carolina

ASM
PRESS

WASHINGTON, D.C.

Address editorial correspondence to ASM Press, 1752 N St. NW, Washington, DC 20036-2904, USA

Send orders to ASM Press, P.O. Box 605, Herndon, VA 20172, USA
Phone: 800-546-2416; 703-661-1593
Fax: 703-661-1501
E-mail: books@asmusa.org
Online: www.asmpress.org

Library of Congress Cataloging-in-Publication Data

Kreuzer, Helen.
  Biology and biotechnology : science, applications, and issues / Helen Kreuzer, Adrianne Massey.
     p. cm.
  Includes bibliographical references (p.   ) and index.
  ISBN 1-55581-304-6
  1. Biotechnology. 2. Biology. I. Massey, Adrianne. II. Title.

TP248.2.K74 2005
660.6—dc22
                                                                    2005043611

Cover and interior design: Susan Brown Schmidler
Art rendering: Patrick Lane

Cover photograph: Fluorescent staining highlights the role of a specific protein during cell division. A red dye binds to DNA, illuminating the chromosomes as they migrate toward what will become new cell nuclei. The green color shows the location of the protein tubulin, vividly revealing its role as the major component of the spindle fibers, filaments that pull two newly duplicated daughter chromosomes apart and separate them into opposite ends of the dividing cell. Researchers are learning more and more about the functions of specific proteins in cellular processes and in observable traits of organisms. This increased understanding enables scientists to use cells and their proteins to make useful products and to solve problems, applications collectively known as biotechnologies. (Photograph courtesy of Nasser Rusan, 2003 Nikon Small World Contest.)

To Anne, Lloyd, and Bob with gratitude and love

About the Authors

Helen Kreuzer is a scientist and an educator. She received her Ph.D. degree in molecular genetics and microbiology from Duke University Medical Center and has taught numerous biology courses to undergraduate students. From 1991 to 1994, she worked with Adrianne Massey at the North Carolina Biotechnology Center, where she designed classroom activities and wrote curriculum materials for teaching about molecular biology and biotechnology in high school. These activities were later published in the book she coauthored with Dr. Massey, *Recombinant DNA and Biotechnology: A Guide for Teachers*, and the accompanying *Guide for Students*. She has taught numerous courses and workshops for high school teachers and developed additional laboratory exercises, models, and videos for molecular biology instruction. Dr. Kreuzer is a member of the biology faculty at the University of Utah, where she researches the use of stable isotope ratios to answer forensic as well as basic biological questions. She and her husband live in Salt Lake City with their basset hounds, TJ and Rosebud, and their dachshunds, Daisy and Pete. In their free time, they perform volunteer work on behalf of homeless dogs and enjoy camping, hiking, and kayaking.

Adrianne Massey has been involved with scientific research, education, and public policy for over 20 years. Her company furthers informed science and technology policy development by mediating consensus-building activities, counseling governments, helping nonscientists understand biology, and training scientists in communications. Prior to founding her company, she was a vice president at the North Carolina Biotechnology Center and a biological sciences faculty member at North Carolina State University. Dr. Massey served as the science advisor for the PBS series *BREAKTHROUGH: Television's Journal of Science and Medicine* and was the original director of the North Carolina Environmental Technology Consortium. She received her Ph.D. in zoology from North Carolina State University and has taught undergraduate and graduate courses in biology and evolutionary ecology. She is currently working with governments in various African and Asian countries to foster responsible economic growth through scientific innovation and technological advance.

Contents

Foreword

Old habits die hard. Perhaps surprisingly, this is nowhere more true than in academia—supposedly a bastion of rational, analytical thought. In the last century, science and technology have utterly transformed our lives. Today, a citizen who does not understand the nature of science and its unique, highly productive way of learning about the world ("science as a way of knowing") cannot claim to be truly educated—any more than someone who cannot write a clear essay. But in our colleges and universities, history is rarely taught to reflect the dramatic ways in which scientific and technological advances have shaped our societies, as well as the course of human events. And most of our science and engineering faculty continue to teach their subject as if it were important only for those who seek to become professionals in their own or a related field.

Even for majors in the subject, we often teach science poorly. Should the purpose of an introductory science course be to commit to memory a large mass of facts that scientists have discovered about the natural world? Life is very different from a quiz show, and anyone with access to the Internet can find almost any fact that he or she wants to know in a few seconds on Google. We need to teach science to all of our students in a way that gives them a clear understanding of how scientific knowledge is accumulated, with its emphasis on logical argument and its insistence on openly presenting evidence—along with details of the methods used to obtain it—so that each scientist's observations can be confirmed (or refuted) by any other scientist. Science is a marvelous community endeavor that enables new knowledge to be built upon old knowledge in ways that have enabled us to gain a tremendous understanding of the physical world. The understanding has in turn enabled humans to manipulate this world in ways that produce great benefits for humanity.

All of those who teach science in our colleges and universities should think carefully about the fact, cited at the beginning of chapter 22, that "only

30 percent of the consumers in the seven largest EU countries know that all crop plants have genes"—or the fact that half of Americans believe in astrology. How can we hope to maintain a rational society when a majority of our citizens are so disconnected from reality? I would contend that, in every science course that is taught, our primary aim should be to encourage the students to engage in the excitement of the subject and to understand deeply the manner in which science is carried out and its relevance for society.

Despite the arguments just presented, we continue to teach biology to college freshmen from massive textbooks that attempt to cover all of biology in a single year. In such courses there is not enough time to delve into any one aspect of biology in enough detail for students to get a feeling for the fabric of science—or to appreciate how powerfully the science connects to technologies that shall continue to change their lives.

Fortunately, *Biology and Biotechnology: Science, Applications, and Issues* is a different type of textbook for a different type of course. By not attempting to explain all of biology and instead focusing on biotechnology and its many applications, the authors have been able to pursue arguments and present case studies in enough depth to give students a real feeling for the nature of science in its modern context. This course also explicitly aims to connect science to the students' lives and the decisions that they will have to make as citizens in a democratic society. The authors display the many connections between science and technology, using carefully chosen, beautifully explained examples. Students can thereby begin to understand the source of many of the changes that they have already experienced during their lifetime.

In closing, I would like to congratulate Helen Kreuzer and Adrianne Massey, as well as their publisher, ASM Press, for having the courage to produce a new kind of textbook. The inertia in academia stems in part from the fact that textbook publishers naturally concentrate on producing books that will be sold in large numbers. They therefore design their new books to match the prevailing large courses. But if all of the attention, energy, and talent are devoted to producing materials for these broad survey courses—a mile wide and an inch deep—there is no hope for a change in academia.

Rarely, a new kind of textbook like this one comes along that creates a "tipping point," triggering a widespread change in how we teach a particular subject. I contend that we are way past the time for a tipping point in the way that we teach introductory biology to college students. Hopefully, this bold new textbook by Kreuzer and Massey will help pave the way for others, thereby catalyzing the spread of many new courses that can inspire students with the wonder and power of modern science.

Bruce Alberts, President
National Academy of Sciences
Washington, D.C.
March 2005

Preface

In the course of our professional lives, we have had many opportunities to talk about aspects of biotechnology with audiences from diverse walks of life: lawyers, farmers, businesspeople, school teachers, students, restaurant chefs, and the general public, to name a few. These people genuinely sought to understand issues that they had heard about in the media and brought a high level of interest, intelligence, and thoughtfulness to the task. Often, their motivation was to become an informed consumer of biotechnology. For example, did they want to purchase and eat genetically modified food, to grow genetically modified crops, to support stem cell research, to favor genetic testing? Sometimes they were merely curious, seeking answers to questions such as how does DNA fingerprinting work? What is cloning? What is recombinant DNA? Our task has been to explain the science and to provide sufficient information about the issues to enable people to formulate their own opinions about them.

This book is an outgrowth of these experiences. Our goal was to create a text that would give readers the foundation they needed for understanding the many inevitable advances in biotechnology that the coming years will bring and a context for making decisions about them as potential consumers. We wrote it for students and readers who have not necessarily chosen biology as their major field of study, although we believe it may offer unique perspectives to biology students as well. The book is self-contained. Readers do not need to have taken a college biology course prior to using it, although since its focus is quite different from that of a typical introductory biology course, the information presented will not be redundant to those who have. Because of its blending of science, consumer applications, regulatory information, and social issues, we believe that the book will be of interest to students and other readers from many disciplines.

We begin our text with a perspective on the interrelationship between science, technology, and society. These three realms have reciprocal effects on one another. Science leads to technologies that provide new tools for doing science. At the same time, technologies are evaluated by society, often in the form of market forces, influencing the future direction of scientific research and technological development. The interweaving of these threads is particularly well illustrated in the case of biotechnology because the social debate about biotechnology products is so public and because the links between scientific and technological advances are so immediate. At the same time, the use by human beings of biology and biological organisms to make products and improve their environment is ancient, giving us a historical context in which to view modern biotechnologies. The perspective offered in the first part of the book infuses the discussion of biotechnology applications presented later on.

Before we discuss applications of biotechnology, we provide the scientific foundation necessary to understand them in the second part of the book, "The Foundational Science." Modern biotechnology concerns the use and manipulation of cells and their subcomponents to make products and solve problems, and the biotechnological manipulation of whole organisms starts with the manipulation of individual cells. Thus, the first subpart of this part is a primer on cell biology titled "From Atoms to Organisms," in which we look at the life processes of cells and how those processes translate to the organism level.

Cell biology alone, however, is not sufficient for understanding many of the applications of biotechnology, which can go beyond individual cells and organisms to affect the progeny of the manipulated organisms. Biotechnologies such as genetic testing look at parents and offspring in context. Agricultural and environmental biotechnologies evoke ecological and evolutionary questions concerning how the characteristics of one member of a community can affect its other members both immediately and over time. The second subpart of this part of the book, titled "From Organisms to Ecosystems," introduces genetics, ecology, and evolution: the transmission of genes from one generation to the next, the interaction of genes and environment to produce traits, the interaction of organisms in communities and ecosystems, and finally, the evolutionary response of organisms and ecosystems to environmental changes.

Throughout "The Foundational Science," we present examples of how the scientific knowledge being explained has been translated into technologies used to solve problems in medical and everyday settings. Also imbedded within this part are a few narratives of the history of particular scientific developments. We include these as examples of how science progresses and to illustrate the impact of society and social context on science.

Having laid this scientific foundation, we can now focus on biotechnology in the third major part of the book, "Biotechnology Applications and Issues." Although the title of this part would probably cause most people to think of commercial applications, so far the most significant applications of biotechnology have been in the research laboratory, enabling scientists to gain new knowledge about the natural world at an ever-increasing pace. We begin with a look at this impact of technology upon science in a subpart titled "Research Applications," which describes biotechnology techniques and shows how these methods are used to gain new scientific knowledge in scientific fields from archaeology to zoology. The two chapters in this subpart will also address "how do they do that" questions.

The second subpart of this part, "Commercial Applications," looks at biotechnology products in society at large. We begin this subpart with a discussion of issues that arise when scientific advances are moved from the research laboratory into society, the concepts of risk and regulation. We present a framework and process for thinking about risk that readers can use to put new technologies and products into perspective in comparison with older ones.

In the following chapters, we look specifically at biotechnology in medicine, food, agriculture, and the environment. These chapters are not simply laundry lists of products and potential products. Rather, we show the scientific basis of a few specific products, as well as the complexity surrounding their introduction into the market. We also discuss how product introductions are regulated and the process that regulators go through in making decisions.

Biotechnology applications can trigger ethical dilemmas in which there are no easy or perfect answers. Our goal in these chapters is not to provide readers with an opinion but rather to provide them with tools for conducting their own informed, critical evaluations. To that end, we attempt to provide the essential information required for understanding both the science and the issues involved in each chapter's example applications. As we discuss the issues raised by each application, we use the framework presented earlier to analyze them. We hope that these thorough examples will illustrate what kinds of questions need to be asked and how the answers can be put into perspective as the readers think about any new technology.

In the past, technologies were usually adopted without consideration of their potential impact on society and the environment. In recent decades, the impact of technology on society and the environment has become an issue of great concern to many, and biotechnology as a whole is one of the first broad categories of technology to receive public scrutiny before its widespread introduction. Societal decisions about which technologies to adopt may have profound implications for humanity in terms both of what technological options are available to us today and of the impact that our decisions have on the future direction of scientific and technological advances.

Readers, the decisions you make about biotechnology will contribute to our societal decisions as a whole and thus to the future of science, technology, and society. We hope that this book will empower you to evaluate issues independently and critically.

Helen Kreuzer and Adrianne Massey
April 2005

Acknowledgments

Writing this book was a long and difficult task. Many people supported and encouraged us in many different ways during the process, and it is now our pleasure to thank those who helped this book come to be.

First, we thank Jeff Holtmeier, the director of ASM Press. Jeff gently prodded us to write a college-level textbook, based on our first book on biotechnology, which was also published by ASM Press. Early in the process we realized that we wanted to write a textbook that bears little resemblance to our other book, and without hesitation Jeff allowed us to do this. As Bruce Alberts notes in the foreword, publishing a textbook that differs significantly from existing textbooks that are geared to familiar, well-established markets requires courage. Jeff, we thank you for your courage and for your faith in us.

We also thank the ASM Press staff, particularly Susan Birch, Laura Ledbetter, and Jennifer Adelman. In addition to providing highly professional assistance in every way, they also offered emotional support and enthusiasm for the project. We are fortunate to work with such a thoughtful and generous group of people. Thanks also to Elizabeth McGillicuddy, who copyedited the book; Susan Schmidler, who did the cover and interior design; and Patrick Lane, who rendered the art.

Many members of the scientific community improved the content and visual quality of the book. A number of anonymous reviewers read the first draft of the entire book, and other scientists, including Harold Coble, Fred Gould, Karyn Hede, and Ron Kuhr, read selected chapters. Reviewing is a time-consuming task, and we are grateful for your time and effort. Your conscientious reviews led to numerous improvements in the manuscript. Other scientists allowed free use of their photographs. Some exceeded our requests and found additional images that they thought might be useful. We thank all of you for your generosity, especially Steven Baskauf, Carol and Dennis Gonsalves, Nasser Rusan, Kent Schwaegerle, Hans and Petra Sommer, George Seidel, and Michael Vernon.

We thank Thomas Martin, a computer graphics expert and professional photographer, for his multifaceted support throughout this project. He handled countless image and file transfer issues, took several photographs specifically for the book, volunteered to be a nonscientist reviewer for most of the chapters, and served as a computer consultant throughout the project. And he did all of this at no charge! Tommy, we can't thank you enough for the many tangible and intangible contributions you made to this book.

A special note of thanks goes to Bruce Alberts, president of the National Academy of Sciences, for his willingness to write the foreword. We are honored that you found the time to read the book and gratified that you saw so clearly the heart of our intentions in writing it.

Finally, we could not have finished this project without the moral support and encouragement of our friends and loved ones. You know who you are. Thanks, y'all.

Helen Kreuzer and Adrianne Massey
April 2005

P A R T I | Perspective

chapter 1 | Science, Technology, and Society

Take a minute and try to imagine a world without technology—and by technology we don't mean computers and cell phones but the broader sense of the word. (In popular parlance, the word "technology" seems to have become synonymous with telecommunications, computers, and other advanced technologies based on microelectronics. In this chapter, we use Webster's definition of technology: "the totality of means employed to provide objects necessary for human sustenance and comfort.") You will soon find that it's virtually impossible, because technology is woven so tightly into the fabric of your life and your self that you cannot extricate yourself from its influences. Trying to assess the impact of technology on your life resembles a fish stepping back and objectively observing water. Not only does the fish's total dependence on water make this impossible, but also the essence of "fishness" is inseparable from the properties of water.

You may resist the idea that technology has penetrated every tiny facet of your life so deeply that you have become technology dependent and are as incapable of living in the natural world as a fish is of living on land. If you're reluctant to accept this assessment, then pay attention to your actions during the rest of the day and notice the many ways technology, both simple and sophisticated, has shaped your world. Some impacts—televisions, cars, and microwave ovens—are immediately obvious. Others are so subtle and pervasive you might not think of them as technology but instead take them for granted as a "natural" part of life. The clothes you're wearing, the chair you're sitting in, the pen you may be holding, the paper this book is printed on, the food you had for lunch are all products of technology.

As you are paying attention to the ways technology impinges on your life, look at the flip side of the coin, too. How often do you come face-to-face with nature? Probably not very often, because humans rejected the idea of living with nature thousands of years ago. Look around the room to see if you can find anything that's natural, such as a spider building its web in a corner or mold on the cheese you wanted for lunch. How do you feel about these calling cards from nature? If you're like most people, you're not pleased they invaded your life. This distaste for the adversities nature indifferently dispenses has driven society's constant quest for new technologies. People use all technologies for the same purpose: to change the environment so that the natural world suits us better.

Technology and society: it's a two-way street

Technologies not only alter the environment, they change people. People's views of the world, the questions they ask, the way they think, their sense of values, hopes, and fears are all products of life in a world shaped by technology. Through its effects on people, technology changes society as a whole, always in unpredictable ways.

Even though technological innovations develop as solutions to very specific problems and well-defined needs, they also provide doors to unforeseen innovations, novel applications, and unexpected societal changes. As a result, once a technology is introduced to society, predicting the nature and speed of all subsequent technological developments and societal effects is impossible. How could the inventors of the automobile ever guess it would lead to technologies for extracting and refining oil and, ultimately, to the international petroleum industry with all of its wealth and thousands of products (Figure 1.1)? Nor could they have possibly known that the automobile's social impacts would include fragmented family units; the creation of suburbia, shopping malls, and industrial parks; and, ultimately, the economic decline of some major urban areas.

Beneficial technologies that meet legitimate societal needs can also have negative societal impacts, some of which are also unpredictable. To compli-

Figure 1.1 Refining crude oil. Engineers had to develop a large-scale process to produce gasoline from crude oil when the sudden popularity of automobiles created a surge in the demand for gasoline. The process, known as distillation, uses heat to separate the hundreds of compounds found in crude oil from each other. The different compounds in heated oil turn from gas back to liquid form at different temperatures. As crude oil vapors rise in the distillation column, they gradually cool, and different sets of compounds, or fractions, are collected as they become liquid again. At first, oil refiners saved only the gasoline fraction and threw away the others. Later, they realized that all of the fractions had uses.

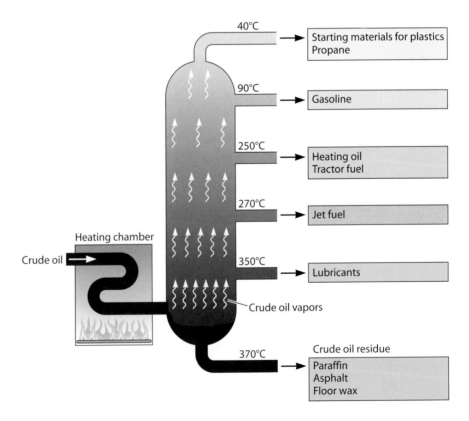

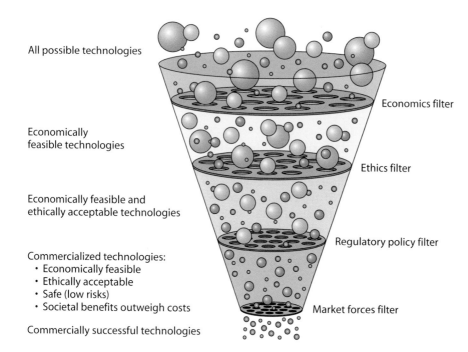

All possible technologies

Economics filter

Economically
feasible technologies

Ethics filter

Economically feasible and
ethically acceptable technologies

Regulatory policy filter

Commercialized technologies:
- Economically feasible
- Ethically acceptable
- Safe (low risks)
- Societal benefits outweigh costs

Market forces filter

Commercially successful technologies

Figure 1.2 Technology development. Not all technologies that are both scientifically and technically possible make it all the way through the development process and become commercialized. They must pass through a series of filters, created by society, before they become reality. The order of the filters shown does not necessarily reflect the actual sequence of barriers that every technology confronts during development, except the final filter, market forces. Only certain technologies that are commercialized are successful in the marketplace.

cate matters further, often the risks associated with a technology become apparent only after it has infused society so completely that lives and economies depend on it and rejecting the technology is not feasible. The automobile developers probably expected its widespread use to carry certain risks: cars colliding with each other, horses, and perhaps even pedestrians. But they never imagined that automobiles would lead to worldwide air and water pollution, as well as concerns about climate change from global warming. Once society finally realized that driving cars causes serious environmental problems, the car was so central to the structure of our society and economy that giving it up was not a viable option. The best society can do when it realizes an essential technology causes problems is devise ways to minimize its negative impacts—which typically means developing another technology, such as the catalytic converter in the case of auto emissions and air pollution.

Even though it may feel like the relationship between technology and society is a one-way street with technology in the driver's seat, each society plays a powerful role in determining its own technological profile. Only certain technologies out of all possible technologies are realized. The agents guiding technology development along certain paths rather than others include economics, ethical values, government policies, market opportunities, consumer preferences, and, in democracies, public opinion (Figure 1.2). Because so many different forces can alter the course of technology development, different segments of society with divergent interests have the power to influence its trajectory. As a result, the benefits of technology are not necessarily distributed equally throughout society, and one group often benefits at another's expense.

Has it always been this way? Yes, in principle. As soon as humans started fashioning tools from stone, technologies began to influence human societies and our species' relationship to the natural world, and on the other hand, societies have always nudged technology development down certain paths but not others. Today, however, technological advances are more pervasive, their effects are more potent, and the societal changes they bring occur at an increasingly rapid pace. As a result, the impact of technology on our lives and societies has greatly intensified. The increased presence, power, and rate of technological change can be traced to the role scientific understanding plays in propelling technology development.

THE RELATIONSHIP BETWEEN SCIENCE AND TECHNOLOGY

Many people equate science with technology, but the two differ in very fundamental ways (Table 1.1). Technologies can exist in the absence of science, because the concept of technology encompasses the practices and products that humans develop to modify and control nature for sustenance and comfort. As such, technology predates the birth of science by many centuries. Early humans created technologies, such as agriculture and tool making, empirically through trial and error and independent of any understanding of the laws of nature—the domain of science. Manual craft skills, such as lens grinding and metal forging, enabled the creation of tools that allowed scientists in the 1600s and 1700s to observe nature more precisely, which increased their understanding of the physical world. Thus, initially, technological tools facilitated scientific progress by opening the door to precise observations and, eventually, scientific experimentation. As science-based understanding of the natural world broadened and deepened, science and technology began to converge, and advances in one began to drive advances in the other.

Equipped with a more sophisticated understanding of the world they wanted to control, people began using scientific understanding to drive technology development. In addition, they replaced their empirical and relatively inefficient approach to developing technologies with the methodical, experimental approach of science. Buttressed by both scientific knowledge and methods, the technologies became more effective. With better technologies available to them, scientists could probe more deeply into the causes and effects of natural phenomena, which in turn gave rise to new, improved technologies. This science-based technology development led to the increasingly complicated modern technologies that people typically envision when they think of technology: software, DVD players, pesticides, antibiotics, and vaccines.

Table 1.1 Fundamental differences between science and technology

Science	Technology
Search for knowledge	Practical application of knowledge
Way of understanding ourselves and the physical world	Way of adapting ourselves to the physical world
Process of asking questions and finding answers, then creating broad generalizations	Process of finding solutions to human problems to make lives easier and better
Looks for order or patterns in the physical world	Looks for ways to control the physical world
Evaluated by how well the facts support the conclusion or theory	Evaluated by how well it works
Limited by the ability to collect relevant facts	Limited by financial costs and safety concerns
Discoveries give rise to technological advances	Advances give rise to scientific discoveries

Science, technology, and society: it's a three-way street

Science-based technology development accelerates economic growth through its effects on industrial productivity. New technologies create new products, stimulate the creation of new companies and even new industries, improve existing products and processes, and lower manufacturing costs. They also provide industrial researchers with tools and techniques for discovering new products.

In the last century, technological knowledge, or in other words, human capital, and physical capital (machines and infrastructure elements, such as roads and widespread electrification) have replaced land, labor, and natural resources as the primary drivers of economic growth. Governments that have invested in scientific research, technology development, and infrastructure requirements; improved their educational systems; and created policies and institutions to encourage science and technology development have profited from their investments. The overall economic well-being of their citizens has improved, as have their longevity, infant mortality rate, and quality of life. The division of the world into industrial nations and developing countries can be explained by the roles science and technology have played in stimulating economic growth in the former, but not the latter.

However, as described above, advances in technology have unexpected negative effects on society. Some people argue that the economic and social benefits that science-based technology development has provided to some of us are not worth the costs, particularly to the environment, paid by all of us. They link environmental degradation directly to modern technology development. While it is true that some technologies have had negative environmental effects, some of which have been very serious, the relationship between technological advance and environmental quality is more complicated than the simplistic view that technology ruins the environment. New technologies also improve environmental quality and prevent environmental problems. Most people in industrialized nations associate the synthetic chemicals used by farmers solely with environmental pollution, but herbicides also decrease soil erosion, the major contributor to water pollution, while fertilizers and pesticides increase yields, and in doing so they prevent deforestation. When it comes to technology and the environment, the situation is never as simple as we would like it to be.

The rate of technological change is increasing

The positive feedback loop between scientific understanding and technological innovation drives a constantly accelerating rate of technological change (Figure 1.3). This does not mean simply that scientific knowledge increases and leads to more technologies—the *rates* at which knowledge accrues and technology changes also increase (Table 1.2). The acceleration of scientific innovation and technological change has important implications

Figure 1.3 In today's world, the relationship between progress in science and technology is circular, not linear. Changes in one lead to changes in the other. In addition, the relationship between the two is reciprocal: science is as important to technology as technology is to science.

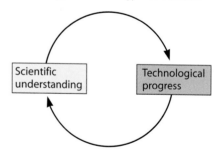

Table 1.2 Time lapse between technology introduction and widespread use

Technology	Infiltration time (yr)[a]
Electric lights	80
Personal computers	16
Internet	4

[a]The speed with which a technology infiltrates a society decreases over time. The measure of infiltration is held constant at 50 million users. (Source, Organization for Economic Cooperation and Development, Paris, France.)

for individuals and their governments. Societies struggle to cope with the present breathtaking rate of technological advance, and because of the circular relationship between scientific understanding and technological innovation, the pace will quicken continuously in the future. Will government policies that are intended to control the social impacts of technology or the direction of scientific research manage to keep up with the galloping rate of change?

Consciously or unconsciously, many people realize they have become increasingly dependent on science and technology as their understanding of both has decreased. This paradoxical circumstance generates uneasiness, at best. Knowledge really is power, so understanding something about two of the primary forces that make today's world go round—science and technology—helps people feel less vulnerable to manipulation. When your car won't start, knowing a few simple things about how internal combustion engines work gives you a bit of protection against unscrupulous auto mechanics, doesn't it? In addition, when a decision involves personal health and well-being, people need to feel they are acting in their own best interest, and increasingly, that means understanding something about science and technology.

But how can someone with very little background in science be expected to stay abreast of scientific and technical developments when the meteoric pace of scientific discovery precludes even scientists in one area of research from understanding discoveries in other areas?

This text tries to address this problem by providing some basic scientific information about a science, biology, and its derivative technology, biotechnology, both of which will have a great impact on your life in the coming years. We want to provide you with enough information about the general principles underlying how organisms work so that you can develop a mental framework that will help you understand new developments in biotechnology as they arise. Having a mental framework that both grounds and guides you should increase your confidence about making informed decisions, as well as help you become an informed participant in discussions about the appropriate uses of biotechnology. But, to be perfectly straight with you, we have to admit that we can never provide you with all the relevant facts about biology you will need to make informed decisions, because they change on a weekly basis.

Another one of our goals is to help you understand how science is done. That may seem like a paltry offering in light of the complex issues you will face due to advances in biology and biotechnology, but understanding how science works may be the most valuable tool we can provide, because that knowledge will *never* be made obsolete by new discoveries. Knowing how science works creates a second framework for interpreting the scientific advances you hear about. Learning how science works also means learning how scientists think. If you learn how to think like a scientist, you will give yourself the best possible defense against misinformation and manipulation.

To think like a scientist, you must be willing to ask questions repeatedly, without being emotionally invested in a certain answer, and then base your conclusions on evidence, not preconceived notions of what ought to be true. You also need to be equally skeptical of all those who are answering your question until the preponderance of evidence makes it clear that one answer is more correct than others. Finally, and probably most surprising to those who think science has all of the answers, you need to be comfortable with ambiguity, uncertainty, and being wrong.

THE NATURE OF SCIENCE

How often do you read a newspaper article about a scientific discovery that contradicts last week's discovery and think, "Why can't scientists make up their minds?" Contradictory reports can be especially maddening when the debate relates to personal choices you make about your health and well-being. As you read these reports and become increasingly frustrated and distrustful, scientists reading the very same contradictory newspaper articles are thinking, "Hmmm, that's interesting." Why the difference? You might assume the scientists' blasé attitude stems from knowing a truckload of scientific facts not included in the story, but rarely is that the reason. They are calmly accepting the contradictory reports because they know how science works.

If you are like most nonscientists, you probably have an idealized view of science that is a natural and almost subliminal outgrowth of the way that you were taught science in school. Given the large amount of material teachers must cover, they need to present science as a set of eternal laws, absolute truths, and uncontested facts that represent the final end point of a linear process in which each discovery led, rationally and obviously, to the next.

If only it were that easy! Science is much, much messier than is implied by a retrospective listing of sequential discoveries building logically, one upon the other, and culminating in a "scientific truth." Your teacher couldn't share the messiness with you, because describing the mess and mistakes would have consumed every minute of class time, leaving no time to explain the discoveries. Unfortunately, learning about the discoveries and not the mess creates a utopian view of the scientific process and a belief that every scientific discovery represents a concrete, immutable stepping-stone in the path of discovery leading to an infallible truth. Contradictory newspaper reports belie this myth. As is always true with utopian concepts, accidentally glimpsing clay feet or learning about the mess in the closet makes people recoil from their idealistic belief with a ferocity that equals the strength of their embrace of it. When it comes to science, neither cynical disillusionment nor blind acceptance is in your best interest, so learning that every scientific finding is always tentative and open to reinterpretation or outright rejection will help you tolerate contradictory results.

Knowing how science works puts scientific facts in perspective

Do you remember when you realized that the quickest and most surefire way to figure out how to get the bunny through the maze to the carrot patch was to start at the carrot patch and go backwards? That is how science is taught (Figure 1.4). Knowing the current thinking about a certain scientific topic, your teacher described the most direct path of discovery that got scientists to that point (the carrot patch).

But that's not how science is conducted. Scientists must enter the maze where the rabbit does, unsure of where they're headed and uncertain how to get there. As a result, the history of science is filled with blind alleys; negations of previous "breakthrough" discoveries; results repudiated in one era that become cornerstones of major scientific theories in another; invalid (even ridiculous) ideas accepted solely because a great scientist proposed them; and seemingly disparate findings made compatible by another finding or a new, improved theory capable of accommodating them all. But don't think of the blind alleys, incorrect theories, and repudiated results as "mistakes" simply because they are not findings that served as stepping-stones on the path to discovery. Even negative results contain useful information; at a minimum, they tell scientists they're on the wrong track.

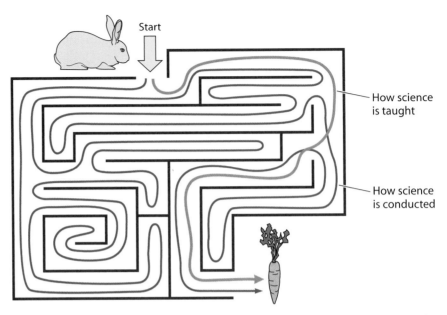

Figure 1.4 The nature of science. The yellow line indicates how science is taught, but the blue lines show how science is conducted.

Doing science is like trying to assemble a jigsaw puzzle without having the picture on the front of the puzzle box. Scientists discover where they were headed only after they get there. It is only once they arrive that they finally know what the picture on the puzzle box looks like. Once there, they turn around to see how they got there, and the path is so obvious it lights up like a yellow brick road—in retrospect. Interestingly, once scientists have arrived where they were headed, they keep testing whether they're where they think they are, just to be sure.

Scientists report their findings in scientific journals, which are of little interest to most nonscientists. The reported findings include the blind alleys, repudiated results, and invalid ideas described above, though the scientists reporting them don't know their findings are wrong or irrelevant at the time they are reporting them. Recently, however, newspapers have also begun announcing scientific findings that might excite their readers. The reporters reading the journal articles and writing the stories are ill equipped to understand where a certain finding fits in the scheme of things. Is it a blind alley in the maze or a key stepping-stone in the path to a breakthrough discovery? A piece that belongs in a different puzzle box? Two puzzle pieces from the correct box that don't seem to fit together? Or is it that magical puzzle piece that makes all of the other pieces lock into place?

The messiness of the scientific process that used to occur outside of the public's view is now splashed across the front pages of major newspapers. It's the scientific equivalent of having neighbors watch a family that is airing its dirty laundry so the family members can understand each other better.

Science is a process, not a list of discoveries

Science is not simply a collection of facts, findings, and theories; it is an ongoing, methodical, iterative process with a specific goal: explaining and understanding the physical world *by way of* facts, findings, and theories (Figure 1.5). The first step in the scientific process is to gather facts, but only

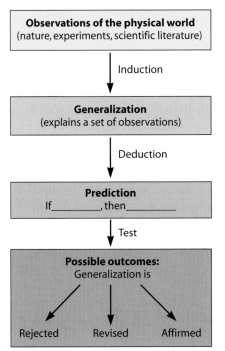

Figure 1.5 The scientific process. Science always begins with observations of the physical world. Those observations may come from observing nature, conducting an experiment, or reading the scientific literature. First, the scientist uses inductive reasoning (from specific to general) to create a generalization to explain a set of observations, which are often referred to as facts. Then, the process is turned on its head, and the scientist uses deductive reasoning (from general to specific) to make a prediction based on the generalization. The scientist then tests the prediction to see if it is correct. If the prediction is not accurate, the generalization is either revised or rejected. If the prediction is accurate, the generalization is accepted. Scientists, however, can never prove that a generalization is accurate.

those facts that are relevant to the question or problem being studied are gathered. To a scientist, facts are observations of events or material objects in the physical world. Once the facts are gathered, the scientist then attempts to order them by fitting them together and making a general statement that encompasses all of the facts. For example, a generalization that would provide order to a set of your observations of the natural world is "All birds have feathers and can fly."

The next step in the scientific process is to assume that the generalization is valid and make a prediction—an "if-then" statement—based on the generalization. For example, "If sparrows, eagles, and doves are birds, then they have feathers and can fly." You check and see that indeed they do have feathers and can fly, so your generalization is strengthened by your additional observations. Then one day you come upon a nest of newly hatched birds. Not only are they featherless, they also can't fly, so you revise your generalization accordingly. For example, your new generalization might be "All adult birds have feathers and can fly." Later, you take a trip to Australia and New Zealand and see two new bird species, emus and kiwis, neither of which can fly even though both have feathers. You modify your generalization once again, and now it might read "Most adult birds have feathers and can fly," or "All adult birds in North America have feathers and can fly." But as a scientist, you would know that this generalization, like the ones before it, is always open to new observations and endless revisions.

After reading this explanation of the scientific process you probably realize that you follow the very same thought process in your personal life. You order some of your observations into a generalization and work from the generalization until it's proven false. Sometimes you may even test your generalization to make sure you've got your head on straight. If facts don't fit into your generalization, you may revise your generalization slightly to accommodate the new observations or throw it out completely and start again. But now you have many more facts to incorporate into a generalization, so your new generalization will probably be more accurate than the first one. Thus, the scientific process is not magical, nor is it so demanding that only a few geniuses can master it.

If you think this way every day, does that make you a scientist? If not, then what distinguishes science from your thought processes? One distinguishing trait is the type of information that can be used to create the generalization. In your personal life, you might let hopes, prejudices, or politics creep into your generalization in addition to facts. Also, you might ignore certain facts that don't fit into your generalization, because you're just sure it's right or you want very badly for it to be true. The only facts a scientist can use in constructing a generalization are observations of the physical world. Scientific observations cannot be subjective, because they must be verifiable by others. Scientists also are not free to ignore facts because they don't fit into their generalization.

The nature of the generalization may differ as well. A generalization drawn from your personal life can be subjective and might include a value judgment. In addition, when all of your observations support your generalization, you probably decide it's valid and don't continually revisit the issue. In science, the generalization must be amenable to testing, using the if-then process; it also must be open to contradiction by anyone, anywhere, at any time. That is, it must be refutable.

These essential attributes of scientific generalizations are key to understanding what science can and cannot prove. Science can prove a generalization is *not* true, but it can never prove a generalization *is* true. Why? Because scientists know they will never be able to observe all of the relevant facts. In the example above, if you had never gone to Australia and seen emus, you would continue to think your generalization that all adult birds have feathers and can fly was valid. Therefore, the phrase "scientifically proven to be safe" is inherently false. In addition, if someone says society should never approve the use of a new technology until it is proven to be safe, they are actually saying society should never approve any new technology.

Science accepts a generalization as valid when observation after observation supports it, but that acceptance is tentative and conditional, so nonscientists should not interpret acceptance as proof. This explains why scientists who are quoted in the newspapers are always qualifying their statements. They understand the limitations of science; they also understand that the public does not.

Societal values influence science and technology

Despite certain essential differences, science and technology also converge in subtle ways that may not be apparent to people outside the scientific and engineering communities. Most people would readily agree that decisions to develop certain technologies and not others are often driven by societal factors, such as politics, economics, and cultural values. In other words, the development of certain technologies instead of others is value laden. Fundamental, value-based assumptions drive the quest for more and better technologies spawned by ever-increasing scientific knowledge. One assumption is that science and technology result in progress. A second is that we should dominate nature.

What most people are not aware of is that science is influenced by these same societal factors of politics, economics, and cultural values. The most obvious influence is the source of research funding. Scientists need money to conduct their research. Most of their funding comes from government grants, but significant research monies are also provided by companies, non-profit organizations, special interest groups, and research institutions. All funding sources favor certain research questions over others, and scientists will choose questions that have the greatest chance of being funded. Often, politicians decide that a certain research area deserves more funding, and scientists shift their research accordingly. The politicians may have made their decision based on the needs of all citizens or only a handful. Thus, just as only some technologies are developed from the universe of possible technologies, only certain scientific questions, out of all possible questions, are asked.

Nonscientists view science as an objective search for the truth, and in many ways it is. Honoring the scientific process we just described—systematic observation, generalization construction, and repeated testing—researchers attempt to establish their generalizations objectively and gather facts through detached, accurate observation and careful, methodical experimentation. The scientific community respects the integrity of the scientific process and, on the whole, does its very best to keep that part of the research process value free.

But scientists were raised in society, and that society has shaped the way they view the world. The scientific process cannot shield scientists from the social context in which they conduct research. Society has molded their sense of values and goals since the day they were born. The narrower social context of the scientific world in which they function imposes pressures and erects powerful filters for selectively viewing the world. Facts and findings are interpreted through these filters, and they also play a key role in determining which questions are asked. Consequently, even if scientists could fund their own research, only certain questions from the universe of available questions would be asked. So, while the methodology that scientists use to answer questions may be as objective as possible, the questions asked and the interpretation of the results are not value free.

THE BIOLOGY CENTURY

In the past two centuries, the primary scientific drivers of technology development were physics and chemistry. Technological innovations rooted in these sciences were key components of the Industrial Revolution, Information Age, and Green Revolution, all of which transformed the nature of our economies and the structure of our societies. Society now finds itself on the threshold of an era dubbed the Biology Century by a number of authors because breakthroughs in biology will be the predominant force transforming our lives, driving economies, and perhaps even provoking people to reconsider the role of the human species on the planet.

The Biology Century will be fueled by biotechnology, and all aspects of society will be affected. Biotechnology is propelling a breathtaking rate of scientific research and discovery, forcing governments and judicial systems to reevaluate their laws and policies in light of new discoveries. You already face choices made possible by biotechnology, and the list of possible options will grow exponentially, tracking the rapid pace of scientific discovery and technological innovation.

If you are not quite sure you understand what biotechnology is, rest assured you are not alone. A great deal of confusion surrounds the word "biotechnology," because different people have used that term to designate different things, much as Americans and Britons have different games in mind when they say "football." Having productive discussions about the technical, complex, and often emotional topic of biotechnology can be difficult in and of itself without the added difficulty of using the same word to mean different things.

What is biotechnology?

Defining biotechnology is actually quite easy. Break the word into its roots, and you have:

> **Biotechnology** the use of living organisms or life processes to solve problems or make useful products.

After reading that definition, you can immediately see one source of confusion. Living organisms have *always* met our need for sustenance and comfort by providing us with food, shelter, clothing, and fuel. In truth, Stone Age farmers kicked off the biotechnology revolution over 10,000 years ago when they domesticated plants and animals, because domestication is inseparable from genetic modification (Figure 1.6). They extended their use of living organisms to microorganisms around 8,000 years ago, when they began exploiting microbial biochemical processes (unknowingly, of course) to convert grapes into wine, milk into yogurt and cheese, and grains into beer and raised breads. So even though biotechnology is quite ancient, the phrase "biotechnology revolution" is bandied about today as if it were brand new, which can be confusing.

What is it about today's biotechnology that sets it apart from ancient biotechnology and elevates it to the status of a revolution? Scientific understanding distinguishes new and old biotechnology. With earlier biotechnology, when people used organisms and attempted to change them to better meet their needs, they did not understand the mechanics underlying the life processes they wanted to control and improve. In fact, they did not even know that microorganisms were responsible for the fermentation processes that turned grapes into wine and milk into cheese until 150 years ago—a mere 7,850 years after they began using them. In the absence of understanding, their exploitations and manipulations of plants, animals, and microorganisms were trial and error ventures.

True to the relationship between science and technology we described earlier, over the centuries, the biological sciences provided insights into the inner workings of living organisms, and people gradually became more proficient at using and improving them—controlling nature to meet their needs. Scientific progress in the last half of the 20th century ultimately led to an understanding of organisms at their most basic levels, the cellular and molecular levels. This deep and rich understanding of the fundamental mechanics of the processes of life and its biomolecular components (Table 1.3) has given

Figure 1.6 Crop genetic modification. The wild ancestor of corn, teosinte, bears little resemblance to modern corn. Ancient farmers in Central America used genetic modification through seed selection to convert teosinte, which had been a wild gathered plant, to corn.

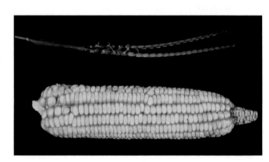

Table 1.3 The four classes of biological molecules[a]

Type of biological molecule	Familiar examples
Lipid	Fats, steroid hormones
Protein	Enzymes, collagen
Nucleic acid	DNA, RNA
Carbohydrate	Starch, glucose

[a]Biological molecules, or biomolecules, are the large molecules that are unique to living organisms.

birth to today's biotechnology. Therefore, a more appropriate definition of the biotechnology we will be discussing in this book is:

> **Modern biotechnology** the use of cells and biological molecules or cellular and biomolecular processes to solve problems and make useful products.

If today's biotechnology marks the next step in a continuum of technologies based on living organisms, does it deserve the moniker of revolution? Time will tell. So far, the primary changes associated with commercial applications of biotechnology seem primarily to be in *how* people do things but not in *what* they do. If any biotechnology revolution has occurred, it has been in the research laboratory. As you recall, technological innovation plays a key role in facilitating scientific discovery. Biotechnology has provided researchers with an amazing set of research tools that are responsible for the breakneck speed of discovery in biology today. The biotechnologies are allowing researchers to answer longstanding questions and prompting them to ask questions that would not have occurred to them if these technologies did not exist. It is in that sense that they have been revolutionary.

Biotechnology is a collection of technologies used by many industries

Biotechnology is not a single technology but a collection of technologies (Table 1.4). Some have been essential technological components of industries for many decades, such as bioprocessing and plant cell culture; others, such as genetic engineering and antisense technology, owe their existence to scientific discoveries made in the 1970s and 1980s. The common thread uniting these technologies is their foundation: they are based on living cells and biological molecules. They capitalize on the cell's capacity to reproduce itself, manufacture biological molecules precisely and repeatedly, degrade a variety of substances, and respond to environmental factors.

These are the very same capabilities humans have relied on for centuries, unknowingly, and that have provided the foundation for many well-established industrial sectors, such as agricultural production, pharmaceutical manufacturing, and waste treatment. The success of these industries has always hinged on effectively expropriating the biochemical processes of living organisms and shaping them to specific purposes. With today's biotechnologies, scientists co-opt these same cellular capabilities knowingly and purposefully.

All of the biotechnologies listed in Table 1.4 can be used by the many industrial sectors listed in Table 1.5 to conduct basic and applied research, improve processes, and create new products and services. Consequently, biotechnology, like all technologies developed in the last two centuries, will stimulate economic growth by increasing industrial productivity.

Table 1.4 Biotechnologies[a]

Technology description	Examples of current applications
Monoclonal antibody technology Uses immune system cells that make proteins called antibodies. Antibodies bind to substances with extraordinary specificity.	Diagnose infectious diseases Treat autoimmune diseases Detect harmful microorganisms in food Locate and measure environmental pollutants Distinguish cancer cells from normal cells
Bioprocessing technology Uses living cells, such as bacteria, yeast, and mammalian cells, or their enzymes, to manufacture useful products, break down molecules, or generate energy.	Clean up toxic waste sites Produce energy from agricultural refuse Manufacture therapeutic compounds and vaccines Produce fermented foods and nutritional additives Manufacture industrial enzymes and feedstock chemicals
Cell culture technology Is the growing of cells in appropriate nutrients in laboratory containers or in bioreactors in manufacturing facilities.	Increase use of biocontrol in agriculture Replace animal testing with cell testing Treat certain medical problems by replacing malfunctioning or injured cells with healthy cells Produce naturally occurring plant therapeutics
Biosensor technology Consists of a biological component, such as an enzyme, linked to a tiny transducer that produces an electrical or optical signal when the biological component binds to another molecule.	Measure blood glucose levels Monitor industrial processes in real time Provide physicians with instant test results Locate and measure environmental pollutants Measure the nutritional value and safety of food
Recombinant DNA technology (genetic engineering) Uses molecular techniques to join, or recombine, DNA molecules from different sources.	Treat certain genetic diseases Improve food nutritional value Develop biodegradable plastics Provide new and improved vaccines Enhance biocontrol agents in agriculture Decrease allergenicity of certain foods Increase crop yields and decrease production costs
Microarray technology Allows analysis of thousands of genes, proteins, or other molecules simultaneously.	Detect genetic mutations Tailor drug treatment to patient Assess potential toxicity of drug Identify stage of disease progression Find microbes for cleaning up pollution
Protein engineering technology Improves existing proteins, such as enzymes and antibodies, and creates proteins not found in nature.	Create novel enzymes Improve catalytic ability of enzymes Develop sustainable industrial processes Improve proteins responsible for bread rising
Antisense technology Decrease the production of specific proteins by blocking the genes encoding them.	Slow food spoilage Control viral diseases Engineer metabolic pathways in crops Treat diseases such as asthma and certain cancers

[a]Biotechnology is a collection of technologies, all of which utilize certain unique properties of cells and the molecules within them. This list includes only some of the biotechnologies and focuses on commercial applications and not the uses of biotechnology in basic research.

Table 1.5 Examples of industrial sectors affected by biotechnologies[a]

Human health care

Knowing the molecular basis of health and disease can lead to improved and novel methods for diagnosing, treating, and preventing diseases. Biotechnology products already on the market include detection tests for many infectious organisms, certain cancers, hormone levels, and genetic diseases; therapeutic compounds for rheumatoid arthritis, diabetes, cystic fibrosis and other genetic diseases, multiple sclerosis, cardiovascular diseases, and many cancers; and vaccines for hepatitis B, meningitis, and whooping cough.

Agricultural production

The agricultural production industry uses biotechnology to increase yields, decrease production costs, diagnose plant and animal disease, enhance pest resistance, improve the nutritional quality of animal feed, broaden the use of biological control agents, and provide alternative uses for agricultural crops. Currently marketed products include insect- and disease-resistant crops, herbicide-tolerant crops, healthier oilseed crops, and crops that provide renewable sources of raw materials for soaps, detergents, and cosmetics.

Food and beverages

Food processing, brewing, and wine making have always relied on biotechnology to enhance the nutritional quality and processing characteristics of their starting materials—grains, fruits, and vegetables—as well as improve the microorganisms that are essential to these industries. All fermented foods and beverages depend on the action of microorganisms, which also serve as the source of many food-processing aids, preservatives, texturing agents, flavorings, and nutritional additives, such as amino acids and vitamins. In addition, biotechnology-based diagnostic tests improve food safety by identifying microbial contamination and food allergens.

Enzyme industry

The enzyme industry and its products are essential to the operations of many of the other industrial sectors, such as food processing, textiles, and brewing. Microorganisms have been the essential manufacturing work force of this industry, and their impact will increase in the future as genetic engineering gives new manufacturing capabilities to standard production microorganisms and improves manufacturing process efficiency and production economics.

Forestry/pulp and paper

Biotechnology is being used to create trees that are resistant to diseases and insects and to improve the efficiency with which trees convert solar energy to wood production. Extensive research is being conducted on microbes and their enzymes for pretreating and softening wood chips prior to pulping, removing pine pitch from pulp to improve the efficiency of paper making, enzymatically bleaching pulp rather than using chlorine, and deinking recycled paper.

Textiles

Many textiles, such as cotton, wool, and silk, occur naturally, while others are derived from natural substances, such as wood pulp. Biotechnology should have an indirect impact on the textile industry by improving the source materials, as well as a direct impact. Enzymes are currently used in natural-fiber preparation and value-added finishing of the final product, such as stonewashed denim jeans. Leather manufacturers use enzymes to remove hair and fat from skins and to make leather pliable. Genetically engineered microbes have produced textile dyes, such as indigo, and the protein found in spider silk.

Chemical manufacturing

Biotechnology can provide cleaner, more efficient ways of manufacturing chemicals than do current methods. Microbes have been used for decades to convert biological materials, such as corn, into feedstock chemicals. Public and private institutions are conducting research on increasing the use of plant biomass and microbial enzymes in chemical manufacturing, because using both is likely to generate fewer toxic waste products.

Energy

Before fossil fuels can be used for energy production, sulfur must be removed, and biodesulfurization relies on microbes and their metabolic enzymes. Microbes have also been used to enhance oil recovery from in-ground crude-oil formations for more than 30 years. In the future, as fossil fuels become depleted and oil prices increase, society will need to establish alternative energy sources, such as biomass-based fuels, like the ethanol that is currently added to gasoline. Advances in biotechnology are making production of ethanol more attractive economically. Other potential areas of energy production include genetically engineered microbes to generate methane from agricultural or municipal wastes or photosynthetic microbes for hydrogen production. However, both will require a number of decades of research before they become economically viable.

Waste treatment

Microbes have always been essential for degrading organic wastes, whether the waste is generated by humans or agricultural and industrial operations. As the human population increases, supplies of potable water decrease, and the standard of living of people in developing countries improves, society will need to apply biotechnology to improving the efficiency of natural microbial degradation processes. In addition to utilizing microbes to break down wastes, environmental engineers are also turning to microbes to clean up soils and water that have become contaminated with environmental pollutants.

[a]Even though the investment community refers to biotechnology as an industry, biotechnology is not an industry but a set of enabling technologies used by a wide variety of industrial sectors. This summary is not intended to be comprehensive but only suggestive of the potential roles biotechnology will play in these industries. The roles are discussed in detail later in the text.

Some biotechnology applications tackle problems that to date have evaded technological solutions, while others represent new approaches to problems people have addressed with previous technologies. For example, for years physicians have successfully treated diabetes with injections of pig insulin purified from the pancreatic tissue of butchered hogs. Now, diabetics have the option of using a pure form of human insulin manufactured by genetically engineered microorganisms rather than the insulin from pig pancreatic tissue, thus alleviating potential problems, such as contamination with disease-causing organisms or reaction to foreign proteins. This new solution to the problem of treating diabetes saves both lives and money.

Biotechnology capitalizes on cell properties

The fundamental problems confronting people today are essentially the same problems they have faced for centuries—growing crops, curing disease, and getting energy—but the technological tools brought to bear on these problems have improved, especially during the last century. In spite of the improvements, these technologies have sometimes created other problems. Will the solutions provided by today's biotechnology represent improvements over previous technologies? Equally important, will they solve problems without creating a host of new ones?

By using cells and biological molecules as the foundations of a technology, companies can develop products that capitalize on innate properties of life at those levels: specificity, unity, and reproducibility. These intrinsic characteristics give rise to technologies that, by their very nature, have the potential to generate new and improved solutions. The jury is still out on how much of that potential will be realized.

Specificity, precision, and predictability

Cells and molecules exhibit extraordinary specificity in their interactions, so the tools and techniques of biotechnology are quite precise and can be tailored to operate in known, predictable ways. As a result, the products of biotechnology could represent improvements over earlier, comparable technologies by being better targeted to solving specific problems, generating less severe side effects, and having fewer unintended consequences. The specificity of cells and molecules also enables biotechnology-based detection techniques to identify substances that occur in minuscule amounts and, once identified, to measure them faster and with great accuracy.

Unity and flexibility

Cells and molecules from very diverse organisms display remarkable similarity. Because all cells (1) work with essentially the same set of molecular building blocks, (2) have similar manufacturing processes, and (3) are able to read and implement the genetic instructions from virtually any other cell, the technologies based on cells and biomolecules allow great flexibility in developing products and solving problems. By working at the level of cells and molecules, all of nature's immense diversity becomes accessible, providing an unprecedented number of options for designing technological solutions to specific problems. The essential questions that remain unanswered are which problems will be given the highest priority and who will be involved in establishing those priorities. The answers will determine the groups that will be the greatest beneficiaries of biotechnology's potential.

Reproduction and renewable resources

Many human activities rely on petroleum, a nonrenewable resource and the major contributor to pollution and solid-waste generation. Because organisms contain molecules similar to those in petroleum, they can be used as energy sources and material inputs in manufacturing processes. As a result, biotechnology could help replace petroleum with renewable resources and, in doing so, facilitate the quest for sustainability and decrease manufacturing costs. In addition to being renewable, these resources could lead to products and processes that generate less solid waste or pollution.

However, this same capacity for reproduction could also cause problems in certain situations. Some living organisms released into the environment, either intentionally or unintentionally, could increase in numbers. We will discuss this issue at length later, but we raise the point here to reiterate a fundamental truth about technological innovation: beneficial applications are accompanied by some costs. How can society act wisely in determining how best to use the new capabilities biotechnology provides?

Using these technologies in ways that maximize benefits and minimize costs is more likely if decisions are informed by past mistakes made with earlier technologies. Luckily, humans have a very long history with biotechnology that is directly pertinent to today's applications.

Biotechnology is a continuum of technologies

As mentioned above, biotechnologies did not arise without antecedents but represent the next step in a continuum of technological change. Each of the biotechnologies listed in Table 1.4 has a technical and conceptual heritage characterized by an increasing dependence on scientific understanding rather than intuition or experience. Below, we illustrate this concept of an ever-evolving continuum of biotechnologies by describing the historical development of genetic-modification techniques, culminating in the modern biotechnology technique of genetic engineering. The concept of a continuum of genetic-modification techniques is so essential to understanding many of the biotechnology applications and societal issues discussed later in the text that it merits a detailed explanation. We focus on genetic modification of crop plants but could draw very similar pictures for the history of genetically modifying virtually all of the organisms people use, such as domesticated animals, yeasts that make bread and wine, or microbes that produce antibiotics.

THE HISTORY OF CROP GENETIC MODIFICATION

For 99% of human history, people lived as small bands of nomadic hunters and gatherers, but with the advent of agriculture approximately 10,000 years ago, they settled down into permanent, self-supporting communities. The last 1% of human history bears little resemblance to the previous 99%, because agriculture, the first technological revolution, affected virtually every aspect of life. All technological, economic, and social progress associated with human civilization presupposes and depends on having an adequate food supply provided by agriculture.

Until approximately 8500 BC, all the plant material people used came from wild gathered plants. Around that time, they began to save seeds and intentionally plant them rather than relying on the plant's mechanisms of random seed dispersal. Domestication of rice, corn, barley, and wheat was complete around 7000 BC, and by 5000 BC, virtually all of the major crops people rely on today were fully domesticated.

Stage 1 was genetic modification through seed selection

As soon as humans started planting certain seeds and not others, they unknowingly began to alter the genetic makeup of the wild plants they had been gathering. At first, seed selection was not goal oriented and may have been based on traits that had little or nothing to do with better crop performance. For example, perhaps they gathered seeds that accidentally fell to the ground during harvesting and that were easy to see because they were large and light colored. Once they recognized that plant offspring resemble their parents, they began to save seeds from plants with traits they valued. Purposeful seed selection led to the complete domestication of wild plants to crop plants (Figure 1.6).

Whether or not the choice of seeds was goal oriented, the effect was the same: genetic modification. Consistent discrimination among different inherited characteristics changes the genetic makeup of populations. In the example just given, the seeds of that crop plant would have been larger and lighter than those of its ancestral wild plant. Our ancestors genetically transformed wild gathered plants to domesticated crops by trial and error, relying only on the minimal understanding they had amassed through experience.

Therefore, the first stage in the 10,000-year history of crop genetic modification was one of unwitting exploitation of certain life processes, such as plant reproduction and heredity. Genetically changing crops may have progressed from haphazard selection to using empirical know-how to intentionally create crops with certain traits, but humans were stuck in the "artisan seed selection" stage for thousands of years because they understood nothing about the biological processes they were relying on: seed production and trait inheritance. They had learned one method for creating better crops—planting seeds from the best plants—but were limited to selecting the best seeds from whatever nature provided. Reaching the next level of crop improvement meant exerting control over the type of seeds nature produced.

Stage 2 was genetic modification through plant breeding and selection

A technological advance, the invention of the microscope, permitted the next developmental leap in genetic-modification techniques (Figure 1.7). Through this new window on the invisible world of cells and microbes, scientists learned that reproduction in animals involved the fusion of egg and sperm and identified the flower as a plant's reproductive structure. In plant reproduction, pollen produced by the male portion of the flower fertilizes the eggs, or ovules, hidden deep within the female part of the flower, triggering the formation of fruit and seeds (Figure 1.8).

Now that they understood how plants reproduce, farmers in the 1700s began incorporating desirable traits into crops by intentionally cross-pollinating certain plants by hand. By controlling plant reproduction, they were no longer restricted to selecting seeds produced by nature's random mating process. However, for reasons we discuss later, these controlled crosses produced plants with unexpected characteristics, some desirable, some not. Like the ancient agriculturists before them, farmers selected seeds from certain plants for the next round of reproduction and discarded others, gradually increasing the proportion of plants with desirable characteristics generation by generation.

Hand pollination, which limited seed production to the best plants, along with selection of only certain offspring from that cross as seed sources for the next year's planting, greatly accelerated crop genetic improvement. Yields increased as farmers developed crop varieties specifically adapted to

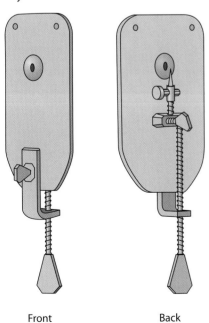

Figure 1.7 The first microscope. In the 16th century, Zacharias Johanssen, a theatrical producer whose hobby was lens grinding, invented the first microscope. Scientists would look through the lens opening on one side of the metal plate to study objects mounted on the point of a pin on the other side. They would bring objects into focus by moving the pin with a system of screws.

Front   Back

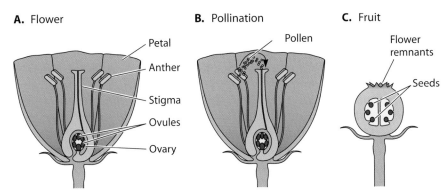

Figure 1.8 Flowering plant reproduction. A single flower usually contains both male and female reproductive organs and gametes from both sexes. The male organ, the anther, releases pollen, which fertilizes ovules (eggs) and leads to the production of fruit and seeds. Seeds contain the embryos from that fertilization event. Many plants can self-pollinate in addition to cross-pollinating.

local conditions. Taking control of plant reproduction also allowed them to create crops that would never have occurred in nature, because they began overriding some of nature's restrictions on cross-pollination.

Understanding plant reproduction sparked a flurry of progress, but crop improvement reached another plateau because the hereditary mechanism responsible for transmitting traits from one generation to the next remained mysterious. Simple observation did little to paint a complete and accurate picture of heredity, because inheritance patterns can be so unpredictable. Offspring might exhibit traits not seen in either parent. Parental traits disappear in one generation and reappear in another, sometimes many generations later. Some observable traits might never reappear, because (scientists now know) they were caused by specific environmental factors, such as soil conditions, and so they could not be passed between generations.

For centuries farmers, and even some scientists, overwhelmed by their confusing set of observations about biological inheritance, resorted to invoking spiritual forces and supernatural powers to explain phenomena that seemed inexplicable and beyond understanding. A few scientists persisted in trying to understand, and society owes the remarkable progress that has culminated in modern biotechnology to them. As you read about their discoveries, you will see that the forces that propel one area of science forward include technological tools, findings in other areas of science, and just plain old luck. You will also see that as science progresses, sometimes the next essential discovery for moving the field ahead is obvious, and scientists race each other so they can be the one that makes the breakthrough discovery. In such cases, if one researcher doesn't make the discovery, another will within a few months, or sometimes only days or weeks. At other times, important discoveries are ignored because the scientific community is not able to hear what is being said.

Why are scientists sometimes unable to hear the correct answer to a problem they have been trying to solve? Because the generalizations the scientific community accepts at a point in time often do more than organize observations. Sometimes they act like mental filters that let in only those observations that are consistent with the generalizations and block those that are not. The field of genetics was born at such a time. The Father of Genetics, Gregor Mendel, described a set of experiments about the nature of inheritance to a scientific community that could not grasp the extraordinary significance of what he was saying until 35 years later.

The hereditary material occurs as discrete information packages

One evening in February 1865, a monk named Gregor Mendel presented a lecture to the local scientific society in Brno, Czechoslovakia. He described in great detail the data and conclusions from 8 years of work intentionally crossbreeding and counting thousands of peas. The results he shared that night rank among the most important findings in the entire history of biology, but none of the scientists in the audience understood their significance. Although he never used the word "gene," Mendel's work revealed the inherent nature of genes; gave birth to a new branch of biology, genetics, that influenced researchers in every other branch of biology; and ultimately transformed the way biologists view the natural world by providing them with an accurate model of heredity.

Models are mental frameworks people construct to make sense of what they observe. They provide a way of organizing observations into a coherent whole, rather than viewing them as individual events unconnected to each other. To explain how the natural world works, scientists create models by assembling a set of currently accepted generalizations, just as they gathered together a set of observations to create a generalization. Thus, in the scientific world, models are supposed to order and explain a very large number of observations. Like generalizations, models are subjected to repeated testing. If the model holds up under all of the scrutiny it receives, it becomes elevated to the status of a **theory**. The word theory has a special meaning in science that is very different from the way most people use the word. Scientists use theory as defined in Webster's dictionary: "a plausible or scientifically acceptable principle or body of principles that explain natural phenomena." Scientists do not mean Webster's other definition: " a hypothesis assumed for the sake of argument." A scientific theory is similar in spirit to a scientific law, not to the sense of speculation and conjecture most people associate with the idea of a theory.

The fluid-blending model of heredity

When Mendel conducted his research, the prevailing model of inheritance involved fluid hereditary material that blended together when egg and sperm fused, much like mixing two colors of paint. While the fluid-blending model of heredity may have organized one set of observations, that both parents contribute hereditary material to the offspring, it ignored another set of observations: nature's extraordinary variation. Even though the same two parents contribute hereditary material to their offspring, no two offspring, with the rare exception of identical twins, are exactly alike, and offspring have characteristics not seen in either parent.

Let's use an everyday example to help explain why the fluid-blending model of inheritance could never be compatible with the variation that exists in nature. Imagine pouring together milk (maternal hereditary material) and chocolate syrup (paternal hereditary material) in various proportions to make glasses of chocolate milk (offspring) that vary along a lighter-to-darker continuum of more or less chocolate. Now, divide each glass into two portions, to represent eggs or sperm, and mix each with an equal amount of a different shade of brown to mimic the joining of eggs and sperm in fertilization. You can see that even though you created variable offspring in the first round of reproduction, in very few "generations," all of the offspring would be the same 50-50 shade of brown. Also, no matter how many times you repeated this dividing and mixing, crossing offspring to produce the next generation, you would never again have a glass of pure chocolate or a glass of

Figure 1.9 Models of inheritance. Shown are graphic representations of two models of inheritance, fluid blending and discrete particle.

pure milk (Figure 1.9). In other words, attributes of the parents that disappeared in the first generation would never reappear in subsequent generations.

The discrete-particle model of heredity

Mendel's research proved that the fluid-blending model of inheritance was inaccurate by demonstrating that the hereditary material that passes from one generation to the next is organized as discrete packets of information. When egg and sperm (or pollen) fuse during fertilization, the maternal and paternal hereditary materials do not blend together like paints. Instead, the hereditary material from each parent retains its distinct identity. Bundling up hereditary material into chunks that exist independently of each other and therefore can be separated from each other provides a constant means of generating genetically variable offspring.

Before we explain a few basic findings of Mendel's research, we first want to show you how the discrete-particle model of inheritance explains nature's variation. Use objects that have distinct identities that can be separated from each, such as different-color jelly beans, marbles, or index cards, to symbolize genes. We'll use jelly beans in this example. Each individual has two genes for every trait, one from its mother and one from its father. Begin with two jelly beans of the same color to represent the genes for a trait, such as flower color, that each parent has and could contribute to the next generation, such as two purple jelly beans in the father and two white ones in the mother (Figure 1.10).

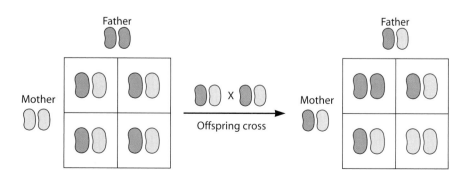

Figure 1.10 Discrete-particle inheritance. Mendel's theory of heredity provided an explanation for the observation that nature contains a great deal of diversity.

Each parent contributes only one of its genes for the flower color trait to its offspring, so all of the first-generation offspring will have one purple gene and one white gene. No other combination is possible. Now use the purple-white offspring as parents for the next generation, and you will see that three types of offspring are possible: purple-purple, purple-white, and white-white; genetically variable offspring.

Mendel organized his observations into generalizations

We will describe Mendel's research at length in chapter 11. At this point, we focus on two of Mendel's theoretical breakthroughs that propelled rapid changes in crop genetic modification. Mendel studied the inheritance patterns of seven different traits in the garden pea (Box 1.1). He found the same inheritance patterns for all traits; we will use the one just described, flower color.

BOX 1.1 *Mendel's methods and observations*

Until the mid-1800s, the science of biology essentially involved observation, classification, and description. Nature did the experimenting, and people simply observed the results and tried to make sense of their observations. The generalizations created from observations of nature were either supported or refuted by additional observations from nature. Influenced by the experimental approaches pursued by scientists in other disciplines, 19th century biologists began to adopt experimentation as a method for testing hypotheses. Mendel's work epitomized this new approach to doing science. Experimentation, in and of itself, is not necessarily a superior way of conducting research because poorly designed experiments can obscure more than they illuminate. Here are a few aspects of Mendel's experimental design that demonstrate that not only did he adopt a new approach to doing research in biology, his work exemplifies research methodology that scientists refer to as "good science."

He chose appropriate research material

Mendel's research material, the garden pea, had two necessary characteristics for any research organism in genetics: short generation time and high productivity. In addition, peas produce fertile offspring after self-pollinating or cross-pollinating, so Mendel could observe the results of all possible crosses. Because of the pea's unique flower struc-

ture, accidental cross-pollinations are virtually nonexistent, and Mendel could be certain the results he observed were due to hand pollinations that he could control (see Figure 11.3). An additional element that made peas attractive research material is essential to all biological research, not simply genetics: expediency. Pea seeds were cheap, plentiful, and readily available commercially.

He picked the appropriate variables to measure

Mendel picked seven traits that could be studied one at a time and, most important, were "discontinuous," that is, each trait had clearly distinct types and was not represented in the population along a more-or-less continuum, as variables such as weight and height are. Focusing on discontinuous traits allowed Mendel to assign his experimental results (trait of offspring) to clear, unambiguous categories, of which there were only two for all seven traits he studied.

Finally, environmental conditions did not affect the presence, absence, or quality of these traits.

He generated a large amount of data and mathematically analyzed his results

Subjecting experimental results to mathematical analysis is such an essential property of science today that it is difficult to conceive of a time when this was not so. Mendel's mathematical approach to biological research was new but not at all sophisticated: he counted peas. Influenced by the statistical way of thinking recently adopted by physicists studying phenomena that could be understood only with probability and statistics, such as the behavior of individual molecules when heated, Mendel adopted a statistician's mindset. This way of thinking allowed him to use information about the population as a whole to understand what was happening in an individual.

Pea characteristics studied by Mendel

Trait	Type
Seed shape	Round or wrinkled
Seed color	Green or yellow
Ripe-pod shape	Inflated or pinched
Unripe-pod color	Green or yellow
Stem length	Long (5 to 6 ft) or short (9 to 18 in.)
Flower color	Purple or white
Flower position	Along the stem (axial) or bunched at the top (terminal)

Mendel always began a series of experiments by crossing two different purebred lines for the trait of interest. When he crossed plants with purple flowers and plants with white flowers, all of the offspring had purple flowers. Mendel allowed the offspring to self-pollinate, and surprisingly, some of the offspring had purple flowers and some had white flowers. A trait that disappeared in the first generation, white flowers, reappeared in the second; so the first generation offspring must have had the potential to produce either parental type even though they displayed only one of those types. Mendel described the relationship between the hereditary contribution made by these two parents as **dominant** and **recessive**. The gene for a dominant trait overpowers a recessive trait but does not subsume it, because the capacity to produce the recessive type remains (Figure 1.11).

To explain his observations, Mendel proposed that the hereditary material for a trait is packaged as a discrete particle, or "factor." Each offspring receives one factor for each trait from each parent, so the factors occur in pairs. During the production of reproductive cells (e.g., eggs, sperm, or pollen), members of a pair separate from each other before cell division, so a reproductive cell has only a single factor for each trait. Scientists now call those factors genes.

Mendel's next theoretical leap was even more creative. He proposed that the outward appearance of an organism, which biologists refer to as its **phenotype**, may or may not directly reflect its **genotype**, which is the genes that are present in the organism. In other words, what you see is not necessarily what you get. Plants with identical phenotypes, such as purple flowers, can differ genetically. In one plant, the gene pair for flower color consists of two genes for purple flowers, while another contains one gene for purple flowers and one gene for white flowers. Therefore, when scientists want to understand the details of an organism's genetic makeup, sometimes its outward appearance, or phenotype, helps them know its genotype and other times it can be very misleading. Having a firm grasp of the fundamental ambiguity in the relationship of genotype to phenotype is so essential to understanding many biotechnology applications that we devote an entire chapter to the topic.

Figure 1.11 Inheritance of flower color. Mendel ordered his experimental results (observations) into a number of generalizations that ultimately led to his theory of inheritance. Two of the most important generalizations he constructed are the concept of dominant and recessive genes and the idea that organisms that look the same (have the same phenotype) can have different genotypes. Purple flower color is dominant to white flower color.

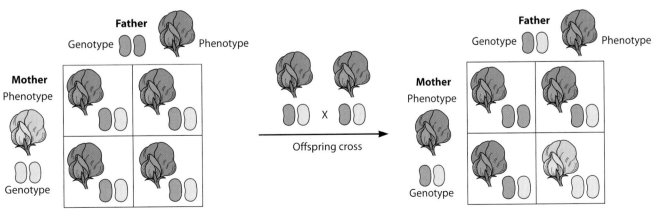

It would be difficult to overstate the importance of Mendel's findings to the present understanding of the workings of biological systems at every level from DNA to evolution, but no one was impressed the night he presented his results or for 35 years afterward. In 1900 and 1901, at least three biologists in different disciplines unearthed Mendel's work simultaneously and confirmed his results. The simultaneous nature of their discoveries tells us that the time was now right for Mendel's Theory of Inheritance; finally, the scientific community could hear what he was saying. Mendel triggered an intellectual revolution in biology 16 years after he died. Science works like that more often than you might think.

Stage 3 was science-based plant breeding

The intellectual revolution catalyzed by Mendel's work precipitated a transformation in agriculture. Over the centuries, farmers had used controlled pollination to develop **cultivars**, which are varieties of the same crops having genetic makeups that differ slightly from each other. These cultivars had been genetically modified so that they were well suited to their specific locales. Armed with a fundamental understanding of genetic mechanisms provided by Mendel's work, farmers and plant breeders now knew they could cross well-adapted cultivars to another plant with a desirable trait, no matter how inferior or poorly adapted, without fear of losing the genes in the superior, well-adapted line and replacing them with the inferior plant's genes. After creating a hybrid plant whose genetic makeup consisted of half superior and half inferior genes, including those for the desirable trait, they could cross the hybrid to the superior line for a number of generations, each time selecting only those hybrid plants having the desirable trait. Through repeated crossings of offspring having the new, desirable trait to the superior parental cultivar, they could gradually replace the inferior plant's genes with the superior plant's genes. After many generations of crossing hybrid offspring with the superior parent, the superior cultivar would have the new, desirable trait, but other genes from the inferior plant would have been removed.

Plant breeders began to look for desirable traits in cultivars from all over the world and in the ancestral wild plant of the crop in question and its relatives. They developed a number of laboratory techniques that allowed them to hybridize crops and plants that would never have been able to interbreed in nature. Many of these "wide crosses" involved plants belonging to different species (**interspecific hybridization**). In some cases, the amount of genetic difference between the crop and other plant was even greater, because breeders began crossing plants from different genera (**intergeneric hybridization**), the next level of genetic difference. For example, bread wheat has been crossbred with more than 10 different species in four different genera (Table 1.6). The wheat you have eaten all of your life is not "natural" in that it would not exist without the extensive genetic manipulation of plant breeders.

By picking and choosing useful genes from a wide variety of plants, plant breeders developed crop cultivars that were resistant to many diseases, insects, and other pests; had greater yield potential; and were drought resistant and easier to harvest. These developments led to dramatic increases in the amount of food produced throughout the world in the 20th century. According to the U.S. Department of Agriculture, the combined production of 17 crops in the United States increased almost 300% from 1900 through 1990, while the amount of land under cultivation did not increase. Improvements in cereal crops, such as wheat, corn, and rice, led to U.S. yield in-

Table 1.6 Examples of genes and traits transferred to bread wheat from plants in different species but the same genus (interspecific cross) and different genera (intergeneric cross)[a]

Source of gene	Desirable trait
Interspecific crosses	
Triticum monococcum	Stem rust resistance
Triticum timopheevi	Powdery mildew resistance
Triticum turgidum	Hessian fly resistance
Intergeneric crosses	
Aegilops comosa	Stripe rust resistance
Aegilops ovata	High kernel protein
Aegilops speltoides	Stem rust resistance
Aegilops umbellulata	Leaf rust resistance
Agropyron elongatum	Drought tolerance
Secale cereale	Yellow rust resistance
	Winter hardiness
Elymus scabrous	Green bug resistance

[a]The scientific name of bread wheat is *Triticum aestivum*. Organisms in different species but the same genus have the same "first name," which is *Triticum* in this case, but different "last names." Organisms in different genera have different first and last names. The rusts are a type of fungi that cause a variety of devastating diseases in wheat and related crops.

creases that were much more impressive than a mere 300% and staved off the widespread famine predicted for the developing world in the last half of the 20th century (Table 1.7).

At this point, plant breeders reached another plateau in crop genetic improvement. They had pushed the breeding envelope about as far as they could. If they were to make additional significant progress in crop improvement, plant breeders needed new ways to find useful genes and introduce them into our crops. The birth of recombinant DNA technologies in the 1970s opened the door to all of nature's genetic diversity.

Studies of gene structure and function led to new understandings of heredity

Mendel's theory of heredity led to the proliferation of a number of subdisciplines in biology as scientists followed a variety of investigatory paths to probe these newfound hereditary particles from different angles. Some researchers began asking questions about gene action: what do genes do; how

Table 1.7 Increasing wheat production[a]

Yr	Production (millions of tons)	No. of adults provided with needed carbohydrates[b]
1966	11.39	3
1971	23.83	94
1976	24.10	96
1981	36.50	186

[a]Wheat production in India increased significantly after the introduction of high-yield varieties developed in the 1960s by Norman Borlaug, the father of the Green Revolution, and other plant breeders. Borlaug was awarded the Nobel Peace Prize for his work. (Data from the Indian National Wheat Program and Norman Borlaug, *Science* **219**:689–693, 1983.)

[b]Based on 65% of the calories in a 2,350-calorie diet being supplied by carbohydrates, or 375 g of wheat/person/day.

do genes work together to coordinate body functions; how is genetic information copied so that it can be transmitted from one generation to the next; how does the information in genes become elaborated into an organism during its development from a fertilized egg? To answer these questions, they needed to know the chemical nature of the genetic material.

What are genes made of?

One of the first clues came from a British public health official, Frederick Griffith, who was trying to develop a vaccine for the pneumonia-causing bacterium. Only certain strains of this bacterium cause pneumonia (virulent strains), while other strains do not (nonvirulent strains). In 1928, he mixed a dead virulent strain with a live nonvirulent strain in hopes of making an effective vaccine. He injected the mixture into mice, and instead of becoming immunized, they caught pneumonia and died. How could virulent, but dead, bacteria cause a disease? Griffith reasoned that a substance must have transferred from the dead virulent bacteria to the live nonvirulent bacteria and transformed them into a virulent form (Figure 1.12). We now know that the substance was DNA, the genetic material, but because Griffith was not a geneticist, he did not think in those terms and called the substance simply a "transforming factor."

In 1943, O. T. Avery and his colleagues at the Rockefeller Institute, New York, N.Y., all of whom were physicians studying microbial diseases, determined that the transforming factor was the nucleic acid, DNA. Many scientists doubted their findings because DNA is an extraordinarily simple molecule. They questioned how something so simple could carry enough information to turn a fertilized egg into a human being. They believed that proteins were the only biological molecules capable of carrying a sufficient amount of information to create an entire organism.

Figure 1.12 Discovery of the transforming factor by Frederick Griffith. The virulent strain of a pneumonia-causing bacterium has a gene for a protective outer covering that gives it a smooth appearance. The nonvirulent (rough) strain does not have this gene and is therefore recognized and attacked by the immune system. When Griffith mixed nonvirulent live bacteria with heat-killed virulent bacteria, the gene for the protective coat moved from the dead smooth bacteria into the rough forms and transformed them into virulent bacteria.

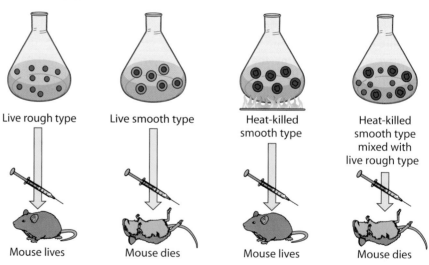

Live rough type — Mouse lives

Live smooth type — Mouse dies

Heat-killed smooth type — Mouse lives

Heat-killed smooth type mixed with live rough type — Mouse dies

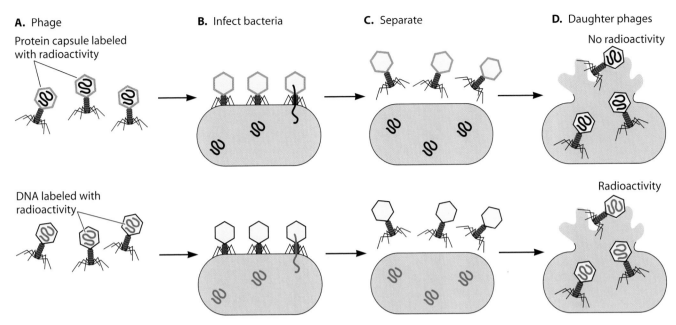

Figure 1.13 Experiments of Alfred Hershey and Martha Chase. One group of bacteriophage viruses containing protein labeled with the radioactive isotope ^{35}S and a second group containing DNA labeled with the radioactive isotope ^{32}P infected bacteria by injecting their genetic material into the host's cells. Hershey and Chase separated the viral coats from the bacterial cells and found ^{35}S in the viral coats and ^{32}P within the bacterial cells. Viral progeny that resulted from the infection also contained ^{32}P, confirming that DNA is the genetic material because it had been transmitted from one generation of viruses to the next.

The DNA-versus-protein debate was resolved by definitive experiments conducted in 1952 by Alfred Hershey and Martha Chase using viruses that infect bacteria, the **bacteriophages**. Their work depended on a technological innovation that had nothing to do with science. Work on the atomic bomb in World War II gave scientists access to the radioactive isotopes that were crucial to the Hershey-Chase series of experiments.

Viruses consist of a protein coat with a small amount of genetic material inside, and they infect a cell by injecting the genetic material into the host cell, while the coat protein remains outside. Hershey and Chase exploited this bit of viral molecular biology to settle the debate on the molecular nature of the genetic material. Proteins contain sulfur but no phosphorus; DNA contains phosphorus but no sulfur. They grew viruses labeled with the radioactive isotopes of sulfur and phosphorus, ^{35}S and ^{32}P, and infected bacteria with them. They separated the viral coat proteins from the infected bacteria and found they were rich in ^{35}S, while the infected bacterial cells contained ^{32}P (Figure 1.13). Hershey and Chase concluded that DNA was the genetic material because the viruses injected their DNA into bacteria while leaving their protein coats outside.

What is the structure of DNA?

As soon as the DNA-versus-protein debate was settled, another question was immediately raised: what was the structure of the DNA molecule? Scientists wanted to know how genes worked, and they knew that the structure of DNA would reveal facts about gene function. They also knew that whoever described the structure would be guaranteed an important place in history.

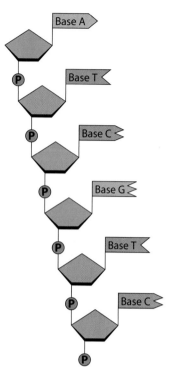

Figure 1.14 Schematic representation of DNA. Long before Watson and Crick determined the three-dimensional structure of DNA, biochemists had proven that it was a very long molecule composed of a subunit that repeated many times. The subunit consisted of a phosphate (P) attached to a sugar, which was also attached to one of four nitrogenous bases. The four bases are represented by the familiar letters A, T, C, and G that people associate with DNA.

Long before Hershey and Chase proved conclusively that DNA was the genetic material, scientists knew that DNA was a large molecule made of a few subunits repeated thousands of times; they also knew that each subunit consisted of three components—a phosphate, a sugar (deoxyribose), and one of four nitrogenous bases (Figure 1.14).

Before biologists knew that DNA was the genetic material, they also had established an essential fact about gene function. They knew that genes work through proteins. In other words, the information in genes is converted into proteins that are responsible for carrying out specific functions dictated by genes. They also realized that the DNA molecule must be capable of replication.

So what else was there to know about the chemistry of this molecule? The three-dimensional structure. A key to determining the underlying mechanics of two primary DNA functions, replication and translation of its information into proteins, resided in its three-dimensional structure.

From 1948 to 1951, findings from a number of branches of science coalesced and provided sufficient information to uncover the next piece of the puzzle.

- A physicist and chemist, Linus Pauling, described the rules governing the formation of chemical bonds and a novel chemical structure, the helix, and introduced the tactic of building physical models to determine molecular structures.
- The physicist Maurice Wilkins and the chemist and crystallographer Rosalind Franklin provided physical images of the DNA molecule created by X-ray diffraction analysis of DNA crystals, a technology developed in the early 1920s.
- The chemist Erwin Chargaff discovered that DNA molecules always have equal amounts of adenine and thymine and equal amounts of cytosine and guanine.

At Cambridge University, Frances Crick, a graduate student in physics, and James Watson, a geneticist on a postdoctoral fellowship, were determined to be the ones who unearthed the structure of DNA, and they succeeded in 1953. Unlike Mendel's discovery, which went unnoticed for a generation, the time was ripe for Watson and Crick's. Had they not proposed the structure of DNA, another scientist would have in a matter of weeks or months.

Stage 4 was plant genetic engineering

Understanding that genes are made of DNA molecules that contain instructions for making proteins informed the work of plant breeders, but this knowledge did not allow them to *do* something new. They now understood that inherited traits, such as disease resistance, could be traced to one or more proteins, yet knowing this did not significantly expand their capacity to identify disease resistance genes and incorporate them into crops.

The next developmental leap in genetic-modification techniques came from an unexpected line of investigation and provides a classic example of scientific serendipity changing the course of scientific research and technological innovation.

As mentioned above, confirming that genes are made of DNA opened the door to a variety of questions about gene function on the molecular level. To answer these questions, scientists turned to the simplest of organisms, bacteria, and the even simpler viruses, the bacteriophages, that infect bacteria. Bacteria reproduce by dividing in two, so the resulting cells are genetically identical to the parental cell, except for rare occasions when a mutation occurs naturally. However, researchers noticed genetically pure bacterial strains acquiring new capabilities, and therefore new genes, when mixed with certain other strains. This observation reconfirmed Griffith's finding. In addition, sometimes bacteriophages carried bacterial genes from one bacterial cell to another. Microbiologists wanted to determine exactly how this transfer of genetic information occurred.

During their investigations, they noticed that some bacteria would destroy new genes from other bacteria, as well as genetic material from the infecting phage virus, by secreting chemicals that specifically degraded foreign DNA but had no effect on their own. Researchers isolated these chemicals and determined that they were enzymes that cut DNA into small pieces. Bacteria use these enzymes, known as restriction enzymes, to defend themselves against other organisms. You will learn a great deal about these enzymes in subsequent chapters, because they are indispensable tools of genetic engineering.

Restriction enzymes, in conjunction with other enzymes, allow researchers to move single genes between organisms. Because of the unity of life described earlier, all organisms can read the DNA instruction book of any other organism and convert the genetic information into the appropriate protein. Plant breeders now have access to all of nature's rich genetic diversity. Later in this text we discuss at length how they are using their new capabilities to genetically modify crop plants.

This newest genetic-modification technique is often referred to as recombinant DNA technology because it allows researchers to recombine genes from different organisms. While this may sound like a frightening prospect, later in the text you will learn that nature recombines DNA from other organisms on a regular basis. So once again, people are following the pattern they established thousands of years ago: exploiting natural processes to create technologies to meet human needs for comfort and sustenance.

In summary, as scientific understanding of life processes has broadened and deepened over the past centuries, humans have used that knowledge to create a variety of genetic-modification technologies. Paralleling the increasing importance of science in developing these technologies, there are two other evolutionary trends.

1. The technologies become more specific and precise because people understand more about the processes they are trying to control or change.
2. The technologies allow scientists to do things that are increasingly "unnatural."

SUMMARY POINTS

Science and technology are significant forces shaping society. While this has been true for the past two centuries, the pressures exerted by these forces are increasing.

The relationship between science and technology is circular. Scientific discoveries drive technological innovations, and technology drives science. As a result, the rate at which science and technology change society is accelerating.

The relationship among science, technology, and society is also circular. Not only do science and technology change society, society also influences the conduct of science and the development of new technologies. Only certain scientific questions are asked; only certain technologies are developed and commercialized. The societal forces altering science and technology include economics, ethics, government policies, and public opinion.

Biotechnology is a collection of technologies based on cells and molecules. It represents the next step in a continuum of technological change that results from an increased understanding of biological systems. This parallel evolution of scientific understanding and technological advance can be seen in the evolution of crop genetic modification techniques used in agriculture.

Many diverse industrial sectors use biotechnology to conduct research, develop new products, and improve processes. Therefore, like all technology, biotechnology will drive economic growth by stimulating industrial productivity.

Biotechnology capitalizes on a number of cell properties that, if used wisely, could offer inherent advantages over earlier technologies. Society can use the tools of biotechnology to answer questions and solve problems. The issue before everyone is to determine which questions to ask and which problems to solve. How society addresses this issue will ultimately determine whether it will witness a true biotechnology revolution.

KEY TERMS

Bacteriophage	Genotype	Interspecific hybridization	Recessive
Cultivar	Intergeneric hybridization	Phenotype	Theory
Dominant			

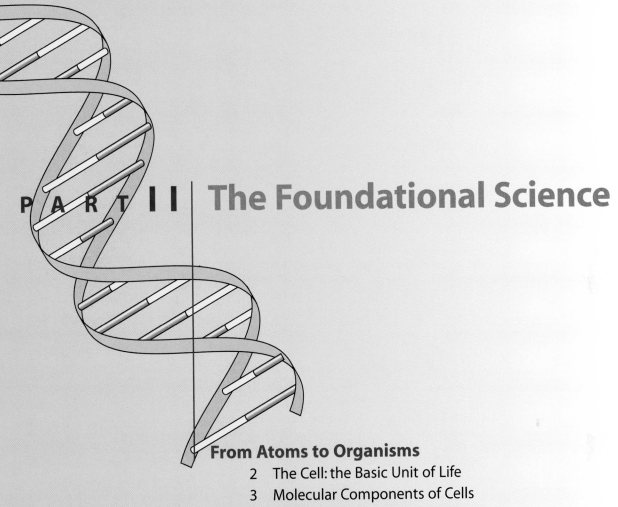

PART II | The Foundational Science

From Atoms to Organisms

From Organisms to Ecosystems

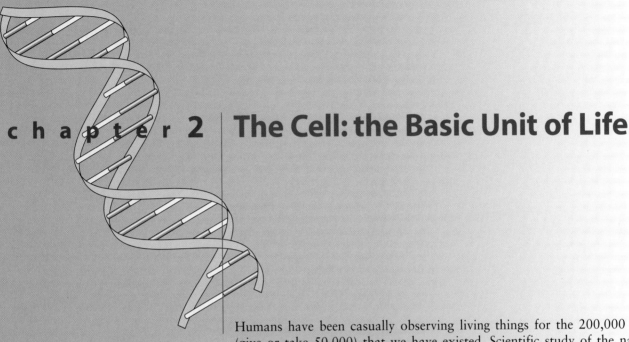

chapter 2 | The Cell: the Basic Unit of Life

Humans have been casually observing living things for the 200,000 years (give or take 50,000) that we have existed. Scientific study of the natural world, in the sense of making careful observations, asking questions, and conducting research, is a more recent development of the past 2,000 years. From the Greeks to the Middle Ages, scientific inquiry focused on the living things that we could see. With the invention of the microscope in the late 1500s, scientists began observing and studying aspects of the natural world that are invisible to the naked eye.

During the 17th century, a British scientist named Robert Hooke sliced off a thin piece of cork and examined it under a microscope. He saw something that looked a bit like a honeycomb—row upon row of tiny pores surrounded by walls. Hooke called the tiny pores **cells** because they reminded him of the cells in monasteries in which monks lived (Figure 2.1). Hooke's name for them stuck.

You may have had the opportunity to use a microscope yourself at some point. If you swabbed the inside of your cheek, collected a drop of blood, or peeled off a layer of onion skin and examined it under the microscope, you probably saw individual cells. Since the time that Robert Hooke first named them, we have learned that cells constitute the smallest, most basic unit of all living things.

This chapter is an introduction to the basics of cells. In it we discuss:

- essential functions of cells
- common features of cells
- the chemical basis of cell organization
- the two major classes of cells

These very basic concepts about what cells do and how they are put together will form a framework for the rest of this section of the book. The next eight

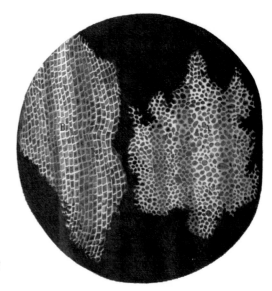

Figure 2.1 Hooke's drawing of cork cells.

chapters will take closer looks at cellular components and the essential functions touched on here. As we look more closely at the various features or functions of cells, we will also illustrate how they can be used to solve problems or make useful products, the definition of biotechnology.

CELLS

During the 18th and 19th centuries, dramatic technological improvements in the microscope allowed scientists to see the newly discovered microscopic world in increasingly fine detail. The advances in microscopy coincided with rapid and significant advances in the science of chemistry. As a result, scientists interested in living things accumulated a great deal of information about both their structures and chemical natures.

Biologists examining the microscopic structures of microbes, plants, and animals found that every tissue or organism was made of cells, each enclosed by a membrane separating it from surrounding cells and the external environment. Chemical analysis revealed that even though various cell types looked very different physically, their chemical makeups were remarkably similar (Figure 2.2). In addition, chemists discovered that cells contain four classes of molecules found only in living things: proteins, carbohydrates, fats, and nucleic acids, such as DNA.

The work of various scientists in the 18th and 19th centuries established a fundamental principle of biology: cells are the basic unit of life. Single cells live and reproduce, but nothing less than a cell can reproduce on its own. All organisms are composed of one or more cells, and those cells:

- have the same essential properties
- are made of the same types of chemicals
- use the same basic design and building materials
- carry out essentially the same processes in very similar ways
- use DNA to store and transmit information for directing their own activities and for reproducing themselves
- use many different kinds of proteins to carry out their many processes

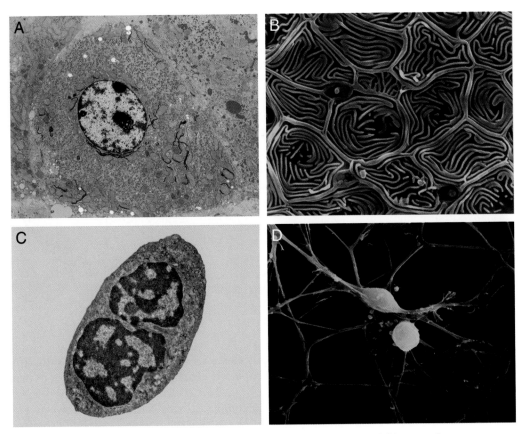

Figure 2.2 Different cell types. **(A)** Liver cell. **(B)** Goldfish skin. **(C)** Neutrophil (a type of white blood cell). **(D)** Neurons. (Photographs copyright Dennis Kunkel Microscopy, Inc.)

Living cells carry out several essential functions

What features distinguish living and nonliving things? You can probably come up with some examples by giving the question a little thought, but some of the defining attributes of living things are less obvious.

Cells grow

Over time, cells can increase in size, and this growth requires building materials and energy. Conveniently, cells get both of these supplies from the same source, the large molecules that make up food: proteins, carbohydrates, fats, and nucleic acids. Cells break down large, complex food molecules to get smaller, simpler building block molecules, and in the process, the energy in food is released. Cells then recombine the simple molecular raw materials into the large molecules they need in order to grow, build and rebuild their structures, and carry out activities (Figure 2.3).

This process of breaking down large molecules for building blocks and energy and then channeling both materials and energy into molecular manufacturing is known as **metabolism**.

Cells reproduce themselves

Just as new animals and plants are generated by the reproduction of animals and plants, new cells are generated by the reproduction of cells. Cells grow to a certain size and then divide in two. Obviously this process requires careful management. During the growth phase, the cell manufactures molecules

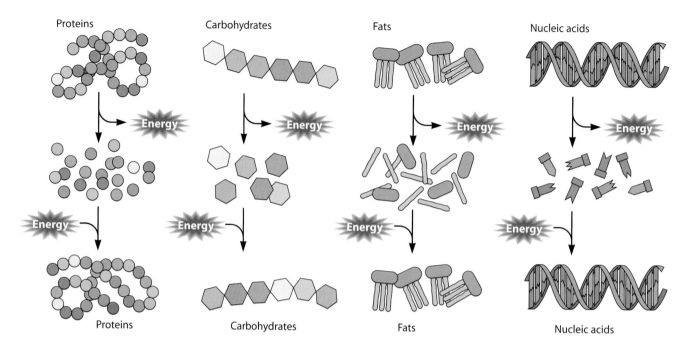

Proteins	Carbohydrates	Fats	Nucleic acids

Energy   Energy   Energy   Energy

Energy   Energy   Energy   Energy

Proteins	Carbohydrates	Fats	Nucleic acids

Figure 2.3 Cell metabolism. The biological molecules in food are broken down for energy and building block molecules and then reassembled according to the needs of the cell.

and cell structures. When the cell senses its supply of materials is sufficient for two fully functional cells, it copies its genetic material and divides in two (Figure 2.4).

For each daughter cell to be a complete, living cell, it must receive an entire correct copy of the genetic material from the parental cell and also must have a complete complement of the cell parts it needs to carry out its activities. Once separated, the two daughter cells grow to a certain size and then divide. This cyclical process of growth and division is known as the **cell cycle** (Figure 2.5).

Cells maintain their internal environments

If you compare the chemical makeup of a cell to that of the environment surrounding it, you will find that the identities and amounts of molecules are quite different. Cells manufacture and keep inside themselves many unique kinds of molecules not found in their immediate environments. Cells also contain molecules readily found in their external environments, such as water, salts, and energy sources like sugars, but the concentrations of these substances inside the cell are different from their concentrations outside the cell.

Figure 2.4 Dividing cells. (Photograph copyright Dennis Kunkel Microscopy, Inc.)

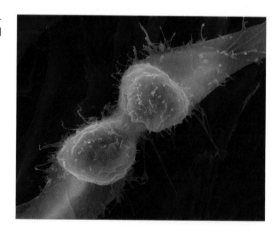

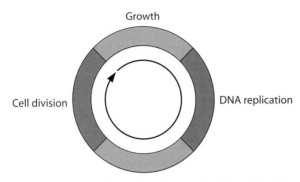

Figure 2.5 The cell cycle. Cells go through cyclic periods of growth, DNA replication, and cell division as they reproduce.

Cells must expend energy to maintain differences in the internal and external concentrations of molecules. In addition, while the external concentration of a molecule such as salt might change, the concentration inside the cells stays constant (Figure 2.6).

Cells respond to their external environments

As alluded to above, cells sense and respond to their environments. When the concentration of salt or water in the external environment changes, cells sense the increase or decrease in salinity, and they make changes to keep from becoming too salty or too watery inside. One-celled organisms like bacteria move toward food and away from toxins; under harsh environmental conditions, many enclose themselves in tough coverings and become inactive. A plant cell responds to an insect eating it by releasing protective, insect-killing chemicals. The presence of food causes stomach cells to release digestive juices.

To be capable of these and a variety of other kinds of responses, cells must be able to (1) sense a change in their environment and (2) respond to it (Figure 2.7).

Figure 2.6 The cytoplasm of this amoeba is much more concentrated than the water in which it is living. (Photograph copyright Micrographia, Inc.)

Figure 2.7 The sensitive plant responds to being touched by curling up its leaves. **(A)** Before stimulus. **(B)** After stimulus. (Photographs courtesy of Kent Schwaegerle, University of Alaska.)

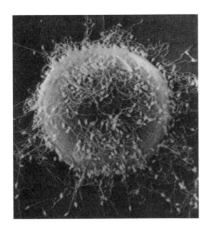

Figure 2.8 Egg and sperm exchange chemical signals during the process of fertilization. (Photograph courtesy of Gerald Schatten, University of Pittsburgh.)

Cells communicate with each other

Cells in multicelled organisms communicate among themselves to coordinate their activities. Nerve cells give muscle cells the signal to contract. Cells in one organ manufacture and secrete hormones that affect molecule manufacturing in the cells of a target organ far from the source of the hormone. A cell in a tissue senses when it is touching a neighbor cell and stops growing. One-celled organisms also communicate with each other. The two sexes of yeast cells use chemical signals to find one another and fuse. Certain disease-causing bacteria signal to each other to determine when their population size is sufficient to launch an attack (Figure 2.8).

In multicelled organisms, cells specialize

During the development of a multicellular organism from a single fertilized egg, cells grow, divide, and become specialized to perform certain tasks and not others (Figure 2.2). The cells organize themselves into different groups; each group is composed of cells committed to a certain function. The cell specialization process is known as **differentiation** (Figure 2.9).

Certain features are common to all cell processes

In addition to sharing a basic set of processes, cells carry out these processes in very similar ways. Here are a few of the overarching principles underlying and uniting all cell processes.

Cell processes require a constant supply of energy

It is clear from the discussion above that cells are very active and highly organized. They can do many different things simultaneously: break down mol-

Figure 2.9 A single fertilized egg divides thousands of times, and the descendent cells differentiate into many different cell types in the process of becoming a baby. **(A)** Four-cell human embryo. (Photograph courtesy of Michael Vernon, West Virginia Center for Reproductive Medicine.) **(B)** A newborn. (Photograph courtesy of Anastasia Lott.)

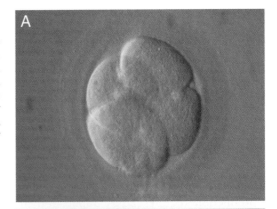

ecules, sense, respond, communicate, and grow, to name a few. In addition, they build their own intricate internal structures from many types of simple, building block molecules. They constantly ensure that their internal chemical compositions differ from that of the external environment.

Carrying out activities, building and rebuilding molecules and structures, and maintaining an internal environment that differs from its surroundings all require energy, and cells obtain it from a variety of sources. Virtually all plants, and a few other organisms, have cells that are able to capture the energy in sunlight and convert it to the type of energy found in food molecules. Others cannot utilize sunlight energy and must obtain energy from breaking down food molecules or from other types of chemical reactions.

Cell processes consist of a set of chemical reactions

Breaking down large food molecules into their component molecules and ultimately atoms provides an obvious example of a set of chemical reactions that constitute a cell process. But all of the cell processes described above can ultimately be reduced to a set of chemical reactions. Cells are tiny chemical factories that perform thousands of chemical reactions simultaneously. Through these chemical reactions, molecules inside the cell join together, break in two, and exchange atoms with each other.

The chemical reactions that occur in cells would occur very slowly, if at all, if we poured the cell's molecules into a test tube. To drive these reactions, cells manufacture protein catalysts known as **enzymes**. Each enzyme is responsible for making a specific chemical reaction occur (Figure 2.10). By manufacturing some enzymes but not others, cells can control which chemical reactions occur and which do not.

Cell processes occur in a series of small steps

The set of chemical reactions that constitute a certain cell process do not occur independently of each other but are linked into a sequence of consecutive reactions. Each small step in a cell process is usually carried out by a specific enzyme that carries out that step and no other. Structuring processes as a stepwise series of small, sequential changes, each requiring a specific enzyme, provides a cell with many points to control a process. A **pathway** is a process consisting of a series of steps leading from a starting material to an end product.

Cells regulate their processes

Cells carry out their life processes in an orderly, regulated way. They control which process occurs, when it occurs, and where it occurs. For example, cell reproduction requires DNA duplication and cell division into two equal

Figure 2.10 An enzyme catalyzes one specific chemical reaction, during which molecules fit together precisely.

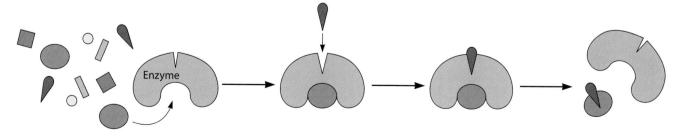

halves, each with enough parts and molecules to enable self-sufficiency. DNA duplication and cell division must be regulated to ensure that:

- they occur only after the cell reaches a certain size and has enough parts for two free-living cells
- DNA duplication happens before cell division happens
- the ratio of DNA duplication events to cell division events is one to one

You will learn more about cell division, how it is regulated, and how failure to regulate it properly can give rise to cancer in chapter 9. Other examples of regulation, such as of blood sugar levels, body water balance, and blood pressure, will also be presented as you learn more about how cells work. For now, the bottom line is this: *cells have a number of tactics for regulating their activities, and all involve the interaction of molecules, especially proteins, with each other.*

CELL ORGANIZATION

If all cell processes consist of nothing more than thousands of simultaneous chemical reactions between molecules, how are cells able to carry out their activities in such an organized and tightly regulated fashion? After all, you probably find it difficult to get—and stay—organized and schedule your activities, and you have a brain. We can trace the remarkable organization of cells to a few simple rules governing molecular interactions. Below, we focus on two of the most important rules and describe how adherence to these rules leads inevitably to a certain degree of organization.

The interaction of different molecules with water is key to cell organization

The degree to which molecules interact comfortably with water determines to a large extent how they are organized within a cell. Cells live in and are designed for a watery environment. Single-celled organisms usually live in water or in very moist environments. You may have read that your own body is 70% water; human bodies are typical of animal bodies in that regard. Animal body cells are bathed in watery body fluids, as are living plant cells. The internal environment of cells, too, is watery.

From your own experience, you know that some chemical substances interact readily with water and others do not. If asked what happens when you stir sugar into tea, coffee, or water, you would probably say the sugar dissolves in the watery fluid. But what does dissolving in water actually mean? The sugar granules dissolve as the sugar molecules interact with water molecules and become evenly dispersed throughout the watery fluid. Even though you can no longer see the sugar, you know it is still there because you can taste it. The same is true for salt or vinegar.

But what happens if you pour cooking oil into water? The cooking oil does not disperse evenly throughout the water but coalesces into a single unit as the oil molecules congregate together. If you stir the oil-water, you can break the oil up into small droplets, but as soon as you stop stirring, the droplets will come back together into one large drop of oil.

Hydrophilic and hydrophobic molecules self-organize

Sugar and cooking oil illustrate two fundamentally different types of interaction with water. Sugar is **hydrophilic** (hydro, water; philic, loving); it dissolves easily in water and disperses evenly throughout a watery liquid. Oil is **hydrophobic** (phobic, fearing); it does not dissolve in water. The interaction between water molecules and oil molecules is unfavorable, so the water molecules keep to themselves and the oil molecules keep to themselves.

Therefore, the first rule governing molecular interactions is that hydrophilic molecules dissolve in water and hydrophobic molecules do not. As a result, when hydrophobic and hydrophilic molecules are mixed together, they spontaneously segregate themselves into two groups, each composed of the same type of molecules.

This tendency for hydrophilic molecules, like sugar, salt, and water, to congregate together and for hydrophobic molecules, like oils, to congregate together is an organizing force inside cells. You can see in the oil-water example that the oil and the water create two compartments inside their container: an oily, hydrophobic compartment and a watery, hydrophilic compartment (Figure 2.11A). If you stirred some salt into the oil-water mixture, all the salt would dissolve in the water. If you added a different oil, it would mix into the first oil. In fact, you can go home and try this. If you start with light-colored canola oil in water and add greenish olive oil, you will see that the olive oil mixes completely with the canola oil and the oil mixture stays separate from the water. Stir in some salt quite thoroughly and then taste the oil; it will not be salty (Figure 2.11B and C).

Molecules can have both hydrophilic and hydrophobic regions

We need to add one more piece of information to the picture before moving on to the second principle underlying molecular interactions. In the examples above, we designated entire molecules either hydrophobic or hydrophilic. While that is true for some molecules, like sugar or oil, it doesn't have to be. Some molecules have parts that are hydrophobic and parts that are hydrophilic. These molecules follow the same pattern as do oil and sugar: the hydrophobic parts congregate (if possible) with other hydrophobic molecules (or parts of molecules), and the hydrophilic parts congregate (if possible) with other hydrophilic molecules or parts of molecules. Figure 2.12 illustrates several ways such molecules might arrange themselves.

Figure 2.11 Hydrophobic and hydrophilic molecules tend to congregate and form compartments. **(A)** Oil, a hydrophobic substance, and water form two separate compartments. **(B)** A hydrophilic substance like salt will dissolve in the water but not in the hydrophobic oil. **(C)** A hydrophobic substance will dissolve in the oil but not in the water.

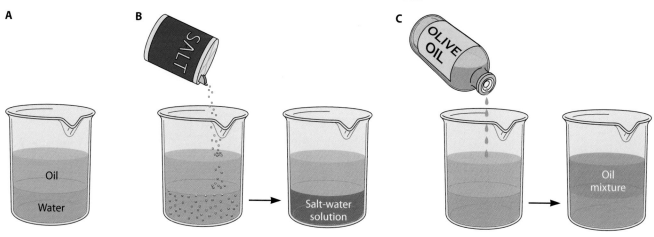

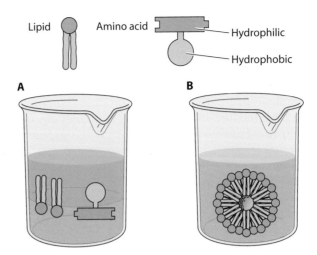

Figure 2.12 Some molecules have both hydrophobic and hydrophilic parts. **(A)** The most favorable arrangement for them is to keep the hydrophobic parts in hydrophobic environments and the hydrophilic parts in hydrophilic environments. **(B)** Here, the molecules form a sphere with the hydrophobic parts on the inside. You are looking at a cross section.

Molecules have distinct shapes and must fit together precisely to react with each other

In Figures 2.10 to 2.12, we have drawn molecules as indistinct blobs, color coded for their hydrophobic or hydrophilic nature. In reality, molecular structure has very fine detail. An individual chemical bond holds two atoms together in a specific geometry. If you connect many atoms via bonds of specific geometry, you can get very complex shapes indeed. Molecules have overall geometry, and their surfaces may have protrusions, nooks, pockets, and crannies (Figure 2.13). These variable shapes are key to organizing molecular interactions and, through these interactions, cell processes.

In cells, molecules make things happen by fitting together. Scientists refer to molecules fitting together as **binding**. When two molecules have shapes that fit together precisely (think of a lock and key), a chemical reaction might occur, a signal might be created, or one of the molecules might be moved through the cell membrane, to give a few examples. The protein molecules that drive cell processes are so specific and intricately shaped that most can bind to only one molecule. As a result, even though a molecule may collide

Figure 2.13 Three-dimensional structure of the protein flavodoxin. (Structure courtesy of Antonio Romero and Javier Sancho.)

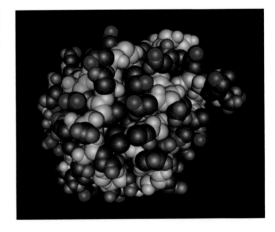

with thousands of other molecules in a cell, a chemical reaction occurs only when it binds to a molecule with the correct shape.

Fitting together depends not only on shape but also on the chemical natures of the parts, especially whether the molecules are electrically neutral or have areas that are partially charged. For example, a hydrophobic pocket on one molecule would have to fit with a hydrophobic protrusion of the right size and shape. A correctly shaped hydrophilic protrusion would not be compatible with a hydrophobic pocket.

Remember this principle of molecular interactions requiring molecules with complementary shapes and compatible chemistries as you read through this book. You will encounter many examples of molecules binding or not binding with each other in the sections and chapters to come. Virtually all healthy cell processes depend on the right molecules binding together.

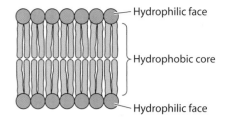

Figure 2.14 Cell membranes are a two-molecule-thick layer with a hydrophobic core and hydrophilic faces. The cell membrane surrounds the cell like the membrane of a balloon. A cross section is shown.

Cellular membranes provide structural features that help cells organize their activities

Cells are separated from their environments by membranes, and cells are divided into subcompartments by additional membranes. Membranes are obviously a crucial structural element of cells. Cellular membranes are made of molecules with one hydrophobic end (a long tail) and one hydrophilic end, arranged as shown in Figure 2.14. When placed in a watery environment, these bipolar molecules adhere to the molecular interaction rule described above and spontaneously orient themselves so that the hydrophobic parts align with each other and avoid water while the hydrophilic components face the watery environments inside and outside of the cell. This two-layer arrangement creates a hydrophobic barrier between the watery material inside the cell, the **cytoplasm**, and the watery external environment (Figure 2.15).

Cellular membranes contribute to the highly organized and tightly controlled activities we observe in cells in a number of ways. First of all, cells maintain themselves as distinct units, each with a cellular membrane, the **plasma** or **cell membrane**, that delineates the inside and outside of the cell. As discussed below, this membrane controls which molecules enter and leave the cell.

Most cells also use a series of internal membranes to create small internal compartments and to cordon off essential structural elements within the cytoplasm, such as the cell **nucleus**, which houses the cell's genetic material.

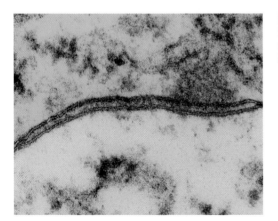

Figure 2.15 Micrograph of cell membrane. (Photograph copyright Dennis Kunkel Microscopy, Inc.)

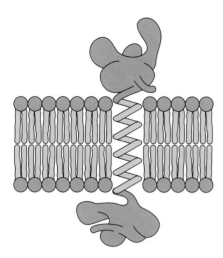

Figure 2.16 A protein embedded in a membrane. The portion of the protein inside the hydrophobic core of the membrane is also hydrophobic, while the external portions are hydrophilic.

Like the plasma membrane, the internal membranes help keep things organized by determining which molecules enter and exit the structure or internal compartment they surround.

This intracellular compartmentalization of certain molecules in specific locations helps cells keep all of the steps in a chemical reaction pathway together. In fact, often the enzymes involved in a pathway are physically lined up along a membrane in the correct sequence. These internal membranes also create compartments that separate chemical reaction pathways from each other, which is essential for pathways that share certain molecules.

In addition, an extensive network of internal membranes creates interconnecting sacs, tubules, and channels. Two of these sets of membrane-bound intracellular spaces, the endoplasmic reticulum and the Golgi apparatus, sort and process the molecules they manufacture and efficiently transport the correct molecule to the appropriate location (see Figure 2.17).

Cellular membranes control which molecules enter and leave the space or structure they surround

The contribution of cell membranes to cellular organization extends beyond the structural organization they provide. Both the plasma and internal membranes are actively involved in selecting which molecules will be allowed to pass through. Membranes make these "decisions" using the two principles of molecular interaction described above: hydrophobic-hydrophilic interactions and precise binding of one molecule to another.

Only a few molecules can move easily through the hydrophobic barrier the cell membrane establishes between the cell and its environment. Those that can are small, electrically neutral molecules, such as carbon dioxide and oxygen, and small, very slightly charged molecules, such as water. However, many of the biological molecules essential to cell processes are large molecules with partially charged regions, such as glucose, or small ions that are highly charged, such as calcium and potassium. Because many cell processes depend on these types of molecules, cells must have developed some way to get these molecules across the hydrophobic barriers that cell membranes create.

In Figure 2.14, you can see that the cell membrane consists of two layers of molecules, each with a hydrophobic and a hydrophilic end. Embedded in that hydrophobic bilayer are various protein molecules with portions that are hydrophobic and portions that are hydrophilic (Figure 2.16). These proteins serve a wide variety of functions that will be discussed in later chapters. In every case, however, the proteins bind and interact with only one or a very few molecules. This extraordinary specificity allows cells to control very precisely the molecules that enter and leave the cell or move into or out of a membrane-bound compartment within the cell. How various substances get into and out of cells will be discussed in detail in chapter 7.

TWO FUNDAMENTAL CELL TYPES

Up until now, this chapter introducing you to cells has focused on the remarkable similarity of cells, whether the cell is a one-celled organism, such as a bacterium, or a highly specialized cell in a multicellular organism. All cells share certain basic features: they have molecular machinery for duplicating DNA and breaking down and synthesizing molecules, they reproduce by dividing in two, they use the same molecular building blocks, and they are enclosed by a hydrophobic membrane that separates the cell from its surroundings.

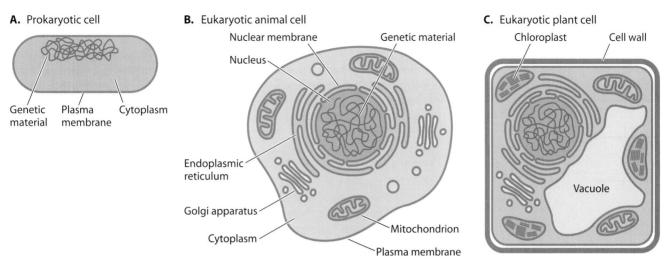

A. Prokaryotic cell

Genetic material   Plasma membrane   Cytoplasm

B. Eukaryotic animal cell

Nuclear membrane   Genetic material
Nucleus
Endoplasmic reticulum
Golgi apparatus
Cytoplasm
Mitochondrion
Plasma membrane

C. Eukaryotic plant cell

Chloroplast   Cell wall
Vacuole

Figure 2.17 Typical prokaryotic and eukaryotic cells. These simplified representations incorporate features from typical cells, but there is great variety in cell shape and content (for examples, see Figure 2.2). **(A)** A prokaryotic cell has no internal membranes. **(B)** Animal and other eukaryotic cells contain many membrane-bound organelles that are the sites at which specific processes occur. Here, we show some of them. **(C)** Plant cells are surrounded by supportive walls. In addition to the organelles found in animal cells, they contain membrane-bound chloroplasts, where solar energy is converted to chemical energy, and a storage compartment called a vacuole.

But only certain cells have the sort of internal compartmentalization we described in the previous section. This distinction, visible with a powerful microscope, led biologists to divide cells into two different types, **prokaryotic** and **eukaryotic** cells (Figure 2.17). The organisms made of these two cell types are called prokaryotes and eukaryotes. As it turns out, the differences made visible with sophisticated microscopy are mirrored in the invisible details of the molecular biology underlying many cell processes. Prokaryotes and eukaryotes carry out the same cell processes, but they often do so in very different ways.

Some of the differences between prokaryotes and eukaryotes play significant roles in biotechnology, so you will need to remember which organisms are prokaryotes and which are eukaryotes, which is actually quite easy.

Prokaryotic cells are quite simple

Prokaryotic cells are much simpler and usually significantly smaller than eukaryotic cells. As described above, the most obvious difference between prokaryotic and eukaryotic cells is the absence of internal membranes in prokaryotes. The lack of a nuclear membrane surrounding the genetic material is the origin of the name "prokaryote" (pro, before; karyon, kernel or nucleus). Because most prokaryotic cells are so small, they are able to function efficiently without subcellular compartments.

All prokaryotes are single-celled organisms, but not all single-celled organisms are prokaryotes. Some prokaryotes may live in colonies, but there are no true multicellular prokaryotes in which cells become specialized for certain tasks.

The two major groups of prokaryotes are eubacteria and archaea. Eubacteria include the familiar bacteria, such as *Escherichia coli*, and blue-green algae (sometimes called cyanobacteria). Archaea (also called archaebacteria) are a fascinating group of organisms that often inhabit extreme environments such as very hot springs or deep ocean vents. They are genetically and biochemically similar to one another but are as different from eubacteria as plants and animals are. When we talk about prokaryotes or bacteria in this book, we are referring to eubacteria.

Because bacteria grow rapidly, are easy to grow in large quantities, and have much less genetic material than eukaryotic cells, they are a favorite tool of biotechnologists.

Eukaryotic cells are larger and more complex

Eukaryotic cells are present in most organisms with which you are familiar, such as plants, animals, and fungi. While most plants, animals, and fungi are multicellular organisms that contain many specialized tissues and cell types, some eukaryotes are single-celled organisms such as yeast, green algae, or amoebae.

Like prokaryotes, eukaryotic cells are surrounded by a hydrophobic membrane, but unlike prokaryotes, they also have internal membranes. These internal membranes divide their cytoplasm into specialized compartments collectively known as **organelles**. Organelles perform specific functions within eukaryotic cells. You have already learned about two organelles found in eukaryotic cells, the endoplasmic reticulum and the Golgi apparatus. All eukaryotic cells also contain organelles called **mitochondria** that play a crucial role in energy production (Figure 2.17B). In plant cells, organelles called

---

BOX 2.1 *Viruses*

Viruses cause a number of familiar infectious diseases, such as influenza, chicken pox, and AIDS, but viruses are very different from other disease-causing organisms. Viruses are not cells. Most viruses consist of nothing more than genetic material enclosed in a protein covering called a capsid (figure). The molecules of which the capsid are made interact through their molecular

A virus, in its simplest form, consists of genetic material enclosed in a protein capsid.

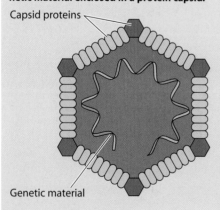

Capsid proteins

Genetic material

shapes and chemical natures with the surfaces of cells, and by so doing, allow the viral material to enter the host cell. The viral genetic material then takes over the host's molecular machinery to produce new virus particles. Because viruses are not capable of independent reproduction or metabolism, they are not considered to be alive per se.

Viruses infect all types of cells, both prokaryotic and eukaryotic, but most viruses are extremely specific in the host cells they are able to infect. Not only does a specific virus usually infect only one type of organism (such as humans or cats), viruses are also limited to certain cell types within the organism. For example, a "cold virus" infects only the cells of the upper respiratory tract, while a "stomach virus" infects only the cells of the digestive tract. However, a few viruses, such as rabies virus, normally infect several different host species, and once in a while, a virus acquires the ability to infect a new host.

When physicians are certain a patient has been infected by a virus, they will

not, and should not, prescribe antibiotics, because antibiotics are ineffective against viruses. Antibiotics specifically target various components of the prokaryotic biochemical machinery found in bacteria so that the antibiotics don't harm your eukaryotic cells. Antibiotics don't work against viral infections because the virus is using your own eukaryotic machinery against you. Taking antibiotics to treat a viral infection is not only ineffective, it is counterproductive. The antibiotic will not kill the virus that is making you ill, but it will contribute to antibiotic resistance in bacteria, which we discuss in detail in chapter 13.

In addition to viruses that infect humans or other animals, there are plant viruses specific to certain plant tissues and bacterial viruses specific to certain bacterial strains. Bacterial viruses have played an important role in the development of molecular biology and have been given a special name: **bacteriophages.**

chloroplasts are responsible for capturing the energy in sunlight and converting it to chemical energy.

Trying to understand minute differences in the molecular mechanisms of cell processes in prokaryotes and eukaryotes may seem like an arcane scientific exercise, but these differences are very important in biotechnology. Sometimes the differences can be exploited to accomplish a certain goal. For example, antibiotics kill bacteria without harming your own cells because they are targeted quite specifically to prokaryotic molecules and mechanisms (Box 2.1). At other times, the differences between prokaryotic and eukaryotic cells complicate matters, forcing researchers to work harder (and spend more money) to reach their goal.

SUMMARY POINTS

All living things are made of cells. Cells break down and manufacture molecules, reproduce, maintain an internal environment different from the external environment, receive and respond to environmental signals, regulate their activities, and obtain and use energy.

Cells carry out their processes through chemical reactions that occur in many small steps. Each step is mediated by a protein catalyst called an enzyme.

All cells are made of the same kinds of molecules and are organized in essentially the same way. Cells contain four kinds of molecules that are found only in living things: proteins, fats, carbohydrates, and nucleic acids.

Plant cells and certain others can obtain energy from sunlight. Animal cells and other cells get energy from breaking down large molecules. Pieces of these large molecules can also be used to synthesize molecules and structures the cell needs.

One of the principles underlying cellular organization is that hydrophobic and hydrophilic substances self-segregate. The tendency of hydrophobic substances to separate themselves from water allows the creation of compartments inside cells.

Hydrophobic membranes separate the cytoplasm of the cell from the external environment, and other membranes create subcompartments inside cells.

Activities inside cells take place as a result of the precise fitting together of molecules as a function of their shapes and chemical natures. Most cellular processes are carried out by proteins.

There are two major classes of cells: prokaryotic, which have outer membranes but no internal membranes, and eukaryotic, which have outer membranes but also many internal membranes that define compartments called organelles. Organelles are associated with specific functions, such as the nucleus with information storage, the chloroplasts with photosynthesis, and the mitochondria with energy production.

Viruses are not cells and do not carry out the life processes that cells do. Viruses consist of genetic material packaged in a capsule and can reproduce only by infecting a host cell. The infection process involves precise molecular interactions between the virus capsule and the cell surface, and viruses are therefore very specific in terms of the kinds of cells they can infect.

KEY TERMS

Bacteriophage	**Cytoplasm**	**Hydrophobic**	**Organelle**
Binding	**Differentiation**	**Metabolism**	**Pathway**
Cell	**Enzyme**	**Mitochondrion (plural,**	**Plasma or cell membrane**
Cell cycle	**Eukaryotic**	**mitochondria)**	**Prokaryotic**
Chloroplast	**Hydrophilic**	**Nucleus**	**Virus**

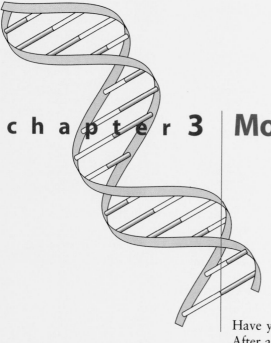

chapter 3 | Molecular Components of Cells

Have you ever wondered why you cannot live on a diet of soil and rocks? After all, there is a lot more of that kind of material on earth than there are plants and animals. Part of the answer is that your cells don't contain enzymes that catalyze reactions between the chemical components of soil and rocks. But even if you did have those enzymes, you'd still have a problem. Soil and rocks are largely composed of different elements than you are, so you could not easily retrieve from soil and rocks the elements you need to build muscle, blood, or other tissues.

The chemistry of living things (often called biochemistry) is different from that of nonliving things. Living things are largely composed of just four elements, carbon, hydrogen, oxygen, and nitrogen, with everything from a scoop to a sprinkling of many other elements added in. As earlier scientists studied cells, many of them separated the cells into specific chemical components to learn what kinds of substances they are made of. They discovered that cells from all kinds of organisms, bacteria to elephants, are chemically quite similar. Cells contain many small molecules and ions, but their major structural and functional components are formed by four types of large molecules unique to living organisms:

- lipids
- carbohydrates
- proteins
- nucleic acids (**DNA**, short for **deoxyribonucleic acid**, and **RNA**, short for **ribonucleic acid**)

This chapter introduces you to these molecules. We begin by laying some groundwork for discussing and drawing molecules, including defining some terms we've already been using (such as molecule). Having done that, we will look at the four major classes of large molecules and the roles they play in cellular structure and function. In later chapters, you will encounter them again and again as we look at how cells get energy, reproduce, control their environments, and communicate.

ATOMS, IONS, AND MOLECULES

As we said above, the most common elements in biological molecules are carbon, hydrogen, nitrogen, and oxygen. An **element** is a substance that cannot be broken down into anything else—it is a pure ingredient. For example, gold is gold; it is not made up of anything else. Gold is an element. Table salt, on the other hand, is made up of sodium and chlorine and is also called sodium chloride. Likewise, water is composed of hydrogen and oxygen (the familiar formula H_2O). Water and table salt are not elements.

Elements are composed of atoms

The smallest piece of an element still recognizable as that element is an **atom**. An atom of a given element is made up of a specific number of protons, some electrons, and some neutrons, which we will ignore in this discussion (Table 3.1). Protons carry a positive charge, and electrons carry a negative charge. In general, an atom isolated from other atoms carries no charge because it has equal numbers of protons and electrons.

Electrically charged atoms are called ions

If an atom gains an electron, it becomes negatively charged; if an atom loses an electron, it carries a positive charge (Table 3.2). An atom that is charged because the numbers of its protons and electrons are not equal is called an **ion**. An atom of a given element usually loses or gains a specific number of electrons and thus always forms an ion of the same charge.

Ions are very important in biology. Their movement moves the electrical charges they carry, so ion movement can constitute an electrical impulse. You may have heard that nerve impulses are electrical; the means by which a nerve impulse is generated is through the concerted movement of ions (chapter 7). A substance composed of a positively charged ion and a negatively charged ion (or ions) is called a **salt**, and when you hear that a salt is important to maintaining human health, what that really means is that the ions making up that salt are important. We will look at how the body regulates its salt and water balance in chapter 8.

Atoms bond together by sharing electrons, forming molecules

Atoms can form **bonds** (also called **chemical bonds**) with each other by sharing electrons. The term **molecule**, which we have been using freely, refers to two or more atoms bonded together. Atoms of individual elements form specific numbers of bonds. Each hydrogen atom can form one bond, carbon forms four, nitrogen usually forms three, and oxygen forms two.

Table 3.1 Different elements are made of atoms with various numbers of protons

Element	No. of protons (+ charge)	No. of electrons (− charge)
Hydrogen	1	1
Carbon	6	6
Nitrogen	7	7
Oxygen	8	8

Table 3.2 Some biologically important ions

Ion	No. of protons	No. of electrons	Charge
Calcium	20	18	+2
Sodium	11	10	+1
Potassium	19	18	+1
Chlorine (chloride ion)	17	18	−1

In the last chapter, we discussed the importance of the hydrophobic or hydrophilic nature of molecules to the organization of cells. Molecules are hydrophobic, hydrophilic, or some of each, because of the characteristics of the bonds between their atoms, which are a function of the identity of the atoms participating in the bonds. We will come back to this subject in chapter 5 when we look in detail at the structure of proteins.

Drawing biological molecules

Throughout this book, we will be depicting biological molecules in figures so that you can see important features or look at changes they undergo during biological processes. Sometimes, when the individual atoms in the molecule aren't particularly important, we will portray molecules as blobs. However, at other times we will want to emphasize or at least acknowledge the individual atoms that comprise a molecule.

When we want to show the atoms that comprise biological molecules, we will show the molecules as ball-and-stick figures, usually two dimensional. In these figures, we will use the chemical symbol of the element (which conveniently enough happens to be the first letter of the names of hydrogen, carbon, nitrogen, and oxygen) with lines representing the bonds it is forming. Table 3.3 lists the chemical abbreviations of elements you will encounter in this book. Some examples of small molecules are shown in Figure 3.1, drawn in ball-and-stick form.

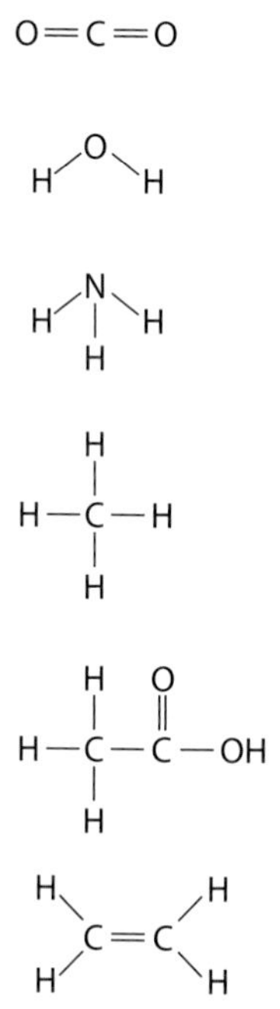

Table 3.3 Chemical symbols for biologically important elements

Element	Symbol[a]
Hydrogen	H
Carbon	C
Nitrogen	N
Oxygen	O
Sulfur	S
Phosphorus	P
Sodium	Na
Potassium	K
Chlorine	Cl
Calcium	Ca

[a]The symbols for sodium (Na) and potassium (K) come from their Latin names, natrium and kalium.

O=C=O   Carbon dioxide (CO_2)

Water (H_2O)

Ammonia (NH_3)

Methane (CH_4)

Acetic acid (vinegar; CH_3COOH)

Ethylene (C_2H_4)

Figure 3.1 Examples of small molecules drawn in ball-and-stick style.

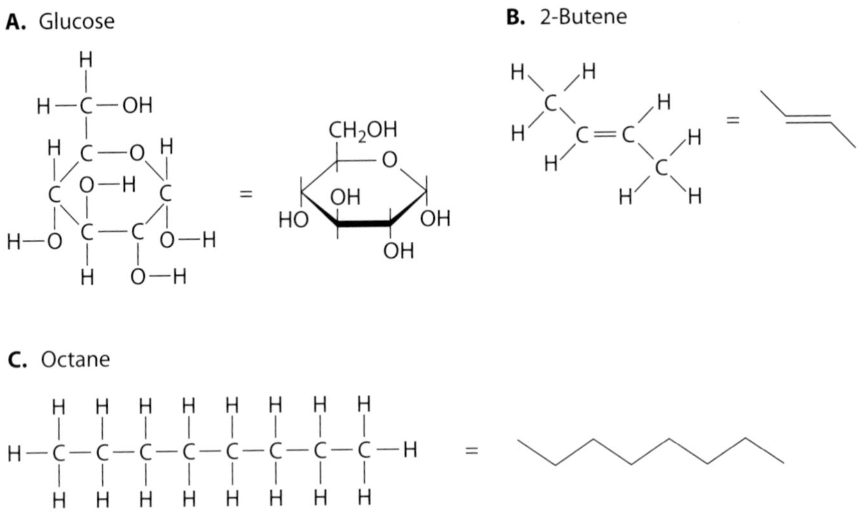

A. Glucose

B. 2-Butene

C. Octane

Figure 3.2 Skeleton representations of molecules. Unlabeled corners are carbon atoms. Hydrogen atoms bound to carbon are usually not shown and are assumed to be present.

When larger molecules need to be represented, drawing all the individual carbon and hydrogen molecules can get very messy and can actually interfere with getting the point across. In these situations, we may use an abbreviated system in which carbons are understood to be at the corners of a skeleton and hydrogens are understood to be filling the rest of the carbon's possible bonds. If another type of atom (say, oxygen) is present, it will be shown specifically (Figure 3.2).

Now that we've explained the various ways that we are going to present molecules visually, let's look at the four classes of biological molecules. As you read the following descriptions of these large molecules, you'll notice that many of them are made of small repeating units (Table 3.4). This is an important theme, because when organisms eat, they break the big molecules down into their repeating units, which can be recycled into new big molecules or broken down further for energy (see Figure 2.3).

LIPIDS

Lipids are a diverse collection of biological molecules that share the property of not being soluble (capable of dissolving) in water. Lipids are hydrophobic: fats, oils, cholesterol, and related molecules. Another characteristic of lipids is that their chemical bonds, which are almost solely carbon-hydrogen and carbon-carbon bonds, contain a lot of energy. That means lipids are a good energy storage molecule; a great deal of energy can be packed into a single molecule.

Table 3.4 Subunits of biological molecules

Class of molecule	Examples	Smallest repeating unit
Lipid	Fats, oils	Glycerol, fatty acids
Carbohydrate	Sugars, starch, cellulose	Simple sugars
Nucleic acid	DNA, RNA	Nucleotides
Protein	Enzymes	Amino acids

Animals use excess energy to manufacture lipids and then store those lipids until the energy is needed. Some plants, too, store energy in the form of lipids, usually in seeds to provide energy for the new plant (think of sunflower seeds, walnuts, and almonds, all high in oils). If an animal or a developing plant needs energy, it can break down the lipids and release the energy stored in their chemical bonds. Lipids offer organisms a way to pack a lot of energy into a small space. The flip side of this, as any dieter knows, is that you have to expend lots of energy to burn up a single pound of unwanted fat.

A modified form of lipids plays the starring role in cellular membranes. These modified lipids have a hydrophilic chemical group attached to them, creating bipolar molecules that organize themselves into membranes as described in chapter 2. Lipids are thus essential structural components of cells, in addition to being energy storage molecules. Let us now look at some specific types of lipids.

Fats are made from glycerol and fatty acids

Fats and oils have a common structure. They are made up of glycerol, a three-carbon molecule, combined with **fatty acids** (Figure 3.3). Fatty acids are long chains of carbon and hydrogen with an acidic group at one end. The acidic groups of three fatty acids react with the glycerol to form a **fat**. If the fat is liquid at room temperature, it is called an oil.

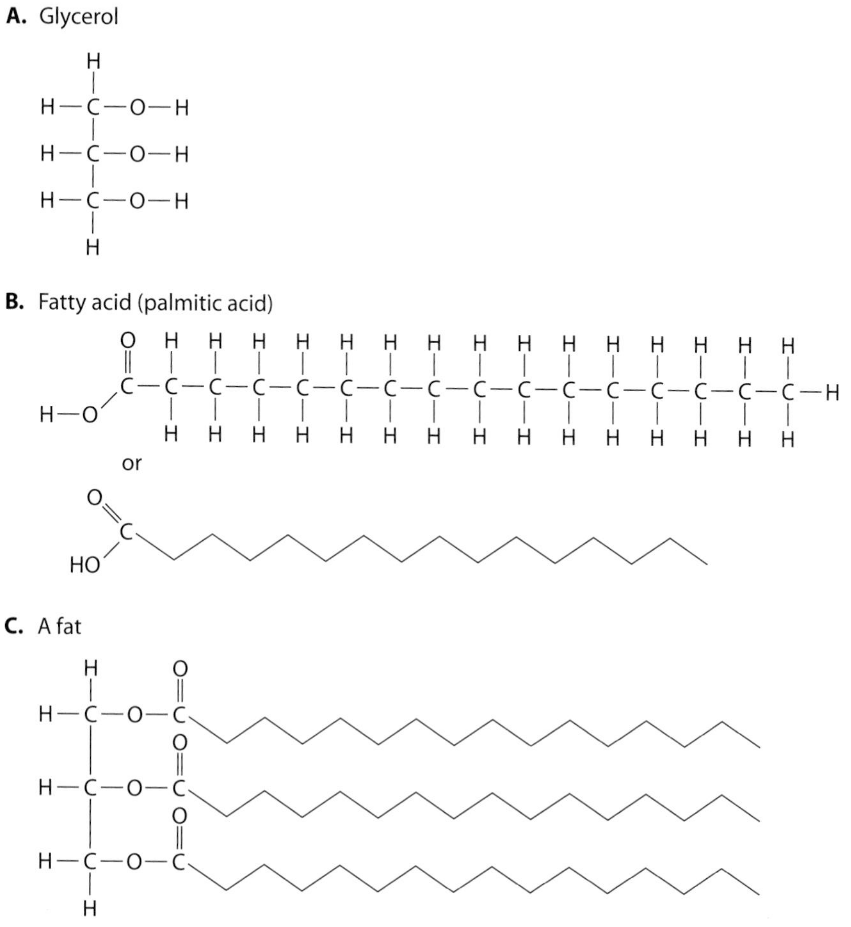

A. Glycerol

B. Fatty acid (palmitic acid)

C. A fat

Figure 3.3 Fats and oils are composed of a glycerol molecule joined to three fatty acids.

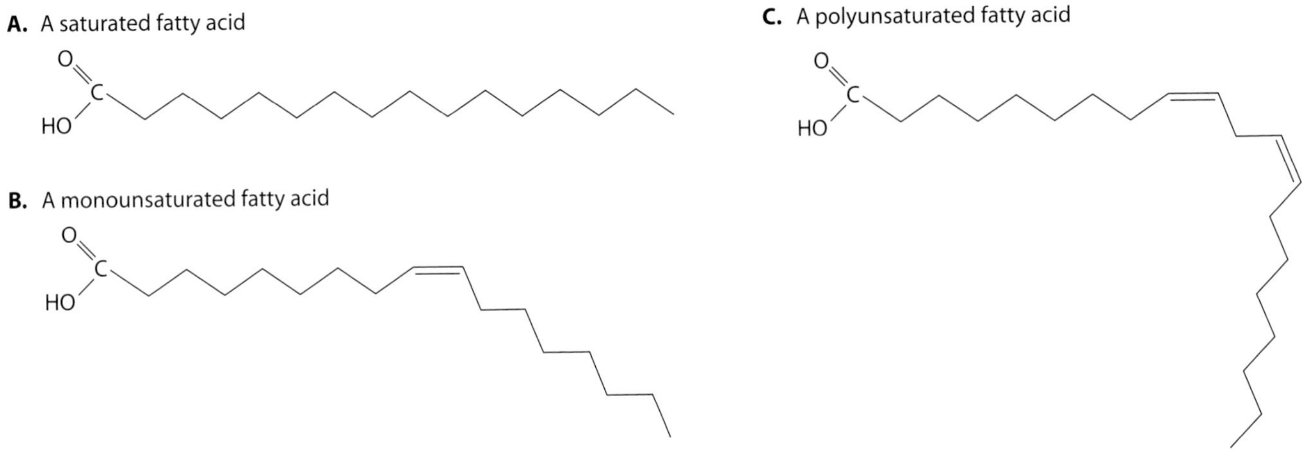

A. A saturated fatty acid

C. A polyunsaturated fatty acid

B. A monounsaturated fatty acid

Figure 3.4 Saturated and unsaturated fatty acids.

You may have heard of fatty acids, and if you have, you probably know that there are saturated and unsaturated fatty acids (which form saturated and unsaturated fats, respectively). A saturated fatty acid contains all the hydrogen atoms it can possibly hold, as shown in Figure 3.4. An unsaturated fatty acid, in contrast, does not. Instead, it has a so-called double bond between two of its carbon atoms. A fatty acid can have more than one double bond, in which case it is called a polyunsaturated fatty acid.

Saturated fatty acids form straight chains because of the geometry of the bonds between the carbon and hydrogen atoms. In the fat shown in Figure 3.3 (a saturated fat), the straight saturated fatty acid chains pack neatly side by side. Because of this physical arrangement, saturated fats are solid at room temperature. Look at the unsaturated fatty acid in Figure 3.4 and note how the double bond changes the geometry of the chain (the distortion is real and not just how we happened to draw the picture). Unsaturated fatty acid chains don't pack together neatly like saturated fatty acids, and the more double bonds a fat has, the more likely it is to be a liquid at room temperature.

The major components of cell membranes are phospholipids

In the last chapter, we stated that cellular membranes are made of a bilayer of molecules with one hydrophobic end and one hydrophilic end. These molecules are **phospholipids**. Most of these phospholipids are like fats or oils in that they have a glycerol backbone with fatty acids attached, but instead of three fatty acids as in fats and oils, they have only two. The third carbon of the glycerol backbone is instead bonded to a charged phosphate group, which may be bonded to an additional small polar molecule (Figure 3.5).

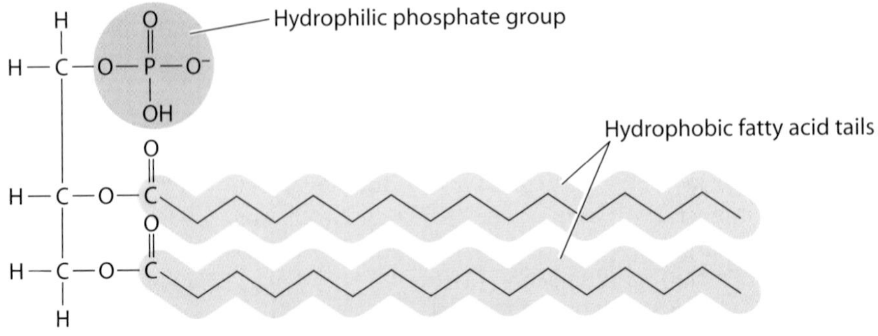

Figure 3.5 A phospholipid. Phospholipids are major components of membranes. The charged phosphate group is hydrophilic, while the long fatty acid tails are hydrophobic.

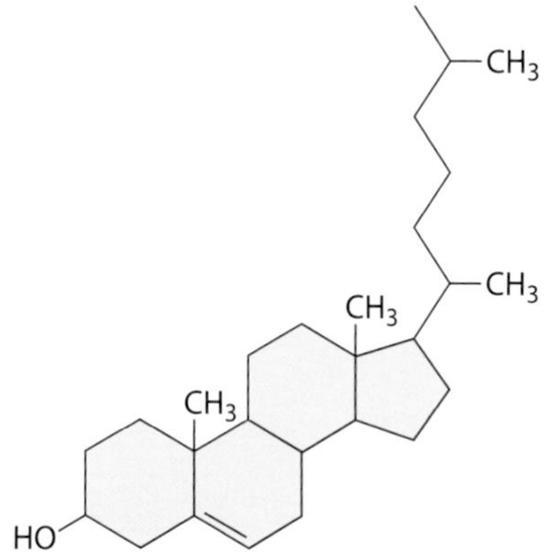

Figure 3.6 Cholesterol, a sterol lipid, is composed almost entirely of carbon and hydrogen.

This structure provides long hydrophobic tails from the fatty acids and a polar head group from the phosphate and attached polar molecule. In a membrane, the fatty acid tails point toward the interior while the polar head groups point outward to the extracellular fluid or the cytoplasm.

Sterols have a completely different structure

Another important class of biological lipids is the **sterols**. These lipids are made of ring-shaped assemblies of carbon with associated hydrogens. Figure 3.6 shows cholesterol, an example of this group. While we hear a lot of negative publicity about cholesterol, it is actually an essential part of your body. Cholesterol is a component of animal cell membranes, including yours, and if you don't eat enough, your body will manufacture it. Plant cell membranes don't contain cholesterol, which is why food that comes from plants is cholesterol free, but they do make other types of sterols for their membranes. Cholesterol helps keep the fluidity of your cell membranes at the right level, and it is also the starting material for synthesis of many hormones and bile, a substance essential for the digestion of fats. The hormones synthesized from cholesterol are called **steroid hormones**. Among them are estrogen, testosterone, and cortisone (Figure 3.7).

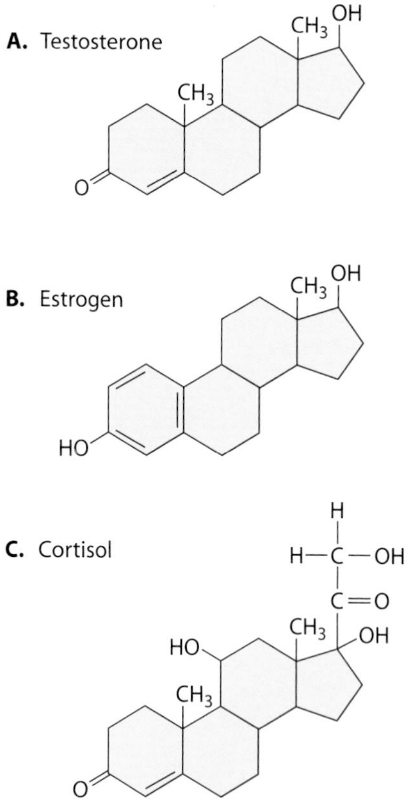

CARBOHYDRATES

Carbohydrates are defined by their overall chemical formulas: they contain a ratio of one carbon atom to two hydrogen atoms to one oxygen atom (C, H_2, O). Nineteenth century chemists recognized that this composition was the same thing as one carbon atom and one molecule of water and coined the term carbohydrates to describe this class of chemicals. Examples of carbohydrates are sugars, starches, and related molecules.

The smallest carbohydrates are the simple sugars. Simple sugars can be chemically hooked together to form additional carbohydrate molecules. For example, one molecule each of the simple sugars **glucose** and fructose can be combined to form our familiar table sugar, **sucrose** (Figure 3.8). One molecule of glucose and one molecule of a sugar called galactose combine to form the milk sugar **lactose**. A large number of glucose molecules can be connected to form pectin, starch, or cellulose, and other simple sugars are combined by organisms such as marine algae to make other large carbohydrates, like agar (a gelling agent) and carrageenan (a thickener; look for it in food ingredient labels on products such as ice cream or artificial whipped topping).

Figure 3.7 Examples of steroid hormones.

A. Testosterone

B. Estrogen

C. Cortisol

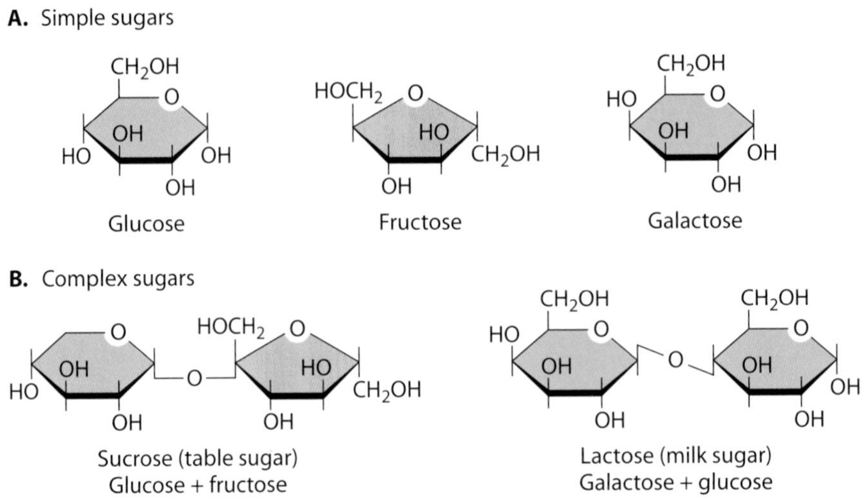

A. Simple sugars

Glucose   Fructose   Galactose

B. Complex sugars

Sucrose (table sugar)
Glucose + fructose

Lactose (milk sugar)
Galactose + glucose

Figure 3.8 Simple sugars can combine to make more complex sugars.

As you can see, carbohydrates range in size from small molecules to gigantic assemblies of thousands of these small molecules. The gigantic assemblies are often referred to collectively as **polysaccharides** (poly, many; saccharides, sugars), while those intermediate in size are called oligosaccharides (oligo, several). Carbohydrates are hydrophilic, though the large ones are so big that they do not dissolve readily in water.

Carbohydrates play key roles in energy metabolism

One important role of carbohydrates is in energy metabolism. Plants use energy from sunlight to synthesize the simple sugar glucose in the process known as **photosynthesis** (Figure 3.9). The plants then break down the glucose to recover the energy and use it for growth and other life processes. Plants store excess glucose by linking the glucose molecules together in a long chain that we call **starch** (Figure 3.10). When a plant needs to use the stored energy, it breaks the starch molecule back down into its component glucose molecules. Animals such as humans also use glucose for energy, but instead of making it, they obtain it by eating plants or other animals. Glucose is the nutrient your body circulates to cells for energy. If more glucose is available than is needed, animal cells can also connect glucose molecules together into a different form of starch called **glycogen** for storage, or they can break down the chemical bonds in the glucose and use the released energy to make lipids.

Plant cells also use glucose molecules to make **cellulose**, the strong structural material in plants. It is a major component of wood and is the substance that is extracted from wood to make paper. Like starch, cellulose is a chain of glucose molecules, but they are connected together in a different way.

Figure 3.9 Plants use the energy in sunlight to power the synthesis of glucose from carbon dioxide and water (photosynthesis).

$$6\,CO_2 + 6\,H_2O + energy \longrightarrow C_6H_{12}O_6 + 6\,O_2$$

Carbon + water + energy $\longrightarrow$ glucose + oxygen
dioxide

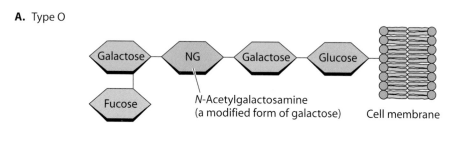

Figure 3.10 Plants store glucose as starch.

Carbohydrates can form part of molecular recognition molecules

Carbohydrates are often found connected to other molecules on the outsides of cells. In these positions, carbohydrates play roles in cellular recognition, cell signaling, and cell adhesion. In fact, the A, B, and O blood groups are defined by carbohydrates on the outside of red blood cells. One particular chain of sugars attached to the plasma membrane of the red blood cell is defined as type O. If one additional sugar, called *N*-acetylgalactosamine (a modified form of galactose), is attached to the O chain, it is designated type A; if simple galactose is attached to the O chain, it is designated type B (Figure 3.11). A person with type AB blood has both the A and B oligosaccharides on their cell surfaces. The presence of a foreign blood cell signal (such as type A blood cells in a type O person) activates the body's immune system, causing it to attack the type A cells and destroy them.

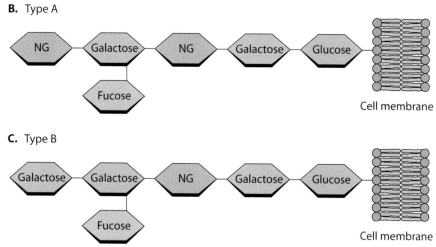

Figure 3.11 Carbohydrates on the outside of red blood cells determine A, B, O, and AB blood types.

PROTEINS

Thus far, we have seen that lipids and carbohydrates play roles as structural molecules, in creating membranes, as energy stores, and as recognition molecules. What is missing up to this point is action: molecules that recognize other molecules, that create other molecules, that break them down, and that make things happen. These molecules are the **proteins**.

Proteins supply much of an organism's structure and carry out its functions

It would be reasonable to say that proteins carry out nearly every function necessary for life. As you read in chapter 2, a class of proteins called enzymes makes and breaks chemical bonds, carrying out cellular chemical reactions as diverse as digesting nutrients, synthesizing fats, replicating DNA, and repairing damaged components. Proteins called receptors sit in the cell membrane and bind to specific signaling molecules in the extracellular environment. Binding of the specific signal to the receptor protein is analogous to turning a key in a lock; it triggers changes in the structure of the protein molecule. Those changes in turn signal other molecules on the inside of the cell, triggering yet other actions (Figure 3.12).

Proteins of the immune system called antibodies recognize which cells belong in the body and which do not. The binding of antibodies to foreign cells or other foreign matter signals the immune system to destroy the invaders. Transport proteins bind specifically to certain molecules and carry them to distant parts of the body, as hemoglobin carries oxygen. Some proteins have a structural role: hair and nails are made of proteins called keratins. Muscle tissue is composed of the proteins actin and myosin, which contract in response to signals from nerves. Thus, proteins provide the structure and carry out the functions of organisms, whether that organism is a goldfish, an oak tree, a fruit fly, a bacterium, a human being, or any other life form on earth.

It is not simply the nature of an organism's proteins that give the organism a particular identity. After all, the proteins in different animals can be pretty similar—muscle proteins, hemoglobin, enzymes to replicate DNA, and

Figure 3.12 A receptor protein binds to a specific chemical signal and undergoes a change. The change triggers further reactions inside the cell, transmitting the signal from outside to inside.

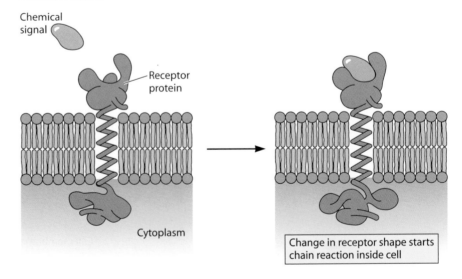

so on. The proteins of closely related organisms are very much alike, which is why diabetic humans who cannot make the protein insulin can be treated with insulin obtained from pigs. Even organisms as different as animals and plants have many types of proteins in common.

Much of the diversity in nature is due to the organization of proteins within an organism, particularly structural proteins and those that synthesize additional structural body components. In some ways, the organization of similar proteins into different body plans is analogous to using a pile of bricks in construction. You could assemble them into the wall of a building or into a barbecue pit, or even a sidewalk. That analogy is an oversimplification, of course, since proteins do differ between organisms, and the more distantly related the organisms, the more they differ. The organization of protein synthesis during the development of an organism leads to its unique form, whether it is a fish, a tree, or a human. An organism does what it does and looks like what it is because of the nature of its proteins and how they are organized during development.

Proteins are chains of amino acids

Proteins are chains of smaller molecules called **amino acids**. Amino acids have a common hydrophilic backbone and usually one of 20 different side chains (Figure 3.13). These side chains are quite variable in their shape and chemical nature; some are hydrophobic, and some are hydrophilic. Proteins are made by connecting from a few to thousands of these amino acids in various orders to make chains. For a simplistic analogy, think of making a chain of pop-beads from an assortment of 20 different kinds of colored beads. But a protein is more than a straight chain of beads. You must also imagine that the chain is folded and coiled into a specific three-dimensional shape (Figure 3.14).

A protein's function depends on its three-dimensional structure

Proteins perform their functions essentially by fitting specifically and precisely, or binding, to another molecule or molecules and causing something to happen. The "something" might be the breaking of a chemical bond, the forming of a chemical bond, the rearrangement of a different part of the protein itself,

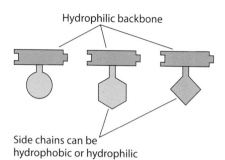

Figure 3.13 Amino acids have a hydrophilic backbone that can form chains and one of 20 different side chains that can be either hydrophobic or hydrophilic.

Figure 3.14 A protein is a chain of amino acids that folds into a specific three-dimensional shape.

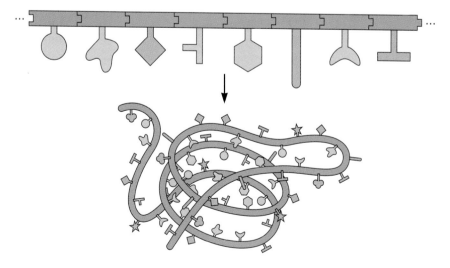

or a combination of these things. Through specific binding and subsequent actions, proteins carry out a remarkable variety of functions.

Picture a single receptor protein embedded in the outer membrane of a cell. Our imaginary receptor protein looks like an irregularly shaped glob with interesting nooks and crannies. The shapes of the nooks and crannies are absolutely critical to the protein's function: one cranny on the outside of the cell membrane is the place where a growth hormone molecule must fit exactly to signal the cell to grow. Other nooks and crannies on the inside are sites that fit precisely to other molecules for communication with the rest of the cell. Through interaction at these sites, the receptor protein can tell the cell that the hormone signal has arrived.

Our imaginary receptor illustrates a crucial point about protein function. *A protein depends on its ability to fit with or bind to other molecules (sometimes other proteins) to carry out its functions. That ability is determined by its three-dimensional structure.*

The three-dimensional structure is a consequence of amino acid sequence

What determines the three-dimensional structure of a protein? When amino acids are assembled into a protein chain, that chain immediately folds back upon itself to assume the most "comfortable," or energetically stable, shape. The most energetically stable shape is determined by the interactions of the individual amino acids that make up the protein. Therefore, the identities of the component amino acids and the order in which they occur in the chain govern the final three-dimensional structure of the protein. The order of the amino acids in the chain is thus extremely important to the function of the protein. As you can imagine, the possibilities for constructing different and unique protein chains are almost limitless. (Imagine how many unique chains of pop-beads you could make using 20 different colors of beads.) This variety is fortunate, considering the many and varied functions that proteins must perform. In fact, the forms and functions of proteins are so central to biology that chapter 5 is devoted to them.

Your body contains thousands of different proteins with different structures and functions. Two that illustrate some of the variety of these proteins are keratin and amylase. Amylase is a digestive enzyme found in your saliva and also secreted by your pancreas. It binds to starch molecules and breaks them down into their glucose components. Amylase is soluble in water (in saliva, too), because the amino acids on its surface are hydrophilic.

Keratin is the structural protein that makes up your hair and fingernails (in slightly different versions). The three-dimensional structure of keratin is fibrous, and the keratin fibers bind strongly to one another. Is keratin soluble in water? (Think about it—does your hair dissolve when you wash it?) The reason it is not is that, in contrast to amylase, the amino acids on the surface of the keratin fibers are hydrophobic. Since keratin fibers interact so strongly with each other, they don't dissolve in oil either, luckily for us.

NUCLEIC ACIDS

The genetic material of all life on this planet is made of DNA. When scientists first discovered DNA, the consensus was that it was far too simple a molecule to contain the information responsible for the extraordinary diversity of life forms on our planet. Yet it does, with elegant simplicity and effectiveness.

DNA stores genetic information in a remarkably simple structure

DNA is made of only six components. These components are a sugar molecule (deoxyribose), a phosphate group, and four different nitrogen-containing bases: adenine (A), guanine (G), cytosine (C), and thymine (T). The essential building block of the DNA molecule is called a **nucleotide**, which consists of a deoxyribose molecule with a phosphate attached at one place and one of the four bases attached at another (Figure 3.15). The carbon atoms of the deoxyribose sugar portion of a nucleotide are always numbered in the same way. The base is always attached to carbon 1, and the phosphate group is always attached to carbon 5.

Nucleotide terminology

Some nucleotide terminology is potentially confusing, and we want to clear it up before going any further. The bases themselves are called adenine, guanine, cytosine, and thymine. When they are incorporated into nucleotides, they are bound to sugar and phosphate groups, and these nucleotides have slightly different names to distinguish them from the free bases. The name of a nucleotide tells you the identity of the base and how many phosphate groups are present. If a nucleotide is incorporated into DNA, the part about the number of phosphates is unnecessary: it is one. However, when a nucleotide is free in the cell, enzymes can add and remove phosphate groups, and the molecule may have from one to three phosphate groups in a row. If the base in the nucleotide is adenine, these forms would be called adenosine monophosphate (AMP), adenosine diphosphate (ADP), and adenosine triphosphate (ATP) (Figure 3.16). For the other bases, substitute guanosine, cytidine, or thymidine for adenosine. Certain free nucleotides, particularly

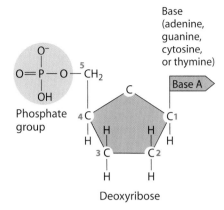

Figure 3.15 A nucleotide, the building block of DNA, is composed of deoxyribose sugar, a phosphate group, and one of four bases.

Figure 3.16 A free nucleotide can have one, two, or three phosphate groups.

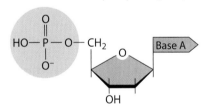

A. Adenosine monophosphate (AMP)

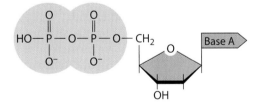

B. Adenosine diphosphate (ADP)

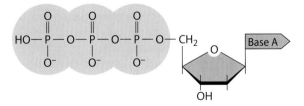

C. Adenosine triphosphate (ATP)

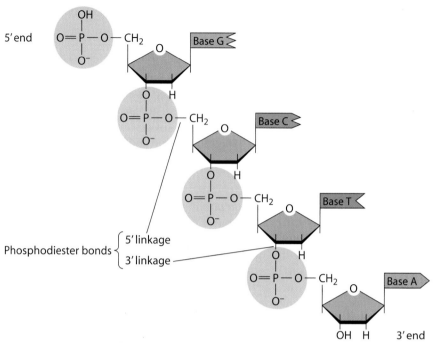

Figure 3.17 Nucleotides form chains through phosphodiester bonds between the phosphate group of the number 5 carbon of one nucleotide and the OH group on the number 3 carbon of another.

those containing A and G, play roles in cellular signaling and energy metabolism, and you'll encounter this terminology when we look at those cellular functions, but here we focus on them as components of nucleic acids.

Nucleotide chains

In a DNA molecule, thousands to millions of nucleotides are strung together in a chain by connecting the phosphate group on the number 5 carbon of one deoxyribose molecule to the number 3 carbon of a second deoxyribose molecule (a free water molecule is created in this process). Figure 3.17 shows an example in which three nucleotides are connected. Because the nucleotides are held together by bonds between their sugar and phosphate entities, DNA is often said to have a **sugar-phosphate backbone**. Notice that the ends of the molecule in Figure 3.17 are labeled 5' and 3' for the carbon atoms that would form the next links in the chain at either end.

Base pairing

The sugar-phosphate backbone of DNA is an important structural element, but it is the same all along the length of the DNA. The information DNA carries is contained in the four bases. The key to the transmission of genetic information lies in a characteristic of these bases: that adenine and thymine together form a stable chemical pair and cytosine and guanine form a second stable pair. The pairs are formed through weak chemical interactions called hydrogen bonds. These two pairs, adenine-thymine and cytosine-guanine, are called **complementary base pairs** (Figure 3.18).

In a DNA molecule, two sugar-phosphate backbones lie side by side, one arranged from the 5' end to the 3' end, and the opposite strand arranged from the 3' end to the 5' end. These oppositely oriented strands are described as **antiparallel**. The bases attached to one strand are paired with their part-

Figure 3.18 Complementary base pairs in DNA.

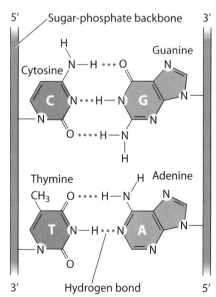

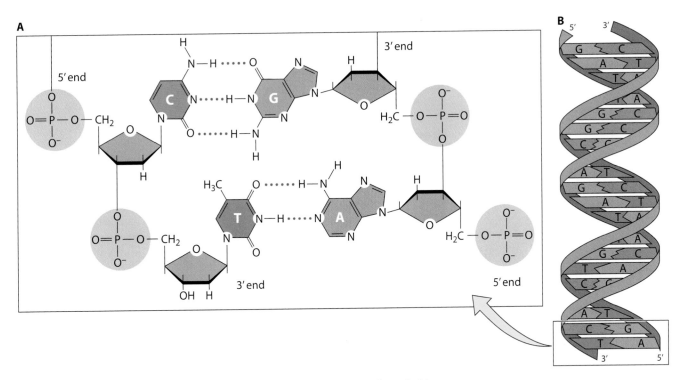

Figure 3.19 DNA structure. **(A)** In DNA, two antiparallel nucleotide strands are held together by complementary base pairing. **(B)** DNA is a double helix with the base pairs on the inside like rungs of a ladder and the sugar-phosphate backbones spiraling around the outside of the molecule.

ner bases attached to the opposite strand (Figure 3.19A). Thus, the order, also called the sequence, of the specific bases on one strand is perfectly reflected in the order of the complementary bases on the other strand. Knowing the sequence of bases on one strand allows us to deduce the base sequence on the complementary strand.

DNA is often drawn as a flat molecule, because that shape is easy both to draw and to look at. In reality, each of the two sugar-phosphate backbones is wrapped around the other in a conformation called a double helix (Figure 3.19B). The base pairs are on the inside of the helix, like the rungs of a ladder, while the sugar-phosphate backbones are on the outside. For ease of representation and viewing, DNA can be presented by using a model in which each backbone is represented as a thin ribbon and the base pairs are represented schematically, as in Figure 3.19B. In fact, since the nucleotide sequence of one strand specifies the nucleotide sequence of the other strand, a DNA molecule or region of a molecule is very often represented by the sequence of only one strand, always written in the 5'-to-3' direction.

The double-helix structure of DNA is consistent with the tendency of hydrophilic molecules to be found in watery environments and the tendency of hydrophobic molecules to segregate themselves. The sugar-phosphate backbone of DNA is hydrophilic, so it is comfortable exposed to the cellular environment. The DNA bases, however, are hydrophobic and so are comfortable together inside the helix.

Chromosomes

As you might guess from looking at the figures showing its structure, DNA is a long, slender molecule. Just how long varies with the organism, but the

DNA inside a cell is always many, many times longer than the cell itself. You can imagine that packing a fantastically long DNA molecule inside a cell could present an organization problem. The solution to this potential problem is that DNA molecules are wrapped around certain proteins called histones, rather like thread wrapped around a spool. The complex of one DNA molecule and its associated protein scaffold is called a **chromosome**.

Proteins can interact specifically with DNA

From the ladderlike representation of DNA we have shown you, it might seem like DNA molecules would all be identical in shape. In a large sense, they are. They all have the double-helix conformation with the sugar-phosphate backbones around the outside and the base pairs on the inside. In fact, some proteins (like the histones that make up chromosomes) interact with the parts of DNA molecules that are common to all of them. However, many cellular processes depend on interactions between proteins and specific base sequences within a DNA molecule.

Even though the base pairs are on the inside of the helix, the "edges" of the base pairs are available to contact cellular proteins between the spiraling strands of the sugar-phosphate backbone. The edge of each base presents a different chemical nature and shape to the outside, so a protein that interacts directly with DNA can "read" the base sequence. Thus, a DNA-binding protein can interact with the edges of a specific series of DNA bases in the same lock-and-key way that a hormone receptor protein can interact with its hormone target. You will encounter several examples of proteins binding specifically to DNA in the following chapters. This kind of interaction is crucial to the carrying out and regulating of the two major cellular functions involving DNA.

DNA must be duplicated and its information expressed

We have said that DNA contains the genetic information of organisms. If you think of DNA as a library of instructions, it's clear that two different kinds of things have to happen to it during the life of an organism. First, the information encoded by its base sequence has to be translated into the actual "stuff" of an organism—its form and functions. Second, when a cell divides, the DNA has to be accurately copied so that the new cell can have its own instruction library. Now that we have looked at some detail at the structure of DNA, let's see how that structure meshes with these two cellular functions.

DNA replication

The double-stranded structure of DNA immediately suggests how DNA can be faithfully replicated, or copied. You can see that either of the two strands of DNA can be used as a template, or pattern, to reproduce the opposite strand by using the rules of complementary base pairing. When a cell is ready to replicate its genetic material, the two opposite strands are gradually "unzipped" to expose the individual bases. Each exposed strand is used as a template to form two new strands (Figure 3.20). The result is two daughter DNA molecules, each composed of one parental strand and one newly synthesized strand, and each identical to the parent.

Expression of genetic information

Although the structure of DNA immediately suggests how the molecule can fulfill the requirement for faithful transmission through duplication, it is not

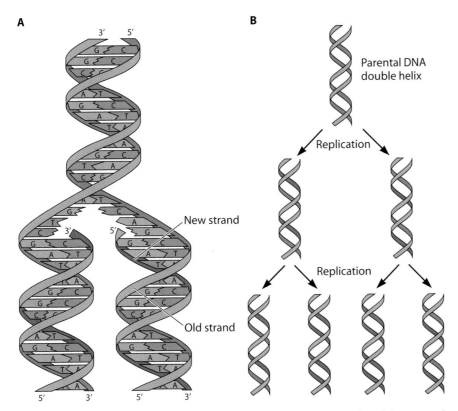

A

B

Parental DNA
double helix

Replication

New strand

Replication

Old strand

Figure 3.20 DNA replication. **(A)** An existing strand of DNA is unzipped, and the exposed bases on each strand serve as a pattern, or template, to guide the synthesis of a new strand. **(B)** DNA replication results in two identical daughter molecules, each with one new and one old strand of nucleotides.

so obvious how such a simple molecule can determine the development of creatures as complex and varied as a goldfish or an oak tree. The key to the expression of information encoded in DNA is proteins: those molecules that provide much of the structure and carry out nearly all the functions within living cells.

We have seen that an organism's structure and functions depend on the proteins present in its cells. We have also seen that a protein's function depends upon its three-dimensional structure. Since the three-dimensional structure of a protein depends upon the sequence of amino acids in the protein chain, the key to what a cell can do and how it is structured is found in the amino acid sequences of each of its proteins. *DNA determines the characteristics of an organism because it determines the amino acid sequences of all the proteins in that organism.* The next chapter describes how DNA fulfills this critical function.

RNA

Cells contain a second type of nucleic acid that plays key roles in the expression of genetic information. This second nucleic acid is called RNA. Like DNA, RNA is made up of nucleotides composed of a sugar, a phosphate, and one of four different organic bases. However, there are three important differences between DNA and RNA, two of them chemical and one of them structural.

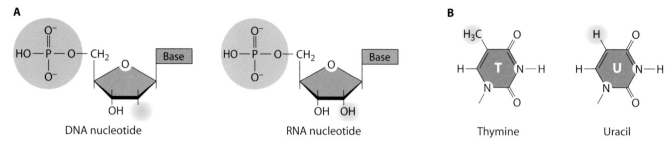

Figure 3.21 Chemical differences between DNA and RNA. **(A)** The sugar in the DNA nucleotide is deoxyribose. In RNA nucleotides, the sugar is ribose. (The arrows indicate the points of difference.) **(B)** RNA contains the base uracil instead of thymine. (The yellow highlights indicate the points of difference.) Uracil can form a base pair with adenine, as thymine does.

The chemical differences are that (1) instead of the sugar deoxyribose, RNA contains the sugar ribose (hence the name ribonucleic acid) and (2) instead of the base thymine, RNA contains the base uracil (Figure 3.21). In RNA, uracil can form a complementary base pair with adenine. The sugar-phosphate backbone of RNA is linked together like the DNA backbone, and the bases are attached to the number 1 carbon, as in DNA.

The important structural difference is that, although RNA bases can also form complementary pairs, RNA is usually composed of only a single strand of sugar-phosphate backbone and bases. It does not have the base-paired double-helix structure of DNA (Figure 3.22), although it is capable of pairing with other single strands of DNA or RNA. As we shall see in the following chapter, the single-stranded structure of RNA is ideally suited to its task of transferring information.

BIOCHEMISTRY AND BIOTECHNOLOGY

All cells, whether prokaryotic or eukaryotic, contain lipids, carbohydrates, nucleic acids, and proteins. These large molecules play similar roles in all cells. The similarity of cellular biochemistry is a manifestation of the relatedness of different kinds of cells and organisms—we are made of the same things and, at a cellular level, work in the same ways.

If you stop to think about this, you can see that there are some important implications. One of them is that different kinds of organisms can eat each other and use molecules or parts of molecules in their food to make new cellular components for themselves. In fact, it may have already occurred to you that many of the terms for the biochemicals we've been discussing—lipids, fats, oils, carbohydrates, and proteins—are things you're quite familiar with from discussions of diet. If so, you're right on target. The protein you eat is the same protein we discussed above: the folded chains of amino acids that do so many different things. In your body, the proteins you eat are digested back into amino acids that can be either recycled into new proteins or broken down further for energy. Carbohydrates are broken down into glucose; lipids are broken down into their subcomponents.

Recalling the definition of biotechnology from chapter 1, using cells or parts of cells to solve problems or make useful products, you may begin to see that the similarity of cells opens up possibilities for using one type of cell, such as a bacterium, to make something useful for another type of cell or organism, such as human insulin. Since proteins are the movers and shakers inside cells, biotechnology is usually about manipulating proteins, either to get

Figure 3.22 Although RNA bases can form complementary pairs, RNA is usually single stranded.

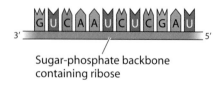

Sugar-phosphate backbone containing ribose

the proteins themselves or to get the products the proteins make. Since DNA encodes the instructions for making proteins, biotechnology requires manipulating DNA.

In the next chapter, as we discuss how the information in DNA is expressed in proteins, you'll begin to see how the instructions for making a protein in one cell could work in another cell. Getting a cell to produce a new protein, whether the new protein is something to be harvested and used by itself, such as human insulin, or a protein that gives the host cell slightly different properties, such as making it more resistant to freezing, is one of the major ways that biotechnologists use cells to make products or solve problems.

SUMMARY POINTS

All cells contain four major classes of large molecules that play essential roles in the structure and function of the cell. The four classes are lipids, carbohydrates, nucleic acids, and proteins.

Lipids are hydrophobic molecules. Lipids create membranes that separate the cytoplasm of cells from the external environment and that, in the case of eukaryotic cells, separate organelles from the cytoplasm. Lipids also play an important role in energy storage.

One important class of lipids is fats and oils. These lipids are made from a glycerol molecule and three fatty acids, which can be saturated or unsaturated. Phospholipids, which resemble fats except that a polar head group containing a phosphate group is substituted for a fatty acid, form the structural backbone of cellular membranes. Another important class of lipids is the sterols, which include cholesterol and many steroid hormones.

Carbohydrates are hydrophilic molecules ranging from simple sugars, like glucose and fructose, to giant molecules, like cellulose and starch. Carbohydrates are important in energy metabolism and storage (sugars and starch), structure (cellulose in plants), and recognition.

The nucleic acids DNA and RNA are the molecules that store and transmit genetic information. They are structurally simple, composed of repeating units called nucleotides. A single nucleotide contains one of four bases: adenine, thymine, cytosine, or guanine. These bases form complementary pairs: A with T (U in RNA) and C with G.

DNA consists of two complementary chains of nucleotides. Since they are complementary, the base sequence of one chain can be used as a template for synthesizing the other chain, which is how DNA is replicated.

Inside cells, long DNA molecules are wrapped around certain proteins, forming structures called chromosomes.

Proteins carry out essentially all the active functions in a cell, and some have structural roles. Proteins are chains of amino acids, which are chemically quite variable. These chains fold up in energetically favorable configurations that give each type of protein a unique shape. The unique shape of a protein is essential to its function.

DNA determines the characteristics of an organism by determining the amino acid sequences of all its proteins.

KEY TERMS

Amino acid
Antiparallel
Atom
Carbohydrate
Cellulose
Chemical bond
Chromosome
Complementary base pair

Deoxyribonucleic acid (DNA)
Element
Fat
Fatty acid
Glucose
Glycogen
Ion

Lactose
Lipid
Molecule
Nucleotide
Phospholipid
Photosynthesis
Polysaccharide
Protein

Ribonucleic acid (RNA)
Salt
Starch
Steroid hormone
Sterol
Sucrose
Sugar-phosphate backbone

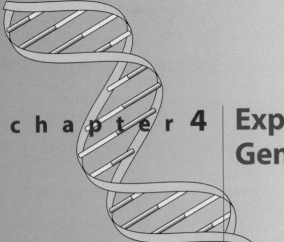

chapter 4 | Expression of Genetic Information

The word "gene" has become part of our everyday language. "I got my mother's red hair gene." "The corn in that field has been genetically engineered." "My dog is a cross between a Labrador retriever and a basset hound. He sure got the short legs gene."

Schools, movies, and the media in general have done a great job conveying the notion that genes are behind many observable traits. Unfortunately, the information often stops there and the big question of how genes contribute to traits is ignored or glossed over. There's a good reason for the information gap: "genes determine traits" is essentially a catchphrase, but looking at *how* they do it requires quite a lot of information and explanation.

In the last chapter, we stated that the structure and function of a cell are determined by what proteins the cell contains. At the end of the chapter, we stated that genes determine the characteristics of cells by encoding the instructions for making every protein that a cell can make.

This chapter focuses on how the information in DNA is translated into proteins and how changes in DNA and subsequent changes in proteins can affect their functions. Specifically, we will look at:

- the genetic code that links DNA base sequences to proteins
- the process by which a DNA base sequence is translated into a protein sequence
- what happens to proteins once they are synthesized
- what happens if the DNA base sequence undergoes a change

In future chapters, you'll encounter many different kinds of proteins as we look at how cells carry out their essential life processes.

THE GENETIC CODE AND THE GENOME

Let us begin this discussion of how cells get from DNA to proteins by focusing for a moment on the goal: a protein. Recall that proteins are chains of amino acids. These amino acid chains coil and fold upon themselves to assume the most energetically comfortable configuration in the cellular environment. This final folded three-dimensional shape is intrinsic to the protein's function: its shape and chemical nature determine how it interacts with other molecules.

Since a protein's three-dimensional shape is determined by its amino acid sequence, the amino acid sequences of all of the proteins in a cell in essence determine what that cell has the potential to do. The amino acid sequences of proteins are the link in the chain leading from the base sequence of DNA to the characteristics of an organism. *DNA determines the characteristics of a cell or an organism by specifying the amino acid sequences of all of its proteins.*

The genetic code

How does a simple molecule like DNA, which is composed of only four different nucleotides, determine an amino acid sequence that can involve 20 different amino acids? Pretty much in the same way that Morse code can represent the alphabet using only two symbols: dots and dashes. DNA contains a **genetic code** for amino acids in which each amino acid is represented by a sequence of three DNA bases (Table 4.1). These triplets of bases are called

Table 4.1 The genetic code

First base in DNA triplet	Second base in DNA triplet	Choices for third base in DNA triplet	Amino acid encoded
A	A	A, G	Lysine
		C, T	Asparagine
	G	A, G	Arginine
		C, T	Serine
	C	A, C, G, T	Threonine
	T	A	Isoleucine
		G	Methionine
		C, T	Isoleucine
G	A	A, G	Glutamic acid
		C, T	Aspartic acid
	G	A, C, G, T	Glycine
	C	A, C, G, T	Alanine
	T	A, C, G, T	Valine
C	A	A, G	Glutamine
		C, T	Histidine
	G	A, C, G, T	Arginine
	C	A, C, G, T	Proline
	T	A, C, G, T	Leucine
T	A	A, G	Stop
		C, T	Tyrosine
	G	A	Stop
		G	Tryptophan
		C, T	Cysteine
	C	A, C, G, T	Serine
	T	A, G	Leucine
		C, T	Phenylalanine

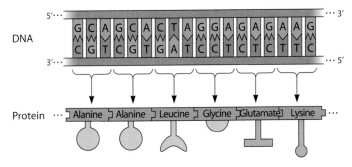

Figure 4.1 The base sequence of DNA determines the amino acid sequences of proteins.

codons. The order of the codons in a DNA sequence is reflected in the order of the amino acids assembled in a protein chain (Figure 4.1). The complete stretch of DNA needed to determine the amino acid sequence of a single protein is called a **gene**, and the complete set of genetic material in an organism is called its **genome**.

The human genome

In 1990, the U.S. Department of Energy and the National Institutes of Health began the Human Genome Project with the goal of determining the sequence of bases in the entire human genome. The project was completed ahead of schedule in 2003. The human genome consists of ~3 billion base pairs of DNA, and over 90% of it does not encode proteins. This **noncoding DNA** is often in the form of repeated sequences, sometimes in large clusters, which appears to be typical of eukaryotic genomes. Our genome contains ~30,000 genes, and those genes encode the proteins that make us what we are.

PROTEIN SYNTHESIS

The process by which proteins are produced from the genetic code has several steps. DNA is essentially a passive repository of information, rather like a reference library. The action of making a protein occurs at special structures in the cell called **ribosomes**. Therefore, the first step in protein synthesis is to relay the information from the DNA to the ribosomes. Imagine you needed to get a recipe from a book in a library out to the site where you were making something but that you could not remove the book from the library. What would you do? You'd make a copy of the directions you needed (not the whole book) and carry the copy out, right?

Exactly the same thing happens in cells. Cellular enzymes synthesize a working copy of a gene to carry its genetic code to the ribosomes. This working copy is an RNA molecule called **messenger RNA (mRNA)**, and the process of synthesizing it is called **transcription** (an elegant term for copying something from one medium to another, such as transcribing a verbal interview into writing).

The mRNA carries the genetic code for a protein to the ribosomes. In the second step of protein synthesis, the codons in the mRNA must be matched to the correct amino acids. This step is carried out by a second type of RNA called **transfer RNA (tRNA)**. Finally, the amino acids must be linked together to make a protein chain. The ribosome (which is made of proteins and yet more RNA) performs this function. When the protein chain is complete, a genetic "stop sign" tells the ribosome to release the new protein into the cell.

Transcription is the synthesis of an mRNA copy of a gene

The first step of protein synthesis is to make mRNA. This process resembles DNA replication in many ways. First, the DNA double helix must be unzipped to reveal the information-containing bases. Then, complementary nucleotides (which contain the sugar ribose, since we are making RNA) are paired with the exposed bases. During the synthesis of RNA, the base uracil substitutes for thymine and pairs with adenine. Bonds are formed between the nucleotides, and the new mRNA contains a base sequence that is exactly complementary to the template DNA strand (Figure 4.2). tRNA and ribosomal RNA molecules are also encoded in DNA and synthesized by transcription, but unlike mRNA, they are not translated into protein.

There are two major differences between transcription and DNA replication (compare Figures 3.20 and 4.2). In DNA replication, both strands are used as templates to generate two new strands for two new helices. In RNA synthesis, only one DNA strand is used as a template and only a single RNA strand is made. The second difference is that the new mRNA molecule is released from the DNA template as it is made. The DNA double helix "zips back up" as the mRNA is released. Newly synthesized DNA remains part of a new DNA helix, paired with its parent strand.

Figure 4.2 Transcription. **(A)** Base pairing between incoming RNA nucleotides and the DNA template guides the formation of a complementary mRNA molecule. A single strand of the DNA molecule is used as a template. **(B)** The DNA template closes behind the transcription site, releasing the RNA molecule and leaving the DNA template intact.

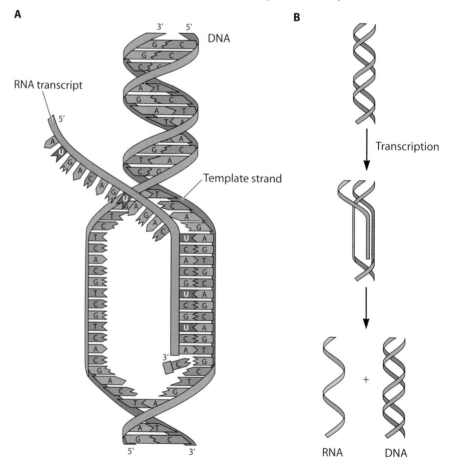

The process of synthesizing RNA, like other synthesis processes inside cells, is carried out by enzymes. The RNA-synthesizing enzyme **RNA polymerase** has an interesting task. Not only must it select the correct complementary nucleotides and link them together (as does its counterpart, DNA polymerase), but it must also decide where a gene is. A DNA helix can contain thousands or millions of base pairs. The RNA polymerase must determine exactly where to start and stop synthesizing RNA so that it will transcribe a complete gene.

A promoter is a sequence of bases where transcription begins

How does RNA polymerase "know" where to start making mRNA? Genetic traffic signals called **promoters** are built into the DNA base sequence to tell it. A promoter is a specific sequence of DNA bases that has the right chemical shape and nature to bind very specifically to RNA polymerase. Although we draw DNA like a ladder with the bases in the middle, the edges of the bases in the DNA helix are exposed between the two strands of the sugar-phosphate backbone. The RNA polymerase protein's three-dimensional shape and the chemical nature of its amino acids fit with the edges of the promoter bases like a lock and key. When that happens, the DNA base pairs separate and the building of mRNA can begin. As you might imagine, other signals tell RNA polymerase to stop synthesizing RNA and leave the DNA template. These signals are called **terminators.**

The notion of sequences of bases in a DNA molecule that provide signals for controlling transcription is an important one. Cells not only must use the information in DNA to synthesize proteins, they must also regulate when a protein is made and how much of it is made. For example, you may have heard of insulin, the hormone that causes cells to absorb the sugar glucose from the blood. What you may not know is that insulin is a protein. When you eat foods that cause your blood sugar level to rise, the hormone insulin is released from your pancreas, and cells within your pancreas begin to synthesize more insulin. Transcription of the insulin gene is regulated to meet your body's needs.

Regulating gene expression usually involves proteins binding to DNA. In later chapters, you'll see some examples of DNA-binding proteins that cause transcription to start and others that bind to DNA and cause transcription to be blocked. For now, just remember that in addition to the protein-coding sequence, genes contain specific sequences of bases that are targets for the binding of proteins—always for RNA polymerase and often for proteins that regulate whether and how much transcription occurs.

In eukaryotes, mRNA undergoes a process called splicing

In the 1970s, scientists were astonished to discover that eukaryotic proteins are not encoded by one long, continuous stretch of DNA. Instead, the coding sequence is interspersed with stretches of DNA bases that do not contribute to the amino acid sequence of the final protein. Comparing the mRNA of a eukaryotic gene to its DNA showed that the noncoding stretches of bases are first transcribed into the RNA molecule but then removed in a process called **splicing** (Figure 4.3). The noncoding stretches of bases were dubbed **introns,** and the coding stretches were called **exons.**

Following transcription, introns are removed from the RNA transcript and the mRNA is ready for the next step in protein synthesis.

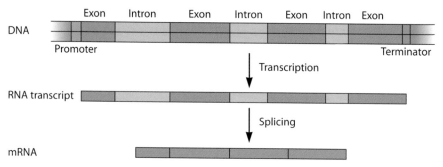

Figure 4.3 Splicing of an RNA transcript in eukaryotes.

Using mRNA to make a protein

After transcription and processing are complete, the mRNA moves to the ribosome, the site of protein synthesis. The ribosome and the mRNA fit together so that the mRNA codons can be read correctly. Look back at Figure 4.1 and think about this for a minute. The task at hand is to translate the DNA base code in the mRNA into amino acids, as shown in the figure. You can translate a DNA base code into amino acids by looking at the genetic code table (Table 4.1). The code table is a key for you, connecting a sequence of bases to a particular amino acid.

Cells have a genetic key, too. The cellular genetic key consists of a series of molecules with one end that fits exactly with one and only one codon and another end that fits with the correct amino acid. Together, these molecules constitute a molecular key for translating each of the codons in Table 4.1 into the correct amino acid. The molecules that make up the molecular key are the tRNA molecules. All cells use tRNA to translate the base code of DNA (at the ribosome in the form of an mRNA copy) into the amino acid sequence of proteins—your cells, bacterial cells, lizard cells, and the mold cells on the old sandwich in your refrigerator. This process, in which the mRNA base sequence is translated into a protein amino acid sequence, is called (amazingly enough) **translation** (Table 4.2).

tRNA molecules are folded in on themselves to resemble cloverleaves. At the tip of one of the lobes is a sequence of three bases called an anticodon (Figure 4.4). This anticodon pairs exactly with one of the codons on the single-stranded mRNA, using the rules of complementary base pairing. At the other end of the tRNA molecule is an amino acid. Each different tRNA is connected to the right amino acid for its anticodon. The result is that, when the anticodon on the tRNA pairs with its codon on the mRNA, the correct

Table 4.2 DNA and RNA process terminology

Term	Literal meaning	Nucleic acid process
Replication	Making an exact duplicate	Using a DNA molecule as a pattern to make two identical daughter molecules
Transcription	Copying from one medium into another, such as from an audiotape into writing	Copying the base sequence of a DNA molecule into an RNA molecule
Translation	Taking information in one language and putting it into another language	Taking the sequence of codons in mRNA and turning it into a chain of amino acids in a protein

amino acid is brought to the ribosome. If you are wondering how the correct amino acids get on the tRNAs in the first place, those bonds are formed by enzymes, one for each tRNA, that fit that and only that tRNA and its correct amino acid.

The ribosome holds the mRNA molecule so the tRNAs pair with their complementary codons one at a time, in order. As the tRNAs bring in the correct amino acids, the ribosomes link the amino acids into a growing protein chain. Once an amino acid has been linked to the chain, the tRNA molecule is separated and released from the mRNA-ribosome complex (Figure 4.5).

Genes contain traffic signals for transcription and translation

Protein synthesis involves a variety of interactions among enzymes, DNA, and RNA. As you saw above with the promoter, the base sequence within a nucleic acid can do more than simply encode amino acids. Certain sequences of bases can provide binding sites for enzymes or other proteins. These binding sites can act as traffic signals for starting, stopping, and regulating protein synthesis, and RNA as well as DNA molecules contain them. In fact, every mRNA molecule contains more than just the codons needed to make a protein. It also contains traffic signals for the ribosome.

Recall that mRNA is held by a ribosome in the correct configuration for protein synthesis to take place. The first signal that is needed in an mRNA molecule is something that will cause it to interact with a ribosome, so mRNA molecules have a sequence of bases whose three-dimensional shape specifically fits a binding site on the ribosome. Once the mRNA molecule has bound to a ribosome, another need arises: the ribosome has to know which three bases in the mRNA comprise the first codon for the new protein. This need is fulfilled by a so-called **initiation codon**. The initiation codon is usually the base sequence AUG, which encodes the amino acid methionine. Protein synthesis starts with the pairing of a methionine tRNA anticodon with the initiation codon on the mRNA and continues with each subsequently encoded amino acid. Additional methionines can be incorporated into the protein according to its genetic instructions.

This brings us to a final question. How does a ribosome know when to stop? Well, of course, a ribosome doesn't know anything, but the stopping point for the synthesis of a particular protein is determined by the presence

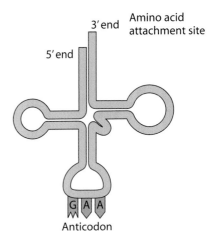

Figure 4.4 A tRNA molecule. Complementary base pairing between different portions of the tRNA molecule maintains its shape. We show only the anticodon bases.

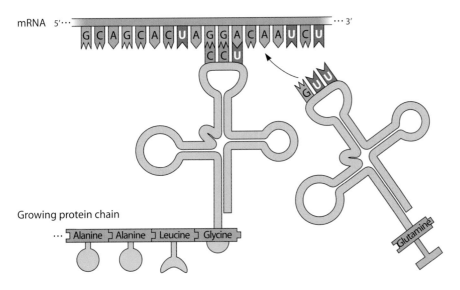

Figure 4.5 Translation. Complementary base pairing between the anticodons of incoming tRNA molecules and the codons of the mRNA guides the formation of the amino acid chain.

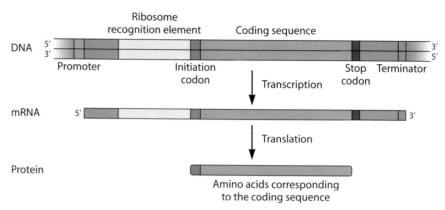

Figure 4.6 Major genetic traffic signals. Promoter, site where transcription begins; ribosome recognition element, bases recognized and bound by ribosome to hold RNA in place for translation; initiation codon, first codon in protein-coding sequence (ATG or AUG for methionine); stop codon, where translation stops; terminator, site where transcription ends.

of a signal called a **stop codon** (Table 4.1). There are no tRNAs that pair with stop codons. Instead, proteins called termination factors bind to stop codons and cause protein synthesis to terminate. Since the ribosome recognition sequence, initiation codon, and stop codon are present in the base sequence of the mRNA, they must also be encoded by the original DNA template that was transcribed to make the mRNA.

Don't be overwhelmed by the thought of all the signals combined with the actual processes of transcription and translation. It's actually very logical. If you walked up to a drawing of a DNA molecule and had to make a copy of the base sequence of part of it, you'd ask, "Where do I start?" and "Where do I stop?" The promoter and terminator for RNA polymerase provide those signals. If you walked up to a drawing of an RNA molecule and had to translate its codons into an amino acid sequence, you'd ask, "Where is the first codon?" and "Where is the last codon?" The initiation codon and stop codons provide these signals.

A summary of the major genetic traffic signals is presented in Figure 4.6. The DNA molecule is an impressive information storehouse. It contains not only the blueprints for the amino acid sequences of every protein an organism makes but also the traffic signals that direct the cell to interpret the information properly. As we shall see in a later chapter, DNA even contains signals that allow a cell to regulate the synthesis of its proteins to meet the requirements of its environment.

CELLULAR FATE OF PROTEINS

Cells need proteins for many different kinds of functions. Some of the functions, such as breaking down molecules for energy (see chapter 6), duplicating DNA (see chapter 9), or synthesizing cell components (see chapter 6), take place inside the cell. The proteins involved in these processes stay inside the cell once they are synthesized. Other functions, such as interacting with the environment and controlling what comes in and goes out of the cell, take place at the cell's plasma membrane. Proteins involved with these processes are inserted into the cell membrane. Still other proteins exert their effects outside the cell entirely—proteins that act as hormones or digestive enzymes, for example—and must be secreted.

Thus, part of the protein synthesis process is analogous to sending a package through a shipping department; as proteins are synthesized, they have to be shipped to their destinations. How does a cell know what to do with a particular protein? It turns out that the genetic code for proteins contains built-in address tags. These tags are short sequences of amino acids at the very front end of the protein-coding sequence. As these amino acids are strung together at the ribosome, they assume a shape that interacts with proteins found in the cell's internal membranes or other structures in such a way as to target the protein to the right place.

The first stage of the cell's shipping department is the ER

Eukaryotic cells have the equivalent of a shipping department. Figure 4.7 outlines the passage of a secreted protein, such as insulin, through it. The first stage of this tiny shipping department is the network of membranous tubes inside the cell called the **endoplasmic reticulum** (ER) (endo, inside;

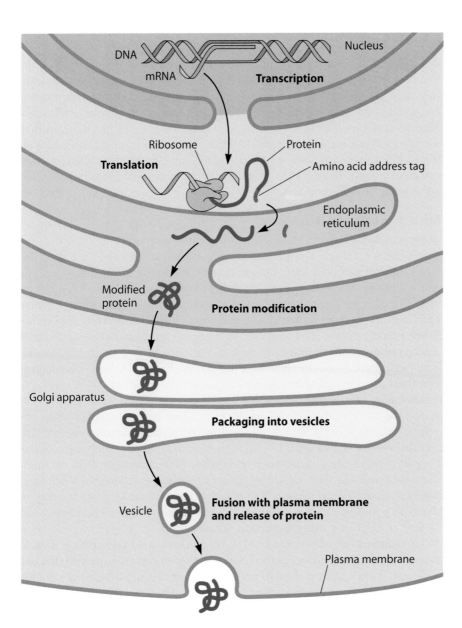

Figure 4.7 Passage of a secreted protein through the cell's shipping department. The newly translated protein has a sequence of amino acids at one end that targets it to the ER. There, the protein is modified and shipped to the Golgi apparatus, where it is packaged into vesicles for secretion.

reticulum, little net). The space inside the membranes of the ER is separate from the cytoplasm. Proteins whose function is outside the cytoplasm (whether outside the cell altogether, in the cell membrane, or elsewhere) are transported into the ER by virtue of their amino acid address tags. Proteins that carry out their functions in the cytoplasm lack the amino acid address tag for the ER and stay in the cytoplasm after translation. Once a new protein is inside the ER, enzymes clip off the amino acid address tag. The protein can then be further modified—carbohydrates can be added to it, or additional pieces may be removed—and it is passed along to the second stage.

The Golgi apparatus is the second stage of the shipping department

When the ER has completed the first stage of processing proteins for export, the proteins are sent to another set of membrane-bound compartments called the **Golgi apparatus**, named after the Italian scientist who first observed them under a microscope. In the Golgi, proteins can be further modified, but the main outcome is that they are wrapped up inside tiny membrane-enclosed packages called **vesicles** (imagine putting something in a box to ship it). The membrane-bound packages even have chemical address tags on them.

The vesicles can then travel to their destination membrane (such as the plasma membrane), fuse with it, and deliver the protein, or they can wait inside the cell until a particular signal causes them to fuse and deliver. In your body, the control of vesicles is an important means of regulating events like absorbing glucose from the blood and concentrating or diluting your urine in response to your water or salt balance. Having proteins waiting in vesicles allows a cell to respond to a signal much more quickly than if it had to make the protein from scratch. We'll look at the regulation of blood glucose and urine water in chapter 8.

Most proteins are short-lived

Most proteins, once synthesized, last for a while and then break down. Some proteins have an average lifetime of only a few minutes, while a very few (such as proteins in the lens of the eye) last the lifetime of the organism. The rest fall somewhere in between. The fact that proteins don't last lets a cell respond flexibly to signals from its environment. For an example, let's return to insulin, the hormone that causes cells to absorb glucose from the blood. When you eat sugary food, the level of sugar in your blood increases. Insulin is secreted into your bloodstream, and in response, your cells absorb the glucose (we'll look more closely at this process in chapter 8). Your blood sugar level falls.

What would happen if insulin lasted forever? Your cells would constantly be in "absorb glucose" mode, even when it was inappropriate, and your blood sugar level would fall too low. Fortunately, insulin doesn't last. As long as your blood sugar level is high, your body makes more of it. When your blood sugar level falls, your body stops making insulin. Without stimulation by insulin, your cells stop absorbing glucose and your blood sugar level stays where it should be. The fact that insulin breaks down is a key to the body's ability to regulate blood glucose levels. Protein breakdown is a similar key to many other important processes.

A consequence of proteins not lasting forever is that cells always need to synthesize at least some proteins. Though cells that are growing and dividing obviously have a greater need to produce proteins, even cells that are not actively growing must synthesize some, too (Box 4.1).

BOX 4.1 *When protein synthesis fails: poisons and pills*

The ability to synthesize proteins is absolutely essential for life. You will never read about a genetic disease in which the patient cannot synthesize proteins, because such a person could not exist. However, there are a number of chemical substances that block protein synthesis. If these substances block protein synthesis in animal cells (like our own), we usually call them poisons because of their toxic effects. If these substances block protein synthesis in bacterial cells without harming our own cells, we call them antibiotics.

Poison or antibiotic: it's the cell type that makes the difference

How can a chemical block protein synthesis in animal cells but not bacteria, or vice versa? The answer to this question comes back to one of our basic themes: the specific fitting together of molecules according to their shapes and chemical natures. Bacteria are prokaryotes, and animals are eukaryotic. Prokaryotic and eukaryotic cells are much more alike than they are different. They are made of the same types of molecules, they use DNA for their genetic material, and they synthesize pro-

teins in the same way. However, there are many differences in the details of how these processes occur and in the structures found inside the two respective cell types. The reason a poison can block protein synthesis in animals but not in bacteria and an antibiotic can block protein synthesis in bacteria but not in animals is that the ribosomes of prokaryotic and eukaryotic cells are different enough that their three-dimensional shapes are not recognized by the same molecules.

Examples: ricin, erythromycin, and tetracycline

Ricin, which is produced by the castor bean plant, is one of the most potent poisons known. It is estimated that 1 mg (0.001 g) of ricin can kill a 70-kg (154-lb) adult human. Ricin is one of a number of highly toxic substances that are of concern as potential bioweapons. Ricin exerts its lethal effect by blocking protein synthesis. It is actually an enzyme that binds to eukaryotic ribosomes and cuts their RNA in a specific place, destroying the activity of the ribosome. Since prokaryotic ribosomes have a different three-dimensional shape and

base sequence, ricin has no effect on bacteria.

The situation is essentially reversed with the antibiotics erythromycin, tetracycline, and their relatives. All of these chemicals bind to bacterial ribosomes and interfere with protein synthesis but do not interact with eukaryotic (e.g., your) ribosomes. Thus, these substances can be used to treat bacterial infections. Erythromycin and its relatives azithromycin, clarithromycin, and others bind to one site on bacterial ribosomes and freeze the protein synthesis process by preventing the addition of amino acids to a growing peptide chain. Tetracycline and its relatives, such as doxycycline and minocycline, bind to a different site on prokaryotic ribosomes and so distort the shape of the ribosome that the anticodon of a tRNA molecule can no longer align properly with a codon of mRNA and protein synthesis cannot take place.

This discussion should make it even clearer why antibiotics do not work against viral infections. Viruses use the host cell's (for example, your) ribosomes to synthesize their proteins.

MUTATIONS

Any change in a DNA sequence is called a **mutation**. It is a fact of life that mutations happen. They result from normal cellular processes and unavoidable environmental hazards. During DNA replication, the DNA polymerase enzyme occasionally makes errors. Environmental factors, such as ultraviolet light or mutagenic (mutation-causing) chemicals, damage DNA regularly. Most of this damage is corrected by DNA repair enzymes, proteins that recognize and repair abnormalities in DNA, but occasionally the repair enzymes miss something. When DNA polymerase attempts to use a damaged base as a template during DNA replication, it often cannot read the base properly and so inserts an incorrect base into the new strand, thus creating a mutation (more information about mutation and repair is presented in chapter 9). Genetic events such as transposition (see chapter 13) result in insertion or loss of segments of DNA, also changing the original sequence. Errors in cell division can lead to a rearrangement of the segments of chromosomes or even a change in chromosome number.

The consequences of mutation depend on many factors

The effect of a mutation on an organism depends on how the mutation affects the expression of genetic information and how the change in gene expression (if any) affects the organism within its environment. For example, a sequence change in a region of DNA that wasn't part of a protein-coding sequence would probably not have any effect on the organism. Many amino acids are encoded by more than one codon; for example, the codons TTT and TTC each encode the amino acid phenylalanine (Table 4.1). A mutation that changed TTT to TTC would not have any effect on the protein. Mutations with no effect on a protein are often called silent mutations.

Even a change in the amino acid sequence of a necessary protein might not alter its function in a significant way. Recalling the example from chapter 3 of the receptor protein in the cell membrane, imagine that a mutation occurred in a region of the protein apart from the specific hormone recognition area. As long as the change did not distort the overall shape of the protein or impair its interaction with the cell, it might not affect the function of the protein.

Obviously some mutations will abolish the functioning of a protein. Mutations in genetic traffic signals, such as promoters and ribosome recognition sequences, can completely shut down the synthesis of a protein. Sometimes a single base change in a coding region can result in an amino acid substitution that severely harms or destroys the protein's ability to perform its function. If a mutation eliminates a protein's function, then the effect of the mutation on the organism will depend on the role of that protein in the organism's environment.

For example, suppose a bacterium acquired a mutation that made it unable to break down the sugar lactose. As long as the bacterium had other food available, the mutation would not affect it. If the bacterium were in an environment where the only food source was lactose, however, it would die. On the other hand, a mutation that blocked an organism from breaking down all sugars would be fatal in all environments.

Still other mutations can knock out the function of a protein that is not essential and so have no effect on survival. For example, the hair of dogs and other mammals is pigmented because cells called melanocytes in the hair follicles synthesize pigments and place them in the hairs. The melanins are a family of black and brown pigments synthesized when receptors in the melanocyte cell membrane bind to a hormone called melanocyte-stimulating hormone 1, or MSH-1 (Figure 4.8).

In yellow Labrador retrievers, the gene for the hormone receptor is mutated and the receptor protein is nonfunctional. Therefore, it cannot receive the hormone signal to synthesize melanins, and so yellow Labs are yellow instead of brown or black. In the habitat of yellow Labs (backyards and houses), being yellow does not put a dog at a disadvantage, and so the lack of the hormone receptor has no effect on the dogs' survival (Figure 4.9).

Once in a great while, a mutation could even be beneficial, again depending on the environment. Changes in the amino acid sequence of a protein might make it more resistant to heat, which could be an advantage if the organism's environment is becoming warmer. An alteration in the shape of another protein might allow it to bind to and break down a different type of sugar, which could be an advantage if its environment contained that sugar. If a change in a protein's function is not immediately fatal, it might actually help the organism under the right environmental conditions. Being yellow rather than black could be beneficial to a Labrador retriever if it lived in a yellowish landscape and had to hide from giant dog-eating birds.

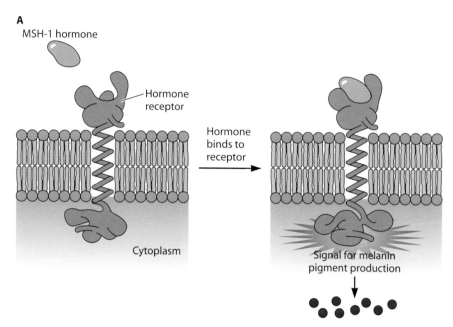

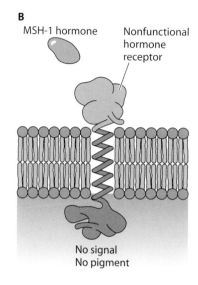

Figure 4.8 Hormone signaling in pigment production. **(A)** In black or chocolate Labrador retrievers, binding of the hormone MSH-1 to its receptor gives the signal for melanin pigment production. **(B)** In yellow Labs, the MSH-1 receptor is nonfunctional, so cells never receive the signal to produce melanin and the dogs are yellow instead of black or brown.

Figure 4.9 The coat color difference in the chocolate Lab and the yellow Lab puppies is caused by the puppies' lack of a functional receptor for MSH-1. (Photographs courtesy of Thomas A. Martin [top] and Donna Morgan [bottom].)

A. Normal hemoglobin DNA and protein sequences

B. Sickle-cell hemoglobin DNA and protein sequences

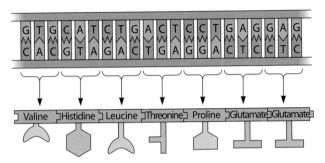

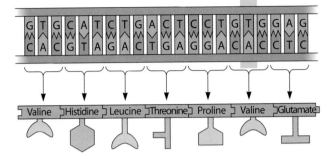

Figure 4.10 A single base change in the hemoglobin gene, highlighted in yellow, gives rise to the disease sickle-cell anemia.

Sickle-cell anemia

An example of a harmful mutation is found in the disease sickle-cell anemia. This disease is the result of one specific mutation in the gene encoding the oxygen-carrying blood protein hemoglobin. In sickle-cell anemia, a single A-to-T mutation changes the sixth codon from GAG to GTG. This mutation changes the sixth amino acid in the 146-amino-acid protein from glutamic acid to valine (Figure 4.10). Glutamic acid is hydrophilic, while valine is hydrophobic. The sickle-cell disease is a result of this change in character in the context of the rest of the hemoglobin molecule. In hemoglobin, the sixth amino acid sits on the surface. The mutant hydrophobic surface valine affects the way hemoglobin molecules interact with one another.

As it happens, the hydrophobic valine side chain on the surface just fits into a hydrophobic pocket that is exposed on the hemoglobin molecule when it is not bound to oxygen. The surface valine is not positioned to fit into its own pocket, but it can fit into the pocket on a second molecule. Since hydrophobic molecules tend to congregate in hydrophobic environments, the hydrophobic valine tends to snuggle into that hydrophobic pocket once it is exposed.

When the mutant hemoglobin molecules give up oxygen in the capillaries, the deoxygenated molecules fit together in a lock-and-key fashion. The surface valine fits into the pocket of another molecule, which itself has a surface valine to fit into another molecule, and so on (Figure 4.11). The mutant hemoglobin molecules aggregate into long fibers, changing the red blood cells' shape to a sickle form and interfering with circulation of the cells through the capillaries. The impaired circulation gives rise to a number of deadly problems in the afflicted individual.

Thus, the fatal disease sickle-cell anemia is a consequence of hydrophobic interactions and altered protein structure. If the hydrophobic valine did not happen to fit into the hydrophobic pocket on the deoxygenated hemoglobin molecule, you would not get polymerization of the mutant hemoglobin, sickling of cells, and impaired circulation. (Figure 4.12).

A beneficial mutation?

An example of a potentially beneficial mutation is the inherited condition benign erythrocytosis. Individuals with this condition have highly elevated levels of red blood cells, which is where the oxygen-carrying protein hemoglobin is found. Far from being ill, these individuals have greatly enhanced stamina because their blood can carry more oxygen. One such person, the

Figure 4.11 Representation of sickle-cell hemoglobin aggregation. **(A)** Normal hemoglobin molecules do not stick together. **(B)** The hydrophobic patch on the surface of sickle-cell hemoglobin caused by the glutamate-to-valine substitution at position 6 fits neatly into a hydrophobic pocket on a second molecule. Thus, sickle-cell hemoglobin molecules can form chains in a head-to-tail arrangement.

A. Normal hemoglobin

B. Sickle-cell hemoglobin

Val-6

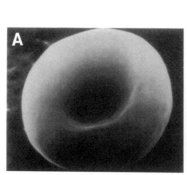

Figure 4.12 Sickle-cell anemia. **(A)** Normal red blood cell. **(B)** Red blood cell in sickle-cell anemia. Chains of mutant hemoglobin molecules distort the shape of the cell, impairing circulation. (Photographs courtesy of Connie Noguchi, National Institute of Diabetes and Digestive and Kidney Diseases.)

Finnish athlete Eero Mäntyranta, won three gold medals for cross-country skiing in the 1964 Winter Olympics. Scientists recently determined the molecular basis of benign erythrocytosis.

Red blood cells arise from progenitor cells called stem cells that are found in bone marrow (see chapter 10 for more information about stem cells). Stem cells are stimulated to mature into red blood cells by the hormone erythropoetin. The hormone communicates with the stem cell through a 550-amino-acid receptor protein embedded in the stem cell's outer membrane. One portion of the protein lies outside the cell, forming a docking site for the hormone. When the hormone binds, the portion of the receptor positioned inside the cell transmits the maturation signal. The cytoplasmic end of the receptor also contains a docking site for a cellular protein that prevents the transmission of the maturation signal. Docking of this cellular protein thus acts as a molecular brake on red blood cell production (Figure 4.13A). The Finnish athlete and other members of his family carry a mutant version of the receptor gene. A G-to-A mutation changes their codon 481 from TGG, for tryptophan, to TAG, a stop codon. This single base change causes the athlete's ribosomes to stop synthesis of the receptor protein 70 amino acids early.

Losing the last 70 amino acids does not disrupt the extracellular hormone-binding region of the protein, its cell membrane-spanning region, or the intracellular region responsible for transmitting the maturation signal. However, loss of those 70 amino acids removes the docking site for the braking protein. Thus, Eero Mäntyranta's red blood cell production has no molecular brakes, and he and other mutation-bearing family members have higher-than-normal levels of red blood cells (Figure 4.13B). Their blood can carry more oxygen than normal, so they have enhanced stamina. It is possible that their elevated red blood cell levels would be detrimental in some circumstances. High concentrations of red blood cells can cause the blood to thicken, leading to the possibility of heart attacks and strokes, particularly during hard exercise, and possibly death. Some reports state that people "afflicted" with benign erythrocytosis have normal life spans, while other reports claim that their average life span is reduced.

Interestingly, the hormone erythropoetin is also a drug of abuse in sports. You may have heard it called EPO. Athletes who cheat with EPO take the

A. Normal receptor protein

B. Mutant receptor protein

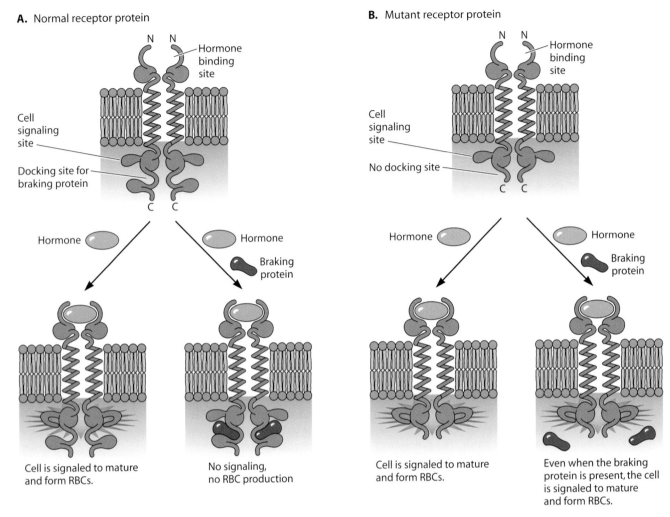

Figure 4.13 Schematic representation of how the loss of 70 amino acids from one end of the erythropoetin receptor protein results in increased red blood cell (RBC) production.

hormone to boost levels of red blood cells in their bodies and thereby achieve added stamina. In the late 1990s, several athletes were disqualified from the Tour de France bicycle race because of EPO use. This form of cheating can be detected because athletes who are using EPO have abnormally high levels of it in their systems, and those levels can be measured. Eero Mäntyranta did not have extra EPO in his body, but because his EPO receptor lacked molecular brakes, his body responded in a manner similar to the way most people would respond to added EPO, by producing additional red blood cells.

Was Mäntyranta's natural advantage unfair? That is a good question for debate. Mäntyranta did not just walk up and win the Olympics; he worked very hard to train, as do other athletes. Was his receptor mutation any less fair than the unusual height that confers advantages on National Basketball Association players? It is likely that as research progresses, the biological bases for many perceived advantages will be uncovered: extra height, faster reflexes, better balance, keener color perception, and better fine motor coordination. It is inevitable that some of our genetic differences will result in characteristics that confer advantages in certain situations. In the future, you may need to make decisions about the acceptability of altering someone's characteristics to give that person a perceived advantage or to eliminate a perceived liability.

Genetic variation is natural

One consequence of mutations is that the chromosomes of all sexually re-producing individuals differ in many ways. Although all humans have similar chromosomes and produce the same sets of essential proteins, the exact DNA sequences in those chromosomes vary. One obvious manifestation of variation is that we look different: one individual has genes resulting in blue eyes, and another has genes resulting in brown eyes. Less obvious is the accumulation of changes in noncoding regions of DNA. Great variety can be present in these regions, particularly in the number of repeated sequences present, with no outwardly observable effects. In fact, it is extremely unlikely that any two individuals (except identical twins) would share the same sequences in all of their noncoding DNA. It is this silent variation that makes DNA fingerprinting possible (see chapter 16).

The accumulating data about DNA and protein sequences indicate that protein structures and functions can often be maintained through many amino acid changes. The keratin proteins that comprise our hair and nails are a good illustration. Their specific amino acid sequences vary, yet they are all recognizable, though different, versions of keratin. The genetic mutations that resulted in different versions of the protein did not destroy structure or function. Another example of this kind is hemoglobin, which has more than 300 known genetic variants in the human population alone. Though you just learned of one very harmful hemoglobin mutation, most of these variants are single amino acid changes with only minor structural and functional effects. Protein structure, it seems, is reasonably robust in the face of many amino acid changes.

Mutation is the ultimate source of genetic variation, which is the raw material for evolution. Once different versions of genes are created by mutations, there are additional ways in which different combinations of genes can be generated. You'll read more about genetic variation, how it is created, and its importance in the natural world in chapters 11 to 13.

SUMMARY POINTS

DNA determines the characteristics of organisms by encoding the amino acid sequences of all of the organism's proteins. The genetic code uses a system of three DNA bases for each amino acid.

Protein synthesis occurs in a multistep process. The first step, transcription, is the synthesis of a working copy of a gene in the form of mRNA. The mRNA molecule moves to the ribosome, where protein synthesis occurs. The ribosome holds the mRNA so that its codons can be bound, in order, by anticodons on tRNA molecules. tRNA molecules carry the amino acid and the anticodon that corresponds to the genetic code for that amino acid. One by one, in order, the amino acids are connected in a chain to form the protein encoded by the gene.

A gene is the portion of a DNA molecule containing the instructions for making one protein. A gene contains more than just codons indicating the order of amino acids in a protein. It also contains built-in traffic signals that, for example, comprise the sites where RNA polymerase binds to begin transcription, where a ribosome recognizes and binds an mRNA, and where regulatory proteins bind.

The protein-coding regions of eukaryotic genes are interspersed with stretches of bases called introns that do not contribute to the final protein sequence. After the entire DNA sequence between the promoter and terminator has been transcribed, the introns are removed in a process called splicing.

SUMMARY POINTS *(continued)*

Proteins whose functions take place outside the cytoplasm have amino acid address tags that cause them to be put inside the ER. There, they can be processed and are moved to the Golgi apparatus, where they may be further processed. In the Golgi, they are packaged into membrane-bound vesicles. The vesicles can either take them to their final destination right away or wait for a signal to deliver the proteins.

Most proteins don't last forever. Protein lifetimes can be just a few minutes or years, but most are broken down relatively quickly. Because most proteins don't last, cells can respond flexibly to environmental signals. In addition, even cells that are not growing must continue to synthesize proteins.

The effect of a mutation depends on its effect on protein function or expression. Mutations can be neutral, harmful, or beneficial.

One consequence of mutations is genetic variation. Genetic variation can be silent or can contribute to observable differences between individuals.

Antibiotics treat bacterial infections by blocking essential bacterial processes, such as protein synthesis. Bacterial cells are prokaryotic, and their ribosomes are different enough from human ribosomes that antibiotics can interfere with bacterial protein synthesis without harming human protein synthesis. Conversely, poisons such as ricin block protein synthesis in animals without harming bacterial protein synthesis.

Antibiotics used to treat bacterial infections are useless against viral infections because viruses use host cell machinery to synthesize proteins and get energy.

KEY TERMS

Codon	Golgi apparatus	Promoter	Transcription
Endoplasmic reticulum	Initiation codon	Ribosome	Transfer RNA (tRNA)
Exon	Intron	RNA polymerase	Translation
Gene	Messenger RNA (mRNA)	Splicing	Vesicle
Genetic code	Mutation	Stop codon	
Genome	Noncoding DNA	Terminator	

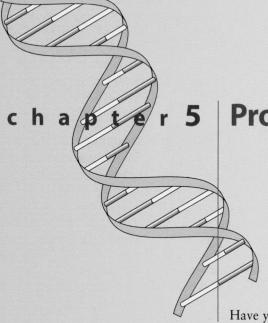

chapter 5 | Protein Structure and Function

Have you ever fried an egg? Had a hair permanent? You probably didn't realize it at the time, but what is going on in both of these activities is the manipulation of protein structure. By the time you get to the end of this chapter, you'll be able to explain just what the smelly hair permanent solution is doing to the hair's protein structure and how hair can be reshaped in the process.

Protein structure and function are at the heart of most biological processes. This chapter will give you a more in-depth look at the structure of proteins and how it relates to their functions. The objective is not to make you an expert on protein structure but rather to demystify proteins a bit. There are a few basic principles underlying their structure, and you've already encountered one of the most important: the tendency of hydrophobic and hydrophilic substances to interact with their own kind and avoid one another. The other principles are just as straightforward.

In this chapter we will look at:

- the characteristic of chemical bonds that makes substances hydrophilic or hydrophobic
- how amino acid chains fold to keep hydrophobic and hydrophilic portions comfortable
- the types of bonds that hold protein structures in place
- the association between regions of proteins and the specific functions they carry out
- examples of specific proteins and how they work (including the hair permanent)

Consider this chapter a sort of fantastic voyage into the inner workings of biological processes.

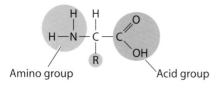

Figure 5.1 General structure of an amino acid. R signifies one of the 20 different side chains shown in Figure 5.4.

AMINO ACIDS AND PRIMARY STRUCTURE

As you learned in previous chapters, proteins are chains of amino acids folded into specific three-dimensional structures that represent the most energetically comfortable three-dimensional shape. The function of a protein depends on that three-dimensional shape and the chemical characteristics of its amino acids—the shape and chemistry must fit in a lock-and-key fashion with the molecules with which the protein interacts. In both chapters 3 and 4, you've seen depictions of proteins with a uniform backbone and varying amino acid side chains portrayed as blobs. We will now take a look at the actual chemical structure of amino acids and a fundamental property that determines how comfortably they interact with each other and with the intracellular environment.

A protein's primary structure is its amino acid sequence

Amino acids have the general chemical structure shown in Figure 5.1. The R in the figure can be any of 20 different so-called side chains. These different side chains give the 20 amino acids their separate identities. When amino acids are joined to make a protein chain, the OH group on one end of the amino acid reacts with the NH_2 group of another amino acid. A water molecule (H_2O) is lost, forming what is called a **peptide bond** (Figure 5.2A).

A protein consists of many amino acids joined together via peptide bonds. In fact, a chain of amino acids is often called a **polypeptide** (poly, many). Like a DNA strand, a protein backbone has a direction. One end has a free NH_2 group, and the other has a free COOH group. These ends are called the N terminus and the C terminus, respectively (Figure 5.2B). The overall effect is that a protein has a uniform **peptide backbone** with various amino acid side chains. The identities and order of the side chains in a pro-

Figure 5.2 Peptide bonds. **(A)** Peptide bonds are formed between the NH_2 group of one amino acid and the COOH group of another, with the formation and loss of a water molecule. R_n, amino acid side chain. **(B)** A protein has a polypeptide backbone with various amino acid side chains.

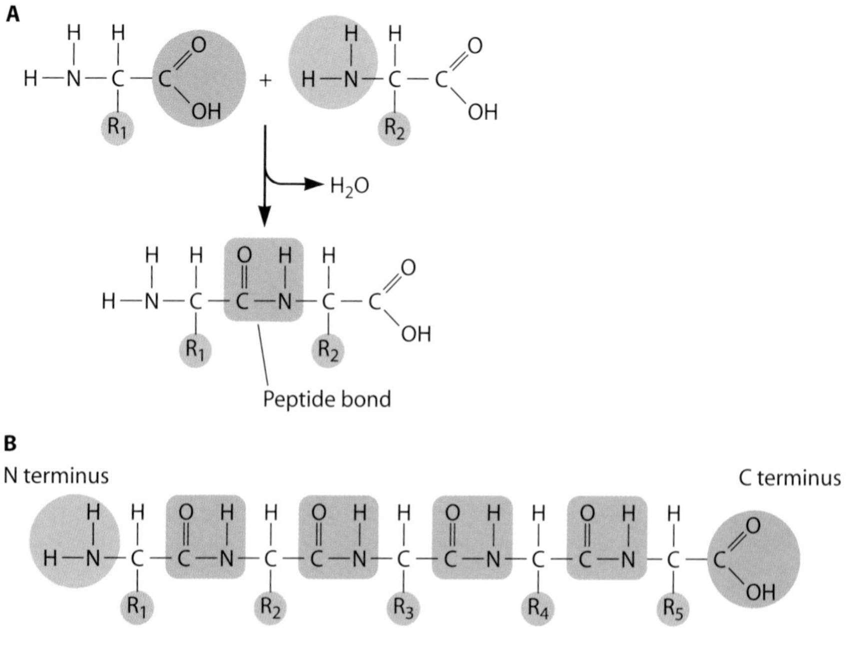

tein are called the **primary structure** of the protein. The primary structure is a direct consequence of the DNA base sequence in the gene encoding that protein.

You have probably already figured out that if all proteins have a uniform peptide backbone, then the nature of the amino acid side chains must be the factor that governs how an individual protein folds into a three-dimensional structure. That is true, and fortunately, one particular property of side chains is the most important for influencing three-dimensional structure. This property is the hydrophobicity or hydrophilicity of the side chain: how comfortably the individual side chains interact with water molecules. We discussed this property of molecules in chapter 2 as a basic organizing force inside cells without explaining what causes it. Now, we will look at that underlying cause, which is a function of chemical bonds.

Electron distribution determines whether a chemical bond is polar or nonpolar

The atoms in a protein molecule are held together by chemical bonds, which consist of electrons shared by two atomic nuclei. The nuclei are positively charged, while the electrons are negatively charged. If the electrons are shared equally by the nuclei, then their negative charge is distributed evenly over the area of the bond and balanced by the positive charges in the nuclei. However, the nuclei of certain elements attract electrons more strongly than other nuclei do. The ability to attract electrons is called **electronegativity**; you can think of this as a fancy term for the tendency to hog electrons.

When a strongly electronegative element forms a covalent bond with a less electronegative element, the electrons tend to be found near the strongly electronegative nucleus. Think of the electronegative nucleus as pulling the bond electrons away from the less electronegative nucleus. This uneven distribution of electrons creates a partial negative charge around the electronegative nucleus and a partial positive charge at the other nucleus (Figure 5.3). Chemical bonds with this type of uneven charge distribution are called **polar bonds**, because they have positive and negative poles.

Amino acid side chains can be polar or nonpolar

Amino acids (and proteins) are made primarily of carbon, hydrogen, oxygen, and nitrogen. Of these four elements, oxygen and nitrogen are strongly electronegative, and carbon and hydrogen are not. Imagine atoms of these four elements as children on a playground: O and N are electron hogs, while C and H are not. You can easily picture what happens when combinations of two of these children get together to share electrons (to form a bond between atoms). Bonds between oxygen and hydrogen, oxygen and carbon, nitrogen and hydrogen, and nitrogen and carbon are polar. Bonds between carbon and hydrogen are not. Whether a large chemical group like the side chain of an amino acid is polar or not depends on its constituent chemical bonds.

The chemical structures of the side chains of all 20 normal amino acids are shown in Figure 5.4. The structures are grouped according to whether the side chains are fully charged (ionized) under typical cellular conditions, are polar (partially charged), or are nonpolar.

We chose to show the structures of all the amino acids in this book to reinforce some basic concepts, not to emphasize chemical formulas or individual structures. Look at the amino acids in Figure 5.4 and note that each one has the same backbone (that of proline is slightly altered, but the basic idea

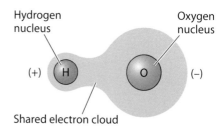

Figure 5.3 A polar chemical bond. Although the oxygen and hydrogen nuclei share electrons, the highly electronegative oxygen nucleus tends to draw them away from the weakly electronegative hydrogen nucleus. As a result, the oxygen end of the bond acquires a partial negative charge, while the hydrogen end is partially positive.

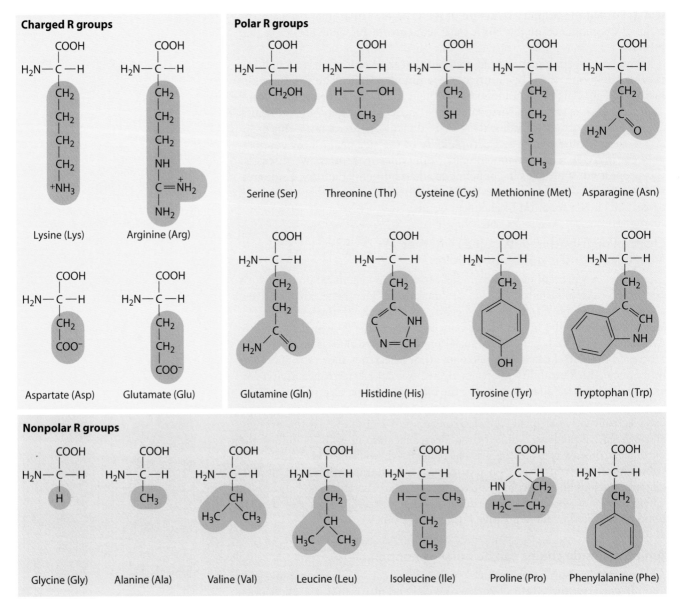

Figure 5.4 Amino acids commonly found in proteins.

is there). Look at the side chains and see that those labeled nonpolar are composed almost solely of carbon and hydrogen. In contrast, the amino acids labeled polar or charged contain oxygen, nitrogen, and sometimes sulfur (which is similar to oxygen in its bonding properties but not as electronegative).

Hydrophobicity and hydrophilicity

The polarity of the amino acid side chains is fundamentally important to protein structure, because polarity determines how stably a side chain interacts with other elements in the protein and in the cellular environment. A compound that has an overall polar character or is fully charged (like an ion) is energetically stable when it associates with other compounds with complementary charges or partial charges. The complementary charges neutralize each other. Nonpolar molecules do not associate with charged or polar molecules in an energetically favorable manner. Instead, they associate comfortably with other nonpolar molecules.

We stated above that the single property of amino acids that most determines three-dimensional protein structure is their ability to interact stably with water. Interactions with water are critical, because the intracellular environment is water based, as are other body fluids. How stably an amino acid side chain interacts with water, therefore, determines how "comfortable" it is when exposed to the intracellular fluid.

Here is the bottom line about water. Water consists of two polar oxygen-hydrogen bonds and is a very polar molecule (Figure 5.5). It thus associates comfortably with other polar or charged molecules. For this reason, molecules that are electrostatically charged or polar are hydrophilic. Since nonpolar molecules do not associate comfortably with water, they are hydrophobic. Hydrophobic amino acid side chains (the nonpolar ones in Figure 5.4) do not associate stably with the intracellular fluid. Hydrophilic amino acid side chains (the charged and polar ones in Figure 5.4) do associate stably because their charges or partial charges can be neutralized by complementary partial charges of polar water molecules.

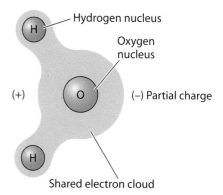

Figure 5.5 Water is a very molecule. The strongly electronegative oxygen nucleus hogs the electrons it shares with the hydrogen nuclei.

The fundamental consideration in protein structure is interaction with water

Now that we have looked at the features of covalent bonds that are important for understanding protein structure, let's see what it all boils down to. A protein molecule is a long peptide backbone with a mixture of charged, polar, and nonpolar amino acid side chains. Cytoplasm is a watery environment, so the charged and polar side chains will be stabilized through interactions with water molecules there. However, the hydrophobic amino acid side chains do not associate stably with water; they are more stable when clustered together away from water.

It appears that the basic rule underlying protein structure is, as much as possible, to fold up hydrophobic amino acid side chains together in the interior of the protein, creating a water-free hydrophobic environment. Hydrophilic side chains, meanwhile, are stable when exposed to the cytoplasm on the surface of the protein molecule. This is not to say that you would never find a hydrophilic amino acid in the interior of a protein or a hydrophobic one on the surface, but in general, the rule holds true. A protein is therefore said to have a hydrophobic core. The three-dimensional structure of each individual protein can be thought of as a solution to the problem of creating a stable hydrophobic core, given that protein's primary structure. However, every protein structure must solve one common problem. As you might guess, solving that problem forms another theme in protein structure.

FOLDING THE AMINO ACID CHAIN

There is one major problem in folding a protein to create a hydrophobic core: the backbone. Look at Figure 5.2. The peptide backbone is full of NH and CO bonds, and both kinds of bonds are highly polar. On the surface of a protein, these partially charged bonds can be readily neutralized through hydrogen bonding with water. However, for a protein structure to be stable, the partial charges of the peptide backbone must also be neutralized inside the protein core, where there is no water. The solution to this problem is a major factor in determining protein structure. So that you'll understand the

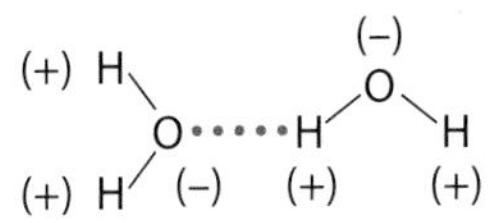

Figure 5.6 A hydrogen bond (dotted line) is a weak attraction between opposite partial charges.

solutions, we must first describe another type of chemical interaction that plays a major role in the structures of proteins and other biologically important molecules.

Highly electronegative elements can form hydrogen bonds

The other important chemical interaction is called a **hydrogen bond**, the same kind of bond found between base pairs in DNA (see Figure 3.18). Hydrogen bonds are not covalent bonds. They are much weaker and form when two highly electronegative nuclei "share" a hydrogen atom that is formally bonded to only one of them. The partial positive charge on the bonded hydrogen is attracted to and neutralizes the partial negative charge on the second electronegative nucleus (Figure 5.6).

Hydrogen bonds form only between pairs of any of the three most electronegative elements: oxygen, nitrogen, and fluorine. Of these three elements, only oxygen and nitrogen are common in biological systems, so you can forget about fluorine when thinking about protein structure. In proteins, the most important groups involved in hydrogen bonding are N, NH, O, OH, and CO groups. Water can form hydrogen bonds with all of them (Figure 5.7). Look at the hydrogen bonds in DNA in Figure 3.18, and you will see the same groups.

Although hydrogen bonds are weaker than covalent bonds, they are stronger than hydrophobic and hydrophilic interactions, and a large number of hydrogen bonds together are very strong (Box 5.1). Imagine them being like wooden toothpicks or broom straws: they are easy to break one at a time but very difficult to break in large bundles. Large numbers of hydrogen bonds hold the two strands of DNA together and anchor many aspects of protein structure.

Now that we've introduced hydrogen bonds, let's get back to the problem of what to do about the hydrophilic protein backbone inside the hydrophobic protein core.

The hydrophilic backbone can self-neutralize in two different types of structure

The fundamental solution to the problem of the peptide backbone in the hydrophobic interior is for the backbone to neutralize its own partial charges. The NH groups can form hydrogen bonds with the CO groups (Figure 5.7), neutralizing both. Since every amino acid contributes one NH group and one CO group to the backbone, this solution is very convenient. However, because of geometric constraints, the NH and CO groups from the same amino acid are not in position to form a hydrogen bond with one another. Instead, the peptide backbone must be carefully arranged so that the NH and CO groups along it are in position to form hydrogen bonds with complementary groups elsewhere along the backbone. Two basic arrangements work well, and these two arrangements form major components of protein structure.

Figure 5.7 Common hydrogen bonds (dotted lines) in biological systems.

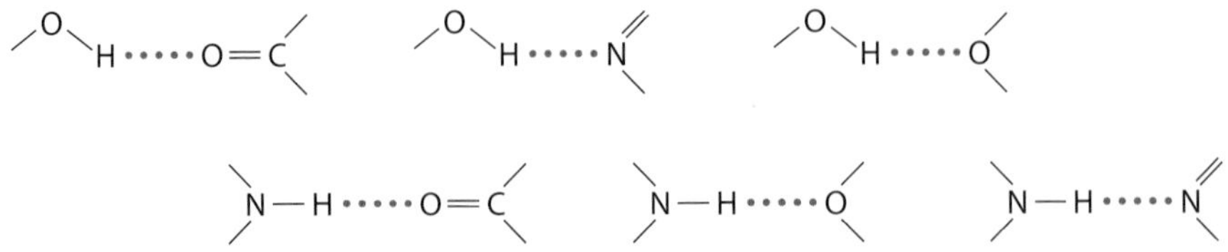

BOX 5.1 *Hydrogen bonding in action*

Hydrogen bonding may sound like yet another obscure chemical phenomenon, but you can readily observe its effects for yourself. Water molecules are highly hydrogen bonded to one another, so they stick together even more than they would from hydrophilic interactions alone. You can see their tendency to stick together in the way that undisturbed water surfaces seem to form a skin.

(Photograph courtesy of larvalbug.com.)

Look at the photograph of the water strider standing on water (figure). At this scale, you can easily see how the surface of the water dimples under its feet, just like the surface of a balloon might. This tendency of the surface of a liquid to stick together like a skin is called **surface tension**. Hydrogen bonding between water molecules makes the surface tension of water especially strong.

You can observe surface tension for yourself by floating a variety of unlikely objects on water: a needle or a paperclip, for example. It may help to make the surface of the object slightly greasy so that the water molecules will not stick to it; wiping it between your fingers will probably be sufficient. One way to get an object like a needle to float is to lay a small square of paper toweling on the surface of the water and lay the object on top of the tissue. When the paper towel piece becomes saturated with water, it will sink, leaving the needle floating (figure).

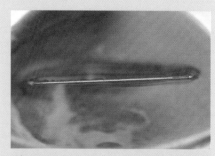

(Photograph courtesy of Thomas A. Martin.)

Another way to observe the effects of surface tension is to fill a glass exactly to the brim with water. Guess how many pennies (or other small objects) you could add to the glass before the water spills out of it; then test your guess. The ability of the water to bulge up over the top of the glass without spilling (again like a balloon) is a result of hydrogen bonding between the water molecules.

The first self-neutralization arrangement for the peptide backbone is for the backbone to form a helical coil, as if it were winding around a pole. The amino acid side chains point outward, away from the imaginary pole. The NH and CO groups along the backbone form hydrogen bonds with complementary groups above or below them on the pole, as shown in Figure 5.8. This arrangement is called an **alpha helix**.

In the second self-neutralization arrangement, stretches of the peptide backbone lie side by side, so that a CO group on one backbone can form a hydrogen bond with an NH group on the adjacent backbone (Figure 5.9). The amino acid side chains point alternately above and below the plane of the backbone. This arrangement is called a **beta sheet**, and the individual stretches of backbone involved in the sheet are called beta strands. Beta sheets are usually not flat, but twisted.

Within a protein molecule, particular stretches of the amino acid chain may assume an alpha helix or beta sheet conformation. Certain amino acid sequences favor the formation of each one, although we cannot predict these structures from the primary structure with perfect accuracy. The regions of alpha helix and beta sheets within a protein are referred to as **secondary structure**.

Disulfide bridges anchor structures in place

One other type of bond that contributes to protein structure forms between two properly aligned side chains of the amino acid cysteine. Find cysteine among the polar amino acids in Figure 5.4 and note that it has an S-H group at the end of its side chain. Under the right conditions, two cysteine side

Figure 5.8 The alpha helix. C_α indicates the carbon atoms with side chains, which are not shown.

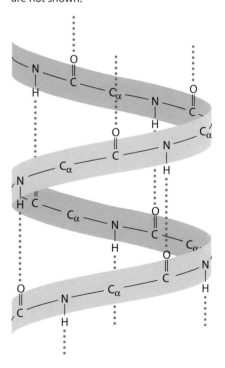

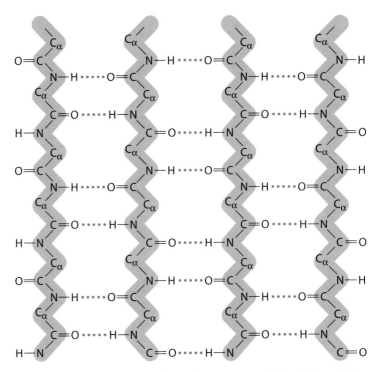

Figure 5.9 A beta sheet. C_α indicates the carbon atoms with side chains, which are not shown.

chains can form a covalent bond between their S-H groups, losing both hydrogens in the process (Figure 5.10). This S-S bond is called a disulfide bond, and the two cysteines are often said to form a **disulfide bridge**. The disulfide bond is covalent and therefore very strong, so disulfide bridges act as anchors in their local regions of the protein.

Unstructured loops of amino acids can change conformation

Within the hydrophobic core of a protein, some segments of the backbone may be found in the alpha helix conformation while other segments may be arranged as beta sheets. (Some proteins are formed entirely from alpha helices; others are formed entirely from beta strands.) These secondary structures are often connected to one another via stretches of amino acids on the surface of the protein, where the partially charged backbone does not need to assume a particular secondary structure because it is neutralized by water in the cellular environment.

These unstructured amino acid loops are free to change conformation; therefore, they often play key roles in a protein's activity. When the protein's lock-and-key partner molecule comes along, an unstructured loop can bind to it and change shape (picture your hand curling around something), triggering additional changes. Since a protein's activity depends on its shape, changing its shape can change its activity. Thus, a protein binding to some other molecule can trigger cellular events like pigment production, as we described for Labrador retrievers in the last chapter (see Figure 4.8).

Below, we will show you an example of an unstructured loop interacting with a second molecule, causing a change in the protein's overall three-dimensional shape and in its activity. Before we do that, we must address the issue of drawing the three-dimensional shapes of proteins.

Figure 5.10 Two cysteine side chains can form a disulfide bridge.

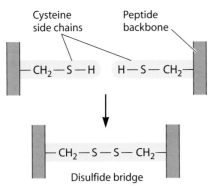

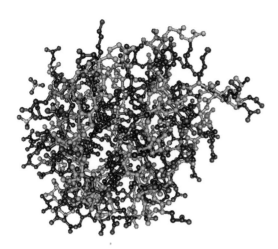

Figure 5.11 A ball-and-stick representation of the protein flavodoxin. (Image courtesy of Antonio Romero and Javier Sancho.)

Our protein drawings emphasize the backbone

If you look at a picture of a protein in which all the atoms are shown, you can't tell much (Figure 5.11). The picture contains too much information and too many atoms to tell what the big picture is. For purposes of understanding the overall structural plan, it is most helpful to look at just the configuration of the backbone. One popular way of showing this configuration is to draw the backbone as a ribbon, with alpha helices coiled and beta strands drawn as arrows. The arrows point toward the C terminus of the amino acid chain. Free loops are uncoiled regions of ribbon without arrowheads. Examples of some protein structures drawn in this manner are shown in Figure 5.12. Don't worry about the names of these proteins; we simply want you to see how some of the structural elements we have just discussed fit together in real proteins.

Examining actual protein structures

Look closely at the drawings in Figure 5.12. Plastocyanin (Figure 5.12A) is composed of beta strands connected by loops. The center part of flavodoxin

Figure 5.12 Ribbon drawings of protein structures. (Drawings courtesy of Jane Richardson.)

A. Plastocyanin   **B.** Flavodoxin   **C.** Triose phosphate isomerase

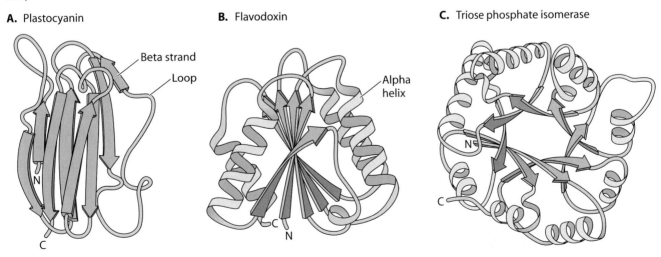

(Figure 5.12B) is a twisted beta sheet. The strands of the beta sheet are connected by regions of alpha helix. The center of triose phosphate isomerase (Figure 5.12C) also consists of beta strands, but they are arranged somewhat differently. Try using your finger to follow the peptide backbones of all three structures from N to C. You can see that adjacent strands in a beta sheet do not have to come from contiguous stretches of the backbone but may be segments that are widely separated in the primary structure.

You can also see from the flavodoxin and triose phosphate isomerase structures that alpha helices are apparently at the surfaces of these proteins. This arrangement is fairly common because of a handy property of alpha helices. Since the amino acid side chains stick out from the center of the imaginary barber pole, one side of the pole can have mostly hydrophilic side chains while the other has mostly hydrophobic ones. With such an arrangement, the helix can sit comfortably at the protein surface, its hydrophobic side buried in the protein core and its hydrophilic side exposed to the cellular fluid. Neat, isn't it?

HIGHER LEVELS OF STRUCTURE

The proteins we showed you in Figure 5.12 are relatively small ones. Their various elements of secondary structure are combined in one stable, compact, three-dimensional shape called a **domain**. The proteins in Figure 5.12 consist of a single domain. Larger proteins may consist of several separate domains, often linked by unstructured loops. The structures of individual domains within a protein and the way multiple domains fit together are called the **tertiary structure** of the protein.

Domains are fundamental units of protein structure and function

Domains are usually formed from continuous stretches of amino acids and therefore are translated from continuous regions of mRNA. In multifunctional proteins, it is not uncommon to find that the protein folds into several domains and that each domain is associated with one function. For example, Figure 4.8 in the last chapter shows the MSH-1 receptor binding to the MSH hormone and stimulating pigment production. The MSH receptor has three domains: the hormone-binding domain outside, the transmembrane domain inserted into the membrane, and the activating domain. Sometimes it is possible to separate the domains of a protein through various chemical or enzymatic treatments, and the separate domains occasionally retain their individual functions, acting like miniproteins all by themselves.

Scientists believe that domains are fundamental units of protein structure and function because similar domains appear in different proteins, sometimes many different proteins. Some of the domains appear to be mostly structural; others are connected with specific functions. For example, a DNA-binding domain called the homeodomain (rhymes with Romeodomain) is found in a large number of gene regulation proteins that interact with a specific type of DNA sequence. (Homeotic genes and their proteins will be discussed in chapter 10.) These proteins activate different sets of genes and are found in a diverse set of organisms, including worms, fruit flies, and humans. All of the homeodomain proteins bind to DNA via their homeodomains, and the amino acid sequences of the various homeodomains are very similar.

A two-domain DNA-binding protein

Another example of a gene regulation protein is the repressor protein of the bacteriophage lambda. This protein regulates viral gene expression by binding to a specific sequence of bases in the virus DNA and blocking transcription of genes located near the binding site. As we have mentioned before, the binding of proteins to DNA is a key activity in the regulation of gene expression. It turns out that many gene regulator proteins have two functions: binding to DNA and interacting with a second protein molecule. You'll read more about proteins like this in chapter 10 when we discuss how an organism develops from a fertilized egg into an adult. The lambda repressor is a well-studied example of such a DNA-binding protein.

The lambda repressor consists of 236 amino acids folded into two domains. The first 92 amino acids fold into the N-terminal domain, and amino acids 132 through 236 fold into the C-terminal domain. The other 40 amino acids connect the two domains (Figure 5.13A). Each domain of the lambda repressor carries out one of the protein's functions.

The N-terminal domain of the lambda repressor binds to DNA. The C-terminal domain binds to the C-terminal domain of a second molecule of the repressor protein, forming what is called a dimer (Figure 5.13B). A protein dimer is a stable association of two copies of the same polypeptide chain (see "A protein can contain more than one polypeptide chain" below). Each of the chains is called a monomer. The binding of the two C-terminal domains to one another stabilizes the interactions of the N-terminal domains with DNA. In fact, the lambda repressor does not work if it cannot form dimers.

Modular proteins

You may be wondering by now if it is possible to put together proteins by connecting domains like modules or Tinkertoys. The answer is yes. Although not all proteins are made in this way, many are, with perhaps a few unique regions thrown in. Figure 5.14 shows some modular protein structures. Again, don't worry about the names of the proteins or the domains. The important point is the implication: these modular structures suggest that many genes did not evolve from scratch (read more about evolution in chapter 13). Rather, it looks as if many genes were patched together from pieces or copies

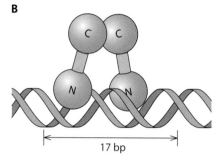

Figure 5.13 Domain structure of the bacteriophage lambda repressor protein. **(A)** The N-terminal domain consists of amino acids 1 to 92, and the C-terminal domain consists of amino acids 132 to 236. **(B)** The repressor forms dimers through interactions between the C-terminal domains. The N-terminal domains bind to a specific DNA base sequence. (From M. Ptashne, *A Genetic Switch*, 2nd ed. [Blackwell Scientific Publishing and Cell Press, Cambridge, Mass., 1992].)

Figure 5.14 Domain structures of some modular proteins. Epidermal growth factor (EGF) is a protein that signals several cell types to divide. The other four proteins are protein-digesting enzymes with a variety of physiological roles.

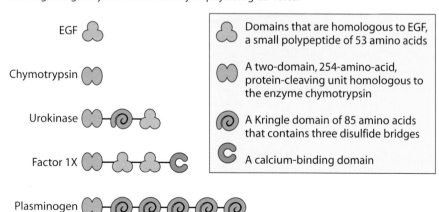

of preexisting genes, resulting in proteins that contain domains common to many other proteins.

The modular structure of some proteins has implications for biotechnology. Nature has produced proteins with many different functions by joining similar domains in different ways over the course of evolution. Using recombinant DNA technology (described in chapter 15), scientists can now swap domains, too, by swapping portions of genes. For example, the portion of the lambda repressor gene encoding the first 92 amino acids (the DNA-binding domain) can be replaced with a similar domain from a different repressor protein. The hybrid protein works as a repressor, but it recognizes the DNA-binding site of the second protein.

A protein can contain more than one polypeptide chain

Many functional proteins consist of a single amino acid chain, but many contain more than one polypeptide chain. These chains can be multiple copies of the same chain, as in the dimeric lambda repressor described above, or they can be assemblies of different polypeptides. *Escherichia coli* RNA polymerase contains five different polypeptide chains encoded by five different genes. The identities and number of the polypeptide chains and how they fit together in the final protein are called the protein's **quaternary structure**. The various levels of protein structure are listed in Table 5.1.

Protein structure can be disrupted

Three-dimensional protein structures are held together largely through the relatively weak chemical interactions of hydrogen bonds and the favorable interactions of hydrophobic side chains in the interior. Anything that disrupts these weak interactions—heat, extremes of pH, organic solvents, or detergents—can alter the folding of the protein (Box 5.2). The most extreme form of alteration is the complete unfolding of the amino acid chain, a process called **denaturation**. Denaturation of a protein is often irreversible. You can observe denaturation by frying an egg. The egg white protein albumin is water soluble in its native state. As you heat it, the albumin denatures and

Table 5.1 Levels of protein structure

Level of structure	Description	Notes
Primary	Amino acid sequence	Encoded directly by the DNA base sequence of the gene; every protein has primary structure.
Secondary	Local areas of alpha helix, beta sheets, or unstructured loops	Much secondary structure is a solution to keeping hydrophobic amino acid side chains together and neutralizing the hydrophilic peptide backbone.
Tertiary	Domains	A protein may contain one or more domains. Domains are compact structures usually encoded by a continuous stretch of DNA and often associated with a specific function.
Quaternary	Association of multiple polypeptide chains	Proteins can contain more than one copy of the same polypeptide chain or be made of associations of different polypeptides encoded by different genes. Proteins made of a single polypeptide chain do not have quaternary structure.

BOX 5.2 *A disease of protein structure?*

In the early 1990s an outbreak of so-called mad cow disease (bovine spongiform encephelopathy) struck the British cattle industry. Symptoms of mad cow disease include erratic behavior and resemble symptoms of the sheep disease scrapie. In fact, the British mad cow disease appeared to have been transmitted to cattle through feed made in part from the remains of scrapie-infected sheep. Scrapie and mad cow disease are two of a group of diseases that affect the brain, including the human diseases Creutzfeldt-Jakob disease and kuru.

Using animal models, scientists were able to isolate an infectious agent that transmits scrapie. The agent passed through filters that would exclude bacteria, so it was assumed to be a virus. However, no one could detect either DNA or RNA in the scrapie agent, and it was unaffected by enzymes that destroy nucleic acids. However, the agent was neutralized by enzymes that degrade proteins. Researchers named the mysterious agent a **prion**. How could an agent that apparently contained no nucleic acid transmit a disease? Many scientists assumed the scrapie researchers must be in error.

After 20 years of research, however, the story is becoming clearer. It appears that prion diseases are diseases of protein structure. Normal individuals manufacture the prion protein, and it assumes its normal configuration. Under some circumstances, however, the prion protein can assume an alternative configuration. Once it is in this alternative configuration, it does not return to normal, and when the altered form of the protein comes into contact with the normal form of the protein, the normal form changes to the altered form. As the altered form becomes more and more prevalent, disease begins. Thus, the aberrant form of the protein can act as a disease agent.

coagulates, forming a white solid that is no longer soluble in water. Cooling the cooked egg does not reverse the process.

Some proteins are harder to denature, or are more stable, than others. One way to quantify stability is to measure the temperature at which a given protein denatures. This is sometimes called the melting temperature (T_m) of the protein. The reason that you can usually purify drinking water by boiling it for several minutes is that at that temperature, at least some essential proteins of nearly all living organisms are irreversibly denatured, so that the organisms are killed.

Some proteins require much higher temperatures to unfold than others do (i.e., they are thermostable). For example, the enzymes of organisms that inhabit hot springs and ocean thermal vents are stable at very high temperatures. No single thing makes these enzymes more thermostable. It appears that many different aspects of their primary and tertiary structures contribute to their heat resistance.

One feature of protein structure, however, makes a significant contribution to stability: disulfide bridges. Disulfide bridges anchor regions of the protein in a specific configuration, stabilizing the structure. Disulfide bridges can form between distant portions of the same domain or between different polypeptide chains within a quaternary structure. Further on in this chapter, you will read about two examples of manipulating disulfide bridges to change protein structure: in hair permanents and in engineering a protein for greater stability.

EXAMPLES OF PROTEIN STRUCTURE AND FUNCTION

Protein structure is an interesting topic in itself, but it is critically important because protein function depends on it. Let's look at a few specific examples: a structural protein and two DNA-binding proteins.

The structure of keratin makes it tough and water insoluble

The **keratins** are a family of similar proteins that make up hair, wool, feathers, nails, claws, scales, hooves, and horns, and they are part of the skin. To

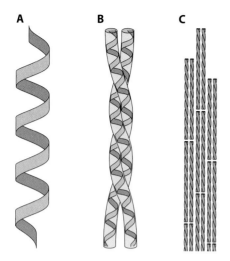

Figure 5.15 Keratin, a structural protein. **(A)** The keratin polypeptide forms an alpha helix with hydrophobic side chains (not shown). **(B)** Two keratin helices wrap tightly around one another. **(C)** The coiled helices lie side by side and end to end, forming fibers.

fulfill their functions, these proteins must be very strong. In addition, they must not be soluble in water (it would be quite unhandy if your hair dissolved when you took a shower), even though most proteins are. Let's look at how the structure of keratin makes strong, water-insoluble fibers possible.

Hydrophobic alpha helices

The amino acid chain of keratin folds into one long alpha helix. Almost all of its amino acids—alanine, isoleucine, valine, methionine, and phenylalanine—are hydrophobic, and they extend outward from the helical backbone. The presence of all these hydrophobic side chains everywhere violates the general rule that hydrophobic side chains must be buried in the protein's core. As you probably expect, this rule violation is important: it means that keratin is not energetically comfortable surrounded by water molecules and therefore is not soluble in water. Imagine, the reason your fingernails don't dissolve when you wash them is all those hydrophobic side chains.

Instead of associating with the watery environment inside cells, keratin molecules associate with each other in large groups. Their structure, a single long helix, lends itself to forming fibers. First, two alpha helices of keratin wind around each other. Their hydrophobic surface side chains interact favorably in this conformation. These two-chain coils lie end to end and side by side with from one to many other coils, forming fibers (Figure 5.15).

Disulfide bridges

These fibers are not only held together by favorable hydrophobic interactions between side chains but are also stabilized by disulfide bridges between the coils. The bridges make the fibers strong and rigid. Different keratin proteins have different amounts of cysteine to make the bridges: the harder the final structure (fingernails versus hair, for example), the more disulfide bridges are present. In the toughest keratins, such as tortoise shells, up to 18% of the amino acids are cysteines involved in disulfide bridges.

A hair and disulfide bridges

If you have ever had a permanent wave in your hair, you have manipulated the disulfide bridges of your keratin hair fibers. Disulfide bridges form only in the right kind of chemical environment. When you get a permanent, the hair stylist first wraps your hair around small rods. Next, a smelly solution is applied to your hair. The smelly solution contains a chemical that changes the environment in your hair and breaks the disulfide bridges holding the keratin fibers side by side (the chemical, called a reducing agent, is the component of the solution with the strong odor, too).

While this is going on, the hair stylist has also arranged for your hair to be warm, either by placing you under a hair dryer or by putting a plastic bag over your hair. The moist heat breaks some hydrogen bonds that keep the keratin alpha helices stiff, allowing them to relax a little. The net effect is that the keratin helices move a little with respect to one another while they are wound around the rods.

After your hair has incubated sufficiently in the warm environment with the reducing agent, the stylist rinses out the reducing agent and applies a neutralizing solution. This solution restores the original hair environment, allowing disulfide bridges to re-form. But here's the catch. Your hair has been relaxed around the curling rods, and many of the cysteine SH groups will form disulfide bridges with new cysteine SH groups. These brand-new SH bonds hold the hair fibers in the conformation they were in around the curl-

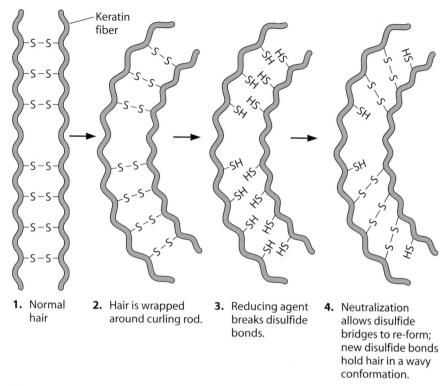

1. Normal hair

2. Hair is wrapped around curling rod.

3. Reducing agent breaks disulfide bonds.

4. Neutralization allows disulfide bridges to re-form; new disulfide bonds hold hair in a wavy conformation.

Figure 5.16 Biochemistry of a permanent hair wave.

ing rods: a permanent wave (Figure 5.16). Rinsing and cooling your hair allow the keratin helix hydrogen bonds to reestablish themselves, returning your hair to normal, except that new disulfide bridges now hold it in a wavy shape.

The lambda repressor binds to a specific DNA sequence

Regulation of gene expression is important for many different cellular and organismal functions. Gene regulation usually involves the binding of proteins to DNA, and we want to give you a close-up look at how a protein can bind to a specific base sequence. We introduced the lambda repressor above in our discussion of protein domains. Now we will show you in more detail how the structure and amino acid sequence of the N-terminal (DNA-binding) domain of the repressor allow it to bind DNA.

The DNA sequence to which the repressor binds is symmetrical; each N-terminal domain of the repressor interacts with identical bases. Figure 5.17A shows how the two N-terminal domains sit on their DNA-binding sites. The amino acid side chains of helix 3 contact specific bases within the DNA-binding sequence (Figure 5.17B). When the protein binds to DNA, helix 3 sits along the DNA molecule so that these specific contacts can occur. Figure 5.17C shows one of these amino acid-base contacts in detail. Note that the amino acid forms hydrogen bonds with the edge of the adenine base.

From this example, you can see that both the overall structure of the lambda repressor and its specific amino acid sequence are important to its function. The helices within the domain are oriented so that helix 3 can sit alongside the DNA molecule. The specific amino acids within helix 3 must contact specific bases for the binding to work, so the protein binds only to its recognition sequence.

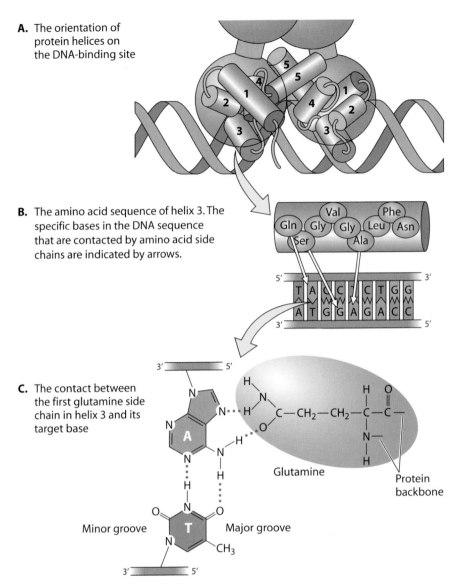

A. The orientation of protein helices on the DNA-binding site

B. The amino acid sequence of helix 3. The specific bases in the DNA sequence that are contacted by amino acid side chains are indicated by arrows.

C. The contact between the first glutamine side chain in helix 3 and its target base

Figure 5.17 Binding of lambda repressor protein to DNA. (From M. Ptashne, *A Genetic Switch*, 2nd ed. [Blackwell Scientific Publishing and Cell Press, Cambridge, Mass., 1992].)

The Trp repressor changes its shape and binds DNA in response to environmental conditions

Another theme in this book is that the binding of a protein to some other molecule can change its shape and thereby its activity. We will now show you an example, again involving gene regulation. Our example protein, the tryptophan (Trp) repressor, is found in *E. coli* and bears some similarity to the lambda repressor.

Like the lambda repressor, the Trp repressor is a dimeric protein that binds to DNA through amino acids that are part of a helix. Binding of the Trp repressor to its DNA target shuts off the transcription of nearby genes. Unlike the lambda repressor, the Trp repressor must also be bound to tryptophan itself in order to be able to bind to DNA. This requirement enables *E. coli* to respond beautifully to its environment.

The genes regulated by the Trp repressor encode enzymes that synthesize the amino acid tryptophan. *E. coli* needs tryptophan for synthesizing its own proteins, so these genes (the Trp genes) are normally "on," being transcribed into mRNA. If an *E. coli* cell finds itself in an environment where tryptophan is plentiful, though, it can simply absorb it from its environment and no longer needs to expend the energy to make it from scratch. Under these circumstances, it makes sense to turn the Trp genes "off." The Trp repressor carries out this function.

As noted above, unstructured amino acid loops are often involved in binding other molecules, and when they bind, they can change conformation and trigger additional shape changes in other regions of the protein. In the Trp repressor, the DNA-binding helices are folded in toward the main body of the protein and aren't positioned to fit alongside the DNA molecule (rather like the folded-in claws of a crab). Here's where tryptophan and the unstructured loop come in. If enough tryptophan is present inside the *E. coli* cell, the dimeric repressor protein binds to two molecules of it (one per monomer). The unstructured loop indicated in Figure 5.18 changes conformation as it binds to the tryptophan.

The tryptophan molecule and the amino acid loop fit into the structure like a wedge between the DNA-binding helix and the body of the protein, forcing the DNA-binding helix to swing out (imagine the crab's claws unfolding outward from its body) (Figure 5.18). This process takes place in both monomers. Now the helices are positioned to bind their DNA recognition sequences. The tryptophan-Trp repressor complex binds to its DNA recognition sequence and blocks further transcription of the Trp genes. The requirement that the Trp repressor be bound to tryptophan so that it can bind to DNA means that the Trp genes are shut off only when tryptophan is plentiful and the *E. coli* cell does not need to spend energy making any more.

When the concentration of tryptophan inside the cell falls, tryptophan dissociates from the Trp repressor, which then can no longer bind its DNA target. The Trp repressor releases the DNA, and transcription of the genes

Figure 5.18 Binding of the amino acid tryptophan to the Trp repressor protein changes the conformation of the repressor so that it can bind to DNA. (Reprinted from *Nature* **327**:591–597, 1987, with permission.)

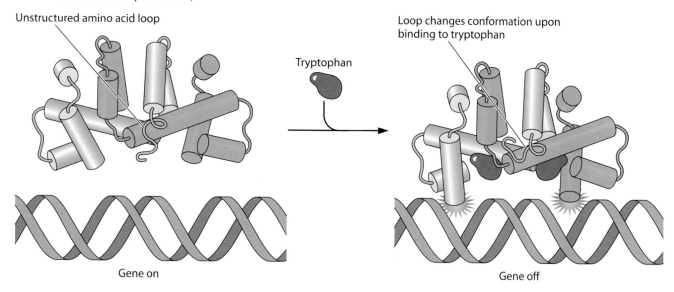

Unstructured amino acid loop

Loop changes conformation upon binding to tryptophan

Tryptophan

Gene on

Gene off

encoding tryptophan-synthesizing enzyme can occur. The enzymes then make tryptophan from scratch, supplying the cell with the amino acid it needs. Thus, the structure of the Trp protein lets *E. coli* respond to its environment appropriately.

PREDICTING PROTEIN STRUCTURE

The determination of a protein's structure is a scientific specialty all its own, requiring expensive instrumentation and highly specialized training. Most molecular biologists who study the functions of genes and proteins do not have the training or the instruments to determine a protein's structure. Even if they did, the methods available at this time do not work on all proteins. Thus, predicting protein structures from DNA (and therefore amino acid) sequences is part of trying to figure out what a protein does and how it does it or what part of a protein is doing a specific thing.

Unfortunately, predicting the three-dimensional structure of a protein from its amino acid sequence is a major unsolved problem in structural biology. Structural biologists have searched and are searching for clues in solved protein structures, looking for amino acid patterns that correlate with specific structures. Structure prediction computer programs based upon these statistical studies have been developed, and they are useful but not perfect.

Since many proteins share common domains, one of the ways scientists try to figure out a protein's likely structure and function is to compare it to other proteins. If you are trying to predict a three-dimensional structure for a protein whose gene you have recently discovered, or to figure out what the function of that protein might be, one of the first things you would probably do is compare the amino acid sequence of your protein (which you can figure out from the gene, using Table 4.1) to every other gene and protein anyone else has analyzed and published. If you found a previously published protein sequence, or domain of a protein, that was very similar to your protein, you'd have a clue about its structure or function. For example, finding a common transmembrane domain in your protein sequence would suggest that your protein sits in a membrane.

Public databases make information easily accessible

Making those comparisons used to be extraordinarily time-consuming, if not impossible. Scientists publish their results in numerous scientific journals, and finding references to all the published information and then getting access to the articles used to be a potentially Herculean task. In the 1980s, the U.S. government, through the National Institutes of Health, provided a resource that renders the task relatively simple and makes a wealth of scientific information readily available to anyone with Internet access.

In 1988, the National Institutes of Health established the National Center for Biotechnology Information (NCBI), which consists of numerous electronic resources, including databases of gene and protein sequences. Scientific journals now require scientists to deposit gene and protein sequence information in these databases as part of publishing the results of their research. All the databases are accessible through the NCBI website, so with a few mouse clicks, you can compare any DNA or protein sequence to all the sequences in the databases. These online databases and the search engines developed to go with them have revolutionized the process of comparing gene and protein sequences (Box 5.3).

BOX 5.3 *Tour some protein structures*

One of the NCBI databases is a protein structure database. We encourage you to go to the website (http://www.ncbi.nlm.nih.gov) and observe a very tiny portion of your tax dollars at work. At the NCBI homepage, search the structure database for plastocyanin. Click on any entry from the results page, and choose to view the structure. You'll be able to download freeware to do this. Play with the structure view. Find the ball-and-stick and space-filling styles, and look at the animations. Try some other proteins, human hemoglobin or keratin, for example.

While you're at the NCBI website, surf around. The website is a resource for scientists and physicians, so much of it is written in highly technical language (we'll be the first to admit we don't understand everything on the website), but there are segments written for the public. If you surf through some of the human genome resources, you'll find that you can even download a copy of the sequence of the human genome.

Testing structure-function predictions

Suppose that you've compared your protein sequence to the databases and found enough similarities to other proteins to make predictions about your protein. What next? Molecular biologists have come up with ways to test predictions about what parts of a protein are involved in specific functions even when they don't know what the protein's structure is. How can they do this? In a way, they imitate nature. They introduce mutations into the protein (by manipulating the DNA sequence of its gene) and determine what the changes do to the protein's function. By carefully designing these mutation experiments, they can often get good evidence that specific amino acids are crucial to specific functions of the protein. In the end, though, the only way to be completely certain of a protein's structure is to use one of the instrumental methods described in chapter 15 for determining it, a long and exacting process.

PROTEIN ENGINEERING

As you read in chapter 1, advances in scientific understanding and new technological capabilities usually nudge each other along. The field of structural biology is no exception. As scientists have learned more about the relationship between protein structure and function and learned how to determine the contribution of individual amino acids to both aspects of a protein, they have also begun to manipulate them. The deliberate manipulation of a protein's amino acid sequence to change its function or properties is called **protein engineering**. Proteins can be manipulated simply to learn more about how they function, but when we say protein engineering, we mean the altering of proteins to make them more useful to humans.

Candidates for protein engineering are proteins that are used outside of the cells in which they were originally made, such as proteins administered to people as therapeutic agents, used in manufacturing, or added to products to enhance them. In the Biotechnology Applications and Issues section of this book, we provide information about many different uses of proteins.

One arena in which protein engineering is potentially very useful is chemical manufacturing. Often, the very same reactions used in the chemical industry are carried out inside cells by enzymes. The same enzymes can often be used outside of cells to carry out the reactions in industrial processes, usually generating far less pollution than traditional chemical manufacturing (see chapter 24). Increasing the use of enzymes in manufacturing is thus an environmentally desirable goal. Protein engineering can make enzymes more convenient and cost-effective in industrial applications.

Proteins can be made more resistant to denaturation

One desirable way in which industrial enzymes could be modified would be to increase their stability so that they would be less vulnerable to changes in temperature or other reaction conditions. Hardier enzymes could be used under a greater variety of industrial conditions, and higher temperatures could mean higher reaction rates. Increasing the stability of a protein means increasing the stability of the protein's tertiary and, possibly, quaternary structures. Therefore, before a scientist can undertake to engineer a protein for increased stability, he or she must know its three-dimensional structure. Once that structure has been determined, the scientist looks for ways to increase stability without altering the protein's function.

The most obvious way to increase a protein's stability is to introduce disulfide bridges to hold the tertiary structure together. These bridges have to be geometrically compatible with the protein's structure and must not alter crucial amino acids. The bacteriophage enzyme lysozyme has been engineered in this way. This enzyme has two domains. Using recombinant DNA technology, scientists introduced codons for cysteines in positions within the gene that were selected after careful study of the protein's structure. The altered gene was reintroduced into cells, which duly produced the protein with the new cysteines. The cysteines combined to form three disulfide bridges that tied the domains together in their appropriate three-dimensional configuration (Figure 5.19). These alterations significantly increased the thermal stability of the enzyme. The engineered protein withstood heating to a temperature 23°C higher than the native protein could endure.

Figure 5.19 Protein stability engineering. Three engineered disulfide bridges (gold) tie the two domains of bacteriophage T4 lysozyme in the proper configuration. The two domains are shown in lavender and green. The numbers indicate the positions of the cysteines in the 164-amino-acid polypeptide. The cysteine at position 54 was changed to a threonine to keep it from interfering with proper formation of the engineered bridges. (Adapted from M. Matsumara, G. Signor, and B. W. Matthews, *Nature* **342:**291, 1989, with permission.)

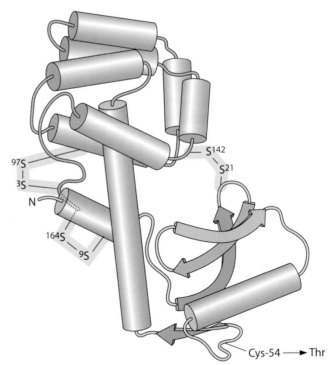

Your laundry detergent may already contain an engineered protein

The bacterial enzyme subtilisin breaks peptide bonds and so digests proteins. Proteins are a common component of laundry stains, and laundry detergent manufacturers have been adding subtilisin and other enzymes to detergent for years. Unfortunately, subtilisin is inactivated by bleach. Scientists studied the inactivation and found that it was caused by a reaction between molecules of bleach and a methionine side chain at position 22 in the protein.

Scientists cloned the gene for subtilisin and altered the codon for methionine-22. They substituted codons for a variety of different amino acids and tested the resulting proteins for their protein-digesting activity and bleach resistance. They got the best results when they substituted an alanine for the methionine. The mutated subtilisin was still active in digesting proteins and was also far more resistant to bleach. Many laundry detergents already contain the cloned, genetically engineered protein—something to think about when you do your laundry.

Scientists are already looking beyond simply modifying existing proteins to creating new ones from scratch. A group from Denver has already produced the first artificial enzyme. Using computer programs that make predictions about protein structure plus their knowledge of existing enzymes, they designed a protein-digesting enzyme from scratch. When it was synthesized, it had the predicted activity. Researchers involved in making artificial enzymes hope that they can be used to usher in an era of "green chemistry," chemical manufacturing that uses renewable resources and doesn't produce toxic wastes. More information about the industrial uses of enzymes is presented in chapter 24.

SUMMARY POINTS

Proteins are chains of amino acids connected by peptide bonds. Amino acids consist of a common chemical backbone, which participates in the peptide bonds, and unique side chains, which give an amino acid its identity and determine its chemical nature.

Some amino acid side chains are hydrophobic, while others are hydrophilic. The backbone is hydrophilic. When a protein folds, it generally segregates the hydrophobic side chains in the interior of the protein, where they interact with one another away from water molecules in the cytoplasm. Segments of the amino acid backbone usually adopt one of two standard arrangements, the alpha helix or the beta sheet, to neutralize their own charged natures.

Amino acid chains tend to fold up into compact three-dimensional structures called domains. A protein can be made of one or many domains. Domains are fundamental units of overall structure and function, and often a particular protein domain is associated with a particular function of that protein. Similar-looking domains can be found in different proteins, and many proteins appear to be modular assemblies of domains found in many other proteins.

Protein structure is held together by hydrogen bonds and disulfide bridges between cysteine residues. Proteins can be denatured by treatments that break these bonds, such as heat or harsh chemicals. If a protein is denatured, it often cannot refold itself even if the denaturing agent is removed.

Proteins carry out functions through molecular interactions between specific side chains and their targets. Scientists can discover which amino acids are necessary for specific functions by systematically altering them via genetic manipulation and seeing if the function is affected.

Scientists also learn about new proteins by comparing their amino acid sequences to the sequences of all other known proteins, using national databases and search engines.

Researchers are using their knowledge of protein structure and function to improve the usefulness of enzymes in industrial and other processes. They do this by manipulating the DNA sequences of genes encoding these proteins, so that the proteins will have altered amino acid sequences and characteristics.

Prion disease agents, such as the causative agent of mad cow disease, appear to be improperly folded proteins that are able to corrupt native, correctly folded proteins. Once the native proteins are corrupted, they cannot return to the correct folding and so lose their functions.

KEY TERMS

Alpha helix	Electronegativity	Polar bond	Quaternary structure
Beta sheet	Hydrogen bond	Polypeptide	Secondary structure
Denaturation	Keratin	Primary structure	Surface tension
Disulfide bridge	Peptide backbone	Prion	Tertiary structure
Domain	Peptide bond	Protein engineering	

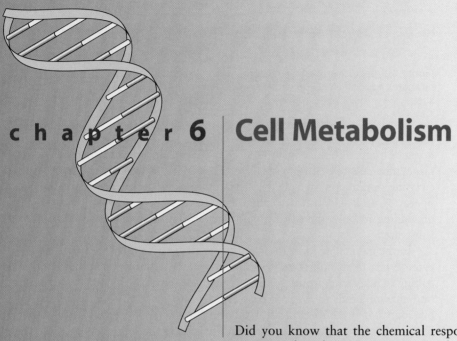

chapter 6 | Cell Metabolism

Did you know that the chemical responsible for the burning sensation in your muscles when you exercise very hard and the alcohol in beer and wine are both breakdown products of glucose, or that the artificial sweetener in diet soda and the hormones in birth control pills are both manufactured with the assistance of microorganisms? Welcome to the world of metabolism, the processes by which cells break down food to get energy and use energy and building materials to manufacture their component parts.

Obtaining and using energy are some of the processes carried out by all living cells: your cells, plant cells, microbial cells, all cells. When you eat, you may think you are filling your stomach, but what you are really doing is supplying your cells with energy. In this chapter, we will discuss how food is converted to energy for your cells and how cells use that energy to make molecules they need. Specifically, we will look at:

- an overview of metabolism at the molecular level
- the consequences of various metabolic deficiencies
- examples of how biotechnologists are using our understanding of metabolism to solve problems and create useful products

As you read this chapter, don't be intimidated by biochemical names, and please don't concern yourself with memorizing them. We had to call the molecules something to distinguish them from each other, and naming them molecule 1 and molecule 2 often ends up being more confusing than calling them by their real names. Our goal is for you to understand the concepts.

METABOLIC PATHWAYS

Cells break down and manufacture molecules in a sequential set of reactions known as **metabolic pathways**. Each step in a metabolic pathway is catalyzed by a specific enzyme. Recall that the function of an enzyme, which is a protein,

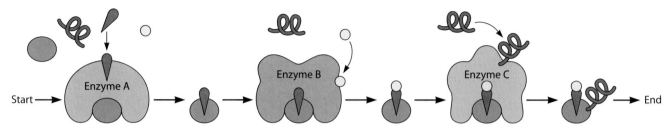

Figure 6.1 Cellular processes are carried out by a sequence of enzymes, each of which promotes a specific reaction with a specific substrate, rather like a robot in an assembly line.

depends on its three-dimensional shape. The enzyme must fit correctly with its target molecule to be able to catalyze a change in the target. The molecules that enzymes operate on are called **substrates**, and the molecules they produce by operating on a substrate are called their **products**. Most enzymes are capable of binding to only one substrate and generating only one product from that substrate.

In the generalized metabolic pathway depicted in Figure 6.1, the first enzyme binds to its substrate, the starting material, and an action occurs, resulting in a specific product that is the substrate for the second step. The second enzyme binds to that molecule, and another action occurs, generating a product that is the substrate for the third and final step, which leads to the pathway's end product. You can think of a metabolic pathway as rather like an assembly line of robots, in which each robot has a shape that allows it to attach to its target and only its target.

Pathways branch, converge, and form networks

Metabolic pathways do not exist in isolation from each other. Sometimes the end product of one pathway is the starting substrate in another pathway, so that the first path feeds into the second. In Figure 6.1, the materials produced by enzymes A and B on the way to the end product are called intermediates. Sometimes the end product of one pathway is an intermediate in another pathway, linking the two pathways into a network.

Metabolic pathways are very often branched because there is more than one possible fate for an intermediate in the pathway. Another way to look at branching pathways is that two processes share many steps, but at a certain intermediate, they diverge. Figure 6.2 illustrates a pathway containing a branch. If oxygen is present, the intermediate at this branch point, pyruvic acid, will be broken down to carbon dioxide and water. If not, yeast cells will convert the pyruvic acid to alcohol, and animal cells will convert it to lactic acid.

Some metabolic pathways converge with each other like two roads merging into one. For example, the pathway that cells use to break down fats to get energy converges with the metabolic pathway for breaking down glucose to get energy.

When an enzyme in a pathway is missing, intermediate products accumulate

If one of the enzymes in a metabolic pathway is missing or doesn't work properly, it is as if one of the robots in an assembly line was broken: it derails the entire assembly line. For example, if the second protein in the

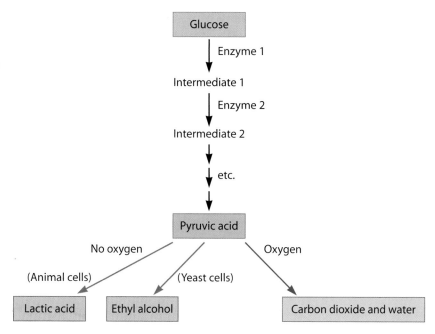

Figure 6.2 A branching metabolic pathway.

process is missing, then its particular operation will never be carried out. The third protein cannot bind to the product of the first step, so the product of the first step in the process will build up (Figure 6.3).

The consequences of having a broken step in a pathway depend on a number of things. How important is the normal end product of the pathway? If the normal end product or any of the intermediates between the broken enzyme and the end product is essential to the survival of the cell, then the cell dies. If none of these molecules is essential for the cell, will the substrate of the defective enzyme build up in the cell? If that molecule is at a branch in the pathway or is the molecular link to another pathway, it may not build up but be channeled into the other pathways. If it does build up in the cell, is its accumulation harmful? If so, how harmful? Does the accumulation kill the cell or simply change its characteristics or functions?

Later in this chapter, we will look at some specific examples of the varied consequences of not having certain enzymes involved in metabolism.

Figure 6.3 If one enzyme in a multistep process is missing or fails to function, the intermediate synthesized in the previous step may build up, and the end product cannot be made.

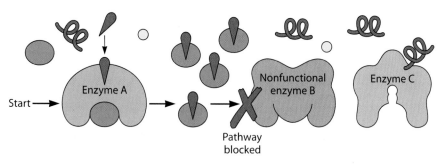

Different organisms use the same or very similar metabolic pathways

One thing scientists have learned as they have collected more and more information about cell and molecular biology is that organisms are fundamentally much more alike than they are different. Nowhere is this statement more obvious than at the level of metabolic pathways. Organisms that seem to be very different, such as bacteria, plants, and humans, share virtually all of the pathways involved in breaking down proteins, fats, and carbohydrates for energy. The more similar the organisms, the more pathways they share. Therefore, scientists can study metabolic pathways in an organism like yeast or the fruit fly and apply the knowledge they gain, with very few changes, to other organisms. Much of the information you will read in this and later chapters was originally learned through studying small, easy-to-use organisms like bacteria and yeast.

CATABOLISM AND ANABOLISM

Cells require a constant supply of energy to stay alive. Plants, algae, and a few bacteria can use the energy in sunlight, but once the sun goes down, even they have to get energy elsewhere. Chemical bonds contain energy, and when bonds are broken, energy is released. Conversely, when chemical bonds are formed, energy is required. Biologists use the word **catabolism** for the breakdown, energy-yielding half of metabolism and **anabolism** for the synthesis, energy-requiring half. All organisms have the ability to break bonds in molecules to get energy, and they all use energy to form chemical bonds in molecular synthesis.

Energy is released when chemical bonds are broken

Actually, nearly all chemical reactions involve both the making and breaking of chemical bonds. Whether energy is released or consumed in the reaction depends on which set of bonds contained more energy—the starting materials or the products. A good example of energy release through bond breaking is the burning of gasoline. You have to put in a little energy (from a match, perhaps) to get the reaction started, but then a huge amount of energy is released in the forms of heat and light.

What's happening at the level of the gasoline molecules? The chemical reaction is shown in Figure 6.4. Gasoline is a mixture of **hydrocarbons**. Hydrocarbons are substances made from hydrogen (hydro) and carbon. Bonds between carbon and hydrogen contain a great deal of energy. When a hydrocarbon like gasoline burns, the carbon-hydrogen and carbon-carbon bonds in the gasoline molecules break, and new bonds between carbon and

Figure 6.4 Burning gasoline breaks the C-H and C-C bonds in gasoline to produce CO_2, H_2O, and lots of energy.

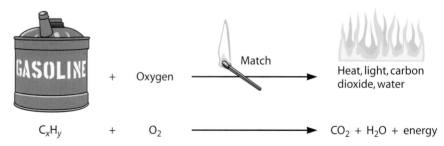

oxygen and hydrogen and oxygen form. The products of the reaction are carbon dioxide and water. The bonds in the product molecules contain far less energy than the bonds in the hydrocarbon, so when a hydrocarbon burns, energy is released.

Believe it or not, the same kind of process takes place in your cells. Bonds between carbon and hydrogen and between carbon and carbon, as well as others, are broken. Your body uses oxygen in the same reactions as the burning gasoline, producing the identical end products, carbon dioxide and water. In the body, too, the most energy-rich bonds are between carbon and hydrogen. Guess which molecules contain the most carbon-hydrogen bonds? Lipids (fats and oils). That is why fats have more calories per gram than other types of food. When your body breaks down a fat, all that lovely energy in those carbon-hydrogen bonds is released. The energy released by breaking bonds in food molecules is used by your body to keep you warm, to power movement, and to drive the synthesis of your own cell's components.

In the cell, chemical bonds are broken and formed by enzymes

It may be difficult to relate the burning of gasoline to the chemical breakdown of fats, since we obviously don't have fire going on inside of us. The difference lies in how the process takes place. When gasoline burns, the reaction requires heat to get it started, and then the heat from the burning gasoline keeps the fire going. In your body, the reaction takes place at body temperature (obviously). The fat-burning reaction can happen at this relatively cool temperature because enzymes are making it happen.

In the previous section, we spoke rather vaguely about enzymes causing "an action to occur." When fats are being broken down, the action occurring is bond breaking. Certain enzymes break chemical bonds within molecules, and other enzymes, which we will discuss later, make chemical bonds between molecules. An enzyme's three-dimensional shape and the side chains of the amino acids within that shape allow it to bind very specifically to its substrate and then to break and/or form bonds to make the product. The product has a shape different from that of the substrate, does not fit with the enzyme, and is released.

The energy balances of enzymatic reactions are the same as those of chemical reactions like burning gasoline. If the bonds in the substrate molecule(s) have more energy than the bonds in the product molecule(s), then energy is released in the reaction. If the bonds in the product molecule(s) have more energy than the bonds in the substrate molecule(s), then energy must be added to the reaction (Figure 6.5).

Catabolism of food begins with the digestive system

In the human body, energy and building materials are obtained from breaking down, or catabolizing, carbohydrates, lipids, and proteins. The first stage of catabolism, in which the large molecules are broken into their component parts—carbohydrates into sugars, proteins into amino acids, and fats into fatty acids and glycerol—occurs in the digestive tract (Figure 6.6). The breakdown of starch, a carbohydrate that is a chain of glucose molecules, begins in the mouth. You may already know that if you take a bite of a saltine cracker or a piece of white bread and hold it in your mouth, you can begin to taste a slight sweetness as an enzyme in your saliva, amylase, breaks the starch down into its building block, glucose. Digestion of proteins into amino acids begins in the stomach, and digestion of fats to fatty acids begins in the small intestine.

Figure 6.5 Energy balance in chemical reactions. If the chemical bonds in the product(s) contain more energy than the bonds in the reactants, energy is consumed in the reaction. In the opposite situation, energy will be released in the reaction.

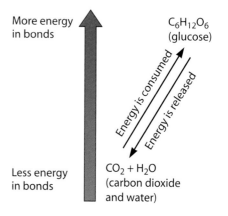

More energy in bonds

$C_6H_{12}O_6$ (glucose)

Energy is consumed

Energy is released

Less energy in bonds

$CO_2 + H_2O$ (carbon dioxide and water)

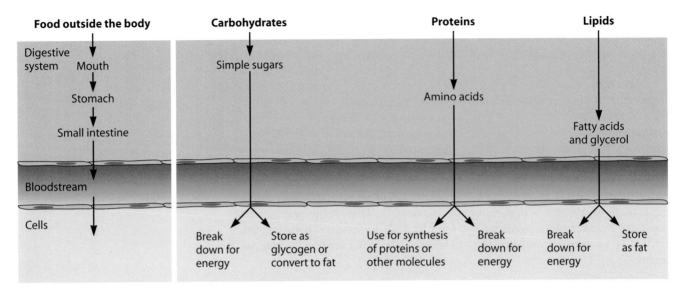

Figure 6.6 Catabolism in animals.

Lipids, carbohydrates, and proteins can be used for energy or as sources of building block molecules

Sugars, fatty acids, and amino acids are absorbed into the bloodstream through the walls of the small intestine. Cells that need a quick shot of energy take up glucose or another sugar from the blood and break it down immediately. If the supply of glucose exceeds the body's demand for quick energy from glucose, enzymes in liver and muscle cells hook the glucose units together into a glycogen molecule, which is stored until the demand for energy from glucose increases again. Fatty acids absorbed into the bloodstream can also be broken down by cells that need energy, such as exercising muscle cells, or if energy is not needed by cells, they can be converted to fat and stored as fat droplets in fat cells. Glycogen and fat are the two main energy storage compounds. Glycogen acts as a short-term storage molecule, and most people have a glycogen supply that could provide cells with energy for about 1 to 2 days. Fats are for long-term energy storage, and a normal-size person has enough fat to power cells for 4 to 6 weeks.

Finally, cells absorb amino acids from the bloodstream and link them together to build proteins, or they remodel them into other amino acids. If necessary, cells can also use amino acids for energy, but that is not their primary fate. Cells would rather not break down amino acids for energy because proteins are responsible for making cells work. The protein inventory of a cell changes from one minute to the next, so cells always need a ready supply of all 20 amino acids for protein manufacturing.

Glucose is broken down for energy via a series of three pathways

Glucose is the major energy molecule for cells from almost all organisms, both prokaryotes and eukaryotes. Cells break down glucose to release the energy in its bonds via three sequential pathways. The first two pathways convert glucose to carbon dioxide and water and capture most of the energy released in temporary storage molecules. In the third pathway, the energy is transferred to a form the cell can use to power other reactions. The overall process of breaking down nutrients (glucose, as well as other molecules) to obtain energy is called **respiration**.

The first pathway, **glycolysis**, is found in bacteria, as well as plants and animals. The presence of this biochemical pathway in such diverse organisms implies that it evolved in ancient times, before the ancestors of these various life forms diverged. In this nine-step pathway, glucose, which contains six carbon atoms, is converted to two molecules of the three-carbon compound pyruvic acid. No oxygen is required to convert glucose to pyruvic acid, so even microorganisms that live in the absence of oxygen can use this pathway. Other sugars, such as fructose, enter the nine-step pathway at the second or third step, so they, too, are converted to pyruvic acid.

If oxygen is present, pyruvic acid is broken into two molecules: carbon dioxide and acetyl coenzyme A (acetyl-CoA). Acetyl-CoA feeds into the second pathway, which is called the **Krebs cycle**, tricarboxylic acid cycle, or citric acid cycle. Krebs is the name of the scientist who worked out the steps in the cycle; tricarboxylic acid and citric acid are intermediates in the cycle. The Krebs cycle requires oxygen to finish the complete breakdown of glucose to carbon dioxide and water. The great majority of the energy released from breaking down glucose is generated in the Krebs cycle. As energy is released from breaking the chemical bonds in glucose, it is captured in temporary storage molecules.

If oxygen is not present, the pyruvic acid is converted to a different end product, depending on the enzymes present in the organism. In yeast, for example, pyruvic acid is converted into ethyl alcohol. People take advantage of this metabolic pathway to brew beer or make wine. For information about how your own cells handle pyruvic acid in the absence of oxygen, see Box 6.1.

How are molecules other than glucose broken down to get energy? All of them are broken down into molecules that are intermediates in the glucose breakdown pathway just described. Fatty acids, another important energy molecule in cells, are broken down via a multistep pathway to acetyl-CoA, which enters the Krebs cycle. The 20 different amino acids have 20 different starting structures. By 20 different, but converging, pathways, all of which first require the removal of nitrogen, amino acids are converted into either pyruvic acid, acetyl-CoA, or one of the other molecules in the Krebs cycle. Like cars merging onto an interstate highway, sooner or later the breakdown pathways for glucose, fatty acids, and amino acids merge into a common pathway (Figure 6.7).

The electron transport pathway stores energy as ATP molecules

As molecules of lipid, carbohydrate, and protein are broken down in glycolysis and the Krebs cycle, the energy released is used to form temporary energy storage molecules. These temporary energy storage molecules then feed into the third major step in catabolism, the **electron transport pathway**. In this pathway, the energy in the temporary storage molecules is transferred to the molecule that forms the cell's major energy currency: **ATP**.

You have already met adenosine: it is a nucleotide with the base adenine. In nucleic acids, nucleotides are associated with one phosphate group. When they are free in the cell, nucleotides can have one, two, or three phosphates connected in a row. ATP stands for **adenosine triphosphate**, the three-phosphate form (see Figure 3.16). The adenosine found in DNA is made with the sugar deoxyribose; the form of ATP used for energy storage is made with ribose. The bond between the second and third phosphate is a very high-energy bond. It takes a lot of energy to make it, and when that bond is broken, a lot of energy is released.

B O X   6 . 1   *The exercise burn and a metabolic branch point*

Have you ever exercised until you felt your muscles burn? The burning is a direct result of a branch point in the metabolic pathway for breaking down glucose and other substrate molecules.

When you exercise, your muscles need a lot of energy. Exercising muscle cells break down lots of glucose—muscles even have their own glycogen stores for just such situations. Breaking down glucose to carbon dioxide and water requires another input: oxygen. When your muscles are working hard, you are breathing harder, and your heart is beating faster to supply your muscles with extra oxygen to power all that glucose catabolism. Sometimes, though, you cannot get enough oxygen to your muscles. You might be putting extraordinary demands on isolated muscles, as when you lift weights, or demanding output from your entire body that exceeds your capacity to supply oxygen via breathing and blood circulation, as when you sprint for a while.

When your muscles can't get enough oxygen, the normal catabolism of glucose to carbon dioxide and water cannot happen. Look at the figure. When oxygen is present, pyruvic acid is converted into acetyl-CoA and on to carbon dioxide and water. This step is actually a

metabolic branch point. If oxygen is not present, your muscle cells do something different with the pyruvic acid. They have another enzyme that converts it into a three-carbon compound called lactic acid. When lactic acid builds up in your muscles, you feel that burning sensation.

The lactic acid burn is a sign that your muscles are working anaerobically (without oxygen). You've probably heard

of aerobic exercise. The word aerobic means with oxygen. Aerobic exercise by definition causes you to breathe hard, but your breathing can keep up with your muscles' need for oxygen. That means you can sustain the exercise for longer periods of time. Because you are breathing harder and your heart is beating faster than usual during aerobic exercise, sustained exercise of this kind can strengthen your heart.

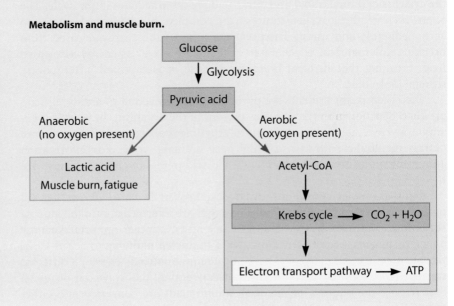

Metabolism and muscle burn.

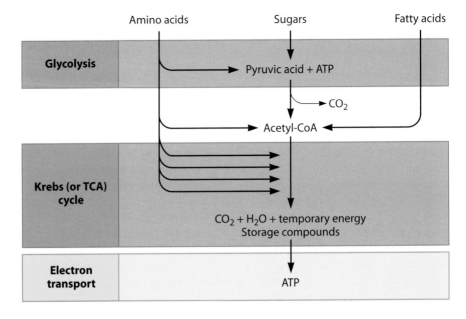

Figure 6.7 Breakdown pathways for sugars, amino acids, and fatty acids converge. TCA, tricarboxylic acid.

In the electron transport pathway, the energy released during the step-by-step breakdown of carbohydrates, lipids, and proteins is harvested from various temporary storage molecules to make ATP by attaching a phosphate group to adenosine diphosphate, or ADP. ATP molecules are like money in the cell's energy bank. When the cell needs energy to drive a process, it can break that high-energy bond and release the energy (Figure 6.8). The body is a fast spender, though; it doesn't store ATP. It is estimated that if ATP production were stopped, you would use up all your ATP in about 5 minutes. ATP is constantly being formed and then broken back down into ADP and phosphate.

If an organism's energy intake and output are balanced, its cells break down sugars and fatty acids to form ATP and then use that ATP to synthesize their own proteins, lipids, carbohydrates, nucleic acids, and other components, as well as to power movement and other processes. If the organism takes in more energy than it needs, it uses excess ATP to drive the synthesis of long-term energy storage molecules, such as starch, glycogen, and fats. If an organism takes in less energy than it needs, it must obtain energy by breaking down its long-term storage molecules and releasing the energy captured in those chemical bonds.

The electron transport pathway is the final pathway in the process of obtaining energy from the food you eat. As part of the electron transport pathway, oxygen from the air you breathe is converted into water molecules inside your cells. Your body has such a constant need for the ATP supplied by the electron transport system that you will die quickly if deprived of oxygen. In fact, certain poisons, like cyanide, kill by blocking electron transport, resulting in quick and unpleasant death. Cyanide binds to one of the enzymes involved in electron transport and inactivates it. The effect on the victim is similar to what would happen if a person were deprived of oxygen. Cellular energy production stops, and cells die. Though we may think of cyanide in the context of murder mysteries, many plants produce compounds that can break down to produce cyanide, and cyanide poisoning sometimes occurs in animals and people who eat these plants.

Figure 6.8 ATP is the cell's major energy storage molecule. It is synthesized from ADP and a phosphate group, with a large energy input. Breaking ATP down into ADP plus phosphate releases the energy.

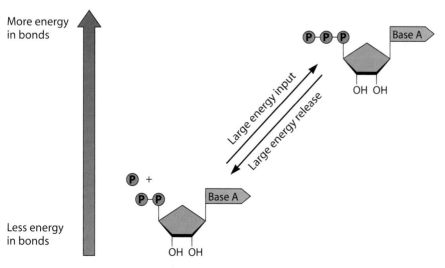

Anabolic processes require energy from ATP and chemical building blocks

Cells synthesize many different molecules for their own purposes. There are hundreds of enzymes inside a cell, and every one is a protein the cell synthesized. The cell membrane is composed of lipids synthesized by the cell. Muscle and liver cells synthesize the carbohydrate glycogen. Plant cells synthesize and store starch. Cells synthesize large molecules by joining together the component building blocks, and they also synthesize the building blocks from scratch.

Many of the intermediates of glucose breakdown are used as starting materials for making amino acids, nucleotides, fatty acids, glycerol, and other small molecules, such as steroids (Figure 6.9). Therefore, sugar molecules like glucose can be used not only for energy but also to make amino acids. Amino acids and fatty acids can also be broken down and reassembled into glucose molecules. The carbon and hydrogen atoms in a fat may end up in DNA after having been part of a glycogen molecule. If there is more energy available than the body needs, the energy and carbon atoms in glucose can be converted into a fatty acid for storage.

The synthesis of biological molecules costs energy. Cells power the energetically costly anabolic reactions by coupling them to energy-yielding reactions, usually the breaking down of ATP to ADP, rather like burning gasoline (an energy-yielding reaction) to heat water (an energy-requiring process). ATP thus serves as a sort of energy currency for the cell—it is created during

Figure 6.9 Many intermediates of glucose breakdown can be used as starting materials in the synthesis of other molecules.

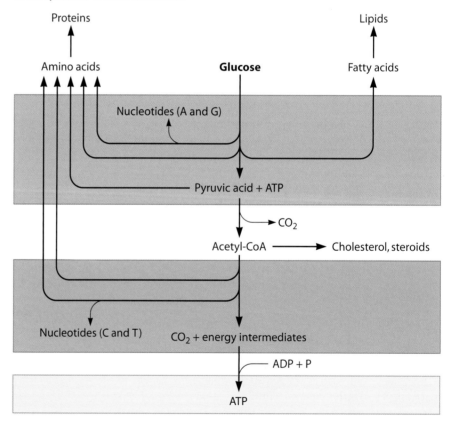

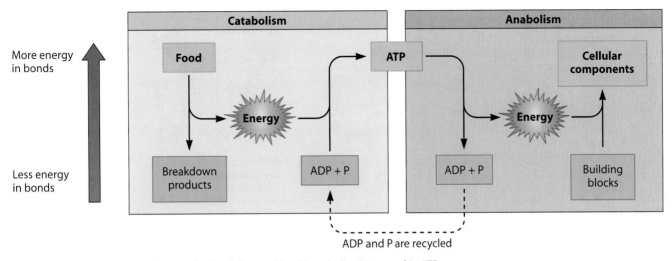

Figure 6.10 Energy released during the breakdown of food (catabolism) is stored in ATP. ATP is then broken down, releasing its stored energy, to power biosynthesis of cellular components (anabolism).

the energy-yielding processes of catabolism and spent in the energy-requiring processes of anabolism (Figure 6.10).

REGULATION OF METABOLISM

In chapter 2, we introduced the concept of regulation. Cells regulate all of their processes in order to ensure that they meet their needs, which change from one moment to the next, as efficiently and effectively as possible. As we mentioned previously, the stepwise nature of pathways provides many possibilities for regulation, since any step can be controlled. One way to modulate a pathway is to alter the activity of an enzyme in the pathway. A second mechanism involves not synthesizing an enzyme in a certain pathway.

In feedback inhibition, the end product shuts off the process

Feedback inhibition is a common mechanism for controlling a process, even those that are not biological. A form of feedback inhibition that is probably familiar to you is the operation of a thermostat that turns off the furnace when the room temperature reaches a predetermined temperature and then turns the furnace back on when the room gets too cold.

In metabolic pathways, a substance produced in the pathway, typically the end product, inhibits the activity of an enzyme at the beginning of the pathway. The inhibitor accomplishes this by binding to the enzyme in such a way that it alters the enzyme's shape so that the enzyme can no longer bind to its substrate. This action disables the pathway; the inhibiting molecule is no longer produced, the enzyme regains its binding capability, and the metabolic pathway is reactivated (Figure 6.11). Using the analogy of the thermostat-controlled heating system, the heat acts as the inhibitor and the furnace plays the role of the enzyme.

Examples of feedback inhibition in amino acid biosynthesis

One of the simplest examples of feedback inhibition in metabolism is the pathway for synthesis of the amino acid isoleucine. The starting substrate in the isoleucine synthesis pathway is another amino acid, threonine. Isoleucine

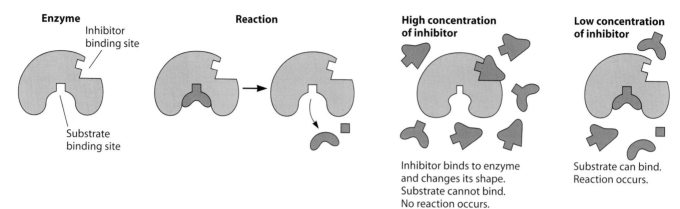

Figure 6.11 Enzyme inhibition.

feeds back and inhibits the enzyme that catalyzes the first step in the pathway. When a surplus amount of isoleucine is available, it blocks its own synthesis. When the amount of isoleucine decreases, the enzyme is released from inhibition and isoleucine synthesis resumes (Figure 6.12).

Using feedback inhibition, cells not only modulate the production of an end product, they also shunt activity down certain routes in branched pathways. In the amino acid synthesis pathway in Figure 6.13, the end product of the first branch, methionine, inhibits the first enzyme in that branch, while the end product of the second branch, threonine, inhibits the first enzyme in that branch.

As we mentioned in chapter 2, one of the benefits of having pathways that consist of many steps catalyzed by specific enzymes is the availability of a variety of control points for regulating the synthesis of different end products. As you can see in Figure 6.14, cells carefully control which amino acid is produced from aspartic acid, an amino acid that serves as the starting material for synthesizing other amino acids.

Cells can control metabolism by synthesizing certain enzymes but not others

Cells can also regulate their metabolic pathways by altering which enzymes are produced. For example, most bacteria have genes encoding enzymes capable of breaking down a variety of sugars they might encounter. However, synthesizing these enzymes would be a waste of the cell's energy if the sugars were unavailable to the cell. Most bacteria can sense which sugars are available in their environment, and they use that information to synthesize the necessary enzyme(s) for breaking down that sugar.

Conversely, most bacteria produce enzymes capable of synthesizing all of the amino acids bacteria need to make their proteins. However, synthesizing these enzymes would be wasteful if the required amino acids were already present in the environment, so most bacteria produce a specific amino acid-synthesizing enzyme only if that amino acid is not available to them from their environment.

Controlling tryptophan biosynthesis in *E. coli*

Escherichia coli requires tryptophan on an ongoing basis for synthesis of its proteins. The bacterium can synthesize tryptophan from scratch via the activities of several enzymes encoded by the *trp* genes. The *trp* genes are lined

Figure 6.12 A simple example of feedback inhibition. When isoleucine is plentiful, it blocks its own biosynthesis by inhibiting enzyme 1.

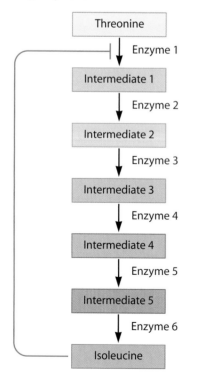

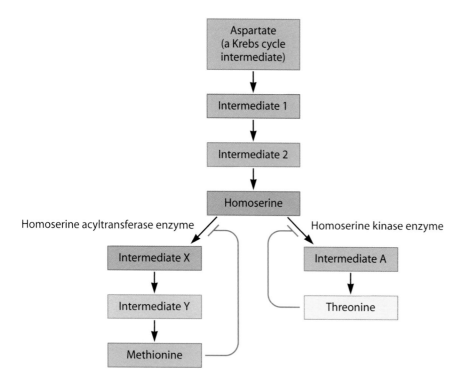

Figure 6.13 Feedback inhibition at metabolic branch points.

Figure 6.14 Feedback inhibition within a network of pathways.

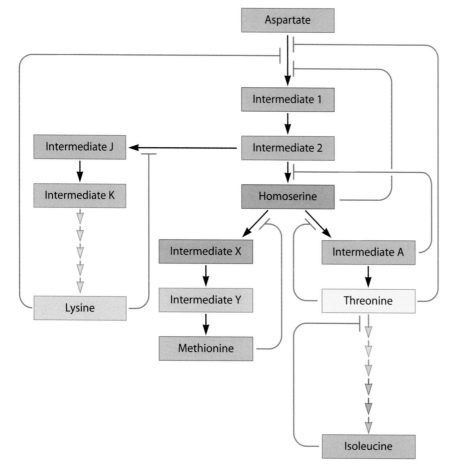

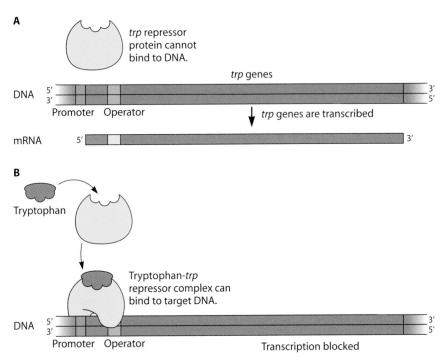

Figure 6.15 Regulation of tryptophan biosynthesis genes. P is the promoter of the *trp* genes. O is the repressor binding site. **(A)** Little tryptophan in cell. The repressor does not bind to target DNA. *trp* genes are transcribed. **(B)** Plentiful tryptophan in cell. Tryptophan-*trp* repressor complex binds to DNA and prevents the transcription of *trp* genes.

up along the *E. coli* chromosomes and are transcribed into a single long mRNA from the *trp* promoter. *E. coli* regulates expression of the *trp* genes through a repressor protein, which we introduced in the last chapter.

The *E. coli trp* repressor alone cannot bind to its target DNA near the *trp* promoter. In the presence of the repressor protein alone, the *trp* genes will be transcribed and the tryptophan biosynthesis enzymes will be produced. Those enzymes will synthesize tryptophan for the cell. However, if the concentration of tryptophan in the *E. coli* cell rises sufficiently, the excess tryptophan forms a complex with the repressor protein. Binding of tryptophan to the repressor alters its shape so that the complex can bind to DNA and shut off transcription of the *trp* genes (see Figure 5.18). Thus, when tryptophan is plentiful, transcription of the *trp* genes is turned off (Figure 6.15) and the cell does not waste energy making enzymes for synthesizing an amino acid it has in excess.

We will look at other mechanisms by which cells control gene expression in later chapters.

Multicellular organisms use hormones to coordinate and regulate some aspects of metabolism

Cells like your own, which exist inside a multicellular body that eats complex foods, don't usually encounter the variety of different kinds of food molecules that a free-living single cell such as a bacterium might. Even though you may eat a variety of foods, from potato chips to salads to ice cream, all of these foods are broken down in your digestive system into lipids, sugars, and amino acids. The problem your body faces is one of allocation and regulation rather than one of adapting to constantly changing substrate molecules.

For your body to function properly, your brain has to have glucose all the time. Your brain cannot run on any other fuel, and it requires a constant supply (even if you occasionally seem to be brain-dead). If there is extra glucose available, your body can store the energy as glycogen or fat. If you've been fasting and your blood glucose levels fall, your body must release some stored energy in the form of glucose so your brain can keep working.

Your body regulates glucose metabolism through the hormones insulin and glucagon, two small protein molecules that signal your body's cells to absorb glucose or to stop absorbing it and signal your liver to store glucose or release it. The mechanisms by which these hormones accomplish this will be discussed in chapter 8.

ERRORS IN METABOLISM

The cell's scheme for breaking down and synthesizing carbohydrates, fats, proteins, and their component building blocks consists of hundreds of steps, many of which link separate pathways into a network. Because a unique enzyme catalyzes each metabolic step, there are hundreds of opportunities for something to go wrong. What happens when a mutation leads to a missing or malfunctioning metabolic enzyme?

Just like any network, sometimes the consequences of a malfunction go unnoticed because other parts of the network compensate for the error. Other times, the error causes a network failure, and the system crashes. The same is true for errors in genes that encode metabolic enzymes. Sometimes the missing or malfunctioning enzyme is so crucial to survival that a fertilized egg with this mutation dies very early in development, so people with these types of devastating mutations do not exist. Other times, mutations for defective or missing metabolic enzymes pass from parents to children, sometimes with serious negative consequences and other times with little or no impact on the child's health.

The effect of a metabolic error depends on several factors

A genetically determined enzyme defect leading to interruption of a metabolic pathway at a specific point is known as an inborn error of metabolism. The defective enzyme can be the catalyst for a catabolic or anabolic reaction. How does an inborn error manifest itself? It depends. Think back to the discussion of metabolic enzymes as robots. We can be sure that two things will happen when a metabolic enzyme is missing completely or seriously malfunctioning.

1. The substrate of the broken enzyme will not be converted into a product.
2. Without this product to serve as the substrate for the next enzyme in the pathway, subsequent products in that pathway cannot be produced.

Does either or both of these biochemical effects lead to serious medical consequences? It depends. Consider consequence number 1, the substrate of the broken enzyme. Will it build up in the cell, or can it be converted into something else? If it builds up, is it harmful? If it is converted into something else, is that substance harmful? Sometime it is, and sometimes it isn't.

Now consider consequence number 2, the missing product. Is that a problem? That, too, depends on a number of factors. Is that product available from another source? If so, then the inability of the defective enzyme to generate the product probably will not pose a problem. If not, is that product, or anything downstream of it in the metabolic pathway, crucial to health? If so, the absence will cause a problem.

The complex relationships between enzyme defects and the variable biochemical effects and medical consequences they can cause are best illustrated by an example. While a number of interesting diseases result from biochemical defects due to inborn errors of metabolism, we are choosing to focus on problems in the pathways involved in the metabolism of two amino acids, phenylalanine and tyrosine. These pathways intersect, which helps to reinforce the concept of metabolism as a network of pathways. In addition, mutations in genes for different enzymes in these pathways illustrate a number of important points about genetic diseases. For example, some genetic defects do not warrant therapeutic intervention, because they do not lead to serious health problems. Others that do cause serious problems can be controlled through environmental factors, even though the cause of the disease is purely genetic. Other genetic diseases do not cause serious health problems but do lead to social problems. Understanding these types of considerations will be important to our discussions of medical applications of biotechnology, such as genetic testing and gene therapy.

Enzyme defects and amino acid metabolism

Unlike plants and most microbes, which can synthesize all 20 amino acids from scratch, humans and other mammals can synthesize only about half of them. The rest, called **essential amino acids**, must be obtained from the diet. Phenylalanine is one of the essential amino acids.

Cells use phenylalanine in two ways: they incorporate it, as is, into protein molecules, or they convert it to another amino acid, tyrosine, which has a number of possible fates. Mutations leading to defective or missing enzymes at different steps in the phenylalanine/tyrosine pathways have various effects.

Phenylketonuria

Phenylalanine is converted to tyrosine by an enzyme called phenylalanine hydroxylase (PAH) (Figure 6.16). If individuals with defective genes for PAH consume more phenylalanine than is needed for its role as a building block for protein molecules, trouble begins. The excess phenylalanine cannot be converted to tyrosine; it accumulates, and some is shunted to an alternate pathway that converts phenylalanine into products called phenylketones. This inborn error of metabolism is known as **phenylketonuria** (PKU), be-

Figure 6.16 Phenylalanine is normally converted to tyrosine by the enzyme PAH. If PAH is not present, phenylalanine is converted to phenylketones, resulting in the condition phenylketonuria.

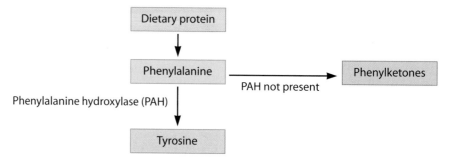

cause the phenylketones and phenylalanine are excreted in the urine (the term phenylketonuria means "phenylketones in the urine").

Excess phenylalanine interferes with the normal development of the nervous system. A baby's diet usually includes more phenylalanine than is required for manufacturing its own proteins, so babies that cannot synthesize PAH are at risk for irreversible mental retardation and early death. When PKU was discovered, about 1% of institutionalized mentally retarded people were found to have the condition. Babies with PKU appear completely normal at birth, but physicians now routinely test for this defect immediately after birth so that babies with PKU can be identified and placed on a diet that is extremely low in phenylalanine. Without excess phenylalanine in their bodies, their nervous systems develop normally.

A great deal of nervous system development has already occurred by the time a baby is born, so how can babies with PKU be normal at birth? A phenylketonuric baby is normal at birth because its mother, who has the enzyme the baby is missing, has been clearing excess phenylalanine from the baby's blood by converting phenylalanine to tyrosine.

Note that even though phenylketonuria is a genetic disease, it can be controlled or "treated" via an environmental factor—appropriate diet. Phenylalanine is an essential amino acid, so the diet must provide enough phenylalanine for protein synthesis that occurs in normal growth and development but not so much that it accumulates and damages the developing nervous system. Because PKU can be treated through diet, people with this condition now lead normal lives (Box 6.2).

Another effect of having a defective form of PAH is the lack of its product: the tyrosine that was supposed to be derived from phenylalanine. Does this lack of a product caused by defective PAH have any adverse effects? Not necessarily. Tyrosine is readily available in dietary proteins. An appropriate diet will supply sufficient tyrosine to meet a person's biochemical needs.

BOX 6.2 *Sweet amino acids*

Do you ever drink diet soda? Or do you know anyone who does? The next time you're around a can of diet drink, look carefully at the label. It will contain this or a very similar warning: "Phenylketonurics: contains phenylalanine."

Aspartame (NutraSweet), the artificial sweetener used in diet sodas and many other reduced-calorie foods, is composed of two amino acids: aspartic acid and phenylalanine. Amino acids contain calories, of course, but since aspartame is 200 times sweeter than sugar, foods can be sweetened with such tiny amounts of it that the calorie addition from aspartame is negligible. When aspartame is digested, it is broken down into its component amino acids, which then undergo digestion or reuse just like other dietary amino acids. A serving of nonfat milk provides about 6 times more phenylalanine and 13 times more aspartic acid than the same amount of beverage sweetened with aspartame. Neither aspartame nor its components accumulate in the body over time.

One other product is produced during aspartame digestion: methanol, a type of alcohol (not beverage alcohol) found in many foods, such as fruits, vegetables, and their juices. In sufficiently large amounts, methanol is poisonous, but in tiny amounts it is not harmful. According to the International Food Information Council (IFIC), a serving of tomato juice contains about six times as much methanol as the same amount of diet soda.

The IFIC calls aspartame one of the most extensively tested food ingredients of all time. Before its initial approval by the U.S. Food and Drug Administration in 1981, it underwent safety testing in more than 100 different studies. Since its approval, groups of people have arisen who claimed it caused various problems, such as headaches, seizures, and allergic reactions. In every case, careful studies proved these claims were baseless. These results make sense, since we eat the components of aspartame every day in larger amounts than we get from drinking even several cans of diet soda.

In fact, the only group to whom aspartame presents a potential health problem is people with phenylketonuria. So unless you have PKU, enjoy your diet soda without concern.

Alkaptonuria

Further down the phenylalanine breakdown process shown in Figure 6.17, the intermediate homogentistate (HG) is converted to another intermediate in the pathway, maleylacetoacetate (MAA). If the enzyme responsible for this conversion is defective or missing, HG accumulates, leading to **alkaptonuria**. Excess HG, unlike excess phenylalanine, causes no serious ill effects, and as a result, alkaptonurics do not need to minimize phenylalanine intake.

The product missing from the breakdown of HG, the molecule MAA, is not a starting material or essential intermediate in any other pathway. In addition, the final end products of MAA breakdown are carbon dioxide and water, which we get plenty of from other pathways. Therefore, the absence of MAA has no serious repercussions.

If no obvious medical problems are associated with alkaptonuria, how do we know this genetic disorder exists? Excess HG, which turns black when it is exposed to air, is secreted in urine. We can identify alkaptonurics because their urine is black.

Comparing the fates of people with enzyme defects at different points in the phenylalanine catabolism pathway resulting in either PKU or alkaptonuria illustrates an important point. Neither person can break down phenylalanine completely, but in one case, a harmful intermediate accumulates, and in the other, a nonharmful intermediate accumulates. Individuals with PKU will be severely mentally retarded unless they are put on a special diet at birth. Individuals with alkaptonuria have black urine but no serious medical problems. The severity of the consequences of these inborn errors of metabolism does not stem from being unable to break down phenylalanine but rather from the precise step where the pathway is blocked and the nature of the intermediate that accumulates.

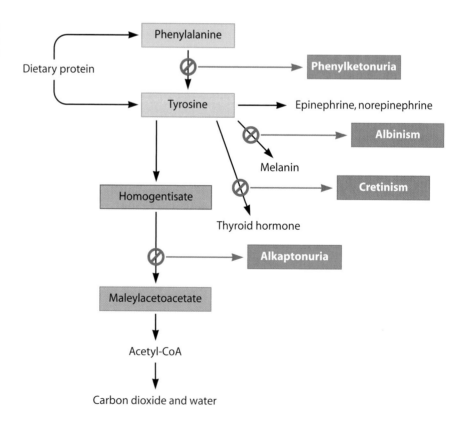

Figure 6.17 Potential disruptions in phenylalanine and tyrosine metabolism. The arrows represent from one to many metabolic steps.

Albinism and cretinism

Refer again to Figure 6.17. Note that tyrosine is a necessary starting material for many important products, including the neurotransmitters epinephrine and norepinephrine, the pigment molecule melanin, and thyroid hormone. People lacking enzymes that convert tyrosine to melanin lack coloration in their skin, hair, and eyes, a condition known as albinism. Some people with albinism have visual disorders, but albinism is not life-threatening nor are its medical consequences serious.

On the other hand, the inability to convert tyrosine to thyroid hormone has very serious medical implications. The lack of thyroid hormone causes cretinism, which is characterized by severely impeded growth and maturation of the skeletal and nervous systems, leading to mental retardation and dwarfism. These effects can be reversed by providing thyroid hormone replacement therapy to infants with enzyme defects that prevent production of thyroid hormone.

There are no inborn errors of metabolism that block production of the neurotransmitters epinephrine or norepinephrine from tyrosine. Why do you suppose this is so? No doubt, the absence of people with enzyme defects for synthesizing neurotransmitters is due to the essential roles these molecules play in nervous system function. Epinephrine and norepinephrine are involved in controlling respiration, heart rate, blood pressure, blood glucose levels, and blood flow to the brain and kidneys. An embryo incapable of synthesizing neurotransmitters from tyrosine would die early in development.

BIOTECHNOLOGY APPLICATIONS

How are people using a detailed understanding of metabolic enzymes and pathways to make useful products and solve problems? We have chosen a few very different examples from a long list of possible examples in order to illustrate the extraordinary breadth of potential applications.

Metabolic disorders can be treated with enzymes and genes

Having just learned about diseases caused by inborn errors of metabolism, you might be wondering whether it is possible to use our understanding of metabolic pathways to treat these diseases. Yes, but only for some of these diseases. When the identity of the defective or missing enzyme responsible for a certain disease is established, physicians can sometimes treat the disease by giving patients functional versions of the missing or defective enzyme. When the gene that encodes that enzyme is also identified, it may become possible to treat patients with correct genes for the defective enzyme.

Gaucher's disease

Gaucher's disease (pronounced go-shay) is caused by a deficiency of the enzyme that breaks down a particular type of lipid (glucocerebroside) found in red and white blood cell membranes. Worn-out blood cells are taken up and digested by specialized cells called macrophages. In Gaucher's disease patients, the macrophages cannot break down the membrane lipid; it accumulates inside the macrophages, and they become enlarged. Enlarged macrophages full of this undigested lipid are called Gaucher cells, and physicians look for them when diagnosing the disease. Gaucher cells most often accumulate in the spleen, liver, and bone marrow, affecting their functions. They can also accumulate in other tissues, including the nervous system.

There are three types of Gaucher's disease, which differ in severity, age of onset, degree of nervous system involvement, and the nature of the mutation causing the disease. In type I Gaucher's disease, the most common and least severe form, the accumulation of lipids leads to enlarged livers and spleens and painful bone lesions but no damage to the nervous system. Type II Gaucher's disease, which is very rare, results in extensive neurological damage in infancy, and patients die when very young, typically by the age of 2 years. In type III Gaucher's disease, lipids accumulate in nerve cells as in type II, but symptoms are not apparent until the child is about 10 years old.

Studies of Gaucher's disease patients have revealed that many different mutations in the gene for the lipid-degrading enzyme can cause the disease. As you might expect, mutations with the most potential to disrupt the three-dimensional structure of the protein are associated with the most severe forms of the disease. For example, one type of mutation actually terminates synthesis of the enzyme after just a few amino acids, so these individuals lack the enzyme completely and suffer from type II Gaucher's disease. By contrast, a mutation associated with type I Gaucher's disease causes a substitution of one amino acid for a similar amino acid, and the function of the mutant enzyme is impaired but not completely lost. A different amino acid substitution significantly alters the shape of the enzyme, leading to the more severe forms of the disease.

Enzyme replacement therapy for Gaucher's disease

In the past, therapeutic options available to Gaucher's disease patients were limited to those that alleviate symptoms, such as the removal of painfully enlarged spleens. While this treatment may alleviate the pain, it does nothing to solve the fundamental problem. Lipids and Gaucher cells continue to accumulate, exacerbating the problems in the other affected organs, the liver and bones.

Once scientists identified a defective lipid-degrading enzyme as the source of the problem, physicians began trying to treat Gaucher's disease by giving patients the missing or malfunctioning enzyme. At first, the enzyme was prepared by extracting it from human placental tissue, but this proved to be prohibitively expensive for all but a few very wealthy patients.

In the mid 1980s, scientists used new biotechnology techniques to identify and locate the gene encoding the lipid-degrading enzyme missing in Gaucher's disease patients. Using recombinant DNA techniques, which we will describe in detail later, scientists developed methods to produce large amounts of the enzyme more economically. Scientists also tinkered a bit with the gene so that the enzyme would be taken up more readily by macrophages, where it is most needed. Patients with Gaucher's disease receive periodic infusions of the engineered form of the lipid-degrading enzyme, which proved more effective as a treatment than did infusions of the unaltered enzyme. This form of treatment comes much closer to addressing the cause of the disease than simply dealing with its symptoms.

Gene therapy for Gaucher's disease patients?

Can we get even closer to correcting the fundamental problem in Gaucher's disease? Rather than giving patients regular injections of the enzyme, why not provide them with correct copies of the gene so their bodies can make the functional form of the enzyme? Since the sequence of the gene that encodes the correct form of the enzyme is now known, treating Gaucher's disease patients with genes rather than the enzyme might be possible. However, identifying the gene and knowing its sequence are necessary but not sufficient.

For any gene therapy to be successful, the introduced gene must (1) become a stable part of the host and (2) produce the correct protein so that it is at the right place at the right time. These are more complicated tasks than determining the sequence of a gene. Researchers are exploring two angles for treating Gaucher's disease. One involves introducing the correct gene into bone marrow, where macrophages arise. The other strategy is to introduce the altered gene for the macrophage-targeted form of the enzyme into muscle cells. Both approaches are still highly experimental.

We will discuss gene therapy further in chapters 19 and 20 on the medical applications of biotechnology.

Microbial metabolism is exploited in many ways

The facts that enzymes are very specific in what they do and that they carry out reactions under mild conditions can be big advantages in manufacturing. Purely chemical processes often produce more than one product from the same starting material. Enzymes, however, carry out one specific activity, so the amount of product obtained from starting material can be much higher. Chemical processes can require high temperatures and harsh solvents, while enzymes work at relatively low temperatures in water-based solutions, potentially saving energy and minimizing or eliminating the production of chemical waste.

Biotransformation

There are two general approaches to using enzymes in manufacturing: use a whole organism to make the desired transformation or isolate the enzyme and use it alone. If the product you want is the endpoint of a microbial or cellular pathway, then you can grow the relevant microbe or cell in the presence of your starting material and collect the end product. Bacteria are particularly useful for these **biotransformations** because of the many different kinds of enzymatic pathways they possess. One particularly advantageous aspect of biotransformation is that the waste generated in the process (dead bacteria and their culture fluids) is biodegradable.

Some of our most ancient biotechnologies are biotransformations. Six thousand years before the existence of microbes was discovered, they were providing people with fermented foods, such as wine, yogurt, beer, and bread. Humans provided these invisible workers with an abundant supply of glucose, such as the starch in flour, sugar in grapes, and lactose, the sugar in milk, and in the process of breaking down these sugars for energy, different microbes synthesized useful by-products: carbon dioxide leavened bread, lactic acid turned milk into yogurt, and acetic acid pickled vegetables.

Pasteur's discovery that microbes were responsible for generating these products opened the door to a new phase of exploiting the metabolic pathways of microbes. Microbiologists began to systematically look for microbes that could make useful products. They discovered many organisms that generated desirable by-products from breaking down glucose (Figure 6.18 and Table 6.1) and from other metabolic pathways used to synthesize and break down biological molecules for growth, reproduction, and energy production (Table 6.2).

Microbes carry out more metabolic functions than just respiration and growth. Different organisms synthesize specialty compounds required for activities such as infecting other organisms, defending themselves against other microbes, and communicating with each other. As luck would have it, many of these specialized products are also useful to humans.

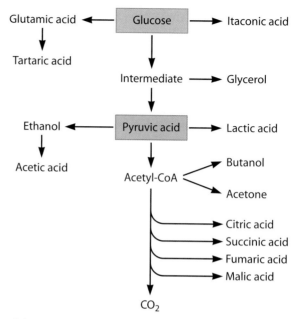

Figure 6.18 Useful products of glucose breakdown provided by various microbes.

Table 6.1 Chemcials currently produced by microbial metabolism of glucose and their industrial applications

Chemical	Microbial source	Industrial uses
Ethanol	*Saccharomyces*	Industrial solvent, fuel, beverages
Acetic acid	*Acetobacter*	Industrial solvent, rubber, plastics, food acidulant[a] (vinegar)
Citric acid	*Aspergillus*	Food, pharmaceuticals, cosmetics, detergents
Gluconic acid	*Aspergillus*	Pharmaceuticals, food, detergent
Glycerol	*Saccharomyces*	Solvent, cosmetic preparations, soaps, antifreezes
Isopropanol	*Clostridium*	Industrial solvent, cosmetic preparations, antifreeze, inks
Acetone	*Clostridium*	Industrial solvent, intermediate in many chemical synthesis reactions
Lactic acid	*Lactobacillus, Streptococcus*	Food acidulant, fruit juice, soft drinks, dyeing, leather treatment, pharmaceuticals, plastics
Butanol	*Clostridium*	Industrial solvent, intermediate in many chemical synthesis reactions
Fumaric acid	*Rhizopus*	Intermediate in synthesis of synthetic resins, dyeing, acidulant, antioxidant
Succinic acid	*Rhizopus*	Manufacture of lacquers, dyes, and esters for perfumes
Malic acid	*Aspergillus*	Perfumes
Tartaric acid	*Acetobacter*	Acidulant, tanning, commercial esters for lacquers, printing
Itaconic acid	*Aspergillus*	Textiles, paper manufacture, paint

[a]An acidulant is a substance added to food or beverages to lower pH and to impart a tart, acid taste.

Table 6.2 Useful products from microbial metabolic pathways other than glucose metabolism

Type of product	Examples	Applications
Amino acid	Glutamic acid, phenylalanine, aspartic acid, lysine	Nutritional supplements, flavor enhancers, sweeteners
Carbohydrate	Dextran, xanthan gum	Food emulsifiers and thickeners, oil recovery
Vitamin	B_{12}, riboflavin, β-carotene	Nutritional supplements, pigments
Metabolic enzyme	Proteases, amylases, lipases	Detergents, sweeteners, brewing, cheese making, textiles, leather softening
Nucleotide	Guanosine, inosine	Flavor enhancers

- Virtually all antibiotics are the very same molecules that microbes manufacture to kill other microbes.
- A microbe that infects rice secretes a molecule to stimulate plant growth, and plant nurseries now use that molecule to do the very same thing.
- The food industry uses the set of enzymes that allow microbes to penetrate the skins of fruit to enzymatically peel fruit before canning.

Microbes, steroids, and birth control pills

Sometimes manufacturers use microbial pathways and enzymes not for the end product but for a specific step in a pathway that confounds chemists attempting to make a specific molecule. One of the most famous examples of using a microbe to carry out one step in a manufacturing process comes from the pharmaceutical industry. In the 1940s, pharmaceutical chemists discovered the anti-inflammatory properties of the steroid hormone cortisone, but they needed over 700 lb of starting material (which had to be extracted from the bile of slaughtered animals) to produce 1 g of cortisone in a 37-step chemical process. They concluded that economical production of cortisone through chemistry was impossible.

The extreme inefficiency of the chemical synthesis process could be traced to one thing: the chemists needed to put an oxygen molecule on one specific carbon atom in a molecule that was a cortisone precursor. They found a microbe with an enzyme specific for this step, converting the 37-step process to one with only 11 steps and cutting the costs of production by 70%. They began commercial production of cortisone soon thereafter. Eventually, they found a number of microbes with enzymes for other steps in the process, and the price of cortisone today is 400 times less than its original price.

The most widely used steroids are the estrogens and progesterone found in birth control pills. Pharmaceutical chemists ran into the same sort of problems when they tried to synthesize sufficient quantities of these hormones at an economical price. Luckily, they found microbes capable of carrying out the specific reactions that were impeding commercial development. Thus, the widespread availability of birth control pills hinges on a specific enzyme that catalyzes one step in a metabolic pathway of a microbe.

Cleaning up environmental pollution

Microbes inhabit an impressive variety of environments, many of which are uninhabitable by most organisms, such as hot sulfur springs, deep ocean vents, and oil slicks. The microbes that live in these extreme environments contain enzymes that allow them to use substances available to them in those environments. For example, microbes found in oily environments use the oil for food. Their enzymes break down oil molecules much as ours break down carbohydrates and proteins.

Over the past 30 years or so, scientists have studied how microbes eat oil, and industrial scientists have used oil-eating microbes in a process that has created a new industry: **bioremediation**. Bioremediation is the use of biological methods to degrade pollutants at contaminated environmental sites. A typical site for bioremediation might be a gas station where an underground storage tank has leaked into the soil and the underlying groundwater, or a coastal area where oil has been spilled. Environmental cleanup workers add oil-eating microbes and extra nutrients to the polluted area

Figure 6.19 Environmental Protection Agency personnel monitor an oil spill bioremediation test site. (Photograph courtesy of Albert Venosa, Environmental Protection Agency.)

(Figure 6.19). The microbes consume the gasoline or oil and multiply, giving off clean water and carbon dioxide gas.

Enzymes play useful roles in manufacturing

Sometimes, rather than use a living microbe for a biotransformation, manufacturers use individual enzymes to carry out a specific reaction. For example, soft-centered chocolates are created by first coating a hard sugar center with chocolate and then injecting a tiny amount of the enzyme invertase, which converts the hard sugar to a soft syrup. Stone-washed jeans now get that well-worn look and feel through treatment with the enzyme cellulase, which partially breaks down the cellulose in cotton fibers. The enzyme amylase, which breaks down starch into glucose, is used by industry in a number of ways. One of our favorite examples is its use in making "lite" beer.

Reduced-calorie beer

Beer is brewed by growing yeast in the absence of oxygen on sugars extracted from sprouting grain, especially barley. Grains are actually plant seeds, and the plants store starch in grain to provide food for the young plant when it first sprouts. When the seed begins to sprout, amylase enzymes in the grain are activated and begin to break down the stored starch into sugars. Brewers soak grain in water to activate the amylase enzymes; then, when it begins to sprout, they dry it to stop the enzymes until they are ready to brew the beer. This process is called malting.

When it is time to brew the beer, the brewer crushes the malted grain in water to reactivate the enzymes and convert more starch to sugars and then boils the sugar solution with hops for flavor and adds yeast. Yeast can take in and break down glucose and maltose (a sugar made of two glucose molecules) but not larger carbohydrates. The malting process, however, leaves much of the starch in the form of carbohydrates, which remain in the beer. Humans are perfectly capable of breaking down the carbohydrates, so they represent additional calories in the beer.

To make lite beer, brewers add amylase enzymes during the brewing process to finish converting the carbohydrates to sugars that the yeast can use. By doing this, they eliminate leftover carbohydrates in the beer, causing the beer to be lower in calories.

Laundry detergents

In addition to being used to create products, enzymes are also sometimes added directly to products and become part of them, as in laundry detergent. Think about what kinds of dirt you get on your clothes—food, grease, sweat, oils? Many of the kinds of molecules that make your clothes dirty are the very same kinds of molecules that cellular enzymes can break down for energy.

Manufacturers of laundry detergents have been adding enzymes to detergents for years. The enzymes help get your clothes cleaner by breaking down large molecules into smaller ones that may not cling to the fabric as well and can be removed more easily. Your laundry detergent probably contains lipases (to break down lipids; what kind of dirt would lipids constitute?) and proteinases (also called proteases; they break down proteins). Read a detergent label some time and see what it says.

You'll find many more examples of uses for enzymes in manufacturing in the Biotechnology Applications and Issues section of this book.

SUMMARY POINTS

Cellular processes occur in pathways made up of series of small steps. Each step of a pathway is carried out by a specific enzyme. Many cellular pathways are branched, meaning that a specific compound participates in more than one. Other pathways converge, as with the pathways that break down various molecules for energy.

Cells break down molecules to obtain energy. Your body uses carbohydrates, lipids, and proteins for energy. The breakdown pathways for all of these molecules converge on three major pathways: glycosis, the Krebs cycle, and the electron transport pathway.

In the electron transport pathway, the energy obtained from food molecules is used to drive the addition of a phosphate group to the molecule ADP to form ATP. When your cells require energy, they break the bond attaching that phosphate group to release the stored energy, a phosphate group, and ADP. The electron transport pathway requires oxygen. It is so essential that blocking it results in death within minutes.

Cellular enzymes also synthesize proteins, lipids, carbohydrates, and other molecules, such as hormones and pigments. These synthesis processes usually start from molecules obtained by breaking down food. Synthesis requires energy.

Cells regulate their metabolism so that an organism can respond appropriately to its environment or the needs of the body in which it is located. In feedback inhibition, the product

of a pathway regulates an early step in the pathway to keep levels of the product within a particular range. Free-living cells, such as bacteria, often regulate the expression of genes encoding enzymes for breaking down specific molecules according to whether those molecules are present in the environment. Multicellular organisms use hormones to coordinate metabolism in all the body's cells.

If an individual cannot make an enzyme, the process for which that enzyme is required is blocked. Some processes are so critical that no individual can develop without them, but other processes can be blocked and an individual can live, with consequences that vary in severity. The consequences depend on factors such as whether there are other pathways to handle the intermediates that could accumulate, whether those intermediates are toxic, whether there are alternate pathways to supply needed end products, and what the role of the normal end product is.

Phenylketonuria results when an individual cannot make an enzyme that converts the amino acid phenylalanine to tyrosine. Excess phenylalanine builds up and can cause mental retardation. Alkaptonuria results when an individual cannot carry out a specific step in tyrosine breakdown. Since the intermediate that builds up in alkaptonuria is harmless, individuals with this condition suffer no ill effects.

Gaucher's disease results when an individual cannot make a fully functional enzyme that breaks down a certain type of

SUMMARY POINTS *(continued)*

lipid. The lipid accumulates and causes damage. The severity of Gaucher's disease depends on how impaired the enzyme is, which depends on the nature of the mutation in the gene for the enzyme.

Biotechnology has provided us with techniques for producing large quantities of enzymes and other proteins. These can be used to treat conditions such as Gaucher's disease and to manufacture products, such as more effective detergents and reduced-calorie beer (among many others). Enzyme-based man-

ufacturing is particularly attractive, since it takes place at low temperatures and without harsh solvents and usually produces biodegradable waste.

Living microorganisms can also be used in manufacturing and in bioremediation. These biotransformation processes take advantage of microbial enzymes to convert a substrate into a useful product. Microbial transformations are key to the commercial production of steroid hormones.

KEY TERMS

Adenosine triphosphate (ATP)

Alkaptonuria

Anabolism

Bioremediation

Biotransformation

Catabolism

Electron transport pathway

Essential amino acids

Feedback inhibition

Gaucher's disease

Glycolysis

Hydrocarbon

Intermediate

Krebs cycle

Metabolic pathway

Phenylketonuria

Product

Respiration

Substrate

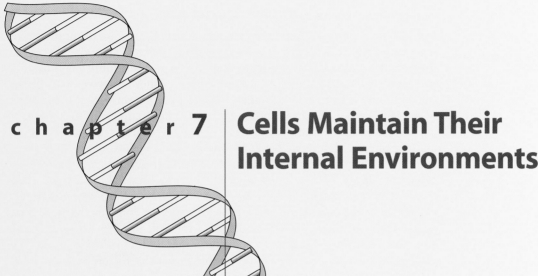

| # Cells Maintain Their Internal Environments

The tide is coming in, and water at the mouth of a coastal river is becoming increasingly salty as ocean water floods it. Suspended in that river water is a bacterium. Though the water around it is becoming increasingly salty, its own internal environment is not. Similarly, your own cells are bathed in fluid that contains about 10 times more sodium than the fluid inside your cells. At the same time, your cells' interiors have about 20 to 40 times more potassium than the extracellular fluid.

Single-celled organisms and the cells in your body maintain their internal environments despite changes in the external environment. Cells also sense changes in their external environments and respond to them. How do cells keep their internal environments different from their external environments, and what benefit(s) do they derive from doing so? And beyond that, how do they sense changes in the environment and respond to them?

The interface between the internal and external environments of cells is the cell membrane. Substances have to cross the membrane to get into or out of a cell. Since the **intracellular** (inside the cell) environment is different from the **extracellular** (outside the cell) environment, cells must be selective about what does get in and out.

In this chapter, we will look at how cells sense and respond to their environments. For our first step, let's look at what kinds of substances get in and out of cells and how they move across the cell membrane.

CROSSING THE CELL MEMBRANE

In chapter 2, we described the hydrophobic membrane that separates the outside environment from the interior of a cell. Recall that a cell membrane is made of two layers of phospholipid molecules that have hydrophilic head groups and long hydrophobic tails. In the membrane, the hydrophilic heads face outward toward the watery exterior environment and inward toward

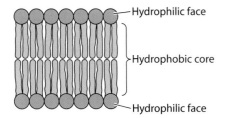

Hydrophilic face

Hydrophobic core

Hydrophilic face

Figure 7.1 The basis of the cell membrane is a lipid bilayer.

the watery cytoplasm, while the tails from both sides point toward one another (Figure 7.1). As we mentioned before, the cell membrane is more than just a lipid bilayer. For example, cholesterol is an essential part of the membranes of animal cells, and other sterols are found in plant cell membranes.

Cell membranes contain embedded proteins

In addition to phospholipids and sterols, cell membranes are also full of proteins with various functions. For example, receptor proteins embedded in the membrane bind to signal molecules outside the cell and transmit those signals inside the cells. Adhesion proteins bind to other molecules outside the cell and cause cells to stick to things. Recognition proteins signal the identity of the cell to other cells in the body. And particularly important for this chapter, various kinds of transport proteins are involved in moving substances in and out of cells. Before we begin to look at how some of the transport proteins work, let's take a quick look at the general structural features of membrane proteins.

Membrane proteins usually have one domain (region) that is actually embedded in the membrane and another domain or domains that are either on the inside of the cell, on the outside of the cell, or some of each. Recall from the discussion of proteins in chapter 5 that a general rule in the folding of proteins is the creation of a hydrophobic core inside the protein with a hydrophilic exterior facing the watery cytoplasm of the cell. Protein domains embedded in cell membranes generally don't follow that rule, because the membrane is a hydrophobic environment.

The membrane-spanning domains of membrane proteins have hydrophobic amino acids pointing outward into the oily membrane. These domains are essentially inside out when you compare them to the more usual arrangement of a hydrophobic protein core with hydrophilic surface amino acids. Along with their hydrophobic exteriors, the membrane-spanning domains of transport proteins have hydrophilic cores, which are ideally suited to their functions. These transport proteins permit hydrophilic molecules, such as sugars, amino acids, and ions, to penetrate through the hydrophobic membrane that would normally exclude them. Their hydrophilic interiors provide an environment that is chemically compatible with the hydrophilic molecules that they transport.

Other proteins can associate with embedded membrane proteins either just inside or just outside the cell. Proteins like these often play key roles in transmitting signals to the inside of the cell. Figure 7.2 shows an artist's conception of the cell membrane, with various proteins embedded in it.

Now that we have a more complete picture of the cell membrane, let's get back to the question of how a substance crosses this barrier.

Hydrophobic substances and very small molecules can cross the membrane unassisted

Hydrophobic molecules usually don't have any problem crossing the cell membrane. These molecules are very comfortable in the oily environment inside the membrane and readily enter it. In fact, if you rubbed a small amount of the hydrophobic substance DMSO (dimethylsulfoxide) on your skin (note we said *if*—don't go do this), it would pass through your skin, cross over cell membranes, and spread throughout your body, and you would almost immediately taste it. Other nonpolar molecules that can cross the membrane are oxygen and carbon dioxide. Many medicines are also hydrophobic molecules.

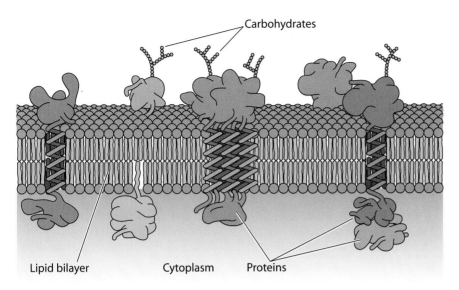

Carbohydrates

Lipid bilayer   Cytoplasm   Proteins

Figure 7.2 The cell membrane is a lipid bilayer with embedded proteins. Many of the external membrane proteins have attached carbohydrate groups.

Small polar molecules, such as water, can also cross the membrane. Ethanol, or drinking alcohol, is another. In fact, ethanol is so good at crossing cell membranes that it can get into your brain, which is protected from most substances by the blood-brain barrier. This very selective barrier consists of specialized cells that block most transport.

With the exception of these kinds of molecules, though, the cell membrane effectively blocks the entry of most other substances. Most of the biologically important substances, such as sugars, amino acids, and ions, cannot get into cells on their own. These substances enter cells with the help of transport proteins embedded in the cell membrane.

Concentration is an issue in transport

Before we get a closer look at how transport proteins work, there is another issue we need to consider: **concentration**. Not the effort you are putting forth as you read this, but rather the amount of a substance that is present inside and outside of a cell. When a solid substance, such as a sugar, a salt, or an amino acid, is dissolved in a liquid, the mixture is called a **solution** and the formerly solid substance is referred to as a **solute**. The concentration of a solution refers to the amount of a solute that is dissolved in a given volume of liquid. A cup of water that has 20 g of sugar dissolved in it is a more concentrated solution than a cup of water that has only 10 g of sugar dissolved in it.

The reason concentration matters when we consider crossing the cell membrane is that it determines whether energy is required to transport something. Imagine a house sitting with its windows open on a completely windless day. All power in the house is off, so no fans are running. If the house is full of smoke, some of that smoke will gradually move out of the house into the outside air. If, instead, the house contained no smoke but the outside air was full of smoke, the smoke would gradually move into the house (Figure 7.3).

Diffusion

This scenario illustrates the process of **diffusion**. Even without forces such as wind or fans, molecules are always moving—vibrating on the atomic scale—

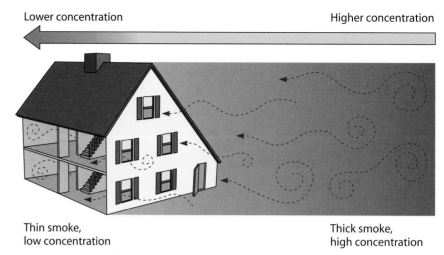

Lower concentration   Higher concentration

Thin smoke,   Thick smoke,
low concentration   high concentration

Figure 7.3 Diffusion happens. If not blocked, substances will move from areas of higher concentration to areas of lower concentration.

and they move from areas of higher concentration (the smoke-filled outside air) to lower concentration (the smoke-free house). If you left the situation alone long enough, eventually the air inside the house would have the same amount of smoke in it as the air outside. In fact, you could make a bumper sticker for your car: Diffusion Happens. That is, it does if the substance in question can move unimpeded from one area to another.

Now imagine that you come home as the smoke is entering your house, and you don't like it. What do you have to do to get the smoke out? You could shut the windows and doors so that no more smoke could get inside, but that wouldn't remove the smoke that had already gotten in. To do that, you might turn on fans, but whatever you did, it would require energy. The smoke would not leave by itself.

Applied to cells, what this scenario means is that if a substance is moving from an area of higher concentration (say, the extracellular fluid) to an area of lower concentration (say, the intracellular fluid), it will move by itself if it can get through the membrane. No energy input will be required. Conversely, if something is to be moved from an area of low concentration to an area of high concentration (like pushing the smoke back outside), energy will be required.

Concentration gradients

It can be helpful to think of concentration in terms of a hill. At the top of the hill is the area of high concentration. Substances will move downhill to areas of lower concentration without added energy; in fact, energy is released. If something is to be moved from an area of low concentration (downhill) to an area of higher concentration (uphill), energy has to be applied. We call going downhill moving **with a concentration gradient** or along a gradient, and we call moving uphill moving **against a gradient** (Figure 7.4).

Thus, there are two issues involved in getting things in and out of cells: (1) a passageway through the membrane and (2) whether energy is required. Even when no passage is required, as with hydrophobic molecules or water, the question of concentration still applies. Molecules will not move on their own from an area of lower concentration to an area of higher concentration.

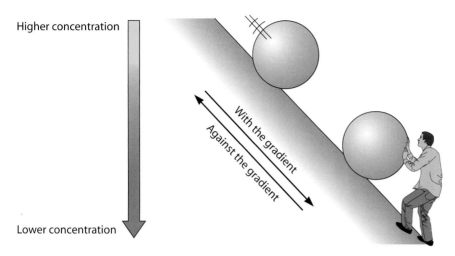

Figure 7.4 Energy is released when substances move from areas of higher concentration to areas of lower concentration. It takes energy to move a substance from an area of lower concentration to an area of higher concentration.

TRANSPORT PROTEINS

We've already mentioned that some of the many proteins embedded in the cell membrane are transport proteins. As you might imagine, there are different types of transport proteins for different substances and situations. Some types of transport proteins use energy to move substances against concentration gradients; others do not.

If the substance in question is moving down a concentration gradient from an area of high concentration to one of lower concentration, simple diffusion will do the trick and no additional energy is required. All that is needed is a passage for the substance in question to get through the membrane or an escort to chaperone it through that oily environment. We'll consider these types of transport proteins first and then look at transport proteins that use energy to move things against a gradient.

Channel proteins allow diffusion when their gates are open

Channel proteins are structured so that they form a little pore, or channel, through the cell membrane that only the desired substance can fit through. Channels provide the means by which ions, like sodium, potassium, chloride, and calcium, enter and leave the cell along their concentration gradients. Even water has a channel protein, called an **aquaporin**. Water can move through its channel much more quickly than it can diffuse across the cell membrane.

Just as you would be unlikely to leave the door to your house open all the time, cells do not usually leave their channels open all the time. Most of the channels are "gated." That is, some part of the channel proteins can change conformation and either block or open the channel (Figure 7.5). Opening and closing channels provides one means by which cells control their internal environments.

Carrier proteins escort substances across the membrane

Another means by which substances can cross the cell membrane is via **carrier proteins**. These proteins act rather like escorts. They bind to their particular substance on the outside of the membrane, where the concentration is higher, and release them on the inside, where the concentration is lower. As

Figure 7.5 A gated channel protein allows its target substance to pass through when the gate is open. Transport is with the concentration gradient.

Closed

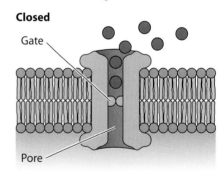

Open

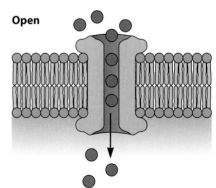

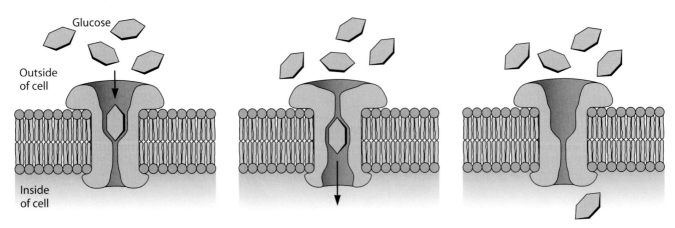

Figure 7.6 The glucose carrier protein binds to glucose and permits it to pass through the cell membrane. Transport is with the concentration gradient.

you might imagine, transporting substances by escort is slower than allowing them to diffuse through channels—think of escorting people one by one into a room versus just opening the door and letting them walk in unescorted. Carrier proteins are used in particular to transport energy substrates and metabolic building blocks, such as glucose, amino acids, and nucleosides, into cells.

The glucose transporter is a particularly important carrier protein. There are actually different versions of glucose transporters found in different cell types, but the glucose transporters all move glucose from an area of high concentration to an area of low concentration. Specifically, the carrier protein binds to glucose and then changes its shape and releases it on the other side of the cell membrane (Figure 7.6)

In the liver, glucose transporters work in both directions. The liver absorbs glucose from the blood when blood sugar is high and stores it in the form of glycogen (see chapters 6 and 8). When the blood sugar concentration falls, liver enzymes break glycogen back down into glucose. The concentration of glucose in the liver cells is then higher than the concentration in the blood, and the glucose transporters work in reverse, moving glucose from the liver back into the blood.

Pumps move substances against gradients

Channels and carrier proteins both allow the movement of substances with their concentration gradients. They do not require energy for their actions. However, cells often move substances against their concentration gradients, and this kind of movement requires proteins that harness energy. One type of protein that uses energy to move substances is called a **pump**. The extracellular fluid bathing cells inside the body is called **interstitial fluid**. If you look at Table 7.1, you'll see that there are vast differences between the concentrations of sodium (Na^+), potassium (K^+), chloride (Cl^-), and calcium (Ca^{2+}) in the interior of the cell and the interstitial fluid outside it. The only way such differences can be maintained is through pumps that spend energy to move these ions against their gradients.

Pumps can get energy from two different sources. One source is ATP, the molecule that acts like the cell's energy currency (see chapter 6). Some pumps get energy by splitting off the third phosphate from ATP as they pump their substance across the membrane. The breaking of the high-energy ATP bonds provides the energy to move the substance against the gradient.

Table 7.1 Approximate concentrations of ions in intracellular and extracellular fluids

Ion[a]	Intracellular concn (mM)	Interstitial concn (mM)
Sodium (Na^+)	10	145
Potassium (K^+)	150	5
Calcium (Ca^{2+})	0	3
Chloride (Cl^-)	5	110

[a]The most abundant ions in interstitial fluid are sodium and chloride ions, which are the components of table salt.

Other pumps move substances using energy inherent in gradients. As we just stated, animal cells maintain a much lower concentration of sodium inside than there is in the extracellular fluid. What this means is that anytime a pathway opens by which sodium can get into a cell, it will do so. The energy of sodium ions squirting into a cell down their steep concentration gradient can be used to pull other molecules into the cell at the same time. Cells that line the small intestine use this type of pump to suck in glucose and amino acids from the intestinal contents, as described below (Figure 7.7).

Sodium/potassium ATPase

The body's most common pump is one that simultaneously moves sodium and potassium. Animal cells maintain a much higher concentration of potassium inside themselves than is present in the extracellular fluid. Simultaneously,

Figure 7.7 The glucose pump uses the energy of sodium ions moving down their concentration gradient to transport glucose against its gradient.

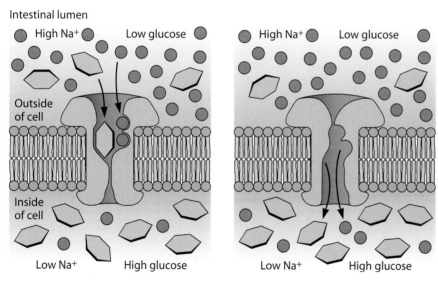

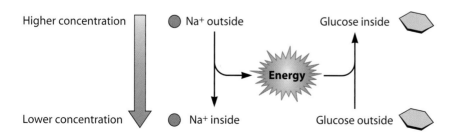

they maintain a much lower concentration of sodium (Table 7.1). They do this by means of a pumping protein that moves one potassium ion (K^+) in and one sodium ion (Na^+) out while breaking down ATP. In fact, this protein is called the **Na^+/K^+ ATPase** after its activity. Maintaining the proper internal concentrations of sodium and potassium is essential for keeping your digestion working, your nerve impulses transmitting, and your heart beating. The activity of the Na^+/K^+ ATPase is so important that some cells spend about 25% of their total ATP on it.

TRANSPORT PROTEINS AT WORK IN YOUR BODY

Transport proteins work in ingenious ways to keep your body working. We are now going to discuss how transport proteins function to transmit your nerve impulses and make your heart beat and how disturbances in ion gradients can disrupt these critical body functions. Next, we will turn to your digestive system and look at how transport proteins get food from your digestive tract to your body's cells, how your body maintains a salt-water balance, and even why lactose intolerance leads to diarrhea.

Ion concentration gradients power nerve impulses and muscle contraction

Nerve impulses are essential to many of the most basic functions in your body, from controlling your breathing, heartbeat, and involuntary muscle contractions to receiving and integrating countless sensory impulses and responding to them. Your brain receives sensory input and issues commands to your body via electrical impulses through the nervous system. What you may not know is that the very ion gradients we have been discussing are what power these nerve impulses.

The key players in transmitting nerve impulses are sodium (Na^+) and potassium (K^+) ions. Both of these are positively charged (note the small plus signs next to their abbreviated names). A nerve impulse is generated when a neuron (nerve cell) is stimulated. In response to the stimulation, sodium ion channels in the neuron's membrane near the stimulus site open. Sodium ions squirt into the cell from the outside because of the concentration gradient. The increased positive charge inside the cell (from the influx of sodium ions) causes the sodium channels to shut and, after a slight delay, potassium channels to open. Then potassium ions squirt out of the cell.

The movement of positive charges into, and then out of, the cell constitutes an electrical impulse. The change in charge at the site of the original stimulus stimulates the region of the membrane nearest it, and in a chain reaction, the electrical impulse is transmitted down the neuron. When the nerve impulse reaches its target cell, a different chain reaction occurs that leads to an outcome such as a muscle contraction.

In the nerve cells, the Na^+/K^+ ATPase pumps quickly restore the proper sodium and potassium ion concentrations inside the cell (high K^+ and low Na^+), and the nerve is ready to fire again.

Muscle contraction is controlled by calcium ions

Muscle contraction, too, is controlled by ions and concentration gradients. Refer back to Table 7.1. One of the ions in that table that we haven't talked about yet is calcium. Calcium ions play a key role in the contraction of your body's muscles, including the heart. Inside muscle cells, calcium ions are packed into a membrane-bound compartment called the sarcoplasmic reticulum (SR). There is essentially no calcium free in the intracellular fluid.

Clench your fist. For this to happen, a nerve impulse, powered by the sodium and potassium gradients, fired down your arm to a variety of muscles in your hand and forearm. When the nerve impulse reached the muscle cells, it triggered calcium channels in the SR to open, flooding your muscle cells with calcium. Inside your muscle cells, the calcium ions bound to a protein called troponin, which normally prevents your muscle cells from contracting by preventing the contractile proteins from interacting. Calcium binding to troponin changes its shape and inactivates it, allowing the contractile proteins to interact and your muscles to contract. When the stimulus to contract your muscle ceases, calcium is pumped back into the SR by an ATP-using ion pump, and your muscle relaxes.

The heart works in a similar way. The involuntary (that is, not under your conscious control) electrical impulse to beat first opens a calcium channel in the outer membrane, letting a little calcium in from the extracellular fluid. This triggers the SR channels to open, flooding the heart cell with calcium and permitting contraction. After the contraction, the calcium ions are pumped back into the SR, and a few are passed through channels in the outer membrane back into the extracellular fluid. This process happens very rapidly. The heart of a person at rest may beat 50 to 80 times per minute, while during intense exercise, the heart may beat at rates approaching 200 times per minute, depending on the individual.

WHEN GRADIENTS FAIL

After reading these sections, you might guess that if a person's cells couldn't maintain ion gradients at least fairly well, that person could not live. You'd be right. There are no examples of people who lack, for example, the Na^+/K^+ ATPase. We do have an example of what happens when potassium gradients are destroyed, again from the gruesome world of poisons. An injection of potassium chloride will rapidly kill a person by interfering with nerve impulses and preventing the heart from contracting. Using what you've just learned about nerve impulses and muscles, why do you think a sudden increase in potassium ions in the blood might do that?

Ion channel defects can cause heart rhythm irregularities

Ion channels are implicated in a defect of the heartbeat rhythm called long QT syndrome, or LQTS. The heart is a muscle, and to pump blood effectively, its cells must contract regularly in a coordinated manner. The signal to contract is normally transmitted in an orderly wave through the heart as an electrical impulse propagates a flow of sodium and potassium ions through channels in the muscle cell membranes. Disturbances in heart rhythm are called arrhythmias (meaning lack of rhythm). The type of arrhythmia seen in LQTS can be triggered by a variety of causes, from diseases to alcoholism to certain prescription drugs, but some individuals have an inherited form. About 1 in 3,000 to 10,000 people is estimated to have this condition.

LQTS is characterized by abnormally long recovery periods before the heart can contract again after one beat. This defect is dangerous not only because the heart cannot beat normally but also because the recovery periods from cell to cell become variable. Normally the electrical impulses that prompt the heart to beat sweep regularly through the heart from top to bottom. In people with this disturbance, the current can get sidetracked, producing dangerous arrhythmias that can cause loss of consciousness or even sudden death in an apparently healthy person.

Since the mid-1990s, several genes that can be involved in hereditary LQTS have been identified. All of them encode ion channel proteins; most are potassium ion channels, but one is a sodium ion channel. The potassium ion channel defects interfere with the normal outflow of potassium ions after sodium flows in. The sodium ion defect allows sodium to continue to flow into the cell after influx would normally have ceased. Either of these problems prolongs the time it takes for the ion concentration to be restored so that the cell can fire again.

The discoveries of various genetic defects leading to hereditary LQTS have led to the investigation of therapies tailored to the specific defect. The drug mexiletine, for example, blocks sodium channels and is being investigated for use in those patients whose sodium channels stay open inappropriately. Other therapies are being tested for patients whose defect is in potassium transport. Since the discovery of these genetic defects is relatively recent, treatments are still being developed.

A defect in calcium pumping can cause hereditary heart failure

Heart failure affects more than 4.5 million Americans and occurs when the heart loses its ability to pump blood. There are thought to be many causes of heart failure, and at present there are no cures. Recently, scientists have begun to pinpoint some molecular malfunctions that can underlie heart failure, and some of these (not surprisingly) involve calcium ion gradients.

Heart failure typically develops later in life, but there are rare instances of inherited early heart failure that develops by the person's mid-20s. In 2003, scientists pinpointed the mutation that causes inherited heart failure in one family. The mutation turned out to be in a protein that regulates the calcium pump on the SR. Recall that calcium is normally stored in the SR of muscle cells and is released as part of muscle contraction. Following a contraction, the pump moves calcium ions out of the muscle cell and back into the membrane-bound compartment, readying the muscle cell for a new contraction.

The newly discovered protein acts like a switch on the calcium pump. In the "off" position, it binds to the pump and inhibits its activity. To turn the pump on, a phosphate group is added to it by a phosphorylating enzyme. The phosphate group changes the shape and electrical charge of the protein, and it can no longer inhibit the pump (Figure 7.8). The pump then transports calcium ions back into the SR.

The family members with hereditary heart failure have an altered form of the regulatory protein with a shape that cannot be phosphorylated. Without the phosphate group, the regulatory protein constantly inhibits the pump. Without the activity of the calcium pump, the proper calcium gradients in the heart muscle cannot be maintained and the heart cannot contract properly.

There are other interesting science stories out there about ion channels and heart failure—for that matter, about muscle proteins and heart failure, too. This is a rapidly moving area of research, and many scientists are studying these problems. If you're interested in learning more, search the Internet for university press releases and science articles targeted to laypeople. Be aware that if you're reading about current research findings, future results may change how the current findings are interpreted.

Regulatory protein inhibits
the calcium pump.

Phosphorylated regulatory protein
allows pump to operate.

Mutant regulatory protein cannot be
phosphorylated; calcium pumping
is blocked.

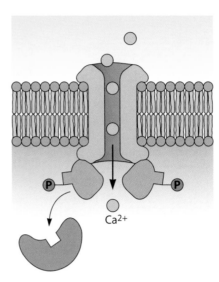

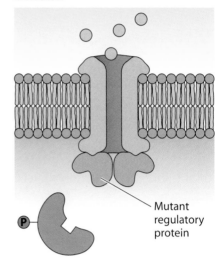

Figure 7.8 Inhibition of calcium pumping in hereditary heart failure.

PUMPS, CARRIERS, AND NUTRIENT DISTRIBUTION

When you eat, food enters your digestive tract. From there, nutrients have to get to your toes, your brain, and everything in between. We already mentioned (chapter 6) that enzymes in the small intestine help break down the large molecules in foods (carbohydrates, complex sugars, fats, and proteins) into their components (simple sugars, fatty acids, and amino acids). These component nutrients must move out of the intestine and into your bloodstream to circulate through your body. The cells lining your digestive tract, also known as the **intestinal epithelium**, play a starring role in this process. These cells have a system of pumps, transporters, and channels, along with a structural organization, that is brilliantly suited to this task. To look at how the intestinal epithelium works to get nutrients to your cells, we need to start with a look at its structure.

Body compartments are separated by strong sheets of cells called epithelia

Cells that cover body surfaces and line internal organs are called **epithelial cells**, and the sheetlike tissues they form are called epithelia (singular, epithelium). These cells form a barrier between the inside and the outside: between the body and the environment, and between an organ and the rest of the body. In a sense, epithelia are the body's version of a cell's membranes, in that epithelia separate the body into compartments and separate the body from the outside world.

An epithelium is a one-cell-thick sheet with an inside and an outside. In the case of the intestinal epithelium, one side faces the contents of the small intestine and looks like little brushes. A multitude of little protrusions called **microvilli** stick out into the intestine so that there is plenty of cell surface area to contact the intestinal contents and absorb nutrients (Figure 7.9). The cell membrane of the microvilli is studded with enzymes and transporters. This is where the enzymes that break down complex sugars, such as the milk

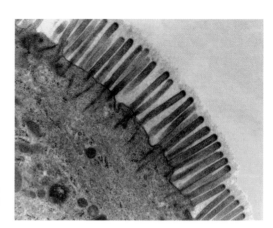

Figure 7.9 Intestinal microvilli. (Photograph copyright Dennis Kunkel Microscopy, Inc.)

sugar lactose and table sugar (sucrose), into simple sugars (like glucose) are located, along with transporters for bringing the simple sugars into the cell.

Consider for a moment that the microvilli of intestinal epithelial cells are their "heads," facing the interior of the small intestine. In this case, the "feet" of the epithelial cells are anchored to a tough network of extracellular proteins and carbohydrates that together make a matrix to support the epithelium, giving it strength. The bodies of the cells are glued together by bands of yet another protein, so that the intestinal contents cannot squeeze between cells (Figure 7.10). The strong connections between the bands of protein are called **tight junctions**. These junctions are tighter in some epithelia than in others. The ones in the intestinal epithelium do let some water and ions through, but no large molecules.

The feet of the epithelial cells and the matrix to which they are anchored are bathed in interstitial fluid, which contains ions (Table 7.1) and other small molecules. The tight junctions in the intestinal epithelium form a barrier between the intestinal contents and the interstitial fluid.

The form of the intestinal epithelium is well suited to its function

Think now about what the intestinal epithelium has to do. It must pull nutrient molecules out of the contents of the small intestine and pass them through to the body, even if there is a higher concentration of nutrients inside the epithelial cell than exists in the contents of the intestine. That means that the transporters on the intestinal side have to use energy to suck in the nutrients: they are pumps. On the other side of the cell, however, the concentration of nutrients in the extracellular body fluid is usually lower than in the epithelial cell.

The structure of the epithelial cell exploits this situation beautifully. The tight-junction band around its middle effectively divides the side of the cell exposed to the intestinal contents (the microvillus side) from the side of the cell anchored to the extracellular matrix. The transporter proteins in the epithelial cell membrane are distributed unequally to take advantage of this arrangement. On the microvillus side, the cell membrane is full of pumps to suck glucose into the epithelial cell against the concentration gradient. On the other side of the tight junctions, the epithelial cell membrane is full of glucose carriers that bind glucose and carry it across the membrane, out of the cell, sort of like pushing the glucose up the concentration hill on the intestine side and letting it roll down the hill out the back on the membrane side (Figure 7.11). The structural keys that make this neat system work are (1) the

Figure 7.10 Structure of the intestinal epithelium.

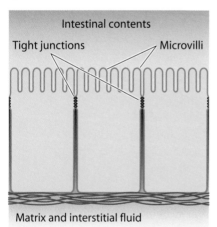

Intestinal contents

Tight junctions   Microvilli

Matrix and interstitial fluid

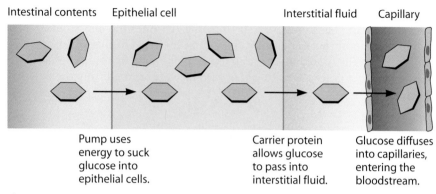

Figure 7.11 Transport of glucose from the small intestine into the bloodstream.

tight-junction band that separates the epithelial cells into two distinct sides and (2) the separate localizations of the pumps on the intestinal side and the carrier proteins on the membrane side.

The glucose pump is powered by the sodium gradient. The pump protein binds two sodium ions and lets them squirt in along with each glucose molecule brought into the cell. The concentration-driven movement of the sodium ions down their steep gradient has enough energy to bring in the glucose. This system depends on the maintenance of the high sodium concentration gradient. On the other side of the epithelial cells are many Na^+/K^+ ATPase pumps that pump the sodium back out of the cell (Figure 7.12).

Glucose and amino acids transported out of the epithelial cells are released into the interstitial fluid behind. Within the fluid in close proximity to your intestinal epithelium is a dense network of capillaries. Capillary walls are designed to let all but the largest molecules (for example, blood proteins)

Figure 7.12 A system of pumps and carrier proteins keeps glucose flowing from the small intestine into the bloodstream.

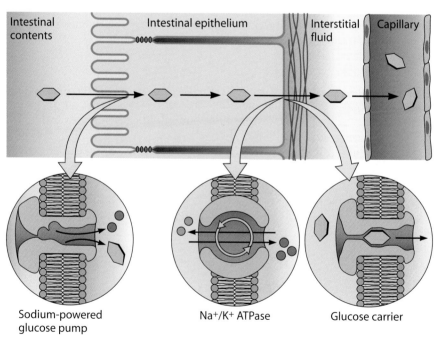

cross over, so the nutrients transported across your intestinal epithelium diffuse into them. The capillaries surrounding the intestine and stomach feed into larger blood vessels that carry blood to the liver. The liver absorbs excess glucose from the blood and stores it as glycogen. One reason it can do this is that is gets first crack, so to speak, at the glucose absorbed from the digestive tract.

CELLS, SALTS, AND WATER BALANCE

We've been discussing the movement of nutrients (glucose and amino acids) and ions (sodium and potassium) across cell membranes. Now, we turn to the movement of water. We've already established that water can enter cells through water channels called aquaporins or by simply crossing the cell membrane. Let us now look at the overall movement of water in and out of cells.

We will be the first to admit that, on the surface, water movement in and out of cells doesn't sound like a fascinating topic. Yet water movement in and out of cells is part of the movement of water in and out of your body. Maintaining the correct water balance in your body is essential for good health and, ultimately, for life. Ingestion of too much ethanol disrupts one of the ways your body maintains its water balance, producing dehydration and those nasty hangover symptoms. If you lose too much water through your intestinal epithelium into your intestinal tract, you have diarrhea. Diarrhea may be nothing more than an uncomfortable nuisance, but under some circumstances the dehydration accompanying diarrhea can be fatal. Diarrhea-induced dehydration is a particular problem for small children who lack access to medical care. Once we have looked at the basics of cellular water balance, we will examine how our understanding has led to life-saving rehydration therapies and commercially available sports drinks.

Water has an effective concentration in the body

We have already seen how dissolved solutes diffuse from areas of higher concentration to areas of lower concentration if they are not stopped by a barrier (like a cell membrane). Water behaves in exactly the same way, except the notion of "concentration" of water is a little different.

Figure 7.13 Water balance in a cell. Water will move in or out of a cell in response to the solute concentration in the environment.

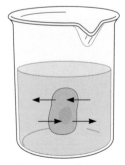

Cell is in osmotic balance with extracellular fluid.

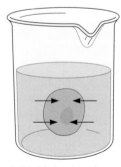

Cell is higher in solutes than is extracellular fluid. Water enters the cell.

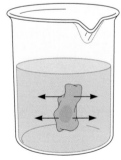

Extracellular fluid is higher in solutes than is the cell. Water leaves the cell.

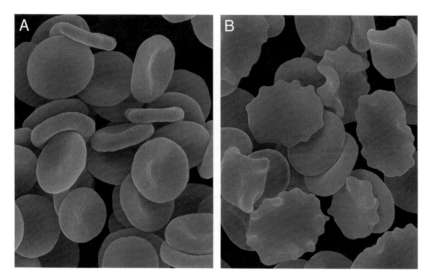

Figure 7.14 Red blood cells illustrate osmosis. **(A)** Blood cells and the environment are in osmotic balance. **(B)** The environment has a higher salt concentration than the cells' cytoplasm. Water has flowed out of the blood cells, leaving them shriveled. (Photograph copyright Dennis Kunkel Microscopy, Inc.)

The total concentration of solutes, that is, the sum of all the dissolved substances (sugars, salts, proteins, amino acids, etc.) dissolved in water, determines how it moves. Water moves to equalize the total concentration of solutes (assuming there is no barrier in the way). If a cell with plenty of water channels is sitting in a solution that has a lower concentration of solutes than does the cytoplasm, water will flow into the cell to try to equalize the concentrations. Similarly, if a cell is sitting in a solution that has a higher concentration of solutes than its cytoplasm, water will flow out of the cell to dilute the surrounding liquid (Figure 7.13). Finally, if a cell is sitting in a solution with a concentration of solutes equal to that of its cytoplasm, no net movement of water will occur. The movement of water across membranes is called **osmosis**, and a system in which no net water movement is occurring (because solute concentrations are equal) is said to be in **osmotic balance**.

This movement of water is often demonstrated with red blood cells. If you put some blood on a microscope slide, add fresh water, and watch, you would see the blood cells swell up as the water enters them. Likewise, if you put some blood on a microscope slide and add very salty water, you would see the blood cells shrivel up as water leaves them (Figure 7.14).

Your cells stay in osmotic balance with your extracellular body fluids

The solutes inside your cells and those in your interstitial fluid are a little different. Your cells contain many proteins, amino acids, and other small molecules. The majority of the solutes in your extracellular fluid, however, are salts, particularly sodium chloride (table salt). In Table 7.1 you can see the concentrations of a few other ions as well.

If you look more closely at Table 7.1 and add up the concentrations of ions inside cells and in the extracellular fluid, you'll notice that they are unequal. The intracellular total is about 165, while the extracellular total is about 260. It might seem like water should flow out of all your cells to dilute your extracellular fluid, and in fact it would, except for one thing.

Table 7.1 doesn't include all the proteins and small molecules inside the cell. These molecules make the effective solute concentration inside the cell equal to that outside the cell. Those ion pumps that spend so much ATP pumping sodium and other ions out of the cell are in fact keeping your cells in osmotic balance with your body fluids.

Water follows salt

A shortcut saying that describes, in a loose way, what happens with water in the body is "water follows salt." A more accurate saying might be "water follows solutes," but it's not as catchy. What this means is that anytime you increase the amount of solutes (salt is the major one) in a body compartment, water will move into that compartment to dilute the salt if it can.

One example of this is your overall body water balance. A 165-lb (75-kg) man has about 45 liters of water in his body. Of this, about 30 liters is intracellular. About 3.75 liters is in his blood plasma (the fluid that comprises blood minus the blood cells), and the remaining 11.25 is in the extracellular fluid bathing his tissues. Since blood capillaries are permeable to water and small molecules, the compositions of his blood plasma and his extracellular fluid are about the same in terms of solutes (note that plasma proteins and red blood cells do not leak out of the capillaries).

Let's say this person is a student studying for an exam. As he studies, he eats a lot of salty pretzels. His intestinal epithelium takes up glucose from the starch, along with sodium, and both sodium and chloride leak through the tight junctions into his extracellular fluid. His extracellular fluid has just become saltier. Two things happen. One is that water from his intestinal contents also leaks through the tight junctions to dilute the salt. Another is that some of the intracellular water may in fact move out of his cells to dilute his extracellular fluid. Water is following the salt.

The student's body will return itself to the proper water-salt balance. Maintaining the right amount of water and the right concentration of ions in the plasma, extracellular fluids, and cells is essential for the proper functioning of your body (especially the nervous system and the circulatory system, and if you don't have those, what have you got?). It is also important in maintaining blood pressure. The more fluid present in your blood vessels, the harder your heart has to work to push it through, and the higher your blood pressure. Consequently, your body has many interconnecting systems for maintaining its water-salt balance, and we will look at some of the major ones in the next chapter. Eating a bag of pretzels while studying is hardly a serious disruption to your water-salt balance, but you might notice a slight increase in weight from water retention if you consume a large amount of salty food at one time.

Why lactose intolerance includes diarrhea as a symptom

Water balance explains why people who are lactose intolerant get diarrhea as a result of eating dairy products. **Lactose intolerance** results from a lack of the enzyme lactase that breaks the complex milk sugar lactose into its simple-sugar components, glucose and galactose. The enzyme normally sits in the cell membrane of the microvilli. Most Caucasians continue to produce the enzyme and metabolize lactose throughout their lives, but many racial groups, including Africans, Asians, and Native Americans, tend to produce much less of the enzyme after early childhood.

People with insufficient lactase in their microvilli are said to be lactose intolerant. If they consume certain dairy products, they get diarrhea and gas.

The diarrhea is a result of water following not salt, but solutes. All that unusable lactose sugar in the intestinal tract attracts water out of the body into the small intestine. More water in the intestine is experienced by the owner of that intestine as diarrhea. The gas, incidentally, is also a result of the lactose. Though the person cannot metabolize the lactose, bacteria in his or her large intestine are happy to do so (the next chapter will explain how intestinal bacteria respond to lactose in their environment). When gut bacteria metabolize the lactose, they produce gas.

The same principle that causes diarrhea in lactose-intolerant individuals is operating in the actions of certain laxatives, such as those based on the magnesium ion. When a person ingests a high-magnesium laxative, the increased concentration of ions in the intestinal tract causes an efflux of water into the intestine. The intended result of the extra water is that the contents of the intestine will be washed out, relieving constipation.

Cystic fibrosis is caused by impaired salt transport

Water-salt imbalance also explains the symptoms of people who have cystic fibrosis, the most common fatal inherited disease of Caucasians. Cystic fibrosis patients produce abnormally thick mucus in several epithelia, particularly those lining the respiratory and gastrointestinal tracts. The thick mucus interferes with the secretion of digestive enzymes from the pancreas into the small intestine and causes digestive problems, but the primary clinical symptoms are respiratory. The airways of victims are frequently blocked by thick mucus. In addition, the lungs cannot clean themselves. In normal lungs, hairlike projections called cilia constantly sweep mucus and embedded particles (including bacteria) up and out. In cystic fibrosis patients, the thick mucus blocks this process, and recurrent respiratory infections result.

The gene that is defective in cystic fibrosis was isolated in 1989. It turned out to encode a chloride ion transporter located on the outside of the affected epithelia. Although the exact mechanism by which the defects in the transporter cause disease isn't yet clear, experiments show that blocking chloride transport in these epithelia results in reduced water secretion (water follows salt, and when salt is blocked, so is water) and thicker mucus.

BIOTECHNOLOGY APPLICATIONS

An understanding of the basics of membrane transport and the roles it plays in so many processes has led to improved therapies for quite a few conditions. We describe just a few here. Although not glamorous, the therapy that has saved and continues to save the most lives is rehydration (Box 7.1).

Rehydration therapy: water follows salt

Diarrhea is estimated to kill over 2 million children every year worldwide. The diarrhea itself may originate, in part, from a lack of sanitary infrastructure, but death from diarrhea is a result of dehydration. Effective treatment for afflicted children must include rehydration.

Simply drinking water is not the most effective rehydration therapy for severely dehydrated patients. A much more effective therapy has been developed based on an understanding of osmosis and the mechanism of glucose absorption from the small intestine. When the epithelium absorbs glucose, it also absorbs sodium. The uptake of these solutes creates an osmotic difference, and water flows from the small intestine into the tissues. Giving

BOX 7 . 1   *Cholera and ion transport*

Cholera is an infectious disease caused by the bacterium *Vibrio cholerae*. Cholera is usually spread by the consumption of food or water that has been contaminated with human fecal material and often occurs as epidemics in crowded areas. The major symptom of cholera is massive diarrhea, in which the stools of patients come to resemble "rice water," or clear fluid with white suspended particles. The particles are actually cells and mucus shed from the victim's gut. The diarrheal fluid also contains *V. cholerae*, which explains why the disease can spread epidemically in unsanitary conditions. Left untreated, cholera has a mortality rate of 25 to 50%. Victims die of dehydration.

Although cholera is now extremely rare in developed countries, it used to be a major health concern. For example, the city of Chicago reported over 4,500 deaths from cholera in 1849 to 1866. The cities of London, Gateshead, and Newcastle, England, reported over 10,500 deaths from cholera in 1853 alone. As cities and countries developed sanitary infrastructures, cholera outbreaks ceased.

Cholera is still a major health concern in the developing world and is one of three diseases that require notification of the World Health Organization. In 2001, 184,311 cases were reported in 58 countries, with 2,723 deaths. The World Health Organization estimates that the number of cases reported represents about 5 to 10% of the actual number of cases. In a major outbreak in Rwandan refugee camps during 1994, 48,000 cases occurred in a single month, with 23,800 deaths.

The killer behind this potentially lethal diarrhea is a protein encoded in the *V. cholerae* genome: cholera toxin. Cholera toxin causes diarrhea by disrupting the normal ion balance inside intestinal epithelial cells. The toxin is an enzyme that chemically modifies an intestinal epithelium protein involved in regulating ion transport. Normally, this regulatory protein is in the off position. When it switches to "on," it activates a pump that transports ions out of the epithelium into the small intestine.

The cholera toxin chemically modifies the regulatory protein, turning it on. The modified regulatory protein starts a chain reaction that activates the ion transporter, and the activated transport proteins begin pumping ions out into the small intestine. Because water follows salt, water pours into the intestine as well. Ion transporters and channels on the other side of the epithelium attempt to correct the ion drain by bringing more ions into the epithelial cells from the interstitial fluid, and the new ions are spewed out as well, draining even more water from the patient. Cholera patients can lose up to 1 liter of fluid per hour.

Cholera can be treated by immediate rehydration therapy to replace lost fluid and ions. For 80 to 90% of patients, drinking sufficient volumes of a solution of sugar and salts is all that is needed. In cases of severe dehydration, intravenous fluids may be required. With prompt rehydration, fewer than 1% of cholera patients die.

severely dehydrated children a solution of sugar and salt to drink is more effective than water alone, sugar and water, or salt and water as a rehydration drink.

Many sports drinks are also solutions of sugars, salts, and water, though some contain extras, like caffeine and vitamins. Research has shown that the right proportions of sugar and salt do result in faster absorption of water and that small amounts of carbohydrates in the drinks can provide an energy boost during sustained workouts. However, the general consensus also seems to be that unless you are exercising intensely for over an hour, water is completely adequate as a rehydration fluid. If you use sports drinks, you might want to do some research on their components, making sure you're not just reading marketing material.

Enzyme treatments for lactose intolerance

As described above, lactose intolerance results when a person lacks the enzyme to break lactose down into the simple sugars glucose and galactose. The enzyme is called lactase, or more properly, β-galactosidase. This enzyme is now produced industrially and is used to treat milk and dairy products so that they do not contain lactose (you can find these in the dairy case at the grocery store). The enzyme itself can also be purchased, and lactose-intolerant individuals can take it when they eat dairy products so that the lactose will be broken down and they won't get uncomfortable symptoms.

A related enzyme is marketed as a digestive aid for breaking down complex galactose-containing sugars present in high quantities in beans and other vegetables. Humans lack enzymes for breaking down these sugars, called galactosides, into their simple components and so cannot digest them. Like lactose in lactose-intolerant individuals, the galactosides pass through the small intestine and go on to feed intestinal bacteria in the large intestine. These bacteria do have an enzyme, α-galactosidase, for breaking down the galactosides. One of the by-products of bacterial galactoside breakdown is, you guessed it, gas. Microbial munching on galactosides is said to be responsible for one of the more unfortunate consequences of, say, a bean-rich meal in a Mexican restaurant.

In what has been touted as both a scientific and a social breakthrough, you can buy preparations of α-galactosidase in a drugstore under the trade name Beano. When taken with galactoside-rich foods, the enzyme is said to break down the galactosides into sugars you can digest, leaving the microbes without fuel for gas production.

SUMMARY POINTS

The cell membrane is a barrier between the intracellular and extracellular environments. Only a few small hydrophobic molecules, water, and ethanol can cross the membrane on their own. All other compounds that cross into or out of a cell do so with the assistance of a membrane protein.

The transmembrane domain of membrane proteins is an inside-out version of normal protein structure. Hydrophobic amino acid residues face outward into the oily membrane, and hydrophilic residues are found on the inside. In the case of transport proteins, the hydrophilic residues form a hospitable environment for the transport of hydrophilic substances, such as sugars, amino acids, and ions.

Molecules will move down a concentration gradient if there is no barrier in their way. Gated channel proteins and carrier proteins allow certain molecules to move into or out of a cell along a gradient.

Energy is required to move molecules across the cell membrane against a concentration gradient. Transport proteins that use energy to do this are called pumps. The energy to power a pump can come from ATP or from harnessing another gradient, typically the sodium gradient.

The Na^+/K^+ ATPase pump maintains the relatively high intracellular concentration of potassium and the relatively low intracellular concentration of sodium, using energy from ATP. This ion pump is so important that some cells use 25% of their ATP to power it.

Nerve cell impulses are propagated by the sodium and potassium gradients.

Muscle cell contraction is caused by the release of calcium ions from the SR into the muscle cell cytoplasm. The release of the ions is triggered by nerve impulses to the muscle.

Epithelia are organized sheets of cells with two distinct sides separated by tight junctions between cells. The intestinal epithelium is specifically organized for absorbing nutrients from the small intestine via pumps and releasing the nutrients into the interstitial body fluid via carrier proteins.

Water follows solutes in the body. Water will move across membranes and through aquaporins, when possible, to maintain osmotic balance in various cell compartments.

The profuse diarrhea characteristic of cholera is caused by the cholera toxin protein. This toxin activates an ion transport channel in the intestinal epithelium, causing it to pump ions out of the body into the intestinal tract. Water follows the ions, resulting in diarrhea.

KEY TERMS

Aquaporin	Epithelial cell	Microvilli	Solution
Carrier protein	Extracellular	Na^+/K^+ ATPase	Tight junction
Channel protein	Interstitial fluid	Osmosis	With or against a
Cholera	Intestinal epithelium	Osmotic balance	concentration gradient
Concentration	Intracellular	Pump	
Diffusion	Lactose intolerance	Solute	

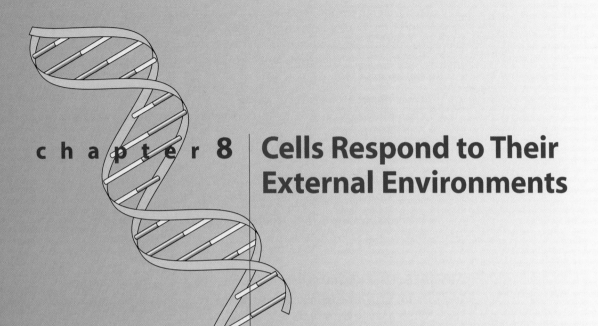

chapter 8 | Cells Respond to Their External Environments

A lactose-intolerant student at a party is unable to resist the temptation of a large bowl of homemade ice cream covered with chocolate syrup and sprinkles. When the sugars in the ice cream reach her small intestine, her cells cannot break down the lactose because they lack the enzymes for it, so the milk sugar passes into her large intestine. The sucrose, however, is cleaved to glucose and fructose, which are absorbed. Her blood sugar level climbs quickly but then begins to fall as her cells absorb it. Meanwhile, bacteria in her large intestine sense the influx of lactose. They immediately begin to synthesize enzymes for digesting the sugar and feast on the new nutrient source. Unfortunately, one of their lactose breakdown products is carbon dioxide gas, which builds up and causes the student some discomfort later on.

The student and the bacteria in her gut may not seem to have much in common, but in both cases, their cells sensed and responded to their environments. The bacteria sensed the presence of the new nutrient and responded by synthesizing enzymes to break it down for energy. The student's body sensed the increased concentration of glucose in her bloodstream and responded by releasing insulin, which in turn caused her blood sugar concentration to drop back to normal levels.

These are examples of cells responding to their external environments, but such communication is not a one-way street. Cells also release signals into their environments that can affect the activities of neighboring cells. In the above example, cells in the student's body reacted to the increased glucose by releasing insulin, which acted as a signal to other cells and changed their activities. Sending and receiving signals allow cells to respond to environmental changes and to communicate with one another (Box 8.1).

The ability of cells to respond to their environments is crucial to survival. Single-celled organisms have to be able to sense and respond to changing conditions to find and take advantage of new sources of food, to protect themselves from changes in salinity or the presence of toxins, and even to

157

Chemical warfare in the plant world

Like animal cells, plant cells both receive and produce signals of all kinds. When plants are under attack by herbivorous insects, the damage to their tissues can trigger the release of chemical signals. These signals act like a warning system. When they bind to receptors on other plant cells, they stimulate those cells to synthesize compounds that are poisonous to the insects.

Certain insects, however, have learned to use plants' warning systems to evade the plants' chemical defenses. In the corn earworm, the warning chemicals released by damaged plant cells act as signals to increase the production of detoxifying enzymes in the worms' guts. Exposure to the plants' alarm chemicals causes the worms to gear up to break down the plant poisons.

find mates. For cells in a multicellular organism such as yourself, the environment is the inside of the organism. These cells send and receive signals that allow them to respond to external emergencies (such as being chased by a predator), maintain the internal environment of the body within healthy ranges, meet the body's needs for energy and elimination of waste, and regulate growth and development. All these communication activities are mediated by various kinds of signals and receptors for those signals.

In this chapter we will look at:

- types of signals and receptors
- examples of direct interaction between the environment and single-celled organisms
- how hormones regulate the environment within multicellular organisms
- regulation of glucose concentration in the blood
- regulation of salt and water balance and blood pressure

In later chapters, we will look at examples of the regulation of cell division, growth, and development.

SIGNALS AND RECEPTORS

In order for cells to communicate with and respond to their environment, three things have to happen: (1) there must be a signal, (2) cells must be able to detect the signal, and (3) detecting the signal has to induce changes in the cell that form a response to the signal. The variety of kinds of signals cells can detect is remarkable, including chemicals, light, sound, electrical impulses, solute concentration, and pressure.

The kinds of cellular changes that can be induced in response to signals include, but are not limited to:

- activation of enzyme activity
- suppression of enzyme activity
- activation of transcription and synthesis of new proteins
- suppression of synthesis of specific proteins
- changes in the permeability of the cell to certain substances through alterations in the activities of channel proteins
- release of stored proteins

These immediate changes can be part of a process that leads to responses, such as generating a nerve impulse, metabolizing a nutrient, moving, migrating, growing and dividing, differentiating, or even dying. You have already encountered examples of some of these responses in the previous chapter, for example, nerve impulses stimulating the release of calcium ions, leading to muscle contraction.

Specialized receptors detect many different kinds of signals

Different kinds of receptors are required to detect different kinds of signals. For example, photoreceptors contain molecules called pigments that respond to specific wavelengths of light. The pigments absorb energy from the light, setting off a chain reaction of events inside the cell leading to a response. Some signals, like pressure, exert a direct effect on the structure of the cell and stimulate a response. Specialized cells called **baroreceptors** (baro, pressure, as in a barometer) respond to touch, stretching, or pressure by opening

Table 8.1 Receptors and the five senses

Type of receptor	Activating stimulus	Cellular response	Brain's interpretation of nerve impulse
Photoreceptor	Light	Change in membrane channels	Vision
Auditory receptors	Vibration	Release of stored neurotransmitters	Sound
Olfactory receptors	Various molecules in the air	Change in membrane channels	Smell
Taste receptors for sweet and bitter	Various dissolved molecules	Change in membrane channels	Sweet or bitter taste
Taste receptors	Na^+, Cl^-, K^+ (salty) H^+ (sour)	Release of stored neurotransmitters	Salty or sour taste
Baroreceptor	Deformation of cell	Change in membrane channels	Touch, pressure

ion channels in their membranes and generating a nerve impulse that stimulates action in other cells (for example, the release of a hormone). Similarly, **osmoreceptors** respond to changes in salt concentration in your body fluids. When the salt concentration in your body fluids increases, these cells shrink (because of water loss from the cell), and the size change affects their geometry, again opening ion channels in their membranes and generating a nerve impulse.

Our five senses involve receiving direct input from the outside environment, and the various sensory stimuli are received by different kinds of receptors that generate nerve impulses that travel to the brain. The brain, in turn, interprets the nerve impulses as sights, sounds, tastes, smells, or touch. Receptors involved in our senses are described in Table 8.1.

The most common signals inside your body are various kinds of molecules. Chemical signals are detected by receptor proteins whose shapes allow them to bind specifically to particular signal molecules. Receptors for molecular signals can be in the cell membrane facing out, or they can be inside the cell, depending on whether the signal can cross over the cell membrane. A signal can cross if it is the right sort of molecule (small and hydrophobic) or if there are transport proteins or channels in the cell membrane that let it in. You'll see some examples of signals that get into cells below. If a signal cannot cross the cell membrane, then it must interact with a receptor on the cell surface. You'll see examples of signals with surface receptors below, too.

When the signal binds to its receptor protein, it causes a change in the shape of the protein and therefore changes its function (Figure 8.1). The receptor-signal complex is thus able to perform a function the receptor alone could not do or it can no longer perform an activity the unbound receptor was doing. Either way, the receptor protein's activity changes.

Responding to signals often involves a chain reaction

You have already learned that cell metabolic processes usually involve many steps, each one carried out by a specific protein. The same is true for responding to signals. Although some response pathways involve just one or two proteins, most are more like chain reactions, in which the initial activation of the receptor starts a cascade of changes in other proteins that eventually leads to the end response (Figure 8.2).

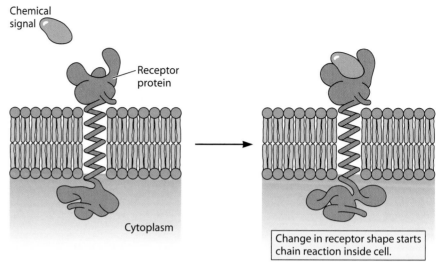

Figure 8.1 When a chemical signal binds to its receptor protein, the receptor's shape and activity change.

A key concept for this chapter is that in order for a cell (or, by extension, a multicellular organism) to respond to signals from the environment, all the steps from the signal to the effect must be in place. If a receptor is not present or doesn't work, then the cell cannot detect the signal. If a step in the chain reaction between the receptor and the effect is short-circuited by a missing or defective protein, then the effect will not take place. We will discuss some examples of cascades below as we consider some of the ways our bodies regulate salt and water balance. First, however, we will look at an example from the bacterium *Escherichia coli* to illustrate some basics of signaling and response.

Figure 8.2 The pathway leading from a signal to a response often involves a cascade of changes.

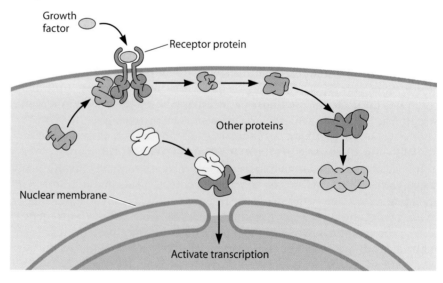

Single-celled organisms respond directly to the outside environment

One way cells can make changes in response to environmental signals is through regulation of the synthesis of proteins. Turning on or shutting off the synthesis of a particular protein changes the cell's functioning. In the chapter introduction, we used the example of a student's gut bacteria responding to the milk sugar lactose by synthesizing new enzymes that allowed them to metabolize it. Here is a closer look at how this process takes place.

Lactose breakdown in *E. coli*

Most bacteria have genes encoding enzymes capable of breaking down quite a variety of sugars for energy. However, synthesizing these enzymes would be a waste of the cell's energy if the sugars were not available to the cell, so most bacteria synthesize an enzyme that breaks down a particular sugar only if that sugar is present in the environment.

The sugar lactose can be used by the gut bacterium *E. coli* and many other bacteria as an energy source. *E. coli*'s lactose utilization genes (called the *lac* genes) are lined up in a row along its chromosome and are transcribed from a single promoter into one long mRNA. The genes include the lactase enzyme, also called β-galactosidase, which cleaves lactose into the simple sugars glucose and galactose. These simple sugars are then broken down for energy, and carbon dioxide gas is one of the end products. (Recall from chapter 6 that carbon dioxide gas is a normal breakdown product of sugar metabolism in your body, too.) The lactase enzyme made by *E. coli* is the same enzyme, though a slightly different variation, that is missing in lactose-intolerant individuals, as described in the previous chapter. When lactose is present in the environment, *E. coli*'s *lac* genes are transcribed, lactase is made, and *E. coli* derives energy from the sugar. (See chapter 4 if you need a quick review of gene expression.) When no lactose is present, these proteins are not synthesized.

How does *E. coli* achieve this appropriate regulation of its *lac* genes? The following description is somewhat simplified but gives the basic idea.

Regulation of the *lac* genes

E. coli normally synthesizes a repressor protein that prevents transcription of the *lac* genes. The repressor prevents transcription because its three-dimensional shape and chemical nature cause it to bind to the *E. coli* chromosome at a unique sequence of DNA bases (called the operator) near the promoter of the *lac* genes. When the repressor protein is bound to its target base sequence, it blocks RNA polymerase from binding to the promoter and transcribing the *lac* genes. Consequently, the bacterium does not waste energy making lactose-utilizing enzymes when there is no lactose in the cell.

If lactose is present, however, the bacterium needs enzymes for using it. The *lac* regulation system allows the cell to respond beautifully to this new need. When lactose enters the cell through transport proteins, it binds to a special site on the *lac* repressor protein. This interaction changes the shape of the protein and renders the repressor unable to bind to its site on the *E. coli* DNA. The repressor protein releases the DNA, leaving the gene free to be transcribed. The cell then can make the lactase enzyme and take advantage of the new energy source (Figure 8.3).

The system for regulating lactose breakdown is very similar to the one that regulates tryptophan biosynthesis, described in chapters 5 and 6. In both cases, a small molecule (lactose or tryptophan) binds to a repressor protein and changes its shape. In the case of lactose, the shape change renders the

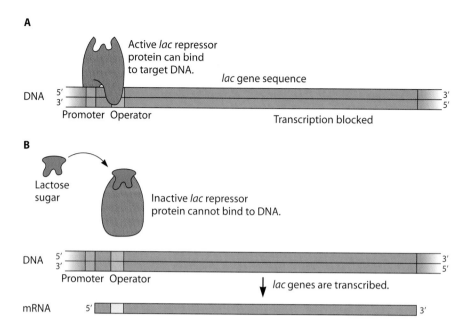

A

Active *lac* repressor protein can bind to target DNA.

lac gene sequence

DNA 5′ 3′

Promoter Operator

Transcription blocked

3′ 5′

B

Lactose sugar

Inactive *lac* repressor protein cannot bind to DNA.

DNA 5′ 3′

Promoter Operator

lac genes are transcribed.

3′ 5′

mRNA 5′

3′

Figure 8.3 Regulation of lactose utilization genes in *E. coli.* **(A)** No lactose in cells; active repressor prevents transcription. **(B)** Lactose in cells; inactive lactose-repressor complex cannot bind to DNA. Transcription occurs.

Figure 8.4 Slime mold amoebas swarm toward a chemical signal, piling onto one another at the focus. (Photograph courtesy of Danton O'Day, University of Toronto at Mississauga, Mississauga, Canada.)

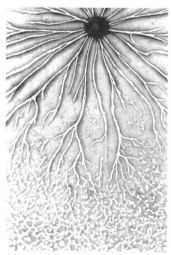

repressor unable to bind DNA and allows transcription of the *lac* genes and breakdown of the sugar. In tryptophan biosynthesis, the shape change causes the repressor to bind to DNA, shutting off transcription of the *trp* genes so that the cell does not waste energy synthesizing tryptophan if there is already plenty of it in the cell.

The *lac* and *trp* genes of *E. coli* are two well-understood examples of how interactions between substances in the environment and receptors change gene expression and allow cells to respond appropriately to changing conditions. There are a multitude of other examples along these lines, for example:

- yeast cells changing gene expression in response to chemical signals secreted by other yeast cells so that they can mate
- *E. coli* responding to increasing salt in its environment by closing channels in its cell membrane to exclude ions
- slime mold cells responding to chemical signals by migrating toward one another (Figure 8.4)
- plant-pathogenic bacteria secreting substances that, when they reach a critical concentration, activate transcription of genes required to cause disease, so that no attack occurs until a critical population size of bacteria is present

At this point, rather than examining more examples like these, we will turn to discussing how your own multicellular body uses chemical signaling to coordinate the responses of many cells so that your internal environment stays within constant and healthy ranges.

Your body uses hormones to coordinate cell responses

For single-celled organisms, interaction with the environment is straightforward. The cell senses changes in its environment via membrane or cytoplasmic proteins and responds. For multicellular organisms with different tissues and compartments (like you), the task is more complex. The change in the environment must be sensed, whether it is the increasing glucose concentration in your blood from the candy bar you just ate or the angry grizzly bear

Figure 8.5 Imagine you saw this grizzly bear near you while you were hiking in a remote area. Can you feel changes in your body just thinking about it? If you received this signal for real, a multitude of coordinated responses would prepare you to run for your life. (Photograph copyright Ross Warner.)

you just saw walk into your campsite, and then your cells have to respond in a coordinated manner so that your body can react appropriately (Figure 8.5).

In a multicellular body, specialized cells typically detect the environmental signal and then pass the signal on. The signal may be passed on as a nerve impulse or as a chemical signal. Whatever the initial translation, coordinating your body's response to it usually involves signaling molecules such as **hormones**. Hormones are made in various glands in your body and secreted into your bloodstream (Table 8.2 lists examples). Other signaling molecules are not secreted into your bloodstream, which differentiates them from hormones, but they also act as chemical signals. They are often named for what they do: **growth factors** signal cells to grow and divide; **neurotransmitters** transmit nerve impulses across gaps between nerve cells.

Table 8.2 Examples of human hormones

Hormone	Where secreted	Target(s)	Primary effect(s)
Thyroxine	Thyroid	Many tissues	Stimulates and maintains metabolism; necessary for normal growth and development
Growth hormone	Anterior pituitary	Bones, liver, muscle	Stimulates protein synthesis and growth
Follicle-stimulating hormone	Anterior pituitary	Gonads	Stimulates growth and maturation of eggs in females; stimulates sperm production in males
Melanocyte-stimulating hormone	Anterior pituitary	Melanocytes	Controls pigmentation
Insulin	Pancreas	Muscles, liver, fat	Stimulates uptake and metabolism of glucose; increases glycogen and fat synthesis; reduces blood sugar
Glucagon	Pancreas	Liver	Stimulates breakdown of glycogen; raises blood sugar
Somatostatin	Pancreas	Digestive tract, pancreas	Inhibits release of insulin and glucagon; decreases activity in the digestive tract
ADH	Posterior pituitary	Kidneys	Stimulates water resorption and raises blood pressure
ANH	Heart	Kidneys	Increases sodium ion excretion; lowers blood pressure
Aldosterone	Adrenal cortex	Kidneys	Stimulates excretion of potassium and resorption of sodium ions
Estrogens	Ovaries	Breast, uterus, and other tissues	Stimulate development and maintenance of female sexual characteristics; necessary for proper bone development in males and females; proper seminal fluid formation in males
Androgens	Testes	Various tissues	Stimulate development and maintenance of male sexual characteristics

Hormone receptors

Whatever the signal is called, it must bind to its receptor for a response to occur. Hormone receptors can be on the exterior or within the interior of the cell, depending on whether the signal can cross the membrane. Many hormones are small proteins, and their receptors are typically membrane proteins whose extracellular domains recognize and bind to the hormone. Hormone binding changes the shape of the protein in a way that affects its intracellular domain, so the signal is transmitted to the interior of the cell.

One major class of hormones your body manufactures is the steroid hormones. As described in chapter 3, they are derived from cholesterol and are hydrophobic, so they can cross the cell membrane and enter cells. Receptors for these hormones are located inside cells that respond to the hormones. Steroid hormones bind to their protein receptors, causing a shape change in the receptor. The receptor-hormone complex becomes capable of binding to target DNA sequences, rather like the *trp* repressor. The specific DNA target depends on the identity of the complex. The complex binds to its target DNA sequence, and by doing so, it can either repress or stimulate transcription, depending on how it interacts with other proteins at that site.

Estrogen

Estrogen is an example of a steroid hormone. When estrogen binds to its receptor protein, it causes the protein to change shape and assume the right one to bind to a specific sequence of DNA bases. The estrogen receptor-estrogen complex binding to DNA usually enables the transcription of associated genes rather than blocking it.

Estrogen exerts many effects in the body, and many genes respond to the presence of the estrogen receptor-estrogen complex. Scientists do not yet understand all, or even the majority, of the details of which genes are turned on where in the body in response to estrogen and how the products of those genes go on to exert physiological effects, but here are a few examples of fairly recent research.

As you probably know, estrogen is a dominant hormone in females. High levels of estrogen are present in the first part of the menstrual cycle, when the lining of the uterus is thickening to receive a fertilized egg. In the past few years, it has been shown that estrogen stimulates the transcription in the uterus of genes encoding two different proteins known to be involved in the generation of new blood vessels. In addition, estrogen also stimulates the transcription of the gene for lactoferrin, a protein found in breast milk.

By the way, you may think of estrogen as a "female hormone," but estrogen is also vital to men. Men who lack estrogen or cannot respond to it because they lack the estrogen receptor don't produce seminal fluid correctly. Their semen is dilute, and their fertility may be impaired. In addition, their skeletons do not develop properly, and their bones are thin and brittle. You'll read more about sex hormones and sex differentiation in chapter 10.

The idea of a signal binding to a receptor protein is a theme you are going to encounter over and over as you read more about hormones, cell reproduction (see chapter 9), and cell differentiation (see chapter 10). The signal and its receptor protein must both be present and active for a cell to respond. As you read this and the following two chapters, you'll see example after example of what happens when some component of the signal-receptor system malfunctions and the kinds of conditions that result, from diabetes to cancer to inappropriate sexual development. For now, we will return to our theme of metabolism and look at how glucose levels in your body are regulated.

BOX 8.2 *Insulin in biological science*

Insulin has played a starring role in the development of the fields of biochemistry and biotechnology. It was one of the first proteins to be crystallized in pure form in 1926, and in 1955, it became the first protein to have its amino acid sequence determined. In 1963, it was the first protein to be chemically synthesized in a laboratory (researchers could not produce enough of it for pharmaceutical use), and in 1978, it became the first human protein to be manufactured through biotechnology.

Insulin consists of two peptide chains, the A chain, with 21 amino acids, and the B chain, with 30 amino acids. The two chains are held together by two disulfide bridges, and the A chain has one internal disulfide bridge. The amino acid sequences of insulins from animals are very similar to that of human insulin. In fact, there is only a single amino acid difference between human and porcine (pig) insulins, and only three between human and bovine (cow) insulins.

REGULATION OF BLOOD GLUCOSE CONCENTRATION

Your brain uses the sugar glucose as its only fuel, so it is critically important that there be enough glucose in the bloodstream. If your blood sugar concentration falls too low, you will lose consciousness and can lapse into a coma and die. On the other hand, too much glucose in the bloodstream can cause mental confusion, dehydration, and a number of other problems. Obviously the body needs to control how much glucose is in the bloodstream, and it needs to do this even in the bodies of people who may, at a moment's notice, flood their bodies with glucose by drinking a huge glass of sugary soda and eating a candy bar or two.

The body regulates blood glucose levels through the actions of two hormones: **insulin** (Box 8.2) and **glucagon**. Both insulin and glucagon are small proteins made in specialized cells of the pancreas (Figure 8.6). They alter the ability of cells to take up glucose, as well as the activities of enzymes that store or release glucose and fats. Glucose changes cellular activities in your body (and in those of other animals) not by interacting directly with cells throughout the body but instead by affecting the release of these hormones.

Figure 8.6 The three-dimensional structures of insulin and glucagon. **(A)** Insulin. (Structure courtesy of G. G. Dodson.) **(B)** Glucagon. (Structure courtesy of T. Blundell.)

A. Insulin

B. Glucagon

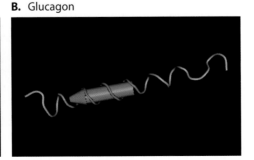

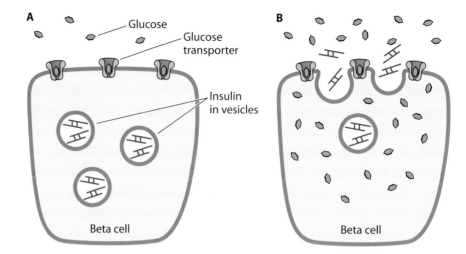

Figure 8.7 An increase in glucose concentration inside beta cells of the pancreas causes release of insulin from vesicles. **(A)** Glucose concentration is low. Insulin remains in vesicles. **(B)** Glucose concentration is high. Insulin is released.

The sugar glucose causes insulin release and turns on gene expression

Insulin is synthesized in the beta cells of the pancreas. Since insulin is made to be secreted, it passes through the cell's shipping department (the endoplasmic reticulum and the Golgi apparatus) and is packaged into vesicles. The membrane-bound vesicles sit in the cytoplasm until they receive a signal that causes them to fuse with the cell membrane and release the insulin.

When you consume high-glucose foods, like sugar and starch, the concentration of glucose in your blood rises as the epithelial cells in your small intestine absorb the sugar from your digestive tract and kick it out into your interstitial fluid so that it can diffuse into your capillaries. Glucose is carried to the pancreas in the bloodstream and enters the beta cells via transport proteins. In a process that is not yet fully understood, the rising concentration of glucose inside the beta cells causes the insulin vesicles to fuse with the cell membrane, releasing the insulin into your bloodstream (Figure 8.7).

Because insulin is sitting in vesicles ready to go, your body can respond quickly to an influx of glucose. Glucose also stimulates the pancreatic cells to transcribe the insulin gene and produce more insulin. Figure 8.8 shows what happens to the blood insulin concentration during a long infusion of glucose. You can see the first quick release of the stored insulin and then the later release of newly made hormone.

Figure 8.8 Blood insulin concentration during an hour-long infusion of glucose. The initial spike (before 10 minutes) is caused by the release of stored insulin from vesicles. The rise in concentration beginning around 15 minutes is due to the release of newly transcribed and translated insulin.

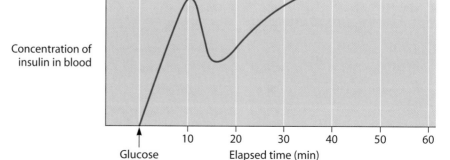

Insulin stimulates transport and modifies enzyme activities to decrease blood glucose levels

Once insulin has been secreted from the beta cells in the pancreas, it travels through the bloodstream. Since it is a protein, it cannot cross over the cell membrane and therefore exerts its effects by binding to protein receptors on the surfaces of cells. When insulin binds to its receptor, the receptor changes shape and adds a phosphate group to itself. Adding the negatively charged phosphate group further changes its shape, as well as its electrical charge, and then the receptor protein adds phosphates to additional proteins. This begins a chain reaction that causes different end effects in different cells, depending on which proteins are present in those cells. Different types of cells have insulin receptors, and their responses to insulin vary.

When insulin binds to receptors on muscle and fat cells, it stimulates the uptake of glucose from the blood. The way insulin stimulates the uptake of glucose is by increasing the number of glucose transporters on the surfaces of these target cells. These cells take up glucose through a glucose transporter called GLUT4. In the absence of insulin, this particular type of glucose transporter is located in vesicles in the cytoplasm of the cells, where it is useless for taking up glucose. The binding of insulin to its receptor on the cell surface, however, triggers changes that cause the vesicles to fuse with the cell membrane, inserting GLUT4 transporters so that they can absorb glucose from the bloodstream. When the concentration of insulin in the blood falls and the insulin is no longer bound to its receptors, the glucose transporters are reisolated into cytoplasmic vesicles (Figure 8.9). Note that liver and brain cells do not require insulin in order to take up glucose; they use a slightly different glucose transport protein (GLUT1) that is not insulin dependent.

Insulin binding to its receptors on muscle and liver cells also leads to the activation of the muscle and liver enzymes that synthesize the storage carbohydrate glycogen. When the liver is saturated with glycogen (about 5% of its weight), insulin also stimulates the synthesis of fatty acids. The new fatty acids are exported in the form of lipoproteins into the bloodstream, where they can be taken up by fat cells for storage as fats or by other cells to be broken down for energy.

Figure 8.9 Effect of insulin on glucose transport in target cells. **(A)** Low insulin concentration in blood. Glucose transporters are stored in vesicles. **(B)** High insulin concentration in blood. Insulin receptors are activated, stimulating insertion of glucose transporters into the cell membrane. The target cell absorbs glucose.

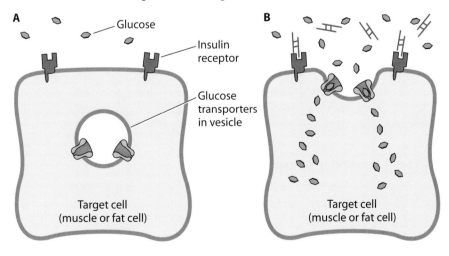

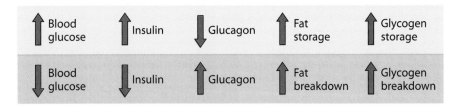

Figure 8.10 Responses to high and low blood glucose levels.

We mentioned above that insulin stimulates fat cells to take up glucose. Fat cells use the glucose to synthesize glycerol. This glycerol is combined with the fatty acids from the liver to make fats. While this fat synthesis is occurring, insulin is simultaneously inhibiting the fat cell enzymes that break down fats.

The net effect of insulin activity is that the liver stores as much glucose as it can in the form of glycogen, muscle and fat cells take up glucose, fat breakdown is inhibited, and if the liver is saturated with glycogen, new fatty acids are released to become fuel or to be used in the synthesis of new fat molecules.

Glucagon exerts a variety of effects to increase blood glucose levels

When the blood glucose concentration falls below a certain level, insulin release slows and the release of the hormone glucagon increases. Glucagon also travels through the bloodstream and binds to receptors on various cells. When glucagon binds to receptors on liver cells, the activity of the enzyme that synthesizes glycogen is inhibited. At the same time, the activity of the enzyme that breaks down glycogen is stimulated. Glycogen is broken down into glucose, which is released into the bloodstream. When glucagon binds to receptors on fat cells, it activates enzymes that break down fats. The fatty acids are exported into the bloodstream, where they can be taken up and used for fuel, sparing glucose for brain cells (Figure 8.10).

The opposing actions of insulin and glucagon keep the fasting blood glucose concentration in a healthy individual somewhere around 80 to 100 mg per 100 ml of plasma.

What happens to people who eat no carbohydrates or sugars?

Glucose is essential for your brain to function, and your cells are prepared to deal with the situation if insufficient glucose is available from the diet. Enzymes in your liver cells can process the carbon skeletons of amino acids and fats and actually make glucose from them in a process called **gluconeogenesis** (gluco, glucose; neo, new; genesis, origin). In animals such as cats, which eat very little glucose, the chief role of glucagon is to stimulate gluconeogenesis. If a person is not eating enough glucose or sufficient fats and proteins to provide the raw chemical material and energy for gluconeogenesis, the body will cannibalize its own fats and proteins (muscle) for gluconeogenesis, because providing the brain with fuel is absolutely critical for survival.

FAILURE OF INSULIN SIGNALING

You may know someone who has diabetes, a disease that afflicts many people. Diabetes is the result of a regulatory failure. Individuals with diabetes cannot properly control blood glucose (often called blood sugar) levels because they do not increase glucose absorption when their blood glucose levels rise. Consequently, their blood sugar levels are too high. They excrete large amounts of glucose in their urine, which causes the kidneys to excrete extra water (remember, water follows solutes, and glucose is a solute). When diabetes was first recognized as a medical condition, it was given the name **diabetes mellitus**. "Diabetes" is a Greek word meaning excessive urination, and "mellitus" is a Latin word meaning honey. Physicians actually used to taste the urine of patients to see if it was sweet. People with diabetes experience excessive thirst and are in danger of dehydration. Because of their metabolic imbalance, acids and other chemicals build up in their blood. Often, eye, kidney, and circulatory problems develop, sometimes so severe that limbs begin to die from lack of blood and must be amputated.

There are two types of diabetes mellitus

Diabetes mellitus can be caused by two entirely different defects in the blood glucose regulatory pathway. In one case, the body cannot produce insulin; in the other, the body can produce it but cannot respond to it. These two defects are associated with different ages of onset of diabetes and also with different levels of severity.

One of the defects that result in diabetes is a loss of the ability to produce insulin. In some people, the body's immune system, which normally attacks and destroys invaders like disease-causing agents, attacks and destroys the beta cells in the pancreas that produce insulin. This happens during childhood. Individuals with this condition develop the symptoms of diabetes when they are quite young. Because they cannot make insulin, they must take insulin for the rest of their lives, monitor their blood sugar carefully, and regulate their diets to complement their insulin intake. Without an outside source of insulin, these patients die young. This type of diabetes is called type I diabetes, juvenile diabetes, or **insulin-dependent diabetes**.

The second type of patient that develops diabetes is usually an overweight, inactive adult or an elderly person. These individuals are perfectly capable of making insulin. Their diabetes develops because their cells lose the ability to respond to the insulin. This type of diabetes is called type II diabetes, insulin-resistant diabetes, or **non-insulin-dependent diabetes**, and accounts for 90 to 95% of all diabetes (Table 8.3). Type II diabetes can sometimes be controlled by weight loss and diet and exercise changes.

Exactly why cells become resistant to insulin is not yet understood and is a subject of current medical research. It appears that there may be several things that can go wrong with insulin responsiveness, which makes sense given that there are several steps between the receptor and the final effects in the cell. There may be fewer insulin receptors on the surfaces of cells, the receptors themselves may still be there but no longer able to make the appropriate changes when insulin binds to them, or the chain reaction that should

Table 8.3 Two types of diabetes mellitus

Type of diabetes mellitus	Life stage of patient at onset	Insulin status	Cause	% of all diabetes mellitus
Type I, insulin dependent	Juvenile	Body does not produce insulin	Immune system destroys insulin-producing cells in pancreas	5
Type II, non-insulin dependent	Adult (usually)	Body produces insulin but does not respond to it	Not known; associated with obesity	95

happen inside the cell once insulin has bound to its receptor may be interrupted. Whatever the molecular mechanism behind non-insulin-dependent diabetes, it is becoming an increasingly serious health problem. As rates of obesity among children climb, doctors are seeing this kind of diabetes developing at earlier and earlier ages.

Other insulin and glucagon diseases

Diseases associated with glucagon are very rare. Cancers of the cells that secrete glucagon can result in excessively high concentrations of the hormone. Individuals with this type of tumor typically experience a wasting syndrome. The insulin-secreting cells of the pancreas, too, can very rarely become cancerous. Cancers of insulin-producing cells produce excessive insulin concentrations in the blood and result in excessively low blood glucose concentrations. The brain becomes starved for glucose, which can lead to insulin shock, a life-threatening condition.

Biotechnology application: recombinant human insulin for diabetics

Insulin and its relationship to diabetes were discovered in 1920 and 1921, and the first human diabetes patient was treated with insulin in 1922. The news of the success of the treatment spread like wildfire, and the drug company Eli Lilly began production of insulin at that time. For the next 60 years, insulin was produced from pig and cow pancreases. Animal insulins are very similar to human insulin and are effective in treating diabetes, but some individuals had allergic reactions to the preparations.

With the development of techniques for manipulating DNA and cloning genes, scientists realized that it would be possible to put genes for human proteins into systems that could produce large amounts of those proteins easily, such as bacteria or yeast. Researchers inserted the DNA coding for human insulin into bacteria and used the bacteria as miniature factories to pump out large amounts of the protein. This recombinant human insulin became widely available to diabetics in the 1980s. Today, almost all diabetics use recombinant human insulin instead of animal insulin. Chapter 15 explains how genes for human proteins (or any other proteins) are cloned and put into different organisms for protein production.

We now want to look at the interconnected system of hormones that regulate your body's blood pressure and salt and water balance.

BLOOD PRESSURE, SALT, AND WATER

In the last chapter, you learned a little about salt and water balance in the body. Now, we are going to look at how body hormones cause changes in cells that regulate the body's water and salt content. Regulating salt and water content is essential not only for proper muscle and nerve cell function (as discussed in the previous chapter), but also for maintaining blood pressure.

Blood circulation provides nutrients to and removes wastes from cells

Your heart is a muscular pump that keeps blood flowing throughout your body. Blood flow brings oxygen from your lungs, nutrients from your liver and small intestine, and hormones from a variety of sources to all your body's cells. These substances get to your cells through the leaky walls of capillaries, the smallest blood vessels. As mentioned previously, capillary walls are permeable to water, ions, and small molecules—only the blood proteins are held in. When blood flows through capillaries, oxygen, nutrients, and other substances diffuse through the capillary walls into the interstitial fluid bathing your tissues and are taken up by your cells. At the same time, waste products eliminated by your cells into the interstitial fluid can pass into the capillaries and be removed. These exchanges work over only very short distances, so you need a lot of capillaries to take care of all your cells. In fact, your body is estimated to have around 60,000 *miles* of capillaries.

Blood pressure

The blood flowing through your blood vessels is under pressure, and maintaining that pressure is important enough that the body has a complex system for regulating it. Why is maintaining blood pressure important? Your blood pressure, like the pressure squirting water from a garden hose, is what keeps blood flowing to your organs. Without an adequate supply of blood and the goodies it supplies, your organs and tissues can't survive. The first one to go would be your brain. In fact, if blood pressure in your brain drops below a certain level you will lose consciousness, or faint. Many people experience dizziness if they've been lying down and stand up quickly. The reason for this is that when a person stands up from lying down, there is a need for greater blood pressure to pump blood up into the head. Although the body quickly adjusts, the person may experience a transient lightheaded feeling or even black out before the blood pressure to his or her brain rises adequately.

Another way to look at the importance of maintaining blood pressure is to observe what happens in people whose blood pressure is chronically high. High blood pressure, or **hypertension**, can cause weakened blood vessels to burst and bleed. If this happens in the brain, it can cause a stroke. High blood pressure is the most important risk factor for stroke. If a blood vessel in the eye breaks, blurred vision or even blindness can result. High blood pressure is also associated with a stiffening (or "hardening") of artery walls. Stiffer arteries cause the heart to work harder to pump blood. High blood pressure is a risk factor for **heart attack**, in which the heart itself does not receive enough oxygen-rich blood to sustain it, and for **heart failure**, in which the heart cannot pump enough blood to supply the body's needs. High blood pressure also affects kidney function (concerning which you are about to read more). Over time, high blood pressure can narrow and thicken the walls of blood vessels in the kidneys. The kidneys filter less fluid, and waste builds up in the body. Eventually, the kidneys may fail altogether. Obviously, keeping blood pressure correctly regulated is very important for health.

Blood pressure and blood volume are interconnected

Now, let's think about blood pressure, blood volume, and artery walls. Blood pressure is measured by determining how much pressure has to be applied to a blood vessel in your arm to stop the flow of blood through it (Figure 8.11). Think about that, and imagine that you could increase the volume of blood in your body. If you had more blood in your body, would you have to press harder or less hard to shut off the flow? Increasing the volume of blood in your system would make it harder to squeeze the artery shut. If the volume of your blood decreased, you could shut off the flow with less pressure. So increasing the volume of blood causes blood pressure to increase.

Your arteries have springy, elastic walls that can contract or relax. Imagine that your arteries contract around your blood. Would you need to apply

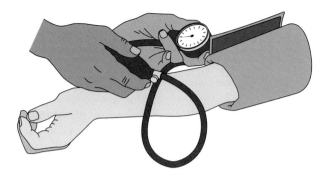

Figure 8.11 A blood pressure cuff measures blood pressure by determining how much pressure is required to stop blood flow.

more or less pressure to shut off the flow? If your artery walls squeeze in around your blood, it would take more pressure to press the artery shut. If your arteries relax, less pressure would do the trick. So constricting your arteries increases blood pressure, too.

Your body's systems for regulating blood pressure involve both the muscle tone in the artery walls and the volume of blood. Quite a few hormones affect arterial muscle tone, and changes in this tone can happen very quickly for fast pressure adjustment, such as when you get out of bed. Regulation of blood volume isn't as fast, but it is interconnected with some mechanisms for regulating arterial constriction.

Blood volume is affected by salt and water balance
In the last chapter you learned that water follows solutes, or salt. In a nutshell, when the amount of salt in your body increases, the amount of water will increase accordingly, and the volume of blood also increases. Our discussion below is focused on how the body maintains salt and water balance, which you now know is also connected with maintaining blood pressure.

Your body contains about 5 liters of blood, and when you are just sitting around (like when you are reading this, unless of course your heart is racing with excitement), your heart pumps somewhere between 3 and 5 liters of blood per minute. That means that if you don't exercise or get excited, your entire volume of blood makes a trip through your body nearly every minute, dispensing oxygen and nutrients and picking up waste. During each trip, it passes through the lungs to give up carbon dioxide and pick up oxygen, and a portion of it passes through the kidneys to give up other waste products.

Your kidneys, through a fabulous system of transport proteins, filter waste products out of your blood, generating urine. The kidneys are also the site where water and salt loss are regulated to maintain ion balance and blood volume. For you to understand how the body regulates water and salt balance and blood pressure, you will need to know a little about how the kidneys work.

How the kidneys work
Your kidneys are essentially a system of tubes. The tubes start out very small and are surrounded by blood vessels. The small tubes feed into larger and larger ones, finally reaching a single tube called the ureter that runs from the kidney to the bladder. As fluid passes into the tiniest tubes, through the system, and into the ureter, it becomes urine.

The tube system of the kidneys starts where capillaries full of blood that needs filtering enter the kidney. The walls of these capillaries are even more leaky than usual, and water and small molecules, including even small proteins, leak out of the capillaries into the extracellular space and then into the tubules, which are just as leaky as the capillaries at that point. The fluid in the tubules at this point is called a filtrate, since it has filtered through the capillary and tubule walls into the tubules.

As the filtrate progresses down the tubule, the tubule walls change and become less leaky, more like the epithelium of the small intestine. At that point, the kidney epithelial cells actively transport material the body needs to retain out of the filtrate. The structure of the kidney epithelium at this point resembles that of the intestinal epithelium. It has microvilli and lots of transporters that suck glucose, proteins, amino acids, and other good things back into the interstitial fluid, where they filter back into capillaries. As in the intestinal epithelium, water follows the solutes out of the filtrate and back into the interstitial fluid, too (Box 8.3). The materials that are left in the filtrate are those that don't have transporters to suck them back into the epithelial cells—normal body waste products, as well as unusual things that may have gotten into the system, such as drugs. Since the epithelium is permeable to water, the filtrate and the extracellular fluid are **isotonic**, another way of saying they are in osmotic balance (Box 8.4).

As the filtrate passes deeper into the kidney, it reaches a section in which the tubule walls are no longer permeable to water. The tight junctions in the epithelium of this section are extremely tight, and there are no aquaporin channels in the epithelial cells. There are, however, many ion transporters that actively absorb sodium, potassium, chloride, and calcium ions out of the filtrate. As the filtrate passes through this part of the system, it becomes less concentrated than the extracellular fluid, though it still contains ions.

Next, the filtrate passes through a region of the tubule that contains variable numbers of aquaporin channels. Water now flows through the channels into the epithelial cells, and the filtrate becomes more concentrated. Here, there are also variable numbers of sodium channels that allow the epithelial

---

BOX 8.3
Diabetes mellitus and the kidney

In diabetes mellitus, the amount of glucose in the bloodstream exceeds the ability of the kidney transporters to reabsorb it. The kidney reabsorbs what it can, and the remainder stays in the filtrate and in the urine, making it sweet (mellitus). The glucose increases the solute concentration, and thus the amount of water, in the filtrate. Excess water is ultimately excreted, resulting in abnormally large volumes of urine (diabetes). Because of their excessive urination, diabetics experience extreme thirst. Diabetes mellitus is hard on the kidneys because of the excess glucose and excessive urination and because of other metabolic imbalances that occur with the disease. Diabetes is a major cause of kidney failure.

BOX 8.4 *Dialysis*

People whose kidneys can no longer filter out waste products build up those toxic products in their blood and eventually die as a result. Kidneys can fail as a result of many conditions, including diabetes mellitus, high blood pressure, infection, hereditary diseases, chronic inflammation, and physical obstructions to urine flow. The treatment of people with kidney failure centers on removing wastes from their blood by a process called **dialysis**. In the classic dialysis procedure, the patient's blood is routed out of the body through a needle inserted into a blood vessel. The blood flows into a tube whose walls are made of a membrane full of tiny pores. The pores are large enough for small molecules to pass through but not proteins or cells. The tubes are immersed in a solution that contains salt and glucose in concentrations that mimic those of healthy blood.

Now, to echo a phrase from chapter 7, diffusion happens. The waste products in the blood diffuse outward into the dialysis fluid, which contains no waste products, and so are drawn out of the blood. Since the dialysis fluid contains salt and glucose, those essential blood components are not lost. Dialysis fluid does not, however, contain amino acids or vitamins, so the patient loses those nutrients during the procedure.

Dialysis effectively removes waste products from the bloodstream, but there is no hormonal control of salt and water balance as there is in a functioning kidney. Dialysis patients therefore must monitor their salt and fluid intake very carefully. Because they lose nutrients during dialysis, they also have to be very careful about getting adequate nutrition. The type of dialysis described here, called hemodialysis, must be performed three times per week to keep a patient healthy. Each session takes 4 to 6 hours.

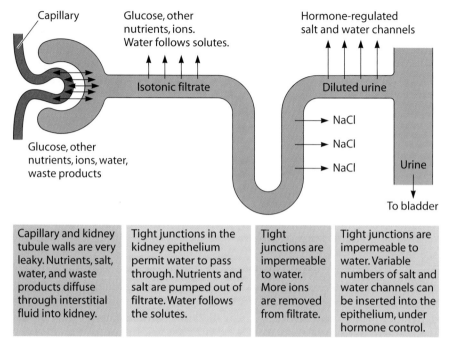

Figure 8.12 Solute transport in the kidney.

cells to absorb some of the remaining sodium. Once the filtrate passes through this segment of the tubule, it is essentially urine. It flows into a larger tube and from there into the ureter and on to the bladder (Figure 8.12).

REGULATION OF BLOOD VOLUME AND SALT BALANCE

As you might have anticipated, the later section of the tubule containing the variable numbers of aquaporins and sodium channels is the place where regulation of blood volume and salt balance occurs. Let's think about the effect of the water and sodium channels on these two things. First, the filtrate is well on its way to being urine by now, but it is quite dilute. Increasing the number of water channels in the epithelial cells would cause water to leave the filtrate, increasing the concentration of the urine, decreasing the volume of the urine, and, by increasing the amount of water reabsorbed, increasing the volume of blood (Figure 8.13). Increasing the volume of blood would increase blood pressure. Similarly, increasing the number of sodium channels

Figure 8.13 Effect of hormone-regulated channels on the salt and water content of urine.

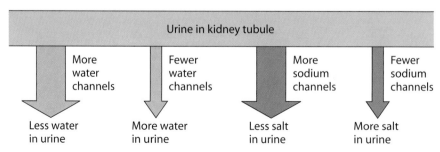

would increase the amount of sodium retained in the body. If the body retained more sodium, it would also retain more water to keep the proper salt concentration, resulting in increased blood volume and pressure.

We are now going to look at the major pathways controlling blood volume and salt balance in your body. In so doing, we will be ignoring many issues, such as potassium concentration and a host of factors that modulate the body's responses, to focus on a few central themes. Please note that there are other, fast mechanisms to adjust blood pressure in situations such as getting out of bed.

Two sets of hormones play key roles in regulating blood pressure

Two interconnected sets of hormones play major roles in regulating blood pressure. One set regulates the muscle tension in artery walls, while the other regulates salt and water balance and therefore blood volume. The signals leading to release of these hormones overlap, and some of them regulate the expression of others.

The pair of hormones that regulate muscle tension in the artery walls is **angiotensin** and **atrial natriuretic hormone** (ANH). Angiotensin causes the muscles to contract, constricting the arteries and increasing blood pressure. ANH causes the muscles to relax, decreasing blood pressure. Both angiotensin and ANH are small proteins.

The hormones that regulate salt and water balance, and thus blood volume, are **aldosterone** and **antidiuretic hormone** (ADH). Aldosterone causes increased retention of salt, and ADH causes increased retention of water (Table 8.4). Water retention increases blood volume and blood pressure. Salt retention can lead to water retention and therefore can also promote increases in blood volume and pressure.

Both angiotensin and ANH have regulatory effects that reinforce their roles in the control of blood pressure. Angiotensin is released via an enzyme cascade that begins in the kidney. The kidney secretes the enzyme renin, which cleaves an inactive blood protein to form angiotensin I. Angiotensin I is cleaved by another enzyme, **angiotensin-converting enzyme** (ACE), to form angiotensin II, the form that causes blood vessels to contract. Angiotensin II also travels to the kidney, where it is converted by yet another enzyme to angiotensin III, which stimulates aldosterone secretion (Figure 8.14). Since the action of aldosterone also promotes an increase in blood pressure, its release reinforces the effects of angiotensin.

ANH, on the other hand, suppresses the release of both aldosterone and renin, and thus of angiotensin. Suppressing aldosterone and angiotensin would tend to further decrease blood pressure, reinforcing the effect of ANH.

Aldosterone and ADH affect channels in the kidney epithelium

Aldosterone and ADH work via channels in the kidney tubules. Aldosterone is a steroid hormone secreted by the adrenal gland of the kidneys. It crosses the cell membrane of kidney epithelial cells and binds to its receptor inside

Table 8.4 Major hormone systems affecting blood pressure

Phenomenon affected	Hormone	Effect
Muscle tension in arteries	Angiotensin	Increases tension
	ANH	Decreases tension
Salt-water balance	Aldosterone	Retain salt
	ADH	Retain water

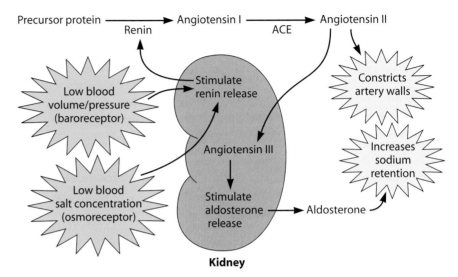

Figure 8.14 The renin-angiotensin cascade.

the cells. As in the case of estrogen, the hormone-receptor complex can then bind to DNA and affect transcription. Aldosterone increases the activity of sodium channels on the side of the epithelium facing the filtrate and also causes increased transcription of a sodium/potassium ion pump on the "back" of the epithelium. The result of aldosterone action is that sodium reabsorption from the filtrate increases. With more sodium in the body, more water is retained, and blood pressure increases.

ADH is a small protein secreted from the pituitary gland in your brain (Table 8.2). ADH binds to a receptor on kidney epithelial cells in the portion of the tubule where aquaporin can be regulated. When ADH binds to its repressor, aquaporin proteins stored in cytoplasmic vesicles are inserted into the membranes of the tubule epithelial cells facing the filtrate, analogous to the insertion of glucose transporters stimulated by insulin. Water can then flow from the dilute filtrate through the aquaporin into the epithelial cells and from there cross into the interstitial fluid. The filtrate becomes more concentrated, and more water is retained by the body (Figure 8.14). Blood pressure increases. In some organisms, ADH also causes blood vessels to constrict, but in humans its primary effect is on the kidney.

Alcohol inhibits the release of ADH, which leads to fewer aquaporin channels in the kidney tubules. Consequently, when people consume alcohol, they excrete a larger volume of more dilute urine than they would without the alcohol. Consumption of excessive amounts of alcohol can lead to mild dehydration, which is thought to worsen hangover symptoms.

Environmental signals are transmitted through osmoreceptors and baroreceptors

The environmental stimuli that trigger the release of these hormones are signals from osmoreceptors responding to the blood sodium concentration and/or baroreceptors (stretch receptors) responding to blood volume (Table 8.5). When blood volume is low, baroreceptors trigger the release of ADH, aldosterone, and angiotensin. Angiotensin increases blood pressure by constricting the arteries, and ADH and aldosterone promote retention of salt and water, increasing blood volume.

Table 8.5 Key hormones involved in regulating salt and water balance and blood pressure

Hormone	Released by	Released in response to	Changes promoted	Effects
ADH	Pituitary gland	Activation of osmoreceptors in response to increased salt in body fluids or to baroreceptors when blood volume is low	More aquaporin channels inserted into kidney tubule epithelium; more water retained	Increased blood volume; increased blood pressure
Aldosterone	Kidney	Activation of osmoreceptors and baroreceptors in kidneys when blood volume or salt concentration is low	Increased transcription of sodium transport genes in kidney tubule; more salt retained	Increased salt concentration; increased blood pressure
Angiotensin	Cleavage of precursor plasma protein (through renin pathway)	Activation of osmoreceptors and stretch receptors in kidneys when blood volume or salt concentration is low	Muscles in walls of blood vessels contract	Increased blood pressure
ANH	Heart	Activation of stretch receptors when blood volume is high	Muscles in walls of blood vessels relax; inhibits secretion of renin and aldosterone; kidneys retain less water and salt	Decreased blood volume and salt concentration; decreased blood pressure

When the blood salt concentration is high, osmoreceptors trigger the release of ADH, which causes increased retention of water, diluting the blood. A different population of osmoreceptors in the brain are thought to stimulate thirst, prompting consumption of fluids. Low blood salt concentration triggers the release of aldosterone, promoting sodium retention.

When blood volume is high, baroreceptors trigger the release of ANH, which relaxes arterial walls and decreases blood pressure. ANH also inhibits the release of aldosterone and renin, causing the kidneys to retain less salt. As salt levels fall, ADH activity decreases and more water is lost, decreasing the blood volume.

Putting the hormone actions together

Now that you have waded through all that information about blood pressure hormones, let's see how their actions are combined when the body responds to two different scenarios.

Salt excess

Imagine that you just ate a large bag of salty snack food—maybe potato chips or pretzels. The concentration of salt in your body fluids will go up. In response to the higher osmolarity of your blood plasma, the osmoreceptors in your hypothalamus fire, generating nerve impulses that cause the pituitary to release ADH and make you thirsty. You drink fluids, and the ADH causes more aquaporin channels to be inserted into your kidney tubules. More water is reabsorbed in the kidneys, and the salt concentration in your body fluids gradually falls to normal as you add water to your system. Because there is more water in your body, blood volume increases.

Simply diluting the extra salt in your body with extra water doesn't return your body to normal; your blood volume and pressure would be too high. However, the increasing blood volume activates stretch receptors in the heart, resulting in the release of ANH. The ANH relaxes your blood vessels, lowering blood pressure, and inhibits the secretion of aldosterone and renin. With less aldosterone around, you excrete more salt, bringing down the osmolarity of your blood plasma and decreasing the stimulus for ADH secretion. As ADH levels fall, you excrete more water, returning your blood volume to normal as well. If you read articles about weight loss, you may read that eating salty foods will cause water retention and an accompanying increase in body weight and that it can take over a day for the excess water to be lost. The actions of ADH and ANH are behind this phenomenon.

Blood loss

Now imagine that you lose a moderate amount of blood—perhaps by donating it to the Red Cross. As your blood volume and blood pressure fall, the baroreceptors in your kidneys stimulate the release of renin, which increases angiotensin and aldosterone levels. Other baroreceptors stimulate the release of ADH. Your blood vessels constrict, increasing blood pressure. Water and salt absorption are increased, reducing the output of urine and conserving both salt and water for blood volume. All of these mechanisms are designed to keep blood pressure adequate to support your heart and brain and to restore blood volume.

Depending on the severity of the initial blood loss and individual response, your capillaries, except for those in your brain and heart, may constrict until they close altogether. This closure conserves the remaining blood for your most vital organs: the heart and brain. The closure of capillaries in your peripheral tissues (close to your skin) can cause you to look pale.

FAILURE OF SALT AND WATER BALANCE

We have just looked at the major hormones involved in blood pressure regulation. Each of these hormones has a receptor through which it exerts its effects. Taking diabetes as an example, you might assume that blood pressure regulation could be impaired by either a failure to make one of the hormones or a failure to respond to it. You would be exactly right.

ADH failure

People with insufficient ADH activity exhibit a symptom you could probably guess from what you know about the action of ADH: they urinate excessively, up to 16 liters (over 4 gallons) per day. As long as these people can drink enough water to replace what they lose through urination, they are all right, though their lives are greatly disrupted. In the absence of sufficient water, the disease is fatal.

To early physicians, people with insufficient ADH activity resembled diabetes mellitus patients in a significant way: both experienced excessive urination. At that time, the easily discernible difference between the two groups of people was that the urine of ADH patients was not sweet. Therefore, the early doctors named the ADH condition **diabetes insipidus**. "Insipidus" is a Latin word meaning tasteless; thus, diabetes insipidus means "lots of tasteless urine."

Like diabetes mellitus, diabetes insipidus can result from either a failure to make ADH or a failure to respond to it. ADH is made in the hypothalamus gland in the brain, and failure to make sufficient ADH is usually the result of head injury or tumors of the hypothalamus. If diabetes insipidus is caused by such a condition, it can be treated by administration of ADH.

Failure to respond to ADH can be caused by kidney disease or by genetic mutations in the response pathway. Some patients have mutations in the ADH receptor and therefore cannot respond to the hormone. Still other patients make ADH and have intact ADH receptors but have mutations in the aquaporin 2 gene. This gene encodes the aquaporin protein that should be inserted into the cell membrane in response to ADH stimulation. These two kinds of mutations are a good example of short-circuiting the response pathway at different points.

Other patients have too much ADH. This condition is called syndrome of inappropriate ADH, or SIADH. It can be the result of a pituitary tumor or other condition of the central nervous system that promotes inappropriate release of ADH from the pituitary or of various cancers that cause ADH to be released from tissues other than the pituitary, or it can be a side effect of various medications, including several anticancer drugs. People with SIADH experience water retention and low blood sodium concentrations. A lower salt concentration in the interstitial fluid causes water to move into cells and can promote brain swelling. Headache, confusion, lethargy, and even coma can result. SIADH is treated by managing salt and fluid intake.

Aldosterone failure

People with insufficient aldosterone activity cannot properly conserve salt. They fail to thrive as infants, have low blood sodium concentrations, and suffer recurrent dehydration (water follows salt). This syndrome, like diabetes insipidus with ADH and diabetes mellitus with insulin, can be caused by either a failure to make sufficient aldosterone or a failure to respond to it.

Aldosterone is made in the kidney and acts on the kidney. Not surprisingly, perhaps, kidney disease can result in aldosterone deficiency, usually in adulthood. Inherited aldosterone deficiency is rare. It can be caused by mutations in the genes for any of several enzymes required to make aldosterone from cholesterol. Patients with aldosterone deficiency can be treated with a synthetic hormone that mimics the action of natural aldosterone.

Other patients have sufficient aldosterone but cannot respond to it. These people are said to be aldosterone resistant. As you might expect, some of the patients have mutations in the gene for the aldosterone receptor. Others have mutations in the gene for the sodium channel protein that allows sodium to be reabsorbed from nascent urine in response to aldosterone (analogous to the aquaporin mutation in diabetes insipidus). Other patients have mutations in other genes involved in the response pathway. Aldosterone-resistant patients cannot be treated with aldosterone. Instead, their condition is treated with salt.

If mutations can block the response pathway of a hormone, can they also activate it? Yes, and there are examples of this kind of activation of the aldosterone pathway. Individuals with mutations that permanently activate the aldosterone response develop severe high blood pressure at an early age, though the levels of aldosterone and renin in their blood are low. Two different types of mutation have been found to cause this condition. One type of mutation affects an enzyme that regulates the aldosterone receptor. The other type of mutation alters the cytoplasmic domain of the sodium channel protein. Apparently, these mutations cause the sodium channel to be always open and eliminate the site on the protein through which aldosterone regulation would normally occur.

Biotechnology application: drugs for high blood pressure

A 2002 briefing by the Director General's office of the World Health Organization reported that worldwide, 10 to 30% of people in almost all countries had high blood pressure. In the United States, a 1999–2000 study estimated that high blood pressure affects over 58 million Americans, or over 28% of the population. As you read above, high blood pressure increases the risk of many significant health problems, including heart disease, stroke, and blindness. Although high blood pressure can often be treated by weight loss, exercise, and dietary changes, many people are not successful with these approaches. It is no wonder that a great deal of research time and money is spent looking for drugs to treat the condition.

Although we outlined some genetic conditions that directly cause high blood pressure, the overwhelming majority of cases of high blood pressure are not directly linked to individual genes. In fact, doctors do not know exactly why most people develop high blood pressure, though it is clear that factors such as obesity greatly increase the risk. Without knowing exactly what is causing high blood pressure in most people, physicians treat the condition with drugs that target salt and water balance and muscle tension in the blood vessels.

ACE inhibitors

One class of high blood pressure drugs is known as **ACE inhibitors**. These drugs inhibit the activity of ACE. This enzyme converts angiotensin I to angiotensin II, a powerful blood vessel constrictor (Figure 8.14). Blocking the ACE enzyme prevents formation of angiotensin II and thus lowers blood pressure. Blocking the formation of angiotensin II also blocks the formation of angiotensin III, which stimulates aldosterone release. Thus, ACE inhibitors also lower the levels of aldosterone in the blood, resulting in more salt excretion, more water secretion, lower blood volume, and lower blood pressure. Some trade names of these drugs are Vasotec, Lotensin, Accupril, and Prinivil.

Angiotensin receptor blockers

Angiotensin receptor blockers are a newer class of drugs that block the receptor to which angiotensin II binds. In doing so, they prevent angiotensin II from causing blood vessels to constrict and thus lower blood pressure. Diovan, Avapro, and Atacand are members of this class.

Diuretics

A **diuretic** is an agent that causes the body to lose water. Diuretic drugs target salt and water secretion. One family of these drugs works by blocking the aldosterone receptor. A widely prescribed example is Aldactone. These drugs prevent aldosterone from triggering salt retention. More salt is excreted, followed by more water, resulting in lower blood volume and pressure.

The other family of widely prescribed antihypertensive diuretics works by blocking the sodium channel through which sodium is reabsorbed. This sodium channel is the same one involved in the genetic mutations described under "Aldosterone failure" above. By blocking sodium reabsorption, these achieve the same result as the other family of diuretics: increased salt and water excretion, lower blood volume, and lower blood pressure. Lasix is an example of a sodium channel blocker.

SUMMARY POINTS

Cells respond to information from their environments through the interaction of extracellular signals with receptors in the cell. The interaction of signal and receptor produces a change in the receptor protein, which initiates a change in the cell's activities, either directly or through a chain reaction of subsequent molecular events. Cells lacking a receptor for a certain signal cannot respond to that signal.

Cells communicate with each other by releasing signals into their environments that interact with receptors on other cells.

Cells fail to respond normally to a signal if the receptor for that signal is missing or nonfunctional or if any step in the chain between the receptor and the final cellular outcome is defective.

Within multicellular organisms, specialized cells receive input from the environment and then coordinate the response of the body through the release of electrical signals (nerve impulses) and/or chemical signals, such as hormones, growth factors, and neurotransmitters.

The reception of a signal can result in any number of alterations in a cell's activities, such as changes in gene expression, changes in enzyme activity, changes in cellular transport, movement, growth, or even death.

In *E. coli*, changes in the environmental concentration of the sugar lactose or the amino acid tryptophan result in changes in gene expression. This happens because the nutrients interact directly with repressor proteins, changing their shapes. In the case of lactose, its presence and effect on the repressor result in the expression of genes for breaking down the sugar. In the case of tryptophan, its presence and effect on the repressor stop the expression of genes for synthesizing the amino acid.

In humans and other mammals, glucose metabolism is regulated by the hormones insulin and glucagon. Increasing concentrations of glucose in the bloodstream stimulate the pancreas to release stored insulin from cytoplasmic vesicles. Insulin travels through the bloodstream and binds to receptors on liver, fat, and muscle cells. Binding to its receptors on these various cells stimulates the synthesis of the storage carbohydrate glycogen in the liver, synthesis of fat in fat cells, and uptake of glucose by muscle cells. Falling blood glucose concentrations stimulate the release of glucagon, which has the opposite effects.

One reason that maintaining a steady level of glucose in the bloodstream is important is that glucose is the only fuel of the brain. If blood glucose levels drop too low, unconsciousness, coma, and death can result.

Failure of the insulin signaling pathway results in diabetes mellitus, or diabetes. People can develop diabetes because they lose the ability to synthesize insulin (type I diabetes) or because their cells lose the ability to respond to insulin (type II diabetes). Type I diabetes is caused by an autoimmune reaction in early life that destroys the insulin-synthesizing cells of the pancreas. Type II diabetes usually develops later in life and is most common in overweight, sedentary adults or the elderly. Its incidence among the young, however, is on the rise.

Type I diabetes is treated through the administration of insulin. In the early days of insulin treatment, the hormone was isolated from animal carcasses. Now, recombinant bacteria engineered with the human insulin gene produce human insulin for diabetics.

The kidneys filter waste products out of the bloodstream by using a combination of transport proteins and changes in the permeability of the kidney epithelium. Salt and water balance is regulated in the kidney through the actions of ADH and aldosterone.

Salt and water balance is important in the regulation of blood pressure, because salt and water balance is related to blood volume. The higher the blood volume, the higher the blood pressure. Blood pressure is regulated through the actions of the hormones angiotensin and ANH, which affect both muscle tension in the blood vessels and the secretion of aldosterone.

ADH causes water retention by causing the insertion of aquaporin channels into the otherwise-impermeable kidney tubules. Aldosterone causes salt retention by activating sodium transport channels in the same tubules.

Aldosterone is secreted in response to a fall in blood volume, sensed by baroreceptors, or a decrease in the blood salt concentration, sensed by osmoreceptors. When these receptors are activated, the kidney secretes the enzyme renin into the bloodstream. Renin cleaves a blood protein to form angiotensin I, which is cleaved by ACE to angiotensin II, a powerful blood vessel constrictor. In the kidney, angiotensin II is converted to angiotensin III, which is the signal for aldosterone release.

Failure of the ADH signaling pathway, from either a lack of hormone or a disruption in the response pathway, results in diabetes insipidus. Symptoms of diabetes insipidus are excessive thirst and urination. Certain drugs, such as alcohol, suppress ADH secretion and cause a temporary increase in urine volume.

Failure of the aldosterone signaling pathway can result in extremely high blood pressure if the pathway is inappropriately active or extremely low blood salt concentrations if the

SUMMARY POINTS *(continued)*

pathway is blocked. Failure can result from a lack of aldosterone or a disruption in the response pathway. Observed response pathway disruptions include both blocking of the response and an inappropriate activation of it. The disruptions can occur at the level of the hormone receptor or the sodium transport channel.

Hypertension, a serious medical condition affecting millions of people, is treated with a variety of drugs that target different hormone pathways. ACE inhibitors inhibit the formation of angiotensin II and thus aldosterone as well. ACE receptor blockers block the blood vessel constriction activity of angiotensin. Diuretics target either the aldosterone receptor or the sodium channel it affects to reduce blood volume.

Kidney failure is a potentially fatal condition because of the buildup of toxic wastes in the blood that accompanies it. Kidney failure is treated with dialysis, in which blood is pumped through porous membrane tubes suspended in fluid containing healthy concentrations of salt and glucose. Waste products are removed from the blood by diffusion into the dialysis fluid.

KEY TERMS

ACE inhibitor

Aldosterone

Angiotensin

Angiotensin-converting enzyme (ACE)

Antidiuretic hormone (ADH)

Atrial natriuretic hormone (ANH)

Baroreceptor

Diabetes insipidus

Diabetes mellitus

Dialysis

Diuretic

Glucagon

Gluconeogenesis

Growth factor

Heart attack

Heart failure

Hormone

Hypertension

Insulin

Insulin-dependent diabetes

Isotonic

Neurotransmitter

Non-insulin-dependent diabetes

Osmoreceptor

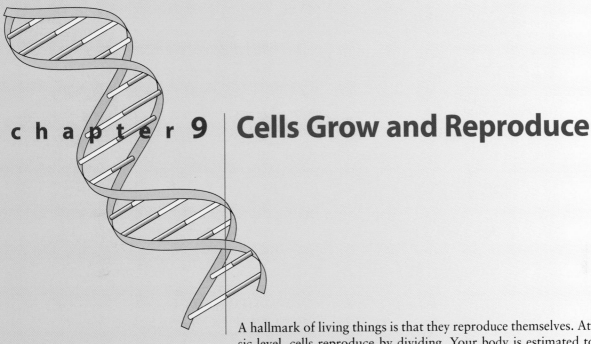

chapter 9 | Cells Grow and Reproduce

A hallmark of living things is that they reproduce themselves. At the most basic level, cells reproduce by dividing. Your body is estimated to contain between 10 and 100 *trillion* cells, and yet you started out as a single fertilized egg cell. It took a lot of cell division to get from that beginning to where you are now, sitting reading this chapter.

When cell division proceeds properly, under control by the body's hormones and growth factors, the result is orderly growth and development of an individual, timely replacement of worn-out cells, and healing of wounds. If the process of cell division escapes the body's regulation mechanism, the result can be uncontrolled and inappropriate cell growth and division, which we know as cancer. In this chapter, we will look more closely at how cell division happens, how it is regulated, how the process can go wrong and produce cancer, and how our understanding of cellular reproduction is being applied through biotechnology.

When you think about it, cells face several different issues when it comes time to divide. For one thing, cells need to grow before they divide if their average size is going to stay the same. For another, the cell reproduction process must ensure that DNA is copied and that each daughter cell gets one copy of everything. Finally, the cell has to split itself in two. Obviously, the timing of these events must be coordinated or disaster would result. As we discuss cell reproduction, we will look at:

- DNA replication
- distribution of chromosomes
- regulation of cell reproduction
- what happens when cell reproduction escapes regulation
- DNA-damaging agents and cellular repair of damage
- how this knowledge is being applied

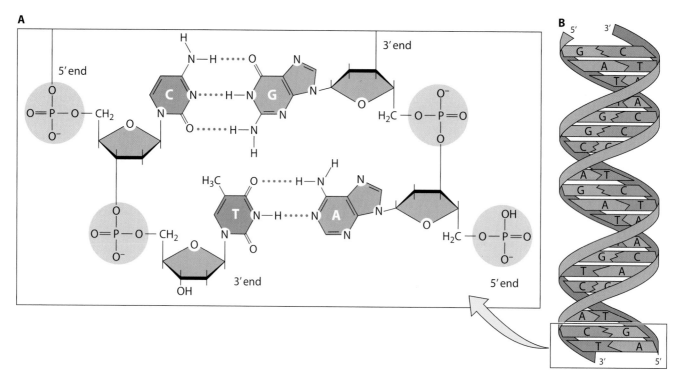

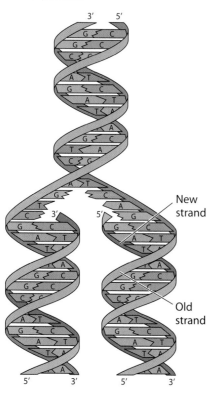

Figure 9.1 DNA structure. DNA consists of two complementary nucleotide strands held together by base-pairing between adenine and thymine or cytosine and guanine.

Figure 9.2 DNA replication. Each strand of the parent DNA molecule is used as a template for synthesizing a new complementary strand.

In this chapter, we make the connection between cell reproduction, DNA mutations, and cancer. We discuss the general kinds of genes involved in cancer and show how two environmental carcinogens work to stimulate its development. Finally, we give some examples of how biotechnology is leading to better cancer treatments.

DNA REPLICATION

In chapter 3, you learned that DNA is made of two complementary chains of nucleotides with a backbone made of repeating units of the sugar deoxyribose and a phosphate group and that each sugar has one of four bases attached (Figure 9.1). The two strands are held together by hydrogen bonds between the complementary base pairs adenine-thymine (A-T) and cytosine-guanine (C-G). The complementary nature of the base pairs means that each strand can be used as a pattern for the other, which is how new DNA molecules are synthesized. The two original strands of the parent molecule separate, and each strand is used as a pattern for synthesis of a daughter strand (Figure 9.2).

Now, let's look a little more closely at the process. DNA replication involves the synthesis of a molecule, and like other cellular synthesis processes, it is carried out by an enzyme. As mentioned in chapter 3, the enzyme that synthesizes DNA is called DNA polymerase. Like other enzymes, DNA polymerase is extremely specific in what it does. DNA polymerase adds a nucleotide to the 3′ end of a growing new strand of DNA, using the complementary base on the opposite (template) strand as a pattern. The enzyme binds to the triphosphate form of a nucleotide (similar to the ATP of energy reactions but made with the sugar deoxyribose instead of ribose), checks to

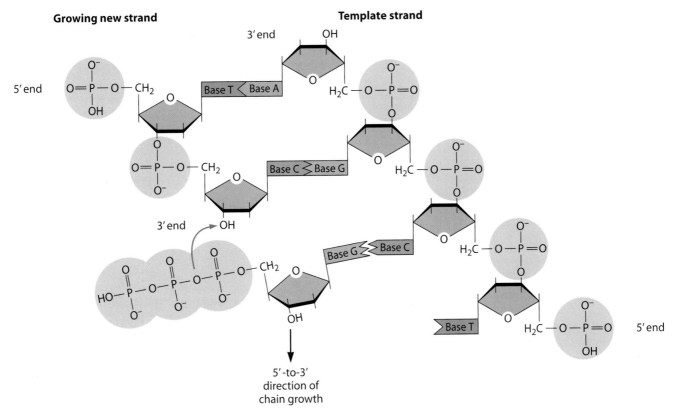

Figure 9.3 In DNA replication, DNA polymerase binds to the nucleotide triphosphate that makes the correct pair with the template base and forms a bond between the 3′ OH group of the growing strand and the 5′ phosphate groups of the incoming nucleotide. A diphosphate is cleaved from the incoming nucleotide when the bond is formed.

see that it forms a complementary pair with the template nucleotide, and if so, forms a bond between the number 1 phosphate on the nucleotide and the free OH group on the number 3 carbon of the last sugar in the template strand, liberating a pair of phosphate groups (Figure 9.3).

To carry out this reaction, DNA polymerase requires several things. First, it must have access to the triphosphate forms of nucleotides. Second, it must have a single-stranded region of DNA (the template strand) with a growing new strand base-paired to part of it. Finally, the growing strand must have a free OH group on the number 3 carbon of the terminal sugar molecule (look at Figure 9.3). How do these things arise?

The nucleotide triphosphates are not a problem; cells synthesize them using amino acids and sugars as starting materials. The single-stranded DNA template with a growing new DNA strand on it is something else again. Cellular DNA is normally a smooth double-stranded molecule. It requires some major changes to go from that form to one that can be used by DNA polymerase for DNA replication.

Getting DNA replication started is a multistep process

The task of replicating cellular DNA begins with the creation of single-stranded DNA with a short complementary **primer** strand base-paired to it. The primer strand serves as a "starter" for the synthesis of a new strand of DNA, providing DNA polymerase with a site to attach a new nucleotide.

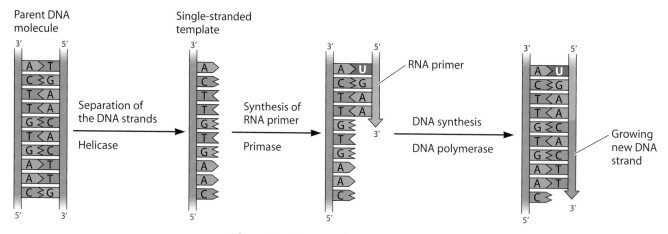

Figure 9.4 Initiation of DNA replication.

When the structure of a single-stranded template strand with a short primer base-paired to it has been created, the DNA polymerase enzyme recognizes it, binds to it, and gets started synthesizing new DNA. This process, up to the point where DNA polymerase gets started, is called **initiation.**

Remember that the cell's mission is to replicate every bit of DNA in the chromosome(s) exactly one time without replicating anything more than once or missing anything. Replication initiation, therefore, is carefully controlled. It begins with the binding of proteins to unique sequences of DNA bases called **replication origins.** One of the proteins that bind early is a helicase enzyme, which separates the two strands of DNA. Once the strands have been separated, further proteins bind to the DNA-protein complex. One of these is a primase enzyme, which synthesizes a short primer using the rules of complementary base-pairing. The primer strand is actually made of RNA, as in transcription (Figure 9.4).

Now we have the DNA structural elements required by DNA polymerase: a single-stranded region of DNA with a primer strand on it. At this point, DNA polymerase can begin synthesizing a new strand of DNA, using the single-stranded template strand as a pattern. DNA polymerase itself works with the help of additional proteins that keep it clamped onto the DNA. In fact, to keep DNA replication going requires the combined actions of quite a few proteins. This coordinated assembly of proteins has been called a protein machine—an apt name.

DNA polymerase proofreads its work to prevent mutations

DNA replication is not a perfect process. Occasionally, DNA polymerase incorporates an incorrect base into the new daughter strand. This can happen in particular with unusual forms of the nucleotide bases that form spontaneously by interconverting with the normal form. For example, the G nucleotide can spontaneously rearrange itself into a rare form that base pairs with T instead of the normal C base (Figure 9.5). DNA polymerase can put the rare form of G opposite a T in a new helix. When the rare form of G spontaneously converts back to the normal form, the bases no longer pair and you have what is called a **mismatch.** Left alone, the mismatch would lead to a mutation: when the new DNA helix was replicated, the template strand with the G would give rise to a G-C pair, while the tem-

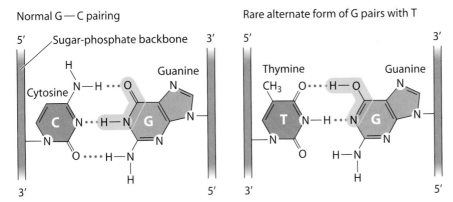

Normal G — C pairing

Rare alternate form of G pairs with T

Figure 9.5 The base guanine can spontaneously assume an alternate form that pairs with T instead of C.

plate strand with the incorrect T would give rise to a mutant T-A base pair (Figure 9.6).

The cell's first line of defense against mismatch-induced mutations is DNA polymerase itself. In addition to adding new nucleotides to a growing chain of DNA, DNA polymerase enzymes proofread their newly synthesized DNA. If the DNA polymerase detects a mismatch before it moves on, it removes the incorrect base and replaces it with the correct one (Figure 9.7).

Sometimes a mispaired base escapes the proofreading function of DNA polymerase, leaving a mismatch. It turns out that both prokaryotic and eukaryotic cells have several systems of proteins that look for and repair not only mismatches, but also other kinds of DNA damage. This is an extremely important capability, because DNA damage is a normal occurrence that can have very harmful consequences. As a final guard against mutation, normal cells have "quality control" checks that block the process of cell division if damaged DNA is present. We will look at more information on DNA damage, repair, and quality control later in the chapter.

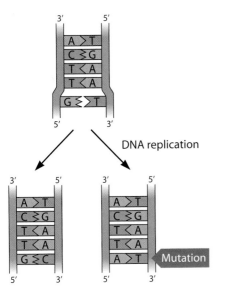

DNA replication

Figure 9.6 Replication of a mismatched base pair leads to a mutation.

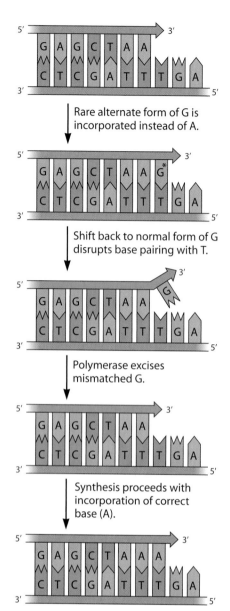

Figure 9.7 Proofreading by DNA polymerase during DNA replication.

MITOSIS

Now that we have looked at how DNA molecules are replicated and mismatches are repaired, let's consider another essential step in cell division: making sure that each daughter cell gets a copy of all of the mother cell's DNA. The processes are different in prokaryotic and eukaryotic cells.

In bacteria, cells grow as they consume energy, and they duplicate their DNA as they grow. Bacterial DNA is connected to the membrane surrounding the bacterial cell, and when DNA is duplicated, the daughter DNA molecule is attached to a different spot than is the parent DNA molecule. The bacterial cell then divides between the two DNA attachment points, and each daughter cell gets a DNA molecule.

In eukaryotes, the cell reproduction process is more complex. In eukaryotic cells, DNA is wrapped around proteins in complexes called chromosomes. The genome of a eukaryote is often spread over several very large DNA molecules, each of which is complexed in a different chromosome. Furthermore, eukaryotic cells usually have two copies of each chromosome. For example, your own cells have two copies of each of 23 different chromosomes. To produce two healthy daughter cells, the division process has to ensure that each daughter gets two copies of each and every chromosome. The elaborate process by which this happens is called **mitosis**.

Although we hate it when biology gets reduced to vocabulary words, we are going to use a few to talk about mitosis. One problem is that there aren't any nontechnical words for some of the structures we have to talk about. Also, you very likely took biology in high school and may already associate some of these words with the process. Finally, when biologists talk about mitosis and cell division, they use these words, and we'd like you to recognize them if you ever see them in a magazine article or hear them in a news report. Please focus on the process more than the terminology. It's really quite beautiful.

Mitosis is an orchestrated set of events

Mitosis is only about the distribution of chromosomes to daughter cells. It is not about DNA replication. Mitosis doesn't begin until DNA replication has already occurred.

Setting the stage

Imagine a cell with one pair of chromosomes undergoing DNA replication. In eukaryotic cells, the two daughter DNA molecules, instead of separating, stay attached to one another via a protein clamp. Picture two long, identical pieces of string side by side, being held together at one point between your thumb and forefinger. The pieces of string represent the two identical daughter DNA molecules, and your thumb and forefinger represent the protein clamp (Figure 9.8).

The place where the two daughter DNA molecules are held together is called the **centromere**. In reality, the centromere consists of unique DNA base sequences to which centromere proteins bind, forming a structure called a **kinetochore**. In mitosis-speak, the two identical DNA molecules held together at the centromere are called **chromatids**.

Mitosis begins after DNA replication, when each chromosome is in the form of a two-chromatid structure held together at the centromere. If you were to look through a microscope at the cell before mitosis starts, you could not see the replicated chromosomes, because they are relaxed and spread out within the nucleus (Box 9.1).

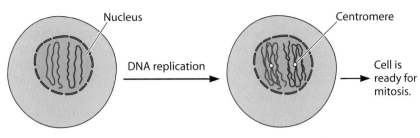

Nucleus

DNA replication

Centromere

Cell is ready for mitosis.

Cell with one pair of chromosomes. The chromosomes are relaxed and spread out in the nucleus.

After DNA replication, the two daughter DNA molecules stay clamped together at their centromeres. The replicated chromosomes are still relaxed and spread out.

Figure 9.8 Getting ready for mitosis.

BOX 9.1 *Mitosis, chromosomes, and amniocentesis*

Chromosomes are ordinarily invisible under the microscope because, in their relaxed state, they are such thin structures. When they condense during prophase, however, it becomes possible to see them (under a microscope), and at metaphase they reach the point of greatest condensation. The reason that human chromosomes look like tiny Xs in photographs is that the pictures are taken when chromosomes are in metaphase, so each "chromosome" is really two replicated chromosomes held together at the centromere (see the figure).

When metaphase chromosomes are stained with various stains or fluorescent dyes, patterns of bands become visible. Each chromosome has a unique pattern that can be used to identify it. In fact, it is possible to detect certain kinds of genetic problems by studying stained metaphase chromosomes and looking for irregularities in the banding patterns.

Pregnant women sometimes undergo a diagnostic procedure called **amniocentesis**, in which technicians examine the banding patterns of chromosomes from fetal cells and screen for certain genetic abnormalities. During amniocentesis, intrauterine fluid is drawn, and fetal cells are isolated from the fluid. These cells are normally present in amniotic fluid and are not forcibly removed from the developing fetus.

The cells are put into a test tube and treated with drugs that halt their divi-

sion process in metaphase, and their chromosomes are stained and photographed. The banding patterns are then used to identify the chromosomes, and the chromosomes are inspected for visible problems. This type of examination does not reveal single-base changes or other small changes in the DNA, but it can show major problems, like chromo-

somal rearrangements, missing chromosomes, or extra chromosomes. It also reveals the gender of the baby (see chapter 10). The staining of chromosomes and examination of the resulting banding patterns is called karyotyping, and the pattern obtained from a given individual is that person's **karyotype**.

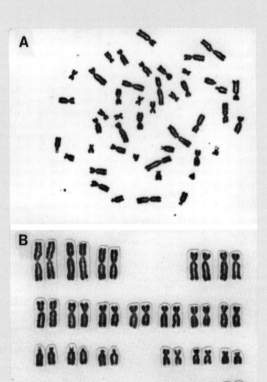

Human metaphase chromosomes. (A) Light micrograph. (B) Paired and sorted. (Photographs copyright Craig Holmes/Biological Photo Service.)

Condensing the chromosomes

The first official event in mitosis is the condensing of chromosomes and their movement to the center of the cell, accompanied by the disappearance of the nuclear membrane. By condensing, we mean that the replicated chromosome coils up tightly into a compact structure, and now you can see it if you look through a microscope. It makes sense to condense the chromosomes before starting to move them around; which would you rather move, a roomful of loose string or the same string twisted into a rope? The shape of the condensed chromosome is rather like an X because of the two chromatids and the centromere (think of your string). This stage of mitosis is called **prophase** (Figure 9.9).

Think about where we are now and what the goal is. We want to get one copy of each chromosome (one chromatid from each pair) into each daughter cell. We have our two replicated chromosomes condensed into their X shapes, held together at the centromere, in the center of the cell. Now, if we could just split each X down the middle and pull one chromatid from each to the opposite side of the cell, we'd be there. That is what happens next.

Assembling the pulling machine

To pull the two chromatids (one from each pair) to opposite sides of the cell, we need to have a pulling mechanism attached to each chromosome. In cells,

Figure 9.9 Mitosis and cytokinesis.

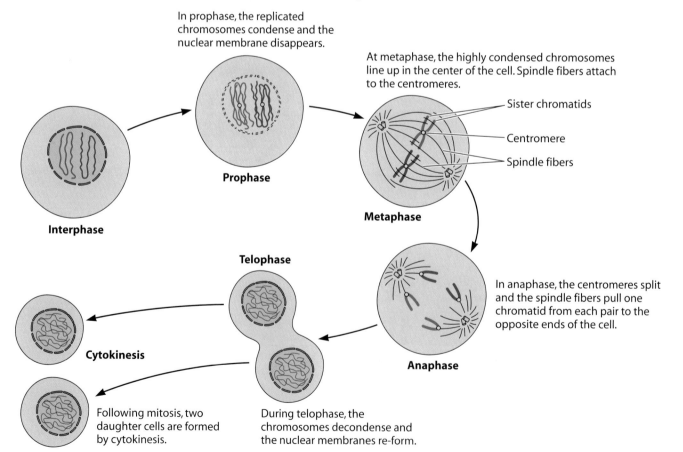

In prophase, the replicated chromosomes condense and the nuclear membrane disappears.

At metaphase, the highly condensed chromosomes line up in the center of the cell. Spindle fibers attach to the centromeres.

— Sister chromatids

— Centromere

— Spindle fibers

Prophase

Metaphase

Interphase

Telophase

In anaphase, the centromeres split and the spindle fibers pull one chromatid from each pair to the opposite ends of the cell.

Cytokinesis

Anaphase

Following mitosis, two daughter cells are formed by cytokinesis.

During telophase, the chromosomes decondense and the nuclear membranes re-form.

the pulling mechanism consists of protein strings called spindle fibers. The fibers are anchored to opposite ends of the cell and reach from those opposite ends to the replicated chromosomes in the middle of the cell. There, they attach to the centromeres of the replicated chromosomes by binding to the kinetochore (Figure 9.10). The phase of mitosis in which the replicated chromosomes are lined up in the middle of the cell with spindle fibers attached to their kinetochores is called **metaphase.**

Pulling the chromatids to opposite sides of the cell

Once the pulling machine has been assembled and every replicated chromosome is attached to it from each side via the kinetochores, the two chromatids in each X-shaped replicated chromosome split apart. Each chromatid still has its own centromere; the two centromeres of the chromatids are just no longer attached. The kinetochore proteins attached to the centromeres pull the two chromatids in opposite directions along the spindle fibers toward opposite poles of the cell (Figure 9.9). This phase of mitosis is called **anaphase.**

Forming two nuclei

Now we have accomplished the goal. There is one copy of each of the chromosomes, identical to the single pair we started with, at each end of the cell. The final events of mitosis put things back in place. The chromosomes decondense, and nuclear membranes form around each set of chromosomes. This phase of mitosis is called **telophase.**

Cell division takes place when mitosis is complete

After mitosis is complete, the cell splits in two in a process called **cytokinesis.** Cytokinesis completes the cell reproduction process, and if everything has gone as it should, the products are two daughter cells that are genetically identical to the original mother cell. Genetically identical means that each daughter cell has exactly one copy of every chromosome that was present in the parent cell (Figure 9.9).

The products of mitosis

This production of daughter cells that are genetically identical to each other and to the mother cell is the point of cell reproduction. It is critical that DNA be replicated accurately and that chromosomes be distributed accurately to maintain the identity and the functioning of the cells themselves. If you remember only one thing about mitosis, remember that it generates daughters identical to each other and to the parent cell.

The time between cell division and the next cycle of mitosis is called, in mitosis-speak, **interphase.** This term may give the impression of an idle period, but as you will see next, the rest of the cell reproduction process takes place in interphase.

THE CELL CYCLE

As mentioned at the beginning of this chapter, cell reproduction has to involve the coordinated, regulated processes of growth, DNA replication, chromosome distribution, and cell division. We've talked about how DNA is replicated and how chromosomes are distributed to daughter cells. Now, we will look at the entire cell reproduction process and how it is coordinated.

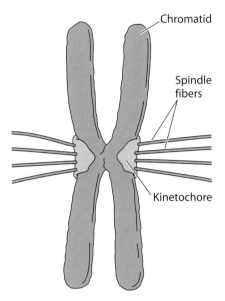

Figure 9.10 Spindle fibers attach to each chromatid by binding to the kinetochore at the centromere.

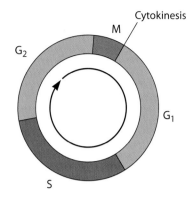

Figure 9.11 The cell cycle.

Scientists observing cells reproducing saw that they went through a repeating cycle of events every time they divided. This repeating cycle was named the **cell cycle** (Figure 9.11). One part of the cycle was DNA synthesis, the replication of all the DNA in the cell. This part of the cycle was called the S phase. Another easily observable part of the cell cycle consisted of mitosis, the orchestrated distribution of chromosomes to daughter cells. This part of the cycle was called the M phase. The splitting of the cell into two daughter cells was called cytokinesis. In between mitosis and the next S phase was a gap during which the cell grew, but no obvious division processes were occurring. This was called gap 1, or G_1. After S phase and before mitosis was another gap during which the cell continued to grow, called gap 2, or G_2. Though the G stands for gap, it is a better memory aid to think of it as G for growth. G_1, S, and G_2 occur in the so-called interphase of mitosis.

Cells divide in response to signals from growth factors

Think about your body throughout your lifetime. When you were young, all parts of you were growing larger. At puberty, in response to sex hormones, certain aspects of your body grew and changed. Now that you are an adult, you have likely stopped growing larger (jokes about girth will be omitted here) and have completed your physical sexual development. However, you synthesize new red blood cells and new intestinal-lining cells continuously, and if you wound yourself, you heal. What does this say about cell division in your body?

What this adds up to is that cell division is very highly regulated. *In fact, your cells are designed not to divide unless they receive signals that cause them to do so.* Late in the G_1 stage of the cell cycle is a point called the **restriction point**. For a cell to pass beyond the restriction point, protein signals called growth factors must be present. Growth factors bind to receptors on a target cell and initiate growth. If the appropriate growth factors are present, the cell passes through G_1 into S phase and is committed to divide. If growth factors are not present, the cell enters a quiescent stage called G_0, in which it can live for a long time without growing or dividing (Figure 9.12). Cells in G_0 are metabolically active, but their rates of protein synthesis are greatly reduced compared to those of actively dividing cells.

The normal state for cells in adult animals is to remain in G_0 until the appropriate growth factors signal to them to divide. For example, skin fibroblasts stay in G_0 and do not grow unless the skin is wounded. When this happens, a growth factor is released by platelets during blood clotting. Platelet-derived growth factor binds to receptors on the surfaces of skin fibroblasts near the site of the wound and signals them to leave G_0 and divide, repairing the wound.

Growth factors and the Ras protein

As you learned in the previous chapter, cells receive and respond to signals through multistep pathways. The first step in the pathway leading to cell division is the binding of the growth factor to its protein receptor on the outside of the cell. When that happens, the shape of the external domain of the receptor changes, and that translates across the cell membrane to a change of shape in the portion inside the cell. That change triggers changes in additional proteins and finally in the replication origin-binding proteins, leading to the initiation of DNA synthesis.

Many growth factor receptors transmit their signals through a protein called **Ras**. The Ras protein sits on the inside face of the cell membrane and

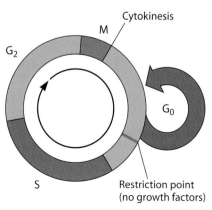

Figure 9.12 Growth factors must be present for a cell to pass the restriction point and commit to divide. Without growth factors, cells enter into G_0.

is activated by a number of growth factor receptors. When the correct growth factor binds to one of these receptors, the activated receptor in turn activates the Ras protein. The activated Ras protein goes on to activate other proteins, leading eventually to the activation of transcription of proteins involved in the initiation of DNA replication and DNA synthesis.

The cell cycle has checkpoints

The restriction point in the cell cycle allows animal cells to respond to signals from the body and thus to coordinate with the body's activities and needs. However, the cell also needs to coordinate with itself in order to progress smoothly through the cell cycle. For example, the cell needs to finish DNA replication before it begins mitosis. If a cell distributed DNA to each of the future daughters before DNA replication was complete, the result would be a disaster because at least one of the daughter cells would not get a complete genome. Likewise, the cell must finish mitosis before it splits in two in the process of cytokinesis.

Finally, the cell division process needs to prevent the replication of damaged DNA. As discussed above, the reason replication of damaged DNA is so harmful is that when a cell tries to replicate damaged DNA, the DNA polymerase enzyme often puts an incorrect base in the daughter molecules. Thus, replication of damaged DNA introduces mutations into cells (Figure 9.6).

Cells regulate their progression through the cell cycle via a series of checkpoints that prevent entry into the next phase of the cycle until the previous phase is successfully completed and that block cell division if cellular DNA is damaged. One of the DNA damage checkpoints occurs at the end of G_1 and is mediated by a protein called **p53**. Damaged DNA activates the expression of the *p53* gene. The p53 protein, in turn, activates the G_1 checkpoint and halts the division cycle before DNA replication begins. This is important, because otherwise the damaged DNA would be copied and mutations would be transmitted to daughter cells. If the cell can successfully repair the damage, then DNA replication and cell division can proceed. If the cell cannot successfully repair the damage, the p53 protein initiates a series of events by which the damaged cell kills itself. This programmed cell death, also known as **apoptosis**, is yet another way the body protects itself against DNA damage.

A similar checkpoint is present at the end of G_2. This checkpoint is sensitive to damaged DNA, but in addition, it is sensitive to unreplicated DNA. As mentioned above, it is essential that the cell completely replicate its DNA before entering mitosis. The G_2 checkpoint ensures that the division process halts until DNA replication is complete. A final major checkpoint comes at the end of M phase. This checkpoint blocks division if mitosis is going awry (Figure 9.13).

UNREGULATED CELL DIVISION: CANCER

When the regulation of cell division fails and a cell divides uncontrollably, producing a mass of descendant cells, we call the condition **cancer**. In the United States, cancer claims the lives of about 500,000 people per year, with cancers of the colon and lung accounting for about 200,000. Over a hundred different types of cancer have been identified. They are usually classified according to the type of cell from which the cancer originated. For example, about 85% of all cancers originate from epithelial cells that line organs or

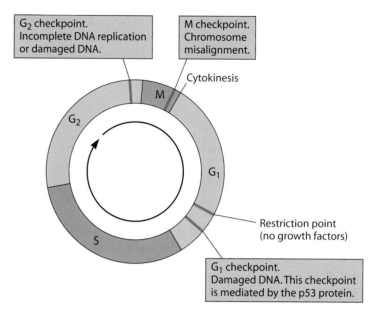

G₂ checkpoint.
Incomplete DNA replication
or damaged DNA.

M checkpoint.
Chromosome
misalignment.

Cytokinesis

M

G₂

G₁

S

Restriction point
(no growth factors)

G₁ checkpoint.
Damaged DNA. This checkpoint
is mediated by the p53 protein.

Figure 9.13 Checkpoints in the cell cycle.

from parts of the skin. These cancers are called carcinomas. Cancers origi-nating from cells of connective tissue, bone, or muscle tissue are called sar-comas. Leukemias arise from white blood cells (leukocytes). Cancers of glan-dular tissue are called adenocarcinomas, and cancers of the nonneuronal cells of the brain are called gliomas and astrocytomas.

A mass of cancer cells descended from a single parent cell is called a **tu-mor**. A tumor that ceases to grow after reaching a certain size, remains con-fined to its original position, and does not invade nearby tissues is said to be **benign**. In contrast, **malignant** tumors continue to grow, invade surrounding tissue, and eventually impair the normal functioning of tissues and organs. Sometimes, cells from a malignant tumor break free of the original tumor, migrate to new sites in the body via the circulatory or lymphatic system, and establish new tumors. This process is called **metastasis**, and it is what makes cancer so lethal. When cancer metastasizes, it may reach sites at which tu-mors can cause severe damage, and it can lodge in so many places that treat-ment by surgery alone is impossible.

The accumulation of mutations in a single cell can lead to cancer

Cancers arise when a single cell acquires the ability to divide in an uncon-trolled manner. To become a cancer cell, a normal cell must accumulate mu-tations in several different genes. These mutations are then transmitted to the cell's cancerous descendants. Since your cells have many mechanisms for re-pairing DNA damage that leads to mutations and normally block the divi-sion of cells with DNA damage, mutations usually accumulate very slowly. That is why most cancers develop late in life.

Research has shown that two general kinds of mutations are necessary to establish cancer, and the findings make sense. Think about what you have just learned about the cell cycle. Cells have to be stimulated by growth fac-tors to divide, and cell division is subject to a number of checkpoints at which the process can be halted.

You can think of the cell division process as being rather like a vehicle with an accelerator and brakes. Depressing the accelerator would be the equivalent of a growth factor signaling the cell to divide, and applying the brakes would be the equivalent of the action of the checkpoints. To get the vehicle to move, you must both step on the accelerator and release the brakes. To get unregulated cell reproduction, the cell must be constantly stimulated to divide, and the checkpoints must be short-circuited. These processes are analogous to the two types of mutations required for cancer to develop.

Oncogenes

The accelerator mutation is a mutation in the signaling pathway that tells cells to divide, so that the cell is always being signaled. Mutant genes that promote cell division are called **oncogenes** (onco, cancer). The most common mutations of this type in human cancers are mutations in the *ras* genes. Cancer-causing mutations in the *ras* genes change the shape of the Ras protein so that it is always activated and therefore always signaling cells to divide. Researchers have identified several different amino acid substitutions (such as valine for glycine as position 12) that cause Ras to become locked in the active form (Figure 9.14). Mutant, permanently activated Ras proteins are involved in about 20% of all human cancers.

Mutant *ras* genes are not the only oncogenes found in human cancers. Any protein in the pathway signaling cell division could potentially mutate to a tumor-promoting form. For example, skin cells are signaled to divide by the binding of platelet-derived growth factor to its receptor. A mutant form of the receptor for platelet-derived growth factor is locked in the "on" position in some cancers. Other oncogenic mutations cause cells to produce growth factors constantly. Still others act further down in the cell division pathway than Ras to stimulate division inappropriately.

Tumor suppressor genes

If the signaling pathway that tells cells to divide is analogous to a car's accelerator, then the cell cycle checkpoints that halt division if damaged or unreplicated DNA is present are the brakes. The oncogene mutation is the accelerator mutation, causing the cell to be constantly stimulated to divide. The brake mutation stops the normal checkpoint processes, allowing cells with damaged DNA to replicate.

Figure 9.14 A single G → T mutation changes the normal Ras protein to a permanently active oncogenic form.

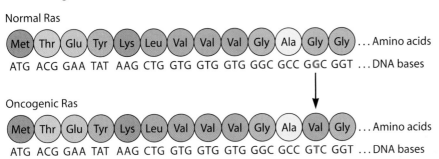

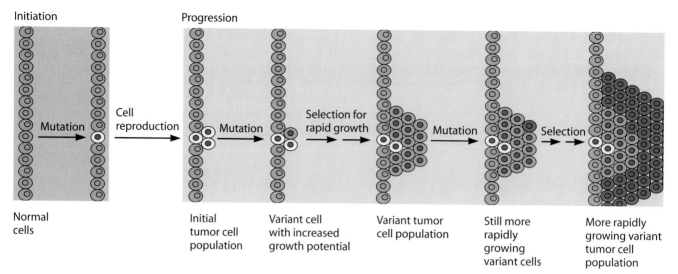

Figure 9.15 The development of full-fledged cancer involves the accumulation of multiple mutations that promote more and more aggressive growth.

Genes encoding proteins that normally halt cell replication are called **tumor suppressor genes.** The most common tumor suppressor mutation in human cancers is inactivation of *p53*. Mutations in *p53* are believed to be involved in up to 50% of all human cancers, including leukemias, brain tumors, and breast, colon, and lung cancer. Other tumor suppressor genes include the *BRCA1* and *BRCA2* genes associated with inherited forms of breast cancer and the *MADR2* and *APC* genes associated with inherited colon cancer.

Base pair mismatches and other forms of DNA damage are a natural result of DNA replication and exposure to the environment. In a normal cell, the damage is usually repaired before DNA replication because of the p53 checkpoint in the cell cycle, so no mutation results. When cells lacking the p53 protein reproduce, there is nothing to stop the replication of damaged DNA.

In cells with the brakes off because they lack p53 or another checkpoint protein, ordinary DNA damage can lead to mutations that change the properties of the cells. If one of these mutations activates an oncogene, the cell in which that mutation occurred will begin to reproduce constantly, forming a tumor. The uncontrolled reproduction provides more opportunities for additional mutations to accumulate in the cancer cells. Some of these mutations will cause more rapid, aggressive growth, and descendants of these cells eventually dominate within the tumor (Figure 9.15). If you combine a permanent signal to divide with a process that allows mutations to accumulate, you've set up a system for cancer to develop.

Accumulation of mutations in colon cancer

The contribution of multiple genetic changes to the development of a specific type of colon cancer has been studied extensively. Colon cancer begins with benign proliferations of cells lining the colon. These proliferations are called polyps, or adenomas. Eventually, mutations accumulate in the adenoma cells, and their descendants become a malignant carcinoma (Figure 9.16). Compare Figure 9.16 to Figure 9.15. Scientists have developed this picture of the accumulation of mutations by analyzing polyps and carcinomas removed

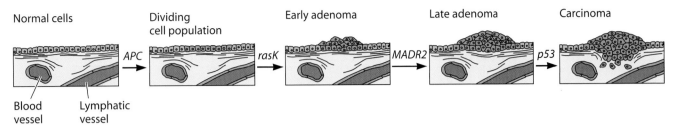

Figure 9.16 Accumulation of mutations in colon cancer.

from cancerous colons. The mutations do not always occur in this order, but this is the most common progression, with *APC* mutations in the early benign polyps and mutations in *p53* and *MADR2* usually appearing in the advanced adenomas and malignant carcinomas. The most important message here, though, is that full-fledged cancer requires the accumulation of multiple mutations in genes that normally regulate cell division.

Inherited mutations in tumor suppressor genes cause a predisposition to develop cancer

We have been discussing cancer arising from mutations in body cells, the scientific term for which is **somatic** (soma, body) cells. Mutations in your somatic cells will not affect your children, because they inherit their DNA only from your reproductive cells: eggs or sperm. Some families, however, carry a mutation in a tumor suppressor gene in their reproductive cells. Children who develop from these cells will have that mutation in their somatic cells and in their own reproductive cells. Because their body cells already carry one mutation needed to develop cancer, their cells do not have to accumulate as many additional mutations for cancer to develop. Families that carry these inherited mutations in tumor suppressor genes have a predisposition to develop cancer.

For example, a few families carry a mutant *p53* gene. Individuals who carry this mutation have a greatly increased risk of developing any of several different kinds of cancer. This multicancer syndrome is called Li-Fraumeni syndrome, for the two doctors who first recognized it. Another few families inherit a tendency to develop the type of colon cancer described above. These families have inherited mutations of the *APC* gene, and individuals who inherit this mutation develop hundreds of polyps that almost always progress to cancer. In another form of inherited colon cancer (a type different from the one described above), families inherit defective genes for mismatch repair. These individuals appear to accumulate mutations at a much higher rate than normal (makes sense, doesn't it?). While colon cancer is the type of cancer most frequently seen in families with these inherited mutations, affected individuals also have higher rates of other types of cancer.

Breast cancer genes

You may be aware that scientists have identified genes that predispose women to develop breast cancer. These genes (*BRCA1* and *BRCA2*) are tumor suppressor genes that normally suppress the development of breast cancer. Research suggests that the protein encoded by *BRCA1* is involved in DNA repair. *BRCA1* is also implicated in ovarian cancer. Some families carry a mutant, nonfunctional form of *BRCA1* or *BRCA2*. If a woman inherits one of these nonfunctional tumor suppressor genes, her chances of developing

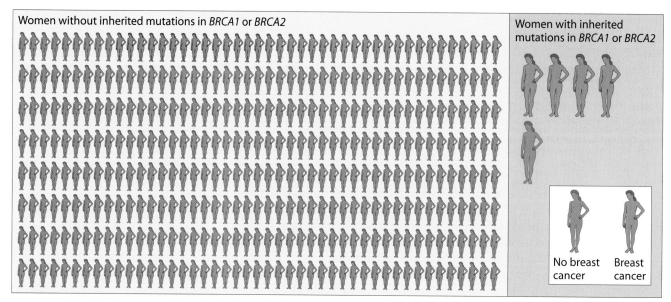

| Women without inherited mutations in *BRCA1* or *BRCA2* | Women with inherited mutations in *BRCA1* or *BRCA2* |

No breast cancer

Breast cancer

Figure 9.17 Breast cancer and inherited mutations. Most, though not all, women with inherited mutations in *BRCA1* or *BRCA2* will develop breast cancer, but most breast cancer is sporadic and develops in women without the inherited mutations.

breast cancer are much greater than those of a woman who has normal *BRCA1* and *BRCA2* genes, and she will likely develop the cancer earlier in her life. A woman who inherits a mutant *BRCA1* gene has a greater than 80% chance of developing breast cancer in her lifetime and about a 40% chance of developing ovarian cancer. Women with normal *BRCA1* and *BRCA2* genes have about a 10% chance of developing breast cancer if they live to age 80.

Many different combinations of mutations can cause cancer

Breast cancer provides a good illustration of the genetic complexity of cancer. Inherited mutations of *BRCA1* and *BRCA2* appear to be involved in only 5 to 10% of all breast cancers; therefore, about 90% of breast cancer cases are sporadic, which means they are not associated with inherited mutations (Figure 9.17). Genetic analysis of sporadic breast tumors shows that they do not seem to be associated with mutations in either *BRCA1* or *BRCA2*, nor are any other mutations consistently associated with them. Instead, alterations in a number of genes encoding growth factors, receptors, and cell division proteins have been found.

Researchers have discovered well over 100 different genes that, when mutated, are involved in cancer. The normal functions of the proteins encoded by these genes have been figured out in some cases, but not yet in others. Table 9.1 lists a few examples of known oncogenes and tumor suppressor genes, along with their roles in the control of cell division.

DNA DAMAGE AND REPAIR

Some environmental agents promote mutations, and high levels of exposure to these agents can increase the likelihood that cancer will develop. Agents that can be shown to promote mutations, usually in tests involving bacteria, are called **mutagens**. If an agent can be shown to increase the rate of cancer development in laboratory animals, it is called a **carcinogen**. Mutagens and carcinogens share the property of being able to damage DNA.

Table 9.1 Some cancer genes and the normal physiological roles of their products

Gene	Normal physiological role
Oncogenes	
sis	Growth factor
erbB, fms, neu	Growth factor receptors
ras, src, abl	Signal transmission within the cell
bcl2	Blocks programmed cell death
myc, fos, myb	Regulators of transcription
Tumor suppressor genes	
rb	Regulation of replication and transcription
p53	Regulation of cell division cycle; stops cells from dividing if their DNA is damaged, allowing time for repair; initiates programmed cell death if DNA is not repaired

As we saw above, DNA mutations arise from the replication of mismatched DNA bases. DNA can acquire mismatched bases through errors in replication, spontaneous chemical rearrangements, or physical damage by environmental chemicals and radiation. Environmental DNA damage can lead to mismatches, because the types of damage that occur distort the helix and block normal replication. If DNA replication cannot proceed normally, replication enzymes can introduce incorrect bases as they attempt to duplicate the DNA base sequence at the damaged sites, resulting in mismatches and the potential for mutation.

UV light and DNA-binding chemicals are environmental carcinogens

One unavoidable environmental DNA-damaging agent is ultraviolet (UV) light. UV light causes adjacent T bases within DNA to form bonds to each other, essentially gluing them together side by side to form what is called a thymine dimer (Figure 9.18). A thymine dimer distorts the structure of the DNA molecule and blocks both transcription and DNA replication. UV-induced DNA damage is harmful: you can kill bacteria and sterilize foods by irradiating them with enough UV light, and exposure to UV radiation is the cause of almost all skin cancer in humans.

Certain chemicals cause DNA damage by forming chemical bonds to DNA molecules. The chemicals normally present in your cells do not bond to DNA, except for the normal binding of proteins and enzymes as part of cellular processes. The chemicals we are talking about here are not a normal part of your cellular chemicals, and because of their nature they can attach themselves to DNA in ways that disrupt normal cellular functions.

Benzopyrene

One such DNA-damaging chemical is **benzopyrene**, a substance found in cigarette and other tobacco smoke; smoke from burning leaves, trash, and wood; diesel exhaust; and other sources. Benzopyrene has been found to promote cancer in laboratory animals and is therefore classified as a carcinogen. Its DNA-damaging ability explains how it can promote the development of cancer.

Benzopyrene binds to G bases within DNA and can lead to mutations if it is not removed. One place along DNA where benzopyrene apparently

Figure 9.18 UV light damages DNA by promoting the formation of thymine dimers.

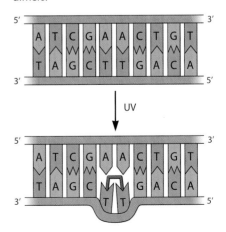

binds exceptionally well is within the *p53* gene. From the information presented above about the role of the p53 protein in the cell cycle, you can understand why a chemical that promotes mutation in the *p53* gene would be especially dangerous.

Lack of DNA repair enzymes increases mutation and cancer rates

As mentioned earlier in this chapter, cells have mechanisms in place for repairing damage. We briefly described the mismatch repair system, which scans newly replicated DNA helices for mismatched bases and corrects the errors. Another system of proteins normally repairs damage that distorts the DNA helix, like the UV-induced thymine dimer or the addition of bulky chemicals, such as benzopyrene. When this system of proteins detects a distorting lesion, enzymes called nucleases clip the damaged DNA strand on either side of the lesion, a helicase enzyme peels away the short piece of DNA containing the lesion, and then a DNA polymerase fills in the single-stranded region with new bases (Figure 9.19). This kind of DNA doctoring is called **excision repair.**

Figure 9.19 The excision repair system removes DNA damage that distorts the DNA helix, such as thymine dimers.

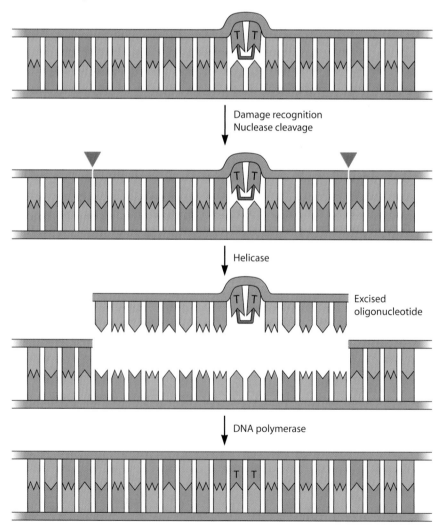

Cells that cannot repair damaged DNA accumulate mutations much faster than those that can. Since cancer results from the accumulation of mutations in specific kinds of genes, people and animals that cannot repair damaged DNA tend to develop cancer much more frequently than those who can. For example, individuals with an inherited defect in the excision repair system have the disease **xeroderma pigmentosum** (XP). The cells of people with XP cannot repair thymine dimers; hence, they are extremely sensitive to UV light. Individuals with XP develop multiple skin cancers wherever their skin is exposed to sunlight. If we did not have such good DNA repair systems, we would all be much more prone to develop cancer.

APPLYING THIS KNOWLEDGE

In the latter half of the 20th century, great strides were made in the medical treatment of cancer. From the mid-1970s to the late 1990s, the rate of 5-year survival for all people diagnosed with invasive cancer climbed from about 50 to about 62%. For some cancers, the news is particularly good: the mortality rate (deaths per some number of the population) for childhood cancers decreased nearly 40% from 1975 to 1998. Despite these successes, people may someday look back on 20th century cancer treatments with the same sort of horror with which we now view historical medical practices such as bleeding or routine battlefield amputation of wounded limbs. The problem with cancer treatment has been that we have known how to attack cancer cells only by targeting their characteristic of rapid division. Our classic anticancer treatments with radiation and chemotherapeutic drugs damage DNA or inhibit DNA replication and thus target rapidly dividing cancer cells. Unfortunately, damaging DNA and blocking DNA replication are toxic to normal cells, too, so these anticancer treatments are essentially poisonous.

A molecular understanding of cancer is leading to targeted, less damaging drugs

Since cancer cells divide more rapidly than normal cells, they are more susceptible to the poisons. The challenge for doctors has been to give patients enough of the treatments to kill the cancer cells without killing the patient, too. The normal cells most susceptible to damage from anticancer treatments are those that divide most frequently: the progenitors of blood cells, cells lining the digestive tract, and hair follicle cells. Thus, the side effects from classic cancer treatments include nausea, hair loss, and anemia.

Our developing understanding of the cellular changes that lead to cancer is producing treatments that are more specifically targeted at cancer cells rather than at dividing cells and to treatments that are tailored to specific cancers. Some improvements in therapy have already taken place. Tamoxifen, one of the older drugs based on a molecular understanding of cancer, has been in use for over 20 years.

Tamoxifen

The hormone estrogen, like other hormones, affects cells that have receptors for it. In women, some of these cells are found in the breast and in the uterus. Each month, estrogen binds to its protein receptor, causing a shape change in the molecule. The estrogen-receptor complex then binds to certain sites on DNA, interacts with other proteins, and in doing so activates the transcription of a number of genes. This sequence of events leads to the growth and division of cells lining the milk ducts and the uterus to prepare for pregnancy.

Some breast cancer cells remain sensitive to estrogen and are stimulated by estrogen to divide. These estrogen-sensitive breast cancers can be treated with the drug tamoxifen. The tamoxifen molecule is similar enough in shape to the estrogen molecule that it can bind to the estrogen receptor. However, tamoxifen binding does not cause the same shape change in the receptor that estrogen binding does, so when an estrogen receptor binds to tamoxifen, it does not activate the transcription of genes and induce cell division.

Herceptin

About 25 to 30% of metastatic breast cancers overexpress the gene that encodes the human epidermal growth factor receptor, called *Her2* or *erbB* (Table 9.1). Metastatic breast cancer cells have many more Her2 receptors on their surfaces than normal cells and thus are abnormally sensitive to stimulation by the growth factor. In 1998, the Food and Drug Administration approved a new drug called Herceptin for use against these cancers. Herceptin specifically binds to the Her2 receptor and blocks its action. Treatment of women with breast cancers that overexpressed the Her2 protein showed that Herceptin significantly reduced tumor growth and prolonged the survival of the patients.

Gleevec

Leukemias are cancers of white blood cells. One type of leukemia, chronic myelogenous leukemia, nearly always involves an abnormal form of the *abl* gene (Table 9.1). The Abl protein is normally involved in the signaling pathway that causes cells to divide. It does so by adding a phosphate group to a target protein, activating it. In its mutant, oncogene form, the *abl* gene produces a protein that is always on. It adds a phosphate group to its target whether or not the cell is receiving a signal to divide.

The drug Gleevec, approved in 2001, is a specific inhibitor of the activity of the Abl protein. In the presence of Gleevec, chronic myelogenous leukemia cells stop dividing in culture. Treatment of patients with Gleevec has been remarkably encouraging as well. In one trial of patients who were not responding to a standard drug, 95% of the patients who took Gleevec were alive after 18 months, and there was no apparent progression of the disease in 89% of them. Gleevec is such a new drug that we do not yet know its effect on the long-term survival rates of patients.

Gleevec is the first anticancer drug that directly turns off the signal of a protein known to cause cancer. It kills leukemia cells without killing normal cells, and treatment with Gleevec involves far fewer side effects than treatment with conventional anticancer drugs or radiation. Industry and academic scientists are working to develop additional drugs to target other cancer-causing molecules. Many efforts are focusing on the Ras protein, and some potential anti-Ras drugs are in evaluation stages.

Genetic testing can provide information helpful for treatment

From reading the information above, you should have gathered that some people have inherited mutations that predispose them to develop certain types of cancer and that similar tumors (such as in breast cancer) can contain different sets of mutations. Genetic testing to determine mutation status can help with treatment in two ways: by facilitating early diagnosis and by indicating the most effective therapies.

First, the best cancer treatment is to remove the cancer early, before it begins to spread. This requires early detection and is the reason you hear so many public service announcements urging people to get regular checkups and screenings. The age at which a particular screening is recommended is based on the statistics showing when people are likely to begin developing the particular type of cancer being discussed. However, people from families with hereditary cancers are likely to develop the cancers much earlier if they inherited the mutation. Genetic testing could show which family members inherited the mutation and which did not. Those who have the mutation would know for a fact that they needed to be screened early and often, and those without it could wait until the age recommended for the general population to begin screening.

Second, the particular mutations present in a tumor can affect how it responds to various treatment options and also indicate how aggressive the cancer is likely to be. By determining the genetic status of the tumor, doctors can make better-informed decisions about how to treat a particular patient. As genetic and cell biology research adds to our knowledge base, we can hope for better, more specific treatments targeted at specific tumor cell types. Keep your eyes and ears open for announcements about new cancer drugs, and see if you can figure out what they actually do from the news stories.

SUMMARY POINTS

The cell cycle is a repeating pattern of events observed in actively dividing cells. It includes growth, DNA replication, the distribution of replicated chromosomes into two new nuclei, and cytokinesis. Cells normally do not divide unless they receive signals called growth factors. Nondividing cells exit the cell cycle into a stage called G_0, in which they are metabolically active but do not grow.

DNA replication is initiated when certain proteins interact with a DNA molecule at a specific sequence of DNA bases called a replication origin. This process is tightly regulated. Once initiation has occurred, DNA polymerase uses the sequence of bases on one strand of DNA as a pattern for synthesizing the complementary strand.

At the end of DNA replication, the two new copies of a DNA molecule are held together at a structure called the centromere, and each DNA molecule together with the proteins it wraps around is called a chromatid. Each chromatid is a copy of its original chromosome.

In the process of mitosis, the chromosomes with two chromatids are lined up, and the chromatids are split apart to form two independent chromosomes and then pulled to opposite ends of the cell. New nuclei form around the chromosomes, and each nucleus contains one copy of every original chromosome. After mitosis is complete, the cell with two nuclei splits into two separate daughter cells. The daughters have the same genetic content as the mother cell.

Cells do not normally divide unless signaled to do so by growth factors. Growth factors bind to receptors on the cell and initiate a chain reaction that ends with the synthesis of cell division proteins and the initiation of DNA replication.

As the cell proceeds through the cell cycle, there are checkpoints at which the cell division process is stopped if the previous task has not been completed or if the cell's DNA is damaged. A key protein in the damage control checkpoint is p53. If a cell contains damaged DNA, the p53 protein is expressed and blocks further progress through the cell cycle.

DNA damage can lead to the formation of mutations. When DNA polymerase attempts to copy a damaged segment, it can insert incorrect bases and change the protein-coding sequence of a gene or a regulatory site for controlling gene expression.

DNA is damaged by environmental agents, such as UV light and chemicals, that attach themselves to DNA. Mismatched bases also occur as a result of uncorrected mistakes in DNA replication or spontaneous chemical reactions inside of cells. DNA damage is normal, and cells have a variety of repair enzymes to correct different kinds of damage. However, the more damage DNA receives, the more likely it is that some will escape the repair enzymes.

Cancer occurs when a cell no longer responds to controls on its division process and begins to divide independently. A cell loses its ability to respond to controls when it accumulates mutations in oncogenes that initiate the

SUMMARY POINTS *(continued)*

division process (such as *ras*) and in tumor suppressor genes that halt the process if it is going awry (such as *p53*).

Some families carry inherited mutations in tumor suppressor genes. Members of these families have a greater risk of developing cancer, because a key mutation in the cancer development process has already happened.

The connection between environmental agents and cancer is that environmental carcinogens damage DNA and thus increase the likelihood of mutation. Benzopyrene, a carcinogen in tobacco smoke, is particularly dangerous because one of the places it frequently damages DNA is in the *p53* gene itself.

Our increasing understanding of the molecular events leading to cancer has given rise to the recent development of cancer therapies that are targeted to the specific cancer a patient has. Older therapies simply attacked dividing cells, which meant they indiscriminately damaged hair follicles, the intestinal lining, and blood cells, causing many side effects. The newer targeted therapies have the promise of being both more effective and safer.

Genetic testing is another new tool for managing cancer and its risks. It can show whether an individual has inherited a specific "cancer gene" and can show precisely which mutations a cancer patient has accumulated.

KEY TERMS

Amniocentesis	**Chromatid**	**Metastasis**	**Replication origin**
Anaphase	**Cytokinesis**	**Mismatch**	**Restriction point**
Apoptosis	**Excision repair**	**Mitosis**	**Somatic cell**
Benign	**Initiation**	**Mutagen**	**Telophase**
Benzopyrene	**Interphase**	**Oncogene**	**Tumor**
Cancer	**Karyotype**	**p53**	**Tumor suppressor gene**
Carcinogen	**Kinetochore**	**Primer**	**Xeroderma pigmentosum**
Cell cycle	**Malignant**	**Prophase**	
Centromere	**Metaphase**	**Ras**	

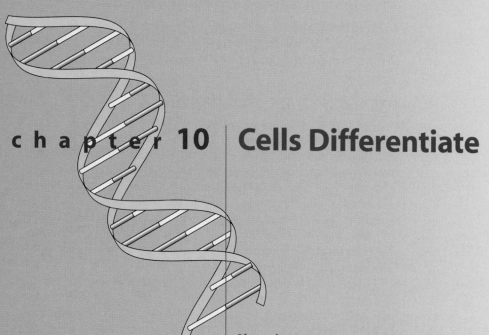

chapter 10 | Cells Differentiate

If you have ever marveled at the emergence of a butterfly from its chrysalis, the transformation of a tadpole into a frog, or the perfection of a newborn baby's tiny hands, you understand what drives scientists to study how intricate body structures develop and change. That baby, like you, was once a single cell, a fertilized egg, spherical in shape, with no nerves, heart, lungs, or limbs. Contained within that fertilized egg was all the information needed for it to develop into a baby, grow into a child, and finally, become a sexually mature adult. Likewise, contained within a fruit fly egg is all the information needed for it to develop into a larva, grow, pupate, and transform into a sexually mature flying insect. The process of transformation from egg to adult is referred to by biologists as development, and the subdiscipline of biology dedicated to understanding the molecular and cellular events that drive these processes is called developmental biology.

This chapter gives you a peek into the world of developmental biology. We need to make clear at the outset that this field, like any other area of science, is a work in progress. We do not have a complete, or even nearly complete, understanding of the molecular mechanisms underlying the development of a multicellular organism, and even if we did, we could not present it to you in a single chapter. Yet this is a heady, exciting era for developmental biologists, and the reason is that the scientific tools provided by the molecular biology-biotechnology revolution have enabled them, for the first time, to begin to tease apart the molecular events that result in the development of a girl instead of a boy, a wing instead of an antenna, a head instead of a tail. We provide here an overview of this fascinating field, including:

- how scientists study development
- model organisms and why scientists use them

- differentiation, morphogenesis, and differential gene expression
- how the body plan is established in *Drosophila*
- early development in mammals
- differentiation of males and females in humans

As you read this chapter, note how the study of these fascinating processes recalls information from earlier chapters in this book, such as gene expression and regulation, protein structure, cell signaling, and receptor proteins.

HOW SCIENTISTS STUDY DEVELOPMENT

Until the 20th century, essentially all that scientists interested in developmental biology could do was observe—and this they did, observing insects and other invertebrates and chicks, frogs, and other vertebrates, using increasingly more advanced microscopes to compile a wealth of observational details. They observed normal developmental processes and occasionally encountered examples of development gone wrong, such as animals with missing or extra limbs. Their work led to great descriptive knowledge of development in many systems—they knew what events would normally happen in what order as an individual developed but nothing about why they happened or what the underlying mechanisms were.

The era of classical genetics began around the turn of the 20th century, as several different scientists were engaged in carefully controlled breeding experiments and the significance of Mendel's work was finally understood. One of the luminaries of this era of experimentation was Thomas Hunt Morgan, a Columbia University professor who studied the fruit fly *Drosophila melanogaster*. One ongoing effort in Morgan's laboratory was to find and breed mutant forms of the fly. Morgan and his students were studying the nature of the gene by observing the inheritance of traits in crosses, and they needed alternate forms of traits (usually found as a result of mutations) for their experiments.

In 1915, a graduate student in Morgan's laboratory, Calvin Bridges, found a mutant fly that had two pairs of wings instead of one. The mutant fly had a second pair of wings instead of the normal tiny appendages called halteres. Bridges called this a **homeotic** mutation, because it changed one part of the body into another (Figure 10.1). The fly was carefully bred, and stocks of it were maintained. In 1934, a graduate student who had a job as a laboratory assistant for a genetics class, which included maintaining stocks of *Drosophila*, spotted a similar mutant fly, which he also carefully bred. Descendants of these mutants, discovered by sharp-eyed students, played key roles in the identification of genes that govern body plan in both flies and other animals, including humans.

Figure 10.1 The fruit fly *D. melanogaster.* **(A)** Normal fly. **(B)** A homeotic mutant with an extra pair of wing segments instead of the normal haltere segment. (Photographs by Ed Lewis, courtesy of the California Institute of Technology Archives.)

Developmental biologists usually study a few specific experimental organisms

The molecular biology revolution, with the accompanying development of methods for manipulating specific pieces of DNA, determining a DNA base sequence, and isolating and studying individual genes (see chapter 15 for descriptions of some of these techniques), gave scientists the tools they needed to begin to tackle underlying mechanisms of development. What they needed was appropriate experimental organisms.

Scientists studying phenomena such as development or DNA replication (rather than studying an organism per se) usually end up using a small number of different kinds of organisms as experimental systems. They call these **model systems**, or models. Many scientists will use the same model system to study a phenomenon, and so it becomes well characterized and more information is available for interpreting experimental results. More specific experimental approaches are developed for that system, too, which adds to its usefulness. Experience has shown that the knowledge obtained from one model system can guide experiments in other systems and that fundamental processes, like DNA replication, are fairly similar from one organism to another.

The fruit fly

The fruit fly was an obvious choice: it is small, so you can keep many of them in a small place; it has a short life cycle, so it develops quickly; it was well characterized, and many mutant strains had been collected over decades of research in many laboratories. Mutants give clues to processes taking place in an organism; if a normal fly has red eyes and a white-eyed fly appears, you know there is a specific pathway to produce the red eye pigment that is altered in the white-eyed fly. Developmental biologists needed mutant strains in which development did not occur properly so they could begin to figure out what was wrong.

It might seem strange to you that scientists chose to study development in a tiny insect rather than in a mammal, such as the mouse. Yet at the time, the only clue scientists had that a particular gene existed was the existence of a mutant (see "Finding Genes" in chapter 16). To find mutants, you need to be able to screen many thousands of individuals, which is impractical even with small animals like mice, and mammals like mice develop inside the mother's body, so you cannot see what is happening.

So developmental biologists turned to biologists' old friend, the fruit fly. In addition to studying the homeotic mutants that had already been isolated, they embarked on a campaign to collect as many developmental mutants as possible. The fruits of their labors have been nothing less than spectacular. Their research has revealed (among many other things):

- a family of genes that control body plan, or what part of a developing embryo becomes what kind of structure, present in organisms as diverse as flies, frogs, mice, chickens, and humans
- how contributions from the mother to the cytoplasm of the egg establish gradients in the fly embryo that determine the body axes (head to tail and front to back)
- genes that specify the axes of the body, versions of which are also found in a multitude of other organisms

The nematode worm

The fruit fly has provided a very productive model system for studying development. Yet it has its quirks: flies are insects, and their developmental

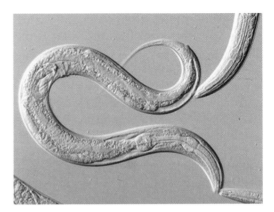

Figure 10.2 The nematode *C. elegans.* (Photograph courtesy of David Gems, University College London.)

pathway through larval stages has some distinct differences from the development of other animals. They are also quite complex organisms. In 1965, the British scientist Sydney Brenner decided to use a simpler model animal for his studies. He chose the tiny roundworm *Caenorhabditis elegans*, a nematode (Figure 10.2).

At the time, people thought Brenner was crazy. No one knew much about the tiny worm back then. Yet in the intervening years, the "worm people" (which is what those who study *C. elegans* call themselves, as opposed to the "fly people") have made invaluable contributions to our understanding of development.

C. elegans adults contain fewer than 1,000 cells, and they are transparent. With the right kind of microscope, it is possible to watch every single cell division that happens as the worm develops from egg to adult. In Brenner's laboratory, John Sulston did just that—he patiently watched nematodes developing until he could trace the history of each and every cell in the worm. The fate of each cell in the worm was predetermined and invariant. Working out the lineage of every cell gave scientists an unprecedented window into the developmental process. Using a tiny laser beam attached to a microscope, scientists in Brenner's laboratory were able to kill individual cells within the developing worm and observe what happened. Instead of revealing what happens when a particular gene product is missing (which is what you see when you study a mutant), they could observe what happens when a particular cell was missing.

The worm people also used genetic approaches. They fed their worms mutagenic chemicals and collected hundreds of mutant strains, identifying many genes associated with specific developmental events or the development of specific tissues. The work of the worm people has been particularly useful in revealing (among other things):

- a process, called apoptosis, by which cells commit suicide, which plays an orderly role in the development of the adult body, as well as in the defense of the body against potential catastrophes, such as cancer
- how a cell's neighbors can contribute to the identity it takes on in a developing organism

Frogs, chickens, fish, and mice

Although findings from research with flies and worms have been spectacular, neither of these organisms can address specific questions having to do with the development of vertebrates, since both are invertebrates. For many years, scientists observed the development of chicken and frog embryos, which develop in eggs outside the mother's body and are therefore readily accessible. They gleaned information about the process by manipulating it in ways such as transplanting tissues from one part of an embryo to another, removing specific tissues, and treating the developing embryos with various substances. A number of years ago, a group of scientists decided to pioneer yet another model system: the zebrafish. These small fish have short generation times, are transparent, and develop outside the mother, so developmental processes can be observed directly (Figure 10.3). Because of their small size and fast growth, it is feasible to mutagenize them and search for developmental abnormalities. Since zebrafish are vertebrates (unlike worms and flies), it is expected that the "fish people" will identify genes specifically related to the development of vertebrates.

Finally, the era of the mouse in development has arrived as well. A set of highly refined tools for making genetically altered mouse strains has been developed and improved over the past several years (see "Genetic engineering of animals" in chapter 16), so that any known gene in a mouse can be specifically mutated. As noted above, it is not practical to try to find random mouse developmental mutants—the space, time, and cost to care for thousands and millions of mice would be prohibitive. But now that work with flies and worms has shown that the same genes govern important processes in these organisms and the mouse, scientists no longer need mouse mutants to point the way to genes. They can use results from flies, worms, and fish to

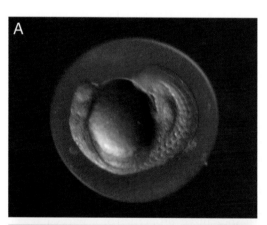

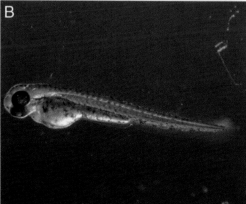

Figure 10.3 The zebrafish *Danio rerio*. **(A)** A 24-hour embryo. **(B)** A newly hatched 3-day fish. (Photographs courtesy of Steven Baskauf [http://bioimages.cas.vanderbilt.edu].)

find the mouse genes and then knock out those genes in mice and study the results. Since mouse and human are 80% similar genetically, what is true in the mouse is assumed to be generally true in humans.

Studying human development

Why not study human development directly? Many scientists do, but they are largely restricted to observing what nature gives them—natural instances of development gone wrong (manifested as miscarriages of pregnancies or as birth defects). Scientists cannot ethically experimentally manipulate human beings, and so they do experiments in other systems and look for evidence that the same is true for people. They do this by comparing normal human gene sequences to those of model organisms, looking for mutations in cases of miscarriage or birth defects, comparing these results to model systems, and observing developmental patterns. This means that the major discoveries are going to be made in the model systems, and discoveries about human development will mostly be made by comparing human cases to the models.

Now you can see why the study of such seemingly obscure creatures as fruit flies, worms, or zebrafish—or even mice—is important for humans.

FUNDAMENTAL DEVELOPMENTAL PROCESSES

The development of an adult organism from a fertilized egg (zygote) involves two basic types of change: differentiation and morphogenesis. **Differentiation** is the generation of different specialized kinds of cells, such as epithelial, liver, or muscle cells, from the **zygote** (the fertilized egg) or other precursor cells. **Morphogenesis** is the creation of form and structure. Differentiation generates blood cells, muscle cells, neurons, and other cell types. Morphogenesis generates the shape of legs, eyes, wings, skin, and other body organs, tissues, and structures.

Differentiation is progressive specialization

A fertilized egg contains all the information needed for it to develop into an adult organism. Descendants of that egg will become all the types of cells found in the adult: neurons, blood cells, muscle cells, sex cells, etc. The fertilized egg is therefore said to be **totipotent** (toti, all). The zygote begins to divide, and at some point, the progeny cells lose their totipotency. In the nematode, this happens at the very first cell division, with each of the two daughter cells giving rise to different parts of the worm. A cell that can give rise to several different, but not all, cell types is called **pluripotent** (pluri, several), or multipotent. A cell with specialized properties of a particular cell type that cannot give rise to any other cell type is called fully or **terminally differentiated**. Terminally differentiated cells do not usually reproduce.

If differentiated cells don't reproduce, how is it we are able to replace skin cells when we are wounded and to replace blood and intestinal epithelial cells on an ongoing basis? Within our bodies, and those of other mammals, are a small but vital number of **stem cells**. Stem cell is a general term for less differentiated cells that can give rise to specific types of differentiated cells. For example, blood stem cells found in the bone marrow divide and give rise to cells that differentiate into all the various blood cell types (Figure 10.4).

Fully differentiated blood cells have limited life spans ranging from less than a day to a few months and must be continually produced. Stem cells in the intestine replace shed epithelial cells, and stem cells in the skin produce

A

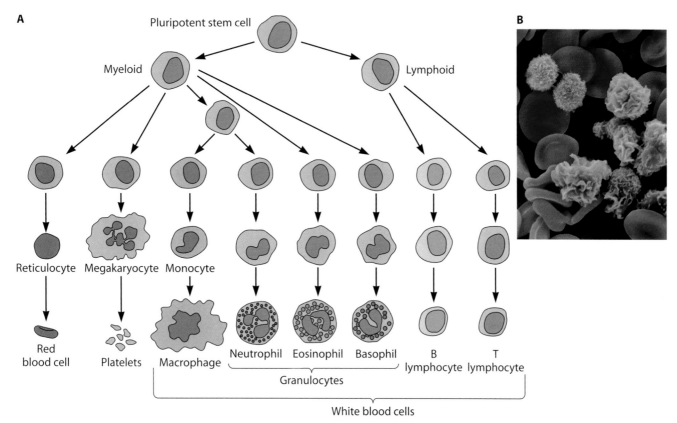

B

Figure 10.4 Differentiation of blood stem cells into blood cells. **(A)** All of the different blood cells develop from stem cells located in the bone marrow. **(B)** Electron micrograph of red blood cells (red), macrophages (blue), and T lymphocytes (pink). The colors were added artificially. (Photograph copyright Dennis Kunkel Microscopy, Inc.)

new skin cells both in response to wounding and to replace dead cells that are normally shed. Stem cells are believed to reproduce in our bodies, unlike terminally differentiated cells. In animals that can regenerate amputated limbs, the process involves dedifferentiation of existing cells and then redifferentiation and development into the missing structure.

Same genes, different expression patterns

All differentiated cells within an organism contain the same set of genes; no genetic material is lost during differentiation. The cells are different because they express different subsets of their genetic material. All of them express the genes necessary to carry out basic cell processes, like protein synthesis (these genes are often called "housekeeping" genes), but each cell type typically expresses characteristic products essential to its specialized function (Table 10.1). The product may be a protein itself, or it may be a substance synthesized by specialized cellular enzymes, such as the pigment melanin.

In a simplistic way, cellular differentiation can be seen as the turning on and off of specific groups of genes within cells so that they become different and express different products. It won't surprise you, then, that many of the genes found to be crucial for defining what part of the body develops from which cells encode proteins or other substances that affect gene transcription. In fact, you've already encountered some examples: the steroid hormones.

Table 10.1 Specialized products of differentiated cell types

Cell type	Specialized product	Specialized function
Keratinocyte (skin cell)	Keratin (protein)	Protection against abrasion and drying out
Erythrocyte (red blood cell)	Hemoglobin (protein)	Transport of oxygen
Melanocyte	Melanin (pigment)	Pigment production
Myocyte (muscle cell)	Actin and myosin (proteins)	Muscle contraction
Pancreatic islet cells	Insulin (peptide)	Regulation of glucose metabolism
Hepatocyte (liver cell)	Numerous enzymes (proteins)	Glycogen storage and breakdown; fatty acid synthesis; gluconeogenesis; other metabolic functions
Neuron (nerve cell)	Neurotransmitters (various)	Transmission of nerve signals

These hydrophobic hormones cross over cell membranes and bind to protein receptors inside the cell. The steroid-receptor protein complex goes on to activate or turn off the transcription of genes. The sex hormones estrogen and testosterone, in addition to playing key roles during puberty and in sexual maturity, are essential to the development of reproductive structures during the development of mammals, as we will discuss below.

Morphogenesis is the development of form and structure

Muscle, bone, and nerve cells are generated from a zygote via differentiation, but the form of an arm, a lung, or a nervous system is created through morphogenesis. Morphogenesis involves the movement, migration, proliferation, and death of cells. One of the earliest morphogenetic events in vertebrate development is the curling up of a flat sheet of cells to form the neural tube, which will eventually become the brain and spinal cord. For a while, cells within the tube proliferate, but eventually they stop dividing and migrate outward. These nerve cell precursors settle into new environments, mature, and begin to form neural circuits. The formation of vertebrate limbs begins with the migration of bone and muscle precursor cells to sites just under the outer layer of the embryo, where they divide and form bulges called limb buds. Cells within the limb buds continue to proliferate, causing the bud to grow outward and form a protrusion that will eventually develop into a limb. The formation of the trachea in *Drosophila* and the lungs in mice begins with the migration of epithelial cells in a branching pattern.

Cell migration and fur pigmentation

The migration of cells associated with the neural tube leaves a visible pattern in furred vertebrates. The cells are the melanocytes, whose specialized function is to produce pigment. Melanocytes in the hair follicles deposit pigment granules into the developing hair shaft, giving it color. Many genes affect the shade of color in the hair, and other genes are believed to affect both the number and migration ability of melanocytes during development. But one bottom line on coat color in vertebrates is this: if there are no melanocytes in the hair follicle, the hair growing from that follicle cannot be pigmented and will be white.

Melanocytes are formed when a population of cells left over at the edge of the developing central nervous system, called the neural crest cells, migrates outward from the region of the spinal cord and differentiates into sev-

Figure 10.5 TJ (left) and Annabelle have identical coloring, but TJ's melanocytes migrated further than Annabelle's during his embryonic development. TJ's white chest and muzzle patches are smaller than Annabelle's, and TJ has a single white toe compared to Annabelle's white socks. (Photograph of TJ courtesy of Thomas A. Martin. Photograph of Annabelle courtesy of John and Jennifer Kleinschmidt.)

eral different types of cells, including melanocytes. Because of the migration pathway of the melanocytes, some of the last regions of the body they reach are the feet and midchest. This migration pattern is the reason why, if an otherwise colored dog has any patches of white on it, the patches are likely to be on the feet or chest.

Although genes (actually the proteins encoded by the genes) appear to affect melanocyte migration, aspects of their progress appear to be somewhat random. A random element in melanocyte migration explains why in a population of piebald spotted animals, none has exactly the same pattern of spots. It also explains why two otherwise identically colored dogs can have varying sizes of white patches on their feet and elsewhere (Figure 10.5)

Apoptosis in morphogenesis

Cell death was recognized as a normal and important part of embryonic development as far back as the 1930s, but it was studies of development in the worm that broke open the mechanisms by which it takes place. The fate of each and every cell of the worm is predetermined and invariant. Along the way to the adult worm's 959 cells, 131 cells are destined to die. The isolation of mutant worms in which appropriate cell death did not occur led to the identification of genes involved in the process. Versions of these same genes have been identified in many other organisms, including birds, mice, and humans.

Programmed cell death, or apoptosis, during development eliminates transitory organs and tissues and sculpts the final form of tissues. The difference between the webbed foot of a duck and the nonwebbed foot of a chicken is that, in the chicken, programmed cell death removes the webbing. Human embryos, too, start out with webbing in between their digits, and occasionally a person is born who still has it. During early development, both *Drosophila* and mammalian embryos have tissues that can develop into male or female sexual organs. As part of normal gender development, programmed cell death eliminates the inappropriate tissues.

Cell migration, proliferation, and programmed cell death are triggered by communication between cells. One cell or group of cells produces a molecular signal and secretes it into the environment. Cells that express a receptor for that signal can receive it and respond to it. The response may be to move, to divide, or even to commit suicide.

DIFFERENTIAL GENE EXPRESSION

Having introduced differentiation and morphogenesis, we find ourselves with a common theme: differential gene expression. Cells become different as a result of expressing different suites of genes within the genome, and they undergo processes that generate shape and form by producing different molecular signals and receptor proteins that allow them to respond to specific signals. Differential gene expression, then, is at the heart of biological development.

In earlier chapters, we touched briefly on the theme of controlling gene expression. When we introduced gene expression in chapter 4, we introduced the promoter as the place where RNA polymerase binds to DNA and begins transcription and said that the binding of other proteins to DNA was involved in regulating this process. In chapter 8, we described how the binding of repressor proteins regulates transcription of the *lac* and *trp* genes of the bacterium *Escherichia coli* in response to environmental conditions. Let's now take a closer look at how gene expression is regulated in eukaryotic organisms.

Control of gene expression in eukaryotes

In eukaryotes, gene expression is controlled primarily at the level of transcription. In our introduction to transcription, we probably left you with the impression that if a promoter was available to the RNA polymerase protein, it would bind to that promoter and begin transcription. In prokaryotes like *E. coli*, that image is accurate enough. Eukaryotes are a completely different story. Eukaryotic RNA polymerase is unable to bind to a promoter and begin transcription without help from quite a few other proteins called transcription factors. Gene expression can be regulated through the presence or absence of specific factors.

Enhancers

Some of the transcription factors are required at the promoter itself to stabilize the binding of RNA polymerase to DNA. In general, these are present in every cell because transcription cannot occur without them. The transcription factors that are involved in cell-specific gene expression usually bind to DNA at base sequences that are not part of the promoter itself. These base sequences, called **enhancers**, are on the same chromosome as the promoter with which they interact but can be several thousand base pairs removed from it.

Enhancers provide the binding sites for specific transcription factors. The enhancer-bound transcription factors have just the right shape and surface characteristics to bind to the RNA polymerase complex at a promoter and alter it so that transcription can begin or so that it begins more efficiently. The enhancer-bound transcription factors are able to reach the RNA polymerase complex at the promoter because the intervening DNA can fold into a loop (Figure 10.6).

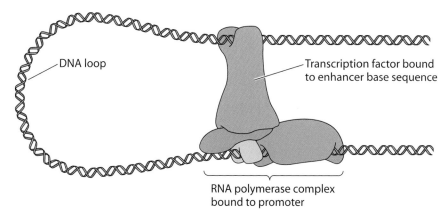

Figure 10.6 Transcription factors bind to DNA base sequences called enhancers and interact with the RNA polymerase complex at the promoter, initiating transcription. The DNA between the promoter and the enhancer folds into a loop.

Enhancer-like DNA sequences and proteins that bind to them can also turn off transcription rather than stimulate it. These DNA sequences are called **silencers**. In addition, some proteins bind to specific transcription factors and block them from binding to enhancer DNA sequences.

Transcription factors

If every gene had to have its own individual transcription factor, then half the genome would consist of genes encoding them—one factor per gene that did something else. Instead, both enhancers and transcription factors have modular structures that make it possible for different combinations of a relatively small number of factors to control many genes.

In chapter 5, we introduced you to protein domains, independent regions of tertiary structure that can have specific functions within a protein. We used the example of the lambda repressor protein (see chapter 5) as a protein that had a DNA-binding domain and a protein-protein interaction domain, each of which contributed to the overall function of the repressor. Eukaryotic transcription factors are generally analogous to the lambda repressor. They have a DNA-binding domain and at least one other domain that interacts with the RNA polymerase complex (Figure 10.7).

Many transcription factors work in pairs, with two proteins binding side by side to enhancer DNA to form the active enhancer complex. Experiments have shown that two different transcription factors can work together on a single enhancer, with each one binding to a different, adjacent sequence of DNA bases. This means that in the very simplest case you could regulate 6 different enhancer sequences with three proteins, 10 different enhancer sequences with four proteins, and so on (Figure 10.8). Of course, nature is not the simplest case, and many enhancers interact with multiple proteins.

Enhancers and transcription factors allow tissue-specific gene expression

The activities of genes can be regulated by the availability of the necessary transcription factors and also by the presence of enhancers near a gene. Cells respond to steroid hormones when the hormones bind to intracellular receptors and, in a complex with those receptor proteins, either stimulate or repress transcription. What happens in this case is that the hormone binding to its receptor changes the shape of that receptor protein so that it becomes a

Figure 10.7 Many eukaryotic transcription factors have one domain that binds to an enhancer DNA base sequence and a second domain that interacts with the RNA polymerase complex.

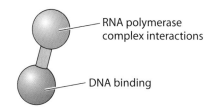

RNA polymerase complex interactions

DNA binding

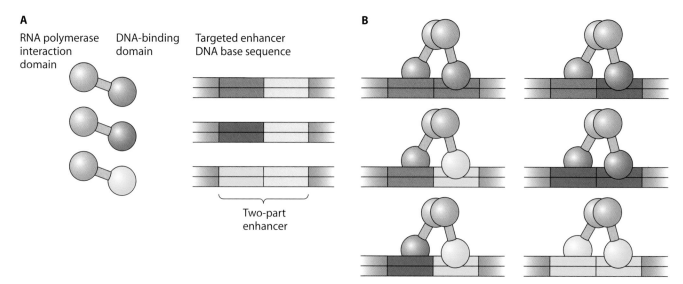

Figure 10.8 Many transcription factors work in pairs, allowing a few factors to activate several different enhancers. **(A)** Three two-domain transcription factors. **(B)** Three factors can activate transcription from a total of six different enhancers. Both enhancer DNA sequence "halves" must be bound by transcription factors for activation to occur.

transcription factor. It binds to specific DNA enhancer sequences and either activates or represses transcription from the associated genes. In this example, the change in gene expression requires the presence of the steroid hormone and its receptor protein.

Another example of gene regulation through enhancers involves the enzyme amylase, which breaks down starch into glucose molecules. You produce amylase in your salivary glands and also in your pancreas, but not in your heart. The reason you produce amylase in your salivary glands is that your amylase genes have an enhancer that binds to proteins produced in the salivary gland. The enhancer-protein complex activates transcription of the associated amylase gene.

Your dog, on the other hand, has amylase genes (how many times have you seen a dog eat starchy foods, like bread?), but your dog does not produce amylase in its saliva. The reason is that your dog's amylase genes lack the salivary-gland-specific enhancer, so no transcription of the amylase gene occurs there and no amylase is produced (Figure 10.9). Your dog's pancreas secretes plenty of amylase, though, because its amylase genes are expressed there.

Enhancers and silencers provide complex on and off switches for genes that respond to the presence of transcription factors made in specific tissues. At one level, the system of enhancers and transcription factors explains how you can have differential expression of genes in different kinds of cells, but it begs the larger question: how did those cells come to make different transcription factors in the first place? In other words, how did the first differentiation event happen?

Searching for how a newly fertilized egg first gives rise to different cell types takes us back to the formation of an embryo. Some of the earliest differentiation events in the development of an embryo are those that define which end of the embryo is "up" or that differentiate one end of an embryo from the other. The past few decades have seen enormous strides in under-

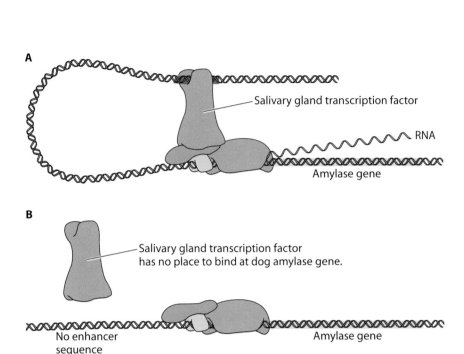

Figure 10.9 Transcription of the amylase gene in the salivary gland is controlled by an enhancer sequence. **(A)** The human amylase gene has a salivary-gland-specific enhancer and is transcribed in the salivary gland. Humans have amylase in their saliva. **(B)** The dog amylase gene does not have a salivary-gland-specific enhancer and is not transcribed there. Dogs do not have amylase in their saliva.

standing how this happens, again with the small, easily manipulated experimental system of *Drosophila* leading the way. We will now summarize some of what is known about the process in *Drosophila* and in mammals.

ESTABLISHING BODY PLAN

A fertilized egg cell divides and divides, growing into an embryo that has a head and a tail. How does a single dividing cell whose progeny are genetically identical give rise to different structures? What determines which end of an embryo will be the head? The tail? The front? The back? In other words, how is the body plan of the embryo established? What are the genes involved in the process, and what do the proteins encoded by body plan genes do?

In the late 1970s, a team of German scientists decided to tackle these questions head-on. To begin their search for body plan genes, they needed mutants in which embryonic development went awry. To find such mutants, they fed adult *Drosophila* flies sugar water laced with a mutagenic chemical. Descendants of these flies showed a variety of defects that eventually allowed the researchers to identify nearly all the genes and proteins that control the overall body plan of the fly.

One type of mutant they discovered produced embryos that had no heads. Instead, the embryos consisted of two tails (abdomens) fused in the middle. They were christened bicaudal (caudal, tailed), and the gene responsible was called bicoid. Another mutant embryo developed a head end but no tail end—it had no abdomen. Since a fruit fly embryo is mostly abdomen, these mutants were very tiny and were christened nanos (from the Greek word for dwarf). Neither bicoid nor nanos embryos could develop into adults.

Maternal genes establish the *Drosophila* body plan and define segments

The discovery of these mutants showed that genes from the mother fly, and not the embryo itself, establish the body plan. Maternal cells surrounding the developing fly egg deposit mRNA into the egg as it is being formed. The bicoid mRNA is trapped near its entry point, establishing an asymmetry in the egg. Upon fertilization, bicoid mRNA is translated into protein that diffuses throughout the embryo, creating a concentration gradient. Meanwhile, from the opposite end of the embryo, the nanos protein, also translated from mRNA deposited by the mother fly's cells, diffuses toward the bicoid end of the embryo, establishing an opposing gradient (Figure 10.10). The bicoid end of the embryo eventually develops into its head, while the nanos end becomes the abdomen.

To test whether the nanos and bicoid proteins alone defined the head and abdomen, the researchers injected bicoid mRNA into the front end of a bicaudal mutant embryo. The injected embryo developed normally. If they injected the mRNA at the "wrong" place in the embryo, the head and thorax developed at the injection site. Similarly, the nanos protein controlled the development of the abdomen. Together, these two proteins establish the **anterior-posterior** (or head-to-tail) **body axis**.

The nanos and bicoid proteins are switches that regulate the expression of other genes according to their concentrations. One such gene, called hunchback, is expressed where bicoid concentration is high and nanos concentration is low. The bicoid protein is required to activate its expression, but nanos acts as a repressor, so hunchback is expressed only in the anterior end of the embryo. The hunchback protein is required for the development of the thorax. Genes turned on in the wake of the bicoid-nanos gradient divide the *Drosophila* embryo into segments from anterior to posterior in the first few hours following fertilization (Figure 10.11).

Bicoid-like genes in vertebrates

Nanos and bicoid were discovered through genetic manipulation of *Drosophila*. Naturally, scientists wondered if similar proteins were present in other organisms. Using sequence information from *Drosophila* as a guide, researchers discovered a homolog of the bicoid protein in the frog and subsequently in chicken, zebrafish, mouse, and human, where they appear to be involved in the development of the head. Understanding the molecular language that specifies the body plan in vertebrates is the goal of much current research.

Figure 10.10 The anterior-posterior body axis in *Drosophila* embryos is established by proteins synthesized from mRNA deposited at opposite ends of the egg by cells in the mother fly.

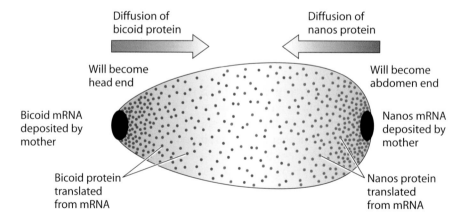

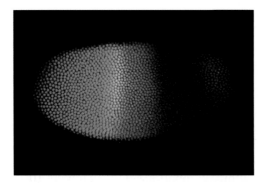

Figure 10.11 Bicoid and nanos proteins regulate the expression of further proteins, such as hunchback (H) and Kruppel (K). These proteins in turn activate others, dividing the embryo into stripes, which become body segments. In this photograph of a *Drosophila* embryo, the hunchback and Kruppel proteins are stained different colors, revealing their locations: green, H; red, K; yellow, both are present. (Photograph courtesy of Jim Langeland, Steve Paddock, and Sean Carroll, Howard Hughes Medical Institute, University of Wisconsin.)

Homeotic genes define the identities of *Drosophila* body segments

The body plan genes of *Drosophila* divide the embryo into segments, but the fates of the individual segments—the structures into which they will develop—are controlled by another family of genes: the **homeotic genes**. As we mentioned above, the first homeotic *Drosophila* mutant, a fly with two pairs of wings instead of one, was discovered in Thomas Hunt Morgan's laboratory near the beginning of the 20th century.

In the 1950s, Ed Lewis at the California Institute of Technology, who studied the descendants of this original mutant plus others discovered since, found that a number of homeotic genes were lined up in a row on one of the *Drosophila* chromosomes. He began to make mutations in these genes with X rays and found he could generate embryos in which the identities of the body segments were transformed: for example, all eight abdominal segments became thorax segments. He named the cluster of genes he found the bithorax complex.

Genes in the bithorax complex controlled the development of the posterior half of the embryo. Amazingly, the genes lay along the chromosome one after another, in the same order as the segments of the fly body they controlled. A second cluster of homeotic genes controlled the development of the anterior half of the embryo. This cluster was named antennapedia (antenna feet), after a gene whose inappropriate expression in the head results in legs where antennae should be (Figure 10.12). The few genes in these two clusters appeared to specify the fate of the embryonic segments.

Homeotic genes in vertebrates

Using the results from *Drosophila* as a guide, researchers looked for similar genes in vertebrates and found them. Not only do vertebrates (including mice and humans) have homeotic genes, theirs are lined up in the same order along the chromosome as the *Drosophila* genes. Furthermore, they appear to control development of the same relative parts of the body: the antennapedia complex controls from the head back to the chest, and the bithorax complex controls from the lower chest to the tail (Figure 10.13). Instead of one bithorax cluster and one antennapedia cluster, however, mice and humans have four copies of each. Researchers are in the process of mutating the mouse

Figure 10.12 Normal (A) and antennapedia (B) fly heads. (Photographs by Rudi Turner, courtesy of Flybase [http://flybase.bio.indiana.edu].)

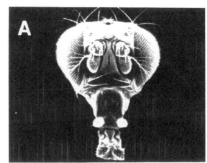

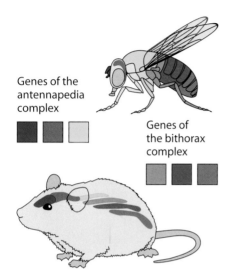

Figure 10.13 Homologous homeotic genes control the development of anterior and posterior body segments in *Drosophila* and the mouse.

homeotic genes via recombinant DNA technology (see "Genetic engineering of animals" in chapter 16) to learn more about the specific functions of the proteins they encode.

How do homeotic proteins determine segment identity? In short, scientists don't really know. That question is one of many at the cutting edge of developmental biology research. Scientists do know that the proteins encoded by homeotic genes are DNA-binding proteins. In fact, the protein products of homeotic genes have similar DNA-binding domains, called the homeodomain. Some of the *Drosophila* homeotic proteins have been shown to be transcription factors, and some have been shown to affect the expression of developmental genes. It makes sense that homeotic genes would specify segment fate by turning different sets of genes on and off, since different cell types are defined by the subsets of genes they express.

EARLY DEVELOPMENT IN MAMMALS

The development of vertebrates proceeds differently from that of *Drosophila*. In general, a fertilized vertebrate egg first undergoes a series of rapid cell divisions to generate a hollow ball of cells called a blastula (not all embryos make a hollow ball, but it's a close approximation). The blastula then undergoes a dramatic rearrangement: **gastrulation**.

In gastrulation, some of the blastula cells migrate through a pore from the outside to the inside of the blastula. Imagine poking part of a balloon inside itself. This cell migration sets up three different **germ layers**: the endoderm, mesoderm, and ectoderm (Figure 10.14). These three layers of cells give rise to different subsets of differentiated cells; thus, their fates are already partially determined. The ectoderm can differentiate into the outer layer of the skin and the nervous tissue; the endoderm will become the inner linings of the digestive organs and circulatory system, and the mesoderm will give rise to everything else, such as muscle, bone, blood, and other internal organs and tissues. Once gastrulation is complete, the cells interact with each other, sending and receiving signals that cause them to further migrate, rearrange, and differentiate into organs and tissues.

By the end of gastrulation, cells have distinct identities as mesoderm, endoderm, and ectoderm and have neighborhoods that may include near neighbors of a type different from themselves. At this point, it is easy to visualize how different cell types could send signals to each other, causing further differentiation. During and after gastrulation, the vertebrate homeotic genes (analogs of the *Drosophila* genes) begin to specify the identities of body segments.

Figure 10.14 Gastrulation, shown as it occurs in the sea urchin.

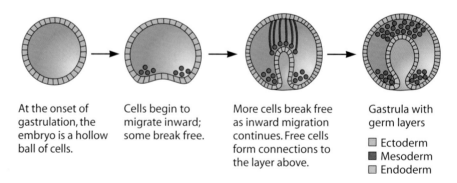

At the onset of gastrulation, the embryo is a hollow ball of cells.

Cells begin to migrate inward; some break free.

More cells break free as inward migration continues. Free cells form connections to the layer above.

Gastrula with germ layers

☐ Ectoderm
■ Mesoderm
☐ Endoderm

Mammalian zygotes give rise to embryonic and extraembryonic tissues

A mammalian offspring develops inside its mother's body, obtaining nutrients directly from her bloodstream through a structure called the placenta, which provides an interface between maternal and fetal circulation. The placenta permits the diffusion of small molecules, such as nutrients and waste products, between the mother's and baby's circulatory systems while keeping the maternal and fetal blood cells separate. The fetal side of the placenta develops from the zygote, while the maternal side develops from the mother. Cells from the zygote also give rise to membranes that surround the fetus during its development. The first differentiation events in mammalian development distinguish what will become embryonic cells from those that will become extraembryonic tissue—placenta and membranes. The following description of early mammalian development is largely derived from studies of mouse embryos, with human examples as noted.

Early cell division and implantation

Fertilized mammalian eggs divide into eight loosely arranged cells (Figure 10.15A) and then undergo rearrangement into a compact ball of cells connected by tight junctions on the outside. Like epithelial tight junctions, they

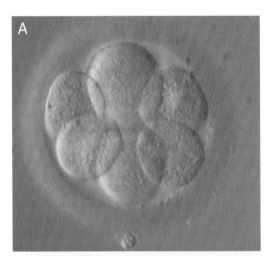
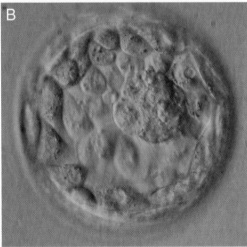

Figure 10.15 Early human embryos. **(A)** An eight-cell human embryo. **(B)** A trophoblast with visible inner cell mass. (Photographs courtesy of Michael Vernon, West Virginia University Center for Reproductive Medicine.)

seal off the inside of the embryo. The 8 cells divide into a 16-cell stage, with a small group of internal cells surrounded by a larger number of external cells.

By the 64-cell stage, the mouse embryo consists of a 13-cell inner cell mass and a separate outer cell layer, the trophoblast. The inner cell mass will give rise to the embryo, its yolk sac, and extraembryonic membranes. The trophoblast cells will form the embryo's portion of the placenta. The trophoblast cells form a fluid-filled ball with the inner cell mass positioned on one side of its interior. This structure (trophoblast and inner cell mass) is called a blastocyst (Figure 10.15B).

At this point, development of the embryo pauses for implantation into the uterus. Trophoblast proteins on the outside of the blastocyst bind to proteins on the uterine lining, allowing the lining and blastocyst to attach to one another. Once in contact with the lining of the uterus, the trophoblast cells secrete proteinase enzymes that digest the extracellular matrix of the lining (which is similar to the matrix underlying the intestinal epithelium), enabling the blastocyst to bury itself in the uterine wall. After implantation, the cells of the inner cell mass undergo gastrulation.

Twins and chimeras

Cells of the inner cell mass of the early embryo are all totipotent; that is, each cell can give rise to a complete embryo. Experiments with animal embryos have shown that the early cell mass can be divided and each portion will give rise to an embryo, even if that portion is a single cell. **Monozygotic** (identical) twins or triplets are formed by the splitting of a single inner cell mass into two or three independent embryos.

Likewise, if cells from the inner cell masses of two embryos are mixed together, a single offspring can be born that contains cells descended from each original embryo. Such offspring are called **chimeras**, after the Greek word for imaginary animals consisting of parts taken from two different ones. Embryo blending can be observed if inner-cell-mass cells from a white mouse embryo and a black mouse embryo are mixed together. The resulting mouse will have patches of black and white fur.

There is evidence that human chimeras can form naturally. Individuals have been observed who have two genetically different cell types within the same body, each with its own genetic traits. The simplest explanation for such an observation is that the uterus originally contained two embryos that fused into one.

ES cells

Experimental evidence thus supports the conclusion that any of the inner-cell-mass cells can differentiate into any adult tissue. These cells can be isolated and grown in culture in such a way that they remain undifferentiated. In this form, they are called **embryonic stem (ES) cells**. A population of ES cells descended from one blastocyst and propagated in culture is called an ES cell line. Mouse ES cells can be genetically manipulated and injected into mouse blastocysts, where they can become part of the developing embryo and ultimately of the adult mouse. The resulting adult mouse is a chimera of the original embryo and the genetically manipulated ES cells. This technology is the basis for making genetically engineered mice (see chapter 16).

ES cells can be induced to develop into different types of cells in culture through the addition of growth factors to the culture fluid. Many researchers and individuals suffering from various diseases hope that someday we will be able to stimulate cultures of stem cells to differentiate into tissues that can be used to replace defective or dead cells within our bodies (Box 10.1). Particularly urgent is the hope that we might be able to correct degenerative neurological disorders, such as Parkinson's or Alzheimer's disease, with an infusion of new, healthy cells. Because the creation of ES cells necessitates the destruction of a blastocyst, however, some people find it ethically unacceptable.

So-called stem cell research is proceeding in animal systems, particularly in mice, where stem cells are readily available. Even when scientists are using human ES cell lines, their research is usually guided by what has been discovered in the models. More information about stem cells and their potential medical uses is presented in chapters 19 and 20.

BOX 10.1   *Stem cell therapy in action*

We already use stem cell therapy in the form of transplants to treat a variety of illnesses involving blood cells: inherited conditions, such as sickle-cell anemia or severe combined immune deficiency, in which a mutation renders the individual unable to make lymphocytes, and cancers of the blood, such as leukemia and lymphoma. The development of all types of blood cells begins in the bone marrow (Figure 10.4). When patients undergo a stem cell transplant, their own marrow is first destroyed, and then they are provided with healthy stem cells from a donor. The stem cells differentiate and generate a population of healthy new blood cells in the patients' body.

Traditionally, donor stem cells were obtained from the bone marrow of donors. Now, stem cells can also be isolated from blood, a procedure that is less painful for the donor. An even newer source of stem cells for transplants is cord blood, placental blood cells that can be harvested when a baby is born and would otherwise be discarded.

Stem cell transplants are also a treatment option in some cancers other than those originating from blood cells. The reason that stem cell transplants can be important in these cancers is a reflection of the crude nature of most of our cancer treatments. Although we have outlined for you a few new approaches (such as Herceptin; see chapter 9) that are based on the biology of specific cancer cells, most cancer treatments are designed to kill cells that are dividing. Cancer is uncontrolled cell division, so the treatments do indeed target cancer cells. Bone marrow stem cells also undergo frequent division as they differentiate to replace worn-out blood cells, and cancer treatments can kill bone marrow stem cells along with cancer cells.

The cure rates for many cancers increase with the amount of chemotherapy given to the patient. Unfortunately, higher doses of chemotherapy destroy more bone marrow cells, too. Bone marrow is the source of red blood cells that transport oxygen, platelets necessary for clotting, and white blood cells for fighting infection, and without sufficient active marrow, a person becomes anemic and immune deficient and ultimately will die. Thus, a dose of anticancer drugs sufficient to cure the cancer might also kill the patient. Bone marrow transplants offer a way out of this dilemma.

To perform a transplant, marrow or stem cells are harvested from a genetically matched donor or, if the marrow of the patient is not diseased, from the patient. The patient is then treated intensively with high doses of chemotherapy and/or radiation that will destroy the bone marrow and (it is hoped) kill the cancer. Afterwards, the healthy marrow or stem cells are infused into the patient's bloodstream. It is expected that the donor stem cells will migrate to the bones, set up housekeeping, and begin repopulating the patient's blood cells.

A bone marrow transplant is a debilitating experience for a patient. The pretransplantation treatment, in which the patient's own bone marrow is destroyed, would kill the patient without the transplant. After the transplantation, the patient may experience nausea, vomiting, bleeding, fatigue, loss of appetite, mouth sores, hair loss, and skin reactions. It takes several weeks for the transplanted marrow to repopulate the patient's body with both red and white blood cells. During that time, the patient is extremely vulnerable to infections because he or she has very little capacity to fight disease. Blood transfusions are necessary because the patient lacks red blood cells to carry oxygen and platelets for clotting. The patient must stay in an isolation room in a hospital for the entire time, and extreme precautions must be taken to prevent infection.

An additional danger from bone marrow transplantation is the possibility that the infection-fighting white blood cells produced by donor marrow could respond to the patient's body cells as if they were foreign invaders and attack them. This is called graft-versus-host (GVH) disease and is very serious. It is not an issue if doctors can use the patient's own marrow, but that is frequently impossible because of the patient's disease. When outside marrow must be used, it is imperative that there be a good genetic match between the donor and the recipient, as a good genetic match decreases the likelihood of a GVH reaction.

Even with the dangers, bone marrow transplants now save thousands of lives each year and can successfully treat once-incurable conditions. The success rate for bone marrow transplants is 50 to 90%, depending on the condition being treated, with success being defined as cure of the underlying disease.

We now conclude this brief look at differentiation and development with an overview of another fascinating process: the differentiation of male and female humans during development. Much of the detailed information was obtained by studying sex differentiation in the mouse, while some of it came from observing natural variations in the process. Note again the recurring themes of gene expression, molecular signals, and receptors.

SEX DIFFERENTIATION

Scientists distinguish two levels of sex determination: primary and secondary. Primary sex determination is the determination of the gonads: whether ovaries or testes form. Primary sex determination can be genetic or environmental. In reptiles, the temperature at which eggs are incubated determines whether the developing offspring will be males or females. In mammals, primary sex determination is chromosomal.

Secondary sex determination applies to the sexual phenotype outside the gonads: male mammals have a penis, seminal vesicles, and a prostate gland. Female mammals have a vagina, a cervix, a uterus, oviducts, and mammary glands. In many mammalian species, each sex also has a specific size range, musculature, and vocal cartilage. Development of both primary and secondary sexual characteristics requires the interplay of many genes and their encoded proteins and is also influenced by environmental factors.

It is beyond the scope of this book to look at variations in patterns of gender and sex differentiation in nature, but we would like to alert you that humans represent just one pattern among many. Even our sex chromosome scheme, human males having two different sex chromosomes and females having two of the same type, is just one scheme. Birds do it the opposite way—males are the same-chromosome gender, while females have two different ones. As stated above, sex differentiation in reptiles is a function of temperature, not chromosomes. Furthermore, many species of lizards reproduce parthenogenetically, without males at all. Quite a few species of fish are able to switch genders as adults. It is very easy to think of the human pattern as "normal" because it is the one we know and experience, but nature holds great variety in terms of gender development and gender roles.

Primary sex determination is a function of the sex chromosomes

Human cells contain 23 pairs of chromosomes, of which 22 pairs are identical in males and females. These 22 pairs of chromosomes are called **autosomes**. The two members of a pair are also called **homologous** chromosomes, or homologs, because they contain the same genes in the same order, though each member of a chromosome pair may contain a slightly different version of any given gene. The 23rd pair is different in males and females. These are the **sex chromosomes**. The 22 pairs of autosomes are numbered from 1 to 22, while the sex chromosomes are named X and Y. A human female's cells contain a pair of X chromosomes, while a human male's cells contain one X and one Y chromosome.

The X and Y chromosomes are very different. The X chromosome is large and contains about 1,500 genes, most of which have nothing to do with gender per se. For example, genes on the X chromosomes encode proteins that have to do with color discrimination in the retina of the eye and proteins that are needed for correct clotting of the blood. Thus, mutations on the X chromosome can result in color blindness and in hemophilia. A human must have at least one X chromosome to survive.

The Y chromosome, in contrast, is quite small and contains far fewer genes than does X (thus far, fewer than 100 genes have been found on Y). A person does not need a Y chromosome to survive, as evidenced by the billions of XX individuals alive today, as well as the individuals born with a single X and no other sex chromosome. What the Y chromosome does is confer maleness upon its bearer. An individual with many normal X chromosomes and one normal Y would still be male, while a person lacking a Y chromosome will develop as a female, although two copies of the X chromosome are required for the complete development of female sexual organs.

As we mentioned in chapter 9, fetal cells from amniotic fluid can be collected and their chromosomes arrested in metaphase, stained, and photographed. The karyotype of the baby reveals the identities of its sex chromosomes. Figure 10.16 shows a sketch of the banding patterns of human chromosomes, including X and Y.

Figure 10.16 Graphic portrayal of human chromosomes showing the 22 autosomes and the X and Y chromosomes. The chromosomes are drawn as X shaped because they are visible at metaphase, when they consist of two chromatids connected at the centromere.

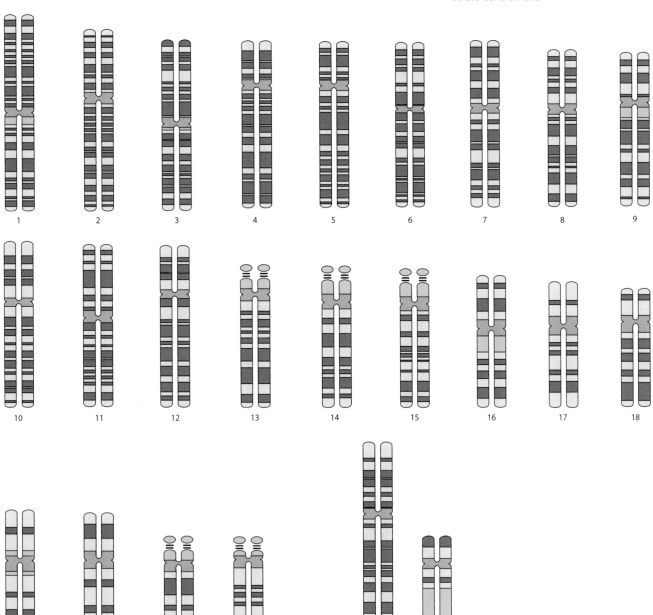

The "maleness gene"

At this point, you might be wondering what gene or genes on the Y chromosome cause a developing baby to be male. Clues to this mystery presented themselves in the form of individuals whose actual sex was different from their chromosomal patterns: XY females and XX males. Some of the XY females were found to be missing parts of the Y chromosome, while some of the XX males were found to have a segment of the Y chromosome grafted into other parts of their genomes. By carefully comparing the regions of the Y chromosome that were missing in the XY females to the regions of the Y chromosome that were present in the XX males, scientists were able to identify a specific portion of the Y chromosome as being responsible for the development of maleness.

In the 1990s, scientists published these and other findings suggesting that they had found the maleness gene. Called the **SRY gene**, for sex-determining region of the Y chromosome, the protein it encodes has a DNA-binding domain and causes DNA to bend when it binds to it. These characteristics are those of a transcription factor. It makes sense that the starting point in a developmental cascade would be a protein that activated additional genes.

Research has supported the notion that SRY is the initial gene acting in a cascade that results in the development of maleness. It is expressed at the right time and place in fetal development. Newly fertilized female mouse zygotes injected with the mouse version of the SRY gene develop penises and testes. Female XY mice lack the mouse SRY protein; male XX mice have it.

Though the preceding paragraph makes it sound like all you need to be male is a functional SRY gene, that is not true. The development of complete male or female sex organs and characteristics requires a number of steps. The molecular mechanisms of sex differentiation are still under active investigation. In the last few years, some light has been shed on the process, and we will now present a brief overview of what is known and look at some of those steps.

Sex differentiation requires the interplay of numerous genes and proteins

Until week 7 of gestation, male and female embryos are indistinguishable in structure. Before sex differentiation begins, a primordial gonadal structure develops that contains the germ cells that will eventually mature into eggs or sperm, along with two ducts called the **Müllerian** and **Wolffian ducts**. This so-called **bipotential gonad** (sometimes called the indifferent gonad) can differentiate into either ovaries or testes, depending on what proteins are expressed in the tissue. In male sexual development, the gonads become testes; the Wolffian duct becomes the epididymis, vas deferens, and seminal tubules; the Müllerian duct degenerates; and the germ cells eventually become sperm. In female development, the gonads become ovaries; the Müllerian duct becomes the oviducts, uterus, and cervix; the Wolffian duct degenerates; and the germ cells become eggs (Figure 10.17).

In chromosomally male embryos, SRY expression can be detected in the gonadal region around week 7 of development. SRY is thought to activate or work with additional gene products to stimulate the gonads to become testes. Two types of cells within the testes then begin to secrete hormones: anti-Müllerian hormone (AMH; also called Müllerian-inhibiting substance) and testosterone. AMH inhibits the growth of the Müllerian ducts (precursors to

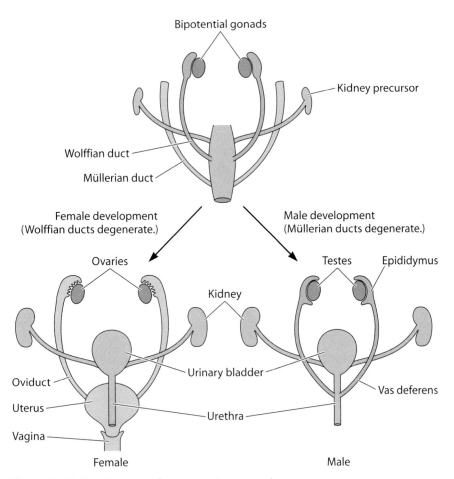

Figure 10.17 Development of sex organs in mammals.

the uterus and oviducts). The Müllerian ducts eventually degenerate, presumably through apoptosis. Testosterone stimulates the Wolffian ducts to develop into the epididymis, vas deferens, and seminal vesicles and causes the external genital structures to grow and fuse so that they become a penis and scrotum.

In the absence of the SRY gene, the bipotential gonads develop into ovaries. This process is thought to require the presence of ovary-determining proteins, but their identities are not clear at this time. Estrogen secreted by the fetal ovaries stimulates the Müllerian duct to develop into the uterus, oviducts, and cervix. Without testes to produce testosterone, the Wolffian ducts degenerate (presumably by apoptosis), and the external genital structures develop into a clitoris, vaginal opening, and labia (Figure 10.18).

Two copies of the X chromosome are needed for complete female sexual development. Babies born with a single X chromosome and no Y develop female external genitalia, but their internal sexual development is incomplete and they are sterile. With no ovaries, they do not produce the surge of estrogen that triggers the development of female secondary sex characteristics at puberty. Since the X chromosome contains many other genes, having just a single one and no Y results in a few other phenotypes, none severe, which together are called Turner's syndrome.

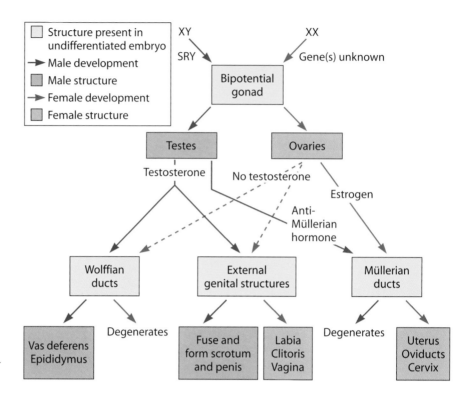

Figure 10.18 The sex differentiation cascade.

Estrogen is required for normal development of males and females

As described in chapter 3, the sex hormones are steroid hormones and are very similar in their chemical structures. In the early 20th century, it was firmly believed that there was a male hormone (testosterone) and a female hormone (estrogen) and never the twain should meet. Scientific reports of significant levels of female hormones in men and stallions were greeted with disquiet. Eventually, however, so much evidence accumulated that the notion of a strict gender-based hormone dichotomy had to be abandoned. It is now known that both sexes synthesize and respond to both estrogen and testosterone.

The female hormone estradiol is responsible for the growth spurts of boys and girls at puberty. Boys' testosterone is converted to estradiol in the bones, where the growth occurs. In females, testosterone is made in the adrenal glands of the kidney and in the ovaries. In female rats, testosterone stimulates the growth of mammary glands, the uterus, and the clitoris.

Female reproductive organs develop normally in the absence of androgens (male hormones), but male reproductive development requires estrogen. Estrogen is secreted by the adrenal glands in both males and females, though the amounts secreted by the ovaries are much higher. Estrogen is required for the complete development of the Wolffian ducts, and in adult males, for fertility. The fertility function of estrogen involves control of water resorption during semen formation.

When semen first forms, it is dilute. In the tubules within the testis, water is resorbed, concentrating the sperm and giving them a longer life span. This resorption of water is controlled by estrogen. If estrogen or its receptor is absent in mice, the water is not resorbed and the mouse is sterile. While concentrations of estrogen in the blood are higher in females than in males, the concentration of estrogen in this part of the testis is even higher than in female blood.

Many variations can occur during sex differentiation

As you can see from the discussion above, sex differentiation involves the co-operation of quite a number of genes and their proteins. Not surprisingly, the program doesn't always proceed in the same way. We present just a few examples of variations in the sex differentiation pathway that result from genetic variations. Although we touch on it only as a consequence of a genetic condition (see the discussion of **congenital adrenal hyperplasia** [CAH] below), an altered uterine environment, whether caused by maternal hormone imbalances or environmental substances, can also affect sex differentiation and other aspects of development.

Androgen insensitivity: XY females

The presence of the SRY gene does not guarantee what we consider normal male sexual development, nor does its absence guarantee complete female development. As should be apparent from the discussion above, the presence of many additional genes is required for the development of males or females.

Testosterone is absolutely required for male secondary sexual differentiation, and its activity depends on its binding to its receptor. The gene for the testosterone receptor is, interestingly enough, on the X chromosome. Some XY individuals have mutations in the gene for the testosterone receptor and therefore cannot respond to testosterone. This condition is called **androgen** (male hormone) **insensitivity**.

During week 7 of these individuals' lives, the SRY gene product initiates testis formation. The developing testes produce AMH, which leads to the destruction of the Müllerian ducts, and testosterone. Since the embryo's cells have no testosterone receptors, they cannot respond to the hormone. The Wolffian ducts behave as if no testosterone were present and eventually degenerate. The external genitalia, too, behave as if no testosterone were present and form a clitoris, vagina, and labia.

At birth, these individuals appear female. Inside their bodies are fully functional testes, but they cannot respond to the testosterone their testes make. Their adrenal glands produce estrogen, and so at puberty they develop further female secondary sexual characteristics. They develop as completely normal women, except that they are sterile and do not menstruate. Instead of ovaries and a uterus (these would have developed from the Müllerian duct, which was inhibited by their AMH), they have testes in their abdomens. These XY women and their families typically have no idea that anything is unusual about their sexual development until they fail to start menstruating.

DHT deficiency

In the brief description of male sex differentiation above, we stated that the developing testis produces testosterone, which stimulates the development of the Wolffian ducts and promotes the development of external male genitalia. Actually, a modified form of testosterone, **5α-dihydrotestosterone** (DHT), is the androgen that masculinizes the external genitalia. Testosterone is converted into DHT in the fetal external genitalia but not in the Wolffian duct.

Some individuals carry mutations in the gene for the testosterone-converting enzyme, which is located on chromosome 2. In males with inactive genes for the enzyme, male external genitalia do not form during fetal development. Although the male infants have testes, they are born with a blind vaginal pouch and a clitoris and appear to be female. At puberty, the higher concentrations of testosterone poured into their bloodstreams from their testes are sufficient to stimulate the external genitalia to masculinize. The penis enlarges, the scrotum descends, and the child becomes a young man. This condition is common in a certain population in the Caribbean and has been observed in other locales as well.

In adult men, DHT is thought to promote enlargement of the prostate and male-pattern baldness. The drug finasteride blocks the action of the enzyme that converts testosterone to DHT and is being sold under two different names at two different dosages to combat male-pattern baldness (Propecia) and to shrink enlarged prostates that are not cancerous (Proscar). Because finasteride does not interfere with testosterone or its receptor, it does not affect male sexual function.

CAH

Sometimes a developing fetus lacks a functional gene for an enzyme needed to produce the steroid hormone cortisol. Cortisol is normally synthesized from the same steroid precursor as are androgens. If the cortisol-synthesizing enzyme is missing, the body's feedback loops sense the lack of cortisol. The adrenal gland responds to the signaling by pumping out more hormone precursors, but since the precursors cannot be converted to cortisol, the end result is overproduction of testosterone and other androgens. This condition is called CAH. If the developing fetus is female, the testosterone from her adrenal glands can affect the development of her external genitalia, enlarging the clitoris and sometimes causing partial fusion of the labia into a more scrotum-like structure. At birth the genitalia of these babies may be ambiguous in appearance or they may resemble those of boys more than girls.

CAH can be diagnosed by measuring hormone levels, and the baby's XX genotype can be confirmed with a karyotype. The appearance of her genitals can be surgically altered to make them more typically female (though this is not always necessary). Cortisol is needed by the body, and CAH is treated by the administration of hormones. When cortisol levels are normalized by treatment, the adrenal glands cease pumping out excess testosterone.

Sex hormones appear to influence gender identity

Development of gender identity, the internal identification of yourself as male or female, appears to occur later in development and separately from the development of external genitalia. The formation of external genitalia is influenced by the concentration of testosterone and DHT, mostly during the middle trimester of human prenatal development. At this time in development, there are no sex hormone receptors in the brain of the developing fetus.

To the best of current understanding, the key player in the development of sexual identity appears to be testosterone. Males experience three especially intense pulses of testosterone during their development, one during the middle trimester associated with external-genital formation, one during the third trimester of development, and one during puberty. The third-trimester pulse appears to be associated with the development of male gender identity. By this time, there are hormone receptors in the brain. Male gender identity is dependent on testosterone and not DHT, and the DHT-deficient children discussed above have a clear sense of themselves as male despite the lack of typical male genitalia at birth.

Conversely, the XY androgen-insensitive females have a clear sense of themselves as female. Though their bodies produce testosterone, they have no receptors for it and do not respond to it. CAH girls also have clear female identity, despite the appearance of their genitalia at birth. Although their adrenal glands produce excess testosterone, the amount is not nearly as high as that produced by the testes of third-trimester males, and the signal is apparently not strong enough to convince the brain that it is male.

Sex differentiation, genes, and traits

Sex differentiation is an example of a developmental cascade from start to finish. At the beginning, a few genes encoding transcription factors (such as SRY) activate a variety of other genes that stimulate morphogenesis and additional gene expression. Some of those genes encode hormone-synthesizing enzymes that produce testosterone, DHT, and estrogen. The hormones bind to their receptors and exert effects in their target cells. For everything to work, all the proteins have to be present: transcription factors, enzymes, hormone receptors, and other protein targets. For these proteins to be functional, their genes must be intact.

Sex differentiation illustrates important points you will encounter in the next chapter about genes in general. For example, genes do not act in isolation: an XY individual with a functioning SRY gene can still develop as a female. As you go into the next chapter, remember that a gene is a set of instructions for making a protein and that the ability of that protein to exert its effect may depend on many circumstances beyond its own genetic code.

SUMMARY POINTS

Much of our understanding of both differentiation and development comes from the study of model organisms. Studies of the fruit fly and the nematode, because they are small enough to be housed by the multiple thousands and because they can be easily mutagenized, led to the isolation of developmental mutants and thus to the identification of many genes important in development. Now we can take that knowledge and find and mutate those specific genes in vertebrates (including the mammalian model, the mouse) to determine their roles in those organisms. Studies of human development are mostly limited to observing what nature gives us and experimenting with ES cells.

Development is the process by which an adult is generated from a zygote. Development involves differentiation, the process through which cells become specialized in form and function, and morphogenesis, the combination of cell migration, proliferation, differentiation, and death that produces the final shape of the body.

Cells of an early mammalian embryo are said to be totipotent because they can differentiate into all types of adult cells. As the embryo develops, the totipotent embryonic cells give rise to cells with more and more restricted possibilities for their ultimate fates, until most of the body is composed of fully (or terminally) differentiated cells, which usually do not reproduce. A small number of pluripotent stem cells remain in the adult body to give rise to replacement cells in some tissues.

Because stem cells can give rise to replacement cells, many hope that research on these cells will lead to therapies for many degenerative illnesses. Stem cell replacement is already used in the form of bone marrow transplants to repopulate a patient's body with blood cells.

The various differentiated cells within an organism have the same genetic content, but they express different sets of genes and thus make different specialized protein products.

Differential gene expression is usually controlled at the level of transcription. Various protein factors bind to sites along the DNA called enhancers or silencers, resulting in the activation or repression of associated genes. Multiple proteins bind to these DNA sites, and they cause the DNA to fold over so that the bound proteins can interact with proteins at the promoter.

During development, cells signal to one another via proteins or other chemicals. These signals bind to receptors on target cells and elicit changes in the targets. The changes can be the turning on or off of genes, movement, proliferation, or even death.

Differentiation begins in a *Drosophila* embryo because the maternal cells surrounding the egg deposited mRNA and proteins at specific places in the egg. After fertilization, proteins translated from these maternal mRNAs diffuse through the embryo, creating gradients. These gradients set up differences within the embryo that signal the expression of different genes in different places, establishing the body axes and dividing the embryo into segments. Later, the differential expression of a suite of homeotic genes specifies the identities of the segments.

Homeotic genes very similar to the *Drosophila* genes are present in vertebrates (including humans) and appear to control body segment identity in these systems. Although we do not yet know exactly how the homeotic-gene products work, they are DNA-binding proteins that probably control the expression of many genes.

A key event in early vertebrate development is gastrulation, in which some cells of the early embryo (often a hollow ball of cells at this point) dive toward the interior of the embryo. This sets up the different germ layers. Germ layer cells are already restricted in their potential fates: ectoderm cells can become neurons or epidermis, endoderm cells become the inner linings of the digestive and circulatory systems, and mesoderm cells become essentially everything else.

In twinning, the inner cell mass is divided and gives rise to two embryos. Even a single cell from the inner cell mass of an early embryo can develop into a complete offspring. If cells from the inner cell masses of two different embryos are mixed together early in development, a single chimeric offspring made of a mixture of descendants of the two types of cells can result. Human chimeras have been observed, suggesting that this process sometimes happens naturally.

Usually, an XY genotype results in an individual being male while an XX genotype results in an individual being female. Sex differentiation in humans begins with the expression of the SRY gene on the Y chromosome. The cascade started by the SRY gene causes the bipotential embryonic gonads to differentiate into testes and begin producing two hormones: testosterone and anti-Müllerian hormone. Testosterone causes the Wolffian glands to develop into male sexual structures, such as the epididymus and vas deferens. AMH causes the Müllerian ducts to degenerate (probably through programmed cell death). A derivative of testosterone, DHT, induces the primitive external genital structure to form a penis and scrotum.

In the absence of the SRY gene and its cascade, a different genetic pathway induces the bipotential gonad to differentiate into ovaries. In the absence of testosterone, the Müllerian duct forms female sexual structures, such as the oviducts, uterus, and cervix, and the Wolffian ducts degenerate. Two X chromosomes are required for the complete development of these internal structures. Without DHT, the primitive external genital structure forms a vaginal opening and other female structures.

The formation of sexual identity appears to be biologically separate from the formation of genitals, though both are influenced by the presence of testosterone. Sexual identity seems to be influenced by a pulse of testosterone in the third trimester of development.

Many variations in gender development occur because of variations in genes involved in the sex differentiation pathway. One example is an individual who lacks a functional testosterone receptor because of mutations in its gene. An XY individual with this mutation forms testes and loses the Müllerian ducts, but because the body cannot respond to testosterone or DHT, the individual develops female external genitalia and secondary sexual characteristics. CAH results in elevated testosterone concentrations in utero and consequent masculinization of the genitals of XX fetuses.

K E Y   T E R M S

5α-dihydrotestosterone	Differentiation	Model system	Terminally differentiated
Androgen insensitivity	Embryonic stem cell	Monozygotic twins	Totipotent
Anterior-posterior body axis	Enhancer	Morphogenesis	Wolffian duct
Autosomes	Gastrulation	Müllerian duct	Zygote
Bipotential gonad	Germ layers	Pluripotent	
Chimera	Homeotic gene	Sex chromosomes	
Congenital adrenal hyperplasia	Homeotic mutation	Silencer	
	Homologous chromosomes	SRY gene	
		Stem cell	

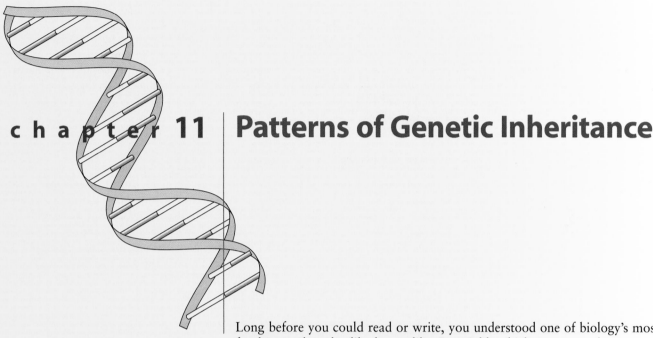

chapter 11 | Patterns of Genetic Inheritance

Long before you could read or write, you understood one of biology's most fundamental truths: like begets like. Invariably, chicken eggs crack open and out pop baby chicks, not ducklings; dogs give birth to puppies, not kittens. The perfect predictability of like begetting like tends to lull all of us into complacency about this truly astonishing bit of biology. It's important not to confuse the familiar with the ordinary, for the regularity of like begetting like is awe inspiring.

Heredity's mysteries have perplexed humans for millennia. Only in the past 200 years have scientists begun uncovering various pieces of the biological processes underlying the continuity of life. In multicellular organisms, the concept of "begetting" encompasses two complex processes: reproduction and the development of one cell into an organism. The begotten organism is "like" its begetter thanks to an invisible hereditary mechanism that ensures that parents transmit traits to their offspring. To really grasp the truism "like begets like," scientists had to discover at least some of the details of all three processes, none of which they could observe directly.

Heredity is the transmission of traits from parents to offspring. Scientists now know that traits pass from parents to offspring because genes encoding these traits are transmitted. In this chapter, we explain basic aspects of heredity, such as:

- the physical basis of heredity
- the mechanics of transmission
- the basic laws governing genetic inheritance

As we mentioned in chapter 1, our goals for this text are twofold: providing scientific facts and familiarizing you with the nature of science. The newness of many biological discoveries described in this text foreshortens our historical view of them, making it difficult to get perspective on the twists, turns, and blind alleys in the scientific maze that ultimately led to their discovery.

Virtually all of the essential pieces of the heredity puzzle locked into place a century ago. We describe some of the experiments and thought processes that uncovered the mysteries of heredity so you can see how science progresses.

The star of this chapter is Gregor Mendel, the father of genetics. We introduced you to Mendel briefly in chapter 1, though we suspect you had already met him in high school. In this chapter, we explain his experimental design and results in more detail and also describe how he organized these results, or observations, into a set of generalizations about heredity. When scientists unearthed his work in 1900, they tested and retested the generalizations he proposed in 1865. His generalizations held up to scrutiny so well that they are now known as Mendel's Laws of Heredity or Mendel's Theory of Inheritance.

EARLY MODELS OF HEREDITY

When you first learned about Mendel's conclusions, you probably were not particularly impressed. They seemed so simple and obvious, didn't they? However, you may have noticed that when biologists talk about Mendel, their tone borders on reverential. Why are they awestruck by something so obvious? Here's why. When Mendel analyzed his experimental results and interpreted them for the scientific community in 1865, his conclusions were revolutionary. Even the greatest scientists of that time could not grasp his ideas, which are the same concepts that are so transparently obvious to you today.

Reproduction, heredity, and development are intertwined

To truly appreciate Mendel's brilliance, try to put yourself in the mindset of the time. Forget what you know about reproduction, embryonic development, and genetic inheritance. Scientists had tried to explain heredity for centuries but were stumped. A valid explanation of heredity had to encompass an incalculable, but remarkably consistent, number of observations that demonstrated the following facts.

- Offspring resemble both parents but often have traits found in neither parent.
- A trait found in both parents may not be seen in their offspring.
- Parents produce variable offspring that nonetheless share many features among themselves.

In other words, scientists had to account for both the similarity and variation between parents and their offspring and among offspring.

In the 1800s, scientists were still grappling with the most basic questions that can be asked about the biological processes responsible for the observation that like begets like: reproduction, development, and heredity. The idea that reproduction involved the fusion of two cells into one had only recently begun to find acceptance in the scientific community. Until the development of microscopes powerful enough to permit visualization of sperms and eggs, most people believed human and animal offspring were coagulation products of semen mixed with menstrual blood.

Once scientists accepted that gametes, not fluids, were involved in reproduction, they also assumed that gametes contained the hereditary material, because gametes were the sole link between generations. However, they attributed the faithfulness with which like begets like to the presence of a fully formed, but miniscule, replica of a parent in its gametes (Figure 11.1). Development of a fertilized cell into a complete organism, it was thought,

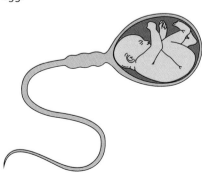

Figure 11.1 Preformationist view of heredity and reproduction. The preformationists believed a tiny replica of a person was contained in either the sperm or the egg.

consisted of the preformed individual simply expanding, like the tiny sponge replicas of dinosaurs that grow 500 times larger when placed in water.

According to this model of reproduction and inheritance, appropriately known as preformation, only one parent could contribute hereditary material to an offspring. Clinging to this view meant ignoring centuries of observations of offspring inheriting characteristics of both parents and not being carbon copies of either. Perhaps scientists supporting this model blinded themselves to these thousands of exceptions because preformationism explained, quite tidily, all heredity phenomena in need of explanation: reproduction, development, and trait inheritance. In the early 19th century, the scientific community finally rejected the preformationist model of heredity when plant breeders clearly and repeatedly demonstrated that both parents contribute hereditary material to their offspring. But what was the something that was being contributed?

Scientists favored the fluid-blending model of inheritance

When Mendel conducted his research in the mid-1800s, most scientists accepted the fluid-blending model of inheritance, which we explained in chapter 1. Unlike preformation, a fluid-blending mode of heredity made sense of innumerable observations of offspring inheriting traits from both parents while simultaneously differing in some ways from both. Even now, that model is intuitively satisfying, because the physical appearance of an offspring often blends parental characteristics and is an intermediate of both, much like red and white paints blending to become pink.

However, as you recall from the chocolate milk analogy in chapter 1, fluid blending is incompatible with centuries of observations of inherited variation (Figure 11.2). In the fluid-blending model, any variation among offspring is quickly diluted, and in a few generations the trait is uniformly expressed in those having it. In addition, a parental trait that disappears in one generation can never appear again in its pure form. Faced with the seemingly contradictory observations of inherited similarity and inherited variation, most scientists took the easy way out and separated the two. They proposed that similarity was inherited but that the variation they observed was due solely to *current* environmental factors that affected the organism's appearance. For example, they knew that soil quality affects plant appearance and yields.

Figure 11.2 Graphic representation of two different models of inheritance: fluid blending and discrete particle.

Fluid blending   Discrete particle

Charles Darwin proposed an invalid model of heredity

Biologists also speak about Charles Darwin reverentially, but as you learned in chapter 1, even the most brilliant scientists are sometimes wrong. Darwin proposed a plausible model of inheritance that explained a huge amount of information but that was nevertheless inaccurate. Like virtually all scientists of his day, Darwin accepted the fluid-blending mechanism of inheritance, even though he recognized that it posed a serious problem for his model of biological evolution. Darwin's explanation of evolution relied heavily on the existence of a hereditary mechanism that cranked out a continual supply of variable offspring, which is precluded in fluid-blending inheritance.

Darwin spent years gathering information on trait inheritance from plant and animal breeding records, reading various theories of heredity, and reviewing his own voluminous natural history observations. The amount of information he amassed was so large that publishing it required two large volumes. He integrated these observations and proposed a theory of inheritance that owed much to the 2,000-year-old writings of Hippocrates and Aristotle. According to Darwin's model, every cell type produced seeds—eye seeds, liver seeds, muscle seeds, and so on. Seeds shed by the cells found their way via the circulatory system to the male and female reproductive organs and ultimately entered the sperm and eggs located there.

This model explained similarity, but how did Darwin account for variation? He could not throw up his hands and claim that all variation was caused by an organism's current environment, because his model of evolution required *inherited* variation. He proposed that the organs giving rise to the organ seeds were altered by environmental conditions and that the seeds carried those new variations to the gametes so that they were inherited. This model of heredity, known as inheritance of acquired characteristics, was rejected by most biologists then and virtually all biologists now.

MENDELIAN GENETICS

To develop a theory of heredity, Mendel, like Darwin, gathered together thousands of observations into a series of generalizations, but the similarity between their approaches ends there. The sources and natures of their observations differed, as did their research methodologies.

Most scientists studying heredity at that time based their ideas on observations of the natural world. They naturally assumed that the more observations they amassed, the more accurate the model would be. It's hard to argue with that way of thinking, when you think of science as a process of connecting the dots. However, using all observable outcomes of the hereditary process to construct a model created more confusion than clarity, because as scientists now know, phenotypic manifestations of genotypes often mislead (see Figure 11.8 for definitions and graphic presentation of some basic terms in genetics).

Mendel's approach was diametrically opposed to the Darwinian method of gathering as much data on as many organisms as possible. Rather than attempting to amalgamate all observations of all traits in all species into a coherent model, he picked the simplest system he could find and methodically focused on one trait at a time. Mendel had studied and taught physics, and the experimental approach to research had infiltrated physics and chemistry, but not biology. Mendel felt that heredity had remained a conundrum because the approach scientists had used to unlock its mysteries had been so slipshod. In his own words,

> Among all the [previous] experiments, not one has been carried out to such an extent and in such a way as to make it possible to determine the number of different forms under which the offspring of hybrids appear, or to arrange these forms with certainty according to their separate generations, or definitely to ascertain their statistical relationships.

In contrast, Mendel's approach to understanding heredity was methodical, systematic, and mathematical. You may want to refresh your memory by rereading Box 1.1, "Mendel's methods and observations." Recall that Mendel selected the familiar garden pea, *Pisum sativum*, as his research organism for both pragmatic and scientific reasons. He also made a number of decisions about his experimental design that seem obvious now but were not common practice then.

- The flower structure of the pea allowed Mendel to use hand pollination to control the crosses (Figure 11.3).
- He focused on a single trait at a time, rather than the organism's overall appearance. This approach, atypical then but routine now, is analogous to the standard practice of using a control when conducting an experiment or controlling all of the variables that may contribute to a result. Controls help researchers determine the role a single variable plays in the phenomenon they are investigating.
- He counted the results, which was possible because he intentionally selected characteristics that were clearly distinguishable from each other, or discontinuous. Thus, he could easily assign a result to one of two categories (Figure 11.4).
- He studied seven different traits; this allowed him to test his predictions seven times in one growing season.

Figure 11.3 Hand pollination of pea flowers. In all flowers, fusion of male and female gametes occurs when mature pollen, produced by the male reproductive organ (anther), lands on the stigma and makes its way down to the eggs. Fertilization leads to fruit and seed formation. **(A)** In peas, the male and female organs are encased in a special structure, the keel, and pea flowers usually self-pollinate before they open. **(B)** Keel cut away to show the reproductive organs. **(C)** Mendel prevented self-pollination by prying open the keel and removing a flower's anthers before its pollen had matured. He cross-pollinated two plants by placing mature pollen from another plant (white flower) on the stigma of the flower lacking anthers. **(D)** He saved the seeds produced by the cross, planted them the next year, and recorded the results—all offspring had purple flowers. For all seven traits, Mendel also did the reciprocal cross (purple flower as male, white as female) and got the same results.

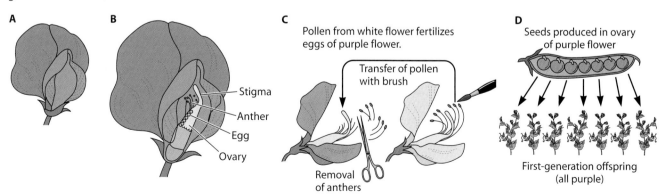

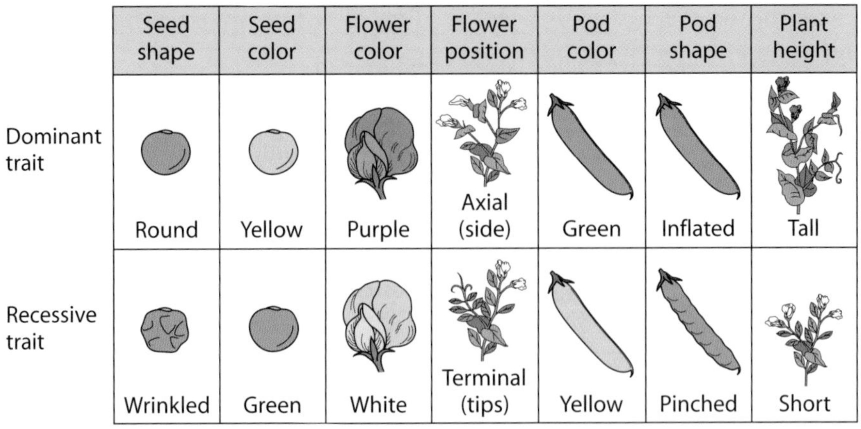

	Seed shape	Seed color	Flower color	Flower position	Pod color	Pod shape	Plant height
Dominant trait	Round	Yellow	Purple	Axial (side)	Green	Inflated	Tall
Recessive trait	Wrinkled	Green	White	Terminal (tips)	Yellow	Pinched	Short

Figure 11.4 Inheritance in peas. Mendel's experiments provided data on seven traits that occurred in two very distinct forms.

In addition, the whole concept of experimental design was relatively new to the biological sciences. Notice that in the description of Charles Darwin's process for developing a model of inheritance the word "experiment" never appears.

Mendel analyzed his results mathematically

For each of the seven traits he studied, Mendel always began a series of experiments by crossing two different **purebred lines** for the trait of interest. A purebred line is one that always produces the same results no matter for how many generations they are crossed with each other.

In every instance, the first-generation offspring always looked identical to each other and to only one parent for each trait. For example, when he crossed plants with purple flowers and plants with white flowers, all of the offspring were purple. Crosses between plants with yellow seeds and plants with green seeds produced only offspring with yellow seeds; those between plants with round seeds and plants with wrinkled seeds produced only offspring with round seeds.

The crosses that shed the most light on the mechanism of heredity were the subsequent crosses (Figure 11.5). Mendel allowed the first-generation offspring to self-pollinate, and surprisingly, offspring in the second generation had characteristics found in both of the original parents: white and purple flowers, green and yellow seeds, and round and wrinkled seeds. Characteristics that disappeared in the first generation reappeared in the second for all seven traits. Clearly, these results were not consistent with the fluid-blending view of inheritance, because the first-generation offspring retained the potential to produce either parental type even though they displayed only one of those types. Mendel described the relationship between the hereditary contributions made by two parents as dominant and recessive rather than blended.

Letting the second generation of pea plants self-pollinate produced another revealing set of results. Plants displaying the recessive characteristic bred true, producing only plants with the recessive phenotype, while those displaying the dominant trait produced both types of plants. For example, the offspring of white-flowered plants always had white flowers, but self-pollination of the purple-flowered plants produced both white-flowered and purple-flowered plants.

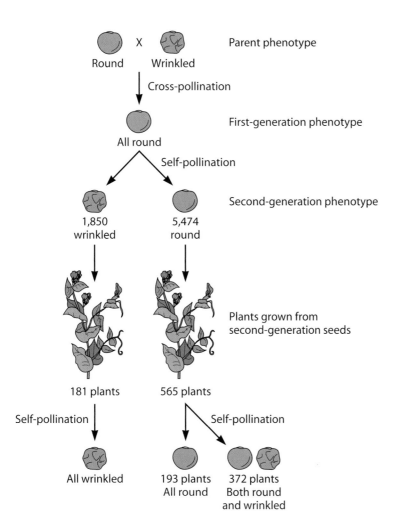

Figure 11.5 Mendel's experimental results. Mendel obtained similar results for each of the seven traits he studied. We use seed shape in the example below. Crossing purebred parents with different phenotypes led to the production of offspring with only one phenotype, which for the seed shape trait was round. Mendel described the phenotype that disappeared as recessive and the phenotype displayed by all first-generation offspring as dominant. Plants grown from the first-generation all-round offspring self-pollinated, and their seeds displayed both parental types, round and wrinkled, in a 3:1 ratio. When the second generation self-pollinated, all plants with wrinkled seeds produced only wrinkled seeds, and one-third of the plants grown from round seeds had only round seeds. Two-thirds of those grown from round seeds had both round and wrinkled seeds. Within their pods, the seeds occurred in a 3:1 ratio, round to wrinkled.

Mendel used his experimental results to propose a model of heredity

The reappearance of traits in the second generation always occurred in a 3:1 ratio for each of the seven traits (Table 11.1). Three-fourths of the plants displayed the dominant form of a trait, and one-fourth had the recessive characteristic. Mendel recognized that such consistent results could not be explained by chance alone, nor could his observation that self-pollination by the second-generation plants always produced consistent results: all plants with the recessive characteristic bred true, as did one-third of the plants that

Table 11.1 Mendel's results

Parental types	First generation	Second generation
Round × wrinkled seeds	All round	5,474 round; 1,850 wrinkled (2.96:1)
Yellow × green seeds	All yellow	6,022 yellow; 2,001 green (3.01:1)
Inflated × pinched pods	All inflated	882 inflated; 299 pinched (2.95:1)
Green × yellow pods	All green	428 green; 152 yellow (2.82:1)
Tall × short stems	All tall	787 tall; 277 short (2.84:1)
Purple × white flowers	All purple	705 purple; 224 white (3.15:1)
Axial × terminal flowers	All axial	651 axial; 207 terminal (3.14:1)

displayed the dominant phenotype. However, two-thirds of the plants with the dominant characteristic produced plants with both traits in a 3:1 dominant-to-recessive ratio. For example, Figure 11.5 shows that when Mendel allowed 565 second-generation plants with round seeds to self-fertilize, 193 bred true and produced only round seeds; 372 produced round and wrinkled seeds in a 3:1 ratio.

To explain his observations, Mendel constructed a number of generalizations and tested them experimentally.

- Each offspring plant inherits two "factors" for each trait from its parents.
- Each parent passes one factor for each trait to an offspring; therefore, each parental gamete contains only one factor for each trait.
- The members of a factor pair are separated from each other in gamete formation.
- When parental gametes fuse in fertilization, the factors do not fuse but retain their distinct identities. Therefore, the first-generation offspring had both a dominant and a recessive factor for each of the seven traits, even though all exhibited the dominant trait (Figure 11.6).

Figure 11. 6 Mendel's Law of Segregation. Mendel's experimental results led him to the conclusion that the hereditary material is packaged as discrete particles, or factors, that occur in pairs for each trait (e.g., RR and rr). Each gamete contains only one member of the pair. The gametes fuse in fertilization, and the resulting offspring once again has a pair of factors for each trait. The best way to visualize Mendel's conclusions is with a matrix, or Punnett square, using a letter to represent a gene for a trait. The letter chosen to signify a trait is usually the first letter of the dominant trait. Uppercase signifies a dominant allele, while the recessive allele is represented by the same letter in lowercase. In the case shown, R refers to round seed shape and r refers to wrinkled seed shape. **(Top)** Production of first generation of 100% round seeds. **(Bottom)** Production of second generation in 3:1 ratio.

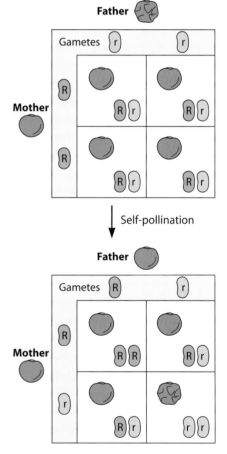

Phenotypic ratio – 3:1 or 5,474(round):1,850(wrinkled)

Genotypic ratio – 1:2:1 or 1,812(RR):3,662(Rr):1,850(rr)

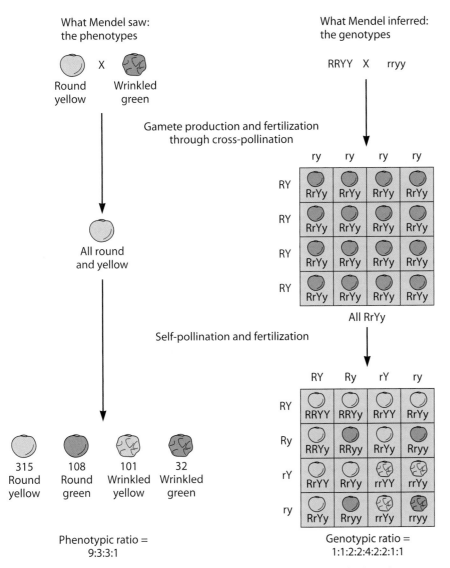

Figure 11.7 Mendel's Law of Independent Assortment. By observing the hereditary patterns of two different traits simultaneously, such as seed shape and color, Mendel inferred that during gamete formation the maternal and paternal genes for one trait, such as seed shape, segregate independently of the maternal and paternal genes for a different trait, such as seed color. Subsequent research by others showed that this law applies only to genes on different chromosomes or genes located very far from each other on the same chromosome.

Mendel conducted additional experiments to determine the relationship of factors for *different* traits to one another. In other words, are different characteristics, such as purple flowers and round seeds, inherited as a unit? His results indicated that the behavior of one factor pair has no effect on other factor pairs, or that when the members of a factor pair separate from each other during gamete formation, they do so independently (Figure 11.7).

Mendel's conclusions were right on target. They have been refined and slightly revised over the years as scientists collected more observations, and today's geneticists use different words, but Mendel's core principles remain intact. His research deduced the fundamental laws of heredity that had eluded researchers for centuries. Mendel's conclusions are summarized in Table 11.2.

Table 11.2 Mendel's conclusions

Inherited traits are determined by discrete factors that do not blend together like paints when gametes fuse in fertilization.

Factors occur in pairs.

One factor of the pair is inherited from the mother, the other from the father.

The factors that constitute a pair separate from each other during gamete formation, so each gamete contains only one factor from each pair. (Principle of Segregation)

The factors from different pairs are distributed randomly among the gametes. In other words, when the factors separate from each other during gamete formation, they do so independently of each other. As a result, a gamete contains some maternal factors and some paternal factors. (Principle of Independent Assortment)

Mendel's conclusions are now the fundamental laws of heredity

Because Mendel's conclusions have withstood the test of time, his conclusions have become laws that, when joined together, constitute Mendel's Theory of Inheritance. (As we explained in chapter 1, the use of the word theory in science is antithetical to the way most people use the word. Scientific theories are not hypothetical conjecture but, quite the contrary, are bodies of principles that explain the physical world.) Here's how today's scientists describe Mendel's laws (Figure 11.8). Geneticists use the word homologous

Figure 11.8 The language of genetics. Genetics has many unique terms that beginning students often have a difficult time remembering. Some of those terms are defined and graphically presented here. We have shown three pairs of alleles on chromosome 1, two heterozygous pairs and one homozygous pair. Chromosome 2, which has genes for traits not found on chromosome 1, has two pairs of alleles, one heterozygous and one homozygous.

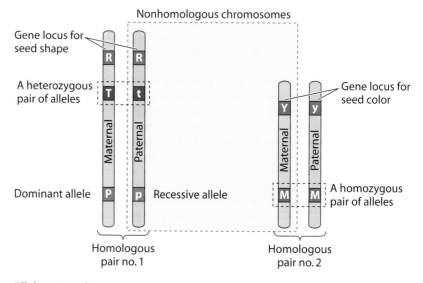

Allele   One of several alternative forms of a specific gene that occupies a certain locus on a chromosome.

Dominant allele   An allele that is phenotypically expressed in the same way in individuals that are homozygous or heterozygous for the allele.

Genotype   The specific genetic makeup of an organism, as opposed to its actual characteristics (its phenotype).

Homologous pair   Two chromosomes that carry genes for the same traits. One member of a homologous pair is donated by the mother, the other by the father. Nonhomologous chromosomes carry genes for different traits.

Heterozygote   An individual with different alleles at a locus.

Homozygote   An individual with identical alleles at a locus.

Locus   The position a gene occupies on a chromosome.

Phenotype   The observable characteristics of an organism as opposed to the set of genes it possesses (its genotype). The phenotype an organism manifests is a result of both genetic and environmental factors.

for members of a gene pair, just as they describe a pair of chromosomes with genes for the same traits as homologous. Therefore, members of a homologous gene pair separate from each other prior to gamete formation. The separation of homologous genes prior to gamete production is known as **Mendel's Law of Segregation**.

For all traits Mendel investigated, the gene occurred in two forms, or **alleles**, one of which was dominant to the other. For example, the gene for flower color has two alleles, white and purple, and purple is dominant to white; the gene for seed shape also has two alleles, round and wrinkled, and round is dominant.

The pea plants inherited a single gene for each of the seven traits from each parent. When the purebred lines were crossed, all offspring inherited a dominant allele from one parent and a recessive allele from the other. Because the alleles within the gene pair are not the same, the gene pair is described as a **heterozygous** pair and the individual is said to be heterozygous for the trait in question. In the purebred parental lines, the gene pair for each of the traits Mendel studied contained identical alleles, which is described as **homozygous**. In Figures 11.5 and 11.6, all of the first-generation offspring are heterozygous for seed shape; in the second generation, half are heterozygous and half are homozygous. Of those that are homozygous, one (RR) is homozygous dominant and the other (rr) is homozygous recessive.

Mendel's second set of experiments looked at the inheritance patterns of nonhomologous genes and asked if they were independent events. Because Mendel studied traits located on different chromosomes, he was able to establish a second principle, known as **Mendel's Law of Independent Assortment**: genes for different traits, or nonhomologous genes, segregate independently of each other during gamete formation. Later in this chapter you will learn about some exceptions to the rule.

Mendel's contemporaries did not grasp the implications of his research

As we described in chapter 1, Mendel's results and conclusions explained the conflicting and confusing body of data on heredity that had frustrated humans for centuries, but no one was impressed when he presented his results to a scientific society in 1865. The publication of his methods, results, and conclusions had no impact on the scientific community for 35 years, even though they were published in a journal found in libraries throughout Europe. Why were Mendel's contemporaries incapable of grasping the significance of his work?

One thing we know for sure, it was not for lack of interest. Scientists of that time should have eagerly seized on any logical explanation of heredity. First of all, understanding how traits are inherited would have improved the efficiency and effectiveness of crop and livestock breeding. Second, the validity of the two overarching theories of biology that had captured the imagination and attention of biologists at that time depended upon the mechanics of heredity.

- The Cell Theory states that all living things are composed of cells, all cells come from previously existing cells, and each cell is potentially self-sufficient because it can carry out all essential life processes. If every cell derives from preexisting cells and is self-sufficient, at least some of the elements of the hereditary program must pass from parent cells to daughter cells. By the mid-1800s, biologists also knew that reproduction of multicellular organisms began with the fusion of two cells, both of which contributed hereditary material to the future offspring.

- The Theory of Evolution, proposed by Darwin, hinged on a system of heredity capable of generating a constant supply of inherited variation. The absence of a valid explanation of the mechanisms of heredity constituted a major weakness in his theory. Mendel's model of heredity provided the constant source of variation Darwin's theory needed, because its basis was particulate inheritance.

If the two overarching theories of biology guiding much of the research of the second half of the 1800s would have been buttressed by a sound theory of inheritance, why didn't biologists of Mendel's day understand and applaud his well-supported, rational, and accurate view of heredity? Perhaps their obtuseness can be explained by Mendel's unfamiliar approach to biological research, which involved mathematics and experimentation. At that time, the universal mode of conducting research in biology entailed observing, describing, recording, and cataloging natural phenomena.

In addition, most biologists of his day probably had difficulty getting their minds around Mendel's conclusions, the crux of which depended upon a theoretical construct (the gene). Even though Mendel had followed the well-trodden research path established by those who preceded him—observing the traits of parents and their offspring—he went a step further. He tried to explain the events occurring inside the black box of inheritance that links parents to their offspring. He extrapolated from his observations and inferred the existence of discrete, invisible packages of hereditary information, or "factors." For biologists, whose fundamental modus operandi consisted of recording events they could observe, explaining a natural phenomenon with an imaginary particle called a factor must have seemed more like practicing black magic than explaining the innards of a black box.

They may have been incapable of understanding Mendel's theory because they could not relate it to something they could *see*. Rather than being satisfied by observing parents, inferring an abstract hereditary mechanism, and observing offspring, perhaps Mendel's contemporaries needed to be able to observe all three: parental traits, a hereditary mechanism, and the offspring. Luckily, during the 35 years that Mendel's work gathered dust on library shelves, technical advances in microscopy allowed scientists to transcend the limitations of human vision. Using these new technologies, they observed chromosomes and watched the unique cell division process that produces gametes, the bearers of hereditary information. When Mendel's work was unearthed in early 1900, biologists had physical images on which to hang his abstract concepts.

RELATED DISCOVERIES IN CYTOLOGY

After Theodor Schwann, Matthias Schleiden, and Rudolf Virchow put forth the Cell Theory in the mid-1800s, cells became a primary focal point for biological research, giving birth to a new subdiscipline within biology: cell biology. Virchow described cells as the "seat of life," because life's essential processes—metabolism, reproduction, and differentiation—are played out on a microscopic level by cells. Biologists realized that if they wanted to understand life, they needed to understand cells.

In its earliest incarnation, cell biology consisted essentially of studying cells by observing and describing cell structures, or **cytology**. Technological progress in two areas of microscopy drove research in cytology. Advances in optics led to the development of light microscopes with the maximum possible resolving power. Scientists also discovered dyes that would preferentially stain different parts of cells and, in doing so, reveal cell structures. The discovery of dyes that react specifically with nucleic acids permitted the visualization of a circular structure consistently seen in most cell types, the nucleus. Under most conditions, the material within the nucleus appeared uniformly granular, but periodically scientists would observe the grainy material condense into individual threadlike structures they called chromosomes.

Cytology informs genetics

In 1873, three scientists independently observed and described the process of mitosis. As you recall from chapter 9, in mitosis, double-stranded chromosomes align along the cell's central axis, and invisible forces pull the strands apart as the cell divides. Each daughter cell receives the same number of chromosomes, now single stranded, present in the parental cell. The chromosomal material then diffuses, reestablishing the granular appearance of the nucleus.

Even though everything about the chromosome's mitotic behavior seemed to advertise its role in heredity, none of the scientists who described mitosis recognized or acknowledged the possible role chromosomes might play in inheritance. Their intense and tightly focused pursuit of their goal—describing cell structure—acted like blinders. Scientists published more than 200 papers describing mitosis in a wide variety of plants and animals before they began to see the relationship among mitosis, chromosomes, and heredity. By the turn of the century, most biologists generally thought that the nucleus and its chromosomes had something to do with inheritance, but they weren't sure what.

Observing gamete formation leads to the chromosomal theory of inheritance

During the 1880s, cytologists studying gamete formation observed a unique chromosomal dance that occurs immediately before the progenitor cells of gametes divide. The process they observed, which is known as **meiosis**, resembles mitosis but also has distinctive features. Meiosis is a unique type of cell division known as reduction division, because the number of chromosomes is halved. In the discussion below, we touch on the aspects of the meiotic process that are relevant to the discussion at hand. Because a solid grasp of both the specific details and the overall process of meiosis is requisite for understanding topics as varied as Mendelian inheritance, evolution, and diagnosis of genetic diseases, we also highlight it in Box 11.1.

The scientists who chose the most appropriate research material, whether intentionally or serendipitously, are those who are credited with describing meiosis and linking it to heredity. In Germany, Theodor Boveri elected to use the parasitic roundworm *Ascaris* to study meiosis and fertilization, because its cells had only four large, easy-to-observe chromosomes. Boveri could see that, unlike mitosis, during gamete production, the four double-stranded chromosomes did not line up along a single axis in the progenitor cell prior to cell division. Instead, the chromosomes formed pairs, and invisible forces pulled the members of a pair away from each other,

BOX 11.1 *Meiosis: a special type of cell division*

In eukaryotic organisms that reproduce sexually, a cell destined to become a gamete undergoes a unique type of cell division that halves its complement of chromosomes. Without this reduction division, known as meiosis, the fusion of gametes in fertilization would double the chromosome number every generation. Meiosis is carefully choreographed to ensure that each gamete receives one member of each homologous chromosome pair. Scientists call any cell containing only one set of homologous chromosomes **haploid** (D and E) and a cell with both members of a homologous pair diploid. Therefore, the goal of meiosis is to produce haploid gametes (E) from a diploid progenitor cell (A, B, and C). To accomplish this, a single progenitor cell undergoes two sequential cell divisions (D and E), creating four haploid gametes. These haploid gametes fuse in fertilization to create a diploid organism. The figure summarizes the key steps in meiosis of an organism with three pairs of homologous chromosomes. The blue member of a pair is of maternal origin; the red member is paternal.

(A) Immediately before meiosis begins, the single-stranded chromosomes condense. The diploid number of this parental cell is six. (B) At the beginning of meiosis, the DNA in each chromosome replicates, producing a two-stranded chromosome. Each strand is known as a chromatid. Even though the amount of DNA has doubled, the cell remains diploid because it contains only six chromosomes. (C) The homologous chromosomes align themselves opposite one another before separating. (D) The cell divides, and the homologous chromosomes separate, producing two haploid cells. Each nucleus contains one member of each homologous pair, but both maternal and paternal chromosomes are present in each nucleus because the chromosomes "assorted independently." (E) The final cell division is like mitosis. The double-stranded chromosomes line up along the center of each cell, and the chromatids separate as the cells divide, creating four haploid nuclei.

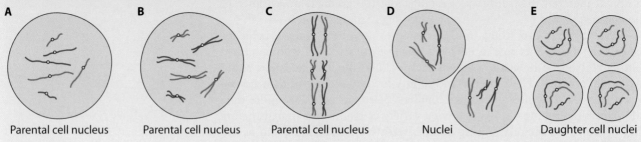

yielding two daughter cells each having only two chromosomes (Figure 11.9). The fusion of two gametes in fertilization restored the number of chromosomes found in parental cells. However, because the four chromosomes looked very similar, Boveri was unable to distinguish one from another, which is key to understanding meiosis fully (Box 11.1).

Walter Sutton, a Columbia University graduate student, was studying gamete formation in grasshoppers (*Brachystola* sp.) at the same time as Boveri. Like roundworms, grasshoppers had large, easy-to-see chromosomes, but they offered an additional advantage. These chromosomes were different sizes and shapes, so Sutton could distinguish them from each other. He could see that the grasshopper's chromosomes occurred in 11 matching pairs; each chromosome had a homolog, a mate of the same size and shape. During gamete formation, when the chromosome pairs aligned in the center, pairing was not random. Each chromosome paired with its homolog.

In summary, Sutton observed the following:

- Chromosomes exist as discrete units.
- Chromosomes come in homologous pairs.
- In gamete formation, homologous pairs line up in an orderly fashion, and the homologs move away from each other when the cell divides.
- Each gamete has half the number of chromosomes found in the parental cell. Most important, each gamete has one member of each homologous pair of chromosomes.

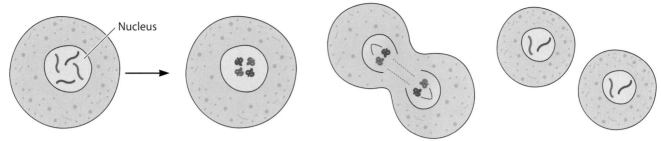

Gamete progenitor cell   Chromosome pair formation   Pair separation   Each gamete has two chromosomes.

Figure 11.9 Schematic of meiosis in roundworms. These cells, which are redrawn from Boveri's 1887 sketches of his observations, show the level of detail microscopes provided to researchers at that time. Even though we have used colors to indicate homologous pairs, this degree of discernment was impossible for Boveri. Note that the progenitor cell of the gametes contains four chromosomes. During meiosis, homologous chromosomes align, and the members of a pair separate as the cell divides, halving the number of chromosomes from four to two. In the last drawing, each gamete contains both a maternal and a paternal chromosome, but in roundworms, it is equally likely that both chromosomes in a gamete are of paternal or maternal origin.

- When gametes fuse in fertilization, the resulting cell has the same number of chromosomes as the original parental cells. These chromosomes occur as homologous pairs, with each parent contributing one member of each pair to the offspring.

Take a moment to compare Sutton's observations to Mendel's conclusions in Table 11.2, and you will see that his observations grounded Mendel's theory in a physical, observable reality. Mendel's work had been rediscovered 2 years earlier; Sutton was familiar with it and recognized the relationship between his observations and Mendel's abstract "factors." In his paper describing his results, Sutton stated:

> Finally, I call attention to the probability that the association of paternal and maternal chromosomes in pairs and their subsequent separation during the reducing division may constitute the physical basis of the Mendelian law of heredity.

Even though the relationship between his observations and Mendel's conclusions was obvious, when Sutton explained this to his mentor, the brilliant biologist E. B. Wilson, Wilson could not make the connection. According to Wilson:

> I well remember when in the early spring of 1902, Sutton first brought his main conclusion to my attention. I also clearly recall that at the time I did not fully comprehend his conception or realize its entire weight.

Isn't it interesting that Wilson, one of the foremost cell biologists of his day, did not understand a discovery that seems so obvious to us now? His admission demonstrates the degree to which mental models make it difficult to see new scientific insights, even when the model exists in the brain of a brilliant scientist. We discuss this in more detail in Box 11.2, "Scientific models: gateways or barriers?"

In 1902, Sutton and Boveri set forth the hypothesis that chromosomes contain Mendel's factors. Sutton immediately noticed a conflict between this hypothesis and Mendel's principle of independent assortment. Roundworms

B O X   1 1 . 2   *Scientific models: gateways or barriers?*

If you think the resistance of the scientific community to Mendel's discoveries was unique to that time, think again. The story of another brilliant geneticist, Barbara McClintock, parallels Mendel's so closely, it is eerie. McClintock is the scientist who discovered "jumping genes," or transposons, which we discuss in detail in chapter 13.

Like Mendel, she studied an organism so familiar it was almost dull—corn. Like Mendel, she was blessed with great powers of observation, an extraordinary intellect, and an uncommon amount of patience and persistence. For over 30 years, she carried out the slow, laborious work of cross-pollinating Indian corn plants and, months later, observing the results of her experimental crosses. By simply using her uncanny ability to find patterns in the color of corn kernels, she developed a theory of inheritance that called into question the fundamental nature of genes and eventually revolutionized our understanding of genetics.

McClintock proposed the existence of jumping genes to the scientific community in the 1950s. Unlike Mendel's audience, McClintock's was impressed—but not positively. In short, some thought a brilliant geneticist, who was an elected member of the prestigious National Academy of Sciences, had gotten a bit confused, to say the least. What she proposed was heresy to the existing view of genes and chromosomes, and so they concluded she was out of touch with reality, not that their model was wrong.

Often scientists forget that the models they develop are supposed to be useful tools for understanding and interpreting the workings of the system under study. Instead, they become wedded to the model and are unable to see and hear information that does not support it. Their model acts like a filter, letting expected information pass through to their brains but blocking the information they don't expect.

This surely was the case during the 20 years McClintock tried to explain to her colleagues what she saw in those corn kernels and was met with complete resistance. Like Mendel, she saw patterns where everyone else saw disarray. She open-mindedly asked what the patterns she saw revealed about the behavior of genes and chromosomes. She did not aim for a specific answer to support an existing theory. Instead, she sought true and deep understanding.

Other scientists were blind to the patterns and deaf to her reasoning. When they couldn't understand her message, they chose to ignore it because it did not fit with their preconceived notions of what was "supposed" to happen. Eventually, after other researchers discovered evidence of transposons in bacteria, yeasts, and fruit flies, her fellow scientists could finally hear what she had to say. In 1983, McClintock received the Nobel Prize for her work on transposons.

and grasshoppers had many more traits than could be explained by either 2 or 11 pairs of chromosomes, respectively. Therefore, Sutton hypothesized, each chromosome must carry many factors, and those factors cannot be inherited independently. They are **linked** and should be inherited as a unit.

Gene linkage supports the chromosomal theory of heredity

One of the first demonstrations of gene linkage comes from Thomas Hunt Morgan's work using the fruit fly (*Drosophila melanogaster*), an organism many people associate with the early days of genetics research. Morgan did not set out to prove or disprove genetic linkage. He was studying the newly discovered phenomenon of gene mutation and was quite excited when a naturally occurring mutant, a male with white eyes, finally appeared in his colony of fruit flies. Morgan crossed the white-eyed mutant with a normal red-eyed female and, just like Mendel's peas, all of the first-generation offspring exhibited the same trait: red eyes. Morgan, therefore, knew that the allele for red eyes was dominant to the white-eye allele.

When he crossed the first-generation offspring with each other, the results departed slightly from those predicted by Mendel (Figure 11.10A). Of the 4,252 second-generation offspring produced, Morgan observed 3,470 red-eyed and 782 white-eyed flies. On further inspection, Morgan noted a more striking oddity: none of the females had white eyes, and he wondered if it was possible for females to have white eyes. Today's geneticists might frame Morgan's question in these terms: is there some sort of biochemical interaction among proteins encoded by genes for femaleness and eye color that prevents females from having white eyes?

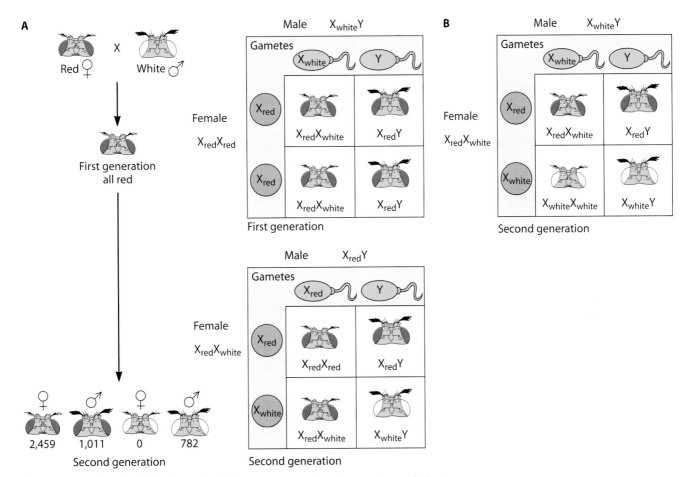

Figure 11.10 Sex-linked inheritance. T. H. Morgan observed a unique pattern of inheritance for eye color in fruit flies. **(A)** Morgan mated a red-eyed female from a purebred line to a white-eyed male. The first-generation offspring displayed the dominant phenotype, red eyes, as predicted by Mendel. When the first-generation offspring mated with each other, all of the flies having white eyes were males. This suggested that a relationship existed between the inheritance of eye color and the inheritance of sex. Morgan reasoned that the gene for eye color was carried on the X chromosome but that no corresponding gene locus for eye color occurred on the much smaller Y chromosome. **(B)** He confirmed that prediction by crossing the original white-eyed male with a red-eyed (heterozygous) female from the first generation.

He knew the red-eyed females from the original cross were heterozygous, possessing a gene for red eyes and a gene for white eyes. He crossed the white-eyed male with a heterozygous red-eyed female and obtained the following ratios (Figure 11.10B):

129 red-eyed females

132 red-eyed males

88 white-eyed females

86 white-eyed males

Being female was clearly not incompatible with having white eyes, so why were white-eyed females absent from the second generation?

Once again, hints for answering a genetics question came from cytology. In 1905, Nettie Stevens, another of E. B. Wilson's graduate students who studied meiosis in insects, saw the commonly observed meiotic pairing of matched chromosomes during egg, but not sperm, formation. When the chromosomes paired during meiosis in the cells destined to be sperm, she could easily see that one pair consisted of two chromosomes that did not resemble each other. One homolog, which Stevens designated X, seemed identical to chromosomes she observed in egg formation, but its partner, the Y chromosome, was much smaller. The XY chromosome combination always occurred in males but never in females, which were always XX. Stevens reasoned that the genes that determine the sex of the insect must be located on these chromosomes. As you learned in chapter 10, the pattern of sex determination Stevens observed for insects also occurs in mammals, including humans.

Some traits are sex linked

Stevens' observations provided evidence to support the Sutton-Boveri contention that chromosomes contained genes and also helped Morgan explain his unexpected results on eye color inheritance in fruit flies. Because the genetic results paralleled the behavior of the X chromosomes, Morgan hypothesized that the gene for eye color was associated with the X chromosome. A male with a single gene for white eyes would always exhibit that trait because the small Y chromosome lacked many genes, including one for eye color. Therefore, no dominant allele could mask the recessive white allele. All subsequent crosses he performed supported this hypothesis, and in 1910, Morgan concluded that the gene for eye color was on the X chromosome and therefore was a **sex-linked trait**.

Those of you who are especially observant may have noticed that the numbers of flies with red and white eyes deviate slightly from the expected ratio of 3:1 in the second generation and the 1:1:1:1 ratio expected when white-eyed males mate with heterozygous red-eyed females. The biochemical defect that leads to the white-eye mutation also leads to higher mortality rates, so there are fewer white-eyed flies than expected.

During meiosis, homologous chromosomes exchange genetic material

A few years after Morgan furnished the first evidence of linked genes, a student in his laboratory, Hermann Muller, discovered that X-rays induced gene mutations in fruit flies. Their work on linkage proceeded rapidly after this technological development, because they no longer had to wait for mutations to occur spontaneously. By creating over 100 easily observable mutations, Morgan and his students demonstrated linkage relationships in many traits. For example, when they crossed heterozygous red-eyed, normal-winged flies with purple-eyed, vestigial-winged flies, they almost never obtained the red-eyed, vestigial-winged or purple-eyed, normal-winged flies that would be expected given Mendel's Law of Independent Assortment (Table 11.3). Morgan concluded, as Sutton had predicted, that the genes for these two traits could not assort themselves independently because they were located on the same chromosome (Figure 11.11).

Perhaps the phrase "almost never" in the last paragraph caught your attention, because from your knowledge of meiosis, you probably assume that genes on the same chromosome are always inherited together. But for every combination of traits that demonstrated a linkage relationship, Morgan and his students almost always observed a certain percentage of offspring that disproved the linkage hypothesis (Table 11.4).

Table 11.3 Genetic linkage[a]

Phenotype (eye/wing)	Morgan's results for offspring of cross red eye/normal wing (RrNn) × purple eye/vestigial wing (rrnn)		
	No.	% Expected (if no linkage)	% Observed
Purple/vestigial	1,196	25	44
Red/normal	1,239	25	45
Red/vestigial	151	25	5.5
Purple/normal	154	25	5.5
Total	2,740		

[a]Geneticists determine if genes are linked by crossing individuals that are heterozygous for two traits with those that are homozygous recessive for both traits. If the genes are not linked, they assort independently, giving a phenotypic ratio of 1:1:1:1. The results in the table show that the genes for eye color (red or purple) and wing shape (normal or vestigial) are linked. The details of the cross are illustrated in Figure 11.11.

Figure 11.11 Linkage. In the example shown, the two traits being investigated for linkage are eye color and wing shape. The two eye color alleles are red and purple. The red-eye trait (R) is dominant to purple eyes (r). The two alleles for wing shape are normal and vestigial. Normal wing shape (N) is dominant to vestigial wings (n). Morgan crossed red-eyed, normal-winged individuals, who were heterozygous at both loci, with flies that had purple eyes and vestigial wings. The results indicated that the genes for eye color and wing shape are linked, because the numbers Morgan observed deviated significantly from the 1:1:1:1 ratio Morgan expected. Had the genes for eye color and wing shape been on separate chromosomes, the numbers would have been distributed equally among the four phenotypes.

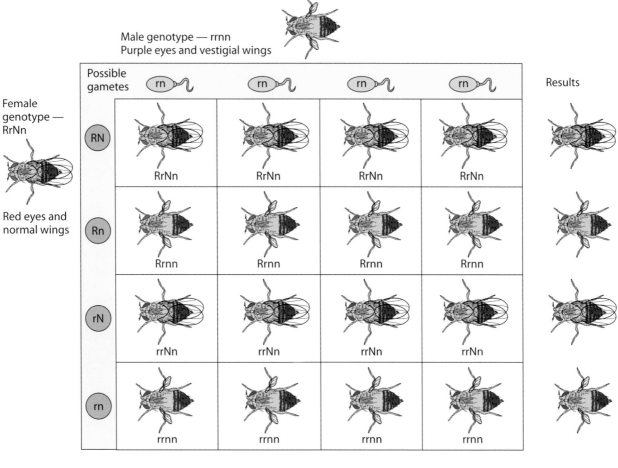

Table 11.4 Genetic linkage of traits[a]

Red eyes/normal wings (RrNn) × purple eyes/vestigial wings (rrnn)		
Parental	Purple eyes/vestigial wings	44%
	Red eyes/normal wings	45%
Recombinant	Red eyes/vestigial wings	5.5%
	Purple eyes/normal wings	5.5%
Tan body/normal wings (TtNn) × black body/vestigial wings (ttnn)		
Parental	Black body/vestigial wing	41.5%
	Tan body/normal wing	41.5%
Recombinant	Black body/normal wing	8.5%
	Tan body/vestigial wing	8.5%
Normal wing/long leg (NnLl) × vestigial wing/short leg (nnll)		
Parental	Normal wing/long leg	35%
	Vestigial wing/short leg	33%
Recombinant	Normal wing/short leg	16%
	Vestigial wing/long leg	16%

[a]If genes for two traits are not linked, equal numbers of the four possible phenotypes are expected when heterozygotes are crossed with individuals that are homozygous recessive for both traits. However, if the genes are linked, the results of the cross deviate significantly from the expected 1:1:1:1, as shown in Figure 11.11. The results of the crosses shown, obtained by Morgan and his students, indicate gene linkage for these pairs of traits. When genes are not linked, the expected percentage is 25% for each of the four possible phenotypes. If linkage is complete, the percentage of recombinants should be 0%, because the loci for the two traits are so close that crossing over is prohibited. Therefore, as the distance between two loci increases, the percentage of organisms exhibiting the recombinant phenotypes for that cross increases. These data were used to construct the linkage map in Figure 11.14.

By the time Morgan discovered linkage, he was convinced that genes were located on chromosomes. He wanted to develop a hypothesis, consistent with the chromosomal theory of inheritance, capable of explaining these aberrant results. He proposed that during meiosis, paired homologous chromosomes sometimes exchange portions of themselves in a process called **crossing over**. He looked to cytology in the hope of anchoring his abstract idea in a visible reality and found just what he was looking for. In 1909, cytologists had observed chromosomes behaving in a way that provided visible corroboration for Morgan's hypothesis (Figure 11.12).

In meiosis, when the double-stranded homologous chromosomes align as pairs along a cell's central axis, their arms cross over each other, and they exchange equal segments of DNA by breaking and rejoining at precisely the same point along the chromosome (Figure 11.13). Some genes from the maternal chromosome come to reside on the paternal chromosome and vice

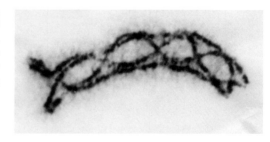

Figure 11.12 Crossing over. The micrograph shows chromatids of homologous chromosomes crossing over each other when they align in meiosis. (Photograph by Bernard John, courtesy of John Kimball.)

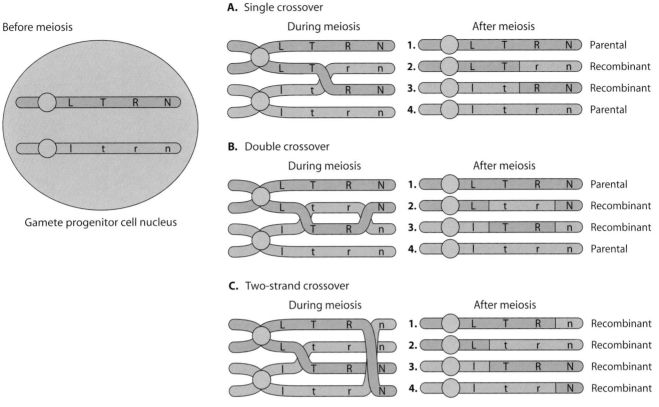

Figure 11.13 Crossing over between homologous chromosomes. Early in meiosis, after DNA replication, homologous chromosomes align so that genes for the same trait are located opposite each other. While paired, the double-stranded chromosomes exchange segments quite precisely. **(A and B)** If the gene-for-gene exchange involves only one chromosome arm, two of the four possible gametes, numbered 1 to 4, contain the chromosome found in the parental cell and two contain a chromosome, known as a recombinant, containing a mixture of maternal and paternal alleles. **(A)** In a single crossover, one chromatid exchanges genes with its homolog at a single point. **(B)** One chromatid can also cross over its homolog at a number of loci. A single chromatid in each chromosome crosses at two points in what is known as a double crossover. **(C)** Other times, both arms of a chromosome are exchanged. If the exchange occurs at different places along the sister chromatids, all four gametes will contain recombinant chromosomes; none contains a chromosome that is genetically identical to the parental chromosome. The gametes shown correspond to the data in Table 11.4.

versa. The chromosome arms that cross over and exchange genetic material are known as **recombinants**, because the genetic material within that arm contains both maternal and paternal genes formerly located on separate chromosomes. Thus, crossing over is one of nature's methods for recombining genetic material and creating genetic variation.

As you can see from the different proportions of recombinants associated with different traits in Table 11.4, the frequency of crossing over depends on the specific combination of traits under investigation. An undergraduate student working in Morgan's laboratory at that time, Alfred Sturtevant, realized that the differences in crossover frequencies, as reflected in the different proportions of recombinants, indicated the linear distances between the genes (Figure 11.14). The greater the distance between two genes, the more likely that crossing over between them will occur and, therefore, the higher the proportion of recombinants. He used the percentages of recombinants to map

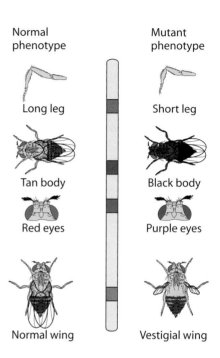

Figure 11.14 Linkage map. This genetic linkage map is constructed from the genetic results shown in Table 11.4. Linkage maps show the relative positions of different gene loci on a chromosome. A large number of recombinants indicates a high frequency of crossing over between two loci. Therefore, in creating a map from results of genetic crosses, the higher the percentage of individuals with the recombinant phenotypes, the greater the distance between the two loci. This map corresponds to the chromosomes in Figure 11.13 that illustrate crossing over.

the positions of genes relative to one another. These **linkage maps** remain an essential tool of geneticists and will be discussed at length when we describe how scientists identify genes.

As a result of the work of Thomas Hunt Morgan and his students, geneticists began to realize that not only did chromosomes carry genes, but each gene occupied a specific, predictable place, or **locus**, on the chromosome. This concept is essential for understanding the process geneticists use to identify genes associated with diseases.

SUMMARY POINTS

Hereditary information is packaged in discrete units, which we now call genes.

In sexually reproducing organisms, genes and chromosomes occur in maternal-paternal, or homologous, pairs. Only one member of the pair is transmitted from each parent to its offspring.

Meiosis is the cell division process that creates gametes with only one member of each homologous pair.

Mendel's Law of Segregation describes the behavior of homologous chromosomes: in meiosis, they separate prior to gamete formation.

Mendel's Law of Independent Assortment describes the behavior of nonhomologous chromosomes: when homologous chromosomes separate in meiosis, they do so independently of the other chromosomes. Thus, the resulting gamete contains chromosomes inherited from both parents.

Using fruit flies as their experimental organism and relying on findings in cytology, T. H. Morgan and his students proved that genes are located at very specific sites, or loci, on chromosomes.

When two gene loci are very close, the genes are described as linked genes because they are inherited together.

During meiosis, homologous chromosomes exchange genetic material. Chromatids cross over each other, break, and rejoin quite specifically. This is known as recombination, because the genetic material of the two chromosomes has been recombined.

The frequency of crossing over allows scientists to determine the relative distance between two genes on a chromosome. When distances among many genes on a chromosome have been determined, scientists are able to create a linkage map that shows the relative positions of genes on a chromosome.

KEY TERMS

Alleles	Homozygous	Mendel's Law of Independent Assortment	Recombinants
Crossing over	Linkage maps		Sex-linked trait
Cytology	Linked genes	Mendel's Law of Segregation	
Haploid	Locus		
Heredity	Meiosis	Purebred lines	
Heterozygous			

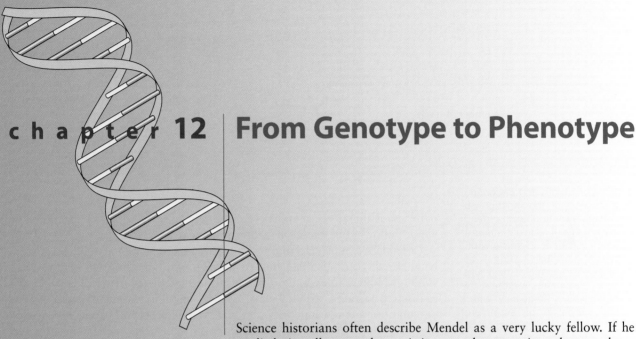

chapter 12 | From Genotype to Phenotype

Science historians often describe Mendel as a very lucky fellow. If he had studied virtually any other trait in any other organism, they say, he would have failed to uncover the invisible hereditary mechanism underlying visible characteristics. In order to elucidate the laws governing heredity, a researcher needed to make a number of improbable choices, and Mendel, lucky fellow that he was, chose correctly every time.

First of all, in deciding which traits to follow from one generation to the next, the researcher would need to select, unknowingly of course, those visible traits that happened to be governed by a single gene. The odds of randomly choosing one such trait, not to mention seven, are quite long, because in all organisms many genes contribute to almost all visible traits. From that small set of one-gene traits, it would be very helpful if the researcher selected traits that occurred in only two allelic forms, for example, round or wrinkled, yellow or green seeds. Of those, the researcher would need to choose those traits, fewer still, in which one of the two alleles was completely dominant to the other. In addition, the unambiguous dominance of one allele should be reflected in unambiguous traits that are so different that the experimental results can be assigned to completely distinct categories. Finally, added to all these lucky happenstances, the researcher should select traits that occurred on separate chromosomes or were at opposite ends of the same chromosome and therefore assorted independently.

Nobody is that lucky! Mendel purposefully chose the seven traits he studied. First, he eliminated most pea characteristics from consideration because they exhibit continuous variation and, as he explained, would be "of a 'more or less' nature that would be difficult to define." Instead, he considered the few characteristics having distinct differences, or discontinuous variation, because those traits would provide unambiguous data that he could place in one of two categories (Figure 12.1). Before beginning his experimental crosses, he identified which of those traits had clear, predictable

A. Continuous variation

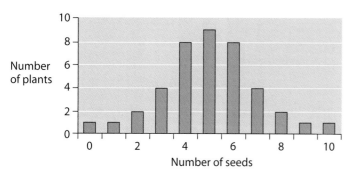

B. Discontinuous variation

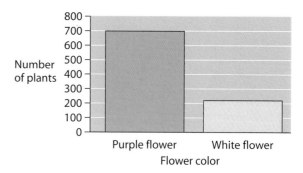

C. Continuous variation

Figure 12.1 Continuous and discontinuous variation. Most of the variation observed in nature is continuous. In order to determine if a trait exhibits continuous or discontinuous variation, scientists count the number of individuals having a certain trait and create a graph of the results. **(A)** Frequency histograms represent the number of individuals (frequency) with a certain trait as a bar. In this example of continuous variation, the phenotype being studied is the number of seeds per plant, and the measurement categories range from 0 to 10 seeds/plant. The number of plants with each phenotype determines the height of each bar. For example, one plant had no seeds, one plant had nine seeds, and eight plants had four seeds. **(B)** A frequency histogram showing discontinuous variation, such as the results Mendel obtained when he crossed pea plants that were heterozygous for flower colors, exhibits no gradation of phenotypes. Offspring occurred in a 3:1 ratio of the two parental types, either purple or white flowers. None of the flowers of the offspring had the lighter or darker shades of purple shown in panel **C**.

patterns of inheritance, and he deliberately selected those as the traits he would follow from one generation to the next.

Mendel's careful selection of traits revealed the fundamental laws underlying heredity and enabled him to develop two concepts that changed the study of biology forever. One, the idea that hereditary material is packaged as discrete particles, which we discussed at length in chapter 11, resolves the paradox of inherited similarity and inherited differences. His second conceptual breakthrough distinguishes genotype from phenotype. In this chapter, you will see why this distinction opened the door to a much deeper understand of heredity, because many factors influence the phenotypic expression of a certain gene, such as:

- interactions between homologous alleles
- interactions with nonhomologous genes
- environmental factors

Before discussing examples of these influences, we need to alert you to some typical misconceptions most nonscientists have about the relationship between genotypes and phenotypes. If we do not correct these misconceptions quite explicitly from the outset, your mental model might make it difficult for you to understand the material in this chapter.

MISCONCEPTIONS ABOUT GENES AND TRAITS

The primary misconception that pervades the public's understanding of genetics and colors its interpretation of genetic information is an implicit belief in a one-to-one relationship between a gene and an observable phenotypic trait that is simple, direct, and inevitable. In other words, most nonscientists believe the following.

- One gene is responsible for one phenotypic trait.
- A phenotypic trait can be traced to the activity of a single gene.
- A certain genotype invariably prescribes a corresponding phenotype.

In the public's mind, an aura of unambiguous predestination surrounds genes. But, as you will learn again and again in this chapter, genes represent potentialities, not inevitabilities. A gene is not the final arbiter of a trait but only one of many factors that determine whether that gene's potential becomes realized as the phenotype commonly associated with it. Virtually every observable trait represents the cumulative effect of the actions of many genes, some of which negate the others. Environmental factors throughout an organism's life influence the phenotypic expression of genetic information, as well. Consequently, genotypes are not predictable from seeing phenotypes, and phenotypes are rarely predictable from knowing the genotype.

What is the source of this pervasive misunderstanding about genes and traits? We are not certain, but we think it can be traced to Mendel and his peas. Most people's educational exposure to the genetic basis of trait inheritance begins and ends with Mendel. As a result, their understanding of the relationship between genotypes and phenotypes reflects Mendel's accurate, but very simplified, explanation of inheritance. Their fundamental misconceptions are then reinforced by the language scientists and nonscientists use when discussing the relationship between hereditary information and observable traits.

Mendel studied traits with very simple inheritance patterns

Mendel wanted to develop a set of generalizations about the transmission of traits from parents to offspring, test his generalizations with experimental observations, refine the generalizations accordingly, and determine if the generalizations applied to other organisms or were unique to peas. His training in physics taught him that successfully teasing apart the details of any complex function, such as the mechanics of heredity, depended upon studying its clearest, simplest examples. To that end, Mendel intentionally chose distinct traits with clear, predictable, and simple inheritance patterns.

The pea characteristics Mendel selected are inherited as one-gene, two-allele traits in which one allele is completely dominant to the other. The traits also seem to be relatively impervious to environmental influences, or Mendel made certain that the pea plants grew in virtually identical environments. In other words, Mendel's traits exhibited the cleanest possible relationship between the invisible genotype and its phenotypic expression at the level of the organism. Very few observable traits in any organism are governed by a single gene, occurring in only two allelic forms with clear dominance relationships that manifest as a single characteristic occurring in only two distinct forms. Figure 12.2 graphically illustrates the complex relationships among genes and observable traits. The path from a gene to its primary protein product to a visible trait is indirect and complex; many genetic factors other than the single gene influence the phenotypic expression of that gene. Overlay on these complex genetic relationships environmental influences that may be affecting each gene's production of its encoded protein, and you can easily see why the relationship between genotypes and phenotypes is rarely as simple as Mendel's work implies.

Language can create and reinforce misconceptions

In a typical conversation, you might hear someone say "she inherited musical ability from her father" or "he inherited sickle-cell anemia from both of his parents." But traits are not inherited; genes are inherited. Traits are the ultimate manifestation of a many-stepped process affected by both genetic and environmental factors. Equally misleading are media reports of scientists discovering "the gene for cystic fibrosis" or "the schizophrenia gene." There is no such thing as *the* cystic fibrosis gene or *the* schizophrenia gene. There are genes that encode proteins that are associated, in some way, with these

Figure 12.2 Relationships among genes and observable traits. A single gene affects many traits, and any one trait results from the actions of many genes. Although this diagram may seem complex, it omits many intermediate products and feedback loops. In addition, it is limited to genetic interactions and ignores the influence of environmental factors on observable traits.

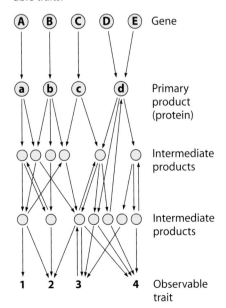

diseases. Unfortunately, these misstatements, which are used by scientists and nonscientists alike, further entrench the widespread misconception of a one-to-one, predetermined relationship between having a gene and exhibiting a certain trait.

The public lexicon about genetics contains these sorts of phrases for valid historical reasons. Imprecise language reflects thousands of years of imprecise understanding. People knew that traits passed between generations, but they had no idea how this happened. Careful observation revealed predictable inheritance of a few traits. Ancient scriptures of many of the world's religions recognize that some diseases are inherited in predictable patterns. Hindu writings warn men not to marry women with diseases or their children and grandchildren will suffer the same fate. The Talmud relaxes the requirement to have male children circumcised if the sons of sisters have died from blood loss, indicating that the authors understood something about the inheritance pattern of hemophilia, a sex-linked genetic disorder.

After the discovery of Mendel's work in 1900, **pedigree analysis**, which is a way to illustrate inheritance patterns of a trait graphically, allowed medical geneticists to infer the genetic basis of a disease, provided the disease, like Mendel's pea traits, was inherited as a single-gene trait (Figure 12.3). However, pedigree analysis did not provide any clues about how information implicit in the hereditary material became explicit as a noticeable trait. For almost 50 years after they discovered Mendel's work and coined the word "gene," scientists had no idea what occurred in the black box linking inherited material to observable traits. Saying "he inherited hemophilia from his mother" was, quite simply, the best they could do.

Little by little, scientists began to tease apart the inner workings of the black box. In the 1940s, they discovered that genes ultimately manifested as traits because they encoded information for making proteins; in the 1960s, they elucidated the code cells use to translate genetic information into functional proteins. However, as Figure 12.2 illustrates, identifying a gene's encoded protein is rarely equivalent to uncovering the relationship between that gene and a visible or measurable trait. Scientists have uncovered the

Figure 12.3 Pedigree analysis for hemophilia. By diagramming the individuals in a family with and without the hemophilia phenotype, medical geneticists determined that its inheritance was associated with the sex of the individual. Therefore, hemophilia is a sex-linked trait, and the gene is located on the X chromosome.

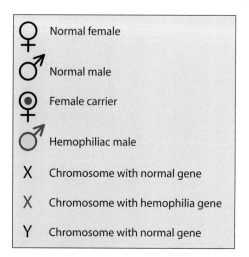

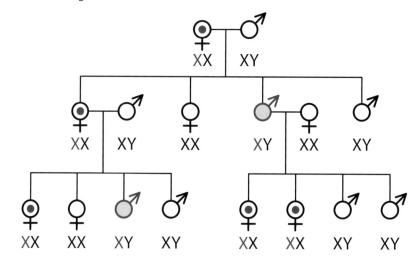

steps linking a gene, an encoded protein, the protein's function, and, ultimately, an observable or measurable phenotype for relatively few traits. Sickle-cell disease, which you learned about in chapter 4, and cystic fibrosis are two genetic disorders for which physicians now understand the sequence of events that culminates in the clinical symptoms they observe (Figure 12.4).

Figure 12.4 From gene mutation to clinical symptoms. Scientists have detailed most of the steps in the chain from the specific nature of the genetic mutation to the encoded protein, the cellular malfunction, and, ultimately, the phenotype for the disease cystic fibrosis. **(A)** In 75% of individuals with cystic fibrosis, the mutation is a three-nucleotide deletion that leads to the loss of a single amino acid, phenylalanine, from the membrane protein that controls chloride movement in and out of epithelial cells. **(B)** The channel protein remains closed, and chloride ions build up within the epithelial cells. This leads to osmotic imbalances that cells try to rectify by taking up more water and preventing the influx of sodium. As a result, thick mucus accumulates in the air passageways in the lungs (bronchioles) and in secretory ducts of organs, such as the pancreas and liver, all of which are lined with epithelial cells. The mucus in the lungs encourages the growth of microorganisms; people with cystic fibrosis have difficulty breathing and are susceptible to frequent respiratory infections. Mucus clogging the secretory ducts prevents the release of digestive juices, so people with cystic fibrosis cannot digest their food and their bodies are deprived of nutrients. They have a difficult time gaining weight, irrespective of their food intake.

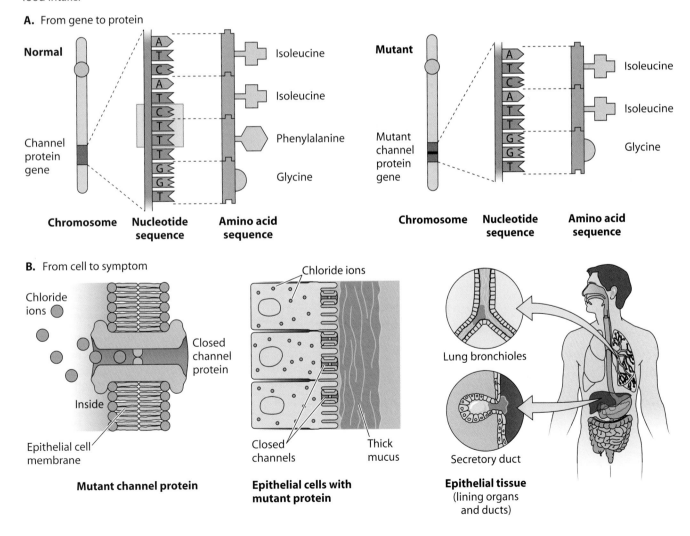

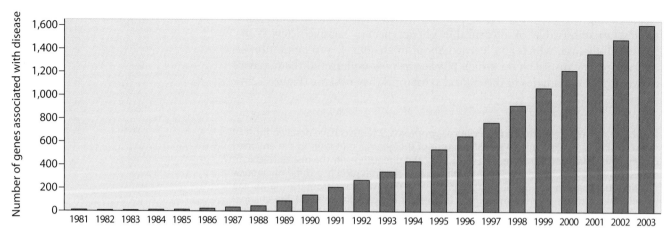

Figure 12.5 Cumulative pace of gene discovery. New technologies are contributing to an exponential increase in the number of genes identified every year. Minimum estimated values are shown.

Rather than saying a child with sickle-cell anemia has inherited sickle-cell anemia from his parents, the correct (and nonmisleading) description of sickle-cell anemia inheritance is as follows.

> He inherited a mutated form of the gene that encodes the hemoglobin protein from both of his parents. This mutation leads to the production of a hemoglobin molecule that, under the environmental influence of low oxygen levels, polymerizes. The polymerization causes red blood cells to sickle and burst, leading to anemia.

Scientists can also replace the imprecise phrase "the cystic fibrosis gene" with more precise language.

> Cystic fibrosis is caused by a gene mutation that encodes a malfunctioning chloride channel protein in epithelial tissue, including the epithelial tissue lining the respiratory and gastrointestinal systems. This leads to the production of thick mucus that blocks pulmonary airways, accumulates in the lungs, and clogs secretory ducts of the pancreas, causing the set of respiratory and digestive problems associated with cystic fibrosis.

Thanks to new research tools available to today's scientists, they are identifying genes and gene mutations at a rapid pace (Figure 12.5). But, even today, filling in all of the steps in the gene–protein–cell–visible-phenotype chain remains a daunting task. Scientists will continue to use phrases like the "gene for curly hair" and the "breast cancer gene" because, very often, that's the best they can do. The media will also continue to use these same misleading phrases, even when the molecular basis of a disease is known, for good reason. The misstatements are more succinct and, paradoxically, they communicate more information than the correct descriptions, which most people would find quite confusing.

Misconceptions about genes and traits can affect your personal life

Replacing the misconception that one gene equals one trait is the most important goal of this chapter. Having an appreciation of the complicated path through which genetic information becomes realized as a phenotypic trait will help you make decisions about the safety of eating genetically modified foods and assess statements about the environmental risks of genetically modified plants, animals, and microbes. Understanding the relationship between genes and traits also has important implications for your mental and physical well-being. In the future, you will hear a great deal about scientists discovering genes associated with certain diseases. You will need to decide if you want to be tested to determine whether you have the normal or mutant forms of those genes. If you ask your physician to conduct the test, you will need to understand the test results to determine your propensity, or lack of one, to inherit, transmit, or develop certain diseases. If you use a "one gene equals one trait" model to interpret the test results, you may well decide you are destined to develop a disease you will never have or, on the other hand, that you are *not* at risk of a disease that you will develop.

Here's an example of how misunderstanding the relationship between genotypes and phenotypes might lead someone to make poor choices about health care. Earlier, you learned about an exception to the one gene equals one trait relationship: *BRCA1* and *BRCA2*, commonly known as the breast cancer gene. Reading about the breast cancer gene, a woman with the one gene equals one trait mindset would have a tendency to interpret the discovery of this gene in the following way. If a woman has the breast cancer gene, she will get breast cancer; if she doesn't, she won't.

Knowing the strong tendency of this gene to cause breast cancer in certain families, she would likely assess her chances of having the breast cancer gene by asking about the incidence of breast cancer in her female relatives (Figure 12.6). If she discovered that no one in her family had had breast cancer, the one gene equals one trait mindset would encourage her to think she

Figure 12.6 Pedigree analysis for *BRCA1* and *BRCA2*. Mutations in the tumor suppressor genes *BRCA1* and *BRCA2* significantly increase the risk of developing hereditary breast or ovarian cancer. Women with a mutation in either gene are approximately three to seven times more likely to develop hereditary breast cancer than those lacking a *BRCA1* or *BRCA2* mutation. Not every woman with a mutation in *BRCA1* or *BRCA2* will develop breast or ovarian cancer. Only 5 to 10% of breast cancers are hereditary breast cancers associated with these two gene loci.

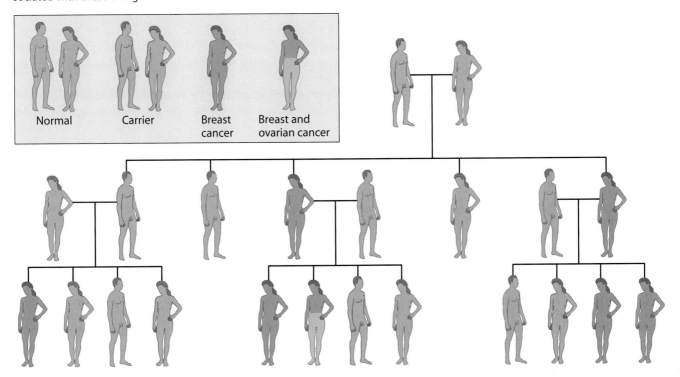

Table 12.1 Risk of developing breast or ovarian cancer[a]

Cancer type	% *BRCA1* and *BRCA2* gene loci	
	Normal	**Mutant**
Breast	13	36–85
Ovarian	1.7	16–60

[a]A mutation in either tumor suppressor gene, *BRCA1* or *BRCA2*, significantly increases a woman's risk of developing breast or ovarian cancer. More than 500 mutations have been discovered for each gene locus. Presumably, some mutations are worse than others. This contributes to the wide range in the percentages of women developing each type of cancer.

would not develop breast cancer. She might decide to forego regular checkups and mammograms. Unfortunately, only 5 to 10% of breast cancers, known as hereditary breast cancer, are associated with having the breast cancer gene; 90 to 95% of breast cancers are nonhereditary. Her risk of developing a nonhereditary form of breast cancer is 13% (Table 12.1).

On the other hand, if she discovered that many of the women in her family had developed breast or ovarian cancer, she might overreact. On learning of the discovery of the breast cancer gene, some women with a family history of breast cancer decided to have mastectomies even though they did not have any symptoms of breast cancer. Unfortunately, not all of these women had a mutation at the *BRCA1* or *BRCA2* locus, and they would not have developed hereditary breast or ovarian cancer. In addition, having a mutation at this locus does not inevitably lead to hereditary breast or ovarian cancer; it does, however, increase a woman's risk of developing both diseases.

In the remainder of the chapter, we will attempt to broaden your understanding of the many factors that affect the relationship between genotypes and phenotypes. Whenever possible, we will use examples from human genetics, but most of our knowledge of genetics does not come from research on humans (for very good reason). Therefore, when you are reading an explanation of the genetic basis of coat color in dogs, please don't think the information is irrelevant to human genetics. It is simply the clearest example we could find to illustrate one of the factors affecting the translation of genetic information into a phenotypic trait. The genetic principles described below apply to all organisms, including you.

Misconception: a gene affects one trait

Most genes affect many different phenotypic traits, not a single trait, a phenomenon known as **pleiotropy**. When you picture the many interrelated biochemical pathways described in the previous chapters, it should come as no surprise that one gene can have many effects. A single enzyme responsible for catalyzing a reaction that produces a molecule with roles in a number of pathways has an impact on all of those pathways. Pleiotropic effects result not only from many different biochemical pathways sharing the same molecules, but also from the interdependence of cells and organs in multicellular organisms. When you stop to think about it, pleiotropy should be the rule rather than the exception.

We have already provided a number of examples of single-gene mutations manifesting themselves as a set of traits, even though we may not have used those words.

- Fruit flies with the white-eye mutation also have higher mortality rates, so the mutation, visible to scientists as the phenotypic trait of white eyes, also has biochemical or physiological effects that are much more harmful.
- The hemoglobin gene mutation associated with sickle-cell anemia leads to the following cluster of phenotypic traits: red blood cells that sickle when oxygen levels are low, anemia caused by lysis of red blood cells, painful joints and organs, jaundice, vascular obtruction, and heart and kidney failure.
- In Gaucher's disease, a single-gene mutation affecting a single enzyme manifests as an enlarged spleen; fewer platelets, white blood cells, and red blood cells; anemia; abdominal pain; and fragile bones.

- A mutation in the gene encoding the enzyme that breaks down the sugar galactose causes phenotypic traits as diverse as gastrointestinal problems, cirrhosis of the liver, and cataracts.

Not only are these familiar examples of pleiotropy, but they also make it clear that knowing the specific identity of the protein encoded by the gene does not automatically explain all of its pleiotropic effects. Why does inability to degrade galactose cause cataracts?

Be careful not to confuse pleoitropy with linkage. Both lead to the appearance of a specific set of phenotypic traits in an individual, but the underlying mechanisms differ. One is based on chromosomal geography, the other on biochemistry. With linkage, traits occur together because the genes associated with the traits are situated near each other on a chromosome and inherited as a unit. Pleiotropic traits co-occur because they can be traced to the presence or absence of a single gene product that has a number of effects.

Misconception: a phenotypic trait can always be traced to a certain genotype

A number of different genotypes can lead to the same observable phenotype. The most straightforward example of this occurs in the one-gene, two-allele traits of Mendel's peas. The homozygous dominant or heterozygous genotypes both produced the same phenotypes—purple flowers, round seeds, and so forth—because one allele was completely dominant to the second. However, the same phenotype can also be traced to different genetic loci. Under these circumstances, the phenomenon is known as **genetic heterogeneity**.

Once again, this relationship of different genotypes to the same phenotype should come as no surprise to you, given your new knowledge of cell biology and biochemical pathways. If different enzymes in a single pathway are mutated, the pathway's end product will not be produced. Expression of phenotypic traits requiring the presence of that end product will be compromised, no matter which step in the pathway is malfunctioning. Think back to the discussion of cell division and cancer in chapter 9. Even though you received a *very* abbreviated description of the factors controlling cell division, you could list at least six mutations of nonhomologous genes, working at different parts of the cell division or DNA repair process, that could lead to colon cancer.

Another example of a genetically heterogeneous trait in humans is retinitis pigmentosa, a disease characterized by degeneration of the retina that follows a very specific course: reduced night and peripheral vision, followed by rod, then cone, degeneration and blood vessel constriction. Scientists have tracked this disease to mutations at more than 15 separate loci, some of which are autosomal while some are sex linked. Some of the new, mutated alleles are dominant to the normal allele, while others are recessive.

Misconception: individuals with the same genotype always display the same phenotype

Individuals with the same genotype at a particular locus can have different phenotypes. Even after scientists have determined the phenotypes that typically derive from specific genotypes, for certain traits they cannot predict the phenotype based on the genotype. This uncoupling of genotype from phenotype may assume two forms, which geneticists characterize as genotype penetrance and genotype expressivity.

Penetrance is the percentage of the population with a certain genotype exhibiting the trait associated with that genotype. If 100% of the organisms with a certain genotype at a locus display the expected phenotype, the gene is **completely penetrant**. From what you know of Mendel's work, you would describe the allele encoding purple flowers as completely penetrant. When some organisms have the appropriate genotype for expressing a trait but do not display the trait, the trait exhibits reduced or **incomplete penetrance**.

Think of penetrance as the probability that an individual with a particular mutation will display the mutant phenotype, assuming the mutant allele occurs in a combination (heterozygous or homozygous) that would permit expression. For example, because cystic fibrosis is a recessive trait, **carriers**, who are individuals having only one recessive allele for a recessive disorder, are not expected to have the disease (Figure 12.7). Because they are heterozygous for a recessive mutation, they are excluded from calculations of penetrance of the cystic fibrosis mutation.

Geneticists have now identified over 900 different mutations in the gene that encodes the chloride channel protein associated with cystic fibrosis! All people who display clinical symptoms of cystic fibrosis have a mutation in that gene, but not all people with a mutation in that gene have clinical indications of the mutation. Therefore, the degree of penetrance of the cystic fibrosis mutation varies with the nature of the mutation (Figure 12.8). Certain

Figure 12.7 Cystic fibrosis pattern of inheritance. Cystic fibrosis is an autosomal recessive trait. One in 2,500 people of European descent has cystic fibrosis, and 1 in 25 is a carrier. **(A)** Cystic fibrosis pedigree analysis. **(B)** Cystic fibrosis Punnett square.

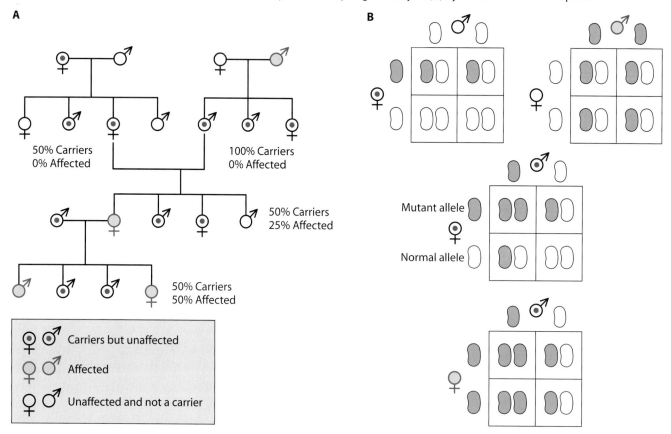

mutations in the gene encoding the chloride channel protein are completely penetrant, because all who are homozygous recessive for that mutation exhibit some of the symptoms associated with cystic fibrosis. Other mutations seem to be completely impenetrant; people with two alleles for these mutations have lived to be 60 years old without developing any symptoms. They happened to discover they have two mutant alleles at the chloride channel protein locus only because physicians now have access to tests for identifying carriers of certain mutations at that locus.

Expressivity refers to the range of phenotypes associated with a certain genotype and is related to the degree to which a gene is expressed in an individual. If a trait takes different forms in different individuals having the same genotype, it exhibits **variable expressivity**. For example, a dominant mutation at a single locus causes the white part of the human eye, known as the sclera, to turn blue. The allele's expressivity varies, ranging from a barely detectable light blue to a blue that is so dark it is almost black. In addition, because the blue-sclera allele is dominant to the normal allele, any individual with a copy of the mutated allele would be expected to have blue sclera. However, only 90% of the people with a gene for blue sclera display the expected phenotype because the blue-sclera allele is also incompletely penetrant.

Cystic fibrosis is another trait that exhibits variable expressivity. Some children develop severe symptoms early in life, others later, and the severity of the symptoms also varies. For years, physicians have recognized this variation, and now molecular evidence showing that the gene has many mutant allelic forms may explain at least some of the variation in phenotypic expression. Interestingly, the severity of the mutation on the genetic level is sometimes not mirrored in the severity of clinical symptoms. People with mutations that lead to total absence of the chloride channel protein often have few, if any, cystic fibrosis symptoms; those with a mutation that leads to the loss of a single amino acid suffer from severe cystic fibrosis and usually die before they are 25 years old.

Another example of a trait that exhibits both incomplete penetrance and variable expressivity is **polydactyly**, the presence of extra digits (Figure 12.9). The gene for extra digits is dominant to the normal allele, but some individuals known to carry the polydactyly allele appear normal, having five digits on both hands and both feet (incomplete penetrance). Some who express the polydactyly allele have only one extra rudimentary toe on one foot, while others with the same genotype have six digits on both hands and both feet (variable expressivity).

Single-gene traits are often associated with predictable phenotypic expression. As a result, you might assume that all diseases caused by a single gene would exhibit no variation in expressivity and complete penetrance. However, as you have just seen, that assumption is invalid. We have described two traits caused by mutations in a single gene, cystic fibrosis and polydactyly, both of which show variable expressivity and incomplete penetrance. In addition, individuals who are homozygous for the sickle-cell trait may have few or all of the symptoms described above. Some die as young children because of the disease; others live to old age.

What is the molecular basis of incomplete penetrance and variable expressivity? With cystic fibrosis, some of the variation is explained by different mutations in the same gene, with variable phenotypic effects. Incomplete penetrance and variable expressivity also result from other factors that exert

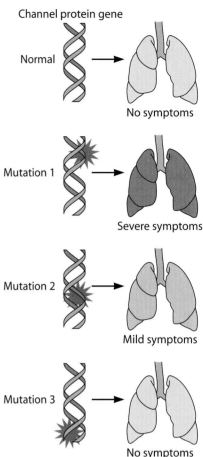

Gene changes in cystic fibrosis

Figure 12.8 Penetrance and expressivity. Different mutations in the gene encoding the chloride channel protein have different phenotypic effects.

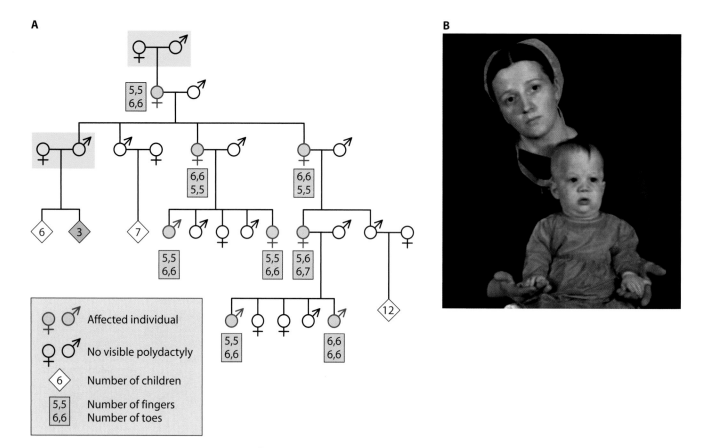

Figure 12.9 Polydactyly penetrance and expressivity. Polydactylism is an autosomal dominant disorder. **(A)** At least one member of the highlighted couples has the polydactyly allele but does not express it. Of those that express it, the degree of expression varies, as evidenced by the variable number of digits among individuals with the same allele. In addition, expressivity is variable within a single individual. **(B)** Polydactyly can occur as an isolated phenotypic effect or as one of many pleiotropic effects of genetic disorders of the skeletal system. The child in the photograph suffers from both dwarfism and polydactylism. A mutation in a single gene is responsible for this autosomal recessive disorder. (Photograph courtesy of Victor McKusick.)

influences along the path from genotype to phenotype, especially genetic interactions and environmental influences. No gene acts in isolation from other genes or the external environment.

GENETIC INTERACTIONS AND PHENOTYPIC EXPRESSION

The action of any gene can be fully understood only in terms of its **genetic background,** which is the overall genetic makeup of an organism in which a gene occurs. Two categories of genetic interaction affect phenotypic expression: interactions between homologous alleles and those among nonhomologous genes.

Interactions between homologous alleles affect phenotypes

Another misconception conveyed by Mendel's work is the idea that alleles of homologous genes exhibit clear-cut dominance relationships. For example, one would never know that a pea plant heterozygous for flower color or seed shape had recessive alleles for those traits because the flowers and seeds of

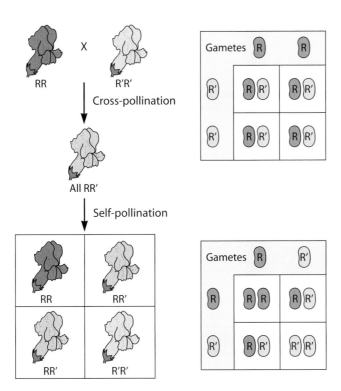

Figure 12.10 Incomplete dominance. When homozygous red (RR) and white (R′R′) snapdragons are crossed, all of the first-generation offspring have pink flowers. When these offspring self-pollinate, all three phenotypes appear. These results reaffirm Mendel's theory of discrete particle inheritance in spite of the "blended" appearance of the first-generation flowers.

such plants appear identical to the flowers and seeds of a plant that is homozygous for the dominant trait. The dominant allele overpowers the recessive allele and masks its presence.

Many alleles do not have that sort of relationship. Some display **incomplete dominance** because the heterozygote has a phenotype intermediate between the two homozygotes. For example, if snapdragons homozygous for red or white flowers are crossed, as Mendel predicted, all of the first-generation flowers will be a single color. But that color is neither red nor white but pink (Figure 12.10). Intermediate dominance, rather than incomplete dominance, might be a more descriptive label for this type of interaction between homologous alleles.

If both alleles are expressed in the phenotype, the relationship between homologous alleles is described as **codominance**. A familiar example of codominance is the ABO blood group. As you learned earlier, the ABO blood types—A, B, AB, and O—result from the presence of different carbohydrates on the surfaces of red blood cells. A single gene existing in three allelic forms, A, B, and O, determines which carbohydrate is present. The A and B alleles exhibit codominance, because people with the AB genotype have both the A and B carbohydrates on their red blood cells. Both A and B alleles are dominant to the O allele, because AO and BO genotypes are expressed as A and B blood types, respectively (Table 12.2).

ABO blood groups are also an example of multiple alleles. Typically, geneticists talk about pairs of alleles because each individual has only two

Table 12.2 Relationship of genotype to phenotype[a]

Phenotype	Genotype
A	AA, AO
B	BB, BO
AB	AB
O	OO

[a]The ABO blood groups provide a simple example of different genotypes leading to the same phenotype.

copies of a gene. This can lead to the misconception that each gene occurs in only two forms, but most genes have many alleles.

Learning the genetic and biochemical basis of the ABO blood groups provides another opportunity for you to connect the dots among a gene, its encoded protein, and a phenotype. The relationship of genotype to phenotype is assessed, not by using a visible trait, but by observing the response of a person's blood to each carbohydrate associated with the ABO blood groups. For example, type A blood reacts immunologically to type B blood, because the B carbohydrate is foreign, or "nonself," to type A red blood cells. Adding type B blood to type A blood causes the blood to form clumps, and vice versa. Neither type A nor type B blood reacts immunologically to the O carbohydrate, but type O blood forms clumps when exposed to either type A or type B blood (Figure 12.11).

Biochemical and genetic bases of ABO blood types

On the surfaces of red blood cells, virtually everyone has many copies of the H substance, a molecule made of five sugars linked in a chain (Figure 12.12). The A allele encodes an enzyme, glycosyltransferase, that adds an additional sugar to the H substance. Individuals with the B allele synthesize a modified version of glycosyltransferase, which adds a different, but very similar, sugar to H. AB individuals have alleles that encode both forms of the enzyme; half of the H substance molecules on their red blood cells terminate in the A sugar, and the other half terminate in the B sugar. Individuals with the O allele produce a nonfunctional form of the enzyme incapable of adding any sugar to the H substance. As a result, type O red blood cells have the five-sugar H substance on their surfaces. Neither type A nor type B blood reacts

Figure 12.11 ABO typing. The laboratory technique of blood typing is based on an immunological reaction of blood to molecules it sees as foreign, or nonself. To determine a person's blood type, a small sample of the blood is divided into four parts, and each part is exposed to one of the other blood types. Blood having one type of carbohydrate (A or B) forms clumps with blood having a different carbohydrate, unless that carbohydrate is the H substance (type O). Synthesis of the H substance is common to all blood types (with the rare exception of the Bombay phenotype), so the H substance is viewed as self by all blood types. As a result, people with type O blood are called universal donors. However, type O blood reacts to all other blood types, because both the A and B carbohydrates are foreign. Type AB blood is the universal recipient, because all three carbohydrates, A, B, and H(O), are self to blood with both the A and B alleles. Type AB blood added to any other blood type causes clumping, because at least one of its carbohydrates, A or B, is foreign to all other types.

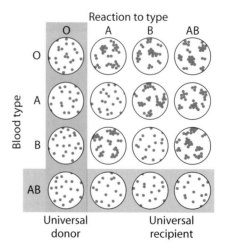

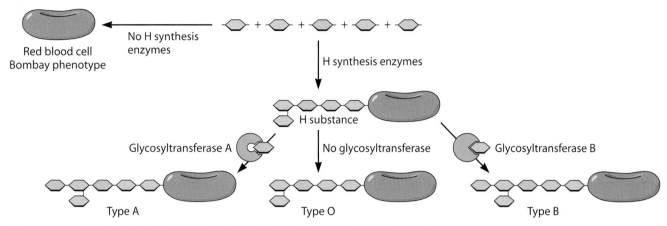

Figure 12.12 Molecular basis of ABO blood types. Three different carbohydrate molecules are responsible for the ABO blood groups. All three can be traced to the activity of a single gene that occurs in three allelic forms.

immunologically to the O carbohydrate, which is the H substance, because red blood cells synthesize H and view it as "self."

What is the source of these allelic forms of glycosyltransferase? Mutation. When scientists compared the A and B alleles, they found four nucleotide substitutions. Those substitutions led to only a slight change in the enzyme's active site, because the A and B forms bind to similar molecules. Comparison of the O allele with either the A or B allele reveals an even smaller difference, but that minor gene change translated into major changes in the protein. A one-nucleotide deletion distinguishes the O allele from both A and B. This single deletion resulted in a frameshift mutation creating a stop codon that terminates translation prematurely and produces a nonfunctional enzyme.

A small percentage of people have a mutation in a nonhomologous gene encoding an enzyme in the pathway leading to the H carbohydrate, and they are incapable of synthesizing the H substance. If their blood is typed, they are classified as type O. In fact, they may have the A or B allele, or both, and if so, their red blood cells produce A or B enzymes. However, those enzymes have no substrate (the H substance) on which to add either a B sugar or an A sugar. Therefore, the phenotype, blood type, is not correlated with the genotype. If, however, the phenotypic measure had been the presence or absence of specific enzymes involved in the synthesis of the A, B, and H carbohydrates, the phenotype would have been a direct reflection of the genotype. This phenomenon, which is known as the Bombay phenotype, illustrates beautifully the ambiguous relationship of the genotype to a specific phenotype and leads us to the topic of interactions among nonhomologous genes (Box 12.1).

Genetic interactions among nonhomologous genes affect phenotypes

An organism's phenotype depends even more upon interactions among alleles of nonhomologous genes than it does on the interaction between alleles of one gene. There are a number of different categories of nonhomologous-gene interactions, but we will discuss only two types of phenotypic expression that result from the many different ways in which nonhomologous genes interact: epistasis and polygenes.

BOX 12.1 *Clarifying the concept of dominance*

No gene is inherently dominant or recessive. Note that whenever we say gene X is dominant, we always add the words "to gene Y." The concept of dominance describes a relationship based upon the interaction between two specific alleles occurring in a heterozygous state. To clarify the distinction between inherent and relative properties, think of yourself. You are inherently a male or female, but you are a sister or brother only in relationship to someone else.

The terms dominant and recessive do not explain anything; they simply describe an observation. As described in the text, to understand why the A and B alleles are dominant to the O allele in blood types, scientists must have information about the proteins encoded by those alleles.

Dominance does not carry with it the concept of "better than." The allele for six fingers is dominant to the gene for five fingers, and the genes for Huntington's disease and early-onset Alzheimer's disease are dominant to normal alleles.

Dominance also does not necessarily mean "more than." Because individuals with dominant alleles always outnumbered the homozygous recessives in Mendel's studies, you might logically

(but incorrectly) assume that in natural populations of organisms, the number of individuals with the dominant phenotype, or the number of dominant alleles in the gene pool, is always much greater than that of recessive phenotypes or recessive alleles. Many factors other than dominance affect the frequency of individual alleles, genotypes, and phenotypes in populations; you will learn about them in detail in the next chapter.

Finally, and perhaps most important, the concept of dominance can get a bit slippery when you alter the specific phenotypic trait being discussed. Sickle-cell anemia provides an excellent example of the arbitrariness of the concept of dominance.

In chapter 4, you learned that sickle-cell anemia is a one-gene, two-allele trait caused by a simple mutation in the gene encoding the hemoglobin protein.

Depending on the aspect of the phenotype under observation—the disease symptom (anemia), the shape of the red blood cells (RBC), or the type of hemoglobin molecule found in the blood (normal [Hb] or mutant [mHb])—conclusions about the dominance relationship between the normal (N) and sickle (S) alleles would vary (see the table).

At the level of the organism, the normal allele is completely dominant to the sickle allele, because heterozygotes (NS) are not anemic. At the cellular level, the relationship could be described as incomplete dominance, because the heterozygote displays a sort of intermediate phenotype, or codominance, because both cell types are present. Finally, at the molecular level, the relationship between the alleles is codominance, because both types of hemoglobin are expressed.

Genotype	Phenotype		
	Anemia	RBC shape	Hb molecule
NN	No	Normal	Hb only
NS	No	Some sickle	Hb and mHb
SS	Yes	All sickle	mHb only

Epistatic interactions

In **epistasis,** the phenotypic expression of alleles at one gene locus masks the expression of genes at another locus, much like dominant alleles mask homologous recessive alleles. The Bombay phenotype described above is an example. Certain alleles for the gene encoding an enzyme required for production of the H substance obscure the alleles at the ABO gene locus when the phenotype being observed is blood type. Another simple but excellent example of epistatic interactions of nonhomologous alleles is seen in a familiar and much-loved animal, the Labrador retriever. Labs occur in three colors, brown, black, and yellow, and scientists now understand the molecular basis of these different phenotypes (Figure 12.13).

Pigmentation in dogs and other mammals, including humans, results from the types and relative amounts of two classes of pigment molecules: eumelanin (black and brown) and phaeomelanin (red and yellow). Both are synthesized from the amino acid tyrosine in pigment-producing cells, the melanocytes. Figure 12.14 shows the enzyme tyrosinase converting tyrosine to a molecule, dopaquinone, that is the precursor of both types of pigments.

To understand why some Labs are yellow, you first need to understand the molecular genetics of brown and black Labs. The enzyme tyrosine-related

Figure 12.13 Molecular basis of coat color. The three coat colors of Labrador retrievers result from an epistatic interaction between two nonhomologous genes, each having two homologous alleles, one of which is completely dominant to the other. (Photograph of black Lab courtesy of the American Kennel Club, photograph of brown [chocolate] Lab courtesy of Thomas A. Martin, photograph of yellow Lab with black nose courtesy of Mary Lynn D'Aubin, and photograph of yellow Lab with brown nose courtesy of Donna Morgan.)

protein 2 (TRP-2) converts dopaquinone to brown pigment molecules. If the enzyme TRP-1 is present, it converts the brown pigments to black pigment molecules. Chocolate Labs lack a functional copy of the enzyme TRP-1; therefore, the pathway terminates after brown-pigment synthesis. If you cross a homozygous black Lab with a homozygous brown Lab, all of the puppies express the black phenotype; therefore, the normal TRP-1 is dominant to the allele encoding the nonfunctional version of TRP-1. The heterozygous puppies carry one copy of the normal allele, which leads to the production of a sufficient amount of TRP-1 to convert enough brown pigment to black, leading to the phenotypic expression of the black coat color.

This is another example of the arbitrary nature of the concept of dominant-recessive discussed in Box 12.1. If dominance were assessed by looking

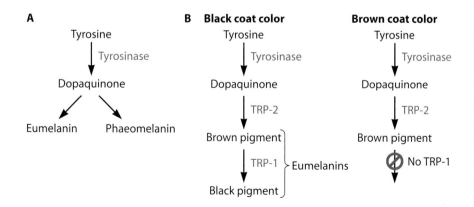

Figure 12.14 Pigment synthesis. **(A)** In mammalian pigment-producing cells (melanocytes), the enzyme tyrosinase converts the amino acid tyrosine to dopaquinone. The pigment synthesis pathway then splits; some dopaquinone is converted to red and yellow pigments (phaeomelanin), and some is converted to black and brown pigments (eumelanin). **(B)** The enzyme TRP-2 converts dopaquinone to brown pigment molecules, and a second enzyme, TRP-1, converts brown pigment molecules to black pigment molecules. Brown (chocolate) Labrador retrievers cannot convert brown pigments to black pigments because they have a mutant form of TRP-1.

at coat color, the relationship would be described as simple dominance of the normal TRP-1 allele to the mutant allele encoding the nonfunctional enzyme. If, however, the amount of functional TRP-1 produced by melanocytes is used as the measure, the relationship would be described as codominance.

The molecular genetics of the coat color phenotypes gets much more interesting when we look at yellow Labs, because epistasis is operating. The nonhomologous allele that masks or negates the TRP-1 and TRP-2 alleles encodes a membrane receptor protein that regulates the synthesis of these two enzymes (Figure 12.15). When the receptor is functioning properly, a hormone secreted by the brain, melanocyte-stimulating hormone (MSH), binds to the surfaces of the melanocytes in hair follicles. Binding to the receptor stimulates the production of the TRP-1 and TRP-2 enzymes, leading to brown and black pigments. Yellow Labs have an allele that encodes a faulty receptor that is incapable of receiving the signal from MSH. Therefore, the hair follicle melanocytes never turn on the eumelanin pathway branch, and

Figure 12.15 Epistatic interaction and coat color. Epistasis, which means "standing on," is a type of interaction in which one gene prevents the phenotypic expression of a non-homologous gene. In this case, a mutation in a gene encoding a receptor "stands on" the gene encoding enzymes responsible for brown and black pigment synthesis. Because of a mutation that changes the shape of the MSH-R protein, MSH cannot bind to the cells that synthesize pigments, the melanocytes. The signaling molecule that activates the expression of TRP-1 and TRP-2 is not released, and all of the dopaquinone is converted to phaelomelanin, the precursor for yellow and red pigments. **(A)** Normal MSH-R protein permits expression of black or brown coat color genotype. **(B)** Mutant MSH-R protein inhibits expression of black or brown coat color genotype.

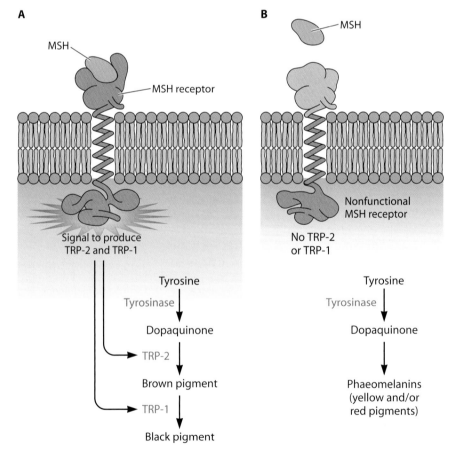

all of the dopaquinone is converted to phaeomelanin pigments, which in Labs happens to be yellow.

The MSH receptor (MSH-R) gene is epistatic to the TRP genes, because even if functional copies of TRP-1 or TRP-2 are present, the dog's coat color will never be brown or black. In order to display the yellow phenotype, the dog must be homozygous for the malfunctioning MSH-R allele. Heterozygotes produce a sufficient number of functional MSH-R proteins to activate the eumelanin branch. Therefore, yellow coat color in Labrador retrievers is an example of recessive epistasis, but dominant epistasis does exist in certain traits not discussed here.

By the way, you can tell whether a yellow Lab would have been brown or black by looking at the color of its nose and lips, because the malfunctioning MSH-R allele affects only the melanocytes in hair follicles. In Figure 12.13, the yellow Lab puppy on the left is genotypically a black Lab; those on the right have the nonfunctional form of TRP-1 and would have been brown if they had inherited one copy of the normal allele encoding the MSH-R protein.

Polygenic traits

In the above discussion of genetic interactions, we focused on phenotypic traits that display discontinuous variation; brown, yellow, or black Labrador retrievers; A,B,O, or AB blood types; normal hemoglobin or abnormal hemoglobin. Mendel's work may tempt you to equate discontinuous traits and one-gene traits. Single genes often govern discontinuous traits (ABO blood types), and other times, more than one gene locus is responsible for discontinuous variation (Lab coat color).

Most phenotypic traits exhibit continuous variation between the phenotypic extremes and cannot be classified into categories that are clearly distinct from each other (refer to Figure 12.1). Height, weight, hair and eye color, metabolic rate, and crop yield and seed size are all examples of traits with continuous variation (Figure 12.16). Classifying people as having a short or tall phenotype is impossible, except at the height extremes. Where would you draw the line for delineating the two categories? Traits exhibiting continuous variation are usually controlled by a number of genes at different loci that interact in a cumulative, or additive, way. Geneticists describe this type of inheritance as **polygenic**, or **quantitative**.

The term "polygenic" can be misleading, because it is easy to confuse polygenic inheritance with the more general concept that phenotypic traits result from the actions of many genes. Geneticists use polygenic to refer to a very specific type of genetic interaction. Polygenic traits represent the combined, or additive, effects of many loci interacting quantitatively. Each contributing locus has **additive alleles** that contribute to the phenotypic expression of a trait and nonadditive alleles that contribute nothing. The contribution of a single additive allele to the phenotype may be quite small, but when the contributions accumulate, a significant amount of phenotypic variation appears.

The genetic determination of kernel color in wheat provides one of the clearest and simplest examples of polygenic inheritance (Figure 12.17). Two pairs of genes, acting in an additive way, determine wheat kernel color, which varies from dark red to white. Dark-red kernels have an additive allele at each locus (four additive alleles), while white kernels have no additive alleles. Crossing dark-red and white kernels gives kernels with an intermediate red color, because all of the offspring are heterozygous at both loci, that is, they

Figure 12.16 Polygenic inheritance. Humans express a wide range of variation in hair and eye color, and that variation is continuous. Traits exhibiting continuous phenotypic variation are usually polygenic traits. (Photograph in top left corner courtesy of Hannah Vaughan; all others courtesy of Thomas A. Martin.)

have a total of two additive alleles. The second-generation kernels from this cross have colors ranging gradually from dark red to white, because the offspring may have zero, one, two, three, or four additive alleles. Creating a frequency histogram of the number of offspring with each trait produces the pattern shown in Figure 12.18, which is the type of frequency distribution exhibited by polygenic traits governed by any number of additive alleles.

Mendelian or polygenic inheritance?

Pretend you are a plant breeder who is trying to determine the genetic basis of wheat kernel color. You, like Mendel, would begin with two purebred lines, red and white, that breed true because each is homozygous for the trait being studied. The results of the parental cross are not very informative, be-

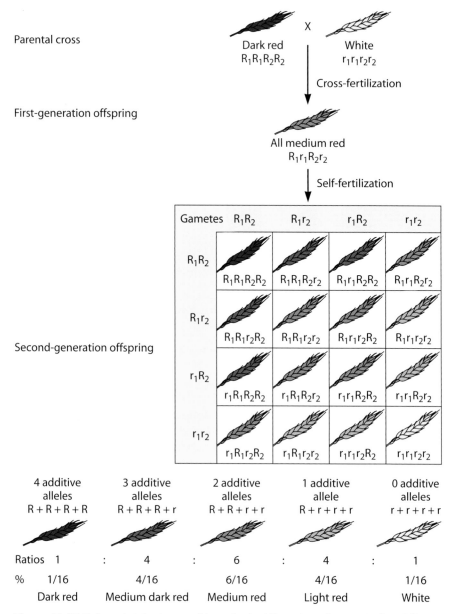

Figure 12.17 Polygenic inheritance of kernel color. Wheat kernels occur in five different colors that range gradually from white to dark red. Two genes are responsible for this trait, and each has both an additive allele (R_1 and R_2) and a nonadditive allele (r_1 and r_2). Each of the additive alleles contributes a small amount of red coloration to the kernel.

cause all offspring have kernels that are an intermediate shade of red. This phenotype can be explained by assuming that the kernel color is either a polygenic trait or a **Mendelian trait** governed by two alleles having incomplete dominance. A Mendelian trait is any trait having inheritance patterns that are consistent with Mendel's ratios.

As was true of Mendel's experiments, the next cross, in which offspring breed with each other, is the cross that reveals much more about the genetic basis of wheat kernel color. If the trait is Mendelian, the kernels will occur in

A. Wheat kernel color

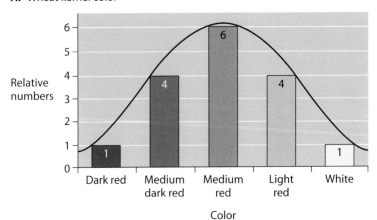

B. Plant height

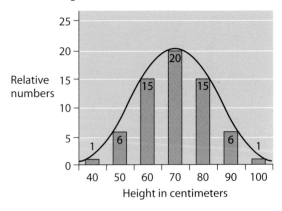

Figure 12.18 Polygenic inheritance and continuous variation. If a trait is polygenic, off-spring of a cross of two individuals that are heterozygous at all contributing loci occur as a range of different types that gradually change along a continuum. **(A)** Because the wheat kernel trait is governed by two genes that have one additive and one nonadditive allele, there are five distinct phenotypes that occur in a 1:4:6:4:1 ratio. **(B)** If three gene loci, each having one additive and one nonadditive allele, contribute to a trait, the offspring of a heterozygous cross would occur as seven phenotypes in a 1:6:15:20:15:6:1 ratio. In this example, the polygenic trait is plant height.

three distinct phenotypes (red, intermediate, and white) in a 1:2:1 ratio. Instead, there are five phenotypes, and graphing the results produces a histogram akin to the familiar bell-shaped curve (Figure 12.19), which is a diagnostic characteristic of polygenic traits. In addition, geneticists can determine how many loci contribute to the trait by noting the ratio of traits to each other. In the wheat kernel cross, some of the phenotypic ratios occur in sixteenths, just like Mendel's two-locus cross of round or wrinkled and yellow or green seeds. This ratio tells geneticists that even though the inheritance pattern is non-Mendelian, the trait is probably governed by two loci that segregate independently.

Kernel color in wheat provides an excellent example for introducing you to polygenic inheritance. The trait involves only two loci, each locus has one additive and one nonadditive allele, and each additive allele's contribution to the trait is small and equal to all the other alleles' contributions. Most polygenic traits involve more than two loci, and the additive alleles have various effects. Some make significant contributions to the phenotypic trait, while the contributions of others are minor (Figure 12.20).

A vast number of traits, including most human diseases and traits important to agricultural scientists, such as size and yield of grains, milk production, and body weight, are under polygenic control. In subsequent discussions of biotechnology applications, most of the traits we discuss are polygenic.

ENVIRONMENTAL IMPACTS ON PHENOTYPIC EXPRESSION

The phenotypic expression of a gene is determined not only by other homologous and nonhomologous genes, but also by environmental factors. The degree to which a certain trait is affected by environmental factors is known as its **phenotypic plasticity.**

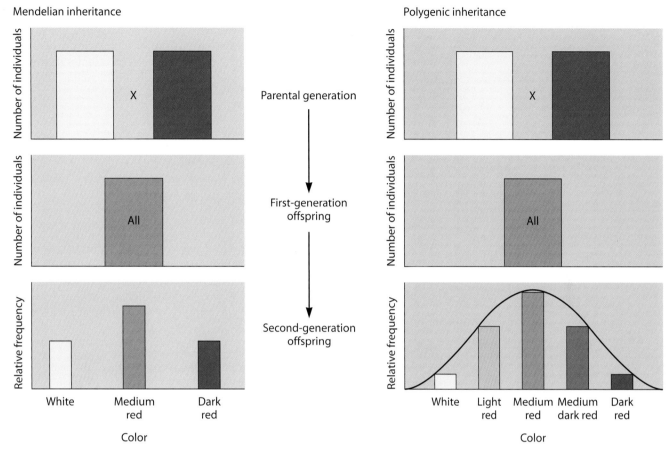

Mendelian inheritance

Polygenic inheritance

Figure 12.19 Mendelian or polygenic inheritance. Geneticists determine the genetic basis of traits by crossing individuals that are homozygous for the trait of interest, recording the numbers of offspring with certain phenotypes, and then crossing those offspring with each other. If wheat kernel color had been a one-gene, two-allele trait and one allele was completely dominant to the other, all of the first-generation offspring would have been the same color as one of the parents. Instead, all were an intermediate color, and this result is consistent with both a polygenic trait and a Mendelian trait governed by alleles with incomplete dominance. Crossing the first-generation offspring allows geneticists to distinguish between these two possibilities.

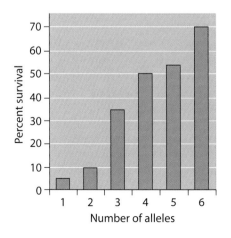

Figure 12.20 Resistance to insecticides. Insects can evolve resistance to insecticides that are used frequently. The genetic basis of this resistance is often additive. Insects with the greatest number of additive alleles are most likely to survive when exposed to the insecticide. Note, however, that each allele does not contribute equally to the resistance trait.

Environmental factors affect phenotypes

You are already familiar with the idea that environmental factors affect phenotypes. For example, you know that poor nutrition can prevent people from reaching their full genetically possible height. The list of examples of the environment affecting phenotypic expression is virtually limitless and includes the following types of phenomena.

Nutrition during development

- Maternal nutrition during pregnancy affects the coat color of genetically identical mice.
- Honeybee larvae develop into workers, drones, or queens according to their diets.

Seasonal effects

- Birds and mammals living in arctic environments, such as snowshoe hares and arctic foxes, turn white during the winter. Decreasing temperatures trigger this change (Figure 12.21).
- Increasing or decreasing day length leads to higher hormone levels in animals that breed seasonally, causing changes in behavior and coloration.

Figure 12.21 A temperature-dependent phenotype. As spring arrives, the arctic fox replaces its white winter coat with a brown one. Both coats help the fox blend into its surroundings so it can approach its prey unnoticed. (Photographs by Keith Morehouse [white fox] and Brian Anderson [brown fox], courtesy of the U.S. Fish and Wildlife Service.)

Ecological effects

- Genetically identical plant cuttings develop large, thin leaves if grown in shade and smaller, thicker leaves when grown in full sun.
- A water buttercup plant produces dramatically different leaves on the stems that grow above and below water (Figure 12.22).
- Fruit flies with the vestigial-wing mutation develop normal wings if raised at high temperatures; conversely, flies with the normal-wing genotype will display a different mutant wing phenotype if suddenly exposed to a high temperature for a brief time.
- Some reptiles and fish change sex if the ambient temperature increases or decreases significantly.

Environmental factors influence phenotypic expression through different avenues

Sometimes a phenotype associated with a particular genotype is not observed simply because the environment does not supply the requisite materials. For example, individuals with the gene mutation for phenylketonuria (PKU) will not develop the PKU phenotype if the amount of phenylalanine in their diets is kept to a minimum. In addition, human embryos with the PKU mutation develop normally because their environment, specifically, the mother's circulatory system, clears their blood of the excess phenylalanine that would lead to developmental abnormalities.

Other times, the environment affects phenotypic expression of a visible trait through direct effects on proteins encoded by certain genes. For example, the seasonal change in coloration we observe in animals living in environments with long, snowy winters occurs because the ambient temperature alters the functionality of one of the enzymes responsible for pigment synthesis.

At other times, the environment's impact on a phenotypic trait is mediated through its effect on gene expression. Recall that bacteria respond to the level of nutrients in the environment by synthesizing certain enzymes and not others. The nutrients interact directly with the regulatory proteins that bind to the DNA molecule and turn genes on and off. A plant that is genetically capable of synthesizing all of the enzymes necessary for manufacturing chlorophyll will not turn green when kept in the dark, even if all of the necessary building block molecules for chlorophyll synthesis are abundant. Light triggers a series of biochemical events that culminate in chloroplast DNA turning on chlorophyll synthesis genes, because chlorophyll plays a key role in photosynthesis.

PHENOTYPES, GENOTYPES, AND THE ENVIRONMENT

As you now see, the relationship between genotype and phenotype is not as straightforward as Mendel's results imply. For the great majority of traits, knowing the phenotype does not allow scientists to predict the genotype. Phenotypes, in turn, are rarely predictable from knowing the genotype. We separated our discussions of genetic and environmental factors responsible for the unpredictable relationship of genotypes and phenotypes, but in the real world, genetic and environmental factors do not act in isolation from each other. Teasing apart the relative contributions genes and environment make to phenotypes can be quite difficult even under the best of circumstances.

Figure 12.22 Environmental effects on phenotypes. The water buttercup lives at the edges of ponds where the water level can change frequently. The submerged parts of the plants produce finely divided leaves, and plant stems above water produce large, undivided leaves.

Genetic and environmental factors interact in unexpected ways

Scientists attempt to determine the effects of genetic and environmental factors on the phenotypic expression of a trait by holding one set of variables constant while changing, in a controlled way, the other variable. If possible, they study genetically similar (ideally, identical) organisms in different environments and genetically different organisms in the same environment. Below, we describe three studies that illustrate how complicated the relationship between genotypes and phenotypes can be.

Eye size in fruit flies

To demonstrate the interaction of genetic and environmental factors, scientists allowed fruit flies with three different genotypes to develop from egg to adult at five different temperatures. The phenotype they studied was eye size, which they measured by counting the individual units of the eye, called facets.

For flies with the normal genotype, as temperature increases, eye size decreases (Figure 12.23). Scientists compared these results to similar results for two mutant genotypes, both of which lead to smaller-than-normal eyes. Figure 12.23 shows that only one generalization can be made about the relationship among temperature, the gene for eye size, and the number of facets: the eye size of flies with the normal genotype is always greater than that of either of the two mutants, irrespective of temperature during development.

Given the results shown in Figure 12.23, if scientists collect fruit flies from nature, what predictions can they make about a fly's eye size genotype by observing its eye size phenotype, and vice versa? Only one. If a fly has 800 facets/eye, they know its genotype is normal. If a fly has 180 facets/eye, they know it has a mutant genotype, but they do not know which one. For all three genotypes, knowing the genotype does not allow them to predict the number of facets, unless they also know the temperature during development, because the number of facets always varies with the temperature during development.

For small and ultrasmall mutants raised at low temperatures, knowing two of the variables (phenotype and temperature or genotype and temperature) does not allow scientists to predict the third variable, nor can they make any generalizations about the effect of temperature during development on eye size, because the effect depends on the genotype. For the normal and ultrasmall genotypes, as temperature during development increases, eye size decreases; for the small genotype, the opposite is true. In addition, the degree of phenotypic plasticity of the eye size trait varies with the genotype; the normal and small genotypes display less plasticity than the ultrasmall genotype.

As we mentioned above, at lower temperatures, scientists cannot differentiate the two mutant genotypes from each other, because the phenotypic variation *within* one genotypic strain is greater than the phenotypic difference *between* the two strains. Flies with the same genotype raised at the same temperature can have different numbers of facets because the genotype is embedded in different genetic backgrounds. Therefore, whenever possible, scientists try to use genetically identical organisms when they investigate the degree to which genetic and environmental factors affect phenotypic expression. This allows them to minimize the amount of within-genotype variation so that they can more easily identify differences between genotypes. With laboratory animals, they create inbred lines that are homozygous at all loci except the genotype being tested. Plants present more opportunity for creating populations of genetically identical individuals, because scientists can root a number of cuttings taken from each plant.

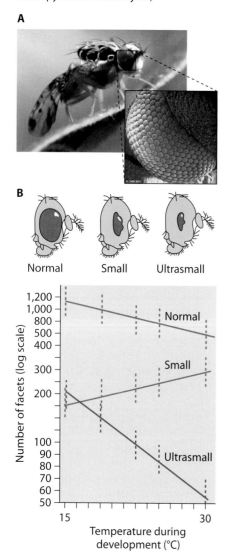

Figure 12.23 Environmental and genetic effects on phenotypes. **(A)** Fruit flies have compound eyes made up of a number of individual units, or facets. **(B)** Temperature during development affects the eye size of fruit flies, which is determined by counting the facets. For each temperature, there is variation in the number of facets (represented by dotted vertical lines); therefore, the solid line represents the average number of facets at a given temperature. (Photograph of Mediterranean fruit fly by Scott Bauer, courtesy of Agricultural Research Service, U.S. Department of Agriculture. Scanning electron micrograph courtesy of David Waddell, University of Queensland Center for Microscopy and Microanalysis.)

Plant height in yarrow

Scientists collected genetically different yarrow plants from a number of locations in the Sierra Nevada Mountains. They rooted cuttings from each
plant to create a set of genetically identical plants for each genotype. They
then grew each genotype at three different elevations and measured plant
height. The picture that emerges about the relationship among genotype,
phenotype, and the environment is even more complicated than the experiment described above (Figure 12.24). Unlike the situation with eye size and
temperature during development, one genotype is not consistently larger or
smaller than all others, irrespective of environmental influences, nor is any
environment always better or worse. In general, plants are shorter if grown
at middle elevations than at low or high elevations, but this is not true for all
genotypes. If scientists could make any generalizations about genetic and environmental effects on this phenotype, would their conclusions change if they
used other phenotypic attributes, such as the number of flowers or stems?

Studies of twins

Dissecting phenotypes into their genetic and environmental components is
not simple, but geneticists are able to get some idea of their relative contributions by controlling the organism's genetic makeup, environmental conditions, or both. Unfortunately, geneticists interested in the genetic and environmental determinants of human traits cannot control either. In addition,
humans who are genetically similar—siblings or offspring and parents—also
share the same environment. As a result, observing that a trait "runs in families" says nothing about whether the trait is due to shared genes or shared

Figure 12.24 Genetic and environmental effects on phenotypes. **(A)** Heights of seven
different lines of yarrow plants grown at three elevations. All of the plants within a single
line are genetically identical. Therefore, any differences within one of the seven plant lines
are due to environmental factors related to elevation. **(B)** When the heights of the seven
lines at three elevations are graphed, it is clear that no generalizations can be made about
the effect of elevation on the height of yarrow plants.

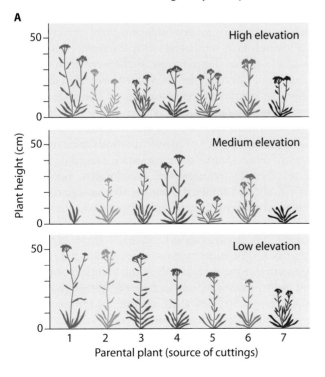

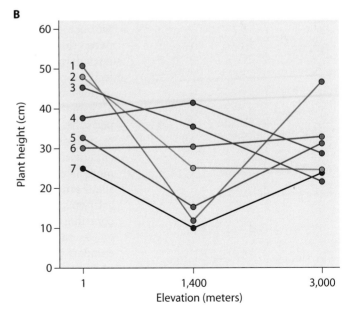

environments. For example, early 20th century researchers became convinced that pellagra, a disease that causes skin lesions and mental disorders, was not an infectious disease but an inherited disorder because its occurrence was strongly correlated in parents and offspring. They concluded, mistakenly and unfortunately, that pellagra was genetically based and that nothing could be done to prevent it. In fact, pellagra is neither a genetic disorder nor an infectious disease. A dietary lack of the vitamin niacin causes pellagra, and adding niacin to the diet cures the disease.

Both genetic and environmental factors contribute to most human diseases, and as the pellagra example shows, knowing if a disease is linked to differences in genes or environments can significantly improve disease prevention and management. Since they cannot control either genes or environment when researching human populations, scientists have been forced to develop other strategies for getting a handle on the relative contributions the two factors make to a disease. Below, we describe two such methods, both of which have limited usefulness because they are unavoidably based on invalid assumptions. In addition, neither method provides any insights into which genes are involved, what the genes do, or how they are inherited. They do, however, shed some light on the degree to which population differences in human diseases, and other traits, can be explained by genetic or environmental differences.

The existence of genetically identical twins allows scientists to control the genetic variable. Most twins, however, are raised in similar environments, so separating genetic similarity from environmental similarity in order to explain phenotypic similarities is quite difficult. A small percentage of identical twins have been raised in different settings, and they serve as important resources for investigating gene-versus-environment questions. If a trait were due entirely to genetic factors, 100% of the identical twins raised in different environments would display the same phenotype. For example, all pairs of identical twins should have the same blood type even if every aspect of their environments differed.

Is the opposite true? That is, if 100% of the identical twins raised in different environments share a trait, is the trait due *solely* to genetic factors? No, because that conclusion presupposes zero environmental similarity between the twins, which is invalid. First of all, the twins experienced identical intrauterine environments, and prenatal environment can have profound effects on phenotypes. In addition, although the twins were raised in different settings, a researcher cannot be certain that every aspect of the environment affecting the phenotype differed, because they have no idea what those aspects are. The inability to control all relevant aspects of the environment is made more significant because the number of identical twins raised in different environments is so small. When sample sizes are small, biased results are more likely, and biased results lead to inaccurate conclusions. Because of small sample size and environmental similarity, studies that compare identical twins raised in different environments tend to overestimate the genetic contribution to the trait being investigated.

In an attempt to control for environmental factors, human geneticists compare the percent similarity for identical twins to that of fraternal twins. If a trait is due entirely to genetic factors, then 100% of the identical twins, but only 50% of the fraternal twins, share the trait (because fraternal twins, like all full siblings, share on average 50% of their genes). As we mentioned above, whenever people see a high percentage of similarity in identical twins, they tend to jump to the conclusion that the phenotypic similarity is due to

Table 12.3 Assessing genetic and environmental contributions[a]

Trait	% Similarity	
	Identical	Fraternal
Blood type	100	66
Eye color	99	28
Mental retardation	97	37
Fingerprint ridge count	95	49
Measles	95	87
Schizophrenia	58	13
Type I diabetes	65	18
Tuberculosis	57	23
Cleft lip	42	5
Congenital hip dislocation	40	3
Club foot	32	3
Breast cancer	6	3
Congenital heart defect	5	5

[a]Comparing the percent similarity of traits in identical twins to the percent similarity in fraternal twins can provide information on the relative contributions of genetic and environmental factors to phenotypic traits in humans.

genetic similarity. However, as is true in the example above, interpreting high similarity percentages can be tricky. Comparing the percent similarity results for identical twins to those for fraternal twins provides a useful reference point. For example, in Table 12.3, the percent similarity in identical twins displaying the measles phenotype is very high. If the researcher did not know that measles was an infectious disease, the high degree of similarity in identical twins might tempt him to conclude that the "measles phenotype" had a strong genetic contribution. However, the percent similarity in fraternal twins, which is significantly greater than the expected 50%, tells the researcher that environmental factors exert strong influences on this trait.

While this approach has its benefits, including a significant increase in sample size, and therefore in the accuracy of the estimate, it also suffers from an invalid assumption. We mentioned above that this method attempts to take environmental similarity out of the equation. Implicit in this approach is the assumption that identical twins have been raised in environments, both pre- and postnatal, that are as similar as the environments of fraternal twins. You know this is an inaccurate assumption. In the first place, fraternal twins are not necessarily the same sex. Attempting to eliminate this variable by using same-sex fraternal twins does not solve the problem, because the amount of environmental variation differs for the two types of twins. People tend to emphasize the similarities between identical twins: they dress them the same, give them similar names, and tend to notice, and remark on, their similarities rather than their differences. The inaccurate assumption that environmental similarities can be eliminated in explaining phenotypic similarity once again leads to an overemphasis on genetic factors.

Determining the relative contributions of genes and environmental factors to human traits has greatly improved because of recent technological developments. Biotechnology has provided new methods for answering questions about how genetic and environmental differences account for phenotypic differences. Using molecular markers or identified genes, researchers can actually see the percentage of individuals with similar genes that show similar phenotypes. We discuss this in greater detail later in the text.

S U M M A R Y   P O I N T S

Mendel studied traits that show simple inheritance patterns, now known as Mendelian: one-gene traits with distinct, observable phenotypic differences and clear dominance relationships. Very few visible phenotypic traits have this type of genetic basis.

Many people have many misconceptions about the relationship between genes and traits. The primary misconception is that one gene leads to one trait and that any one trait can be traced to a single gene. However, a gene usually affects many visible phenotypic characteristics, which is known as pleiotropy. On the other hand, a visible phenotypic trait almost always results from the activities of many genes interacting in different ways, such as additive or epistatic interactions. In addition, the same visible phenotypic trait may be traced to mutations in completely different genes, which is known as genetic heterogeneity.

The other misconception is related to the power of genes to determine a trait. Many people assume that someone who has a gene for a trait will definitely have that trait and that someone lacking that gene will not have the trait. However, a gene never acts alone; interactions among many genes, as well as environmental factors, create phenotypic traits. As a result, a person with a gene associated with a trait may not exhibit that trait, while someone without that gene may have the phenotype.

Many types of environmental factors influence the phenotypic expression of genetic information. In addition, different genotypes are affected by environmental factors to a greater or lesser degree. Therefore, not only genetic factors and environmental factors influence phenotypic expression, but also the interaction of genetic and environmental factors.

Because many genes and environmental factors contribute to the phenotypic expression of a trait, more often than not scientists cannot predict the phenotype from knowing the genotype, and vice versa. Scientists try to determine the relative contributions of genetic and environmental factors to a phenotypic trait by studying genetically identical organisms in different environments and genetically different organisms in the same environment.

Studies of human twins can shed a little light on the relative contributions of genes and environments to human traits, but in general, they tend to overestimate the amount of genetic contribution.

K E Y   T E R M S

Additive alleles	**Expressivity**	**Mendelian trait**	**Polydactyly**
Carrier	**Genetic background**	**Pedigree analysis**	**Polygenic trait**
Codominance	**Genetic heterogeneity**	**Penetrance**	**Quantitative inheritance**
Completely penetrant	**Incomplete dominance**	**Phenotypic plasticity**	**Variable expressivity**
Epistasis	**Incomplete penetrance**	**Pleiotropy**	

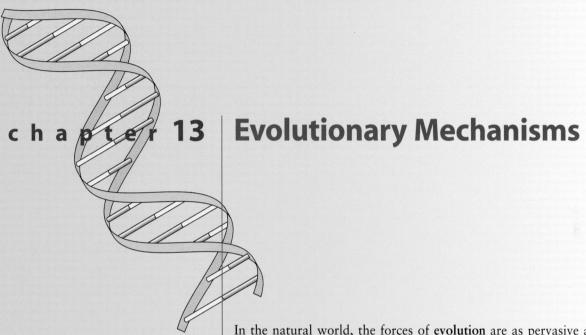

chapter 13 | Evolutionary Mechanisms

In the natural world, the forces of **evolution** are as pervasive and unavoidable as gravitational forces. Evolution's timescale makes its forces less obvious to a casual observer than those of gravity, but it is no less real (Figure 13.1). Living organisms cannot evade biological evolution any more than they can sidestep the laws of chemistry and physics: hydrophobic molecules avoid hydrophilic ones, living cells need energy, and populations evolve (Box 13.1).

When humans try to control or change the natural world but blithely ignore evolution's inevitability, Mother Nature quickly reminds them that such denial is foolish. Attempts to control crop pests and bacterial infections invariably lead to the evolution of resistance to pesticides and antibiotics. Consequently, biological evolution plays a role in determining which biotechnology product ideas are both scientifically possible and technically feasible. For example, will a microbe being developed for environmental remediation be able to compete with indigenous species? Will the pathogenic fungus overcome the transgenic crop's defense mechanism in 1 growing season or 10?

An understanding of biological evolution is useful for assessing its role in biotechnology product development decisions. Companies must factor evolutionary forces into the commercial equation to ensure long-term product viability. In addition, understanding evolution is absolutely essential for objectively analyzing the ecological concerns often raised in discussions of the risks of biotechnology. At their core, questions about the conservation of biodiversity, the impact of genetically modified corn on monarch butterflies, and gene flow from transgenic organisms to native species are questions about evolutionary biology.

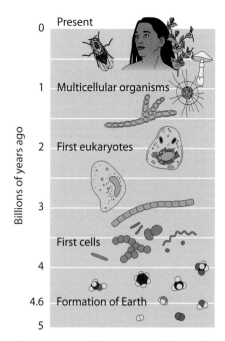

Figure 13.1 Evolutionary time frame. The timescale indicates some of the major events in the evolution of living organisms.

BOX 13.1 *A word about the word "theory"*

Our absolute statements about the inevitability of evolution might surprise you. Because biologists use the phrase "evolutionary theory," nonscientists often have the erroneous impression that biologists are still debating whether evolution occurs.

As you learned in chapter 1, scientific theories are laws, not conjectures or hypotheses. Not only do biologists accept the undeniable fact of evolution, evolutionary theory unites all of biology's subdisciplines under one roof.

The famous geneticist Theodosius Dobzhansky once said, "Nothing in biology makes sense except in the light of evolution." Evolution provides the theoretical framework for organizing all of biology, from protein interactions to the structure of ecosystems. Any biological observation can be viewed from an evolutionary perspective, and that perspective can be summed up in one question: how does this improve survival and reproductive success?

In this chapter, we will introduce you to some of the key concepts of modern evolutionary thought, including:

- the importance of phenotypic and genetic variation
- how genetic variation is generated
- the forces that change the genetic makeup of populations, especially natural selection

We will focus on the evolutionary mechanisms that operate at the level of populations and will not discuss broader evolutionary concepts, such as the origin of new species.

AN INTRODUCTION TO EVOLUTION

Even though the basic premises of evolutionary biology are straightforward and devoid of any emotional or inflammatory language, its ideas have been misused and abused. As a result, in spite of the simplicity and clarity of its fundamental concepts, biological evolution is often misunderstood by nonbiologists.

The first step in understanding evolution is determining what it is and is not. Biological evolution is simply a change in the frequencies of certain genes in a species' gene pool. A **species** is a group of actually or potentially interbreeding organisms.

This straightforward definition lends itself to mathematical analysis and scientific experimentation. In addition, it is devoid of any mention of the object (other than genes), mechanism, or goal of evolutionary forces. This definition of evolution might surprise you, because often nonscientists equate evolution with "the evolution of humans from monkeys."

Darwin and Wallace jointly proposed a theory of evolution

The name Charles Darwin has become synonymous with evolutionary theory, but another English naturalist, Alfred Wallace, deserves credit as well. Wallace and Darwin independently developed similar theories of evolution around the same time. Darwin's name is more commonly associated with evolution because his classic book, *On the Origin of the Species by Means of Natural Selection*, was so popular it became a best seller. Wallace, on the other hand, published his theory in a 20-page scientific manuscript.

The key elements of the Darwin-Wallace theory of evolution by **natural selection** are as follows.

- In every generation, the number of offspring produced exceeds the number that survive.
- Every group of organisms contains individuals that differ from one another in a number of traits.
- Some of these traits contribute to greater rates of survival and reproductive success in certain environments.
- Some of these traits are inherited.
- Those traits that are both favorable and inherited will persist and become more common over time.

Evolution by natural selection is analogous to the process, described in chapter 1, that plant breeders use to change the characteristics of crops over time. In both cases, genetic variation is produced *randomly* by natural processes. With natural selection, the environment replaces the breeder as the selector. Human-mediated selection is known as **artificial selection**, which is the force that domesticated plants and animals and created the diverse breeds of organisms—flowers, crops, cattle, or dogs—that we enjoy today (Figure 13.2).

Therefore, whether biological evolution is driven by natural or artificial selection, evolutionary change depends on a two-step circular process:

1. creating populations of genetically diverse organisms
2. culling certain individuals from that population, which is known in popular parlance as "survival of the fittest"

The individuals left standing after the selection process then reproduce, creating the next generation of genetically diverse organisms subjected to the next round of culling, and so on ad infinitum.

Figure 13.2 Artificial selection. A single, short-legged species known as *Tomarctus* gave rise to various species of jackals, coyotes, and wolves. The wolf (*Canis lupus*) is the immediate ancestor of the domesticated dog (*Canis familiaris*). The original ancestor of wolves and dogs contained a great deal of hidden genetic variation that, under selective pressures imposed by humans, has been molded into more than 100 breeds recognized by the American Kennel Club. (Photographs of dogs 1 to 4 courtesy of the American Kennel Club. Dog 5, Rosie the basset hound, is a close friend of the author.)

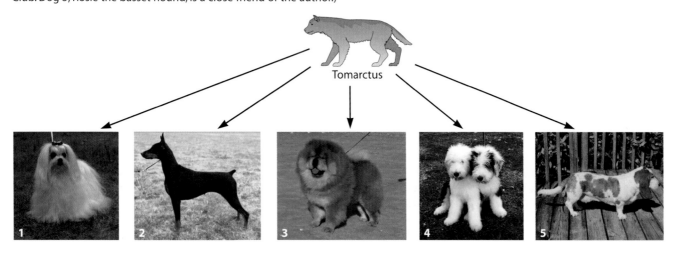

Although the title of Darwin's book refers to the origin of species, he did not provide a detailed mechanism through which natural selection acting on populations (**microevolution**) would create new species (**macroevolution**). According to his vague description, **speciation**, which is the creation of a new species, would occur when enough differences among individuals in a group accumulated to differentiate that group from other groups.

Combining Darwin's and Mendel's theories led to today's views of evolution

The main weakness of the Darwin-Wallace theory at the time of its publication in the mid-1800s and for at least 40 years thereafter stemmed from the scientific community's ignorance of the basic mechanics of heredity. Inheritable genetic variation serves as the linchpin of the Darwin-Wallace theory, but they did not know how traits are inherited or how genetic variation is created and maintained.

When Mendel's work was rediscovered and corroborated at the turn of the century, his theory of heredity and the theory of evolution seemed to contradict each other. Mendel specifically chose to study traits with discontinuous variation. Single genes controlled those traits, which, as a result, showed the classic Mendelian inheritance pattern. In formulating the theory of evolution through gradual, small changes, Darwin and Wallace focused on continuous variation in traits, which is much more common in nature. Traits with continuous variation are controlled by a number of genes and do not adhere to Mendelian patterns of inheritance.

Once geneticists learned more about the additive genetic nature of polygenic traits, they reconciled the theory of evolution through natural selection with Mendel's theory of inheritance. This merger grounded the Darwin-Wallace theory firmly in genetics and provided the modern theory of evolution, which is also known as the **synthetic theory of evolution** because it synthesizes two bodies of knowledge.

Variation provides the raw material for evolution

One of the most important, but subtle, contributions Darwin and Wallace made to biology was to change the way biologists view variation (Figure 13.3). Prior to Darwin and Wallace, scientists thought of variation as deviation from some sort of ideal type. Darwin and Wallace opened the door to viewing variation in a more positive light, because evolution depends upon:

- variation existing among organisms in a population
- some of that variation being inherited
- some of the inherited variation contributing to differential survival and reproduction

Variation is the rule, not the exception, in biological systems, irrespective of the level of biological organization. Whether the focus is DNA sequences, protein molecules, morphological structures, behaviors, or physiological functions, biologists expect to see **polymorphism**, which literally means many forms. By definition, observable variation is phenotypic variation, and some phenotypic variation is attributable to genetic variation and some to environmental factors. Because only genetic variation is the grist of the evolutionary mill, one of the first tasks of the evolutionary biologist is to determine whether the observable phenotypic variation has a genetic basis.

Figure 13.3 Normal frequency distribution. Most characteristics of organisms in nature have a continuous frequency distribution, which is sometimes a normal frequency distribution that can be represented by a bell-shaped curve. In a normal frequency distribution, most, but not all, organisms have a certain characteristic, which statisticians refer to as the mean or average. Before Darwin and Wallace, biologists thought of the mean as the "correct" form of the characteristic, and organisms lacking that form of the trait were considered deviant. Darwin and Wallace shifted attention from the ideal form to the variation in form, or in statistical terms, from the mean to the variance.

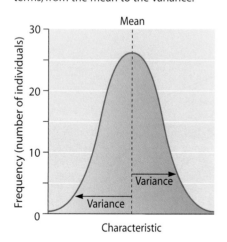

Evolutionary biologists have always studied genetic variation, even if they did not know it. Darwin, who knew nothing about the genetic basis of inherited variation, based his theory of evolution on polymorphisms visible to the naked eye: birds' beaks, iguanas' feeding behavior, and the shapes of tortoises' shells. He focused on phenotypic variation and assumed that some of the variation he could see mirrored invisible inherited variation. Later evolutionary biologists, understanding the relationship between genes and proteins, used various techniques to look at protein variation as a reflection of genetic variation. Today's evolutionary biologists are able to study genetic variation by looking directly at DNA variation, much of which may not be visible as a phenotypic trait. We discuss the use of DNA variation in evolutionary analysis at length in chapter 16.

Irrespective of the method used, when they ask questions about the amount and type of genetic variation, evolutionary biologists are constantly changing viewpoints:

- How much variation is there within a population?
- How much variation is there between populations (within a species)?
- How much variation is there between species?

They also ask the flip sides of these questions: how much genetic similarity is there at each of these levels?

All species are both genetically similar to and genetically different from one another (Figure 13.4). Both genetic unity and genetic diversity characterize all of the Earth's species. Within a species, the same is true: all individuals of a species (or a population) are genetically alike and genetically diverse.

Because a continuum of genetic variation and similarity exists within a species and extends outside of a species to related species, determining where one species stops and another begins can often be difficult and, at times, arbitrary. The point at which one chooses to demarcate a species boundary is flexible and, more often than you might expect, subject to lively debate. In Box 13.2, we discuss a very dynamic branch of evolutionary biology, taxonomy, which is devoted to the question "Where should we draw the line?"

Figure 13.4 Darwin's finches. On the Galapagos Islands, Darwin observed at least 14 different species of finch with different beak shapes, only 8 of which are shown here. He hypothesized that the 14 species evolved from a single ancestor. Competition for food provided the selective pressure that led to the different beak shapes.

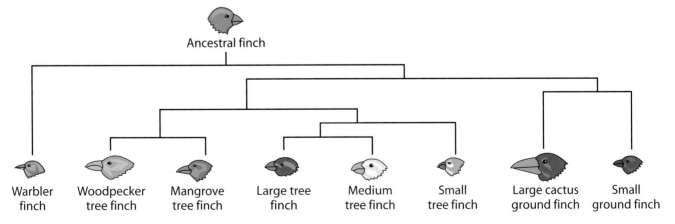

BOX 13.2 *Taxonomy: the science of classification*

Taxonomy, the oldest of the biology sub-disciplines, is the systematic classification of living organisms into categories, or taxons. The Earth supports millions of species, many of which are clearly visible to the human eye. One of the first steps in developing an understanding of any set of diverse observations is to place them into an organizing framework. To bring some sort of order to biological diversity, the ancient Greeks developed one of the first systems for classifying the Earth's organisms based on their phenotypic similarities.

If you were assigned the task of organizing various organisms into sets of similar types, there are many different classification schemes you could use. For example, you could classify animals according to whether they are able to fly, and the "can-fly" category would contain many insects, almost all birds, bats, and perhaps flying squirrels and flying fish, depending upon how you defined "can fly." Faced with the diversity of plants, you might place them in categories according to the colors of their flowers.

While these may be valid ways of sorting the Earth's organisms and assigning them to groups, how informative are these classification schemes? If an organism is identified as a member of a category, does knowing that give you any additional information about it? Learning that an animal is a member of the can-fly category tells you that it has some sort of structural adaptation that allows it to stay airborne but little else.

The categorization system biologists use to order the diversity of the Earth's species is richly informative, because it is a hierarchical classification scheme. Knowing that an organism is in a certain category enables biologists to know much more about it through extrapolation, because all individuals in one category automatically belong to every broader category in the hierarchy. For example, if your school's student population is categorized according to home town, knowing that a classmate belongs to the Seattle category also tells you he

is from Washington state, the Pacific Northwest, the West Coast, the United States, and the North American continent. These are nested categories; all individuals in one category fit inside the broader category, much like the nested beakers toddlers play with.

Taxonomists use the following set of nested categories to classify organisms:

Kingdom
  Phylum
    Class
      Order
        Family
          Genus
            Species

The broadest category in biological classification is the highest taxonomic level, kingdom. At present, biologists recognize five kingdoms: Monera (the bacteria or prokaryotes), Protista (most one-celled eukaryotes, such as amoebas and unicellular algae), Fungi (mushrooms, molds, and yeasts, which are unicellular), Plantae, and Animalia (see the first figure).

The smallest taxonomic unit in biology is the species, which is defined as a group of interbreeding organisms, so taxonomists place all organisms in the same species beaker if they can interbreed. If two populations of similar organisms cannot interbreed because they are separated geographically or temporally, taxonomists make very educated guesses about their potential to interbreed if they coexisted.

If the ability to interbreed is the criterion for belonging to the same species, what factors determine membership in the higher taxonomic categories, such as genus, family, and order? The hierarchical classification system for biological organisms is based on their evolutionary relationships and therefore their genetic similarity. Biologists place organisms in the same category because, after having looked at many different characteristics, they decide that the organisms share a common ancestor. Some characteristics biologists use to determine genetic similarity are phenotypic measures, such as physical features and the

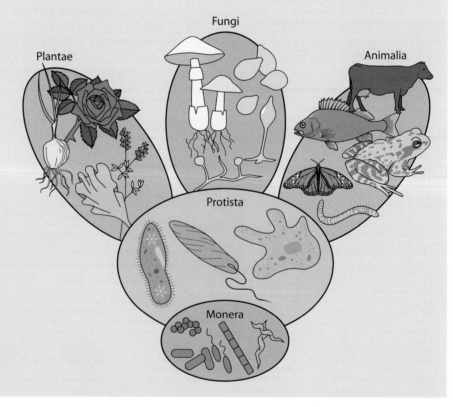

BOX 13.2 *(continued)*

Kingdom

Phylum

Class

Order

Family

Genus

amino acid sequences of proteins, as well as direct assessments of DNA similarity, which we discuss in chapter 16.

Organisms in any one category are more similar genetically to each other than they are to other organisms in a different category at the same taxonomic level. For example, using the table and the second figure, you can see that cattle (Bovidae) and deer (Cervidae) share more genes with each other than deer or cattle share with mice. Therefore, cattle and deer are in the same order (Artiodactyla) and mice are in a different order (Rodentia). Similarly, deer, cattle, and mice are more similar geneti-

cally to each other than they are to frogs.

These relationships are clear from their visible phenotypes. Often, however, visible phenotypes are so similar that they can obscure genetic differ-

ences. In this example, it would be easy to assume that the two deer species are in the same genus when they are not. In these types of cases, measurements of similarity in proteins and DNA can be essential for correct classification.

Animal	Taxonomic relationships of the fallow deer, *Dama dama*					
	Kingdom	**Phylum**	**Class**	**Order**	**Family**	**Genus**
Butterfly	Animalia	Arthropoda				
Frog	Animalia	Chordata	Amphibia			
Mouse	Animalia	Chordata	Mammalia	Rodentia		
Cattle	Animalia	Chordata	Mammalia	Artiodactyla	Bovidae	
Fallow deer	Animalia	Chordata	Mammalia	Artiodactyla	Cervidae	*Dama*
Red deer	Animalia	Chordata	Mammalia	Artiodactyla	Cervidae	*Cervus*

THE SOURCES OF GENETIC VARIATION

Biological evolution is the genetic change of a population over time. Even more specifically, it is the change in the relative proportions of alleles in a population. Therefore, in addition to being the repositories of information for synthesizing proteins, managers of cell processes, and contributors to phenotypic traits, genes are the units of evolutionary change.

Most of our discussion of genes until now has dealt with genes in individuals, or in other words, the mechanisms of gene action and inheritance: how genes are expressed, interact with each other, contribute to phenotypic traits, and pass from one generation to the next. In thinking about evolutionary genetics, step back from the individual and its genes and focus on populations of organisms, not individuals, and extend the timescale from one or two generations to many.

Biologists define a **population** as organisms in the same species that live in the same place at the same time and therefore can interbreed. The combined set of genes within a population is its **gene pool**; therefore, the gene pool contains all of the allelic forms of all genes in that population. Allele pool would be a more appropriate term. Population genetics, the child of the Darwin-Wallace and Mendel merger, is the branch of evolutionary biology that investigates changes in the genetic makeup of populations over space and time.

Nature has a number of methods for continually creating the genetic variation that serves as the raw material for evolution. Biologists group these mechanisms for generating genetically diverse organisms into two broad categories: mutation and recombination.

Mutations create variation by altering genetic material

Mutations are the ultimate generators of the genetic variants that feed into the evolutionary process, because they change a cell's genetic information by altering the DNA molecule. Geneticists recognize two classes of mutations based on the amount of genetic information changed by the mutation: gene mutations and chromosome mutations. Some mutations are induced by external factors, such as ultraviolet radiation, while others result spontaneously from endogenous foul-ups in various cell processes. In either case, they occur randomly and are copied every time the cell divides.

Biologists classify mutations in a number of different ways, some of which are listed in Table 13.1. Mutations can occur in body (somatic) cells or in germ cells (gametes). Only changes that are propagated from one generation to the next are subjected to evolutionary forces, so in this discussion we will focus solely on mutations in gametes, or **germinal mutations**. Initially, they all must be viewed as "mistakes," but with time evolutionary forces convert some of the mistakes in one generation into advantages in another.

In earlier chapters, you learned that the effects of genetic changes vary: some mutations are beneficial, others are devastating, and many are neutral. Whether a mutation is benign or devastating does not *necessarily* depend on whether it involves a small change or a large one. Recall from chapter 4 that changing a single nucleotide in the gene encoding the protein hemoglobin sets in motion a series of events that culminates in the devastating effects of sickle-cell anemia. In some of the examples described below, however, major changes in an organism's genome have no deleterious effects and have

Table 13.1 Categories and types of mutations[a]

Category and type	Description
Cell type	
Germinal	Occurs in the cell line that will become the gametes; therefore, all are inherited mutations.
Somatic	Occurs in any cell other than germline cells. If it occurs in a single cell, it is virtually impossible to detect in fully developed organisms, but if it occurs in a single cell in developing tissues, a patch of cells exhibits the mutation. The great majority are not inherited. If a plant cutting consists primarily of cells with a somatic mutation, some will develop into germinal tissue and the mutation will be inherited.
Cause of mutation	
Spontaneous	Caused by endogenous factors, such as uncorrected errors in DNA replication or the insertion of transposable genetic elements.
Induced	Triggered by an environmental factor, such as the ultraviolet radiation in sunlight and some chemicals.
Nature of point mutation	
Substitution	One nucleotide is substituted for another.
Deletion or duplication	A nucleotide is added or lost. No effect if it occurs in a part of the DNA that is removed from the messenger RNA prior to translation into a protein.
Frameshift	Addition or loss of nucleotide(s) shifts reading frame for protein synthesis. Usually leads to nonfunctional protein.
Effect of mutation	
Silent	No amino acid change.
Replacement	Amino acid change may or may not affect protein structure. Creates new alleles.
Nonsense	Creates a termination codon, leading to a very short, nonfunctional protein.
Neutral	May affect protein structure or function but has no effect on fitness.
Conditional	Mutant phenotype appears only in certain environments.

[a]These categories are not mutually exclusive.

provided excellent raw material for evolution. For example, as explained below, a complete doubling of genetic information has been very important in the evolution of many plants. In general, however, the larger the amount of genetic information that mutates, the more likely the mutation will be harmful.

Gene mutations

Mutations within single genes are called **micromutations**, because a relatively small amount of DNA is changed. **Point mutations**, the ultimate micromutations, are single-nucleotide alterations in a gene that occur randomly and then escape DNA repair.

You have already learned a great deal about how small changes in the nucleotide sequence of a gene might affect the amino acid sequence of a protein and how that change might affect protein function. Some point mutations, **silent mutations**, have no effect on the encoded protein because they do not lead to a change in its amino acid sequence. Those leading to changes in the amino acid sequence (**replacement mutations**) can be beneficial, neutral, or deleterious depending on whether they affect the protein's ability to carry out its function. Because relatively few amino acids are critical to a protein's function, it is reasonable to assume that the majority of replacement mutations are benign because they do not alter the active site.

In studying the relevance of a mutation to evolution, however, we must distinguish between the impact of a mutation on the protein molecule and its impact on the organism. Significantly altering a protein's structure and function may not affect an organism's ability to survive and reproduce. Still other mutations could lead to total loss of gene function without any detrimental effect on the organism. Examples of this phenomenon include the O blood type and coat colors in Labrador retrievers. Such mutations are irrelevant to evolution by natural selection, which is concerned only with mutations that improve or hinder survival and reproduction. No matter what effect a gene mutation may have on its encoded protein, mutations with no impact on survival and reproduction are called **neutral mutations** by evolutionary biologists.

Gene mutations usually create new forms of existing genes, or as an evolutionary biologist would say, they add new alleles to the existing gene pool. Whether or not these new alleles become established in the population as alternate forms of a gene depends primarily upon the effect they have on their owner's survival and reproduction, which is highly dependent on environmental factors.

Chromosome mutations

Changes in more than one gene are called **macromutations**, or **chromosomal aberrations**, and they (1) increase or decrease the total amount of genetic information in a cell or chromosome or (2) rearrange portions of chromosomes, thereby altering the positions of the genes relative to one another or changing the number of chromosomes in a cell without increasing or decreasing the amount of genetic material in the cell.

Unlike point mutations, these mutations can affect chromosome pairing and segregation during meiosis, which often leads to significant loss of gamete viability. As a result, the probability of propagating chromosome mutations from one generation to the next, and therefore the degree to which they provide genetic variability on which evolution can act, depends not only on their phenotypic effect but also on whether they disrupt meiosis.

Many chromosome mutations are lethal or highly deleterious. Most involve chromosomal breakage, and those breaks may occur in the middle of an important gene. In other chromosome mutations, a sizable number of genes may be lost, and loss of a significant amount of genetic information is almost always lethal to both gametes and organisms. However, even though macromutations by their very nature involve major alterations in a cell's genetic material, their effects are not always detrimental to the well-being of either the gamete or the organism carrying the chromosome mutation. Because the impacts of chromosomal aberrations on the evolutionary process vary in both type and significance according to the exact change that has occurred, we must discuss each type individually.

Some macromutations change internal chromosome structure

Chromosome mutations that lead to structural alterations in chromosomes include the following (Figure 13.5).

Deletion One or many genes are lost from the chromosome.

Duplication One or more genes are added to a chromosome.

Inversion A segment of chromosome breaks, rotates 180°, and reinserts.

Translocation Nonhomologous chromosomes break and exchange pieces.

Duplications and deletions

Most gene duplications and some deletions are artifacts of a poorly managed crossing-over event during meiosis (Figure 13.6). Slightly imperfect alignment of homologous genes can lead to a redundant gene in one gamete and a lost gene in another gamete. If the addition or loss of a gene is not lethal to the gamete, then these abnormal gametes will probably unite with normal gametes, giving rise to an organism heterozygous for the duplication or the deletion. The cells that will become gametes will also be heterozygous for the mutation, so the gene duplication or deletion can be passed to the next generation.

Of these two types of chromosome mutations, gene duplication interests evolutionary biologists more than deletions, because the extra gene may not

Figure 13.5 Macromutations. In these diagrams of four types of chromosomal aberrations, the letters represent genetic loci.

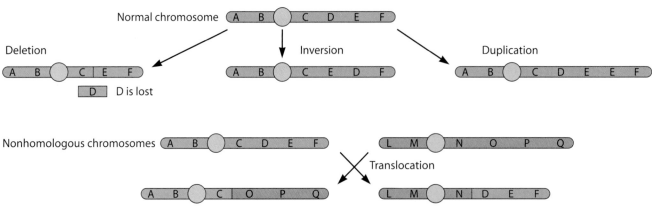

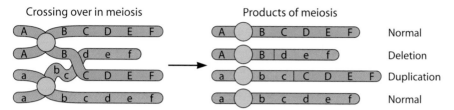

Figure 13.6 Gene deletion and duplication. The letters indicate genetic loci, and the lowercase and uppercase letters represent different alleles for the same trait. In meiosis, if homologous chromosomes are not aligned perfectly when crossing over occurs, certain genes, indicated by the letters, will be lost from some chromosomes and added to others. As a result, some gametes have gene duplications and others have gene deletions.

only alter the chances for a heterozygote's survival but also lead to the appearance of a new gene in subsequent generations. On the other hand, some gene deletions may be tolerated, at best, by an organism heterozygous for the deletion, but the odds are heavily against a beneficial gene deletion occurring and becoming an established component of the gene pool.

Some gene duplications have immediate evolutionary implications because they improve an organism's ability to survive and reproduce by increasing production of the encoded protein. For example, yeast cells with a gene duplication for a certain enzyme are better at extracting phosphorus from their environment. In environments where phosphorus is a growth-limiting factor, a yeast with the gene duplication outcompetes a yeast lacking the duplication.

Gene duplications also play an important role in evolution by serving as a reservoir from which new genes can be created. Because the original gene continues to produce a product, the redundant gene is free to mutate into new forms without jeopardizing the original gene's function. The first few mutations in the duplicate gene might lead to new allelic forms of the existing gene, but as mutations freely accumulate, the duplicate gene could become a new gene locus, not simply a new allele for an existing locus.

Certain groups of related genes appear to have originated via gene duplications. For example, the gene for myoglobin and the hemoglobin gene family, which consists of alpha-globin, beta-globin, epsilon-globin, and zeta-globin, encode a number of similar proteins involved in oxygen transport. The globin proteins are encoded by genes with considerable amounts of sequence homology, suggesting they all are derived from the same parental DNA sequence. Different globin proteins have different survival values. In adult humans, hemoglobin consists of two alpha-globin molecules and two beta-globin molecules, but the hemoglobin found in human embryos contains zeta- and epsilon-globins and has a higher affinity for oxygen than adult hemoglobin, facilitating the transfer of oxygen from mother to embryo.

Inversions and translocations

Because inversions and translocations involve changing the positions of genetic loci relative to one another, they deter appropriate pairing and crossing over in meiosis. Most gametes produced by individuals heterozygous for translocations and inversions do not contain a full complement of the genes (Figure 13.7). Consequently, macromutations that rearrange the positions of genes on chromosomes are not a ready source of genetic variation usable for driving evolutionary change. However, in certain cases, translocations and inversions have become established components of a species' gene pool, so

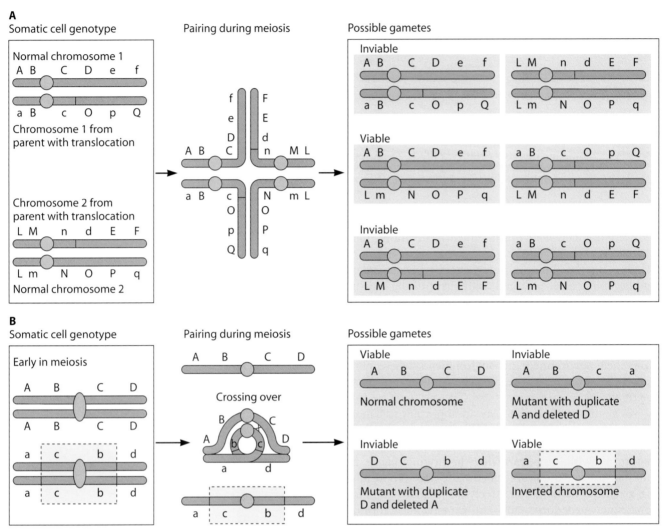

Figure 13.7 Disruptive meiotic pairing in chromosomes with translocations or inversions. **(A) Translocation mutants.** The organism has inherited one set of normal nonhomologous chromosomes 1 and 2 from one parent (solid-colored chromosomes). Its other parent had a reciprocal translocation between chromosomes 1 and 2 (two-colored chromosomes). Under these conditions, pairing of homologous chromosomes in meiosis involves all four chromosomes instead of two members of a homologous pair. Four of the six possible gamete types are inviable. **(B) Inversion mutation.** When a normal chromosome pairs with its homolog, if one member of a pair has an inversion mutation (highlighted), crossing over can lead to gametes with gene deletions and gene duplications. Many of these will not be viable; viable gametes contain either a normal or an inverted chromosome. The chromatids with inversions that do *not* cross over survive in the population, and those that do cross over are selected against. Therefore, the genes in the inverted segment tend to be inherited as a unit.

they are not universally deleterious. Of the two, chromosomal inversions are more easily tolerated by a species and have even proven to be advantageous under certain conditions.

Unlike deletions and duplications, the amount of DNA does not change with a chromosome inversion. Because no genes are lost or gained, inversions are not highly deleterious to the organism unless the break occurs within a gene. Individuals homozygous for inversions are found in many species with no apparent impact on survival and reproduction. However, in some cases, the physical location of a gene in relation to other genetic material may influence the organism's phenotype. For example, if a gene in an inversion is relocated to an area of the chromosome near DNA sequences that are not expressed, its expression may be significantly reduced. Such **position effects** have been described in a wide variety of species.

The significant impact of inversions comes not from the effects they have on an organism's survival but from their effects on gamete viability by disrupting meiosis. In individuals that are heterozygous for an inversion, locus-to-locus alignment during homologous chromosome pairing is achieved through contorted arrangement, so successful crossing over is rare. The chromatids that attempt to cross over usually lose or gain a number of genes, and gametes containing these chromatids are not viable. Because crossing over of at least one chromatid is almost universal in meiosis, inversion heterozygotes suffer from decreased fecundity due to gamete loss. The only viable gametes they are capable of producing contain the chromatids that did not cross over, so the genes within an inversion are essentially locked together and inherited as a unit.

Surprisingly, in spite of the decreased fecundity suffered by organisms heterozygous for inversions, some inversions become established in some populations, indicating a degree of positive survival value to offset the decrease in reproductive output. The answer appears to lie in genes becoming tightly linked and being transmitted as a single unit. Sometimes these linked sets of genes, or supergenes, are important evolutionarily because they represent specific gene combinations that are advantageous.

Some macromutations change the number of chromosomes

In meiosis, homologous chromosomes pair and then separate when the cell divides. If the chromosomes do not separate, one of the resulting cells contains two copies of a chromosome (both members of the homologous pair), while the other has none. The odds are that gametes with abnormal numbers of chromosomes will be fertilized by normal gametes and, if the chromosome addition or deletion is not lethal to the gamete, the resulting offspring will either have too many or too few chromosomes. The failure to separate can affect one or a few chromosome pairs (**aneuploidy**) or the entire chromosomal content (**polypoloidy**). In the latter case, the genetic content of the offspring formed from the fusion of an unreduced gamete (diploid [2N]) and a normal haploid (1N) gamete is triploid (3N), because its cells contain three copies of every chromosome. Triploid organisms typically exhibit poor fertility, because the odd number of chromosomes precludes normal pairing in meiosis.

You are already familiar with examples of abnormal separation of single chromosomes described in previous chapters: Down syndrome (three copies of chromosome 21), Turner's syndrome (XO), and Klinefelter's syndrome (XXY). All of these chromosomal aberrations lead to early death, sterility, or reduced fecundity and therefore play no role in creating genetic variation that can serve as the foundation for evolution.

You are also very familiar with a polyploid organism: the banana, which is a triploid. The normal diploid banana has relatively large, hard seeds, but the bananas people eat have only the tiny dark brown specks that are rudimentary seeds and, as expected in triploid organisms, are therefore sterile. When triploidy occurs spontaneously in wild bananas, the plants are selected against by evolutionary forces because they cannot reproduce sexually. So where do all of the millions of bananas people consume come from? Asexual propagation. The banana provides an excellent example of human interference with natural evolutionary forces and exploitation of mistakes that make people's lives better. We discuss many more examples of human-induced polyploidy for improving crops in chapter 22. However, our interest here is in natural polyploidy and evolution.

Multiplication of the entire chromosome complement has actually been an essential source of variation in the evolution of many flowering plants and ferns, so plants will be the primary focus of our discussion of polyploidy. Although it occurs infrequently, we also find polyploidy in animal species that are self-compatible hermaphrodites (earthworms) or those that can reproduce through parthenogenesis, in which females produce offspring without being fertilized by males (some shrimp, frogs, salamanders, fish, lizards, and insects). In animals, as in plants, polyploidy has played a role in the evolution of new species. For example, species in the salmon family, which includes trout and salmon, have exactly twice the amount of DNA found in their ancestor, suggesting that they appeared originally as tetraploids (4N).

How do tetraploids originate? When a plant (or animal) with unreduced gametes self-fertilizes, the resulting offspring contains two complete sets of chromosomes. Because the tetraploid plant contains an even number of chromosomes, normal pairing in meiosis can sometimes occur, leading to viable gametes. The tetraploid offspring can also self-fertilize or cross with sibling tetraploids, so a sizable tetraploid population can become established fairly quickly. But in order to be an important evolutionary factor, as suggested by its prevalence in plants, tetraploidy must also offer survival advantages, and it does. Tetraploid plants tend to be larger and more robust than their diploid ancestors, and they also have larger fruits and flowers, all of which would be selected for because they improve reproductive success.

But how does polyploidy, increasing the number of chromosomes, increase genetic variation? Just as in gene duplications, because the plant has a full complement of normal genes, the extra genes are free to mutate without jeopardizing the plant's capacity to survive. However, a different form of polyploidy actually creates recombinant offspring that are so genetically distinct from their parents they are incapable of interbreeding.

Some plants hybridize easily with plants in other species. When this occurs, viable seeds from this cross grow into hybrid plants that are, for all practical purposes, sterile because normal meiotic pairing is often compromised and few viable gametes are produced. Nonetheless, as is true of other plants, accidental chromosome doubling occurs infrequently, and each chromosome now has a twin with which it can pair at meiosis. The mistake of spontaneous doubling in a hybrid plant leads to viable gametes and fertile offspring containing a complete complement of the chromosomes found in the original two parent species. Although these events occur rarely compared to normal within-species plant reproduction, the form of polyploidy that combines the genomes of two species appears to have been a major driving force of speciation in plants, including many crop plants.

Transposable genetic elements are a special case

Transposable genetic elements are pieces of DNA that move from one genome site to another. The genome alterations they cause differ conceptually from mutations and chromosome aberrations because the genetic changes that occur do not result from mistakes or malfunctions in normal cell processes. In the 1950s, the corn geneticist Barbara McClintock observed an inheritance pattern for kernel color that could only be explained by assuming that certain genes move from one location to another in the genome. Such an idea challenged the prevailing mindset that envisioned a gene as permanently fixed at the same site on the same chromosome. Transposable genetic elements have now been found in virtually all types of organisms, from viruses to humans.

Transposable genetic elements, which are sometimes referred to as mobile genetic elements, come in a variety of types and operate using different mechanisms, but all are capable of mobilizing themselves. They range in size from a gene fragment to a few genes. As a group, they can move within chromosomes, between chromosomes, between plasmids, and from plasmids to chromosomes. A **plasmid** is a small circular DNA molecule that is not part of the cell's chromosomes and replicates independently of them. Some simply jump from one location to another, but most copy themselves before jumping, so each transposition event leads to an increase in the number of transposable genetic elements in the genome (Figure 13.8). With rare exceptions (discussed below), transposable genetic elements are not beneficial to the organism, because most contain only the gene sequences required for a transposition event. Therefore, until scientists learn more about them, they are best thought of as a sort of genome parasite that seems to exist for the sole purpose of replicating itself.

As a transposable genetic element goes about its business of sending a copy of itself to a different site within the genome, it can cause problems for the genome. If it inserts itself into the middle of a gene, it can inactivate that gene and disrupt gene expression in other, downstream genes. Many generate a high rate of gene deletion in the vicinity of their insertion, while others are associated with chromosomal rearrangements, accompanied by an inability to pair correctly at meiosis. At present, scientists are uncertain how to reconcile the many costs to the host with the widespread occurrence of transposable genetic elements in host genomes.

However, in one case, the value of harboring a transposable genetic element is obvious. In bacteria, a type of transposable genetic element known as a **transposon** contains not only the genes for proteins involved in transposition, but also genes for resistance to antibiotics. This imparts superior survival value to the host under the appropriate conditions. We discuss this at length in "Evolution in Action: the Spread of Antibiotic Resistance" below. In addition, in some cases, when the transposable genetic element replicates itself prior to mobilizing, it also replicates a portion of the host DNA. Therefore, the sort of selective advantages discussed in gene duplication would also apply in this case.

Recombination creates variation by redistributing existing genes

The second mechanism for generating genetic variation is recombination, the joining of genetic information from two sources to produce new genetic combinations. Recombination differs from mutation because the genetic material is not changed but simply reassorted. The precise mechanisms for recombining genetic material from two sources vary with the mode of reproduction of the organism: sexual or asexual.

Sexual reproduction and recombination

A graphical summary of the relationship between sexual reproduction and recombination is provided in Figure 13.9. In sexually reproducing organisms, recombination of genetic material occurs during the production of gametes and their fusion in fertilization. Let's start with the most straightforward example of recombination: fertilization.

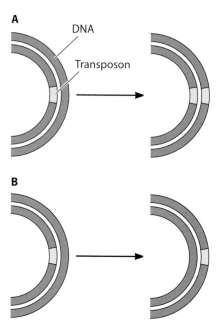

Figure 13.8 Transposable genetic elements. **(A)** Transposable genetic elements that move by replicative transposition are duplicated in the process of jumping: a copy-and-paste mechanism. **(B)** Nonreplicative transposable genetic elements move in a cut-and-paste manner. No copy is made.

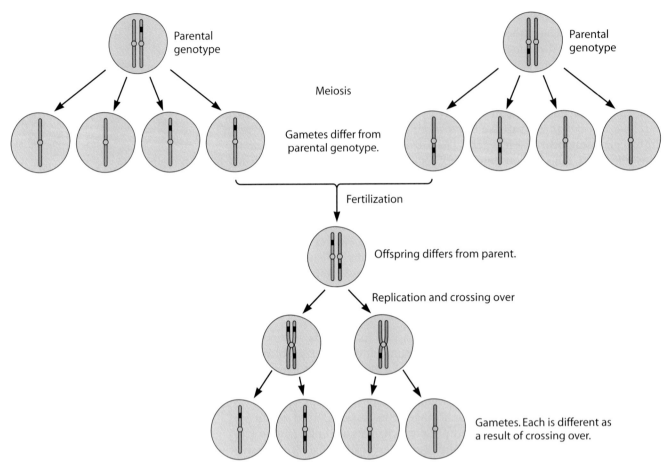

Figure 13.9 Sexual reproduction, genetic variation, and recombination summary. The simplified schematic shows the three ways genetic variation is created during sexual reproduction. (1) A single pair of homologous chromosomes is shown. Gametes produced in meiosis differ from the parental genotype because they contain half the amount of DNA and a random assortment of nonhomologous chromosomes (not shown); some had been inherited from the mother and some from the father of the gamete producer. (2) During fertilization, genetic material from two sources is combined, creating an offspring that differs genetically from both parents. (3) Finally, during gamete production in that individual, crossing over between its maternal and paternal (homologous) chromosomes occurs, creating within-chromosome genetic variation. Although crossing over (3) and independent assortment (1) are discussed sequentially, both occur in each round of gamete production.

When the gametes fuse, the resulting zygote's genetic information is a new combination of genetic information from the two parents. Thus, the offspring that develops from the zygote contains genetic information from two sources and is genetically novel: it is not genetically identical to either parent. All offspring of sexual reproduction, including you, are genetic recombinants.

Recombination also occurs during the production of these gametes through meiosis. Increased genetic variation in populations is the result of two associated meiotic events: the independent assortment and segregation of nonhomologous chromosomes and the crossing over that occurs between homologous chromosomes.

As a result of independent assortment and segregation of nonhomologous chromosomes in meiosis, a gamete differs genetically from the cell that gave rise to it by having one-half as much genetic material (segregation) and a random assortment of the maternal and paternal genetic material (independent assortment of nonhomologous chromosomes). Thus, each gamete contains genetic material from two sources (maternal and paternal) in new combinations. The number of possible combinations of nonhomologous chromosomes that could be generated in the production of each human gamete is 2^{23}, or over 8,000,000.

Independent assortment and segregation alone are rich sources of genetically variable gametes, but the exchange of genetic material between maternal and paternal chromosomes that occurs when homologous chromosomes pair during meiosis generates even more genetic variability. Crossing over produces a chromosome that is a recombinant: a hybrid containing genes of both maternal and paternal origin, or a novel combination of genetic material from two sources.

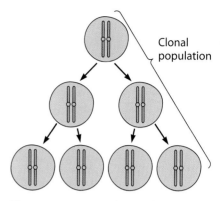

Figure 13.10 Asexual reproduction. Offspring are genetically identical to the parent and, as such, are clones of the parent and each other. Any genetic variation in a clonal population is due to mutation.

Asexual reproduction and recombination

A typical example of asexual reproduction is the division of a bacterium to produce two bacterial cells. These two daughter cells then divide, the four progeny divide, and so on until a large **clonal population**, which consists of genetically identical cells, is generated (Figure 13.10). Bacteria are not the only organisms that reproduce asexually. Yeasts and some other fungi, protozoans, and algae can reproduce by simple fission or by sexual reproduction. A few species of animals reproduce asexually. Many plants can propagate themselves asexually; humans take advantage of this to duplicate desirable plants by rooting leaves and cuttings.

In single-parent, or asexual, reproduction, a copy of the genome of the parent is passed along in its entirety to the offspring, which in the absence of mutation are genetically identical to the parent organism and each other. You can see from comparing Figures 13.9 and 13.10 that populations of asexually reproducing organisms would be far less genetically diverse than sexually reproducing populations if they had no mechanism other than mutation for generating genetically variable individuals. Because of the centrality of genetic diversity to evolutionary success, it should come as no surprise that asexually reproducing organisms have developed novel methods for generating genetically diverse populations. For example, many eukaryotes alternate between sexual and asexual reproductive cycles.

Microbes evolved other mechanisms for introducing genetic diversity into their populations. Although it might be tempting to think that brilliant scientists invented the sophisticated gene transfer techniques they commonly use in modern biotechnology laboratories, microorganisms have routinely been transferring genes among themselves and between species for billions of years. In fact, bacteria have three natural mechanisms by which genetic materials from two sources can be combined: conjugation, transformation, and transduction, all of which have been coopted by biotechnology researchers. Each of these processes not only is important in providing genetic variation for the evolution of bacteria but also plays an important role in human interactions with microbes.

Conjugation.   Conjugation (Figure 13.11) is a process by which one bacterium transmits a copy of some of its DNA directly to a recipient bacterium it is physically contacting. Conjugation is often called bacterial sex, because sex, in a strict biological sense, is the mixing of genetic information.

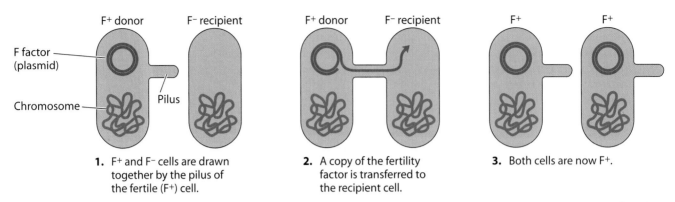

1. F+ and F− cells are drawn together by the pilus of the fertile (F+) cell.

2. A copy of the fertility factor is transferred to the recipient cell.

3. Both cells are now F+.

Figure 13.11 Bacterial conjugation. Genetic material is exchanged between F^+ and F^- cells, creating two F^+ cells.

In conjugation, the genetic mixing occurs within the recipient cell and does not result in the production of offspring. After genetic exchange, a bacterium undergoes asexual reproduction as usual and passes its new genetic information to its offspring.

In conjugation, the donor, or male, cell donates genetic information to the recipient, or female, cell it is contacting. What makes a bacterium male or female? Male cells contain special plasmids called **conjugative plasmids** that contain several genes required for their conjugational transfer. Plasmids do not contain any genes that are essential to the life of the organism. However, they often (but not always) contain genetic material that gives the cell survival advantages under certain conditions, such as genes for synthesizing molecules that kill other bacteria and genes for resistance to antibiotics and harmful heavy metals. Only some plasmids are involved in conjugation.

Scientists called the first conjugative plasmid they discovered the F (for fertility) plasmid. Any cell that contains the F plasmid can synthesize all the proteins needed for conjugation (from the F genes) and so is considered fertile. Fertile cells synthesize a **pilus**, a tubelike appendage that protrudes from the outer membrane. The pilus binds to a recipient cell, which lacks the F plasmid, and brings the pair together. The F plasmid is then copied and transferred to the recipient cell. The recipient thus becomes fertile, and the donor remains so.

Additional genes can be incorporated into the F plasmid by natural recombination processes. F plasmids that contain other genes are often called F′ (pronounced F-prime) plasmids. These bacterial genes are then transferred during conjugation. Finally, the F plasmid occasionally recombines into a bacterial chromosome. When these fertile cells begin conjugation, they attempt to replicate and transmit the entire circular chromosome. (They can actually transmit the whole thing if they remain paired with the recipient long enough.)

Many plasmids besides F encode proteins that allow them to be transmitted by conjugation in a manner similar to that of F. These conjugative plasmids also occasionally allow the transmission of nonconjugative plasmids, so virtually any plasmid can be transferred at some frequency. Some plasmids can replicate in only a few types of host and promote conjugation only between those hosts. Others, however, have a very broad host range and promote conjugation in hundreds of bacterial species.

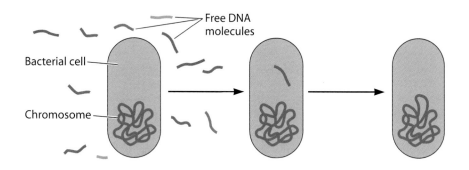

Figure 13.12 Transformation. A cell takes up free DNA from its environment, integrates it into the chromosome, and expresses the encoded products.

Transformation. **Transformation** occurs when cells take up free DNA molecules from the environment and express the information encoded in this DNA (Figure 13.12). Transformation was first observed by Frederick Griffith, who found that when a living unencapsulated (rough) strain of pneumococcus (*Streptococcus pneumoniae*) was mixed with dead cells of an encapsulated (smooth) strain, the rough-colony strain was transformed into the smooth-colony form (see chapter 1).

Bacterial cells in a state that allows them to be transformed are said to be competent. Only some bacterial strains are naturally competent. *Escherichia coli* is one of the many bacterial strains that do not undergo transformation naturally. Naturally competent cells contain proteins dedicated to the process of transformation. Proteins on the outsides of these cells bind DNA and transport it inside. Internal proteins then compare the base sequence of the new DNA with the genome of the organism. If sufficient similarity (homology) is found, the new DNA is recombined into the genome and expressed. As part of the genome, this new DNA is replicated and passed on to daughter cells.

If the new DNA is not similar in sequence to the genome of the organism, it is not incorporated into the genome and is lost. In this way, naturally competent cells have access to genetic variability through transformation but do not waste their time expressing completely irrelevant genes.

Transduction. In bacterial **transduction** (Figure 13.13), a virus intermediary (bacteriophage) carries bacterial DNA from one bacterium to another. At one time, scientists thought that transduction in nature was fairly rare; however, natural transduction has been observed in soil, on plant surfaces, and in lakes, oceans, rivers, and sewage treatment plants. In addition, even though

Figure 13.13 Transduction. A virus serves as a conveyer of genetic material from one organism to another. In this example, the source and recipient of the transferred DNA are both bacterial cells, and the virus is a bacteriophage.

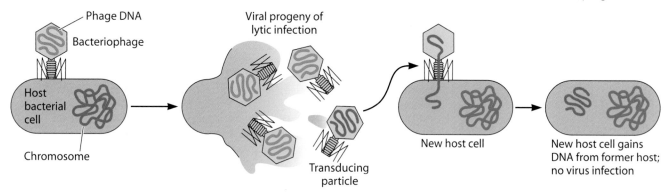

our discussion is limited to bacterial transduction via phage viruses, transduction is not limited to microorganisms and has been observed in organisms such as shellfish and mice.

Many different bacteriophages are capable of transduction. Most infect only one bacterial species, but a few can infect many different species. The details of the transduction mechanism depend on whether a particular phage's life cycle is **lytic** or **lysogenic** (Box 13.3).

During a typical virus infection, the virus injects its nucleic acid into the host cell. In a lytic cycle, the phage hijacks the bacterial enzymes to replicate its nucleic acid and synthesize its protein coat, properly called a **capsid**. Near the end of the lytic cycle, the viral nucleic acid is packaged into the capsid to create a new virus particle, which is then released from the host cell. Some bacteriophages occasionally make an error and package a piece of the host cell's DNA instead. This event is usually random, so any bacterial gene could end up inside a virus particle.

Virus particles containing bacterial DNA instead of viral genetic material are completely capable of attaching to a new host cell and injecting the bacterial DNA, because those functions depend on capsid protein and are independent of its contents. Once inside the new cell, the bacterial DNA can recombine with the resident genome and be expressed and transmitted to future generations. When this happens, transduction has occurred.

BOX 13.3 *Lytic and latent infections*

Viruses recognize their host cells through molecular interactions on the cell and virus surfaces. The proper interaction triggers changes that cause the viral genetic material to be injected into the host cell. At that point, essentially two different types of infection may take place, depending on the specific virus.

Lytic and latent viral infections

After a lytic virus infects its host cell, the virus takes over the cellular machinery. Normal cellular metabolism slows or stops. Cellular enzymes are diverted to make many new copies of the viral genetic material and many viral proteins. As the cell is filled with viral components, new virus particles assemble. Finally, the host cell dies, releasing the progeny virus into the environment. Sometimes, this release of progeny is a gradual process; in other cases, the infected cell bursts (lyses), releasing all of the new virus particles at once.

The course of a latent infection is very different. After a latent virus injects its genetic material into the host cell, it does not hijack the cellular metabolism.

Instead, a few viral proteins that direct the incorporation of viral DNA into the host chromosome are produced. If the host cell is a bacterium, the latently infected host is called a lysogen. (Bacteriophages that set up latent infections are called lysogenic bacteriophages.) The viral DNA lies dormant in the host chromosome until a signal directs it to begin an active (often lytic) infection cycle. In most cases, the nature of that signal is unknown. During the active cycle, viral genetic material is reproduced and viral proteins are made. New virus particles are assembled and released.

There are variations on the lytic- and latent-infection themes. For example, the varicella virus infects its human host and causes the disease chicken pox. During the disease, the virus goes through active infection cycles. However, when the patient recovers, the virus is not gone. Copies of the viral DNA remain integrated into the chromosomes of certain cells as a latent infection. This viral DNA may remain dormant for the rest of the patient's life, or it may reactivate, causing a second disease known as shingles.

In lysogenic infections, the phage genetic material becomes incorporated into the host chromosome, where it can remain for a number of host cell generations. When the phage DNA leaves the host chromosome to initiate the lytic stage of its cycle, it occasionally brings adjacent pieces of host DNA with it. This host DNA then acts like part of the viral genome and is injected into new bacterial cells, where it inserts into host chromosomes. It may remain there when the phage DNA leaves again and become permanently incorporated into the host. Newly infected cells receive the particular fragment of original host DNA that is linked to viral DNA instead of the random fragments described above.

What are the agents that change the genetic makeup of populations?

The above discussion focused on the factors that create genetically variable populations because genetic variation provides the raw material for evolutionary change. Evolution is defined as a change in gene frequencies in a population over time, so what are the factors that can alter the genetic makeup of populations?

The genetic makeup of a population and therefore the amount of genetic variation in a population can be affected by the following factors:

- the rate of mutation
- natural selection
- **gene flow**
- **genetic drift**
- nonrandom mating

Some increase the amount of genetic variation in populations; others decrease it. Some affect the frequencies of genotypes without changing the allele frequency; others alter the frequencies of genes or alleles. The impacts these factors have on the genetic makeup of populations are described briefly in Table 13.2.

NATURAL SELECTION

Evolutionary biologists view natural selection as the primary force changing allele frequencies and charting the evolutionary paths of most species, so the remainder of this chapter focuses on the role natural selection plays in biological evolution by changing the genetic makeup of a population.

Natural selection is differential reproduction

After genetically variable individuals are created by any of the methods just described, these individuals are exposed to the indifferent force of natural selection. Chance determines which genotypes are produced in each generation, and then natural selection establishes the direction and rate of evolutionary change by acting on individuals. But individuals do not evolve; populations evolve. How does this happen?

Natural selection favors (selects for) certain individuals and rejects (selects against) others, i.e., "survival of the fittest." The genes of the fittest organisms are passed to the next generation in greater numbers than the genes of those that are less fit. In this way, gene frequencies in populations change over time, and this change is the process of biological evolution. Thus, natural selection represents the different rates of reproduction of different genotypes, and this alters the proportion of certain phenotypic traits in a population (Figure 13.14).

Table 13.2 Evolutionary forces that lead to changes in the genetic makeup of populations

Natural selection

Differential survival and reproductive success of certain genotypes relative to others. However, survival is important only in the service of greater reproductive success. Generally viewed as the most significant evolutionary force.

Mutation

Spontaneous changes in the genetic material that can create new alleles and, ultimately, new genes and thus change allele and gene frequencies. The direct impact of mutation on gene frequencies differs conceptually from the role that mutation plays in generating the genetic diversity that is subsequently acted on by evolutionary forces, expecially natural selection. The degree to which mutation affects gene frequencies directly depends upon the spontaneous rate of mutation, which is generally too low to shift gene frequencies significantly. Even though environmental factors can affect the *rate* of mutation, the *nature* of the mutation is independent of the environment. Mutations occur randomly, with no regard to the benefit a new trait might have in a certain environment. The environment determines which of those randomly generated mutations is advantageous. In other words, mutations are not goal oriented.

Gene flow

The movement of alleles or genes from one population to another that occurs when individuals change populations, assuming they ultimately reproduce, or in plants, when pollen moves from one population or species to another. Tends to decrease the genetic differences between populations because they share more genes after gene flow has occurred. We discuss this phenomenon at length in chapter 23.

Nonrandom mating

Recall that sexual reproduction does not create new alleles but only juggles the distribution of alleles, just as dealing new hands of cards does not change the number of aces in a deck. Modern evolutionary biologists have mathematically demonstrated that the frequencies of both the alleles and genotypes (percentages of homozygotes and heterozygotes) in a population will stay constant as long as mating is random. In other words, even though sexual reproduction is very important to evolution because it generates genetic diversity, it alone will not change gene or genotype frequencies in a randomly mating population. (For those who are interested, this is known as the Hardy-Weinberg equilibrium. It is discussed at length in every basic biology textbook, but mathematical proofs are beyond the scope of this book, so you'll just have to trust us on this.)

In nature, however, mating is often not random. Inbreeding is common, especially in plants, where some species self-pollinate, and scientists have observed numerous examples of **assortative mating** in animals. In positive assortative mating, those that are alike in some trait selectively breed with each other; in negative assortative mating, dissimilar organisms prefer each other. Nonrandom mating can change the frequencies of *genotypes* in a population, but it alone does not lead to changes in allele frequencies and therefore does not cause evolution. It does, however, change the phenotypic variation evolutionary forces act on. For example, both inbreeding and positive assortative mating increase the number of individuals who are homozygous for certain traits without changing the relative frequencies of alleles in the gene pool. By altering the frequency of genotypes and increasing the proportion of organisms expressing the homozygous trait, nonrandom mating alters the relative distributions of phenotypes. The factor that determines whether these changes in genotypes and phenotypes lead to changes in gene frequencies is natural selection, including sexual selection.

Genetic drift

Random changes in the allele or gene frequencies in a population that result from pure chance. Typically has a significant effect on gene frequencies when populations are founded by a small number of individuals (the **founder effect**) or when a large population has its numbers significantly reduced, which is known as a population bottleneck. As is true of all events governed by the laws of probability, the smaller the sample, the more likely a sampling error will occur. In other words, you are more likely to get a 1:1 ratio of heads to tails with 100 coin tosses than with 10 tosses. When a small number of individuals starts a new population or emerges from a population bottleneck, certain alleles may not be present in the new population, solely by chance. In addition, a gene may have only one allele in the new population even though it had many alleles in the parent population. Whenever a gene is represented by only one allele in a population, evolutionary biologists say the allele has become *fixed*, because 100% of the individuals in that population have that allele.

What characterizes the individuals favored by natural selection, or in other words, those that are the fittest? Because nonscientists typically hear the word "fittest" used in conjunction with survival, they sometimes envision rounds of dog-eat-dog competition leading to a select set of survivors who, by definition, are the fittest. However, even though better survival skills may sometimes be associated with improved fitness, the two are not equivalent. The fittest organisms are those that produce the most offspring that survive to reproductive maturity, so survival of the fittest relates as much to sexual conquests and good parental care as it does to Tennyson's concept of "nature red in tooth and claw."

Populations contain individuals whose abilities to both survive and reproduce vary. If an organism excels at skills required for survival but leaves

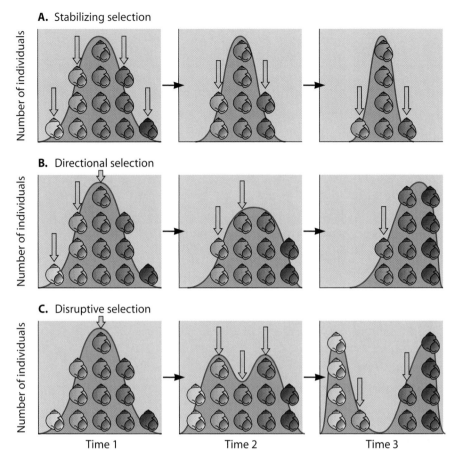

A. Stabilizing selection

B. Directional selection

C. Disruptive selection

Time 1   Time 2   Time 3

Figure 13.14 Natural selection. Selective forces exert different types of pressure on phenotypic traits in nature. In all of the cases shown, the selective force (arrows) is acting on external coloration, a trait that exhibits continuous variation in this population. **(A)** Stabilizing selection selects for the mean color pattern and against those that deviate from the mean, decreasing the variation in this trait. **(B)** Directional selection eliminates one of the extreme types but not the other and shifts the mean to a different phenotype. **(C)** Disruptive selective forces select for the two extremes and select against the mean, leading to a new type of frequency distribution known as a **bimodal distribution** because it has two peaks.

relatively few offspring, it is not fit. On the other hand, traits that have questionable or even negative survival value persist in a population if they are beneficial in obtaining mates. Think of the peacock's tail. It is difficult to imagine how a 5-foot-long tail of brightly colored, iridescent feathers could have positive survival value, but the peahens think those tails are gorgeous; thus, males with long, brightly colored tails are the fittest because they attract the most mates.

Differential ability in acquiring mates constitutes a special subset of natural selection known as **sexual selection**. In the case described above, the selective pressure is provided by members of the opposite sex choosing mates with certain traits. Another form of sexual selection results from members of the same sex, almost always the males, competing with each other for sexual access to the opposite sex (Figure 13.15).

Any trait that helps an organism survive, reproduce, and/or improve its offspring's survival and reproduction is known as an evolutionary **adaptation**, provided that trait has a genetic basis.

Figure 13.15 Sexual selection. In most species, males compete with each other for access to reproductive females and use a wide variety of strategies to entice females to mate with them. **(A)** Visual displays. In *Anolis* lizards, males extend a brightly colored flap under the chin. When this catches the attention of a female, they then do a series of push-ups. (Photograph courtesy of David Crews.) **(B)** Chemical induction of receptivity. In the aquatic animal the red-spotted newt (*Notopthalamus viridescens*), males capture females with their hind legs and rub the female's snout with cheek glandular secretions for as long as 4 to 5 hours. **(C)** Male competition occurs in ways that are much more subtle than physical combat. In this walking stick population in Costa Rica, because males outnumber females two to one, competition for reproductive access to females is intense. Once a male inseminates a female, the pair stays in copulo for 3 to 4 days. By riding piggy-back on the female, he prevents other males from mating with her. **(D)** Male walking sticks (length indicated by blue ×s) are much smaller than females (length indicated by yellow ×s). In many insects, the number of eggs a female produces is proportional to her size, so reproductive success increases with female size. Therefore, competition for access to the largest females is especially intense. This population displayed positive assortative mating; larger females mated with larger males.

Natural selection results from the interaction of the organism with its environment

The evolutionary advantage of having a certain trait depends on a wide variety of environmental factors. These environmental forces acting on species are constantly changing, and a species' survival depends upon its ability to respond with evolutionary changes. When you think of environmental factors as selective agents, you probably envision physical aspects of the environment, like drought and temperature, and yet other living organisms, the **biotic environment**, make up the most important component of a species' environment. Predators, pollinators, and competitors exert powerful selective pressures on populations. Often, the reproductive success and therefore the fitness of two species are so closely linked that when one changes evolutionarily, changes in the second soon follow. This phenomenon is known as **coevolution** (Figure 13.16).

The biological environment of an organism includes not only members of different species but its own as well. Individuals of the same species compete

Figure 13.16 Coevolution. Coevolution can lead to highly complex interactions that often increase not only the fitness of both species but also the dependence of one species on another. In Costa Rica, *Acacia* plants provide three species of *Pseudomyrmex* ants with food and housing. **(A)** *Acacia cornigera* thorns. Ants excavate new, green acacia thorns, creating a safe haven when the thorns mature. The dark circles visible on the tips of some thorns are entrances to hollowed-out thorns. **(B)** Nectaries and Beltian bodies. The ants feed on sugar water secreted by specialized stem structures (nectaries) and on the yellow, protein-rich Beltian bodies on the leaf tips. The plant produces these foods solely for the ants. What do the ants offer the plant in return? **(C)** *Pseudomyrmex ferruginea* ants. The ants defend the plant. In response to the slightest disturbance, ants rush to the site to protect the plant from any animal attempting to eat it. **(D)** Cleared vegetation surrounding acacias. The ants remove competitors. The three shrubs in the center of the photo are acacias. You can see that the ants have removed vegetation from the surrounding area.

for access to resources, such as food, nesting sites, and mates. Some win at others' expense.

In summary, species are continually evolving—changing genetically—in response to the physical environment and to each other. If one species responds to physical environmental changes with evolutionary changes, its evolutionary change becomes an environmental change for a second species. Also, remember that the relationship between organisms and their environment is not linear but circular. The activities of organisms continually alter the Earth's physical environment, often to the detriment of other species. Organism-driven environmental changes also place new selective pressures on individuals.

Many people envision nature prior to human intervention as ideal and ordered: organisms perfectly adapted to unadulterated environments living in balance with each other through time. Such ideas are based on the tacit assumption that Mother Nature would keep a tidy house with a place for everything and everything in its place if humans would stop introducing mess and disorder. While this idyllic view of nature may be appealing, it bears no resemblance to reality. The planet and its species are constantly changing independent of any human presence (Box 13.4).

Natural selection acts on phenotypes

Even though evolution is a change in gene frequencies in a population, natural selection acts on phenotypic traits. Recall that some of the variation observed in populations is exclusively phenotypic variation caused by environmental factors. Such variation may contribute to differences in survival and reproductive success and thus be subject to natural selection, but it is not the raw material for evolution.

Traits that contribute to an increase in fitness will evolve only if they have a genetic basis, because evolution is the response to selective pressures on inherited traits. Therefore, only some of the variation that exists in nature is significant in an evolutionary sense. Some isn't significant now and may or may not be significant in the future, depending on the amount of genetic contribution to the variation and the selective pressures that act on that variation.

Not all traits are adaptive

Some people have faith that through natural selection, evolution has honed all traits in all organisms so that there is a perfect fit between an organism and its environment. Viewing every trait in an organism as an adaptation is appealing but naive. Most traits are present simply because they're not harmful—at that time in that environment. Given another environment or another time, they might be.

At any one time, only some of the traits of an organism are the focus of natural selection. Others are irrelevant to selection and just happen to get passed on with the favored genes. At that time, operating in that environment, natural selection is not "interested" in those traits. As a result, it is not molding a fit between those traits and the environment. Next year, however, those neutral traits might become the focus of natural selection, or a useful attribute one year may be a handicap, or linked to a handicap, the next.

For example, imagine that a mutation produces a color pattern that provides a lizard with better camouflage. If having that trait means the lizard lives longer *and* has more offspring, that allele will be passed to the next generation in greater numbers because the camouflage coloration pattern is

BOX 13.4 *The historical development of evolutionary thinking*

While Darwin rightfully deserves credit for much of the current view of biological evolution, a diverse cast of characters, from French philosophers and scientists to English naturalists and land surveyors, loosened the intellectual soil so that Darwin's theory could begin to take root. Before the scientific world could begin to entertain Darwin's thoughts about evolution, it first had to accept the radical idea that the world had, in fact, *changed* since it was created on "October 26, 4004 BC." (In 1664, James Usher, the Archbishop of the Irish Protestant Church, used a literal interpretation of biblical events to determine that "Heaven and Earth. . . were made in the same instant of time, and clouds full of water and man were created by the Trinity on October 26 4004 BC at 9:00 in the morning.") In the absence of this intellectual revolution, Darwin's theory of evolution would have fallen on deaf ears.

For centuries, the pervasive view of the state of the natural world was stasis, not change. When faced with nature's extraordinary diversity and the remarkable fit between organisms and their environment, scientists and nonscientists offered the same explanation: all species had come into the world simultaneously, fixed in their current forms and specifically designed by the Creator to be perfectly adapted to their roles in nature. Why would species ever change, since they had been slotted into preexisting niches, much like custom cabinetry is built to fit precisely into a space?

In the 1700s, the intellectual climate began to shift slowly, and the idea of a dynamic, changing world began creeping into the worldview. Geologists convinced the public that the Earth had changed since its creation, and they attributed these changes to a series of cataclysms, such as the Great Flood described in Genesis. Accepting geological change did not necessarily threaten the consensus idea of immutable, perfectly adapted organisms. The changes in the Earth's climate may have altered the abundance and distribution of organisms, while their characteristics remained unchanged.

Late in the 18th century, European philosophers and naturalists, including Charles Darwin's grandfather, Erasmus Darwin, proposed the radical idea that living organisms had changed during the Earth's history, which, they stated, was considerably longer than the 5,800 years purported by the church. Land surveyors provided evidence linking geological and biological change, because they found the same fossils in the same rock layers in different locations and different fossils in the different rock layers at a single location.

Even more essential to Darwin's intellectual development was the work of the Scottish geologist James Hutton. In 1785, Hutton proposed that most geological changes could be attributed to slow, continuous, and common processes, such as water and wind erosion, and not to dramatic, catastrophic events. Linking major geological changes to subtle and familiar forces

eased people into the idea that change, not stasis, was normal and constant. Hutton's theory also introduced the concept of small, incremental changes that gradually, over centuries, accumulate and lead to transformations—the basic philosophy that is fundamental to Darwin's theory of evolution. So progress in a very different scientific discipline, geology, led to progress in biological thought.

In 1859, when Darwin proposed his theory of evolution by slight modifications of inherited traits, the concept of a slowly changing world populated by slowly changing organisms was firmly implanted in the mindset of the time. Unlike the reception, or rather the lack of it, that greeted Mendel's theory of inheritance, Darwin's thoughts could at least be heard if not universally accepted. Conditioned by decades of evolutionary thought that had infiltrated philosophy, sociology, geology, and agriculture, his audience was ripe for the ideas expressed in his book, *On the Origin of the Species by Means of Natural Selection*. All of the copies that were printed sold out the first day, and the book's surprising popularity with the general public made it a best seller the year it was published. As an additional testament to the timeliness of Darwin's theory, Alfred Wallace came to the same conclusions as Darwin around the same time. In fact, it was knowledge of Wallace's brief 20-page manuscript that inspired Darwin to finally publish his book about evolution, much of which he had drafted 15 years earlier.

---

adaptive. However, it will drag with it other genes that may have nothing to do with leaving more offspring, and if so, the traits encoded by those genes are, by definition, not adaptations.

On the other hand, the new allele and the resulting color pattern might make the lizard more susceptible to predation. In that case, the gene would be selected against, as would the lizard's neutral genes and perhaps other genes with positive survival value.

Now imagine a situation in which the color pattern makes male lizards more susceptible to predation but is also the color pattern that female lizards prefer when choosing mates. In nature, selective pressures often work in opposition to each other.

In addition, as described above, some of an organism's traits may not depend entirely, if at all, on its genetic makeup. Only traits that have a genetic basis are subject to natural selection. Therefore, if a trait is caused largely by environmental factors, it could not have been honed to perfection by natural selection.

Constant flux characterizes the natural world

Many people fail to realize that disorder, uncertainty, change, and therefore flexibility are common in biological systems, particularly at and above the level of the organism. That understanding is critical for accurately analyzing the environmental issues of biotechnology.

Life on Earth is like a game of chance governed by constantly changing rules. An organism's opponent, natural selection, is also the game's unpredictable referee, rewriting the rulebook on every play. Natural selection is a measure of reproductive success but not necessarily of survival.

For biological organisms, every round of reproduction is like a new hand of cards. Chance determines which genes are combined. These new genotypes give rise to new phenotypes. The path from genotype to phenotype is susceptible to environmental influences, many of which are random occurrences.

These new phenotypes then enter the game uncertain of how they will fare, because some of the rules have changed since the last round. Certain phenotypes might be improvements—in that environment at that time. These phenotypes will be selected for, causing the proportion of certain cards in nature's genetic deck to shift before the next hand is drawn. That same phenotype in a different environment at another time might be the kiss of death.

Given such an uncertain scenario, evolution has selected for biological organisms that hedge their bets through genetic diversity. A species persists from one year to the next by producing genetically variable offspring, some of which manage to survive for another round.

And yet the greater the diversity within a species, the less perfectly adapted to its environment that species is, so organisms attempt to balance diversity—a measure of adaptability—and adaptedness. For any species, a constant and very dynamic tension exists between these two properties.

EVOLUTION IN ACTION: THE SPREAD OF ANTIBIOTIC RESISTANCE

In the 1940s, scientists discovered a microbial enzyme that neutralized the newly discovered antibiotic penicillin and noted that this enzyme might interfere with antibiotic therapy. At the time, they assumed such resistance was the result of a new mutation and, as such, would not present much of a problem due to the low rate of mutation observed in the laboratory. The scientists making these predictions could not have been more wrong.

In the 1950s, an outbreak of dysentery that resisted antibiotic treatment occurred in Japan. The culprit, a *Shigella* species, was resistant to four very different antibiotics. The Japanese clinicians immediately knew something strange was happening because in the laboratory, mutants resistant to the antibiotic streptomycin arose at a rate of about 1 in 10 million (10^7) to 1 in 1 billion (10^9). The odds against four such mutations arising in the same cell would be astronomical: at least one in 10^{28} [$(10^7)^4$]! Even more astounding, *E. coli* cells isolated from the same dysentery patients were resistant to the same four antibiotics. How could resistance to the same four antibiotics arise simultaneously in two different genera of bacteria? The way antibiotic resistance was arising in nature must be different from what had been observed in the earlier laboratory experiments.

To make a long story short, the microbes were sharing genes through conjugation. Virtually all bacteria that cause infections in humans are resistant to one or more antibiotics. In 1952, almost 100% of all *Staphylococcus* strains were susceptible to penicillin, but by 1982, that number dropped to 10%. *Streptococcus*, which causes pneumonia, scarlet fever, and rheumatic fever, could be treated with either penicillin or erythromycin in the 1970s, but by the 1990s, some strains were resistant to nearly every type of antibiotic except vancomycin. Similar evolutionary patterns for resistance to antibiotics have been observed for many other bacteria, such as *Neisseria gonorrhoeae* and *Mycobacterium tuberculosis*, which cause gonorrhea and tuberculosis, respectively.

Antibiotic resistance genes occur naturally

Where did all of these different antibiotic resistance genes come from? To understand the answer to this question, you need to know that antibiotics have been around for eons. Most families of antibiotics were not invented by humans but were discovered from natural sources, such as soil bacteria and other microbes. Microbes produce antibiotics to kill other, nearby microbes and, in doing so, gain an advantage in their local environment. The soil bacteria that through random, spontaneous mutations had genes for antibiotic resistance were favored by natural selection because they were protected from the antibiotic and therefore survived and produced more offspring than those that did not have antibiotic resistance genes.

Most antibiotic resistance genes are transferred on plasmids. These plasmids, often called R (for resistance) plasmids, usually contain several different antibiotic resistance genes close together. Clusters of resistance genes form because most resistance genes are found within transposable genetic elements. Two features of the transposition process make it an excellent vehicle for creating clusters of resistance genes. First, the ends of the genetic element play a key role in the transposition process, acting as molecular handles for moving whatever DNA is between them. Second, when most transposable elements move, they leave a copy of themselves behind in the original location. These two features allow transposable genetic elements to combine into new, larger elements containing a number of genes in addition to the DNA handles. The clusters of resistance genes found on R plasmids are complexes of transposable elements that were presumably generated in this way.

Antibiotic use is a selective force that increases the number of resistant bacteria

The spread of antibiotic resistance requires more than the formation of clusters of resistance genes on R plasmids. Natural selection for the drug resistance abilities conferred by the plasmids must also be operating. Maintenance of a plasmid within a bacterial cell requires energy. In the absence of environmental pressure favoring plasmid-bearing cells, the cell with a plasmid is at a disadvantage compared with plasmid-free cells.

The widespread use of antibiotics has created an environment that heavily favors drug-resistant organisms. In evolutionary terms, the antibiotic is creating selective pressure for drug resistance, which in this case is pressure for maintaining the plasmid. Imagine the microbial population in the intestinal tract of a human being who begins to take an antibiotic for an infection. The antibiotic will kill not just the intended target but also all other susceptible microbes. Any resistant microbes will flourish in the face of reduced competition, creating a larger population with antibiotic resistance genes. In patients taking oral tetracycline, for example, the majority of *E. coli* fecal isolates carry tetracycline-resistant R plasmids within 1 week of the start of drug treatment.

Inappropriate uses of antibiotics have hastened the spread of drug resistance. Antibiotics are prescribed frequently for bacterial infections, which is certainly an appropriate use. In many cases, however, antibiotics have been prescribed for viral illnesses, such as the common cold, which cannot be helped by the antibiotics. According to the Centers for Disease Control and Prevention, 50 million of the 150 million prescriptions for antibiotics written annually in the United States are unnecessary. Additionally, in many parts of the world, people can purchase antibiotics without a prescription. In these regions, people often don't know whether the antibiotic is appropriate or the dose is adequate. Taking antibiotics provides selective pressure for increasing the numbers of any resistant organisms that are present; these organisms would likely disappear from the population in the absence of selection.

In addition, because antibiotics improve the growth of farm animals, livestock and poultry producers add them to animal feed as a routine matter, whether or not the animals are sick. Feeding antibiotics to farm animals increases the spread of drug-resistant plasmids among microbes that infect these animals, and sometimes these plasmids are transferred from microbes within animals to those within people. For example, the use of quinolone antibiotics in poultry has been linked to the appearance of quinolone-resistant *Campylobacter* in people. *Campylobacter* infections are usually caused by handling contaminated poultry or eating undercooked contaminated poultry. In Spain, when quinolones were introduced into poultry and livestock farming in 1989, virtually no *Campylobacter* bacteria infecting humans were resistant to the drug. By the end of 1991, fully 30% of human *Campylobacter* infections in Spain were by resistant bacteria, and in 1999, 80% were resistant. The rapid spread of antibiotic resistance is a consequence of the ability of bacteria to share genes.

Developing new antibiotics takes time and money

As microbes become increasingly resistant to the antibiotics in common use, it is clear that the medical community needs new antibiotics if it intends to lengthen the temporary reprieve from infectious bacterial diseases that society has enjoyed for the past 50 years. Discovering new antibiotics takes a long time, and developing them into commercial products takes lots of money (approximately $500 million), and partly as a result, the newer antibiotics are much more expensive than those that have been on the market for decades. In 1999, the Food and Drug Administration approved the first member of a new category of antibiotic, which is the first new category to appear in the U.S. market in 10 years. The drug, Synercid, works in a different way than other antibiotics, which is why it belongs to a new antibiotic category. Synercid, which is intended to treat hospital patients infected with organisms that are resistant to all other drugs, can only be given intravenously. Injecting it into a small vein is usually so painful that it must be delivered straight into a large vein in the chest or abdomen. The cost of Synercid treatment is approximately $250 to $300 per day, more than 100-fold more expensive than some oral penicillin treatments. Other new antibiotics are under development, but some bacteria will eventually develop resistance to them and share resistance genes with other bacteria. However, society can slow this process by the wise use of antibiotics.

SUMMARY POINTS

Biological evolution, an inevitable force in nature, is a change in the frequencies of certain alleles in a population over time. It consists essentially of a two-step circular process that is repeated over and over again: (1) creating populations of genetically diverse organisms and (2) culling certain individuals, but not others, from that population based on a genetically determined phenotypic trait.

The individuals that survive the culling process reproduce, creating the next generation of genetically diverse organisms. After enough rounds of reproduction and consistent culling of certain types, allele frequencies shift.

Genetic variation is created in two ways: mutation and recombination. Mutation creates new genetic material, such as new genes and new allelic forms of genes; recombination does not create new forms of genes or new genes but simply sorts existing genes into new combinations.

Mutations can involve single nucleotides or entire chromosomes. Some mutations have no effect on the encoded protein; others have no effect on organism structure and function. In both cases, they are viewed as neutral by evolutionary forces, because they have no effect on survival and reproductive success.

Mutations that affect a large piece of a chromosome can be evolutionarily disadvantageous because they decrease the survival and reproductive success of the bearer of the chromosome mutation or affect the viability of the bearer's gametes, or both. A few, such as gene duplications, can be advantageous because one of the duplicates is free to mutate without having a detrimental effect on the organism. The loss or addition of a single chromosome typically has drastic effects on survival and reproductive success; however, doubling the entire complement of chromosomes often has positive effects, because having four homologs of each chromosome can increase the bearer's size and reproductive output.

With transposable genetic elements, mutations are not caused by disruption of a natural process but are part and parcel of the natural process. Transposable genetic elements excise themselves from a location and reinsert elsewhere in the genome, sometimes disrupting gene function.

Sexual reproduction creates genetic variation during gamete production, through independent assortment and crossing over in meiosis, and in fertilization by joining genetic material from two different organisms. Organisms that reproduce asexually also have methods for increasing genetic variation in their populations: conjugation, transduction, and transformation.

Evolution is a change in the genetic makeup of a population of organisms, and the factors that cause the change include mutation, gene flow, genetic drift, nonrandom mating, and natural selection. Some of these factors affect the frequencies of certain genotypes; others affect the frequencies of genes or alleles. Of these, natural selection has the greatest effect on shifting gene frequencies in most populations most of the time.

Natural selection is differential reproductive success. The fittest organisms survive and leave the most offspring that also survive and reproduce. Organisms can improve their reproductive success by surviving longer, if longer life leads to more offspring; by outcompeting members of one sex for reproductive access to the other; and by investing resources in ensuring their offspring's survival.

Inherited traits that improve survival and reproduction are adaptations. Only some of the traits organisms display are adaptations; some are the result of environmental factors, and others may have no effect on fitness or may be detrimental in that environment at that time. Natural selection acts on phenotypes, but only traits that have a genetic basis evolve.

The rise in antibiotic resistance in bacteria is a frightening, real-world example of the evolutionary process. Antibiotic resistance occurs naturally and is inherited. Therefore, exposure to antibiotics acts as a selective pressure, increasing the survival and reproductive success of bacteria resistant to that antibiotic. The evolution of antibiotic resistance occurs at a very high rate because genes for resistance are often located on plasmids that move between bacteria, increasing the spread of the gene through the population. Antibiotic resistance is a serious public health problem that everyone can lessen by using antibiotics for bacterial infections, not viral infections, and, when antibiotics are prescribed, using the complete amount of the drug prescribed.

KEY TERMS

Adaptation	Evolution	Micromutation	Silent mutation
Aneuploidy	Founder effect	Natural selection	Speciation
Artificial selection	Gene flow	Neutral mutation	Species
Assortative mating	Gene pool	Pilus	Synthetic theory of evolution
Bimodal distribution	Genetic drift	Plasmid	Transduction
Biotic environment	Germinal mutation	Point mutation	Transformation
Capsid	Inversion mutation	Polymorphism	Translocation mutant
Chromosomal aberration	Lysogenic	Polyploidy	Transposable genetic element
Clonal population	Lytic	Population	Transposon
Coevolution	Macroevolution	Position effects	
Conjugation	Macromutation	Replacement mutation	
Conjugative plasmid	Microevolution	Sexual selection	

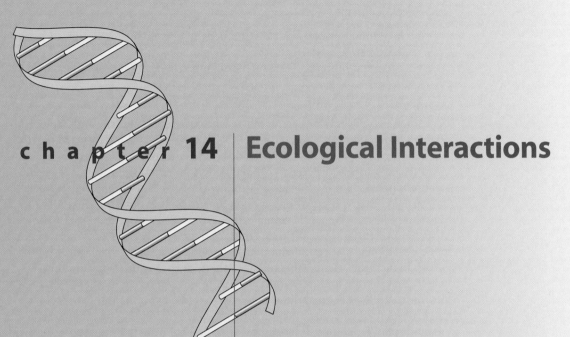

chapter 14 | Ecological Interactions

The study of ecology is inseparable from evolutionary thinking. Ecologists study interactions among organisms and between organisms and the physical environment and, ultimately, the impacts of those interactions. The interactions and impacts that constitute the heart of ecological research *are* the forces of natural selection that lead to adaptations, as described in the last chapter. The concept of the ecological environment encompasses everything, both living and nonliving, that is not an integral part of the organism but that is relevant, both directly and indirectly, to the organism's survival and reproductive success.

Like all biologists, ecologists study patterns in nature, especially the abundance and distribution of species in time and space. The abundance and distribution of organisms represent the results of interactions among organisms and environments and, as such, *are* indicators of past, present, and (perhaps) future evolutionary forces. Ecologists observe a feature or interaction and in the next breath ask the essential evolutionary question: how does (did) this affect survival and reproductive success?

Ecologists investigate interactions occurring at the highest levels of biological organization: the population, community, and ecosystem. Within each level, the spatial and temporal dimensions that ecologists study vary. For example, population ecologists interested in eagles may study adult and immature eagles within a 20-mile radius of a lake in a single summer or all breeding pairs in a forest surrounding the lake in a 30-year period. A community ecologist might study the relationship of the eagle population to the lake, since eagles eat fish; or the population's relation to the lake plus the forest surrounding it, which provides eagles with nesting sites; or the lake and forest plus the meadow on one side of the lake, where flying insects, a fish food, breed and lay eggs. A systems ecologist would study **abiotic factors**, the physical environmental features, such as soil nutrients and rainfall, plus community interactions. For example, fertilizer applications on a nearby golf course create lush putting greens, but if rains wash fertilizer into the lake, they also stimulate the lake's algae to grow, decreasing oxygen levels and killing fish.

In this chapter, we describe some ecological interactions and effects at each level, including:

- the population growth rate and density-dependent regulation of population size
- community interactions, such as competition, predation, and symbiosis
- the flow of energy and materials in ecosystems

Some of you may equate ecology with environmentalism and ecologist with environmentalist, and consequently, you may be looking forward to this chapter because you think we will discuss environmental issues. If so, we need to explain the difference between environmentalism and the scientific subdiscipline of biology known as **ecology** (Box 14.1). Perhaps we also need to warn you that you are about to enter a mathematics zone, because all ecologists rely heavily on math. Some ecologists never venture outdoors to conduct field research. Instead, they conduct their research at computer terminals, developing theoretical mathematical models about how nature should work. In fact, the science of ecology could not progress from natural history, which is observing and describing nature, to an experimental science that relies on testable generalizations derived from observations until mathematicians developed calculus and statistics. We recognize that some of you may be math-phobic, so our descriptions of ecological principles based on mathematics will be as "math-lite" as possible.

B O X   1 4 . 1   *Ecologists and environmentalists*

The word ecology seems to have taken on a new meaning in popular culture. It is often used interchangeably with the word environment, as in "save our ecology," but the two words are not synonymous. Ecology is the scientific study of interactions among organisms and their environments.

Nor is the word ecologist synonymous with environmentalist. An ecologist is a scientist who studies ecological interactions and tries to understand their impacts, especially those affecting the abundance and distribution of species. Environmentalists are people who have made certain value judgments about the impact of human activities on the environment and then try to influence policy or persuade others to agree with their viewpoint.

Both play important, but different, roles in society. A person can be an environmentalist without necessarily having any fundamental understanding of the principles that govern ecosystems, for example, someone who likes to hike

and wants open space and trails preserved. Environmentalism is advocacy based on value judgments that may or may not be grounded in scientific understanding. Ecologists, on the other hand, use scientific approaches, like systematic observation, measurement, model building, and experimentation, to gain a deep understanding of interactions between organisms and their environments, as well as to elucidate the overarching principles that govern these interactions.

The distinction between ecologist and environmentalist can become blurred, because any study of ecology quickly reveals that humanity has managed to evade the natural checks on population size that govern all other creatures on the planet and, because of its population size, is consuming resources and generating wastes at an unprecedented rate. Many ecologists, understanding this situation, make the judgment that humanity needs to change its ways, and they begin to ad-

vocate for change. When they become advocates for a certain position, they also can be labeled environmentalists.

It is important to note at this point that not everyone who understands general ecological principles automatically becomes an environmentalist. Some, rather than judging that humanity needs to change its ways, believe that other interventions, such as colonization of space or improvements in technology, will protect humanity from suffering the ultimate consequences of resource depletion and waste accumulation. Others simply ignore the possible consequences of humanity's collective behavior and focus on understanding ecological interactions in the here and now. The position that one takes regarding the scientific facts of ecology, whether one becomes an environmentalist or adopts a different position, is a consequence of personal values. The overarching principles of ecology, in contrast, are scientifically established and experimentally verifiable.

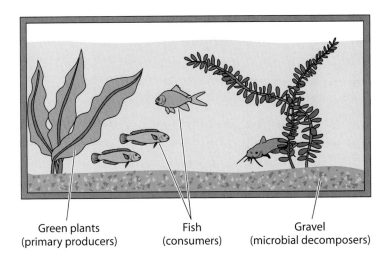

Green plants
(primary producers)

Fish
(consumers)

Gravel
(microbial decomposers)

Figure 14.1 Two ecosystems. (Photographs by Apollo 17 astronauts, courtesy of Johnson Space Center, National Aeronautics and Space Administration.)

SYSTEMS ECOLOGY

An **ecosystem** is a community of organisms linked to one another and to the physical environment by a flow of energy and materials. Ecosystems occur in many different types (for example, deserts, tropical rainforests, and coral reefs) and different sizes. A small aquarium is an ecosystem, but at the other extreme, so is the planet Earth (Figure 14.1). Biological, chemical, and physical processes link different ecosystems across the globe to one another, creating one giant ecosystem. Ecosystems, even those that are the same type, differ greatly in their species compositions and numbers of individuals, as well as their physical features and abiotic components. A Pacific Ocean coral reef has different species in different amounts than an Atlantic Ocean reef, yet both are coral reef ecosystems (Figure 14.2).

Even though species compositions vary dramatically, all ecosystems share certain structural and functional components. They also abide by the same laws governing the production, distribution, and consumption of materials and energy. Below, we describe a few of the most important—and nonnegotiable—laws that all organisms, including ourselves, cannot escape.

All ecosystems require an input of energy

Ecosystems cannot sustain themselves energetically. They are open systems, and like the organisms within them and cells within an organism, if they are to stay alive and well, ecosystems require continual inputs of energy. Chapter 6 explained that organisms depend on high-energy organic molecules (food) for carrying out cell processes. According to **the first law of thermodynamics**, the total supply of energy in the universe does not change, because new energy cannot be created and existing energy cannot be destroyed. If new energy cannot be created from scratch, the energy contained in the chemical bonds of food molecules must have originated elsewhere. The great majority of the time, that origin is the Sun.

Solar energy powers virtually all of the Earth's ecosystems. Multicellular green plants and certain one-celled aquatic organisms known collectively as **phytoplankton** (Box 14.2) capture light energy and, through photosynthesis,

Figure 14.2 Coral reef ecosystems. The species compositions of ecosystems from different geographic locations vary. Even though all of these photos are immediately identifiable as coral reefs, the species compositions are different. **(A)** Two fish species, an angelfish and a sergeant major, from a reef in south Florida. **(B)** Yellow and gray butterfly fish in the Pacific Ocean near Palau. **(C)** A two-banded clown fish hiding in a sea anemone on a coral reef in the Red Sea. (Photographs by Florida Keys National Marine Sanctuary staff [A], James McVey [B], and Mohammed Al Momany [C], courtesy of the U.S. National Oceanic and Atmospheric Administration.)

convert it into a form organisms can use: the chemical energy in glucose and, ultimately, other food molecules. Photosynthetic organisms contain **chloro-phyll**, the pigment molecule that captures the sunlight.

Because they form the foundation of an ecosystem's energy economics, organisms that synthesize large organic molecules from external energy sources are the system's **primary producers**. A few bacteria equipped with fascinating biochemical machinery exploit energy sources other than sunlight to make food molecules from building block molecules. For all practical purposes,

BOX 14.2 *Phytoplankton*

All bodies of water contain vast, but usually invisible, carpets of one-celled photosynthetic organisms known collectively as phytoplankton. Phytoplankton populations are responsible for most of the primary productivity of aquatic systems, and thus, these autotrophs support all other organisms in marine and freshwater habitats. One-celled heterotrophs, such as the protozoan in the photograph, consume phytoplankton; they, in turn, are eaten by secondary consumers and so on up the food chain (see panel A). Phytoplankton populations also make a significant

(A) Three protozoans graze on phytoplankton. (Photograph courtesy of Hans Paerl, Institute of Marine Sciences, University of North Carolina.) (B) Phytoplankton eukaryotes. (Left) Diatoms. (Photograph courtesy of the U.S. National Oceanic and Atmospheric Administration. Photographer, Neil Sullivan.) (Right) Mat of unicellular green algae. (Photograph courtesy of Linda Fisher.) (C) Phytoplankton prokaryotes. (Left) Cyanobacteria, *Anabaena*. (Photograph courtesy of Hans Paerl, Institute of Marine Sciences, University of North Carolina.) (Middle) Cyanobacteria, *Oscillatoria*. (Photograph courtesy of Roger Burks.) (Right) Mats of *Microcystis*. (Photograph courtesy of Hans Paerl, Institute of Marine Sciences, University of North Carolina.) (D) Phytoplankton populations. (Left) The Grand Banks. (Right) Southeastern United States in January. (Photographs by Jeff Schmaltz [left] and Jacques Descloitres [right], courtesy of the MODIS Team, National Aeronautics and Space Administration.)

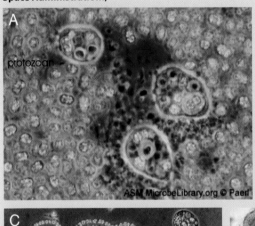

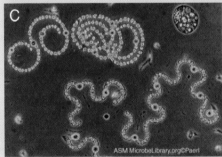

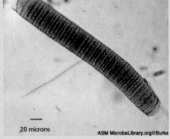

BOX 14.2 *Phytoplankton* (continued)

contribution to global productivity, because 25% of the Earth's total primary productivity can be traced to the photosynthetic activity of phytoplankton.

Phytoplankton communities contain a wide variety of unicellular organisms. All have chlorophyll for capturing sunlight and converting it into chemical-bond energy. Many species in phytoplankton populations are eukaryotes that belong to the kingdom Protista, such as diatoms and one-celled green algae (see panel B, left and right). Diatoms contain large amounts of silicon in their cell walls, and some have a brown pigment molecule that masks the presence of chlorophyll. They are the source of diatomaceous earth, used in toothpaste, metal polishers, and other products. One-celled green algae come in a wide variety of shapes and types and often join together to form large colonies that can be mistaken for a multicellular green plant.

Phytoplankton populations also contain cyanobacteria, which are often referred to as blue-green algae (see the three photographs in panel C). These prokaryotes convert light energy to food; some, such as *Anabaena*, fix nitrogen as well, so they are quite self-sufficient. Cyanobacteria can also occur in colonies so numerous they become visible to the naked eye, such as the stack of *Oscillatoria* cells that join together to form long filaments or the mats of *Microcystis* on the bottoms of shallow streams.

What factors limit the size of phytoplankton populations and define their abundance and distribution? Photosynthesis requires sunlight, so most organisms in phytoplankton occur at the water surface, where light is plentiful. They also need certain nutrients in addition to the food they make through photosynthesis. In addition, temperature affects the rate of photosynthesis, which

in turn determines phytoplankton abundance and distribution.

In the photographs (see panel D, left and right), taken by the NASA Terra satellite, light blue indicates photosynthetic activity. The photograph on the left shows a massive population of phytoplankton in the Grand Banks region of the Atlantic Ocean. The Banks, which are underwater plateaus, disrupt currents, causing deep, nutrient-rich waters to rise to the level of the light-filled shallow waters. These environmental conditions create a bountiful phytoplankton population that supports the fish populations, making the Grand Banks a popular fishing ground.

The photograph on the right, taken in January, clearly shows that the northern, colder waters of the U.S. southeastern coast support smaller phytoplankton populations than the warmer waters off the Florida coast.

however, the Earth runs on Sun power captured by photosynthetic organisms. If ecosystems must have an input of energy from external sources, then *every* ecosystem must have primary producers, which are also known as **autotrophs** (self-nourishers) because they are self-reliant energetically.

Economists develop methods for assessing industrial productivity because economic health depends on it. For similar reasons, ecologists measure the primary productivity of ecosystems. In general, only 1 to 3% of the sunlight reaching photosynthetic organisms is captured and transformed into chemical energy (**gross primary productivity**).

Photosynthetic autotrophs devote 15 to 50% of the chemical energy they create from light energy to powering their own metabolic processes. This energy cost is analogous to what the economist calls the cost of doing business. The energy left over, known as the **net primary productivity**, is the chemical energy available to the rest of the organisms in the ecosystem. As you can see in Table 14.1, the net primary productivities of different types of ecosystems vary greatly from one to another. The net primary productivity of an ecosystem establishes an upper limit for the total amount of energy, and therefore biomass, within an ecosystem. **Biomass**, in this sense, is the combined weight or volume of organisms in a community. As you will learn in this and subsequent chapters, the biomass, both living and dead, provides the energy on which human societies depend.

Energy flows through ecosystems in one direction: downhill

What happens to surplus chemical energy that autotrophs make but do not use in carrying out their life processes? By transforming light energy into chemical-bond energy stored in carbohydrates and other organic molecules,

Table 14.1 Net primary productivities of ecosystems

Type of ecosystem	Net primary productivity[a] (g/m/y)
Aquatic	
Coral reefs and algal beds	2,500
Estuaries	1,800
Lakes and streams	500
Open ocean	125
Terrestrial	
Swamp and marsh	2,500
Tropical forest	2,000
Temperate forest	1,250
Tropical grassland	800
Cultivated cropland	650
Prairie	500
Tundra	140
Desert and semidesert scrubland	90

[a]The values represent the average net primary productivities of different ecosystems. Ecosystems with the greatest net primary productivity can support the greatest numbers of heterotrophs. Nutrient availability is the primary limiting factor of productivity in aquatic systems. In terrestrial ecosystems, productivity is affected by temperature, rainfall, and light availability.

photosynthetic autotrophs feed the ecosystem's other species, the **heterotrophs**, which are all of the organisms (i.e., the nonautotrophs) that use other organisms as sources of energy and building block molecules. The heterotrophs include animals, fungi, most microbes, and some plants, and they are divided into two broad classes according to the nature of the food they consume: living (**consumers**) or nonliving (**detritivores**) (Figure 14.3).

Ecologists divide heterotrophs into additional subcategories. Heterotrophs that consume living autotrophs are primary consumers; they, in turn, are food for secondary consumers, and so on (Figure 14.4). This linear sequence of predator and prey constitutes a **food chain**. Table 14.2 provides additional information on classification schemes biologists use to describe

Figure 14.3 Ecosystem energetics. Energy enters an ecosystem primarily as light energy from the Sun. Some of that light energy is captured by photosynthetic organisms, the autotrophs, which convert it to chemical-bond energy stored in biological molecules. These primary producers use some of that energy to run their own metabolic processes, and the remainder is available to the other organisms in the ecosystem, the heterotrophs, that rely on the primary producers. Note that each time energy is transferred in an ecosystem, some is lost from the ecosystem as heat.

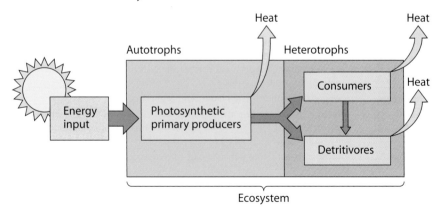

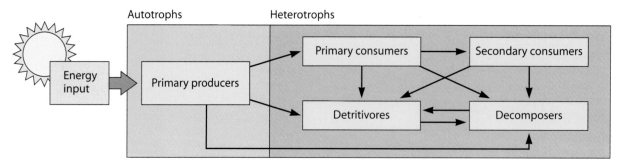

Figure 14.4 Trophic levels. Food chains quickly become food webs when all of the trophic interactions are included.

the relationships of consumers in a food chain. Heterotrophs living on detritus—dead organisms, decaying organic matter, and waste products excreted by living organisms—may be scavengers, such as vultures and earthworms, or decomposers. **Decomposers** are highly specialized detritivores capable of using biological molecules unavailable to other organisms, such as cellulose and nitrogen-containing waste products, as their food source. A few species of bacteria and fungi are the Earth's decomposers (Figure 14.5). Without them, some of the carbon, nitrogen, hydrogen, and oxygen tied up in biomass could never be released and made available to primary producers and, through them, to heterotrophs. We discuss the essential role decomposers play in keeping ecosystems alive over the long term below.

Organizing heterotrophs according to their sequence in the food chain, which is also known as a **trophic level**, has implications beyond the satisfying sense of order classification schemes provide. According to the **second law of thermodynamics**, each time energy is transferred, some is lost, usually as heat. In other words, if energy is "the capacity to do work," then each time a given amount of energy is transferred, less useful energy for doing work is available. Therefore, the total amount of energy at each trophic level is less than at the one before (Figure 14.6). The second law explains why energy flows one way through an ecosystem, making a constant input of new energy necessary.

Table 14.2 Classifyng consumers[a]

Role	Description
Predators	Heterotropic organisms that eat other living organisms, which are known as prey. This class is further subdivided according to trophic level.
Herbivores	Primary consumers that eat primary producers; typically thought of as plant eaters.
Carnivores	Secondary consumers that eat herbivores; also known as primary carnivores. Tertiary consumers that eat other carnivores; also subdivided into secondary carnivores that eat primary carnivores and tertiary carnivores that eat primary and secondary carnivores.
Omnivores	Eat primary producers, herbivores, and carnivores.
Parasites	Heterotrophic organisms that live in or on the organism they eat, which is known as a host, for a significant part of their lives; usually they do not kill their hosts.
Parasitoids	A type of insect parasite that lives in, consumes, and kills the insect that it eats.

[a]Biologists use a variety of terms to describe the roles that different consumers play in ecosystems. All of the organisms listed in this table are consumers because they use living organisms as a source of energy and building block molecules.

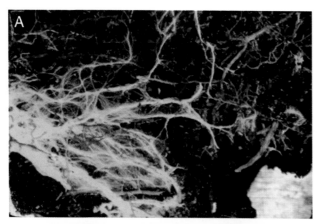

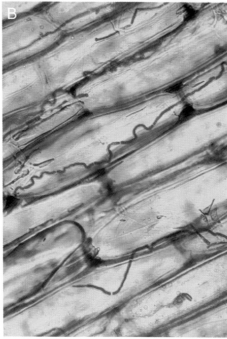

Figure 14.5 Fungi. All fungi are heterotrophs, and many are essential to life on Earth because they permit the recycling of molecular materials. **(A)** Multicellular fungi consist of an intricate mass of filaments known as a **mycelium**. (Photograph courtesy of the U.S. Forest Service. Photographer, Jim Trappe.) **(B)** The dark-blue threadlike structures in the micrograph are individual filaments, or **hyphae**. Decomposition occurs as the hyphae invade tissues and secrete digestive enzymes. (Photograph by Nick Hill, courtesy of the Agricultural Research Service, U.S. Department of Agriculture.)

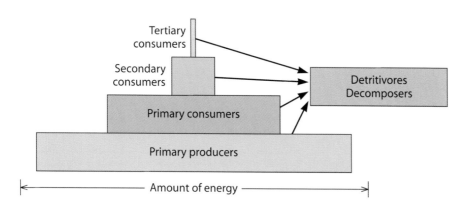

Figure 14.6 Energy pyramid. In every ecosystem, a small portion of energy at one trophic level is transferred to the next trophic level. Some is used by organisms at each level for metabolism, and some is lost as heat.

Within ecosystems, materials cycle

Primary producers not only capture light energy and transform it into a form that will drive the ecosystem, they also secure the molecular building materials for other organisms in the ecosystem. The Earth's water, rocks, soil, and air provide the ultimate source of the chemical elements essential to life. Living organisms must obtain all of their molecular building blocks by drawing from the pool of elements in the abiotic environment.

As you can see in Table 14.3, the relative abundance of the various elements in the Earth's crust does not correspond to their relative abundance in, and therefore importance to, living organisms. Four of the 100 or so elements found on Earth contribute 95% of the constituent elements found in living organisms: carbon, oxygen, hydrogen, and nitrogen. When scavengers and decomposers feed on decaying organisms and waste products so they can meet their own needs for energy and building materials, they also free up the material elements that all living organisms need for constructing themselves.

Compared to energy flow, the material-handling processes ecosystems use are models of efficiency. Atoms making up the food molecules, such as carbon and oxygen, cycle through the ecosystems. They move from the abiotic component to the ecosystem's biotic component and are temporarily stored in living organisms before being released to the abiotic environment.

At a bare minimum, a functioning ecosystem must have decomposers in addition to autotrophs, because they free up the molecular building blocks autotrophs must have to synthesize the food molecules that store the captured energy. Lest we become too arrogant, it is humbling to realize that life on Earth depends on green plants and a few species of bacteria and fungi.

When analyzed on an ecosystem-by-ecosystem basis, these cycles appear inefficient, because ecosystems lose materials. However, materials lost by one ecosystem end up in another, because they are linked into one giant ecosystem. On a global scale, the natural recycling schemes are astoundingly efficient, considering the scale of the operation and the number of interactions. Rather than tracing the cycles of all essential elements, we will highlight two of the most important, carbon and nitrogen. We will also limit our discussion to the biological players in the recycling scheme, even though climatic and geological forces often make important contributions.

The carbon cycle

By now, the importance of carbon to life on Earth should be obvious. All biological molecules are polymers with carbon backbones. Life is carbon based because carbon can form long molecules that provide ample opportunity for creating seemingly unlimited molecular variation.

Table 14.3 Relative abundances of elements

Earth's crust			Humans	Plants	Bacteria
Oxygen	46%	Oxygen	64%	79%	72%
Silicon	28%	Carbon	19%	8%	11%
Aluminum	8%	Hydrogen	9%	9%	9%
Iron	5%	Nitrogen	4%	1%	3%
Calcium	3.5%	Phosphorus	1%	0.5%	0.6%
Sodium	3%	Potassium	0.5%	0.3%	0.3%
Potassium	2%	Sulfur	0.4%	0.1%	0.3%

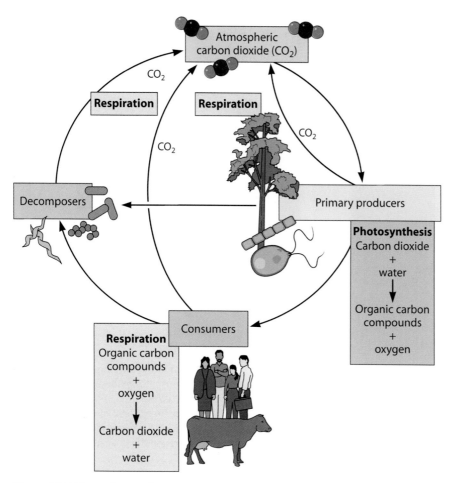

Figure 14.7 The carbon cycle.

The core of the carbon cycle is quite simple. Carbon enters the cycle as an atom in carbon dioxide before assuming its role as a key constituent of organic molecules that are the stuff of living organisms. The carbon atom returns to the gaseous state as carbon dioxide (Figure 14.7). The flow of carbon atoms through ecosystems parallels the energy flows described above, because the two are part and parcel of each other. Photosynthetic organisms, the primary producers, are responsible for the first half of the **carbon cycle**: fixing carbon by moving it from its position as an atom in the low-energy gas molecule carbon dioxide to the solid state as an essential component of high-energy organic molecules, such as glucose. In the process, they generate the oxygen that drives the second half of the cycle (respiration). Through aerobic respiration, organisms break organic molecules apart, freeing the constituent atoms, including carbon as carbon dioxide, and releasing the energy in the bonds holding the atoms together (Box 14.3).

In summary, as plants exhale oxygen during photosynthesis, other organisms (including plants) inhale it so they can access the stored chemical energy that oxygen liberates through cellular respiration. In addition to energy, respiration generates carbon dioxide. As organisms exhale carbon dioxide, green plants and other photosynthetic organisms inhale it.

BOX 14.3 *Photosynthesis: making food from air and water*

Have you ever seen the bumper sticker that asks if you've thanked a green plant today and wondered why you should?

Life on Earth depends on multicellular green plants and certain microbial species that also contain chlorophyll, because they provide all other living organisms with two necessities: food and oxygen.

Through photosynthesis, they capture sunlight and transform it into the chemical energy contained in carbon-based food molecules. These organic molecules provide energy to the photosynthetic organism so it can run its metabolic processes. Because they make more food than they need, their leftovers feed the rest of the organisms on the planet.

Photosynthesis carries with it the ancillary benefit of generating oxygen. Oxygen releases the chemical energy stored in organic molecules, making it available to plant cells so they have power to carry out various processes. In addition to manufacturing a surplus of food molecules, plants also synthesize

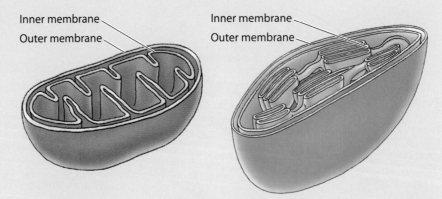

Inner membrane
Outer membrane

Inner membrane
Outer membrane

more oxygen than they need. Other organisms are the beneficiaries of that surplus as well, because oxygen allows them to extract much more energy from a glucose molecule than is possible through anaerobic glucose breakdown.

How do photosynthetic organisms convert light energy to chemical energy so that it can be stored for future use? They take two low-energy molecules, carbon dioxide and water, and convert them to high-energy carbohydrates,

such as glucose. The chemical equation that summarizes photosynthesis is as follows:

$$CO_2 \text{ (carbon dioxide)} + H_2O \text{ (water)} + \text{light energy (sunlight)} \rightarrow CH_2O \text{ (carbohydrate)} + O_2 \text{ (oxygen)}$$

This equation may look familiar to you because it is glucose breakdown in reverse (see Figure 6.5). Recall that the complete breakdown of glucose through cellular respiration occurs in the mitochondria in eukaryotic cells.

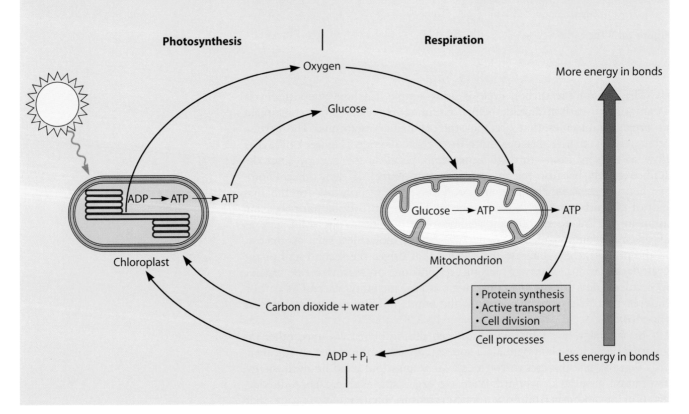

BOX 14.3 *(continued)*

Photosynthesis in green plants occurs in another organelle, the chloroplast, that has many similarities to mitochondria, both structurally and functionally. For example, like mitochondria, chloroplasts have their own DNA and a system of inner membranes (see the first figure).

Chloroplasts contain the pigment molecule chlorophyll, which absorbs sunlight. This energy causes an electron within the chlorophyll molecule to become so energized that it jumps out of the molecule. The energized electron is grabbed by another molecule, which is the lead molecule in an electron transport system very much like the one located in mitochondria that we discussed in chapter 6. The electron is handed from one molecule to the next

in the electron transport chain, generating a molecule of our old friend ATP, the currency of the realm in cell energy economics. So, in photosynthesis, light energy is first converted into electrical energy (remember, electricity is the flow of electrons), which is converted into the chemical energy contained in the ATP molecule.

So, how does the plant or microbe make glucose? The chlorophyll molecule, eager to replace its lost electron, grabs one from a water molecule, which causes it to break into two hydrogen atoms and one oxygen atom. The hydrogen atoms eventually join with carbon dioxide to create a glucose molecule. But as you know, making glucose from carbon dioxide and hydrogen requires

energy, and the ATP molecule generated through the electron transport system provides it. The ATP-driven incorporation of carbon dioxide into an organic molecule is known as carbon fixation (see the second figure).

Under most circumstances, the factor that limits the rate of photosynthesis is access to carbon dioxide, because the percentage of carbon dioxide in the air is very small (0.03%). A 10% increase in the amount of carbon dioxide increases the amount of carbon fixed through photosynthesis by approximately 8 to 9%. We discuss the relationship between increasing carbon dioxide concentration and increased carbon fixation further in "The carbon cycle."

Even though this description of the carbon cycle is completely accurate, it is actually a bit misleading, because it gives the impression that **carbon fixation** through photosynthesis and its subsequent liberation as carbon dioxide is a zero-sum game; that is, the inhaling of carbon dioxide in photosynthesis equals the amount exhaled by all other organisms.

For millions of years, this was the case. The amount of fixed carbon essentially equaled the amount of respired carbon, so the net amount of atmospheric carbon dioxide did not change significantly. During the past 150 years, the ratio of inhaled to exhaled carbon dioxide has shifted dramatically toward exhaled due to human activity, primarily the burning of fossil fuels, such as gas, coal, and oil (Figure 14.8). Millions of years ago, photosynthetic organisms inhaled carbon dioxide, fixed the carbon in high-energy organic molecules, and then died in ways and places that made them inaccessible to decomposers and other detritivores. The fixed carbon could not be liberated from organic molecules as carbon dioxide because detritivores were not using it as a food and energy source.

These high-energy organic molecules, the photosynthetic legacy of organisms that lived long ago, now provide the fuel that runs automobiles, air conditioners, furnaces, and factories. Human societies are accessing the energy that has been stored in chemical bonds for millions of years and, in the process, are instantaneously (in geological time) releasing tremendous amounts of carbon dioxide that had been inhaled gradually over thousands of years.

At first blush, you might think that increasing the amount of carbon dioxide in the atmosphere might be beneficial, because as noted in Box 14.3, primary productivity is limited by the low concentration of carbon dioxide in air. The increase in atmospheric carbon dioxide from burning fossil fuels may well be increasing carbon fixation through photosynthesis. Unfortunately, however, the human activities that lead to tremendous releases of carbon dioxide often go hand-in-hand with activities, such as deforestation, that

Figure 14.8 Fossil fuels and the carbon cycle. Organisms that lived millions of years ago provide fossil fuels, such as coal and oil. **(A)** In 1907, Colorado coal miners extracted coal that contained a fossilized imprint of a palm leaf. **(B)** A leaf from a tropical plant was fossilized in the coal deposits of Antarctica. Both photographs also illustrate the tremendous shifts in species distribution that have occurred during the Earth's history. (Photographs by H. S. Gale [A] and D. L. Schmidt [B], courtesy of the U.S. Geological Survey.)

remove huge numbers of the primary producers that could be absorbing the excess carbon dioxide and fixing it into organic molecules.

What effect does this ever-increasing level of carbon dioxide have? Carbon dioxide in the Earth's atmosphere acts like glass in the windshield of a locked car. It allows sunlight in but prevents heat from leaving. This explains, in part, why many scientists are concerned about global warming. Every year, human societies generate 8 to 10 billion tons of carbon dioxide by burning fossil fuels while simultaneously clearing millions of acres of rainforests and wetlands, two ecosystems with very high net primary productivity. The problem is exacerbated when the cleared vegetation is burned, because this also releases carbon dioxide into the atmosphere (Figure 14.9).

The nitrogen cycle

Photosynthesis provides living organisms with the carbon, hydrogen, and oxygen needed to construct biological molecules, but they also need to synthesize some awfully important nitrogen-containing molecules, such as DNA, ATP, and proteins.

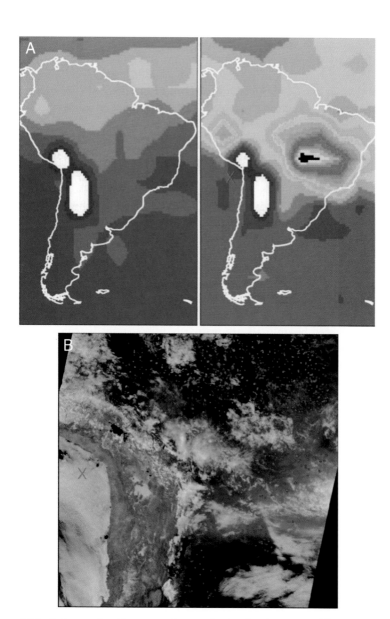

Figure 14.9 Deforestation. Every year, farmers in the developing world prepare to plant crops by clearing and burning millions of acres of vegetation, which releases large amounts of carbon dioxide. Shown are images of Brazil, taken in March and at the beginning of the growing season in September. The Xs mark the same location in the different types of images. **(A)** Created by NASA's MOPITT program, which measures levels of atmospheric pollution, these computer-generated images show the increase in carbon dioxide that occurs when Brazilian farmers clear land for planting crops in September. Blue indicates low levels of carbon dioxide; yellow and orange indicate intermediate levels; purple and black indicate the highest levels. **(B)** This increase in carbon dioxide is due primarily to biomass burning across Amazonia and not to industrial emissions. The red dots indicate fires. (Photographs by David Edwards and John Gille [A] and Jessie Allen, MODIS Team [B]; all images courtesy of the National Aeronautics and Space Administration.)

Of the four elements found most often in biological molecules, nitrogen is usually the one that is most difficult for autotrophs to secure. This may seem odd to you, because you probably know that air is four parts nitrogen and one part oxygen. If primary producers can get their carbon from carbon dioxide, then why can't they use atmospheric nitrogen?

The nitrogen in air consists of two nitrogen atoms bonded to one another (N_2). In chapter 3, you learned that nitrogen typically forms three bonds, so the two nitrogen atoms are linked to one another by a very strong bond known as a triple bond. Only a handful of bacterial species, which we discuss later, have the firepower needed to break this bond and release the nitrogen atoms from one another.

All other organisms must have nitrogen provided to them as atoms bonded to hydrogen in molecules like ammonia (NH_3) or bonded to oxygen as nitrate (NO_3). Only then can they absorb nitrogen and incorporate it into nucleic acids, amino acids, and other biological molecules. Not only is atmospheric nitrogen inaccessible to the great majority of organisms, they also cannot rely on the weathering of rocks to provide a source of usable nitrogen, because the Earth's crust is essentially devoid of nitrogen.

Since the physical environment is of little value when it comes to supplying nitrogen, living organisms have no choice but to use other living organisms as nitrogen sources (Figure 14.10). For autotrophs, a usable form of nitrogen must be located in the soil (or in the water or sediment in aquatic ecosystems) so that it can be absorbed. That's easy to achieve. The waste products excreted by animals are primarily nitrogenous wastes generated by the metabolic breakdown of amino acids and nucleic acids. This nitrogen is excreted in a form that autotrophs can immediately absorb and utilize, primarily ammonia or ammonia-like substances. Bacteria and fungi in the soil also liberate usable nitrogen as ammonia by decomposing dead and decaying autotrophs and heterotrophs. A unique set of bacteria are able to convert soil ammonia into nitrate, the form of nitrogen plants prefer over ammonia, through the process known as **nitrification**.

Autotrophic cells integrate nitrogen into organic molecules via the metabolic pathways that synthesize nucleic acids, amino acids, and other essential

Figure 14.10 The nitrogen cycle.

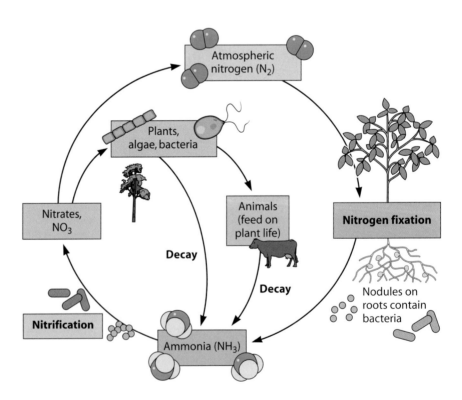

molecules. Heterotrophs then acquire nitrogen by consuming autotrophs, completing the cycle that began with excreted nitrogenous waste products and decomposing organic matter. The **nitrogen cycle** represents the clearest and most literal example of an ecological principle that contributes to ecosystem sustainability: one organism's waste is always some other organism's dinner.

This description gives the impression that the core nitrogen cycle is balanced and self-sustaining, which is not the case. Nitrogen is readily lost from the cycle by a number of natural processes. For example, because ammonia and nitrate are very soluble in water, they are rapidly washed out of the soil into rivers and streams, or they percolate to deeper soil and sediment layers, becoming inaccessible to the autotrophs. Without an ancillary source of nitrogen to continually replace the lost nitrogen, the nitrogen cycle would eventually grind to a halt. What keeps the nitrogen cycle cranking? If you guessed microbes, you're right. Once again, a handful of microbial species, known as nitrogen fixers, save the day and keep the planet chugging along.

Nitrogen fixation. Nitrogen-fixing bacteria are able to tap into the bountiful supply of nitrogen (N_2) in the atmosphere. They have an enzyme, nitrogenase, that catalyzes the combining of one N_2 molecule with six hydrogen atoms to create two molecules of NH_3. Even though the enzyme nitrogenase makes the reaction much easier, breaking the triple bond between the two nitrogen atoms requires so much energy that the bacteria must use 15 to 20 molecules of ATP to make two ammonia molecules.

Some nitrogen-fixing bacteria live freely in the soil, while others have a much easier life because they have developed a symbiotic relationship with one or more plant species. **Symbiosis** is an intimate association between two different species, such as the ant-acacia relationship we described in the previous chapter. Some nitrogen-fixing species of cyanobacteria occur within their host plant's tissues, while other bacterial species are safely ensconced within the roots of certain plants (Figure 14.11). These bacteria infect very young roots and induce the plant to build special structures called **nodules** that house the bacteria. Not only do the plants provide the bacteria with free housing, they feed them glucose. Why do these plants accept and nurture these freeloaders? Because the bacteria provide them with the element that's hardest for plants to find: a usable form of nitrogen. Because both species clearly benefit from their association, their symbiotic relationship is defined as **mutualism**.

Even though nitrogen fixation would seem to solve the problem caused by the continual drain of nitrogen from the cycle by natural processes, it doesn't. Just as human activities upset the balance in the carbon cycle, they also contribute to a chronic scarcity of nitrogen in the soil. Harvesting crops and timber to meet human needs for food, shelter, fuel, and clothing also removes nitrogen from the cycle, because the harvested materials contain amino acids and nucleic acids. These same practices lead to soil erosion, which compounds the problem. Not only does soil erosion contribute substantially to nitrogen loss, it also causes the ecosystem to lose the organisms that could replenish the nitrogen supply, the nitrogen-fixing bacteria.

The present size of the global population demands maximum productivity from the crop acreage under cultivation, so farmers all over the world must add nitrogen to the soil to compensate for the losses this causes. Some of the tactics for supplementing nitrogen mimic natural biological processes, because farmers use animal waste products and decaying organic matter to

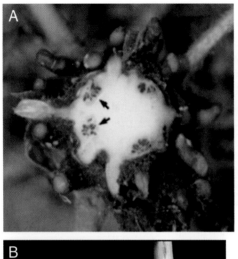

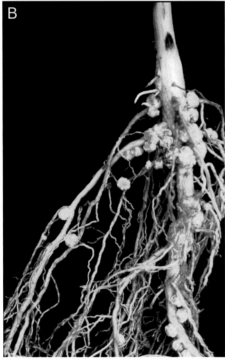

Figure 14.11 Nitrogen-fixing bacteria. **(A)** Cross section (left) of a leaf stem from the tropical plant *Gunnera* (right), with arrows indicating the colonies of nitrogen-fixing cyanobacteria that reside there. (Photograph on left courtesy of David Dalton.) **(B)** Soybean root nodules. (Photograph courtesy of Manuel Becana.)

replenish nitrogen lost from the soil. However, much of the nitrogen used in agriculture is fertilizer that has been chemically synthesized from atmospheric nitrogen.

In trying to get gaseous nitrogen into a form plants can use, people run into the same problems all other organisms face, because the laws of chemistry do not change for humans. Breaking the strong triple bond holding the two nitrogen atoms together always requires a lot of energy, no matter who or what is breaking it. In the manufacturing facilities that chemically synthesize fertilizer from atmospheric nitrogen, fossil fuels provide the energy to drive the conversion of nitrogen gas to ammonia. This process requires so much energy that the amount of energy extracted from food crops is significantly less than the amount of energy required to synthesize fertilizer and get it into the soil. In other words, in nitrogen accounting, the balance sheet shows that humans are deficit spending with regard to energy. We will return to this issue in chapters 22 and 23.

POPULATIONS AND COMMUNITIES

As you can see, energy flow and material recycling in an ecosystem depend upon the composition and structure of its biological community. Because of the intricate interdependence of community members within an ecosystem, any change in the abundance and distribution of one species permeates throughout the system and has implications for other species. If disease reduces the population size of a primary producer, the primary consumers' numbers decrease. If there is redundancy in the ecosystem, that is, if a number of different species assume the same role, not all of the other primary producers in the ecosystem may be susceptible to the disease. If the population size of the primary-producer population remains sufficient to support the primary-consumer population, the primary-consumer population size will not change.

Because the health and stability of ecosystems depend ultimately on the stability and persistence of the populations within them, we now turn our attention to a different level of biological organization: the population.

Populations have remarkable growth potential

The ability of a species to persist over time as an essential component of an ecosystem depends, in part, on its inherent capacity to increase in numbers, which is known as its **reproductive potential**. In developing his theory of evolution by natural selection, Darwin noted that the reproductive potential of all organisms greatly outstrips the environment's capacity to support them. To support his claim, Darwin picked an animal with low reproductive potential because it matures very late and has only one offspring at a time with long intervals between offspring: the elephant (Figure 14.12). He determined that a single pair would produce 19 million elephants in 740 years.

If an elephant can produce enough offspring to cover the globe, then what about organisms with high reproductive potential? A single female

Figure 14.12 Low reproductive potential. During a reproductive life span that can last 40 years, a female Asian elephant produces approximately five or six offspring. Females mature at 12 years and give birth to a single offspring after a 22-month pregnancy. Their lactation period can last 2 years, though some mothers wean their offspring earlier. On average, females give birth every 5 years. (Photograph by Jessie Cohen, courtesy of the Smithsonian Institution National Zoological Park.)

housefly lays enough eggs to create a population of 190,000,000,000,000 flies in 1 year. Under ideal conditions, bacteria divide in two every 20 minutes. A single bacterium could give rise to a colony that would cover the Earth's surface in 2 days; in 3 days, the layer of bacteria blanketing the Earth would be 6 feet deep.

Clearly, the world is not overrun with elephants or buried under a carpet of bacteria. Appreciating the factors that govern the growth of all populations, including human populations, is essential for grasping some fundamental truths about ecosystem structure and stability. The factors that govern population growth fall into two broad categories: intrinsic properties of a species that determine its inherent reproductive potential and extrinsic environmental factors that constrain that potential and hold population sizes within reasonable limits.

Inherent reproductive potential

To assess a species' inherent reproductive potential, scientists first assume that the population inhabits the best of all possible worlds. In other words, resources are unlimited, members of the population do not impede each other's reproductive success or survival, and predators and disease do not exist.

Populations increase in size when the number of individuals added to the population through birth and immigration exceeds the number that die or emigrate. Determining a population's inherent capacity to increase requires knowing the time frame over which this increase occurs, which is defined as the growth rate of the population.

Population growth rate (G) = (number of births + immigrants)/time period − (number of deaths + emigrants)/time period

Mathematical ecologists often eliminate certain variables when they are developing their models. This does not mean that the modelers think the variables are unimportant in the real world. Eliminating variables from equations is equivalent to laboratory scientists holding some variables constant in controlled experiments so that they can measure the impacts of various factors one at a time. Thus, holding a variable constant or eliminating a variable from the equation is actually a way of acknowledging its potential importance.

Population biologists often assume that immigration and emigration are equal, so they can ignore those two factors that affect the population growth rate. Therefore,

G = number of births/time period (B) − number of deaths/time period (D)

Rather than measuring all of the births and deaths in the population, which is difficult for many populations and impossible for others, biologists take a sample from the population and calculate the average number of births and deaths per individual. Subtracting the average death rate per individual (d) from the average birth rate per individual (b) provides the intrinsic rate of increase (or **intrinsic growth rate**) of the population, which biologists symbolize with the letter r.

$$r = b - d$$

Another way to think about r is that it represents the inherent reproductive potential of each individual. The increase in the numbers of individuals in the entire population depends not only on the reproductive potential of each individual but also on the number of individuals in that population (N). A population with an annual intrinsic growth rate (r) of 0.2 increases by 20 if the original population size is 100 and by 200 if it contains 1,000 individuals.

Therefore, under ideal conditions, the change in the number of individuals in a population (G) equals the intrinsic growth rate times the number of individuals in the population, or

$$G = rN$$

Graphing the results of this equation for a number of points in time leads to an **exponential growth curve**, which is also known as a J-shaped curve (Figure 14.13).

You may remember learning about exponential growth curves in high school when you compared the growth rate of savings accounts with simple interest to those with compound interest. If you remember only one thing from that lesson, it is probably this: you make more money if you put your savings in an account with compound interest. Why is that the case? Because the interest that accrues in each time period gets added to the balance. Each time the bank calculates your interest, it uses your new balance, not your original balance, or

Next balance = current balance + (interest rate $\times$ current balance)

Population growth works exactly the same way. Think of r as the interest rate and the number of individuals (N) as the balance.

Next N = current N + ($r \times$ current N)

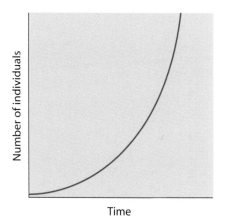

Figure 14.13 Exponential growth curve. This type of growth is seen in populations when resources are abundant.

Implications of the pattern of exponential growth

Two implications of the exponential pattern of population growth are very important but might not be intuitively obvious to you.

1. Even when r, the average reproductive rate per individual, is constant, the **population growth rate** increases. This means that the population size increases faster and faster even though each individual's rate of increase (i.e., the interest rate) doesn't change. When the population is small, the growth rate is small, but as the population size increases, the rate at which it grows increases.
2. No matter how large or small r is, as long as r is positive ($b > d$) and constant, the population increase will be exponential and the growth curve will be J shaped. The only difference will be the shape of the J. Populations with large r values turn the corner faster than those with small r values.

What are some of the characteristics of a species that determine the size of its r?

- age at first reproduction
- number of births per individual
- birth interval
- number of offspring surviving to reproductive age

Which of these factors do you think has the greatest impact on the growth rate of the population? The first rule of investing money—start early—provides a hint. Of all of the factors that influence population growth rates, age at first reproduction has the most powerful effect. Decreasing the age at first reproduction by 1 year has a much greater impact than extending the reproductive life span by 1 year.

Under what circumstances does exponential growth occur in nature? Exponential population growth rates are atypical of most populations. A few species, known as **opportunistic species**, that specialize in colonizing new sites or invading disturbed habitats display exponential growth. Their adaptations, especially those directly related to reproduction, are geared toward rapid and immediate population growth (Figure 14.14). Think of them as sprinters. They're out of the blocks quickly, and they accelerate to top speed almost instantaneously—and like sprinters, they finish not long after start-

Figure 14.14 Opportunistic species. Small rodents and weedy plants are archetypal opportunistic species. **(A)** The house mouse is sexually mature at 1 month. After a 20-day gestation period, a female gives birth to five to seven blind, hairless offspring. Within hours after giving birth, she is impregnated again. Therefore, while she is lactating, another litter is developing in utero. (Left) House mouse adult female. (Right) Three-day-old house mice. **(B)** Weedy plants are designed to maximize their reproductive output after invading disturbed habitats, such as a roadside or a farmer's field. Each knapweed plant produces 25,000 seeds in one growing season and then dies. This leads to large infestations of a single species. (Photographs of knapweed by Norman Rees, Agricultural Research Service, U.S. Department of Agriculture; courtesy of http://www.forestryimages.org.)

ing. Every round of reproduction produces many offspring, and/or birth intervals are very short. After reproducing, they die, move on to another disturbed habitat, or enter a period of dormancy and wait for another disturbance. This reproductive strategy is the antithesis of the elephant's strategy of a 2-year gestational period, 5-year birth intervals, and intensive parental care lavished on a single offspring for a number of years.

Species lacking the adaptations associated with opportunistic species may also experience exponential population growth following some sort of ecological disturbance that disrupts a previously stable ecosystem.

In nature, population size and growth rate are regulated

No matter how cushy an environment may seem when organisms first arrive, at some point the population can no longer sustain an exponential rate of increase. A variety of ecological factors act as governors on runaway population growth by decreasing births, increasing deaths, or both. When they kick in, the rate of increase begins to decline. As the birth rate continues to decrease and the death rate increases, the value of r decreases to 0, which maintains the population at a certain number. In certain situations, r may become negative if deaths exceed births. Because these factors come into play as the population size increases, their regulatory effects are said to be **density dependent**. **Population density** is defined as the number of individuals per unit area.

The actions of density-dependent factors regulating population growth mimic the feedback inhibition of metabolic pathways discussed in chapter 6. An end product (in this case, the numbers of individuals born and surviving) reaches a level that activates a mechanism to prevent more production. Density-dependent regulators of the population growth rate include competition for resources, predation, disease, and habitat degradation, all of which are discussed below.

Other ecological factors, such as changes in temperature, salinity, or rainfall, may also decrease population size. Because there is no feedback between the population size and activation of the mechanism controlling the population size, these factors are said to be **density independent** (though sometimes this distinction is not very clear-cut). As such, they are not true regulators of population growth rates.

Whether they act in a density-dependent or density-independent fashion, physical and biological aspects of the environment that control population size are known collectively as **limiting factors**. A limiting factor is any essential resource that is in short supply or any environmental factor that is too extreme for survival. Over time, limiting factors can change as populations adapt to their environments through evolution.

Limiting factors also operate at the level of the species and delineate a species' geographic range and the types of habitats it can occupy. Therefore, limiting factors are key determinants of the abundance and distribution of species on a global scale. Density-independent limiting factors often serve to establish the maximum possible range of a species. Density-dependent factors, such as predation and parasitism, tend to limit the distribution of a species to a narrower range than the one defined by its physiological adaptations to abiotic factors.

Resource competition and environmental carrying capacity

As population size increases, competition for available resources does as well. An essential resource in short supply becomes a limiting factor for a species' continued population growth and defines the environment's **carrying** capacity for that species. Think of carrying capacity as the maximum number of individuals of the species that the environment can support at a particular point in time.

As the population size (*N*) approaches the environmental carrying capacity (*K*), the population growth rate (*r*) begins to taper off as resources are depleted. Mathematically, ecologists describe this pattern of population growth, known as **logistic growth**, as follows:

$$G = rN \times (K - N)/K$$

Think of the term in parentheses as the availability of the essential resources that limit population growth, such as food, nesting sites, or light in the case of plants.

By exploring the relationships between the different variables you can see that when *N* is small, (*K* − *N*)/*K* is essentially equal to 1 and *G* equals *rN*, the exponential growth rate. As *N* increases, (*K* − *N*)/*K* decreases until, at *N* = *K*, the term (*K* − *N*)/*K* equals 0, and therefore *G* equals 0. If *N* exceeds *K*, the term (*K* − *N*)/*K*, and therefore the population growth rate, becomes negative.

A graph of the logistic-growth equation gives us the S-shaped curve shown in Figure 14.15. This graph represents an idealized form of logistic growth. A single density-dependent regulatory factor—availability of an essential resource—regulates the population size, and when *r* is equal to 0, *N* is equal to *K*. In other words, the zero rate of population growth stabilizes the population size (*N*) at the maximum number of individuals of that species the environment can support at that point in time, its carrying capacity (*K*). In reality, both density-dependent and density-independent factors often keep population sizes of most species below the environment's carrying capacity.

Predation, parasitism, and disease

Every species is dinner for another species. Although it may seem silly to mention something so patently obvious, in some ways the biological profundity of this often escapes us when we become involved in the intricacies of protein structure and function, cell processes, and the mathematics of population growth. The essence of almost every biological phenomenon we've discussed in this textbook can ultimately be viewed as an adaptive mechanism for acquiring food (Figure 14.16) or avoiding being food for another species (Figure 14.17).

Predation, parasitism, and disease are all variations on the same theme: one species uses another species as a source of materials and energy. These

Figure 14.15 Logistic growth curve.

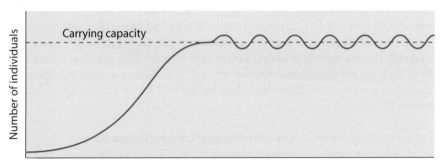

Figure 14.16 Predators and their prey. **(A)** In nature, the predator-prey relationship encompasses interactions much broader than dramatic struggles between large, fierce **carnivores** and helpless prey, such as that depicted here. **(B)** Herbivory. Plant-eating insects are also predators. **(C)** Carnivorous plants. A number of plant species, such as this Venus flytrap, have turned the tables on **herbivores** or other small invertebrates. This one is making a meal of a spider. (Photograph by Barry Rice, courtesy of http://www.sarracenia.com.) **(D)** Predatory fungi. Hyphae from a filamentous fungus lasso a nematode. (Photograph courtesy of George Barron and Nancy Allin.)

interspecific interactions regulate the population size of prey and host species in a density-dependent fashion. The larger the prey or host population size, the easier it is for predators and **parasites** to find their prey or host and for pathogens to spread diseases. As a result, a predator, parasite, or pathogen often keeps the prey or host population size below the environmental carrying capacity (assuming they either kill the prey or host or reduce its reproductive output).

The population size of the predator, parasite, or pathogen often tracks the population size changes of the prey or host, because the availability of the prey or host constitutes a major component of the environmental carrying capacity of predators, parasites, and pathogens (Figure 14.18). As the number of prey or host organisms in an area increases, the population size of the predator, pathogen, or parasite responds accordingly, because r increases due to increased births and fewer deaths. In addition, like most species, predators and parasites tend to congregate where resources are abundant, so immigration into the area also increases the population size in times of prey or host abundance. Competition for limited resources (prey or host), a density-dependent regulatory factor, intensifies as predator, parasite, or pathogen numbers approach the environment's carrying capacity, slowing the population growth rate.

Figure 14.17 Escaping predation. Selective pressures have led to a variety of self-defense strategies in animals. **(A)** Warning coloration. To escape predation, a number of animals synthesize noxious chemicals. They often advertise their distastefulness with bright colors. (Left) The poison arrow frog's skin secretions are so toxic that indigenous people place a few drops on their arrow tips to kill large mammals. (Right) The warning message is reinforced when the animals aggregate, especially when the noxious chemical is sprayed on predators, as these bugs do. **(B)** Mimicry. Some animals hide from predators by having structural adaptations that make them look like something the predator is not interested in eating, such as a leaf (left) or bird droppings (right). **(C)** Protective coloration. A certain pattern of coloration camouflages a Costa Rican ground-nesting bird, the paraque, which is similar to the North American whip-poor-will. The photographs on the left and right were taken from essentially the same spot. The photograph on the left shows two eggs in the paraque's nest; the female is sitting on the nest in the photograph on the right. Can you find her?

Environmental degradation and carrying capacity

As population size increases, individuals within the population may begin to impede the survival of other members, not only by competing with them for limited resources but also because their activities degrade the environmental quality. Accumulation of metabolic wastes can poison the environment for the species producing them and others sharing its environment. Even under

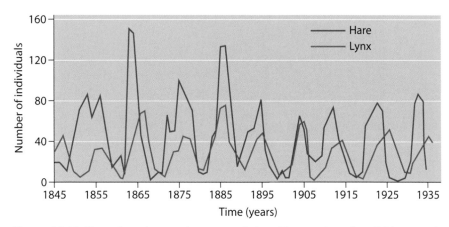

Figure 14.18 Sizes of predator and prey populations. The number of available prey affects the population size of the predators, which in turn affects the population size of the prey. In the classic example of snowshoe hare and lynx populations, an increase in the number of hares consistently led to a corresponding increase in the lynx populations. As the number of lynx increased, greater predation pressure caused a decline in the prey population, causing the lynx population to decrease as well.

conditions of unlimited resources, environmental degradation can lead to a negative population growth rate as deaths exceed births (Figure 14.19). In some cases, the degradation may be so extreme that the environment cannot recover quickly enough and the population goes extinct. Sometimes one species may take another species with it.

Environmental degradation can also redefine an environment's carrying capacity, lowering it to a population size well below a previous carrying capacity. For example, approximately 4,000 deer lived on the Kaibab Plateau in Arizona prior to 1907. In 1907, the government instituted a program to eliminate the wolves and puma that preyed on the deer living on the plateau.

Figure 14.19 Microbial growth curve. When resources are abundant, a population of bacteria exhibits the classic pattern of exponential growth until it reaches the carrying capacity defined by availability of nutrients, space, or both. During the stationary phase, the rate of reproduction equals the death rate. As bacterial waste products accumulate, deaths exceed births, r becomes negative, and the population size decreases.

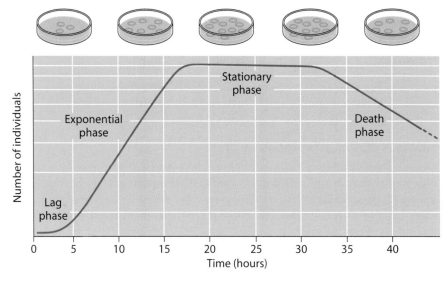

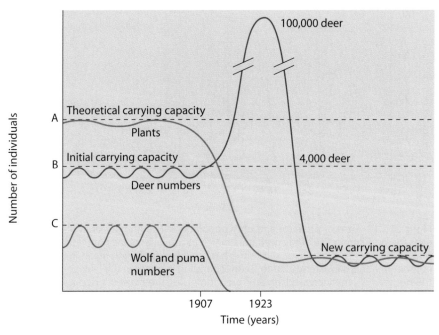

Figure 14.20 Carrying capacity and environmental degradation. In the schematic graph, which is not drawn to scale, lines B and C represent the initial carrying capacities for the deer and predator populations, respectively. Line A is the theoretical carrying capacity for the deer population, based on the number of deer the Kaibab Plateau vegetation could support. Predation pressure is responsible for the difference between the theoretical and actual numbers of deer. The removal of predators caused an exponential increase in the deer population, and the resulting damage inflicted on the island's vegetation caused the deer population to crash. The plant population could never recover to its pre-1907 levels, creating a significantly lower carrying capacity for deer.

By 1923, the deer population had soared to 100,000 individuals, a phenomenon known as **ecological release**, because the factors that had been regulating population size are removed. In the next 2 years, 50% of the deer population starved to death because they had stripped the plateau of vegetation. The disruption was so severe that the plant populations (which had low *r* values compared to that of the deer) could not recover, and eventually the deer population stabilized well below its 1907 level (Figure 14.20).

The carrying capacity of the plateau prior to 1907 was at least 4,000 deer, but it might have been higher if the predators had been maintaining the deer population size below carrying capacity. When the removal of predators released the deer population from regulatory control, the population size increased so rapidly that it overshot the carrying capacity, leading to environmental degradation and lowering the carrying capacity below the 1907 level.

HUMAN POPULATION GROWTH

By now, it should be obvious that humans invest an extraordinary amount of time and energy in devising ways to get around nature's rules. The idea that humans have a right to play by their own rules seems to be one of the defining attributes of the species. Nowhere is that statement more apt than in a discussion of nature's rules governing population growth. Humans have sidestepped certain rules only to be faced with new ones, which they then do their best to sidestep. A historical description of the factors affecting human

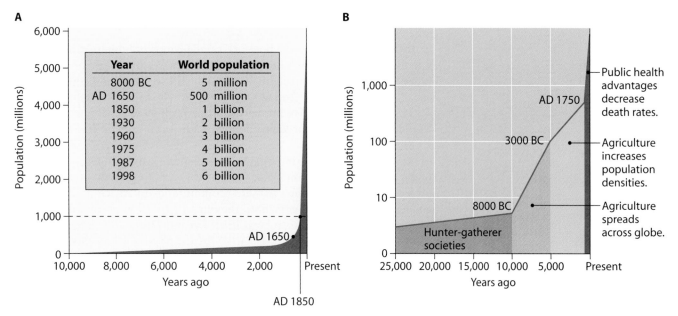

Figure 14.21 Human population growth. Key dates in the graphs are 8000 BC, when people began domesticating plants and animals, and AD 1650 to 1750, when the scientific and industrial revolutions began. **(A)** A lag phase that lasted thousands of years gave way to an exponential growth rate that continues today. **(B)** The graph uses a logarithmic scale for population size instead of the absolute numbers of individuals. It provides information on changes in the *rate* of population growth in addition to changes in the number of people over time.

population growth provides an opportunity to synthesize the various facts you have just learned into a single story. It also sets the stage for subsequent chapters on agriculture.

In other species, populations do not maintain exponential growth rates for very long because increasing population size eventually flips a switch that activates density-dependent regulatory mechanisms, such as resource depletion, environmental degradation, or disease. So far, humans have sidestepped those controls. The human population has been in an exponential phase of population growth for approximately 200 years now (Figure 14.21). At some point, r will begin to decrease as deaths exceed births, causing the population growth rate to decelerate, reach zero, and perhaps become negative. No one knows when this will occur, what the population size will be when it occurs, or how it will happen. But it will happen—assuming a sizable number of people do not emigrate from Earth and establish colonies in outer space, that is.

Early humans made certain limiting factors irrelevant

Two million years ago, the ancestors of today's humans inhabited the grassy savannahs of Africa and ate grains, tubers, and vegetables, supplemented occasionally by small game and meat they scavenged from other predators. Over time, these early humans expanded their geographic range by occupying habitats and climates that had been unavailable to them, because they effectively eliminated the limiting factors that had constrained their distribution. Unlike other species, early humans could mitigate the inhospitable aspects of the environment that had made it uninhabitable by wearing clothes, building fires, and constructing shelters. Making tools that allowed them to hunt cooperatively allowed access to a whole new food supply, large animals, and encouraged the formation of small bands of nomadic hunter-gatherers.

By 40,000 BP (before the present), small groups of hunter-gatherers lived in many different types of habitats around the globe. At that point, the global human population numbered approximately 1.5 to 2 million people. For the next 30,000 years, the population grew slowly but consistently (Figure 14.21B). The r was positive but very small, especially when compared to that of modern humans. What kept r so low, high death rates or low birth rates $(r = b − d)$?

Assuming that the hunter-gatherer societies existing today reflect early human populations, late reproductive maturity and long birth intervals probably kept r small. In hunter-gatherer communities in Africa and Australia, females do not reach puberty until they are 19 to 20 years old, and the interval between births tends to be at least 4 years. The energy demands of the nomadic hunter-gatherer lifestyle may create physiological and hormonal conditions that delay puberty and lengthen birth intervals, such as low body fat accompanied by low estrogen levels.

Agriculture's effects on *r* and *K*

Around 10,000 BP, hunter-gatherer populations at a number of locations across the globe began cultivating the local wild plants they had been gathering. The level of cultivation extended well beyond encouraging the growth of wild plants by providing them with water and removing the weeds that were depleting soil nutrients. Hunter-gatherers began to selectively plant and nurture only those plants from the wild plant population that had desirable characteristics. Mimicking the forces of natural selection described in chapter 13, each hunter-gatherer population guided the evolution of their local, gathered plants into domesticated crops.

Because the hunter-gatherer populations were dispersed across the globe, different hunter-gatherer communities domesticated different crop plants. The crops continued to coexist with their ancestral wild plant populations, but in some cases, the crop plants and wild plants diverged so much genetically that they became different species and could no longer interbreed.

Anthropologists debate which factors—climate change, retreat of the ice sheets, and depletion of local resources, including water—were responsible for three hunter-gatherer populations that lived in very different parts of the world simultaneously (in geological time) starting to domesticate the plants they had been gathering. Whatever initiated the change, slowly the entire structure of human societies began to change. Because agriculture provided food surpluses, communities began to store excess food, which encouraged populations to be less nomadic. The change to a more sedentary lifestyle had a dramatic effect on r, because the birth interval was halved to approximately 2 years and girls began to reach puberty earlier. Figure 14.21 shows that as r increased due to increasing birth rates, the population began growing more rapidly than before.

Accelerating the rate of population growth transformed human society irreversibly by changing not only the abundance but also the distribution of humans. Bands of hunter-gatherers existing today typically contain 20 to 50 members, depending on the productivity of the local environment. When the population reaches a size that exceeds K, a subgroup splinters off and strikes out for a new territory.

The advent of agriculture altered this fissional style of social organization dramatically. As described above, most predators tend to congregate around abundant food supplies, and human "predators" that "prey" on plants are no exception. In nature, as the predator's population density increases, competition for resources also increases. This lowers the r of the predator population, ensuring that the population size stays within bounds. By keeping the predator's numbers within limits, this regulatory mechanism prevents a population from overexploiting its resources or degrading its habitat and causing its own extinction.

These regulatory mechanisms did not kick in with the first agriculturists, which led to the virtual extinction, not of the human species, but of the hunter-gatherer lifestyle. With the rapid increase in r and the natural tendency of predators to concentrate around food supplies, local populations of humans soon reached a size that could not be sustained with anything other than agriculture. In other words, agriculture increased K well beyond the number of individuals that could be supported solely by hunting and gathering. Each hunter-gatherer requires 20 km^2, the same area that can support 6,000 agriculturists. If humans had maintained the hunter-gatherer lifestyle, the carrying capacity of the Earth would have been approximately 20 to 30 million people instead of the 6 billion it now supports (Figure 14.22).

In addition, by clearing forests for cropland and cutting timber for fuel, agricultural communities also degraded the environment that hunter-gatherers depended upon, significantly decreasing the carrying capacity of the environment for that lifestyle. As a result of both of these changes (increase in population size and environmental degradation), the shift from hunting and gathering to agriculture became irreversible. We will return to this point later in this chapter.

High population density increases death rate

Figure 14.21 shows that from 10,000 to 5000 BP the population increased 20-fold, from 5 million to 100 million. From 5000 BP (3000 BC) until AD 1750, the population continued to grow at a constant rate, but that rate was significantly lower than the pace that had been established during the previous 5,000-year period. Even though the population size had increased significantly, it was still well below the new carrying capacity established by agriculture. If resources were not limited, what caused r to decrease between 3000 BC and AD 1750?

In addition to increasing the carrying capacity of the environment, agriculture also increased human population density. Because the carrying capacity of agricultural communities is so large, the settlements spawned by agriculture increased in size. People began living in close proximity to more and more people. Thus, around 3000 BC, density-dependent regulating factors, such as disease and parasitism, began to slow r by increasing the death rate (d).

Lowering d leads to exponential growth

The most dramatic increase in population size began to occur around 1750, and by 1850 the population numbered 1 billion people. It had taken the human race approximately 2 million years to reach the 1 billion mark, but in a mere 80 years (by 1930) it added another billion. From 1930 to 1975, the population doubled again to 4 billion.

The earlier increase in r initiated by agriculture and the more sedentary lifestyle it allowed resulted from increasing the birth rate through earlier puberty and shorter birth intervals. The most recent change in r, however, owes much more to decreasing the death rate than increasing the birth rate. More

Figure 14.22 Modern day hunter-gatherers. In Africa and Australia, a small number of people continue to live as hunters and gatherers. **(A)** This group contained 12 adults and four children. Note the large bow on the left side of the photo. **(B)** The people hunt with primitive weapons, such as the bow and arrow this man is using to kill a young vervet monkey. **(C)** A group member retrieves the vervet monkey from the tree. **(D)** Another group member started a fire with the friction generated from rubbing two sticks together, and they cooked the monkey immediately after removing its intestines.

specifically, it can be traced primarily to better control of infectious diseases, which were the leading cause of death in dense populations.

The first significant improvement in human health began with better public hygiene and improved sanitation and waste disposal. Then and now, controlling the spread of infectious diseases by keeping sewage separate from

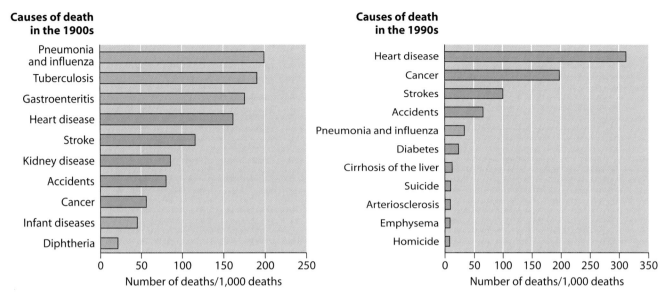

Figure 14.23 Causes of mortality. Immunization and antibiotics have led to significant shifts in the major causes of death in the United States and other industrialized countries in the past century. In the 1900s, infectious diseases (orange bars) were responsible for over 60% of deaths in the United States. Now, noninfectious diseases (blue bars), such as heart disease and cancer, are by far the leading causes of death.

drinking water and preventing infestations of rats, mosquitoes, and other disease vectors profoundly affects the death rate, especially by lowering infant mortality rather than increasing the average life span. However, the most significant factors lowering death rates and contributing to the exponential rate of population growth became operative in the mid- to late 1800s. First of all, the scientific breakthroughs of Louis Pasteur and Robert Koch established the role that bacteria play in causing certain diseases and reinforced the need for improved sanitation and public hygiene. Second, the development of vaccines and antibiotics controlled the spread of many infectious diseases (Figure 14.23).

Agricultural technology slows *r* in the industrialized world

People have long assumed that the early hunter-gatherers abandoned their energy-demanding nomadic lifestyle in favor of agriculture because it gave them more free time and probably improved their diet. In the second half of the 20th century, anthropologists began to test this hypothesis by comparing existing hunter-gatherer societies with groups that practice subsistence agriculture similar to that of early, primitive agriculturists. Their findings surprised everyone.

They began the studies believing that today's hunter-gatherers continue to cling to their lifestyle because they know so little about nature that the idea of cultivating plants has not occurred to them. However, contrary to these preconceived notions and prejudices, the anthropologists discovered that hunter-gatherers understand a great deal about plant growth, development, and ecology. When the anthropologists asked why they didn't purposefully grow plants, the hunter-gatherers replied, "Why should we?" Not only were these tribes well nourished, they also spent less time acquiring food than people who practiced subsistence agriculture.

The !Kung bushmen in Africa's Kalahari Desert obtain approximately 2,400 calories and 96 g of protein per day. The 55 to 65% of the population involved in food procurement typically spend only 2 to 3 days/week hunting and gathering. On average, they spend 600 to 700 hours/person/year in food acquisition compared to 1,000 hours/person/year for the agriculturists. Similar results were obtained for Australian aborigines living in equally harsh environments.

Anthropologists now believe that environmental changes may have forced early humans to take up agriculture. Once the population exceeded a size supportable by hunting and gathering and human fate was forever tied to agriculture, human societies entered a positive feedback loop between increasing the food supply and increasing the number of people in the population. Agriculture is hard work, and children are economic assets in societies based on subsistence agriculture. The more children a family has, the more food it can produce. Children also offer parents security for the day when they are too old to do the hard work agriculture demands. But more children means more mouths to feed, which in turn requires more hands to do the farming. But those hands also come with mouths, and on and on for thousands of years preceding the past two centuries.

The positive feedback linking increasing population size and increasing food production began to weaken when farmers invented ways to uncouple agricultural production from the number of mouths (either human or farm animal) to feed. In the mid-1700s, scientific thinking and technological innovation began to slowly infiltrate agriculture in certain countries. Machines, chemicals, and improved crop varieties allowed farmers to increase crop yields without increasing the supply of farm workers by having more children.

In countries that improved agricultural productivity with science and technology rather than by generating more laborers, r has decreased to virtually zero even though death rates have actually decreased. (In some industrialized countries, r has become negative, which creates its own set of problems.) The population growth rate could decline without sacrificing food production because technological innovations in agriculture made having more children moot (Figure 14.24). Countries that have not integrated science and technology into agriculture continue to be trapped in the positive feedback loop linking the number of mouths to feed and the number of hands to do the work necessary to feed them.

This distinction between countries that use either technology or numbers of workers to improve agricultural productivity has implications that extend well beyond the effect this choice has had on population growth rates. Choosing one path or the other has led to the litany of differences that distinguish industrialized from developing countries.

Around 150 years ago, countries in the industrialized world began using technology to improve agricultural productivity. Over time, machines, chemicals, and increased crop varieties increased crop yields so significantly that people could be released from the agriculture workforce (Table 14.4). These people entered the industrial workforce and became responsible for the economic growth the industrial world has enjoyed during the last century.

High agricultural productivity drove industrial productivity, in part by providing a pool of workers for new industries born during the Industrial Revolution, but also by generating food surpluses. Agricultural workers (farmers) could sell these surpluses to the industrial workers who had left the farm. This provided farmers with money to purchase the goods and services the industrial workers produced—including more and better agricultural technologies. In these societies, a new positive feedback loop began to

A

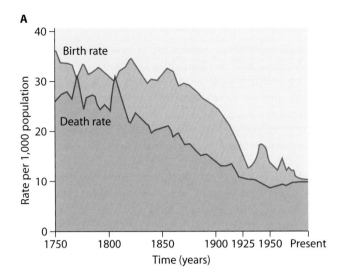

C

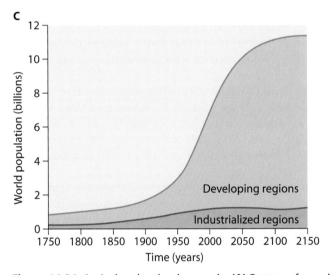

B

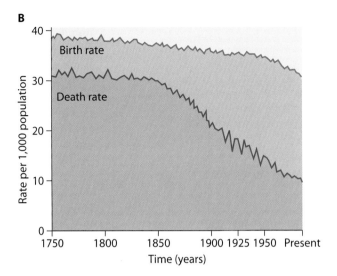

Figure 14.24 Agricultural technology and *r*. **(A)** Pattern of population growth in industrialized countries. In those countries that began incorporating modern agricultural technologies into their farming operations, birth rates began decreasing steadily around 1850, paralleling the decreasing death rate brought on by medical advances. **(B)** Pattern of population growth in developing countries. Those countries that did not use modern agricultural technologies until recently showed no decrease in the birth rate as the death rate decreased. **(C)** Relative numbers of people in developing and industrialized regions. As a result, over the past two centuries, the population growth rate in developing countries has been significantly greater than that in industrialized countries.

Table 14.4 Changes in U.S. agricultural productivity[a]

Parameter	Value in yr:							
	1850	**1880**	**1900**	**1920**	**1940**	**1960**	**1980**	**2000**
% of workforce directly involved with agricultural production	70	55	39	28	20	8	3	1
No. of hours needed to produce a certain amount of food	12	10	8	6	5	2	1	
No. of people supported by one farm worker	1.5	2	5	7	10	20	50	101

[a]Modern technology increased agricultural productivity, which allowed more people to complete their education, leave the farm, join the non-agricultural workforce, and create today's industrial economy. During the last 50 years, one-third as many farms provided 2.5 times as much output with one-fourth as many workers.

operate, linking technological innovation to agricultural productivity to economic growth.

In the developing world, subsistence agriculture is practiced much as it has been for thousands of years, and consequently, as much as 90% of the workforce is involved in agriculture in some developing countries. Because most small-scale subsistence farmers have not incorporated science and technology into their production practices, the relationship between increasing food production and increasing population size, driven by high birth rates, remains intact. Freeing themselves from this cycle is difficult, if not impossible.

In general, crop productivity is low in developing countries, and poor productivity means little saleable surplus is available once the family has been fed. Small-scale farmers who manage to generate surpluses make very little money selling their crops at the local market because most buyers have little disposable income, since they too are subsistence farmers with many mouths to feed and poor crop productivity. People in some developing countries spend as much as 80% of their income on food (for reference, the figure for American consumers is 10 to 12%). Subsistence farmers in developing countries cannot break free of this positive feedback loop, the "cycle of poverty." Technological innovations that broke farmers in the industrialized world out of the cycle cost money, which subsistence farmers don't have.

Many people in today's industrialized societies object to the ecological costs of the agricultural chemicals and petroleum-based machines that replaced farm workers. A surprising number say they want society to return to the time when all agriculture was small scale and chemical free, perhaps because they do not think through the societal and economic ramifications of choosing that option.

Agriculture in today's industrialized societies *is* ecologically expensive, but agriculture in developing countries is just as expensive ecologically, if not more so. Agricultural technologies that rely on chemicals and fossil fuels improve crop productivity, which is yield per acre. Low crop productivity forces farmers in developing countries to improve agricultural output by increasing the number of acres being cultivated through continuous deforestation. In addition to the ecological costs described earlier, deforestation leads to massive soil erosion and loss of biodiversity. Farmers in industrialized countries have been able to increase yields while simultaneously decreasing the number of cultivated acres. Therefore, the ecological costs of modern petroleum-based agriculture should be weighed against the benefit of taking land out of production and allowing it to revert to its natural state.

In addition, by greatly improving crop productivity, fossil fuels and agricultural chemicals provide a food supply that is so plentiful it buys people in industrialized countries their economies, with their goods and services, and therefore, their standard of living. Not until humans could grow more food than they needed could some of them leave the farm, receive an education, enter the workforce, and create modern economies. The ecological cost of today's agriculture buys you the freedom to consider the question, "What do you want to be when you grow up?"

First things must come first, and with all organisms, the first thing is food. Humans can never revert to the hunter-gatherer lifestyle to obtain that food. The enormous size of the human population seals its fate, so agriculture must be the bedrock upon which all economies and societies are built.

The only question is the sort of agriculture that will be practiced in order to produce enough food to feed the population, which now numbers 6 billion.

If citizens in industrialized countries truly want to conserve fossil fuels and stop using chemicals, then more of them will need to join the agricultural workforce. In order to maintain the current level of crop production without using agricultural chemicals, the number of U.S. citizens that would need to return to the farm would cause the economy to collapse. According to the National Council for Food and Agricultural Policy, if U.S. farmers stopped using a single class of chemicals, the weed-killing herbicides, 50 million adults would be forced to leave their jobs and return to the farm to pull weeds in order to maintain the same level of agricultural production.

We will explore this issue further in the chapters on agriculture.

SUMMARY POINTS

Ecology is the study of interactions between organisms and the environment. The ecological environment encompasses everything, both living and nonliving, that is not an integral part of the organism but which affects its evolutionary fitness. Ecologists study interactions within populations of organisms, among different species in a community, and between the community and its physical environment.

An ecosystem is a community of organisms linked to one another and to the physical environment by a flow of energy and materials. All ecosystems need a continuous input of energy, and for most ecosystems the Sun provides that energy. Photosynthetic organisms capture the energy that will support the entire ecosystem by converting light energy into chemical-bond energy.

Within ecosystems, materials cycle. The flow of materials can be cyclical as long as the ecosystem has photosynthetic organisms to fix carbon, nitrogen-fixing bacteria to fix nitrogen, and decomposers to liberate the essential elements that living organisms need to build their biological molecules, primarily carbon, oxygen, nitrogen, and hydrogen. Human activities disrupt nature's balanced cycling of materials, especially carbon and nitrogen.

Populations have remarkable growth potential, but in virtually all cases that potential is not realized because density-dependent regulatory mechanisms control the rate of population growth. A population's growth rate is equal to the number of individuals added to a population through birth and immigration minus the number that are lost to death and emigration over a certain time. A number of factors affect a species' inherent capacity to increase its numbers, including age at first reproduction, number of births per reproductive individual, birth interval, and number of offspring surviving to reproductive age.

The factors that keep population sizes within bounds are limiting factors, such as competition for resources, predation, dis-

ease, and environmental degradation. The necessary resource that limits population size defines the carrying capacity of that environment for the population that needs the resource. When the population size approaches the carrying capacity, the population growth rate decreases and the population size levels off at a certain number.

The human population has used technology to sidestep the normal density-dependent mechanisms that are activated at certain population sizes and retard the population growth rate. Agriculture was the first technology that allowed humans to circumvent factors that govern population growth. Approximately 10,000 years ago, domestication of plants and animals began to make certain limiting factors irrelevant by providing large amounts of food, as well as materials for shelter and clothing to protect early populations from conditions too harsh to support human life.

As human populations congregated around rich agricultural areas, density-dependent factors that decrease population size—disease and parasitism—began to decrease the population growth rate by increasing the death rate. However, humans used environmental technologies (sewers and sewage treatment) and medical technologies to control the spread of infectious diseases, removing that control mechanism.

Modern agricultural technologies have contributed to the slowing of population growth in industrialized countries. As medical technologies decreased the death rate, agricultural technologies simultaneously decreased the birth rate, not directly but indirectly, through social and economic influences. In developing countries, medical technologies also decreased the death rate, but a lower birth rate did not parallel the decreased death rate. The birth rate remained high and constant, because agricultural production continued to rely on manual labor instead of technology.

KEY TERMS

Abiotic factors

Autotroph

Biomass

Carbon cycle

Carbon fixation

Carnivore

Carrying capacity (*K*)

Chlorophyll

Consumer

Decomposer

Density-dependent
regulation

Density-independent
regulation

Detritivore

Ecological release

Ecology

Ecosystem

Exponential growth curve

First law of
thermodynamics

Food chain

Gross primary productivity

Herbivore

Heterotroph

Hyphae

Intrinsic growth rate (*r*)

Limiting factor

Logistic growth

Mutualism

Mycelium

Net primary productivity

Nitrification

Nitrogen cycle

Nitrogen fixation

Nodules

Opportunistic species

Parasite

Phytoplankton

Population density

Population growth rate

Primary producer

Reproductive potential

Second law of
thermodynamics

Symbiosis

Trophic level

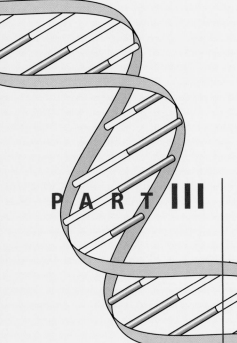

PART III | Biotechnology Applications and Issues

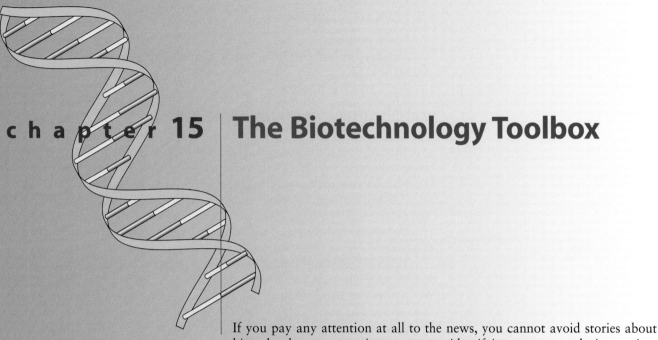

chapter 15 | The Biotechnology Toolbox

If you pay any attention at all to the news, you cannot avoid stories about biotechnology: sequencing a genome, identifying a gene, analyzing ancient DNA, fingerprinting DNA, cloning something, genetic engineering, etc. Have you ever wondered, "How do they do that?"

In this chapter, we will look at some of the tools scientists use in carrying out the research you read about in headlines. The chapter isn't meant to be an exhaustive list (it isn't) or a detailed how-to manual but rather a conceptual guide to basic procedures that are used over and over again in biotechnology. You may find that many of the procedures are familiar to you, because they involve principles and enzymes you have already read about in this book.

As scientists studied basic cellular processes, they often isolated cellular components, such as enzymes or DNA, from whole cells and studied their structures and activities outside the cell. When a process is studied in isolation, it can often be manipulated in ways that are impossible if it is still taking place inside a cell. In the pursuit of knowledge, then, scientists isolated and studied the activities of enzymes that perform various reactions with DNA: for example, copying it (DNA replication) and joining pieces of it (DNA replication, repair, and recombination). They also discovered natural processes in which DNA is transferred from one cell to another (transformation, transduction, and conjugation). As researchers discovered enzymes and processes, they experimented with using them for their own scientific purposes.

A byproduct of the quest for knowledge was that scientists acquired a versatile tool kit for manipulating and analyzing DNA and proteins. Biological knowledge and biotechnology know-how increased hand in glove. Modern biotechnologies harness enzymes and procedures that scientists copied or modified from nature.

In this chapter, we will look at some of the enzymes and other fundamental tools for manipulating DNA and cells that enable us to:

- manipulate and analyze DNA, including determining its sequence
- clone DNA
- analyze proteins

In the next chapter, we will look at how these fundamental techniques are combined to achieve goals, such as finding genes, creating transgenic plants, or constructing a DNA fingerprint.

MANIPULATING AND ANALYZING DNA

The structure of DNA was published in 1953. At that time, scientists weren't sure that DNA was the genetic material, much less aware of how its encoded information was translated into proteins. Yet within 20 years they had constructed the first artificial recombinant DNA molecules and used them to transfer traits in bacteria. Since then, advanced techniques for manipulating and analyzing DNA have continued to accumulate. Some of them have become so routine that high school students often perform them as laboratory exercises.

This veritable explosion of DNA technologies is in part the result of the intense research focus on understanding the function of DNA that began in the 1950s. The other contributing factor is the fact that many of the tools used to manipulate and analyze DNA are natural enzymes and processes. Once scientists discovered and characterized them, they could exploit them.

Natural enzymes allow scientists to cut and paste DNA

One of the challenges facing scientists who wanted to study genes was the fact that in nature, many genes are found on one DNA molecule. Scientists had no way to separate a specific segment of DNA containing a single gene from the rest of a chromosome to study it in isolation. A discovery made by scientists studying a completely unrelated issue made it possible to cut DNA into reproducible pieces and to clone those pieces. This discovery of DNA-cutting enzymes has been credited with making the DNA biotechnologies possible.

Cutting DNA

In the 1960s, scientists studying how certain bacteria resist infection by bacteriophages observed that in these bacteria, the DNA of the infecting virus was cut into segments. Pursuing their observation, they discovered that the phage-resistant bacteria contained enzymes that cut the bacteriophage DNA. Named **restriction endonucleases** or **restriction enzymes**, they recognize specific base sequences in a DNA molecule and cut the DNA at or near the recognition sequence in a consistent way (Figure 15.1). Restriction enzymes can be isolated from the bacteria that produce them. When added to a test tube containing DNA, the enzyme will cut the DNA molecule at every occurrence of its recognition sequence.

The most commonly used restriction enzymes recognize palindromes, sequences in which both strands read the same in the 5′-to-3′ direction. An example of a palindromic sequence is 5′-GAATTC-3′. Remember that the complement of this sequence would read from 3′ to 5′, left to right: 3′-CTTAAG-5′. Now read the complementary sequence from 5′ to 3′. It is identical to the other strand.

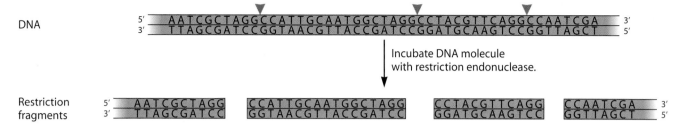

Figure 15.1 Restriction endonucleases recognize and cut specific sites in a DNA molecule. The arrows indicate the cleavage sites of one such endonuclease.

The structure of a palindromic recognition sequence fits with the structure and function of the restriction endonuclease protein. The protein is composed of two identical subunits. Together, they slide along the DNA helix, and when they reach the recognition sequence, each subunit binds to the exact same pattern of DNA bases, one per strand. Next, each subunit cuts the palindrome in exactly the same place (in the example in Figure 15.2, between the G and the first A).

Figure 15.2 shows the products of this cutting action on two different DNA molecules. Note that each of them has a single-stranded protrusion with the base sequence 5'-AATT-3'. This single-stranded region can form correct base pairs with any other 5'-AATT-3' protrusion (remember, one strand must run 3' to 5').

Over 100 different restriction endonucleases have been identified and isolated from many different bacteria. Because they bind to and cut at specific DNA base sequences, they always cut a given DNA molecule in exactly the same way, producing the set of pieces referred to as **restriction fragments**. Scientists use restriction enzymes like precise DNA scissors to cut DNA molecules into reproducible pieces.

Separating mixtures of DNA fragments

To cut a DNA molecule with a restriction enzyme, a solution containing the DNA is typically put into a small test tube and a solution containing the re-

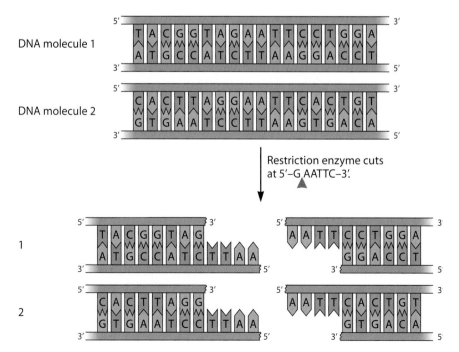

Figure 15.2 The recognition sites of restriction enzymes are usually palindromic. One of the two identical enzyme subunits cleaves at the same position in the palindrome on each strand. The cleavage sites can be opposite one another, as in Figure 15.1. When the cleavage sites are staggered, as shown here, cleavage generates identical complementary protruding single strands at the end of every restriction fragment.

striction enzyme is added. Time is allowed for the enzyme to find its recognition sequences and cleave the DNA molecule. Afterwards, scientists separate the mixture of fragments produced by the enzyme. The standard method used to separate DNA fragments is **electrophoresis** through gels made of agarose or polyacrylamide.

Agarose is a polysaccharide (as are agar and pectin) that dissolves in boiling water and then gels as it cools, like Jell-O. To perform agarose gel electrophoresis of DNA, a slab of gelled agarose is prepared, a solution containing DNA fragments is introduced into small pits in the slab called sample wells, and then an electric current is applied across the gel. Since DNA is highly negatively charged (because of the phosphate groups along its backbones), it is attracted to the positive electrode. To get to the positive electrode, however, the DNA must migrate through the agarose gel.

Smaller DNA fragments can migrate through an agarose gel faster than large fragments. In fact, the rates of migration of linear DNA fragments through agarose are inversely proportional to the $\log_{10}$ of their molecular weights. What it boils down to is that if you apply a mixture of DNA fragments to an agarose gel, start current flowing, wait a little, and then look at the fragments, you will find that the fragments are spread out like runners in a race, with the smallest one closest to the positive electrode, the next smallest following it, and so on (Figure 15.3). Because of the mathematical relationship mentioned above, it is possible to calculate the exact size of a given fragment by comparing how far it migrated to the migration of molecules of known size. After gel electrophoresis, the DNA fragments in the gel are usually stained to render them visible (Figure 15.4). DNA fragments can be isolated and purified from agarose gels, or they can be tested for the presence of specific base sequences (described below). RNA molecules can be separated by electrophoresis as well.

Electrophoresis through polyacrylamide gels works in essentially the same way, except that polyacrylamide makes a tighter mesh than does agarose and so is better for separating smaller DNA molecules and DNA molecules that differ only slightly in size (for example, by 1 nucleotide).

Pasting DNA

Sometimes scientists want to put together fragments of DNA from different sources. This procedure is typically necessary if the goal is to insert a new gene into a microorganism, plant, or animal or to clone a gene as a prelude

Figure 15.3 Gel electrophoresis separates DNA fragments by size.

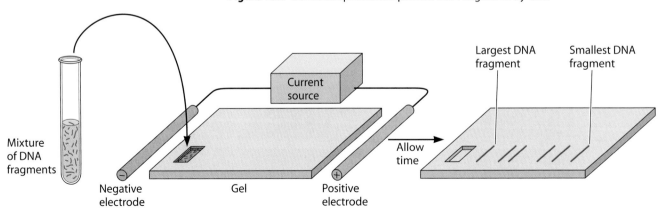

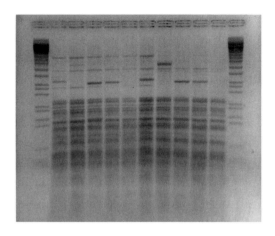

Figure 15.4 Stained agarose gel showing separated DNA fragments. The outlines of the sample wells are visible at the top of the gel. The lanes at the far right and left contain a mixture of fragments of known size.

to characterizing it. We will discuss some of these applications later. The tool used to join pieces of DNA together is another natural enzyme: **DNA ligase**.

DNA ligases join pieces of DNA together by forming new phosphodiester bonds between the pieces (Figure 15.5), a process called ligation. We mentioned DNA ligase when we discussed the repair of DNA damage in chapter 9. Every cell synthesizes ligase to seal gaps in DNA left by replication, repair, or recombination, and DNA ligases have been purified from many kinds of cells.

When two pieces of DNA from different sources are joined together, the result is said to be **recombinant DNA**. The word recombinant means "composed of parts that originally came from two or more sources." You yourself are a genetic recombinant in the sense that half of your chromosomes originally came from your mother's egg and the other half from your father's sperm. You could make a recombinant bicycle by taking apart two bicycles and assembling one from parts taken from each. But in the usual sense, recombinant DNA refers to DNA molecules composed of fragments originally taken from different sources and joined together in a test tube.

The specificity of base pairing is used to detect specific base sequences

Another thing that scientists frequently want to know is whether a specific base sequence is present in a DNA molecule or a DNA-containing sample.

Figure 15.5 DNA ligase joins DNA fragments by forming bonds between the 3′ and 5′ ends of the two backbones.

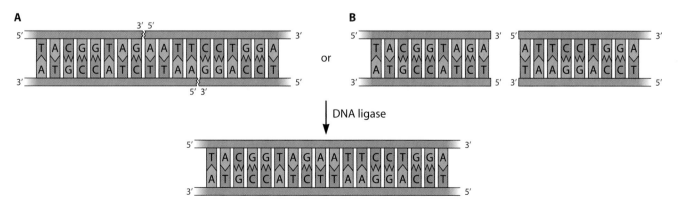

Think about what you've learned about the rules of base pairing in DNA. If you wanted to know whether the base sequence 5'-GGATGCGTCC-3' was present in your sample, one way to find out would be to see whether any DNA in your sample could pair with a strand of DNA having the sequence 3'-CCTACGCAGG-5'. If it could, then the sequence of interest must be present in the sample. This example illustrates **hybridization analysis**, the means by which specific DNA sequences are usually detected.

Hybridization analysis

Hybridization (also called annealing or renaturation) is the term used to describe the process in which two single DNA strands with complementary base sequences stick together to form a correctly base-paired double-stranded molecule (Figure 15.6). Hybridization occurs spontaneously: if two complementary single DNA strands are mixed together and left alone, they will hybridize. The time it takes for hybridization to occur is directly related to the length of the DNA sequences involved; as one might expect, short complementary molecules can line up correctly and form base pairs much faster than long molecules can. Hybridization also works with RNA molecules; complementary RNAs can hybridize to form a double-stranded RNA molecule, and a complementary RNA can hybridize to DNA.

To conduct a hybridization analysis, the double-stranded sample DNA must be separated into single strands. This process, called **denaturation**, is often carried out by heating the sample to 95°C. The heat overcomes the force of the hydrogen bonds holding the base pairs together, and the two strands separate, exposing their unpaired bases.

Once the strands are separated, their bases are free to form pairs with the test sequence. Scientists add what is called a **probe** to the denatured sample DNA. A probe is a piece of single-stranded DNA containing the test sequence of bases. If the sample DNA contains the base sequence complementary to the probe DNA sequence, the probe will form base pairs with the sample DNA, or hybridize to it (Figure 15.7). After time has been allowed for hybridization to take place, the sample is washed to remove unbound probe and then tested to determine whether any hybridized probe is present.

Probes are chemically modified, or labeled, in some way so that their presence can be detected. For example, a probe can be labeled with radioactive phosphorus. If it forms base pairs with the sample DNA, the radioactivity can be detected. Probes can also be labeled with fluorescent dyes or compounds that react with enzymes to make a colored product.

Figure 15.6 Hybridization is the formation of base pairs between two complementary single-stranded nucleic acid molecules. The molecules can be the same or different lengths.

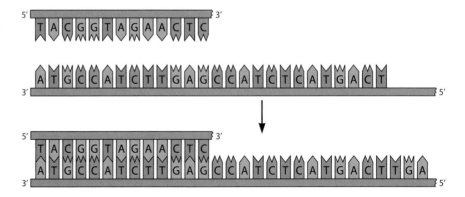

Chemical synthesis of single-stranded DNA

Where can you get a probe for a hybridization analysis? You can isolate naturally occurring DNA, label it, and heat it to denature it, or you can pick up the phone and call in an order for a DNA molecule with a particular sequence (within certain limitations) to any of a number of companies.

Just as biologists studying DNA figured out its role in the cell, chemists studying DNA figured out how to synthesize it in the laboratory. This process has been automated, and computer-controlled DNA synthesis machines loaded with bottles of starting reagents can churn out single-stranded molecules up to about 100 bases long. The user provides the desired base sequence. These relatively short single-stranded molecules are often called **oligonucleotides** (oligo, several). Labels, such as fluorescent dyes, can be added during synthesis, making the use of synthetic oligonucleotides as probes quite convenient.

The term synthetic DNA or **synthetic oligonucleotide** sounds as if the DNA is somehow different from naturally occurring DNA. It is not. "Synthetic" refers to the fact that it was chemically synthesized and not to any differences between it and DNA isolated from an organism. If synthetic DNA is inserted into the genome of an organism, it functions just like any other DNA.

Locating a specific DNA base sequence to a specific restriction fragment

Often it is important to know which DNA fragment in a mixture contains a sequence of interest. For example, your goal might be to clone a particular gene from a given organism. Pursuant to that goal, you plan to isolate DNA from a sample of that organism's cells, cut it into fragments with a restriction enzyme, separate the fragments by electrophoresis, and then isolate the fragment that contains your gene. Obviously, you need to be able to identify which of all the restriction fragments in the mixture contain the gene of interest. Gel electrophoresis and staining alone cannot answer this question, because all DNA fragments look alike when stained. It is possible, however, to combine restriction digestion and hybridization analysis.

Hybridization doesn't work well on DNA fragments embedded in an agarose gel, so if your goal was to separate restriction fragments and then test them, you would perform a procedure called blotting, which is very similar to blotting ink writing. If you write something with a fountain pen and cover the writing carefully with a sheet of blotting paper, the pattern of your writing will be exactly transferred to the blotter. DNA blotting works in the same manner. DNA fragments are separated by agarose gel electrophoresis, and then the gel itself is covered with a membrane and blotting paper. The DNA fragments in the gel transfer to the membrane in exactly the same arrangement they had in the gel. Once the DNA fragments are on the membrane, they can be hybridized to probes to test for the presence of specific sequences (Figure 15.8). Blotting and hybridization analysis can be performed with RNA as well as DNA.

Naturally occurring enzymes allow scientists to make copies of DNA molecules

Yet another challenge for scientists working with DNA can be obtaining enough DNA to analyze. For example, you might have a very tiny sample that you know contains a gene or other DNA sequence you want to study, but the amount of DNA in your sample is too small to be useful. Again, nature provides a solution. DNA is copied inside cells every time a cell divides.

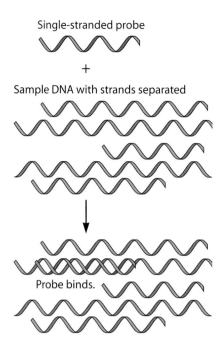

Single-stranded probe

+

Sample DNA with strands separated

Probe binds.

Figure 15.7 Hybridization analysis. A single-stranded probe is added to denatured sample DNA. If the sample DNA contains the base sequence complementary to the probe sequence, the probe will hybridize to the sample and physically stick to it.

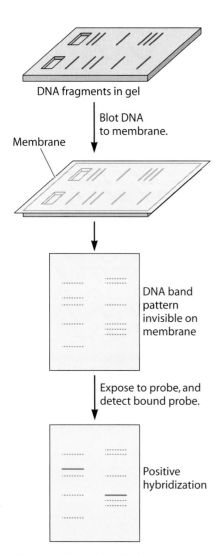

Figure 15.8 A given DNA sequence can be localized to a specific restriction fragment by blotting DNA fragments from an electrophoresis gel to a membrane and conducting hybridization analysis of the membrane.

Scientists take advantage of the DNA replication process to make copies of DNA outside of cells, too.

DNA polymerases

As you have read in previous chapters, DNA polymerases are the cellular enzymes that copy DNA. Using one strand of a parent DNA molecule as a template, they synthesize a complementary strand of DNA by adding nucleotides to the 3′ end of the growing strand (see Figure 9.3). DNA polymerase can be purified from a wide variety of cells and used in test tubes to copy existing DNA molecules.

Within the cell, DNA replication starts with the process of initiation, in which DNA strands are separated and an RNA primer is synthesized (see chapter 9). In a test tube, DNA polymerases also require a single-stranded template with a primer base paired to it. Scientists supply this structure by denaturing the parent DNA as for hybridization analysis (for example, by heating it) and then adding a short, single-stranded DNA molecule that hybridizes to the template and serves as a primer. The first base of the newly synthesized DNA is attached via a phosphodiester bond to the 3′ end of the primer and is complementary to the base on the template strand. Synthesis proceeds as more bases are added to the primer (Figure 15.9).

Figure 15.9 DNA polymerase makes copies of existing DNA molecules (known as the template). The DNA synthesis reaction requires a primer hybridized to single-stranded template DNA, the enzyme, and all four nucleotides. DNA polymerase synthesizes a new complementary DNA strand (shown in red) by adding nucleotides to the 3′ end of the primer. The primer thus becomes part of the new DNA molecule.

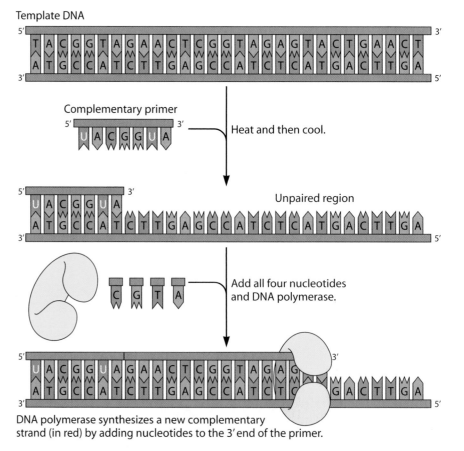

DNA polymerase synthesizes a new complementary strand (in red) by adding nucleotides to the 3′ end of the primer.

The single-stranded primer DNA is the same as a probe in hybridization analysis. It may or may not be labeled for detection. It is called a primer instead of a probe because its purpose is different: to serve as a starting point for DNA synthesis rather than simply to indicate the presence of a DNA sequence by hybridizing to it. Synthetic oligonucleotides are commonly used as primers for DNA polymerase reactions.

Making DNA from an RNA template

Reverse transcriptases read an RNA sequence and synthesize a **complementary DNA** (often abbreviated as **cDNA**) sequence (Figure 15.10). These enzymes are made by RNA viruses that convert their RNA genomes into DNA when they infect a host. Reverse transcriptases allow scientists to synthesize a DNA gene from a messenger RNA (mRNA) molecule. This ability is useful for dealing with eukaryotic genes, since the original genes are often split into many small pieces separated by introns in the chromosome (see chapter 4). The mRNA from these genes has undergone splicing in the eukaryotic cell, and the introns are gone, leaving only the coding sequences. Reverse transcriptase can convert this mRNA into a "prespliced" gene consisting only of protein-coding sequences.

Why would someone want a prespliced gene? Suppose you are interested in moving a eukaryotic gene into a bacterium so that the bacterial cell can produce the protein for you. Since bacteria possess no equipment for splicing, they must be given a prespliced version of a gene if they are to express the correct protein product. Making cDNA from mRNA is important for expressing eukaryotic genes in prokaryotes, as in the production of human

Figure 15.10 Reverse transcriptase uses RNA as a template and synthesizes a cDNA copy. Through additional reactions, the RNA can be removed and replaced with a second DNA strand.

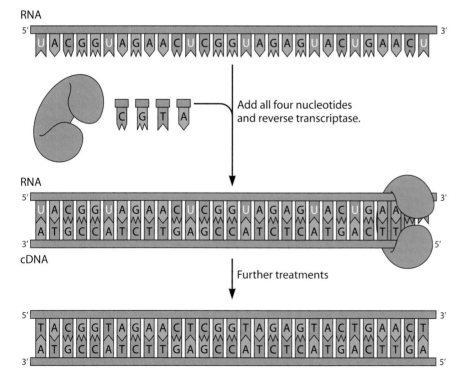

insulin in bacteria. As mentioned in chapter 8, most diabetics now use human insulin made in bacterial cells instead of animal insulin.

PCR makes it easy to produce millions of copies of a target DNA segment

In the 1980s, a scientist driving late at night had a flash of inspiration about how he might use DNA polymerase and primers to create a chain reaction of DNA synthesis, making millions of copies of a target DNA segment. Back in his laboratory, he tested his idea and found that it worked. His invention, the **polymerase chain reaction** (**PCR**), is now used in many different kinds of applications, from disease diagnosis to DNA typing to studying ancient DNA.

PCR requires two primers that hybridize to opposite strands of the template DNA molecule at the boundaries of the segment to be copied. To perform PCR, you would denature the template DNA and allow it to hybridize to both primers, add nucleotides and DNA polymerase, and allow synthesis to begin. DNA polymerase then copies both of the strands, starting at the two primers.

After a few minutes, the reaction mixture is heated to 95°C to separate the product DNA strands from the templates. As the reaction mixture cools, the primers hybridize to the newly synthesized products (as well as the original templates), and synthesis begins again. This cycle is repeated many times, and each cycle doubles the number of DNA molecules, resulting in the synthesis of millions or even billions of copies of the DNA segment that stretches from one primer sequence to the other (Figure 15.11). The process is also called **DNA amplification**.

The 95°C heating step may have caught your attention, since this temperature would be high enough to denature most enzymes and destroy their activity. For PCR, scientists use DNA polymerase enzymes purified from bacteria that live at very high temperatures, such as in hot springs and thermal vents in the ocean floor (Figure 15.12). Since their natural environments are very hot, their DNA polymerases function well at high temperatures. These enzymes can be added to a PCR mixture and will survive repeated heating cycles. Because of this, PCR has been automated. The reaction mixture is assembled and put into a machine that takes it through the repeated temperature cycles.

PCR as a detection method

PCR can be used to amplify scarce DNA so that it can be studied or cloned, but it can also be used as a detection method. The reason it can be used as a detection method is that the primers for PCR must hybridize to the template DNA for amplification to occur. If you added primers to a sample that did not contain the specific template DNA, you wouldn't get the PCR product. In a way, PCR is like hybridization analysis. In hybridization analysis, the hybridized probe is detected directly because of its label. In PCR, hybridization of the primers is detected by the synthesis of the DNA product (Figure 15.13).

PCR is in fact a very good detection method. It is more sensitive than hybridization analysis precisely because it amplifies the target DNA. Regular hybridization analysis is limited by one's ability to detect the hybridized probe. If a sample contained only one copy of the DNA sequence of interest, only one molecule of probe could hybridize to it. In contrast, PCR can theoretically produce slightly over 1 billion copies of a DNA segment from a single starting molecule in 30 rounds of amplification. In fact, PCR has produced a detectable signal from a single copy of a target DNA sequence,

Template DNA plus single-stranded primers (multiple copies)

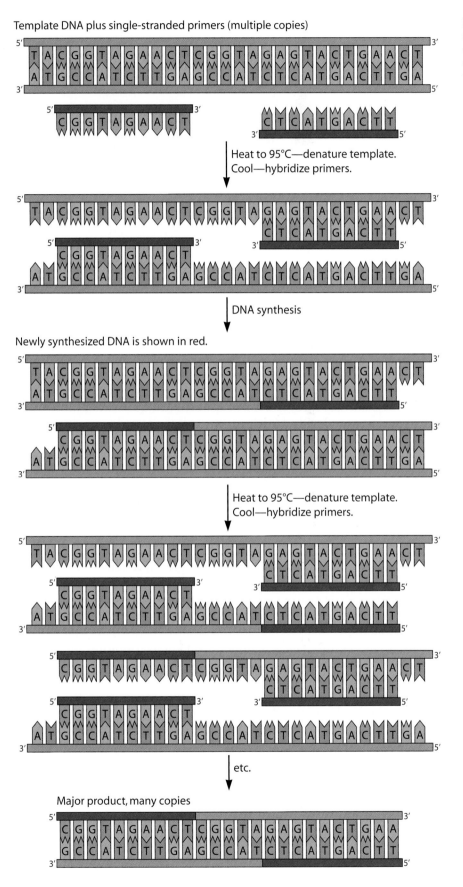

Heat to 95°C—denature template.
Cool—hybridize primers.

DNA synthesis

Newly synthesized DNA is shown in red.

Heat to 95°C—denature template.
Cool—hybridize primers.

etc.

Major product, many copies

Figure 15.11 PCR produces many copies of a DNA segment lying between and including the sequences at which two single-stranded primers hybridize to the template DNA molecule.

Figure 15.12 The vivid yellow, orange, and brown colors at the edges of this steaming hot spring in Yellowstone National Park reflect the presence of heat-tolerant microorganisms. One such bacterium from the park is the source of the DNA polymerase used in PCR. (Photograph courtesy of Thomas A. Martin.)

which is far more sensitive than any method for direct detection of hybridized probe.

Diagnosing disease with PCR

One example of the use of PCR as a detection method is in the diagnosis of disease. The diagnosis and treatment of a particular disease often require identifying a particular pathogenic (disease-causing) microorganism, or **pathogen**. Traditional methods of identification involve culturing these organisms from clinical specimens and performing metabolic and other tests to identify them. The idea behind PCR-based diagnosis of infectious disease is simple: if the pathogen is present in a clinical specimen, its DNA will be present. Its DNA has unique sequences that can be detected by PCR, often using the clinical specimen (for example, blood, stool, spinal fluid, or sputum) itself in the PCR mixture.

DNA technology has proven very useful in the treatment of tuberculosis. Until fairly recently (the 1990s), Americans believed that tuberculosis in their country was a disease of the past. However, the spread of AIDS in the 1980s and 1990s created a population of individuals with almost no ability to fight off disease. Many of these individuals lived in the inner city in dire poverty. Inner-city poverty is associated with crowding, which favors the spread of disease, and malnutrition, which also makes people more vulnerable to infection. Tuberculosis reappeared and spread. To make matters worse, drug-resistant strains arose. In immune-compromised individuals, acute drug-resistant tuberculosis can be fatal in a matter of weeks.

The problem with using traditional methods to diagnose tuberculosis is that the organism that causes the disease, *Mycobacterium tuberculosis*, grows extremely slowly in the laboratory. Traditional culturing, followed by testing for antibiotic resistance, can take months. Patients sometimes died before the test results showing the resistance patterns of the organisms infecting their bodies were in. Once scientists identified the genes responsible for drug resistance in *M. tuberculosis*, they made DNA primers that allowed amplification of the resistance genes. Physicians could obtain information about which drugs could be used to treat a particular patient without waiting for the slow culturing.

Scientists also use DNA polymerase to determine the base sequence of a DNA molecule

The sequence of bases in a gene determines the sequence of amino acids in a protein. The genetic code was worked out in the 1960s, and the first method for determining the base sequence of DNA was also published during that

Figure 15.13 Testing for the presence of a DNA sequence by using PCR.

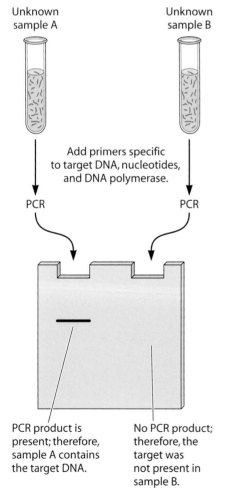

Unknown sample A

Unknown sample B

Add primers specific to target DNA, nucleotides, and DNA polymerase.

PCR   PCR

PCR product is present; therefore, sample A contains the target DNA.

No PCR product; therefore, the target was not present in sample B.

decade. Initially, DNA-sequencing methods were completely chemical approaches and were slow and labor-intensive. As scientists began to work with DNA polymerase enzymes outside the cell, they developed ways to use the enzyme to reveal the sequence of a DNA molecule. Scientists knew that the DNA polymerase enzyme reflects the sequence of a template as it adds complementary bases to a growing new strand of DNA. They figured out how to detect the order in which bases were added, essentially looking over the shoulder of the enzyme as it synthesized DNA.

DNA sequencing with terminators

Enzymatic DNA sequencing employs DNA polymerase and a primer, just as do other DNA synthesis procedures. The DNA polymerase synthesizes a new DNA strand on a single-stranded template, adding new bases one at a time. The key to DNA sequencing is that the nucleotides supplied to the reaction are mixtures of normal nucleoside triphosphates and special nucleotide derivatives. The nucleotide derivatives are forms of adenine (A), guanine (G), cytosine (C), and thymine (T) called **dideoxynucleotides** (ddA, ddG, ddC, and ddT, respectively) that block further base addition because they lack the 3′ OH group that provides the attachment point for another nucleotide (Figure 15.14 and Box 15.1). As DNA polymerase enzymes synthesize DNA,

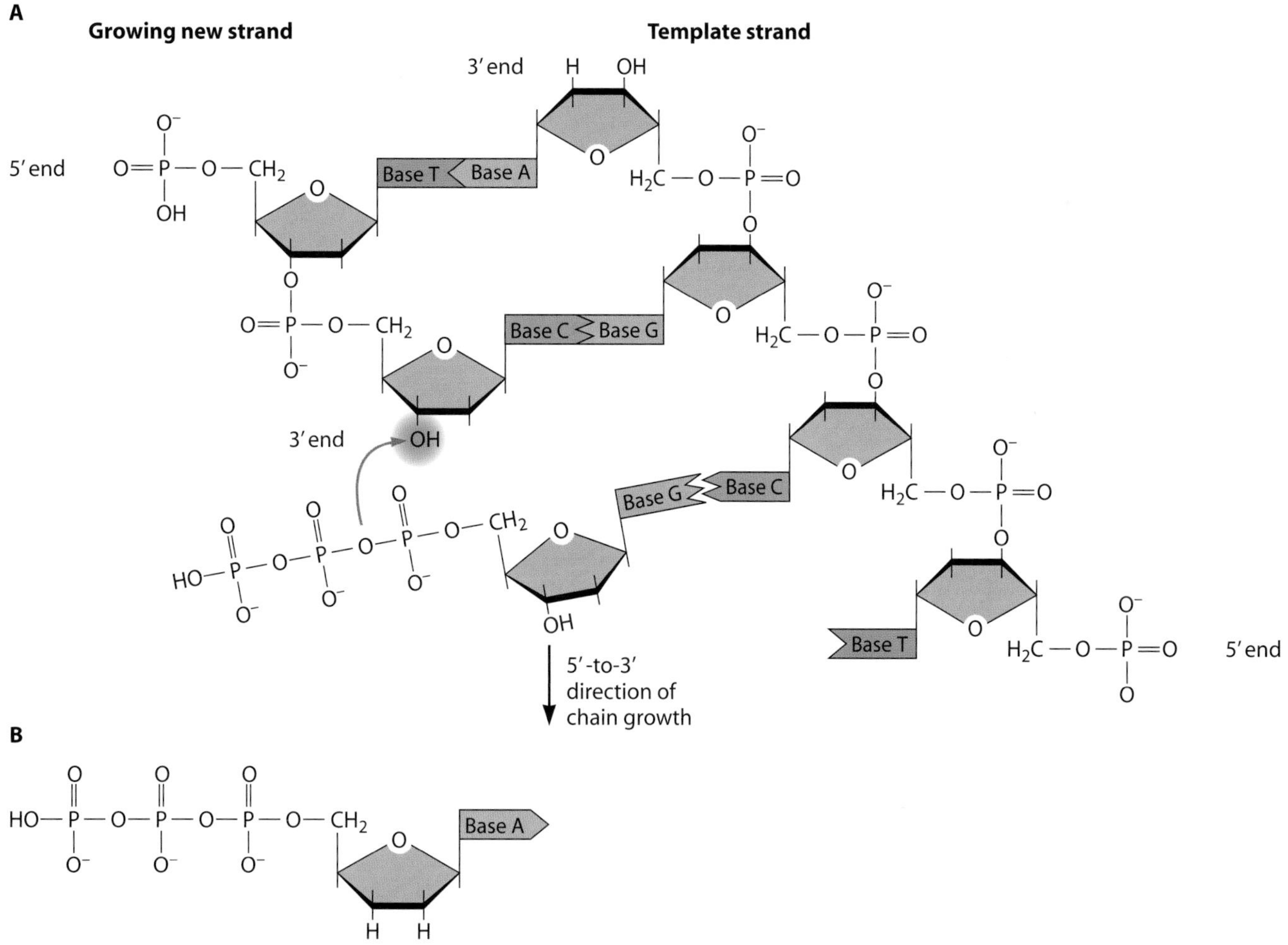

Figure 15.14 Terminator nucleotides. **(A)** A 3′ OH group (highlighted) is required for DNA synthesis, because a new nucleotide is added to the existing strand by the formation of a bond between the OH group and the 5′ phosphate group of the incoming nucleotide. **(B)** A dideoxynucleotide terminates DNA synthesis because it lacks a 3′ OH group.

BOX 15.1 *Replication terminators as antiviral drugs*

Chemicals that were first used as investigative tools in research laboratories often find their way into the pharmacy. Compounds that affect fundamental biological processes are important to basic research and sometimes turn out to be therapeutically useful as well. The chain terminators used in DNA sequencing reactions are a good example of such compounds. Several of the currently available anti-AIDS drugs belong to this class of chemical, as does the best available anti-herpes drug.

These compounds fight AIDS and herpes in the same way they assist in DNA sequencing: by terminating DNA synthesis. For a virus to establish an infection, it must replicate its nucleic acid. Blocking this replication is potentially a very effective way to fight the spread of a virus, but blocking DNA synthesis could be as harmful to the patient as it is to the virus invader. Herpesviruses and the AIDS virus are good candidates for antireplication drugs because they encode their own DNA polymerase enzymes with unique properties that can be exploited.

The AIDS virus (human immunodeficiency virus, or HIV) is an RNA virus. It encodes a reverse transcriptase that synthesizes DNA using the viral RNA as a template. This step is essential to HIV infection. Chain terminators fight the spread of HIV in the body by interfering at this stage. Reverse transcriptase is a good target for chain terminator drugs because it is a sloppy enzyme: it is much more likely to incorporate an incorrect nucleotide or a chain terminator than is the human DNA polymerase. Furthermore, reverse transcriptase lacks the

ability to proofread its work, so it cannot remove an incorrect nucleotide once it is incorporated. This property of reverse transcriptase is also responsible for the high mutation rate of HIV. The drugs are not without toxic side effects, though: although human DNA polymerase is less sensitive to the drugs than is reverse transcriptase, the drugs do affect DNA replication in normal human cells.

There are more issues involved in developing a successful chain terminator drug than simply the sensitivity of the viral replication enzymes. The drug must survive in the body and be absorbed by the proper cells. The active triphosphate form of the chain terminators is not taken up by cells, so the drugs are given in unphosphorylated nucleoside form. Once these compounds enter cells, human enzymes add the phosphate groups (phosphorylate the molecules) to activate them for incorporation into DNA. Different compounds are absorbed and phosphorylated with different efficiencies. Thus, the development of a new drug depends not only on its toxicity to the virus and possible harmfulness to the host but also on how it is metabolized in the human system.

Chain terminators presently employed against HIV include AZT (azidothymidine, or zidovudine), ddC, ddI (dideoxyinosine), 3TC (lamivudine), and d4T (stavudine). AZT, ddC (the same compound that is used in DNA sequencing), and ddI are shown in the figure, along with a normal deoxynucleoside and the antiherpesvirus chain terminator acyclovir. AZT, an analog of thymidine (T), is incorporated into a growing DNA molecule in place of a

normal T. Likewise, ddC is incorporated in place of cytosine. The compound inosine is identical to adenosine (A) except for the absence of one amino group. Cells synthesize inosine and convert it directly to adenosine. When ddI is given as a drug, it enters cells and is rapidly converted to ddA. The ddA is incorporated by reverse transcriptase in place of normal adenosine.

Herpesvirus diseases can be treated with the chain terminator acyclovir (see the figure). The herpesviruses are DNA viruses with relatively large (for a virus) genomes. They encode many of their own DNA replication enzymes, including a DNA polymerase. After the initial infection, herpesviruses remain in the body in an inactive state from which they can be activated and then cause subsequent outbreaks of disease. Herpes simplex virus types 1 and 2 cause fever blisters and genital herpes, respectively. The herpesvirus varicella-zoster virus causes chicken pox when it first infects a person and shingles in subsequent outbreaks.

Acyclovir, marketed under the trade name Zovirax, is relatively nontoxic to humans because human cells take up acyclovir but do not phosphorylate it, so the drug remains inactive. Herpesviruses, however, encode an enzyme that does phosphorylate acyclovir. Therefore, acyclovir becomes an active chain terminator only in herpesvirus-infected cells. Acyclovir is an analog of guanidine (G) and is incorporated opposite cytosine residues in the template. Once incorporated, it terminates further DNA synthesis.

Normal deoxynucleoside shown with chain terminators used as drugs. All are incorporated into DNA from their triphosphate forms.

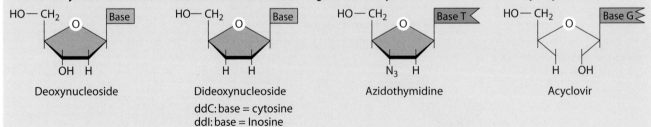

Deoxynucleoside

Dideoxynucleoside
ddC: base = cytosine
ddI: base = Inosine

Azidothymidine

Acyclovir

they may add either a normal or a terminator nucleotide, for example, a normal A or a terminator A.

Each time a terminator nucleotide is added, further synthesis of that molecule is blocked. Separating the synthesis products by electrophoresis and comparing their sizes reveals the order of nucleotide addition and thus the template sequence. One way to do this is to run four separate reactions on the same template, each with a different terminator nucleotide. After synthesis, the reaction mixtures are separated in adjacent gel lanes, and the relative positions of the bands reveal the base sequence of the template (Figure 15.15). Alternatively, the four different terminators (A, G, C, and T) can be labeled with different-colored fluorescent dyes. If synthesis stopped with a terminator C (opposite a template G), that molecule is labeled with the C color. If synthesis stopped with a terminator A (opposite a template T), that molecule is labeled with the A color. These reaction products can be separated in a single gel lane; the fluorescent color associated with each band reveals the identity of the terminator nucleotide at its terminus, and the order of the colored bands reflects the base sequence of the template.

CLONING

The term **cloning** means the production of identical copies of something. In biology, it specifically means genetically identical copies. Asexual reproduction is a natural cloning process, since it generates offspring that are identical to the parent. In fact, an asexually reproducing population is often called a clone, or a clonal population.

Cloning DNA produces identical copies of a DNA segment

When applied to DNA, cloning is usually understood to mean the transferring of a piece of DNA into a cell in such a way that the DNA will be replicated and maintained along with the rest of the cell's DNA. As the cell reproduces, new identical copies of the DNA are produced. Although there are many ways to go about cloning DNA, it usually involves making a recombinant molecule using the fragment of interest and a carrier molecule referred to as a **vector**. A vector facilitates the transfer of DNA into the new host cell and its maintenance within that cell.

Cloning vectors

Many kinds of cloning vectors are available, tailored to different kinds of host cells and applications. The most common kind of vector for cloning DNA into bacterial cells is a **plasmid**. Plasmids are small circular DNA molecules that contain an origin of replication. They are found within the cytoplasm of the bacterial cell and are replicated by the bacterial DNA polymerase. Many types of naturally occurring plasmids have been discovered, and they can be transferred between bacteria via transformation or conjugation (see chapter 13). Scientists typically use laboratory versions of the transformation process to introduce plasmids into host bacteria. Laboratory transformation involves mixing host cells and DNA together and then manipulating the environment of the host cells in such a way that they take up the DNA molecules.

Viruses are also commonly used as vectors, particularly in eukaryotic cells. When a virus is employed as a vector, its natural ability to infect a cell and introduce nucleic acid is exploited, but the natural viral DNA is replaced with the DNA to be cloned. Other types of cloning vectors have themselves

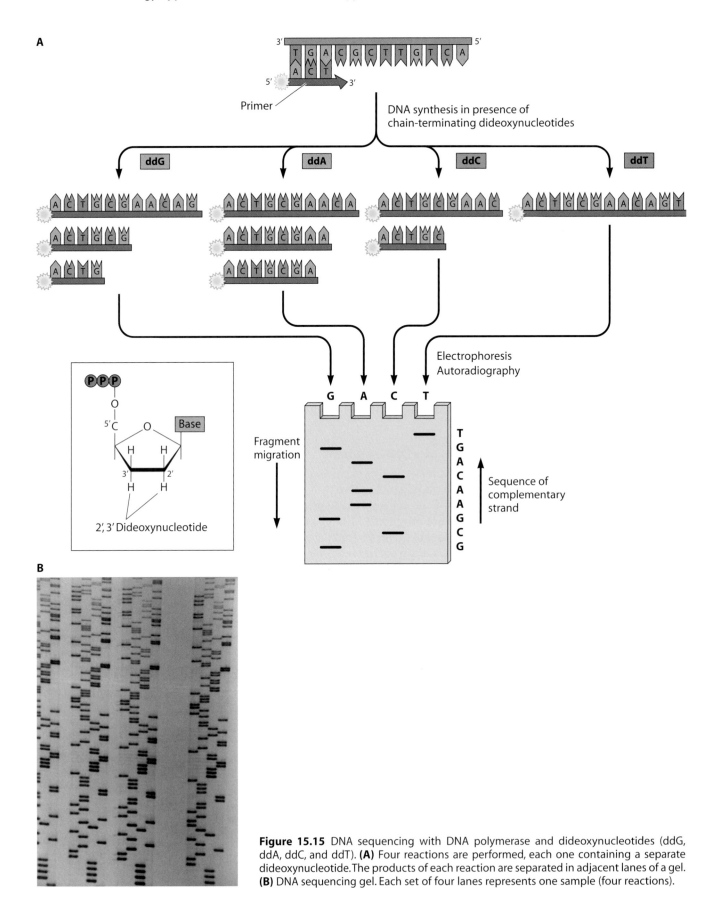

Figure 15.15 DNA sequencing with DNA polymerase and dideoxynucleotides (ddG, ddA, ddC, and ddT). **(A)** Four reactions are performed, each one containing a separate dideoxynucleotide. The products of each reaction are separated in adjacent lanes of a gel. **(B)** DNA sequencing gel. Each set of four lanes represents one sample (four reactions).

been constructed from naturally occurring elements: for example, artificial yeast chromosomes containing replication origins, centromeres, and telomeres have been assembled to serve as cloning vectors for yeast cells.

One particularly useful natural plasmid is the **Ti plasmid** from the plant pathogen *Agrobacterium tumefaciens*. After the bacterium infects a plant, the Ti plasmid transfers itself from the bacterial cell into the invaded plant and inserts part of itself, the so-called T-DNA, into the plant DNA. The T-DNA genes encode proteins that promote tumor growth and the biosynthesis of compounds called opines in the plant. Biotechnologists take advantage of the Ti plasmid by taking out the tumor genes and adding genes they are interested in to the plasmid. The altered Ti plasmid is reintroduced into *A. tumefaciens*, the plasmid-containing bacterium is inoculated into a plant, and the altered Ti plasmid then transfers the new genes to the plant's DNA (Figure 15.16). Unfortunately, only certain plants are susceptible to infection by *A. tumefaciens*.

Just as there are many kinds of cloning vectors, there are many ways to introduce DNA into a host cell. DNA can be injected directly into animal cells with a fine glass needle (**microinjection**) or blasted into plant cells on DNA-coated pellets with a "gene gun." Again, the technology used depends upon the host cell, the vector, and the application.

Marker genes

Regardless of the vector and transfer method used, once the recombinant DNA has been introduced into a batch of host cells, the next task is to identify those cells that took up and are maintaining the recombinant DNA (the transformed cells). DNA transfer procedures are inherently inefficient, so only a fraction, sometimes a very small fraction, of the host cells exposed to recombinant DNA will take it up. Scientists often use **marker genes** to reveal host cells that have successfully received the recombinant DNA.

A marker gene can be any gene that, when expressed, confers a detectable phenotype upon its host. The most common marker genes encode resistance to antibiotics. Cells that take up the recombinant DNA can grow in

Figure 15.16 The Ti plasmid of *A. tumefaciens* is an important cloning vector used in plants. **(A)** When *A. tumefaciens* infects a plant cell, proteins encoded by genes carried on the Ti plasmid transfer the DNA between the T-DNA borders into the plant chromosome. **(B)** Scientists remove the tumor formation genes from the T-DNA and replace it with DNA to be inserted into the plant genome.

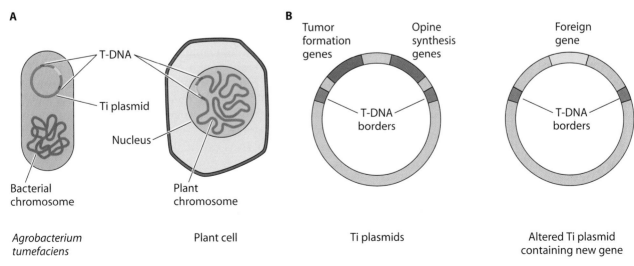

the presence of the antibiotic, while the rest of the host cells cannot. By exposing a mixture of transformed and nontransformed cells to the antibiotic, scientists can kill the nontransformed cells and select the transformed ones. Other kinds of marker genes encode enzymes that cause cells to change color in the presence of specific substrates, relieve nutritional deficiencies by supplying an enzyme the host cells lack, or even make host cells glow in the dark.

Marker genes are part of the vector, and the phenotype conferred by a marker indicates the presence of the vector DNA in the host cell. To determine whether the fragment of interest is present in the vector DNA, additional analysis, such as hybridization to a probe matching the sequence of interest, PCR with appropriate primers, and/or purification and sequencing of the recombinant plasmid, is usually performed.

An illustration of cloning

Figure 15.17 illustrates the cloning of DNA in recombinant bacterial plasmids. First, DNA from the donor organism is isolated and cut into segments with a restriction enzyme (or a combination of enzymes). The selected plasmid vector, carrying an antibiotic resistance gene, is cut with the same enzyme(s). The restriction fragments and the cut vector are mixed with DNA ligase, which forms phosphodiester bonds between the fragments and the vector. Alternatively, large quantities of the fragment to be cloned can be generated by performing PCR using the source DNA as a template. The PCR product is then purified and ligated into a vector.

After time has been allowed for the ligase enzyme to join fragments of DNA, the mixture is introduced into a host cell via transformation. Only some of the host cells will take up and express recombinant plasmids. Following the DNA transfer procedure, the transformation mixture is spread on a solid growth medium to isolate individual cells from one another. The growth medium contains the antibiotic for which the plasmid encodes resistance. Under those conditions, only transformed cells can grow. Each individual transformed cell multiplies, generating a clonal population.

On a solid growth medium, the population takes the form of a little pile of cells called a colony. Colonies look like dots on the growth medium, and each dot is a genetically identical population containing one type of plasmid. The final step in the process of cloning DNA is to isolate plasmid DNA from several colonies, analyze it with restriction enzymes or by sequencing, and determine whether the plasmids from these colonies contain the recombinant plasmid with the cloned fragment that is wanted.

Cloning a particular fragment of DNA gives easy access to it. Once you have a given fragment of DNA in a recombinant plasmid in a bacterial cell, it is easy to produce it in quantity. You simply introduce a few of the bacterial cells into a container of sterile growth medium with antibiotic; allow them to multiply into a large, genetically identical population; harvest the cells; and isolate the plasmid DNA. The techniques described above for manipulating and analyzing DNA, such as restriction analysis and sequencing, require that you start with many identical copies of a given DNA molecule. Cloning a particular piece of DNA is essentially a prerequisite for manipulating it or studying it in detail.

DNA libraries are collections of clones from one organism

A common application of DNA cloning is to produce what is called a **DNA library,** or genetic library. To make a library, the entire genome of an organism is digested with restriction enzymes or otherwise fragmented, and the en-

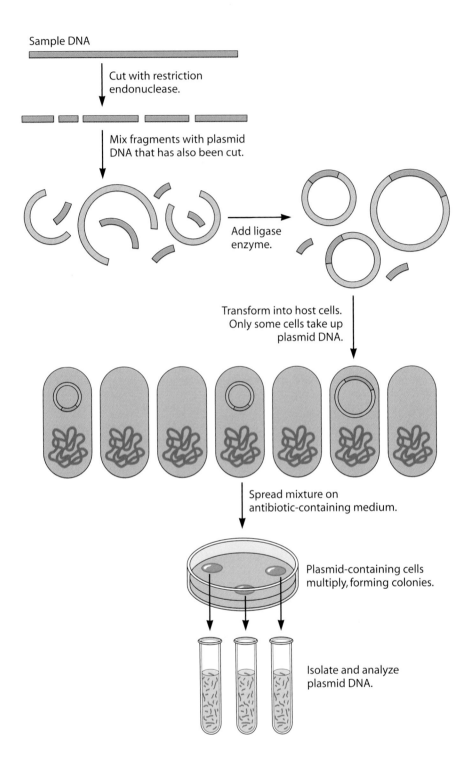

Sample DNA

Cut with restriction endonuclease.

Mix fragments with plasmid DNA that has also been cut.

Add ligase enzyme.

Transform into host cells. Only some cells take up plasmid DNA.

Spread mixture on antibiotic-containing medium.

Plasmid-containing cells multiply, forming colonies.

Isolate and analyze plasmid DNA.

Figure 15.17 Cloning of DNA in recombinant bacterial plasmids.

tire batch of products is mixed with vector DNA and ligated. The resulting mixture of recombinant plasmids is transformed into a host organism. The pool of transformed cells containing the collection of recombinant molecules is called a DNA library of that organism (Figure 15.18A). The idea is that the recombinant molecules in the library will contain fragments covering the whole genome of the organism.

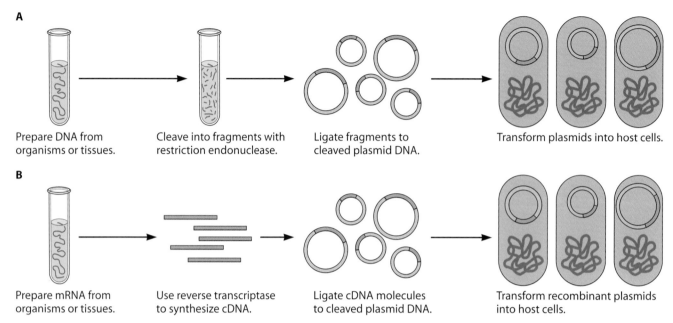

A

Prepare DNA from organisms or tissues. → Cleave into fragments with restriction endonuclease. → Ligate fragments to cleaved plasmid DNA. → Transform plasmids into host cells.

B

Prepare mRNA from organisms or tissues. → Use reverse transcriptase to synthesize cDNA. → Ligate cDNA molecules to cleaved plasmid DNA. → Transform recombinant plasmids into host cells.

Figure 15.18 DNA libraries. **(A)** Genomic DNA library. The insertions in the recombinant plasmids represent the entire DNA content of the organism. **(B)** cDNA library. The insertions in the recombinant plasmids represent genes that were being expressed in the sample.

A DNA library can be used as a resource. For example, if a scientist knows what gene he wishes to study, he can look for it in the library and pull out a cloned version. DNA libraries also serve as starting points for sequencing the entire genome of an organism. The base sequences of the inserted fragments of many, many recombinant molecules are determined, and the regions of overlap are identified. The various sequences are aligned by identifying overlaps until the entire genome is covered.

cDNA libraries

A genomic library should contain clones of all the DNA in an organism, whether the DNA contained genes or was noncoding and whether the genes were being expressed or not. If you were interested in studying only genes that were being expressed in a particular tissue or organism, you could make what is called a **cDNA library**. cDNA libraries are based on mRNA from the starting tissue, so they represent only those genes being actively transcribed. The first step in making one of these libraries is to isolate mRNA from the organism or tissue of interest. Then, reverse transcriptase is used to generate cDNA copies of all the mRNA molecules. These cDNA copies are ligated to vector DNA molecules en masse, and the resulting mixture of recombinant molecules is transformed into host cells, generating a cDNA library. The cDNA library contains DNA with the coding sequences of all the proteins the organism was producing when the mRNA was isolated (Figure 15.18B).

Cloning complex organisms

When the term cloning is applied to plants and animals, it means the production of genetically identical organisms through asexual reproduction. If you have ever taken a cutting of a plant and allowed it to root, you have cloned something. The cutting turned into a new plant that was genetically identical to the original plant.

The production of identical twins is a natural cloning process. Identical twins are formed when a very early embryo splits in two and each portion forms a baby. The two babies are genetically identical clones. Animal embryos formed by artificial insemination can be purposefully split to create identical siblings in a process called twinning.

The kind of cloning that tends to make the news, though, occurs when an adult animal is used as a source of DNA to produce a genetically identical offspring. In general, the way this is done is to remove the nuclei of an adult cell and an egg cell. The adult nucleus is implanted in the egg in place of its own nucleus, and the egg with its new nucleus is implanted in the uterus of an appropriate female. If all goes well, the egg develops normally and the young animal is born.

It sounds simple, but it isn't. Recall that a fertilized egg is totipotent—it has the potential to develop into all the body's tissues. Adult cells, however, are not. Scientists attempting to clone animals manipulate the donor cell to make the DNA more like that of a fertilized egg. Usually, something goes wrong. The success rates of animal cloning procedures are very low.

Note that when the process succeeds, you get a baby animal whose nuclear genes are identical to those of the animal that donated the nucleus. In science fiction movies, "cloning" usually generates an identical, adult copy of the individual being cloned, and frequently the clone has the same memories as the original. This movie image is a far cry from reality. Real cloning generates a baby, and that baby would grow up with its own experiences and memories.

How identical might we expect an adult animal and its clone to be (age difference aside)? One clue comes from looking at identical twins. Even they, who are genetically identical and developed in the same uterine environment at the same time, are not perfectly identical. Some developmental events are heavily influenced by the environment in the uterus, as when congenital adrenal hyperplasia is caused by the hormonal status of the mother (see chapter 10). If identical twins, developing in the same uterus at the same time, are not identical, how much more different might clones that developed in the uteri of different mothers be? In addition, a small portion of eukaryotic DNA is found outside the nucleus. Mitochondria and chloroplasts contain a small amount of DNA (see chapter 16) encoding several genes. When an animal clone is generated by nuclear transplantation, the mitochondria of the cloned offspring come from the egg donor, who could be unrelated to the donor of the nucleus. Taking these differences into account, we would expect a clone generated by nuclear transplant to be less like the donor parent than identical twins are like one another.

ANALYZING PROTEINS

Proteins are the molecules through which genetic information is expressed, the mediators between genotype and phenotype. Learning what role a gene plays in a biological process requires learning what role is played by the protein the gene encodes. Likewise, manipulating the genes of an organism to change its characteristics is really about manipulating some aspect of the expression of proteins in that organism. Analyzing the expression of proteins is therefore an important component of research into the biology of cells, as well as of biotechnology applications.

The proteins in a sample of cells or tissues can readily be extracted by simple chemical treatments. Like mixtures of DNA molecules, mixtures of proteins can be separated by gel electrophoresis (usually through polyacrylamide gels). The details of protein separation are different from the details of DNA separation, but the basic idea and approach are the same. Following electrophoresis, the entire gel can be soaked in a protein stain to reveal the pattern of all the proteins, or the proteins can be blotted to a membrane and tested for the presence of a particular one.

Specific proteins can be detected by antibodies

Specific DNA sequences can be detected by hybridization tests with labeled probes. To detect individual proteins, we need something that will bind to them or react with them with similar specificity. Nature has provided a tool in the form of antibodies.

Antibodies are proteins produced by the B lymphocytes (B cells) of the immune system. Their function is to bind to foreign substances called antigens and tag them for elimination by other components of the immune system. The immune system has the capacity to produce millions of different antibodies. Each mature B cell produces only one kind, which binds to a specific chemical structure. The immune system can produce several different antibodies that bind to the same antigen; each antibody would recognize a different aspect of the antigen's three-dimensional shape and chemical nature and bind to it.

Monoclonal antibodies

Scientists recognized that the most useful protein detection reagents would be pure antibodies of a single type. However, mature B lymphocytes are terminally differentiated cells and do not divide in culture, so it is not possible to culture them as a source of a single pure antibody. In the mid-1970s, a technique was developed that combined the antibody-producing ability of B cells with the ability to divide indefinitely in culture. The complex technology involved the forced fusion of normal B lymphocytes with cancerous ones (myeloma cells), which could divide indefinitely in culture. The result of the forced fusion and subsequent selection procedures was a series of cell lines, each of which could reproduce in culture and synthesize large quantities of one specific antibody molecule. These antibodies are called **monoclonal antibodies** and are a key tool in many biotechnology procedures.

To make monoclonal antibodies to a specific protein or other antigen, researchers inoculate a mouse with that substance. The mouse's immune system responds to the presence of the foreign antigen by making antibodies to it, just as your immune system responds to a foreign invader, like a virus. After the mouse has mounted its immune response, it is killed. B lymphocytes from its spleen are fused with myeloma cells, and the desired antibody-producing cells are identified.

Using antibodies to detect proteins

To determine whether a specific protein is present in a sample, the sample is exposed to an antibody to that protein and then washed and tested for the presence of bound antibody. The process is very analogous to hybridization analysis for a DNA sequence. Antibodies, like DNA probes, can be labeled by the attachment of various chemicals—colored substances, fluorescent dyes, or radioactive elements.

Antibody binding tests can be used to give a yes-or-no answer or to reveal more specific information, such as where in a sample a protein is located. For example, the *Drosophila* embryo shown in Figures 10.11 and 15.19 was exposed to two different fluorescently labeled antibodies. The image in the

chapter 16 | Biotechnology in the Research Laboratory

In chapter 15, we described a collection of biological tools and procedures that can be combined in various ways to answer questions, solve problems, and produce products. Now, we will discuss ways in which these tools are used to accomplish specific goals such as finding genes, analyzing genotypes, generating DNA fingerprints, and genetically engineering both plants and animals. As you read, you'll see that the same kinds of approaches can be used in a wide variety of fields, from basic biological research and forensic science to clinical medicine, agriculture, and beyond.

The tools of biotechnology have revolutionized many areas of scientific research, enabling scientists to conduct experiments, gather data, and answer questions they might only have dreamed about just a few years ago. The new data and findings are spurring the development of even more new technologies and the asking of additional questions. In this chapter, we will look at how some of the techniques introduced in chapter 15 are being used in the research laboratory and elsewhere. We'll focus on three kinds of endeavors:

- finding genes
- analyzing genotypes and genomes
- genetic engineering

These endeavors overlap. For example, if you find a mouse gene that you believe has a particular function, the best way to test your hypothesis is to mutate that gene and see how the mouse is affected. Mutating the gene involves genetic engineering. Testing an engineered mouse to see if its genes have been altered in the way you intended involves analyzing its genotype. The techniques form more of a supportive web than a linear sequence.

Applications of the techniques overlap, too. Besides being used to answer research questions, the techniques we describe have real-world applications. This chapter contains some examples of those as well. In the chapters that follow this one, we will look at commercial applications of biotechnology with a focus on what kinds of products are being developed and the issues they raise rather than on how the products are created.

FINDING GENES

One of the first steps toward understanding the molecular basis of a particular trait is to find the genes that contribute to its expression. Identifying the genes is the beginning point for learning what proteins interact to produce a particular phenotype, how they interact, and how their expression is regulated. The most fundamental tool for finding the gene involved in a specific trait is a mutant organism that expresses some kind of change in that trait. That organism's DNA can be analyzed in search of the causative mutation (which may sound easy but usually is not).

Mutant organisms are invaluable tools for finding specific genes

The notion of using mutant organisms to find genes may sound odd, since the term mutant is so often associated with movies in which mutants are menacing creatures greatly altered from their original state. In science, the term **mutant** is very specific: an organism with an alteration in its genotype. Typically, scientists use the term to refer to organisms with genetic alterations that cause observable phenotypic alterations. Mutants are extraordinarily useful in scientific research, because each different one can provide information about genes and proteins involved in producing a trait. Scientists studying complex processes in an organism often collect as many different mutants with alterations in those processes as they can, because each mutant offers a different window into a process.

Microorganisms

Let us imagine that we would like to find the gene for a specific enzyme involved in the biosynthesis of the amino acid histidine in *Escherichia coli*. What we need is an *E. coli* strain that lacks this enzyme because of a mutation in a histidine biosynthesis gene. To find such a mutant, we could expose a culture of *E. coli* to a mutagen and screen the cells for individuals that had lost the ability to grow on histidine-free culture medium. The mutant we want cannot make histidine and therefore would not be able to live and reproduce unless its medium was supplemented with that amino acid. We could spread our mutagenized *E. coli* bacteria out on a histidine-containing medium, allow colonies to form, transfer some cells from each one to a histidine-free medium, and look for cells that could not grow there. If we identify one that cannot grow on the histidine-free medium, we can go back to the colony on the histidine-containing medium and retrieve cells from that colony for further testing (Figure 16.1). Once we have such a mutant (or a collection of several mutants), we can begin to hunt for the gene.

To find the gene that is defective in our mutant, we might next make a library of genes from a nonmutant *E. coli* strain. We could isolate DNA from the *E. coli* strain and use it to make a DNA library (see chapter 15). The plasmids in this library would contain inserted DNA fragments corresponding to the entire genome of normal *E. coli*, including the histidine biosynthesis genes. We transform a batch of the mutant histidine-requiring *E. coli* bacteria with this mixture of plasmids and look for a transformed *E. coli* strain that has acquired the ability to make histidine. We should be able to find it easily, because it can now grow on histidine-free medium (Figure 16.2).

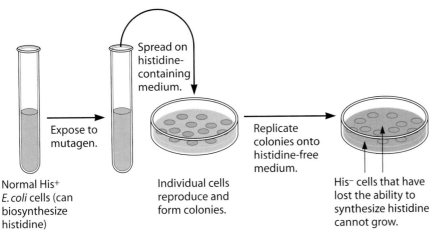

Figure 16.1 Isolating *E. coli* mutants that cannot biosynthesize histidine (His⁻).

The assumption is that any bacterium now able to make histidine must have received the missing gene on its new plasmid. We isolate the plasmid from that bacterium and determine the DNA sequence of the *E. coli* gene it contains, and with luck, we have found the histidine biosynthesis gene we were looking for.

One very important model microorganism is yeast. Yeasts are single celled and can be grown in culture much like *E. coli*. They can be transformed with plasmids, mutated, and in general manipulated much like bacteria. However, yeasts are eukaryotes. As eukaryotes, they are more closely

Figure 16.2 Using an *E. coli* mutant to find an *E. coli* gene.

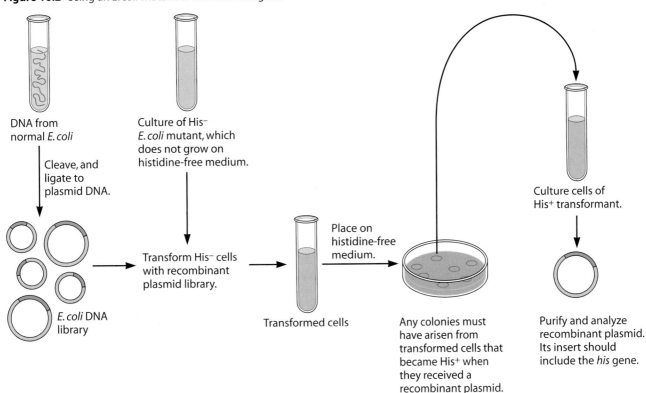

related to organisms like fruit flies, mice, and humans than is the prokaryotic *E. coli*. Yeast provides an easy-to-use system that can be readily manipulated genetically and is thus an excellent system for finding genes that may have close relatives in animals.

Drosophila

The genetic method outlined above is simple and efficient but is applicable only in limited circumstances. Obviously, if you are looking for an animal gene, you cannot collect a batch of mutant animals, transform them with plasmids, and look for restoration of a normal phenotype. A few model systems, however, have well-developed mechanisms for making mutations and finding the genes associated with them.

One of the best is *Drosophila*. A number of years ago, scientists comparing wild and laboratory strains of *Drosophila* discovered that many wild strains had acquired a transposon missing in the laboratory strains. If the transposon was introduced into a transposon-free laboratory fly, the transposon DNA would jump into random locations in the fly genome, causing mutations. This phenomenon has been developed into a system for creating mutant flies. If a scientist finds a transposon-created mutant with a phenotype of interest, he can then take advantage of the transposon again. The DNA sequence of the transposon is known, and so hybridization analysis (see chapter 15) can be used to find DNA segments in which the transposon is located. Since the mutation of interest was caused by the transposon jumping into the middle of a gene, when the transposon is located, the gene is also presumably located.

Drosophila can also be mutagenized by feeding mutagenic chemicals to adults or by exposing them to radiation. The mutagens cause mutations during DNA replication associated with gamete production, and mutations appear in the progeny of the mutagenized flies. It is not as easy to find the site of a mutation created in this way as it is to find the insertion site of a transposon.

Animals

Convenient mutagenesis systems like the *Drosophila* transposon don't exist for mammals. Instead, scientists must study the mutants nature offers them. You might be wondering where or how to find mutant animals, but in fact, many of them are available. Remember that to a scientist, a mutant is simply an organism with a genetic alteration, usually manifesting itself as a phenotypic difference. Humans have bred and collected mutant animals for centuries, perhaps because they liked the way they looked, as in the yellow obese mice of Chinese mouse fanciers (Figure 16.3), or because the variant trait was

Figure 16.3 Analysis of the DNA of genetically obese mice led scientists to genes involved in weight regulation. (Photograph courtesy of Jeffrey Friedman, The Rockefeller Institute.)

Figure 16.4 Purebred dogs are an excellent resource for scientists curious about the genetic basis of many different traits. Scientists in Mexico are analyzing the DNA of short-legged breeds such as the basset hound to determine the molecular basis of their dwarfism. The DNA profile of this basset hound, Champion Rebec's Fuzzy Navel, ROM (aka Fuzzy), is on file with the American Kennel Club. (Photograph courtesy of Reed and Becky Pomeroy, Rebec Bassets.)

useful, as in the short legs of dachshunds and basset hounds, which served hunters' purposes (Figure 16.4). Now scientists are studying the genes of the obese mice to learn about weight regulation and analyzing genes of purebred dogs to find the basis of their characteristic traits, including both anatomy and behavior.

Depending on how you think of it, you could also consider chocolate and yellow Labrador retrievers to be mutants, since each color variant is missing the activity of a protein. In fact, analysis of the genes of coat color variants has led to an understanding of the biological basis of coat color you read about in chapter 12.

In the end, whether you call a yellow Lab a mutant or just a phenotypic variant doesn't matter. The point is that any organism with an observably different phenotype that has a genetic basis can be a guide to finding the genes underlying that phenotype. Scientists have identified many genes for inherited human diseases by studying individuals from families with those diseases.

To find a gene starting from a phenotypic variant, scientists need DNA samples from a large number of related individuals who carry the trait in their family. If they are looking for a human gene, they must find families that fit this description. If they are looking for an animal gene, say, in a mouse, they can breed the mice and make the family. By observing the progeny of breeding crosses or, in the case of humans, by studying the family's medical history, they can deduce which members have the trait and, if the trait is recessive, which members are carriers.

Genetic markers

DNA samples from several individuals are then characterized at many different locations. This can be done by digesting the DNA with many different restriction enzymes and comparing fragment patterns, by amplifying it with many different sets of PCR primers and comparing the sizes of products, or by looking at specific nucleotides that are known to vary in individuals. The goal is to find specific variations among the DNA patterns of individuals in the family tree that are inherited with the trait. A variation can be a unique restriction fragment, a single nucleotide difference, or a unique polymerase chain reaction (PCR) product. Such a distinctive fragment or pattern is called a **genetic marker**. If a marker is found, it is assumed that the region of the chromosome containing the marker is very close to the disease gene.

After a marker for a trait has been identified, a new phase of work begins. Long stretches of DNA (often hundreds of thousands of base pairs) around the marker must be sequenced and searched for sequence patterns that look like genes (with promoters and coding regions). When potential genes are found, their sequences from the normal and mutant individuals must be compared. If the sequences of a putative gene from individuals with the abnormal trait are always different from the sequences of the same putative gene from normal individuals, then the gene is probably the culprit. The human gene *BRCA1*, which is implicated in hereditary breast cancer, was found in this way.

This entire process can take years and is not foolproof, since a marker can be inherited with a trait purely by chance and not because it is close to the gene underlying that trait. Once a candidate for a particular gene has been identified, further study is required to confirm that the identification is correct.

Related organisms usually have similar genes

Finding a gene is usually much easier if you or someone else has already found the gene in a related organism. Given that closely related organisms have very similar DNA sequences, you could use the known gene sequence to make probes and search for similar sequences in your organism of interest. The study of model organisms and the sequencing of their genomes are paying off handsomely in this arena.

It simply is not practical to make tens of thousands of mutant mice to search for specific phenotypes, but it is quite practical to do so in yeast or *Drosophila*. Once a gene is identified in one of these easily manipulated model organisms, that gene sequence can often be used to locate the gene in other organisms. In chapter 10, we described how a research group fed mutagen-laced sugar water to *Drosophila* and identified descendant embryos that failed to develop properly. From these mutants, they identified genes such as *bicoid*. The *Drosophila* genes were subsequently used to find homologs in the frog. By comparing known homeotic sequences of frog and fly to other animals, researchers determined that these genes were present in earthworms, beetles, mice, chickens, cows, and people.

Now that the complete base sequences of the mouse and human genomes have been determined and are available in computer databases, it isn't even always necessary to do physical experiments. Scientists can type base sequences into computerized search engines and ask whether any sequence resembling that one exists in the mouse or human genome. The fact that a similarity exists, of course, does not prove that the correct gene has been found, just as finding a marker that is inherited with a disease does not prove that a nearby gene is the culprit. Further research is needed to absolutely prove the gene's identity. This used to be a very difficult task in animals, but the genetic engineering technique of making so-called knockout mice (inactivating a specific gene in a mouse; see "Knockout mice" below) has given scientists a direct way to test whether a specific gene is involved in a specific phenotype. Finding genes for specific traits still is not an easy task by any means, but biotechnology has provided a whole new set of tools for tackling the project, and scientists are finding and analyzing genes today at a rate that would have seemed incredible just 20 years ago.

Gene sequences or markers can be used for genetic testing

Once a gene has been found and sequenced, the sequence of that gene in an individual can be determined. One application of this kind of genetic testing is to find out whether an individual has a form of a gene associated with an inherited disease. For example, many women with relatives who died at an early age from breast cancer are having themselves tested for the presence of the *BRCA1* and *BRCA2* mutations. In other cases, prospective parents whose families carry a recessive genetic disease have themselves tested to determine if they are carriers of the recessive allele. Additionally, there have been a few instances in which parents who are both carriers of recessive alleles for fatal inherited diseases produced embryos by in vitro fertilization (in which eggs and sperm are united in laboratory dishes) and had the embryos tested before implantation to make sure their future offspring would not suffer and die from the inherited condition.

If the disease gene itself has not been identified but genetic markers have been found, scientists can use the markers as surrogates for specific alleles of the genes themselves. Because markers are usually easily detected features, such as changes in the number or length of restriction fragments or PCR products, they are sometimes used as a screening device even after the actual gene has been defined.

If the marker is present in an individual's DNA, he or she is likely to carry the disease gene. We say likely because genetic markers are not necessarily the disease-causing mutations themselves but may only be unique restriction fragments or PCR products that originate close to those mutations and tend to be inherited with them, so the presence of a marker is not necessarily proof that the mutation is present. Genetic recombination events could separate a marker restriction fragment from the disease mutation. Likewise, the absence of the genetic marker may not be absolute proof that a disease gene is not present. In addition, there may be many different mutations that can cause a given genetic disease. The vast majority of an affected population may carry one particular mutation and its associated genetic marker, but a rare alternative disease-causing mutation may not be associated with the marker restriction fragment. Additionally, the marker might be present in most families but not all of them. In these families, the marker is not a useful diagnostic tool. However, if the limitations are understood, genetic testing for a disease gene can provide valuable information for physicians and potentially affected individuals. Once mutant genes are positively identified, more specific probes for diagnosis can be made.

Testing for SSA

One inherited disease allele that is especially easy to test for is the one associated with sickle-cell anemia (SSA). In SSA, a mutant form of hemoglobin causes red blood cells to assume a sickled shape in low oxygen concentrations, impairing circulation and oxygen delivery (see chapter 4). The mutation that causes SSA alters a restriction site within the hemoglobin (called β-globin) gene. To test for the presence of the mutation, laboratory technicians use PCR to amplify the portion of the genome containing the site where the mutation can occur and then digest the product DNA with the restriction enzyme and look at the pattern (Figure 16.5).

Marker-assisted plant breeding

Genetic markers offer plant breeders shortcuts to identifying plants with traits they wish to propagate. For example, suppose a breeder has noticed

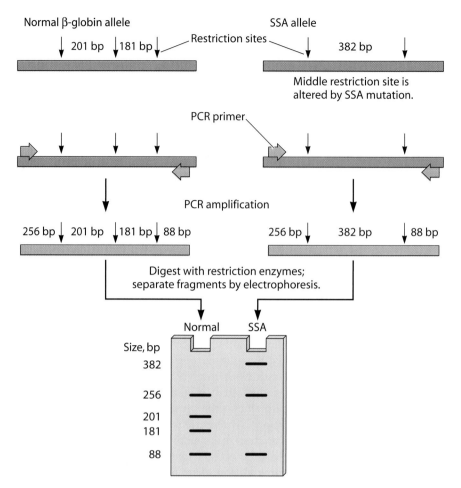

Figure 16.5 Genetic testing for SSA. bp, base pairs.

that certain individuals of an annual crop plant produce especially tasty fruits. When she plants seeds of a tasty-fruit plant, only some of the offspring produce the tasty fruits. She would like to cross two of the tasty-fruit individuals together and eventually make a plant variety that always produces the tasty fruit. The problem is that she must cross the plants before they make fruits, and the tasty-fruit plants and their nontasty siblings look alike. Before the advent of genetic markers, she would have had to cross many plants, wait to see which of them developed tasty fruits, and hope she had happened to cross two of them. This trial-and-error process would have to be repeated for many generations.

If the hypothetical plant breeder could identify DNA markers associated with the tasty-fruit phenotype, she could test leaves from young plants and determine which of them had the tasty-fruit trait before the young plants flowered. She would then know precisely which plants to breed.

In a real application of marker-assisted breeding, a team of crop scientists and Indian farmers sought to increase the drought resistance of a popular type of rice grown in rain-fed rather than irrigated fields in eastern India. Drought resistance is influenced by many genes, as well as environmental factors. Because of the environmental influence, it is difficult to assess the trait in an individual plant in the field. Markers for several relevant genes had been identified in a drought-resistant variety that was less desirable to the farmers because of other characteristics. The team crossed the two rice varieties and screened for drought resistance markers in the progeny. Those with the markers were back-crossed with the desirable rice variety, and those progeny were tested. This process was repeated for several generations until plants were obtained that had the drought resistance markers but the other characteristics of the desirable variety when the farmers tested them in the field (Figure 16.6).

Marker-assisted plant breeding dramatically shortens the time required to breed specific new plants. Because plant breeders can specifically identify the genes of interest in the cross, they can eliminate time-consuming trial-and-error breeding and screening of large numbers of progeny plants.

COMPARING GENOTYPES AND GENOMES

There are many reasons, both academic and practical, to assess the degree of genetic difference between organisms. If we can compare the genotypes or genomes of different species, we can ask how closely related those species might be and draw conclusions about the course of evolution. Shifting our focus to the level of individuals, we can compare genotypes and draw conclusions about whether the individuals are related or whether two samples came from the same individual. In looking at the genotype of a single individual, we can ask about that individual's species, gender, inherited diseases, and more.

Before the advent of DNA technologies, it wasn't possible to analyze genotypes directly. Rather, comparisons had to be based on phenotypic characteristics, that is, some tangible manifestation of the information contained in the DNA molecule. We now know that much of the variation between the DNAs of individuals is found in their noncoding DNA and thus is phenotypically invisible. Furthermore, some identical phenotypes can be encoded by different genotypes (for example, people with blood group genotypes AA and AO both express the phenotype of blood group A, despite the genetic difference), and adaptation to similar environmental circumstances can cause genetically dissimilar organisms to display similar phenotypic characteristics. Now that genotypes can be analyzed directly, variation that would previously have been invisible can be detected and measured.

Genotypes can be compared in several different ways

There are several ways to go about comparing genotypes. One way is to compare whole genomes through hybridization studies. In this approach, genomic DNAs from two species are prepared and then hybridized to each other. The more similar the nucleotide sequences of the two species, the more extensive the regions of base-paired duplex they will form. To measure the extent of the hybridization, the hybrid molecules are heated until the strands separate again. The temperature at which the strands separate (the melting temperature) is a function of the extent of base pairing. The more perfectly matched the two genomes, the higher the temperature required to separate the strands of the hybrid molecules. The melting temperature thus gives a gross measurement of DNA sequence similarity. This kind of assessment has been used to compare the degrees of similarity between the DNAs of different species of organisms.

At the opposite end of the spectrum in terms of the level of detail of the comparison is DNA sequence analysis. This approach gives detailed and

Figure 16.6 The breeding of drought-resistant rice for use in eastern India was facilitated by genetic markers. Rice is a staple crop for billions of people in developed and less-developed countries. (Photograph courtesy of the U.S. Department of Agriculture.)

BOX 16.1 *Solving the mystery of Madagascar's mammals*

Madagascar is the world's fourth-largest island (about the size of California and Oregon combined); it lies about 250 miles to the east of the African continent. Once part of the megacontinent Gondwana, Madagascar broke away from Africa about 165 million years ago (see the first figure).

The mammals living on Madagascar were a mystery for a long time: only four major groups were represented, and the individual species were all unique to the

Madagascar is a large island off the east coast of Africa.

island. How did they get there? The origin of Madagascan mammals was a particularly thorny question because the geology of Madagascar does not favor fossil formation, and there was virtually no fossil record for the last 65 million years. One hypothesis was that ancestors of the mammals had been there when Madagascar broke away from Gondwana and that the present-day species were their descendants. If that hypothesis was true, though, why are there no animals resembling zebra, antelope, lions, elephants, giraffes, or many other African species on the island?

A second hypothesis was that the ancestors of Madagascan mammals crossed to the island 25 to 45 million years ago on a land bridge that might have existed at that time. This hypothesis begs the same question as the first one: why are so few types of mammals present?

A team of scientists from the Field Museum in Chicago, collaborating with the University of Antananarivo (the capital of Madagascar), recently applied molecular analysis to the mystery, focusing on just one group of Madagascan mammals, the carnivora. The carnivorans are members of the order Carnivora,

which includes animals such as dogs, cats, mongooses, hyenas, and bears. Madagascan carnivora include the fossa, a puma-like mammal; the falanouc (see the second figure); the Malagasy striped civet; and four kinds of Malagasy mongoose, which resemble ferrets. The research team sequenced four genes from living Madagascan carnivora and compared the sequences to each other and to those from other living mammals to determine which were more closely related. They also estimated when the animal species separated from each other by calculating the rate of molecular change for each species.

Their results came as a surprise. The gene sequences of all the carnivora of Madagascar resembled one another more closely than any of them resembled any other mammalian sequence. This result indicates that Madagascan carnivora all descended from one common ancestral species after that ancestor arrived on the island. The closest living relative to the Madagascan carnivora is the African mongoose (see the third figure). The number of changes that have accumulated between the Madagascan gene sequences and the

accurate information but is impractical to apply to large stretches of large numbers of samples because it is very time-consuming and relatively expensive. DNA sequence comparisons usually focus on small, specific regions. To give meaningful data, these target regions have to be areas of significant variation at the scale of interest. For example, to compare species that are distantly related, scientists select genes that all the species have in common, which means the genes cannot evolve so rapidly that they have become unrecognizable when one looks from yeast to rice to elephants (see "DNA and protein sequence comparisons" below and Box 16.1).

If the goal was to compare genotypes of individuals within a single species, however, scientists would need to choose DNA that did evolve rapidly so that differences would be found. Comparisons of most human genes would show very few, if any, differences between individuals. For this reason, sequencing studies of individuals of the same or closely related species often focus on **mitochondrial DNA**. Mitochondria have small circular genomes that undergo mutation at a much higher rate than the nuclear genome; thus, mitochondrial DNA sequences display much more variation between individuals than do most nuclear sequences (see Box 16.2).

B O X   1 6 . 1 *(continued)*

The falanouc is one of Madagascar's unique carnivores. (Photograph courtesy of the Field Museum, Chicago, Ill., #CSZ77048.)

African mongoose sequences suggest that the ancestor of today's Madagascan carnivora arrived about 20 million years ago, more recent than the presumed land bridge and far more recent than the splitting of the island from continental Africa. Neither hypothesis about the origin of the mammals was correct.

Then how did the mongoose-like ancestor get to Madagascar? The researchers speculate that it must have floated to the island from the African continent on a raft of driftwood or other vegetation debris. The animal(s) may have been able to survive such a voyage by going into a hibernation-like state, an ability characteristic of many of today's Madagascan carnivora.

Gene sequence comparisons revealed that Madagascar's carnivores descended from one common ancestor whose closest living relative is the mongoose. (Diagram courtesy of John Flynn, Marlene Donnelly, and Anne Yoder, the Field Museum, Chicago, Ill.)

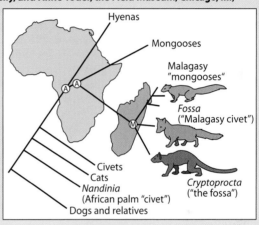

RFLP and AFLP analyses

A shortcut to DNA sequence comparisons is to focus on many small variable regions of the genome. Several different approaches for doing this have been developed. One of the early methods focused on changes to the short base sequences at which restriction enzymes cut the DNA molecule. Changes in a DNA base sequence can create or eliminate these sites or change the distance between two sites. By digesting two DNA molecules with the same restriction enzyme and comparing the lengths of the fragments generated, we can estimate how similar the two molecules are. Two identical molecules will give identical patterns. Two fairly similar molecules will give similar patterns, with perhaps a few differences in fragment size (caused by the addition or removal of cut sites through mutations or changes in the length of fragments due to insertions, deletions, or rearrangements). Scientists estimate the similarity of DNA molecules based on the similarity of their restriction fragment lengths, which is called **restriction fragment length polymorphism** (RFLP) analysis. This approach permits sampling of variation across the entire genome.

The same type of assessment is also conducted with PCR. Genomic DNA from two or more organisms is subjected to PCR using various sets of primers,

B O X   1 6 . 2 *Mitochondrial DNA typing*

Human mitochondria contain a circular DNA genome of about 17,000 base pairs (see the first figure). Mitochondrial genes encode some of the proteins that are associated with the electron transport chain and some of the proteins required for mitochondrial protein synthesis. There are hundreds of mitochondria in a typical cell.

Mitochondrial DNA plays a special role in DNA typing because of its inheritance pattern and its high mutation

Mitochondrial DNA. kb, kilobases.

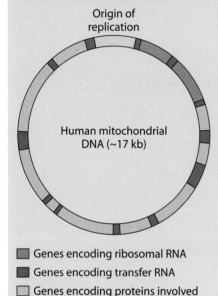

Origin of
replication

Human mitochondrial
DNA (~17 kb)

- ■ Genes encoding ribosomal RNA
- ■ Genes encoding transfer RNA
- ■ Genes encoding proteins involved in electron transport
- □ Hypervariable D-loop region

rate. Except in rare instances, mitochondrial DNA is inherited from the mother. The mitochondria present in the ovum become the mitochondria of the zygote after fertilization. In addition, the region of the mitochondrial genome around the replication origin is highly variable and has a high mutation rate. Therefore, two people with the same DNA sequence in that region are highly likely to share a recent female ancestor.

Taken together, this information means that an analysis of mitochondrial DNA can show very clearly whether two people are related through the female line. Analysis of mitochondrial DNA can be helpful in reuniting families and in identifying the dead. A particularly poignant application of mitochondrial DNA typing occurred in Argentina. During the Argentine military's brutal rule (1976 through 1983), many families were torn apart. Often, parents were murdered and their children were given away or sold. In other cases, parents were dragged away to prison, unwillingly leaving their babies to uncertain fates. When the dictatorship was overthrown, the relatives of these kidnapped or disappeared children tried desperately to find them. Many of the relatives were women whose children had been murdered and who sought their missing grandchildren.

An American scientist, Mary-Claire King, was instrumental in helping Ar-

gentine families find their lost relatives. King used mitochondrial DNA in her analyses. Because the parents of the lost children had often been murdered, DNA from more distant suspected relatives was usually the only evidence available for comparison. Since mitochondrial DNA is passed on through the females in a family lineage, a child's mitochondrial DNA profile exactly matches the profile from her mother's mother, all of her mother's siblings, and the children of her mother's sisters (see the second figure). The approach was successful in identifying many of the missing children.

Mitochondrial DNA typing was also used to identify the skeletal remains of the royal family of Russia, murdered by Bolshevik soldiers in 1918 and buried in an anonymous grave. Because many of the members of various European royal families are related through Queen Victoria, they share mitochondrial DNA sequences. Czar Nicholas's remains were identified through comparison with mitochondrial DNA sequences of his living relatives.

The U.S. armed forces have an ongoing program to identify previously unidentified remains from military conflicts. Although this mission may seem strange at first, to the families of missing soldiers it is very important. In 1999, there were approximately 2,200 soldiers unaccounted for from the Vietnam War,

and the lengths of the product molecules are compared. The assumption is that the more similar or closely related the two organisms are, the more similar their DNAs will be and the fewer **amplified fragment length polymorphisms** (AFLPs; differences in lengths of the amplified PCR products) will be seen.

RFLP and AFLP analyses of genomes are used in evolution and conservation research. More precise forms of genotyping, called DNA typing or fingerprinting, use targeted AFLP or RFLP analysis to zero in on sites in the genome that are known to be polymorphic between individuals (see "DNA typing is based on the uniqueness of every individual's DNA sequence" below). Another type of variation observed to exist between individuals or strains is **single-nucleotide polymorphisms**, or SNPs (differences in single nucleotides); that is, different individuals have different bases at these particular sites in the genome. Genotyping by SNP analysis involves very short specific sequencing reactions at just these sites. RFLP, AFLP, and SNP analyses

BOX 16.2 *(continued)*

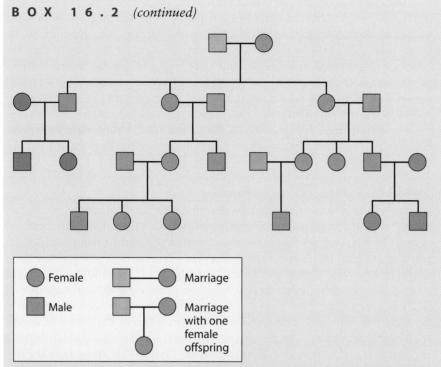

Female

Male

Marriage

Marriage with one female offspring

Inheritance of mitochondrial DNA is matrilineal. That is, the mitochondrial genotype (symbolized here by the colors in the male and female symbols) is inherited from the mother. Note that all family members related through the female line share the same mitochondrial genotype.

8,000 from the Korean War, and 78,000 from World War II. Many of these soldiers' bodies were found but could not be identified and were buried as unknowns.

The U.S. Armed Forces DNA Identification Laboratory uses PCR amplification of mitochondrial DNA to identify human skeletal remains from previous conflicts. Because there are so many copies of mitochondrial DNA per cell, the chances are greater that it can be amplified even from fragmentary remains.

When the laboratory workers attempt to identify a set of remains, they first collect all available information about the remains: where they were found, what unit was fighting there, which soldiers were unaccounted for from that engagement, and so on. This results in a set of possible identities for the remains. Laboratory personnel then contact family members and, with permission, obtain DNA samples for comparison. Under extremely strict conditions designed to minimize contamination, PCR amplifications of mitochondrial DNA are performed, and the products are sequenced. The sequences are compared with sequences from the candidate families. If there is a match, the remains are identified.

In 1998, the remains buried in the Tomb of the Unknown Soldier were brought to the Armed Forces DNA Identification Laboratory under full military honor guard. With great care, the mitochondrial sequence of the remains was determined. The sequence was compared with sequences from seven candidate families, and it matched only one family. The Unknown Soldier was identified as Michael J. Blassie, an Air Force pilot who was shot down in Vietnam in 1972.

The Armed Forces DNA Identification Laboratory now generates a nuclear DNA profile for all soldiers in the hope of eliminating future unknown remains.

are all being explored as methods for rapidly distinguishing strains of bacteria, viruses, or other pathogens.

All of these techniques allow scientists to determine the amounts of similarity or difference between two genomes. DNA comparisons are used in a variety of settings, from evolutionary studies to the courtroom. We will now look at a few applications.

Genotype and genome comparisons shed new light on evolutionary studies

Evolution is a unifying theme in biology, and molecular biology has provided powerful new tools for studying it. Evolutionary biologists seek to discover how individual species arose from earlier forms and to understand the mechanisms of the process of evolution itself. Biotechnology has literally revolutionized both endeavors.

The essential concept of evolution is that, given a variable population of organisms, environmental conditions will favor the survival and reproduction of some and not other individuals (see chapter 13). The genes of the favored individuals will be passed along to offspring more frequently than the genes of less-favored individuals, and over time, the genetic content of the population will shift toward the genotypes of the more successful forms. If two populations of the same organism find themselves in different environments or facing different selection pressures, over time they will become more and more genetically different. Thus, the amount of genetic difference between two species of organisms reflects the amount of time they have been diverging from one another.

DNA and protein sequence comparisons

Formerly, biologists who sought to describe how modern species arose had to rely on morphological, ecological, and behavioral comparisons between various modern species and between modern and fossil forms to deduce degrees of kinship. DNA and protein sequence comparisons have given these scientists an entirely new set of data to consider. To use molecular data in evolutionary studies, biologists first assemble protein amino acid sequence or DNA base sequence data from a specific protein or region of the genome in the group of organisms under study. They then measure the degree of difference in the sequences. On the assumption that changes accumulate slowly and relatively steadily, they construct various "trees" that show how the different sequences could have been generated from a common ancestor or the order in which different groups of organisms may have evolved from one another (see Box 16.1). An evolutionary tree based on the amino acid sequence of the protein cytochrome *c* is shown in Figure 16.7.

Using molecular techniques to probe evolutionary relationships between species is a large and active area of current research. For example, molecular evolutionary analysis found that the kiwi, a flightless New Zealand bird, is more closely related to the flightless birds of Australia than to the moas, the other major group of flightless New Zealand birds. DNA and protein sequence comparisons, as well as hybridization studies, found that humans and chimpanzees are more similar to one another than either is to any other species. These studies suggest that ancestors of modern humans and chimps probably diverged a mere 5 million years ago.

Comparisons of the mitochondrial DNA sequences of 140 dogs of 67 different breeds, 162 wolves from 27 different locations, 5 coyotes, and 12 jackals firmly established that domestic dogs are descended from wolves and not coyotes or jackals (see Box 16.2 for information on mitochondrial DNA). Furthermore, according to estimates of how quickly mutations can accumulate in mitochondrial DNA, it appears that dogs originated about 100,000 years ago. This estimate is in sharp contrast to the appearance of fossilized dog remains, the oldest of which date from about 14,000 years ago. The molecular researchers argue that before that time, dogs may have physically resembled wolves so closely that their fossilized remains are indistinguishable from those of wolves. Molecular evidence gives evolutionary scientists a wealth of new information to argue about and to incorporate into their ideas of the development of species.

Ancient DNA

An interesting twist from the field of molecular evolution is the discovery that we can extract DNA fragments from appropriate ancient samples. DNA

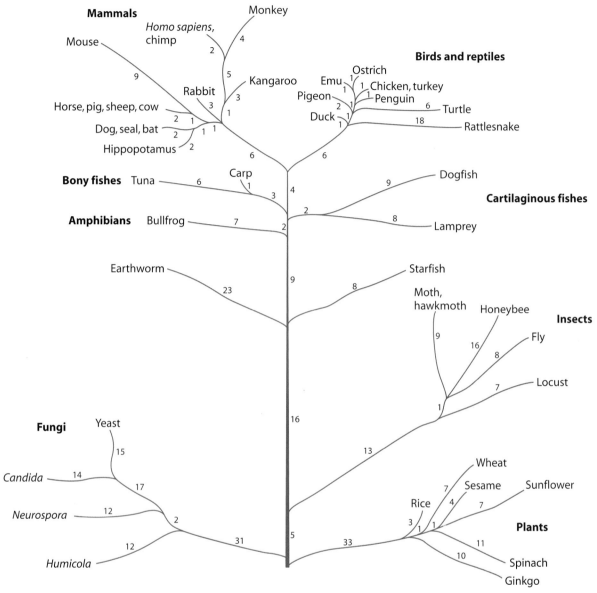

Figure 16.7 Evolutionary tree based on the amino acid sequence of the protein cytochrome *c*. The numbers represent the number of amino acid changes between two nodes of the tree. For example, when the common ancestor of cartilaginous fish diverged from the line leading to bony fish, mammals, etc., it evolved and accumulated two amino acid changes in its cytochrome *c* protein before the lines leading to dogfish and lamprey diverged. The dogfish line accumulated nine more amino acid changes in becoming the modern organism, while the lamprey accumulated eight changes. Thus, the amino acid sequences of cytochrome *c* in modern lampreys and dogfish have 17 differences. (From A. Lehninger, D. Nelson, and M. Cox, *Principles of Biochemistry*, 2nd ed., Worth Publishers, Inc., New York, N.Y., 1993, with permission.)

is rapidly degraded to small fragments after an organism dies, and only special combinations of circumstances will preserve soft tissue cells with DNA. However, samples preserved in bogs, including 17-million-year-old magnolia leaves, have yielded enough DNA for comparison with modern species, as have amber-encased insects (Figure 16.8). It is also possible to recover DNA fragments from bones and teeth, which are more commonly preserved than

Figure 16.8 Scientists obtain information about evolutionary changes by analyzing DNA from ancient specimens, such as insects trapped in amber. This insect is a timber beetle. (Photograph courtesy of Fossilmall.com.)

is soft tissue. PCR is particularly useful in studying ancient DNA because it can amplify the small amounts typically present in specimens.

Early attempts to amplify ancient DNA by PCR were often learning experiences for the scientists involved. What they learned was that it is extremely easy to amplify modern DNA contaminants of the ancient sample. In fact, many attempts to amplify ancient DNA resulted in the amplification of the scientists' own DNA. Through these experiences, laboratory personnel learned that extremely stringent protection against contamination is essential in these kinds of experiments.

In one headline-making study of ancient DNA, a group of scientists painstakingly amplified fragments of mitochondrial DNA from a Neanderthal human fossil. Neanderthals were a race of humans who lived in the Near East and Europe from about 125,000 to 30,000 years ago. Their anatomy was somewhat different from modern humans, and their average brain capacity was slightly larger than ours (Figure 16.9). They cared for one another, as evidenced by fossils showing that individuals survived with physical conditions that would have proven fatal if they had not had help. They also appeared to believe in an afterlife, as they buried their dead in graves that included useful objects and flowers. Anatomically modern humans appeared in Europe around 30,000 years ago, near the time that Neanderthals seem to have died out. For many years, it was assumed that Neanderthals were direct ancestors of modern humans, but other lines of evidence

Figure 16.9 Neanderthals (skull cast on right), with their large brains and anatomy similar to ours, were long believed to be direct ancestors of modern humans (skull model on left). Mitochondrial DNA analysis suggests that modern humans instead arose independently of Neanderthals and displaced them. (Photograph courtesy of SOMSO Models, Coburg, Germany.)

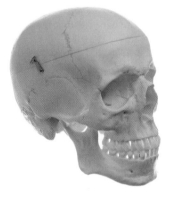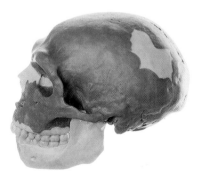

suggested that modern humans had arisen in Africa independent of the Neanderthals, migrated to Europe, and displaced the local Neanderthal population.

The scientists who amplified mitochondrial DNA from the Neanderthal fossil carefully established that the DNA did not come from any of their own cells. They then determined the base sequence of the amplified mitochondrial DNA. The base sequence was very different from the mitochondrial DNA sequences of any known modern human group—so different, in fact, that the researchers concluded that Neanderthals could not be direct ancestors of modern humans.

We stress that the DNA recovered in these processes is extremely fragmentary and is useful only for comparisons with modern sequences to measure the amount of change. So far, experience with ancient DNA suggests that using it to resurrect long-extinct species such as dinosaurs is impossible.

DNA in archaeology

Archaeologists, too, are beginning to use DNA analysis to learn more about bones recovered from ancient graves. DNA analysis gives archaeologists an entirely new way to analyze remains from ancient burials and an entirely new kind of data to consider. For example, a team of scientists studying the Etruscan culture found four skeletons in a tomb in the Monterozzi necropolis near Viterbo, Italy. The skeletons were those of a child, a man estimated to be in his 30s, and two people estimated to be 40 to 45 years old. Because of the state of preservation of the skeletons, the gender of the child could not be determined. One of the two 40- to 45-year-old individuals was identified tentatively as a woman, but the skeleton of the other 40- to 45-year-old individual was too fragmentary for gender assignment. Near the skull of that individual lay bronze earrings, commonly considered a female ornament. When skeletons are too incomplete for definite gender assignment, archaeologists often rely on artifacts associated with burials to draw conclusions about who is buried there, and the earrings suggested the poorly preserved skeleton belonged to a woman. No information was available as to who the buried individuals might have been or whether they might have been related to each other.

To apply DNA analysis to the remains, the team of Italian scientists extracted DNA from bones and teeth. They used PCR with primers complementary to the SRY gene to determine the genders of the individuals. Males have the SRY gene on the Y chromosome; females lack it (see chapter 10). Their results showed that the child, whose gender was previously unknown, was female and confirmed that the 30-something-year-old individual was indeed male. Surprisingly, the 40- to 45-year-old individual tentatively identified as female because the body was buried with earrings tested positive for SRY, showing that the skeleton was that of a man. The other older individual was indeed a woman.

To investigate whether any of the individuals were related through the maternal line, the scientists determined the base sequence of portions of the hypervariable region of mitochondrial DNA from each skeleton (see Box 16.2). The mitochondrial DNA sequences of the woman, the 30-something man, and the child matched, supporting the conclusion that these individuals were related through the maternal line. The scientists hypothesized that the four individuals were a family group, with the older man and the woman being the parents of the younger man and the child. They noted that there was no way to tell whether the individuals had been buried together at the same or different times. Their interpretation of the identities of the skeletons in the tomb paint a different picture of Etruscan culture than the tentative identifications based on the fragmentary skeletons, which would have identified both of the older individuals as female and would not have detected the family relationship. The authors concluded their study by stating that adding molecular data to traditional archaeological data could lead to better understanding of Etruscan culture.

Overarching evolutionary lesson

For scientists who seek to understand the mechanisms behind evolution, molecular biology and biotechnology have provided a gold mine of new information. First, studies of genes and genomes from many organisms have underscored how fundamentally similar we all are. For example, studies using hybridization, RFLP analysis, and sequence comparisons suggest only about a 1.6 to 3% difference between the genomes of humans and chimpanzees. From earlier protein and enzyme studies, it was known that all animals produce about the same complement of enzymes and proteins. Now that genes are being mapped to specific locations on chromosomes, scientists are finding that animals' genes are arranged similarly from species to species. Differences in chromosome numbers and structures seem to have been generated in a process in which chromosomes were cut or broken into large pieces and then put back together in many different ways to produce different final numbers of chromosomes.

Looking at the DNA sequences of individual genes, we see that nature appears to be quite economical: once a protein function that works has evolved, it is often used over and over. Similar domains are found in many different proteins (see chapter 5 for more information). We find proteins that appear to have diverged from two original copies of the same gene, and we find similar proteins playing similar roles in widely disparate species. For example, the proteins that govern body plan in worms, flies, mice, and humans are very similar. The homeotic genes encoding these proteins are even arranged in the same way in the chromosomes of flies, mice, and humans, and they share similar regulatory sequences. Some scientists are now looking at the regulatory sequences associated with homeotic genes as a way to investigate ancestry. The flood of new information about genome structure is giving evolutionary scientists a wealth of fuel for theories about genome evolution.

DNA typing is based on the uniqueness of every individual's DNA sequence

The best-publicized use of genotyping may be its forensic application in criminal cases. Before we had the technologies to look directly at DNA, identifications were made through phenotype: appearance, voice patterns, blood types, dental records, fingerprints, and so on. These methods can be very effective, but they also have significant limitations under certain circumstances. DNA typing, also called DNA fingerprinting, gives both prosecutors and defense attorneys an extremely powerful tool to add to the arsenal of technologies they can use to identify or exonerate an individual suspected of a crime.

It is theoretically possible to identify nearly every individual on earth from his or her DNA sequence, though doing so is not realistic because of the time and money required to sequence an entire genome. It is possible, however, to make very good predictions about individual identity on the basis of a limited and practical examination of DNA with RFLPs or AFLPs.

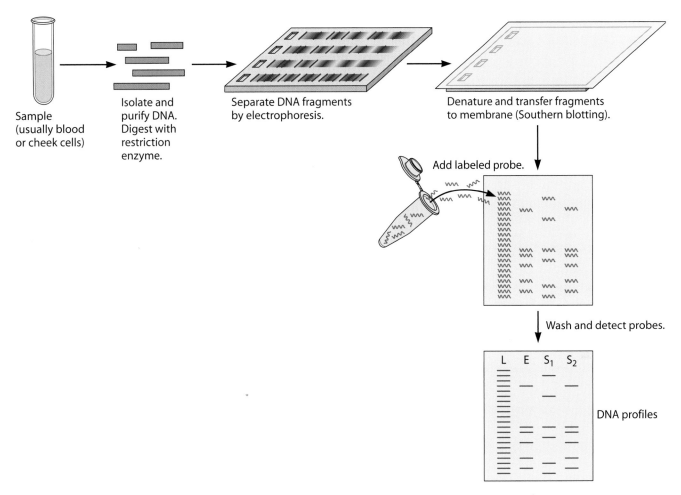

Figure 16.10 RFLP analysis. L, ladder; E, evidence; S$_1$, suspect 1; S$_2$, suspect 2.

To perform an RFLP analysis for DNA typing, DNA must first be extracted and purified from a DNA-containing sample. Following restriction digestion, the DNA fragments are separated by gel electrophoresis. The human genome is so large (about 3 billion base pairs) that digestion with most restriction enzymes generates thousands or even millions of fragments, which overlap in the gel lane, forming a smear. Hybridization analysis permits the visualization of only the regions of the genome known to vary when digested with that enzyme. Following electrophoresis, the DNA is transferred to a membrane by blotting and then hybridized to probes for the sequences of interest. The result is an interpretable pattern of bands (Figure 16.10). In AFLP analysis, primers that amplify only regions of interest are chosen.

The regions of interest in DNA typing are usually areas of the genome that vary greatly from individual to individual in that they contain different numbers of back-to-back repeats of the same DNA sequence. When these regions are cut by a restriction enzyme with sites flanking the repeats or amplified by PCR with primers flanking them, the sizes of the resulting products depend on the number of repeats in the individual's DNA. When the products are separated by electrophoresis, the patterns constitute a genetic fingerprint of the individual (Figure 16.11).

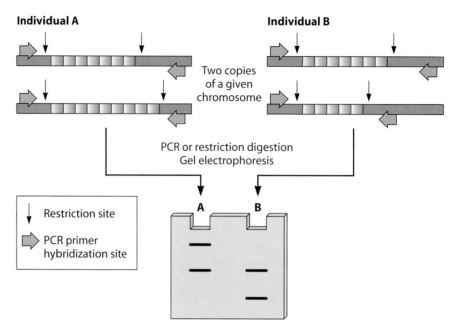

Figure 16.11 Variation in the numbers of tandem (head-to-tail) repeated sequences at many different locations within the genome provides a basis for genetic fingerprinting. To generate a fingerprint, laboratory technicians amplify many different loci using a specific set of primers for each one, or they digest the sample DNA with restriction enzymes known to cut at sites flanking the repeats.

Forensic DNA typing

In criminal cases, DNA typing is used to tie DNA-containing crime scene samples to individual suspects. For example, if blood is found at a crime scene, a DNA fingerprint can be generated from the blood and compared with DNA fingerprints from any suspects. If the crime scene fingerprint does not match a suspect's, that suspect cannot have left the sample and is cleared. In fact, about 30% of the FBI's DNA-typing cases have cleared the prime suspect, and DNA typing has cleared people who were in prison serving time for violent crimes.

If the DNA fingerprint of the sample at the crime scene does match a suspect's DNA fingerprint, the relevant question becomes how likely it is that a person other than the suspect could have left it. Since DNA fingerprinting looks at only a portion of an individual's genome, two people could have very similar fingerprints. Scientists look at databases of DNA fingerprints from many individuals and calculate the frequency of the patterns in the fingerprint in question. They then use those numbers to calculate the odds that another individual will have the same DNA fingerprint. Forensic laboratories typically conduct their testing so thoroughly that random matches are extremely unlikely.

Paternity testing

DNA testing can also be used to establish paternity or maternity. Since an offspring inherits half its chromosomes from its mother and half from its father, its DNA fingerprint should show contributions from both. To establish paternity, fingerprints of the child, mother, and putative father are generated. The child's DNA fingerprint is compared to the mother's, and the bands that match are subtracted. The remainder of the bands in the child's fingerprint should match bands in the father's (Figure 16.12).

The very first use of DNA parentage testing was in fact a maternity case. A young Ghanaian boy arrived in the United Kingdom in 1985 to join his mother. His passport appeared to be forged, and authorities threatened to deport him. The family lawyer contacted a British scientist who had recently made headlines with his newly developed method of DNA fingerprinting and asked if he could help. The scientist believed DNA testing could show maternity and paternity and agreed to try. Test results confirmed that the boy was indeed the woman's son, and authorities dropped the deportation case.

DNA typing is also used to confirm the identity of human remains. U.S. soldiers now deposit samples in a DNA data bank as a backup for the metal dog tags they wear in combat. In the field of human rights, geneticists are using DNA typing to identify the bodies of people slain during recent political upheavals in Guatemala. The work of these scientists corroborates eyewitness reports and sometimes helps bring the killers to justice. DNA typing has also been used to reunite children kidnapped by Argentina's former military dictatorship with their families. In these cases, mitochondrial DNA typing is often the best technology, because the parents of the victim may themselves be dead or missing. Mitochondrial DNA can establish links between any family members related through the maternal line (see Box 16.2).

DNA typing in conservation and ecology

Although DNA typing gets media coverage for its use in forensic cases, the technique is also widely used in conservation biology and ecological research. DNA typing can reveal the degree of kinship of individual animals. This knowledge can be critical to the success of captive breeding programs for endangered species, for which there might be only a very few individuals in captivity. For example, DNA typing has been used on whooping cranes so that biologists could select the most genetically different individuals as breeding pairs.

Knowing degrees of kinship between animals in a living group is essential to behavioral-ecology studies. In many species, it is impossible to determine paternity simply by watching the living group; DNA typing provides a way to solve this problem. The fundamental social group of African lions is the pride, a group of females and their cubs (Figure 16.13). Small groups of male lions take up residence with the pride and father cubs and can be driven from the pride by an invading group of males. Sometimes all of the male lions father cubs (typical for groups of two or three males), and sometimes certain males will help with territorial defense of the pride but not father cubs. Scientists have found that male lions who help out other males in territorial defense but do not get to mate with the females are invariably related to the males who do mate. Thus, their behavior is helping their own kin to reproduce successfully. By contrast, male lions help unrelated males only if they all get to father cubs.

In another kinship analysis, scientists were able to solve the mystery of the Mexican loggerhead turtles. Pacific loggerheads nest in Japan and Australia, not Mexico, yet young loggerheads could always be found off the coast of Baja California. Many biologists did not believe the young turtles could have come the 10,000-mile distance from Japan (Australia is even farther), and so the origin of the Baja turtles was a mystery. Now DNA comparisons have established that the Baja population is made up of turtles from both the Japanese and the Australian groups. Apparently, the young turtles are carried to Mexico by ocean currents, and they then swim back to Japan or Australia to breed. Quite a swim!

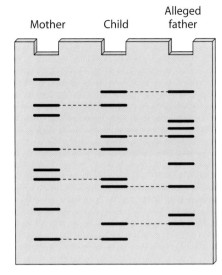

Figure 16.12 Parentage testing by DNA typing. To conduct a test, the DNA fingerprints of the child and the alleged parent are compared. Every band in the child's fingerprint should be present in that of either the mother or father, since a child inherits half its DNA from each parent.

Figure 16.13 Lions are the only cats who live in large social groups, or prides. A pride is made up of 3 to 30 lions, consisting of related adult females and their cubs along with a few males who are not related to the females. DNA typing is revealing new information about social behavior and genetic variation in lions. (Photograph courtesy of the St. Louis Zoo.)

Figure 16.14 This handheld device contains a gene chip with thousands of probes. (Photograph courtesy of Affymetrix, Inc.)

Analysis of genetic variability can provide insight into the status of wild populations. Biologists have assumed that vigorous wild populations contain genetically variable individuals rather than genetically identical ones. DNA typing, together with protein comparisons, provides the means for testing this hypothesis. In a study of lion populations, scientists found that a small population living in an ancient volcanic crater in Tanzania was much less genetically variable than a larger population that roamed the open Serengeti plains. Sperm samples taken from crater lions showed many more abnormalities than comparable samples taken from the plains lions. This result appears to support the greater variability-greater fitness hypothesis.

GENOMICS

One of the newer technologies in the DNA toolbox allows scientists to test for the presence or expression of many genes at once. The key ingredient of this tool is called a **gene chip**, or a **microarray**. A gene chip is a grid of spots of DNA on a tiny glass or silicon slide (Figure 16.14). Each spot contains copies of one specific DNA molecule, either a synthetic oligonucleotide or a fragment of cloned DNA, that acts as a probe for its complementary sequence. The grid itself may contain thousands and thousands of spots. For example, you can obtain chips with DNA representing every single gene in the yeast genome (about 6,000) on a grid that is less than 1 inch square.

Microarrays allow us to ask questions about the presence of many different forms of genes or the expression of many genes at once. Analysis of the entire genome of an organism, or the expression of many genes from an organism, is often called **genomics** or proteomics (as opposed to genetics, which usually refers to the study of one or a few genes at once).

Microarray technology has become possible because scientists have accumulated enough gene sequence information to construct probes for many, many genes from a single organism. The Human Genome Project, an organized international effort to sequence the entire human genome, as well as the genomes of many model organisms, jumpstarted such sequencing efforts. It also sparked the improvement of sequencing technology to the point that the sequencing of the entire genome of a microorganism, once a landmark achievement, is now viewed as relatively routine, though still time-consuming.

Analyzing genotypes

Gene chips can be used to determine which versions of genes (alleles) are present in a given sample or to look at what genes are being expressed in a given cell or tissue. To determine which alleles are present, a gene chip that contains probes specific for different alleles of the gene must be constructed. Constructing this type of chip requires a lot of research into what alleles are present in a population.

To determine an individual's genotype, an appropriate sample is taken and DNA is isolated from it. Next, the DNA is labeled, usually with fluorescent dye molecules, so that it can be detected. Finally, the labeled sample DNA is allowed to hybridize with the DNA on the gene chip. Following hybridization, the sample DNA is washed off to remove any DNA molecules that are not hybridized to the chip DNA, and then the results of the test are visualized. The pattern of hybridization reveals which alleles are present in the person's DNA (Figure 16.15). Because the grids are tiny and densely packed with spots of different probe DNAs, the visualization requires opti-

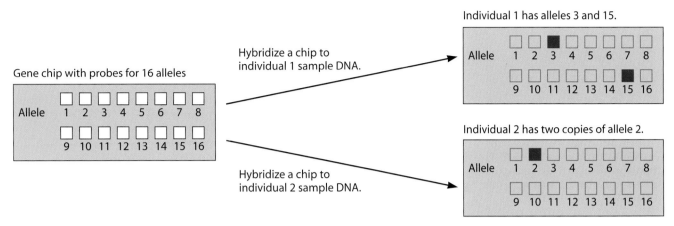

Figure 16.15 DNA chips can reveal an individual's genotype by hybridizing his or her DNA to probes specific to the known alleles of the gene.

cal scanner technology and computer software to interpret the results. An image of a developed gene chip is shown in Figure 16.16.

This kind of test might be used to determine a person's genotype at multiple loci. For example, probes for specific disease-causing mutations could be used to identify carriers of many common inherited diseases. If many different mutations can cause a specific disease, as in the case of cystic fibrosis, arrays of probes to all the different mutations could be assembled on a chip, along with probes for other inherited mutations that cause disease. Since more and more disease genes are being identified, direct examination of those genes is becoming increasingly common as a means of diagnosing genetic disease. For a description of another potential use of genotyping, see Box 16.3.

Analyzing gene expression

Gene chips can also be used to analyze patterns of gene expression to learn what genes are expressed under different conditions in different tissues. To analyze gene expression, messenger RNA (mRNA) is used to probe the gene chip instead of genomic DNA. For example, researchers interested in understanding the total response of yeast cells to different environments (temperature, salinity, and varying food sources, for example) grow yeast in the different environments and prepare mRNA. The mRNA is used to make labeled complementary DNA (cDNA) molecules to test against yeast gene microarrays. One microarray would be incubated with cDNA from control cultures of yeast, and a second microarray would be incubated with cDNA from yeast grown in the experimental environment. The hybridization patterns on the two chips would then be compared.

Researchers interested in learning about the molecular basis of schizophrenia, a poorly understood mental illness, have used microarrays to identify genes that are differentially expressed in schizophrenic and control individuals. To do this, they prepared mRNA from the brains of 12 control and 12 schizophrenic individuals who had died from other natural causes. These two populations of mRNA were tested against a microarray containing probes for over 6,000 human genes to determine whether there were any changes in gene expression associated with schizophrenia and which genes were involved. They observed several specific differences that suggest the development and

Figure 16.16 A developed gene chip must be read by computerized sensors. (Photograph courtesy of Affymetrix, Inc.)

B O X   1 6 . 3 *Prescription drugs and DNA*

One of the first wide uses for genetic fingerprints will likely be for genes involved in the metabolism of medicines. Physicians have long known that different patients respond differently to the same medicines. Drugs that relieve pain in one patient may not help another patient with the very same painful condition. A medication that is not toxic in one person may cause a life-threatening reaction in another.

Research is beginning to reveal the genetic basis for these different responses. Within the body, many medicines are processed by enzymes, and scientists have learned over the past several years that there are different genetic versions of these enzymes in the population. How an individual responds to a given medication depends on what version of the processing enzymes that person has. These differences in responses can determine whether a particular medicine is effective in a given individual and even whether it can cause life-threatening reactions (see the table).

For example, one of the genes involved in processing about a quarter of all medicines, including about 50 of the 100 most commonly prescribed drugs, is called 2D6. About 6 to 10% of Caucasians carry genes for a slow-acting version of this enzyme. When these people take a medication processed by this enzyme, they maintain higher levels of the medicine in their systems much longer than do people with the fast version. Thus, they are more easily overdosed. They are also immune to the pain-killing action of codeine. Codeine is processed by 2D6 into morphine, which exerts most of the pain-killing effect when codeine is taken. People with the slow version of 2D6 do not generate enough morphine from codeine to benefit from it.

This marriage of the study of genetics, genomes, and pharmacology has been named **pharmacogenomics**. Pharmacogenomics is an area of intense scientific research and commercial interest. It is estimated that most commonly used medicines now benefit only 30 to 60% of patients who take them and that adverse drug reactions kill about 100,000 hospitalized patients each year in the United States alone. If the medi-

cine-processing genotypes can be associated with responses to specific medications, physicians can more effectively prescribe for individuals, prevent much misery from adverse reactions, and allow the proper use of medicines that have been withdrawn because they cause severe reactions in a subset of the population.

Gene chips are being developed that contain probes specific to different versions of many genes involved in drug metabolism. These chips can be used in research to identify genotypes associated with adverse reactions to existing and experimental medicines. In the not-too-distant future, a pharmacogenomic fingerprint may become a routine part of health care, so that prescriptions can be tailored to an individual's genotype. People of the 22nd century (or even the late 21st) may look back on our drug prescription practices, where everyone is prescribed the same medications at the same doses, with the same horror with which we now view the early days of blood transfusions, when they were given without blood typing, frequently causing fatal reactions.

Drug (generic name)	Condition(s) prescribed for	Potential life-threatening reaction	Gene involved in reaction
Tofranil (imipramine)	Depression, bed wetting, attention deficit disorder	Heart arrhythmia	CYP2D6
Laniazid (isoniazid)	Tuberculosis	Liver toxicity	NAT2
Coumadin (warfarin)	Harmful blood clots	Internal bleeding	CYP2C9
Adrucil (fluorouracil)	Cancer	Severe immune suppression	DPD
Biaxin (clarithromycin)	Infectious disease (an antibiotic)	Heart arrhythmia	KCNE2
Imuran (azathioprine)	Rheumatoid arthritis	Severe immune suppression	TPMT

function of a particular type of neuron is impaired in schizophrenia. Other genes whose expression was altered in the schizophrenic individuals encoded proteins and peptides involved in neuron signaling pathways.

Once genes that are differently expressed under different conditions are identified, the real work begins. It may take years of research before we finally understand why the genes are differently expressed and just how that expression contributes to the overall phenotype of the organism. In the case of a disease like schizophrenia or cancer, it is hoped that understanding the roles of different genes in the disease process may eventually lead to better treatments.

Microarrays are being used in studies of gene expression in many nondisease systems as well. These include experiments to determine what plant genes are involved in the initiation of flowering, what plant genes change expression in response to changes in the atmospheric carbon dioxide concentration, and what genes are expressed differently in different developmental stages of animals. Before the development of microarrays, it wasn't possible to analyze the expression of thousands of genes in a single experiment. This new technology is opening new frontiers of research into the molecular mechanisms underlying fundamental cellular and organismal processes.

GENETIC ENGINEERING

The term **genetic engineering** refers broadly to the process of directed manipulation of the genome of an organism, usually a specific gene. The goal of genetic engineering, of course, is not to manipulate an organism's DNA per se but to change something about the proteins produced in that organism: to cause it to produce a new protein, to stop producing an old protein, to produce more or less of a protein, and so on. Manipulating the genome is merely the way to influence protein production. An organism that contains a gene originally from another source is often called a **transgenic organism,** and the new gene is sometimes called a **transgene** to distinguish it from the organism's native genes.

Scientists genetically engineer organisms for various reasons. One is to learn about what a given gene does. For example, homologs of the *Drosophila* homeotic genes were identified in mice. To confirm that these genes were important in the development of mice, scientists made genetically engineered mice in which the homeotic genes were nonfunctional and observed the results.

Another reason to make a genetically engineered organism is to obtain large quantities of the protein encoded by the new gene. This kind of approach is used to produce therapeutic proteins, like insulin, from microorganisms. It is also used to produce many of the restriction enzymes and other proteins used as tools in biotechnology. When scientists genetically engineer an organism for protein production, they tinker with the gene to ensure a good output.

No matter what the purpose of genetic engineering is, the gene in question must first be identified and cloned into a plasmid or other vector so that large quantities of a DNA fragment containing the gene can be reproducibly obtained. The cloned gene is manipulated in the laboratory using the tools and techniques described previously to get it ready for insertion into the target organism.

Moving a gene into a target organism is a simple phrase for a process that can be very complex. Techniques for gene transfer, such as transformation, all move DNA into a single cell. If your goal is to move a gene into a plant or animal, you would like to get the gene into many cells, if not every cell, of that organism. That is a more complicated proposition, and we will next look at some methods for doing so.

Genetic engineering of microorganisms

In the last chapter, we described how to clone DNA using bacteria as host cells. Genetic engineering takes the process one step further in that it introduces a specific gene or genes into a bacterial host in such a way that the proteins encoded by the genes will be expressed.

The first step in this process is to clone fragments of DNA that contain the gene of interest. If you are cloning a eukaryotic gene into a prokaryote, you will need to tinker a bit with the gene so that it can be expressed in its new host. For example, most eukaryotic genes have introns that are spliced out following transcription in eukaryotic cells (see chapter 4). Bacterial genes do not contain introns, and bacteria lack the means to splice eukaryotic RNA. Therefore, you would use reverse transcriptase to prepare cDNA from mRNA isolated from donor tissue (see chapter 15). You will also have to add prokaryotic traffic signals, such as a promoter, so that the gene can be transcribed in its new host. Even if the new host is a single-celled eukaryote, such as yeast, you may need to optimize the traffic signals. Finally, you might want to use a vector that has built-in regulatory signals that permit you to control expression of the new gene in its new host. One of the simplest of these is based on the *lac* operon (see chapter 8). Remember that the *lac* genes in *E. coli* are expressed only in the presence of lactose, which binds to the *lac* regulatory protein and prevents it from blocking transcription. If you clone a new gene under the control of the *lac* promoter and regulatory regions, that gene will be expressed only when lactose or a chemical that imitates its activity is present in the growth medium (Figure 16.17).

Once you have finished tinkering with the gene's traffic signals to prepare it for its new host, you would use one of the DNA transfer procedures described in the last chapter to introduce the vector and gene. As a final step, you would verify that the transgenic organisms indeed contained the correct new DNA molecule and that they were producing the desired protein.

Microbes have been engineered to make useful proteins, such as human insulin to treat diabetes. They have also been engineered with new enzymes that allow them to make nonprotein products, such as the dye indigo. Microbes used for industrial production of enzymes such as amylase have had additional copies of their own amylase genes added to their genomes, increasing production of the enzyme.

Genetic engineering of plants

Earlier in this book, we described the historical development of genetic manipulation of crop plants, from seed selection and hybridization to mutagenesis and genetic engineering. Genetic engineering involves the movement of

Figure 16.17 Genes are often cloned so that their expression can be controlled. For example, a gene cloned with the *lac* promoter will be expressed only if lactose is present in the growth medium.

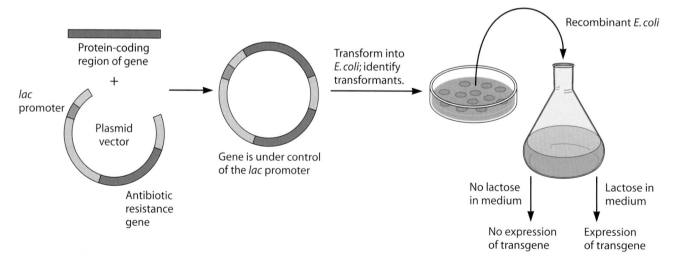

one or a very small number of known genes into a plant to introduce specific characteristics. A few examples of the ways in which genetic engineering has been used are:

- to make crop plants resistant to insect pests
- to make crop plants resistant to viral diseases
- to produce edible vaccine proteins in fruits
- to produce other medicinal proteins in plants
- to make plants frost resistant

The applications of biotechnology in agriculture and food production will be discussed in detail in chapters 21 to 23.

Making a transgenic plant requires the cloning and engineering of the desired new DNA, the introduction of that DNA into plant cells, and finally, the generation of an entire plant from the transformed cells. Cloning and engineering the new DNA and introducing it into plant cells require the DNA manipulation technologies you read about in chapter 15. Once scientists have successfully carried out these steps, they have genetically engineered plant cells, not genetically engineered plants. Producing an entire plant from transformed plant cells requires a set of technologies called **plant tissue culture.**

Plant tissue culture

Plant tissue culture allows the regeneration of an entire plant from a single piece of tissue, or even single cells. If this technology were available for animals, it would be as if you could regenerate a person from a piece of his big toe, liver, or any other tissue. In plant tissue culture, a piece of a plant is surface sterilized and placed in a sterile dish of culture medium containing nutrients and plant growth factors. Under these conditions, the plant piece produces callus cells, an undifferentiated cell type that grows over wounds in plants. Hormone concentrations in the medium can be manipulated to cause the callus to produce shoots, roots, and eventually a tiny complete plant that is genetically identical to the callus tissue (Figure 16.18). The tiny plant can grow to normal size and reproduce.

To make a genetically engineered plant using *Agrobacterium tumefaciens* and the Ti plasmid (see chapter 15) as the vector, the gene of interest is first

Figure 16.18 Plant tissue culture. **(A)** Pieces of plant tissue can be induced to form undifferentiated callus cells (plates on right). Manipulation of hormones in the growth medium causes the callus to differentiate into tiny plantlets (plates on left). **(B)** Close-up of plantlet. Note the fuzzy roots and tiny leaves. (Photographs courtesy of Syngenta.)

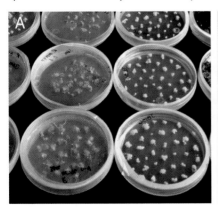
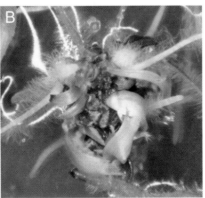

cloned into the plasmid, and the recombinant plasmid is then reintroduced into *A. tumefaciens*. Pieces of the future host plant, usually small pieces of leaf, are surface sterilized and soaked in a solution containing *A. tumefaciens* with the engineered plasmid. After the leaf pieces are exposed to the bacterium, they are transferred to plant tissue culture medium containing an antibiotic. The version of the Ti plasmid used as a cloning vector has an antibiotic resistance marker gene, so the antibiotic allows only those plant cells that received the new DNA to produce callus cells. Therefore, all the plantlets generated from the callus are transgenic.

Resistance to viral diseases

In the 1980s, researchers at the University of Washington in St. Louis, Mo., paved the way for engineering virus resistance into many important crop plants. Plant growers had known for a long time that infection with a mildly virulent virus often protected plants from infection by more dangerous viruses, just as infection by the mild vaccinia (cowpox) virus was found to protect people from infection by the deadly smallpox virus. The researchers showed that plants transgenic for the coat protein gene of the tobacco mosaic virus were resistant to infection by the virus. Following the publication of their work in 1986, other researchers showed that coat proteins from many other important plant viruses provided protection from infection when cloned into a variety of different plants. This technology has been used to protect numerous crops from diseases that normally take a significant toll (see Box 16.4)

Fighting aluminum toxicity

One of the environmental problems affecting the ability of soil to sustain plant life is aluminum toxicity. More than one-third of the world's soil suffers from aluminum toxicity, and the problem is most severe in the humid tropical climates of many developing countries. Aluminum toxicity is also a consequence of soil acidification as a result of acid deposition, seen in sensitive soils of the United States and other countries. Aluminum ions injure plant root cells, thus interfering with root growth and nutrient uptake. In an effort to make crop plants more resistant to aluminum, Mexican scientists transformed corn, rice, and papaya with a bacterial gene for the enzyme citrate synthase. The transformed plants released citric acid, which binds to soil aluminum and prevents it from entering plant roots. The genetically modified plants were able to germinate and develop at aluminum concentrations that were toxic to nontransgenic plants.

Genetic engineering of animals

Making a transgenic animal is similar in concept to making a transgenic plant but quite different in practice. We can take a piece of a plant leaf and manipulate it in culture to regenerate an entire plant, but we cannot take a piece of an animal and regenerate the animal. The only animal cells that can generate entire organisms are fertilized eggs or early embryo cells; therefore, transgenic animals are made by manipulating these cells.

Microinjection of fertilized eggs

To make an animal that expresses a new gene, a recently fertilized egg is microinjected with a DNA fragment containing the desired new gene. The injected egg is implanted into a surrogate mother animal and allowed to de-

BOX 16.4 *Genetic engineering and Hawaii's papaya crop*

Papaya, a tropical fruit rich in vitamins A and C, is Hawaii's second-largest fruit crop. Papaya plants are seriously damaged by papaya ringspot virus (PRSV), which is transmitted from plant to plant by aphid insects. Papaya production on the island of Oahu was virtually eliminated by PRSV in the 1950s, leading to the relocation of papaya farming to the Puna district of the island of Hawaii. The industry thrived there, free of PRSV.

Scientists at the U.S. Department of Agriculture (USDA) and the University of Hawaii feared the situation was too good to last. Knowing that pathogens usually spread, they anticipated that the day would come when PRSV would invade Puna. Thus, in the late 1980s, they set out to create a PRSV-resistant papaya plant for Hawaii's farmers.

Working with the Cornell University scientist who invented the gene gun, the scientists transformed embryos from papaya seeds with the coat protein of a PRSV isolate from Oahu. They used this approach because no tissue culture system existed for the papaya and because the papaya is not infected by *A. tumefaciens*. In 1991, they had promising transgenic papaya plants that produced good fruit and were resistant to PRSV.

In 1992, during a small field trial of the transgenic papaya, the long-feared news arrived: PRSV had invaded the Puna agricultural district. By late 1994, nearly half of Puna's papaya acreage was infected and many farmers were going out of business. Many considered the papaya industry in Hawaii to be doomed. At the same time, field trials of the transgenic papaya were progressing very well (see the figure). Transgenic plants remained resistant to PRSV even

35 months after the trials were initiated. Fruit quality was good. In 1995, efforts to deregulate the plant so that it could be commercially farmed began in consultations with the USDA, the Environmental Protection Agency, and the Food and Drug Administration.

In 1997, the transgenic papaya was released from regulation and approved for commercial production. Due to the severity of the PRSV problem, the Papaya Adminstrative Committee, a USDA marketing group, funded a program to produce seeds of the transgenic papaya. Distribution of the seeds to growers free of charge began in 1998. The genetically engineered papaya is credited with saving Hawaii's papaya farms.

Transgenic papaya has even helped with the cultivation of nontransgenic virus-sensitive strains. The resistant papaya doesn't allow the virus to multiply and spread and so has reduced the amount of virus circulating on the island. Furthermore, stands of resistant papaya appear to provide a protective barrier for nonresistant papaya, since the virus cannot spread through the transgenic plants.

What sounds like a happy ending has become clouded by the same kinds of controversy that commonly accompany the introduction of genetically modified food crops. Japan, a major customer for Hawaiian papaya, has refused to buy the transgenic papaya, citing concerns that they may cause allergic reactions when eaten because of the virus coat protein. Organic farmers, a small minority of the Hawaiian papaya farmers, complain that pollen from the transgenic plants is introducing the new genes into their plants. Groups opposed to genetically engineered foods in general are urging farmers in other countries to reject the plants. Some see the debate as one of large corporations versus small farmers, even though in this case the transgenic papaya was developed by university and government scientists and provided to farmers for free.

How does a reasonable person evaluate whether a transgenic crop poses a threat to human health or ecology? In chapters 21 to 23, we discuss how people attempt to evaluate risks such as allergenicity and gene flow.

A stand of transgenic PRSV-resistant papaya plants surrounded by PRSV-sensitive plants that have been ravaged by the virus. (Photograph courtesy of Dennis Gonsalves, USDA.)

velop into a baby animal. After the baby is born, it is tested to determine whether its genome contains the new gene and whether it is expressing the new protein. Making transgenic animals is partly a numbers game. Some percentage of the microinjected eggs do not survive the injection process. Of those that survive, only a fraction will incorporate and express the new gene. Thus, the procedure has to be repeated with many eggs to increase the chances of success.

Figure 16.19 Researchers at Virginia Tech produced transgenic pigs that secrete human blood-clotting proteins in their milk. Hemophilia patients can be treated with the purified clotting factors without incurring the risk of infection posed by multiple transfusions of human blood. (Photograph courtesy of Virginia Tech Photo.)

Scientists at Virginia Polytechnic Institute and State University (Virginia Tech) have made transgenic pigs that contain the human gene for factor VIII, a protein missing in people suffering from hemophilia A. The blood of hemophilia A patients does not clot properly, and even minor injuries can be life threatening. Hemophilia A patients can be treated with factor VIII protein, but obtaining large amounts of the protein for medical use was problematic. The Virginia Tech scientists engineered the factor VIII gene with a promoter from a milk protein gene that causes the protein to be secreted in the transgenic pigs' milk. The scientists estimate that 300 to 600 milking sows could produce enough factor VIII to meet the world's demand (Figure 16.19).

Gene replacement in ES cells

In the egg microinjection procedure for making transgenic animals, the new genetic material can be inserted anywhere in the genome. Since the point of the procedure is to obtain an animal that makes a new protein, it often does not matter where the new DNA is located. Sometimes, though, rather than introduce a new gene into an animal, scientists want to replace an allele of a gene that is already present with a new allele. This application requires precise replacement of an existing gene segment with a new one.

To replace a specific gene, scientists start with **embryonic stem (ES) cells**. These cells, taken from early embryos and grown independently in culture, retain the ability to differentiate into every kind of cell in the adult animal (see chapter 10). First, the gene of interest is manipulated in the desired way (for example, it may be inactivated by removing a segment or introducing stop codons). The prepared gene is then put into a vector that contains antibiotic resistance genes so that transformed cells can be identified, and the vector with the gene is introduced into ES cells. In this procedure, the vector is linear DNA that cannot be replicated and transmitted independently; it must integrate into the host genome to be stably maintained.

Because the manipulated gene and the host cell's genome contain DNA sequences in common, at some frequency homologous recombination will take place between the newly introduced DNA and the host genome. The recombination event results in the replacement of the intact gene by the inactivated form. Transformed cells are selected by their ability to grow in the presence of the antibiotic and are then tested to make sure the new allele is present at the right place in the genome.

Once an ES cell has been engineered with the new allele and allowed to multiply into a clonal population, some of the cells are injected into early mouse embryos (blastocysts). The blastocysts are implanted into surrogate mothers and allowed to develop into baby mice (Figure 16.20). At some frequency, the ES cells will be incorporated into the embryos and develop as part of the babies.

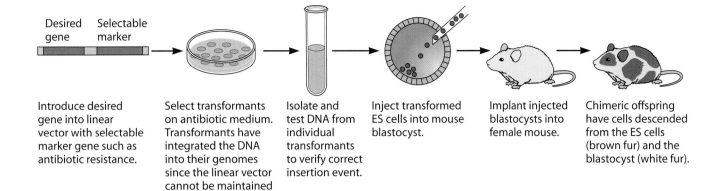

Introduce desired gene into linear vector with selectable marker gene such as antibiotic resistance.

Select transformants on antibiotic medium. Transformants have integrated the DNA into their genomes since the linear vector cannot be maintained independently.

Isolate and test DNA from individual transformants to verify correct insertion event.

Inject transformed ES cells into mouse blastocyst.

Implant injected blastocysts into female mouse.

Chimeric offspring have cells descended from the ES cells (brown fur) and the blastocyst (white fur).

Figure 16.20 Making a transgenic mouse with ES cells.

To help identify baby mice that have incorporated the ES cells, scientists usually use stem cells from brown mice and implant them in embryos from white mice. Babies that have incorporated the ES cells will have patches of brown and white fur. Such mice are called **chimeras**, because they are descended from two sets of parents (the brown mice and the white mice). The parts of the mouse with brown fur are descendants of the engineered ES cells and contain the engineered gene. The parts of the mouse with white fur are descendants of the original embryo (Figure 16.21). To get a mouse that has the new allele in all its cells, chimeric mice whose reproductive tissues are descendants of the engineered cells are bred together.

Knockout mice

One spectacularly useful application of this gene replacement technology has been making so-called **knockout mice**. When scientists identify a gene they believe to be involved in a particular process, the best test of their conclusion is to knock out the gene in a test animal and see if the process is affected. Gene replacement technology is used to inactivate the specific gene in mice by replacing a functional allele with an engineered, nonfunctional one, and the result is called a knockout mouse.

Quite frequently, knockout mice can be used as models for human disease, because the genomes of mice and humans are about 80% similar. For example, in chapter 7 we described a protein implicated in hereditary early heart failure that acts as a switch on the calcium pump on the sarcoplasmic reticulum of heart muscle cells. The Harvard scientists who identified this gene in a family with hereditary early heart failure wanted to test whether the

Figure 16.21 Chimeric mice produced via the fusion of ES cells and a blastocyst. The white fur is derived from blastocyst cells, while the brown fur is derived from the ES cells. (Photograph courtesy of Murinus GmbH, Hamburg, Germany.)

mutant gene alone really was the cause of the early heart failure. To do this, they made a knockout mouse in which the mouse version of the gene was disrupted. The knockout mice developed early heart failure, proving that disrupting just that one gene caused the disease. Knockout mice are also extremely useful for developing and testing new therapies and drugs before they are tested on humans.

Gene expression can be silenced with RNA interference (RNAi)

One of the hottest new techniques in biotechnology provides the ability to silence the expression of specific genes. This technique, like so many others, exploits a natural cellular process, but one that was only recently discovered.

Scientists studying development in the nematode *Caenorhabditis elegans* (see chapter 10) wanted to block the expression of a specific gene. They injected into the nematode an **antisense RNA** molecule, one whose base sequence was exactly complementary to the 5' end of the gene in question. The idea was that the antisense RNA would hybridize to mRNA for the gene and block its binding to ribosomes and translation into protein. This approach had worked in other systems, but not in every case.

Paradoxically, the nematode researchers found that injecting the worms with sense RNA, or exact duplicates of part of the mRNA, which should not hybridize to the mRNA, worked as well. They eventually discovered that double-stranded RNA (dsRNA) whose sequence was homologous to the mRNA was the active agent. Now dsRNA is thought by many to be the true effective agent in cases where antisense RNA has been successful.

Inside the worm cells, the dsRNA was cleaved into short pieces by a cellular enzyme called Dicer. These short pieces of dsRNA triggered the action of specific RNA-cleaving proteins. These proteins bound to the short dsRNA (called small interfering RNA [siRNA]) and denatured them. Using the anti-

Figure 16.22 RNAi. In RNAi, dsRNA homologous to the RNA of the gene to be silenced is used. The cellular enzyme Dicer cleaves the dsRNA into short siRNAs. A second cellular protein uses homology between the siRNA and the target mRNA to locate the target by hybridization and then cleave it.

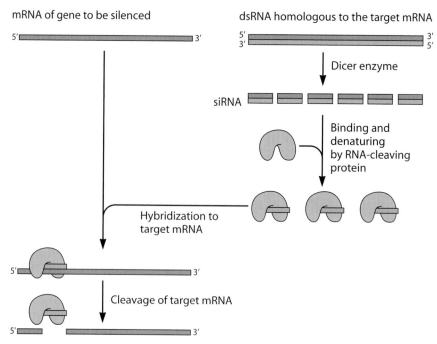

sense portion of the molecule as a guide, the RNA-protein complex bound to the mRNA and cut it (Figure 16.22). The phenomenon of using dsRNA to block the expression of a specific mRNA is called **RNA interference,** or RNAi.

The news of RNAi burst upon the biotechnology scene like fireworks. Scientists quickly realized that this phenomenon had the potential to silence the expression of any gene. They found that supplying cells with siRNA could dampen or block the expression of the homologous gene. siRNA can be provided like a drug or introduced via genetic engineering. RNAi is now being used to study the effects of countless genes by turning off their expression and is being investigated as a means of fighting various diseases.

The related antisense technology (in which the introduced nucleotide is single stranded and complementary to the target mRNA) is not dead. An antisense drug to fight cytomegalovirus-induced blindness, Vitravene (fomivirsen), has been tested and approved and is on the market. Other antisense drugs targeted against cancer, Crohn's disease, psoriasis, colitis, and AIDS, are in various stages of development and clinical trials by several different drug companies. The company that developed Vitravene, Isis Pharmaceuticals, considers siRNA and antisense technology to be one and the same. They believe the new understanding of the mechanism of action of RNA silencing will allow more effective drug development.

SUMMARY POINTS

DNA comparisons can be used on a broad scale to assess the degree of relatedness of two species or on a tightly focused individual level to determine whether two DNA-containing samples could have come from the same individual.

Evolutionary scientists compare the amino acid sequences of proteins and the nucleotide sequences of genes to deduce how species evolved from one another. These analyses are possible because related organisms have similar protein and nucleic acid sequences.

Forensic scientists compare DNA samples by looking at restriction fragment or PCR product sizes from many loci. They analyze enough different sites to be very certain that two matching samples came from the same individual.

Mitochondrial DNA analysis is used to determine relatedness through the female line of descent. It can be used in evolutionary studies of fairly recent events to determine the likely ancestry and time of divergence of species or to identify family members of persons living or dead.

Genes for specific traits are found by using a number of different approaches. Scientists often start by looking for appropriate mutants, organisms in which the trait is altered because of a genetic change. In bacteria and yeast, the gene can often be identified by finding a recombinant plasmid from a genetic library that restores the normal trait to the mutant. Once the gene is known in a related organism, its sequence can be used to make probes to locate the gene in other organisms. If scientists are searching for a gene directly in people or other higher animals, then the genomes of many individuals expressing the trait must be analyzed and compared to those of individuals who do not express it. This is a long and difficult process.

Genetic markers are easily detected features within a chromosome that are inherited with a specific phenotype or genotype and can be used as a surrogate indicator of the presence of a particular allele.

Microarrays, or gene chips, are ordered grids of thousands of nucleotide probes that represent many genes of interest. They are hybridized with sample genomic DNA to reveal genotypes (such as for pharmacogenomic analyses) or with sample mRNA to reveal gene expression patterns.

Genetic engineering is the directed manipulation of an organism's genome. Engineering a plant involves techniques of plant tissue culture, in which an entire plant can be regenerated from engineered cells. Engineering animals involves manipulations of eggs or early embryos, since only those cells can develop into whole organisms.

Knockout mice, in which a specific gene has been inactivated, are invaluable for assessing the functions of genes and as models for developing therapies and testing drugs. Mice are often reasonable models of human genetic conditions because the mouse and human genomes are 80% similar.

The discovery of RNA silencing, related to antisense technology, offers scientists a new way to turn off individual genes to analyze their functions.

KEY TERMS

Amplified fragment length
  polymorphism (AFLP)

Antisense RNA

Chimera

Embryonic stem cell

Gene chip

Genetic engineering

Genetic marker

Genomics

Knockout mouse

Microarray

Mitochondrial DNA

Mutant

Pharmacogenomics

Plant tissue culture

Restriction fragment
  length polymorphism
  (RFLP)

RNA interference (RNAi)

Single-nucleotide
  polymorphism (SNP)

Transgene

Transgenic organism

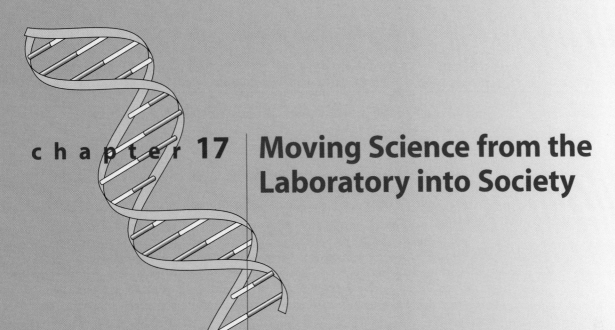

| # Moving Science from the Laboratory into Society

Biotechnology has the potential to improve our lives, but that potential can also be abused. While this thought may be troubling, it is not new, nor is it unique to this technology. Societies have dealt with precisely the same issue ever since humans began to fashion crude tools from stone.

Biotechnology began to generate a great deal of discussion about its potential uses and abuses—from transgenic crops to recombinant microorganisms to gene therapy—at the earliest stages of research, 10 years before any products were available to consumers. The debate about the potential societal impacts of modern biotechnology has continued unabated throughout its development. This discussion is healthy and necessary, because some potential applications of biotechnology raise difficult questions and ethical dilemmas to which there is no right answer. Unfortunately, at times the public discourse about biotechnology has been uninformed, sensationalistic, and, consequently, counterproductive.

Each one of us has a moral obligation to attend to biotechnology's development, but to maximize the good that accrues, that attention must be thoughtful and considered. The primary objective of this book is to equip citizens with information on biology and biotechnology so they can make constructive contributions to discussions and informed decisions in their personal lives. Chapters 2 through 14 provided information on some of the science that must serve as the foundation for productive discussions and informed decisions about appropriate uses of biotechnology. To the same end, in this chapter we provide:

- guidance on identifying and evaluating societal issues derived from biotechnology applications
- a few tips on *how* to think about these issues
- an example of how these critical thinking skills can be applied to analyzing complicated issues that arise from biotechnology

THOUGHTFUL DEVELOPMENT OF BIOTECHNOLOGY

Our society is at a critical stage in deciding how best to use biotechnology's potential, because biotechnology, in general, is still in the first phase of the life cycle pattern modern technologies exhibit (Figure 17.1). Some of the biotechnologies are located farther along the curve in Figure 17.1 than others. Certain industrial sectors have incorporated biotechnology into their research and product development. Some of those, such as the pharmaceutical industry, have commercialized biotechnology-based products. Other industrial sectors that ultimately will be affected by biotechnology development have only recently begun to consider how to use biotechnology's potential to improve their processes and create new products.

The early stages of technology development present some of the richest opportunities for thoughtful, productive discussions about how technology could maximize the benefits offered to the greatest number of people. Many perspectives need to inform this discussion, because like all technologies before and after it, biotechnology will change society in many ways. In any case, no matter how many voices take part in the societal conversation about biotechnology applications, many of its impacts will take society by surprise, and even involving many stakeholders in discussions about maximizing benefits will not protect some segments of society from being harmed by certain biotechnology applications. For example, if biotechnology decreases manufacturing costs in an industrial sector, companies that

Figure 17.1 Technology life cycle. In 1981, Arthur D. Little, Inc., developed a simplified model of the economic impacts of a technology during its life span. During the emerging stage, product and process research and development expenditures are high, as is uncertainty about the technology's eventual success. Product commercialization initiates the growth phase. As the technology-based product or process is adopted by various industrial sectors, product sales increase and process improvements reduce the costs of producing the technology. In the mature phase, the technology is well accepted, sales are stable, and cost reductions from process improvements have reached a plateau. In the final stage of a technology's life span, growth and acceptance of newer technologies (dotted line) displace the older technology.

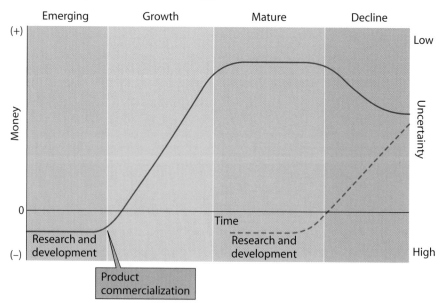

do not integrate biotechnology into their operations may not be able to compete.

Biotechnology's remarkable flexibility generates an enormous number of options for improving existing products and processes and creating new ones. Which products will be developed from the universe of possible products and solutions, and who will be involved in making that decision? Hand in hand with biotechnology's flexibility comes its power. With this power comes a responsibility to make farsighted and informed choices about how these technologies will be used. Where will society draw the line in using biotechnology's power? Will the line be drawn through active choice or passive acceptance? Who will decide where the line is drawn? Just as important, who will draw the line and ensure that society honors it?

In democracies, all of us have a role in technology development

While science and engineering may define the set of potential biotechnology products that are both scientifically possible and economically feasible, a number of factors determine which of those product ideas become realities (Figure 17.2). One of those factors is you.

Your essential role in technology development may come as a surprise to you. Often, people who are not scientists or engineers feel like technology is in charge and they are powerless to influence the direction of technology development. If you feel that way, then you are, in the words of New Age philosophers, "not in touch with your own power." Your power extends well beyond your purchasing power to determine which products on the market become commercial successes. In a democracy, you can exert influence at *every* step along the product development continuum, from basic research through product development, regulatory approval, and finally commercialization (Figure 17.3).

Figure 17.2 Factors that affect technology development.

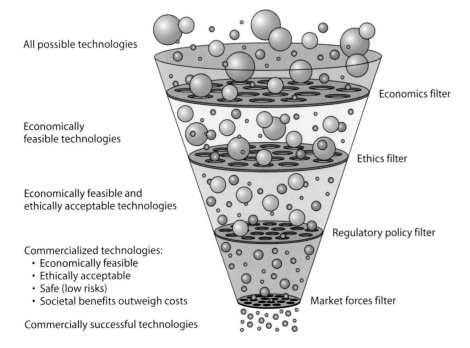

All possible technologies

Economics filter

Economically
feasible technologies

Ethics filter

Economically feasible and
ethically acceptable technologies

Regulatory policy filter

Commercialized technologies:
• Economically feasible
• Ethically acceptable
• Safe (low risks)
• Societal benefits outweigh costs

Market forces filter

Commercially successful technologies

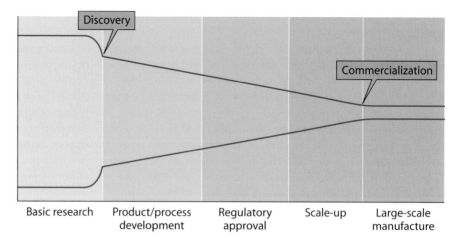

| Basic research | Product/process development | Regulatory approval | Scale-up | Large-scale manufacture |

Figure 17.3 Technology development. Basic scientific research leads to discoveries that give rise to ideas about possible technological solutions to problems. For technical and economic reasons, only a few of those ideas become realized as possible products or processes. Of those feasible products and processes, only some receive government regulatory approval. Additional attrition occurs during the scale-up phase because a sufficient amount of product at a saleable price cannot be produced.

Politics influences the direction of scientific research

If you doubt the power of citizens to influence the earliest stages of product development, you need look no further than the political response to an **emerging infectious disease**, AIDS. An emerging infectious disease is a disease that (1) is new to the human population, (2) occurred in small numbers of people prior to infecting a significant number, or (3) occurred in many people but only recently has been linked to an infectious agent, such as a bacterium or virus. Physicians first described AIDS in 1981; in 1983, researchers identified its causative agent, the human immunodeficiency virus.

By chance, in the United States, AIDS first afflicted the homosexual population. This group had already become politically mobilized around concerns about individual rights and personal freedom. When AIDS struck, they understood the process for effecting political change in this country, and they had a well-established infrastructure for influencing political decisions. They put their political savvy to work and activated their network. In a remarkably short time, they exerted significant influence on which scientific research projects, out of all possible research projects, received federal funding.

The U.S. Congress, the primary funding source for basic academic research in this country, has a limited amount of money to spend. In response to political pressure from a comparatively small but passionate group of advocates, Congress earmarked funds for AIDS research, which decreased the amount of funding available for other areas of scientific research. Even though the great majority of scientists do *not* conduct research on infectious viral diseases, they were ineffectual at protecting their own vested research interests.

Even now, the proportion of federal funding devoted to AIDS research is very impressive when you consider that, unlike many diseases that inflict much higher mortality rates on the U.S. population, scientists know what causes it and people know how to protect themselves from getting it (Figure 17.4). From 1981 to December 2002 in the United States, a total of 501,669 people had died from AIDS-related causes. Every year, approximately

A

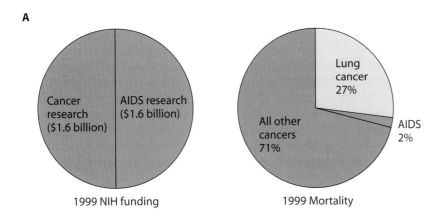

B

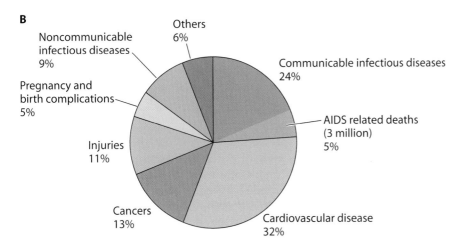

Figure 17.4 Politics and research funding. **(A)** In 1999, the amounts of federal funding earmarked for AIDS research and cancer research in the United States were essentially equal—$1.6 billion. Funding priorities do not reflect the relative impacts of these diseases on mortality statistics. Lung cancer is distinguished from other cancers because, like AIDS, its primary causes are known and the probability of getting the disease can be significantly reduced by altering behaviors. (Sources, National Institutes of Health [NIH] 1999 annual reports and federal budget requests [for funding] and National Cancer Institute and the Centers for Disease Control and Prevention [for mortality statistics].) **(B)** For reference, the 2002 global statistics on causes of death are also provided. Noncommunicable infectious diseases, such as pneumonia, are caused by microbes but are not directly transmitted from one person to another. (Source, World Health Organization, 2003.)

550,000 people in the United States die from cancer. Of those, approximately 62% (340,000) die from cancers that, unlike AIDS, cannot be decreased by altering behaviors, such as avoiding tobacco and using sunscreen. If AIDS in the United States had first afflicted the heterosexual population, would the federal priorities for research funding have shifted as quickly, more quickly, or not at all? Society will never know.

A similar phenomenon is occurring now in the United States with regard to government funding for embryonic stem cell research. A politically mobilized segment of society that understands the political process and has an established infrastructure—the antiabortion lobby—has successfully impeded federal funding for this research, even though it holds great promise for millions of people who suffer from many diseases and spinal cord injuries.

The effectiveness of the political pressure imposed by the antiabortionists has achieved more than simply influencing how federal research monies can be spent. The U.S. House of Representatives recently passed a bill (H.R. 2505) that criminalizes some embryonic stem cell research irrespective of the funding source. In other words, any scientist conducting this type of research could be sent to prison and fined $1 million. The Senate is considering a similar bill that goes even further. A provision in that bill would make it unlawful to reenter the country if a person has left the United States to receive disease treatments in countries where physicians legally use embryonic stem cells to treat or cure diseases.

A successful democracy requires an informed and engaged citizenry

Democratic societies give their citizens the power to influence:

- scientific research priorities
- the stringency of the product regulatory approval process
- postapproval regulatory requirements
- product entry into and removal from the marketplace
- other government-controlled factors that can affect a product's commercial success, such as product labeling

Citizens exert that power directly, by voting on these issues, or indirectly, by electing representatives and pressuring them to vote in certain ways.

Those who created this style of government centuries ago could never have foreseen the technical complexity underlying the decisions politicians and citizens must wrestle with today. In today's high-tech world, absence of understanding can be a tremendous stumbling block to wise action, or worse yet, misinformation can lead to unwise action. Would the inventors of democracy believe in the wisdom and appropriateness of giving all citizens the power to make decisions and influence decision makers today? Our founding fathers believed the solution to the problem of having a misinformed or uninformed public make decisions or exert its political will on decision makers was not to remove the public's power but to provide information.

> I know of no safe depository of the ultimate powers of the society but the people themselves. If we think them not enlightened enough to exercise their control with a wholesome discretion, the remedy is not to take it from them, but to inform their discretion by education.
>
> THOMAS JEFFERSON, 1781

To use the power of biotechnology wisely and to assess its costs and benefits correctly, citizens need access to accurate information. Most people rely on print and broadcast journalism for information on science and technology. Most reporters believe in the importance of balanced reporting and do their best to meet this criterion for good journalism. However, balanced information does not necessarily mean accurate information. In addition, the necessary information for accurately assessing biotechnology applications and issues cannot be crammed into a sound bite. For these and other reasons, the mainstream media has become a poor source of information on science and technology, especially if controversy is involved. The media can alert you to issues, but to formulate thoughtful opinions, you need to seek out other information sources.

> If a nation expects to be ignorant and free, in a state of civilization, it expects what never was and never will be.
>
> THOMAS JEFFERSON, 1799

Society will decide, either actively or passively, the problems and purposes toward which the remarkable power of biotechnology will be directed. Dismissing decisions about the appropriate uses of biotechnology as someone else's concern sets the stage for a small number of individuals to direct the course of scientific investigation and technological change. If a few set the course of biotechnology development, many people will be dissatisfied with the changes it brings, in spite of its extraordinary potential. Some people may even be harmed. If you abdicate your oversight of technological development to a handful of scientists, engineers, and lawyers, or to uninformed but passionate special interest groups, you are ignoring the power, privilege, and responsibility of a democratic way of life. You are also sacrificing a portion of your personal freedom.

> One of the penalties for refusing to participate in the political process is that you end up being governed by your inferiors.
>
> PLATO (428–348 BC)

A MENTAL FRAMEWORK FOR ANALYZING ISSUES

A person does not have to be a scientist to analyze the potential societal costs and benefits of various applications of biotechnology. A productive and user-friendly approach to analyzing any issue associated with biotechnology begins by placing modern biotechnology within the context of centuries of technological change.

As we stressed in chapter 1, modern biotechnology is one of many technologies humans have developed. Medicine, agriculture, and energy production are age-old attempts to improve human health and control the environment with technology. Biotechnology now places at our disposal a new set of technologies for shaping the world to our liking. The tools may be different, but the goals remain the same: to improve human health and alter the environment so that our lives and those of future generations are as long and easy as possible.

The impacts of a technology must be compared to those of similar technologies

For years, people in the industrialized world enjoyed the fruits of technology without fully considering the potential downside. In the second half of the 20th century, they began to recognize that the solutions and gratification that technologies can provide are sometimes accompanied by societal costs. Because many citizens in industrialized societies realize that hidden costs are often associated with technological change, they are focusing much more attention on biotechnology at an earlier developmental stage than was done for any prior technology. Often, however, they view biotechnology in isolation from other technologies and neglect certain key questions that would enrich their analysis.

How do the potential societal impacts of this technology compare to those of existing technologies presently used to accomplish the same goal? For example, how do the environmental and public-health costs of using transgenic microbes to synthesize insulin compare to those of extracting insulin from pancreatic tissue of butchered livestock?

Are the societal issues said to be rooted in biotechnology unique to the new powers these technologies provide, or are they old, unresolved, and sometimes unacknowledged problems associated with earlier technologies? Many of the issues attributed to the development of biotechnology have actually been around in slightly different forms for many years. While it is totally appropriate for the emergence of biotechnology to trigger concerns about the appropriate uses of technology, making biotechnology the scapegoat for the same issues raised with other technologies makes those concerns somewhat hollow.

The costs of technology must be compared to the costs of no technology

While the costs of using a technology are often hidden for many years, more deeply hidden from certain societies are the costs of having *no* new technology. The costs of no new technology are difficult to assess, because those costs are often unknown to people who have lived in a world in which technology has solved certain problems. When comparing the costs of technology and the costs of no technology, they often overestimate the cost of having the technology because they have never experienced the costs of not having it.

For example, getting immunized against infectious diseases carries some risks, including a small chance of dying. During the last decade, the media have focused attention on the risks of immunizations to infants. Some parents have responded by refusing their pediatrician's advice to immunize their infants against whooping cough, or pertussis. The risks of dying from pertussis are *much* greater than those of dying from receiving the pertussis vaccine, but parents in today's industrialized societies have lived in a world in which the mortality risks caused by an infant contracting pertussis are so low they seem nonexistent. A technology—the widespread immunization of infants to pertussis—has made them blind to the risks of not having that technology.

The costs of having no new technology for people with a high standard of living pale in comparison to those for people in the developing world who are trapped in the poverty cycle. As we discussed in chapter 14, although technology development in the industrialized world has had environmental and social costs, its beneficial legacy includes a comparatively healthy, well-fed, educated, and affluent citizenry.

Unfortunately, those people *least* affected by the costs of having no new technology live in countries with the human resources, capital, legal framework, and national infrastructure for feeding and supporting the biotechnology development pipeline. Developing countries that are most affected by the absence of new technologies lack the resources necessary to develop technologies that could improve their standard of living. Many public and private organizations in industrialized countries provide funding and training for scientists from developing countries to enhance their country's technological base for economic development. Even so, the quality-of-life gap remains huge.

Placing biotechnology in historical context can ground analysis

Knowing that today's biotechnologies are the next step in a continuum of technologies has important real-world implications for charting its course. Unfortunately, most public discussions and debates about the societal issues raised by biotechnology do not begin with a clear articulation of relevant historical context. Listeners are often left with the impression that these technologies appeared suddenly, with no historical precedents, and that the issues raised by biotechnology are unique to biotechnology.

A clear description of the technological ancestry that led to biotechnology cannot substitute, by itself, for responsibly discussing and analyzing the issues raised by biotechnology development. However, not incorporating this history into debates, discussions, and decisions perpetuates the misconception that the issues are new and unique. Placing modern biotechnology in a historical context provides reference points that ground assessments. By comparing and contrasting today's biotechnologies to previous, similar technologies from which the biotechnologies evolved, it is possible to:

- delineate how the new biotechnologies resemble and differ from earlier technologies
- get a better handle on the changes biotechnology might bring and determine which changes are significant and why they are significant
- define the set of societal issues specific to the unique capabilities provided by biotechnology and focus attention there
- decrease the uncertainty about potential impacts on the environment and human health
- evaluate the relevance of existing laws and policies precipitated by earlier technologies to the products and societal issues of biotechnology
- use past mistakes with earlier technologies to develop strategies for doing a better job with the new technologies
- compare the safety/risk ratio of these technologies to those of previous technologies for solving the same problem
- assess the social costs and benefits of these technologies compared with those of past activities directed toward the same goal

A PROCESS FOR ANALYZING ISSUES

To be an active and constructive participant in discussions about science and technology in society, you actually don't need to understand a lot of the specific scientific details. Being able to grasp scientific concepts and understanding the idea of a technological continuum are essential, but most important you need to think like a scientist, objective and detached. C. P. Snow, a scientist and author, said it much better than we can: "Science is the refusal to believe on the basis of hope." In assessing the potential impacts of biotechnology, adopt a scientific posture: a willingness to look at all of the data, whether or not they are consistent with your preconceived notions, because you put aside emotional attachment to a certain conclusion.

Wise, judicious, and socially beneficial development of biotechnology depends on informed discussions and critical evaluation of real issues, not hyperbole. All too often, however, discussions about controversial social issues are merely emotional exchanges of opinion, which accomplish nothing, because the participants seem more interested in winning an argument than in achieving greater understanding. Here are some critical steps in objectively evaluating societal issues associated with an application of biotechnology. In general, proceed very methodically as you:

- define and delineate the issue as specifically as possible
- determine whether the application introduces novel issues
- compare the risks and safety of these applications to those of similar ways of solving the problem
- assess the societal benefits and costs of using this application of technology

Often, many of the steps in the process are not sequential but are intertwined and iterative. At each step in the process, you must gather facts; do not rely on hearsay, and be equally skeptical of *all* voices with a vested interest in the debate.

In subsequent chapters, when we describe the applications and issues of biotechnology, we will revisit this process a number of times. Because the methodology described below is generic, some steps may not apply in all cases; in others, more steps may be necessary. The most important thing to take away is the tone of analysis: fact based, objective, methodical, specific, detached, and open to all options. This tone must pervade all analyses of societal issues raised by biotechnology if society hopes to maximize the good that accrues from its use.

Step 1: clarify the issue

Become as clear as possible on the essence of the issue, whether you are analyzing it to determine your position or discussing it with others. Because the issues associated with biotechnology development are multifaceted, you need to make sure everyone is talking about the same thing. This problem is exacerbated when discussion participants use broad, ambiguous terms. Making sure everyone is on the same page may seem tedious and unnecessary, but we cannot tell you how many times we have seen discussions devolve into petty shouting matches just because the people were talking about different things and didn't know it.

Delineate the issue very clearly and precisely

Broad, overarching issues need to be dissected. For example, gene flow from transgenic crops to other plants encompasses at least two issues that differ in almost every aspect relevant to assessing environmental risks and economic costs: gene flow from transgenic crops to wild plants and gene flow from transgenic crops to nontransgenic crops. Ambiguous issues need to be clarified. For example, if someone expresses concern about "genetic engineering," do they mean in plants, animals, microbes, people, or all of the above? If the focus is genetically engineered plants, do they care about the source of the gene (other plants, animals, or microbes) or only about the protein encoded by the gene?

Define the terms quite specifically

In public discussions of issues, if your goal is greater understanding, the value of using discipline in the words you choose cannot be overemphasized. The issues being discussed are almost always technically complex, and frequently they trigger emotional responses. Adding to the confusion by using sloppy language will only make the discussion, or your internal analysis of the issue, less productive.

For example, let's say you attend a debate entitled "The Risks of Biotechnology." Here are a few of the points at which people could be talking about different things and not even know it.

- Biotechnology is an ambiguous term used in a variety of ways by different people. For example, in the 1970s and 1980s, biotechnology meant recombinant DNA technology to American scientists and microbial fermentation to scientists in the United Kingdom. The risks of one technology differ greatly from those of another.
- What type of risk? To some people, the concept of risk relates solely to safety, either to public health or the environment, and is amenable to science-based analysis. Others include social, economic, and ethical concerns when they use the word risk. Science may contribute to

discussions of these concerns, but many factors other than science-based facts affect people's concepts of these types of impacts.

- Which facet of biotechnology development is the target of concern? Biotechnology development occurs along a continuum that begins in a research laboratory and extends through product development to widespread commercial use. The risks of biotechnology laboratory research, small-scale field testing, and large-scale commercial use and manufacturing differ from each other both qualitatively and quantitatively.

Perhaps you feel the discussion should include all of these things. That's fine, but discuss each facet sequentially and methodically, making sure everyone knows exactly which facet is under discussion.

Step 2: put the issue in historical context

Look at the history of using other technologies in similar ways and ask whether the issue is new or unique to biotechnology. Does the concern or issue derive from biotechnology's novel powers that provide brand new capabilities, or does society presently engage in practices that raise similar or even identical issues? If the practice and issues it raises are unique to biotechnology, do all aspects of the issue raise concerns? If not, specifically delineate those areas that trigger concern and focus attention there.

You may determine that the practices that elicit concerns are not at all unique to biotechnology but are accepted as part of life. Does their commonness mean they are not worth analyzing? Absolutely not. Just because people *are* doing something does not mean they *should be*. For example, some people have expressed concern that recombinant DNA technology allows us to genetically modify food by moving genes between different species. But, as you learned in chapter 1, at least a century before the development of recombinant DNA technology, plant breeders genetically modified many food crops by interbreeding different plant species incapable of interbreeding naturally. After learning this, some people may still believe that scientists should not move genes between species, whether they accomplish this through genetic engineering or crossbreeding. In other words, *what* scientists are doing, not *how* they are doing it, concerns them.

Step 3: consider the risks (related to safety)

Although many people discuss risks and benefits in almost the same breath, that juxtaposition can be confusing. Conceptually, these issues need to be separated, because they are fundamentally different considerations (even if they are not sequentially separated in your actual analysis). It is clearer to think of risk vis-à-vis safety, and benefits as societal benefits compared to societal costs.

Assessing risks is a science-based analysis of the probability of causing harm or injury to the environment or human health. We will discuss the formal risk assessment process in detail in the next chapter. In brief, the steps in a risk assessment are to identify the hazard, estimate the probability of exposure to the hazard, and assess the significance of hazard exposure.

For rational risk assessments, you must get facts, not opinions. Obtain objective data that allow you to identify the risks clearly, specifically, and independently. To ground your analysis in sound science, go to primary sources, such as scientific journals or scholarly articles, or to review papers authored by practicing scientists in publications like *Scientific American*. Talk to scientists conducting relevant research. Do not rely on news magazines, popular science magazines, newspapers, television, or hearsay. Be particularly careful of information you gather from the Internet. Some websites are excellent sources of factual information, but others are full of inaccuracies and opinions couched as scientific fact. A general guideline in using the Internet might be to use information from primary sources, academic and government research institutions, and regulatory agencies.

Once you have identified the risks, you must then estimate the probability of the risk occurring. Because much of biotechnology is similar to activities people have been engaged in for many years, there are often abundant data on some products and applications that can be used to estimate general probabilities of specific risks. Others introduce novel risks.

Then, assess the consequences of a particular risk occurring. What is the significance of risk A occurring? If it occurs, will it matter? Under what circumstances does it matter? It may happen on a regular basis anyway without human interference (for example, insects becoming resistant to chemicals produced by plants, or viruses transferring genes across species). Get a sense of the scale of the risk. An adverse effect to an individual organism or a small number of organisms may be environmentally, ecologically, or economically insignificant.

Finally, in discussing the risks of any technology, keep in mind that technologies may be used wisely or unwisely, for good or for bad. The risks of a technology per se must be distinguished from the risks of a technology placed in irresponsible hands. The flaws of a technology and the flaws of human nature are two very different considerations.

Step 4: compare the risks to those of other technologies and no technology

This step formalizes the mental framework described in the previous section. The relevant question to be asked and answered is not "is there a risk" in using this technology or product. Of course there is. There is no such thing as zero risk, because every human activity carries some amount of risk with it. To weigh the value of taking a risk, it needs to be placed in a meaningful context. One way to do this is to compare risks. With biotechnology, that typically means asking how the risks of this new technology compare with those of existing technologies directed toward the same purpose.

Another way to place the risk in a meaningful context is to ask the other essential question: *what are the risks of* no *new technology*. In industrialized countries, people focus almost solely on the risks of a technology. They are free to have that as their singular concern because the risks of *no* new technology are almost always minor, given the quality of life in industrialized countries. As explained above, the same does not hold true for many citizens in developing countries, where the risk (or cost) of no technology may be significantly higher than the risk of the technology.

Step 5: consider benefits and costs

Another aspect of placing risks in a meaningful context involves looking at the application's benefits. Only then can society determine if a risk is worth taking. On the other hand, just because a product is safe, does that mean it should be developed? Does society need this product? Society's resources are limited, and as any economist knows, there are always lost opportunities whenever people make a choice about how those resources—time, people, and capital—should be used. In addition, a safe and legal product may violate one of the fundamental ethical principles.

- Do no harm (nonmaleficence).
- Do good (beneficence).
- Do not violate individual freedom (autonomy).
- Be fair (justice).

The benefits and the costs need to be assessed in equally objective manners, which is difficult, because opinions about what constitutes a benefit vary. In addition, when one group gains, another group often loses, so a benefit for one may be a cost for another.

A CASE STUDY FOR ISSUE ANALYSIS: ANIMAL CLONING

In February 1997, scientists and nonscientists alike were shocked by the news that an adult mammal had been cloned. The responses of these two populations were similar on the surface, but the reasons underlying their responses had little in common. The existence of this clone, a sheep named Dolly (Figure 17.5), unleashed a global media frenzy. This story, which was essentially about a scientific breakthrough, received unprecedented coverage in the mainstream media.

Studying the media's coverage of Dolly offers an excellent opportunity for evaluating how reporters and editors handled the story and whether they fulfilled their responsibility to inform the public. It also illuminates a pervasive problem in the way the mainstream media covers science. Even though the key element that made this story "news" was a scientific breakthrough, the breakthrough that gave the story its newsworthy status was rarely, if ever, explained in the newspaper and television coverage. Many times it was not even mentioned! Instead, the reporters jumped immediately to *possible* applications in medicine and their ethical implications, which cannot be news, since they are purely conjectural and in any case many years away. (For excellent background information on the cloning of Dolly and related topics, visit the website of the Roslin Institute, the research center in Scotland where the cloning of Dolly occurred [http://www.ri.bbsrc.ac.uk].)

The issue of animal cloning presents an opportunity to use the strategy described above to analyze an emotionally charged societal issue raised by a scientific breakthrough related to biotechnology development. The discussion

Figure 17.5 Dolly and her surrogate mother. (Photograph courtesy of Roslin Institute, Edinburgh, United Kingdom.)

below is not meant to be a comprehensive treatment of the issues involved with animal cloning but is included to give you an example of some of the requisite steps in a rational treatment of the issue.

Clarify the issue by defining and delineating

When discussing the "ethics of cloning," be sure to separate the ethics of cloning plants, animals, and people. Some people may feel that the issues are the same irrespective of the organism involved; others may not.

What is a clone?

A clone is a collection of genetically identical individuals, all derived from a single parent. Cloning is the process of propagating those genetically identical individuals. Biologists use the word clone in a number of ways, depending on what is being cloned (Figure 17.6).

In molecular cloning, the cloned "individual" of interest is a gene or a length of a DNA molecule. A clone, in this case, usually refers to both the

Figure 17.6 Cloning. Cloning is the creation of genetically identical copies. The word clone can refer to DNA molecules, cells, or organisms. Identical twins are clones of each other, as are plants that are propagated from a fully differentiated plant part, such as a leaf or stem. **(A)** Molecular cloning. **(B)** Cell cloning. **(C)** Organism cloning.

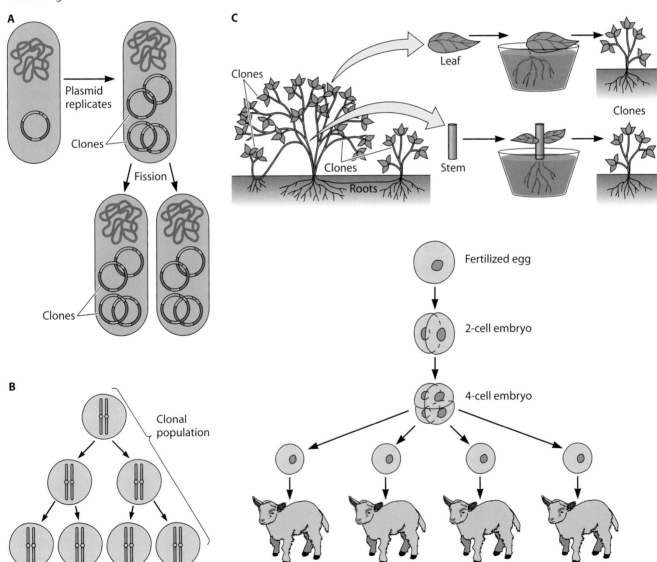

piece of DNA being copied and, by extension, the collection of organisms (usually bacteria) containing the same piece of DNA. In recombinant DNA work, the clone is a recombinant organism, and the piece of DNA is a recombinant molecule.

Cells can also be the cloned entity. Cellular cloning results in cell lines of identical cells. For example, in monoclonal-antibody technology, the immune system B cell that is used to manufacture the antibody of interest is separated from a large population of other B cells. The B cells producing the antibody are clones of each other, as are embryonic stem cells derived from a single fertilized egg.

Finally, sometimes the clone is a multicellular organism. This is the type of clone most nonscientists mean when they use the word. Many plants are propagated through cloning by rooting plant parts taken from adult, fully developed plants. Cloning most animals is more difficult than cloning plants. Except for the simplest animals, such as sponges, scientists cannot simply break off a part of an animal and grow an entirely new organism. In animals, the starting material for cloning a whole organism must be reproductive cells, such as eggs or cells taken from an embryo at the earliest stage of development. The embryo is split into single cells. Each cell that develops into a complete organism is a clone of the others derived from the original embryo.

After reading these different definitions of "clone" and "cloning," it is easy to see why defining terms is essential for an informed discussion. If someone says "clone" and means a mammal that's genetically identical to one of its parents but the listener pictures a petri dish of genetically identical bacteria, misunderstandings will exist from the outset, and the discussion will not be very productive.

For the purposes of this discussion, we will use "clone" to mean a collection of genetically identical multicellular organisms. If not explicitly stated otherwise, assume the organisms are mammals.

What mammalian cloning is not

Cloning mammals is not a matter of creating a copy of an organism instantaneously in the laboratory. If the goal of the work is to produce a whole organism and not simply to maintain cells in culture, the clone must go through normal mammalian gestation and development. The researcher must implant the eggs or embryonic cells into the uterus of a surrogate mother very early in the life of the embryo; then, the organism must go through the gestation period typical for that organism. For example, the embryo that became Dolly was implanted into the uterus of a ewe after having been cultured within sheep oviduct tissue for only 6 days. Dolly was born approximately 6 months later, the typical gestation period for sheep.

A mammalian clone is also not a precise copy of the organism that served as the source of the genetic material. As we discussed earlier, environmental influences, including the uterine environment during gestation in mammals, provide powerful input into the ultimate organism. To assume that clones are perfectly identical copies of one another is to ignore the nurture component of the equation nature + nurture = an organism.

Gather relevant facts

What were some of the facts about Dolly's beginnings?

Dolly was the first animal to be cloned from a differentiated somatic cell taken from an adult animal. Scientists removed 277 cells from the udder of an adult sheep, and then they fused those cells with 277 unfertilized egg cells

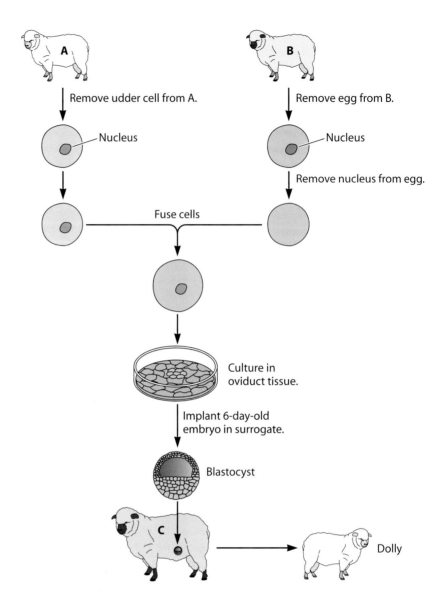

Figure 17.7 Somatic cell nuclear transfer. Dolly was produced by a unique type of cloning. The nucleus in a fully differentiated somatic cell from breed A's udder was inserted into the enucleated egg of breed B by fusing the two cells. The resulting egg contained breed A nuclear genetic material and breed A and B mitochondrial DNAs. The egg developed into a blastocyst in tissue culture, and the blastocyst was inserted into the uterus of sheep C, the surrogate mother of Dolly. Note that Dolly's coloring is identical to that of female A, her genetic mother, and does not exhibit any markings of either her surrogate mother or breed B.

from which the nuclear genetic material had been removed. After culturing the resulting embryos for 6 days, scientists implanted the 29 embryos that appeared to have developed normally into surrogate mothers. Only one produced a live lamb, Dolly, approximately 5.5 months later (Figure 17.7).

What makes Dolly special?

Since the mid-1980s, researchers have used embryonic cells, such as those shown in Figure 17.6, as sources of genetic material to clone sheep, cows, and other mammals. Until Dolly was born, attempts to use adult cells as sources of genetic material for cloning had ended in failure. Scientists assumed that cells containing adult genetic material could not develop into complete organisms because the differentiated cells had become specialized into a certain cell type and could not shed their specific role and "remember" how to give rise to a complete organism. In other words, scientists thought that when certain sections of the cell's DNA "turned off" during cell differentiation, it could not be turned on again. Dolly proved that differentiation was not irreversible and, in being born, challenged a fundamental truth of developmental biology.

That is the scientific achievement that got lost in the hoopla, the scientific breakthrough that made Dolly news: the despecialization of genetic material that had been committed to a special function and its ultimate reprogramming into embryo-like genetic material capable of directing the development of a complete organism. When you read in news reports that scientists are excited about Dolly, it's this scientific discovery that excites them, not the prospect of making genetically identical copies of a multicellular organism. That's old news. Scientists have been doing that for over 50 years.

View cloning in the context of other technological developments: unnatural reproductive technologies

Many people were upset with Dolly's birth because the circumstances surrounding her conception were "not natural." Is this unique to Dolly? No, because unnatural methods of conceiving offspring have been used for over 50 years with animals and 25 years with humans.

Livestock breeding

Animal breeders began to routinely use techniques that interfered with natural methods of reproduction more than 50 years ago. Some of the techniques listed below are not cloning techniques, but all are not natural. Livestock breeders developed these techniques to incorporate desired genetic changes into a herd more quickly. As a result, livestock animals are healthier, and the food derived from them is cheaper and of higher quality (Figure 17.8). In addition, zoos and other conservation organizations use unnatural methods of reproduction to try to save endangered species and to introduce genetic variation into populations.

Artificial insemination. Artificial insemination began to be widely used 45 years ago and is now used routinely in livestock production, thoroughbred horse breeding, and zoo breeding programs. Artificial insemination has allowed U.S. farmers to produce more milk with 10 million cows than they

Figure 17.8 Genetic improvement of livestock. Livestock breeders have incorporated naturally occurring genes that improve productivity, animal health, and product quality into livestock through unnatural reproductive technologies. These technologies increase the speed of genetic improvement of a herd. **(A)** A natural mutation of a gene involved in muscle growth leads to double muscling, which increases the productivity of beef cattle and improves the quality of their meat. **(B)** St. Croix sheep are a hardy breed of sheep that are naturally resistant to certain parasites and have a high tolerance to heat. (Photographs by Keith Weller [A] and U.S. Department of Agriculture [USDA] Agricultural Research Service staff [B], courtesy of the USDA.)

Figure 17.9 Sperm sorter. This machine separates sperm carrying X chromosomes from those with Y chromosomes based on a 4% difference in the amounts of DNA they carry. Scientists use a fluorescent dye that binds to DNA. A laser beam causes the dye to emit light in an amount that is proportional to the amount of DNA in the sperm. (Photograph by Scott Bauer, courtesy of the U.S. Department of Agriculture.)

did in 1945 with 25 million. Semen from the best bulls is frozen and used repeatedly to inseminate cows anywhere in the world. Because of efforts to keep the stud bulls healthy, calves produced through artificial insemination are healthier. In addition, its low cost—approximately $10 per breeding—does not place small farmers at a disadvantage compared to large farmers.

Recently, scientists have developed techniques for separating sperm carrying X or Y chromosomes from each other with 80 to 90% accuracy (Figure 17.9). This allows farmers to predetermine the sex of the offspring, in addition to other economically important genetic traits. Cows on dairy farms are artificially inseminated with semen containing a high percentage of X chromosomes. Beef cattle farmers inseminate cows with a high percentage of Y-chromosome-bearing sperm because male calves grow faster than females.

Multiple-ovulation embryo transfer. Commercial use of **embryo transfer** to genetically improve livestock herds began in the early 1970s (Figure 17.10). To produce livestock by embryo transfer, a valuable female with superior genetic attributes is inseminated after being treated with hormones that stimulate the ovulation of many eggs. The resulting embryos, which are not genetically identical to each other or to either parent, are removed from the mother and placed in surrogate mothers who are less valuable than the genetic mother. The biological mother with superior genetic qualities does not become pregnant, so hormonal treatment can trigger another reproductive cycle. Additional technological advances allow breeders to freeze the embryos and implant them later if the first implantation is aborted. This method of maintaining living tissue is known as **cryopreservation** (Figure 17.11).

Not only has embryo transfer increased the speed with which superior animals can be produced and bred, but it is also an important means for transporting superior genetic material to other countries. Embryos carry fewer diseases than semen or whole organisms, and transportation costs are much lower than for adult animals.

Multiple-ovulation embryo transfer is not a cloning technique.

Figure 17.10 Embryo transfer. In the United States every year, surrogate mothers give birth to 100,000 calves produced from embryos of other, genetically superior cows. Sylvia, the black and white cow, is the biological mother of all of the calves in the photo. Embryos from Sylvia were implanted into the cows on the left, who served as surrogate mothers of Sylvia's calves. (Photograph courtesy of Colorado State University.)

Embryo splitting. **Embryo splitting,** or embryo twinning, has been used in cattle breeding for approximately 20 to 25 years and is similar in concept to embryo transfer. At a very early stage in development, an embryo is split in two, and each half is allowed to develop to the blastocyst stage (Figure 17.12). Then, it is implanted into a surrogate female or cryopreserved.

In this case, the resulting embryos are genetically identical, so embryo splitting is a cloning technique.

Nuclear transfer from embryonic cells. **Nuclear transfer** from embryonic cells is a relatively recent development that bears a resemblance to both embryo splitting and adult cell cloning. It involves taking the nucleus from an embryonic cell of one individual and placing it in the egg (from which the nucleus has been removed [enucleation]) of another individual. In the mid-1980s, a number of research teams successfully produced calves by transferring nuclei

Figure 17.11 Cryopreservation and embryo transfer. **(A)** This sow gave birth to five piglets that had been frozen at the blastocyst stage and surgically implanted in her uterus. **(B)** Prior to implanting cryopreserved embryos, scientists use sophisticated microscopy to ensure the embryos are healthy. (Photographs by Keith Weller, courtesy of the U.S. Department of Agriculture.)

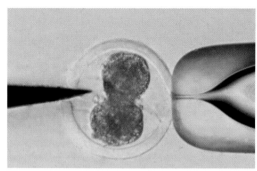

Figure 17.12 Embryo splitting. At early stages of development, scientists use microsurgical blades to separate livestock embryos into separate cells, which they then implant into surrogate females. (Photograph courtesy of George Seidel, Colorado State University.)

from 64-cell-stage embryos into enucleated eggs and implanting these eggs in surrogate females. A variation on this theme involves fusing individual cells taken from a 64-cell-stage embryo with 64 eggs whose nuclei have been removed. These embryos are then implanted into surrogate mothers. This technique is called **nuclear fusion**.

The distinction between nuclear transfer and nuclear fusion, though slight, is relevant to discussions of human reproductive cloning. In nuclear transfer, mitochondrial DNA, which carries approximately 1% of the genetic information, is provided by only one individual, the egg donor. In nuclear fusion, mitochondria from both the embryo cells and the eggs may contribute genetic material to the resulting organisms. Both the nuclear transfer and nuclear fusion techniques are cloning techniques (Figure 17.13).

Figure 17.13 Nuclear transfer and nuclear fusion. In both of these cloning techniques, the nucleus is removed from an unfertilized egg. **(A and B)** Working at a videomicroscope, the scientist holds the egg steady with a pipette **(A)** and then inserts a much smaller pipette into the egg to withdraw the nucleus **(B)**. In nuclear fusion (not shown), the enucleated egg is fused with another cell that still has its nuclear material. **(C)** Nuclei have been removed from other cells, which may be embryonic or fully differentiated somatic cells. **(D)** Each nucleus is injected into an egg cell. Note the nucleus in the pipette in the second photograph. In the final photograph the nucleus is barely visible between the egg cell membrane and cellular cytoplasm. (Photographs courtesy of Roslin Institute, Edinburgh, United Kingdom.)

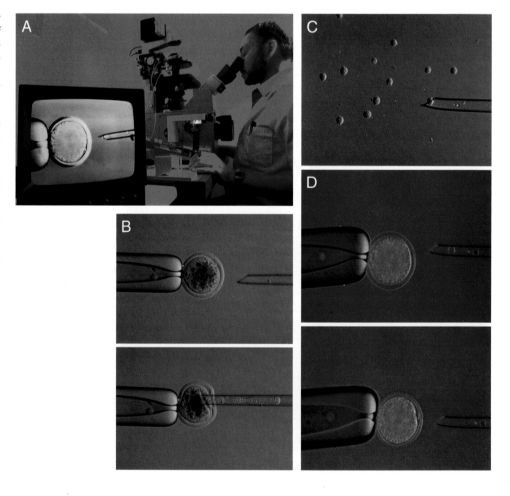

Nuclear transfer from adult cells. The trait that distinguishes nuclear transfer from adult cells from the nuclear transfer technique described above is the source of the genetic material: adult cells versus embryonic cells. Nuclear fusion is also used to transfer adult genetic material into egg cells. This cloning technology was used to produce Dolly and is referred to as **somatic cell nuclear transfer** (Figure 17.6).

Laboratory animals

The application of cloning by nuclear transfer is not unique to mammals. Since the 1950s, developmental biologists have used nuclear transfer from embryonic cells to eggs to study amphibian development.

Humans

In vitro fertilization has been used since 1978 as a method for conceiving children in infertile couples. Eggs and sperm are placed into a test tube, where fertilization occurs. The fertilized eggs are kept in culture for 4 to 5 days until the embryo reaches the blastocyst stage (Figure 17.14). A few of the embryos are then implanted into a female who is usually, but not always, the biological mother. Other embryos are frozen for possible implantation at a later date.

Clearly, in vitro fertilization is not a form of cloning, because genetic material from two sources is combined, creating a genetically unique individual. It is, however, a reproductive technology that is unnatural.

Consider the benefits, risks, and concerns associated with animal cloning

Benefits of animal cloning

Although much media attention has focused on the risks and public concerns about animal cloning, there are a number of benefits.

1. Animal cloning should allow great progress in understanding what turns genes on and off.
2. By using genetically identical animals in experiments, scientists are able to get results more quickly and use fewer animals, because the variation in experimental results due to genetic variation is eliminated.

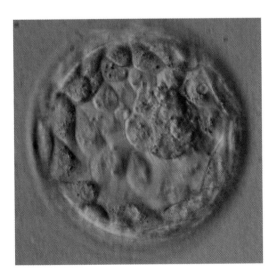

Figure 17.14 Human blastocyst. A fertilized egg develops into a blastocyst in approximately 4 to 5 days whether it is in cell culture or the female reproductive tract. (Photograph courtesy of Michael Vernon, West Virginia University.)

3. Improvements in livestock can be incorporated into herds much more rapidly.

4. Animal cloning could help to save endangered species. Researchers have successfully produced a number of cloned embryos of rare or endangered species, but few have survived.

5. When donor cells are in culture, scientists can introduce desirable genetic changes before the cell is fused with an egg to form the embryonic clone. Therefore, animal clones of consistent quality can be reproduced in large numbers quickly and cheaply. As a result, animal cloning, used in conjunction with recombinant DNA technology, could provide:

   - an alternative and more efficient way of producing transgenic animals that can be used in the production of human therapeutic proteins
   - excellent animal models for studying genetic diseases, aging, and cancer; for discovering new drugs; and for evaluating other forms of therapy, such as gene and cell therapy

Public concerns about cloning

The news of Dolly's birth triggered widespread fear and, in many countries, a call for a total ban on cloning humans. A few have suggested that all animal cloning should be banned. Those who want to ban cloning have justified such bans with vague statements that cloning "poses new ethical problems" or "is contrary to nature." To assess the validity of these claims, you need to determine if cloning does pose new ethical problems and, if so, you should define what they are (see below). Above, we provided information for evaluating the naturalness or unnaturalness of cloning and compared cloning with other widely used unnatural means of producing offspring, both human and nonhuman. If governments ban cloning because it is contrary to nature, then should they consider banning the other unnatural reproductive technologies that couples, farmers, and scientists have used for decades?

Others have sought to ban cloning because of the horrible things they feel could be done if this technology is extended to humans. They have speculated about armies of Hitler clones and brainless cloned duplicates serving as private stocks of replacement parts. Such sensationalist speculation is counterproductive, at best. To assess the costs and benefits of cloning, put concerns in perspective by aligning them with the more immediate and realistic effects of this breakthrough.

Present costs or risks associated with animal cloning research

We will limit our discussion to cloning via somatic cell nuclear transfer.

Using the Dolly research as an example, the success rate for animal cloning is very low (1 in 277). Few implantable embryos result from the fusion of an adult cell with an enucleated egg cell. Of those that are implanted, most have developmental abnormalities and do not survive. Of those that survive, most die early in the pregnancy, and a few die in late pregnancy or soon after birth. Since Dolly's birth, researchers have cloned a number of other species by nuclear transfer from adult cells—mice, cows, goats, and cats—and the success rates have been equally low. The inability to produce healthy clones of endangered species can be explained by these low rates of success.

Scientists are not sure about the long-term health implications of this type of cloning. Because "old" DNA is used to produce the clone, the clone may inherit mutations caused by ultraviolet light or other environmental influences or other genetic problems. This could predispose the clone to premature aging or a higher incidence of environmentally induced cancer.

Other risks related to livestock management may surface. A herd composed of clones would be genetically homogeneous. This lack of genetic variation could result in health problems that affect the entire herd and not just a few individuals. For example, all individuals might be susceptible to the same infectious diseases.

As for ethical issues associated with this research, it is difficult to think of ethical issues that are not associated with animal research in general but are unique to cloning research.

Risks involved in cloning humans

If you are interested in extending this discussion to the question of cloning humans, it is essential to make a clear distinction between cloning cells taken from an embryo that is only a few days old and cloning to make babies (**reproductive cloning**). If the type of cloning being discussed involves implanting an embryo into a woman in hopes of producing a cloned offspring rather than simply growing a mass of undifferentiated mammalian cells in culture, for many people the issues will differ. Problems can arise if that distinction is not clearly delineated.

For example, when the birth of Dolly caused governments to move quickly and ban human cloning, their intent most often was to ban reproductive cloning. However, because they did not consult with scientists before drafting the wording of the laws, some of the laws inadvertently made it illegal to give birth, naturally, to identical twins. Others captured cell cloning, such as monoclonal-antibody production, or other cell culture techniques used in research.

This discussion will focus only on the health risks (not other ethical issues) associated with reproductive cloning via nuclear transfer from adult cells. It is also important to point out that scientists do not know whether human reproductive cloning with adult genetic material taken from differentiated cells is even possible.

Assuming that the problems associated with cloning in sheep would occur in human cloning, then the great majority of the embryos implanted would not survive. For those that survive, environmentally induced mutations and other genetic problems may pose the same risks as they do for animal clones produced through nuclear transfer from adult cells. Even if experience with cloned animals identifies ways of reducing these health risks in animals, is it ethical to perform similar experiments with women to determine whether the risks in humans are significant?

Under what circumstances, if any, might human cloning be acceptable?

Some people believe that there may be situations in the future when nuclear transfer of DNA into an egg cell in hopes of producing a baby could be beneficial.

One example often cited involves people with mitochondrial DNA genetic diseases. Some types of mitochondrial DNA diseases cause blindness and epilepsy. Other evidence suggests that a form of Alzheimer's disease is transmitted via mitochondrial DNA. If the nucleus from an embryo produced by in vitro fertilization is removed and inserted into an egg from a donor, the resulting baby would not have the disorder inherited through mitochondrial DNA, but 99% of its DNA would come from its two parents.

Another example acceptable to some involves cases of infertility that cannot be overcome with in vitro fertilization. The nucleus from an adult cell of one of the parents could be added to an enucleated egg taken from the mother and then implanted. If the donor of the nuclear genetic material is the father, the baby would be the biological offspring of both parents, but 99% of its genes would come from only one parent—the father.

The third example that many believe is an acceptable use of cloning technology does not deal with reproductive cloning but therapeutic cloning. The value of using nuclear transfer from adult cells is that the therapeutic cells would be a perfect match genetically if the nucleus inserted into the enucleated egg was taken from cells of the person receiving the therapy. We will discuss this research and its ethical implications in later chapters.

SUMMARY POINTS

Scientific discoveries can lead to technological innovations. All technologies can be used to benefit people, harm people, or do both simultaneously. It is important to distinguish between problems inherent in a technology and those of human nature.

Certain questions continually resurface in deciding how best to use biotechnology. Where will society draw the line, and why will it draw the line there? How can biotechnology be used to maximize the gains that accrue to the greatest number of people while minimizing its risks and societal costs? Who will be the beneficiaries of biotechnology's potential?

Many factors influence the direction of scientific research, and only certain technologies survive the technology development process. In democratic societies, the public is a potent force shaping scientific research, technology development, and ultimately, society's future. If only some members of society exert pressure on the political system, then scientific research priorities, the nature of the technologies that are developed, and the uses of those technologies will reflect the priorities of special interest groups and not those of society as a whole.

Each individual has a responsibility to attend to biotechnology development to ensure that it is used wisely and safely. However, careful oversight of technology development can sometimes become excessive vigilance that encourages stagnation, and moving cautiously may turn into not moving at all. Sometimes, doing nothing can be costly to a society. To avoid the costs associated with inaction, societies must compare the risks of old technologies and new technologies, weigh the hidden costs of technology against the hidden costs of having no technology, and contrast a technology's risks and societal costs with its societal benefits.

The societal issues raised by advances in biotechnology are often not unique to biotechnology but are the same issues raised by other technologies. However, other issues result solely from the new powers of biotechnology. In charting a course for thoughtful biotechnology development, societies should avoid blaming biotechnology for mistakes of earlier technologies that may or may not apply to biotechnology. Wise decision making is informed by current understanding, as well as past mistakes.

Analyzing the issues of biotechnology within the continuum of technological change provides essential information to help societies and individuals decide how best to use the extraordinary capability biotechnology provides.

Critical analysis of biotechnology applications requires clarity, specificity, and emotional detachment. In discussing biotechnology, participants must define terms and delineate concerns as precisely as possible.

Cloning provides an excellent example of the merits of viewing technological breakthroughs in their historical context, defining terms, gathering facts to delineate specific issues unique to the new development, minimizing the contribution of emotion to a discussion, and recognizing that any technology, no matter how ethically troublesome, also has benefits.

KEY TERMS

Cryopreservation	**Emerging infectious**	**Nuclear transfer**	**Somatic cell nuclear**
Embryo splitting	**disease**	**Reproductive cloning**	**transfer**
Embryo transfer	**Nuclear fusion**		

chapter 18 | Risks and Regulations

Day in and day out, consciously or unconsciously, you take risks. You ride in cars, cross streets, and play sports without thinking twice about the risks involved, because you have "decided," usually without thinking, that the benefits outweigh the risks. Sometimes your choices make good sense and can be supported with factual scientific evidence, but often the level of risk determined through an objective, science-based evaluation and the amount of perceived risk bear little resemblance to each other. People accept huge risks and worry about trivial ones. Cigarette smokers protest nuclear power plants, and people risk pregnancy and AIDS by having unprotected sex and then worry about pesticide residues on their food, yet smoking and unprotected sex have harmed millions more people than accidents at nuclear power plants or pesticides on food.

Choices about acceptable and unacceptable risks are irrational because they are often based on emotion, not facts. As a result, some fears are real, but others are not. In this chapter we:

- compare emotion-based risk perception to science-based risk assessments
- provide an example of the pitfalls of making decisions based on emotions and not facts
- discuss the role that government regulatory agencies play in minimizing the potential risks associated with new products, including those developed using biotechnology

RISK PERCEPTION

Many citizens in industrialized countries seem to suffer great anxiety about the risks involved in simply living. Why is a large segment of society so anxious about risks, when factual evidence shows that people in the industrialized

443

world are living longer, healthier lives in the cleanest environments they have experienced for centuries?

A rational explanation is that new risks do arise. A number of emerging infectious diseases, such as human immunodeficiency virus (HIV) infection-AIDS, Ebola fever, and West Nile virus, have only now begun to cause human health problems. Another rational reason is related to scientific advances that illuminate a risk that has always been there but was previously unknown. For example, viruses transmitted through sexual intercourse were a causative agent of cervical cancer centuries before scientists knew viruses existed. Diets high in salt and low in fiber were contributing to health problems long before physicians established the links.

However, much of the current angst gripping many people is irrational. Society's misplaced anxiety over trivial or nonexistent risks is due to a variety of factors.

- As you learned in the last chapter, many real, high-risk problems have been solved (for now, at least) in the industrialized world. For millennia, microorganisms posed the most serious risks to human health through infectious diseases and microbial contamination of food. Refrigeration, better sanitation, and government-mandated food inspection eased the latter problem, and sewage treatment, antibiotics, and vaccination lessened the former.

- The ability to measure substances that occur in small amounts has improved by many orders of magnitude. People seem to mistake the mere presence of a substance with its level of risk, but measuring techniques can detect quantities so small that they have no real significance for health and safety (Box 18.1). Were vegetables safer before scientists learned that many of them naturally contain small amounts of toxins?

- Risks are much more publicized. Many people are preoccupied with knowing about anything that might possibly cause them harm, and the media indulge their obsession. At times, some reporters may slant the story by presenting the risks in ways that trigger concerns, often unnecessarily. It's not that they lie; they can tell the truth and

---

BOX 18.1 *How much is a billion?*

People living today have become accustomed to hearing the media mention numbers that are so tremendous the human mind is incapable of visualizing them in the abstract: a budget deficit of trillions of dollars, a trade imbalance of billions, 40 million people suffering from AIDS. Putting large numbers in a familiar context can help. A billion seconds ago, World War II ended. A billion minutes ago, St. Paul was writing the Epistles.

Visualizing incredibly small numbers in the abstract presents problems as well. Part of the difficulty in accurately evaluating statements about the risks of substances found in food, water, and air stems from the extraordinary capacity to measure chemical compounds. Highly sensitive methods for detecting and measuring substances can identify quantities so miniscule they have no real significance for health and safety. The quantities are expressed as parts per million, parts per billion, and more recently, parts per trillion. What do these values really mean?

What is one part per million?

1 cent in $10,000

1 ounce in 62,500 pounds

1 inch in 28,000 yards (280 football fields)

1 pancake in a stack 4 miles high

1 drop in 1,000 quarts of water

How much is one part per billion?

1 inch in 16,000 miles

1 minute in 2,000 years

1 cent in $10 million

1 drop in 1 million quarts of water

1 soybean in a silo 50 feet tall and 30 feet wide

How much is one part per trillion?

1 inch in 16,000,000 miles

1 second in 32,000 years

1 needle in a 100,000-ton haystack

mislead simultaneously. For example, a few years ago, major newspapers carried a front-page story about the relationship between hormone replacement therapy and a number of diseases, including breast cancer. All reported that a 10-year study of 20,000 post-menopausal women proved that those who took hormones were 25% more likely to get breast cancer than those who did not. These numbers refer to the **relative risk**, because they compare post-menopausal women who use hormones and those who do not. Few of the media reports provided the actual numbers, or the **absolute risk**. Of the 10,000 women taking hormones, 38 got breast cancer (0.38%); of the 10,000 who did not take hormones, 30 got breast cancer (0.30%). Therefore, approximately one additional woman per year (eight in 10 years) in the hormone replacement group contracted breast cancer. None of the media reports lied, because 38 *is* 25% more than 30. But reporting the actual numbers—0.38% compared to 0.30%, or one more per year—does not sound as scary, does it?

While some believe in the motto "better safe than sorry," being overly cautious is often not in your best interest. Unnecessary fears encourage people to fret over perceived but nonexistent problems, and stress is harmful to health (Figure 18.1). Risk misperceptions also lead people to avoid products that are perfectly safe and spend more money on products with no additional benefits. If a significant proportion of the population or a very vocal minority is unnecessarily fearful of trivial risks, new and improved products are not developed, and government agencies spend their limited resources protecting fearful consumers from small or nonexistent risks. In either case, society as a whole suffers.

What factors skew risk perceptions?

What triggers emotional responses and prevents accurate assessments of real risks? Social scientists have identified a number of psychological elements that skew perceptions of risk, causing people to inflate small risks or trivialize significant risks.

- Voluntary versus involuntary. If people consciously decide to take a risk (a voluntary risk, like smoking), they accept a much higher level of risk than if they feel they had no choice in the matter (an involuntary risk, such as the location of a nuclear power plant).
- Control versus no control. People are more fearful of risky situations over which they have no control. Many people choose to ride in their cars rather than fly in airplanes because they fear flying, even though more people are hurt or killed in auto accidents than in plane crashes. Despite the evidence, people perceive the risks of flying as greater than the risks of driving because they are not in control of the plane.
- Familiar versus unfamiliar. Risk perceptions also vary with the degree of familiarity with the risk. People accept the large but well-known risks involved in driving a car, sunbathing, and drinking alcohol. New risks elicit concern and avoidance, not because the risks are greater but because they are unfamiliar.
- Natural versus synthetic. Another factor affecting risk perceptions is whether the risk is natural or synthetic, i.e., man-made. People tend to view nature as benevolent, despite all the evidence to the contrary.

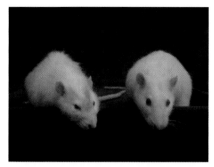

Figure 18.1 Physiological effects of fearfulness. Rats that are chronically fearful age prematurely and die 15% sooner than those that are not. These brother rats are the same age, but the fearful rat on the left has a pituitary gland tumor. Researchers placed novel neutral objects, such as a bowl or brick, in their environment. The fearful rats avoided the objects, became anxious, and maintained chronically high levels of stress hormones after researchers removed the objects. The bold rats investigated the objects, and the surge of stress hormones returned to normal rapidly. (Photograph courtesy of Sonia Cagivelli, University of Chicago.)

Microbes are natural, yet millions of people in developing countries die from diarrhea caused by microbial contamination of water. In developed countries, synthetic products and processes, such as chlorinated water and sewage treatment, prevent this same tragedy from occurring.

Irrational perceptions of the relative risks of natural and synthetic products and processes deserve additional attention, because objectively analyzing the risks of some biotechnology applications requires clearly understanding the pitfalls of equating natural with risk free and synthetic with risk.

Comparing risks: natural versus synthetic

Some of the most instructive examples of problems caused by an unthinking preference for natural products over synthetic products come from issues surrounding food safety. Little has been done to help today's consumer place the risks of synthetic versus natural products in perspective. As a result, many people have become increasingly anxious about the safety of a food supply that is safer than at any other time in history.

Below, we provide a rational discussion of food safety to demonstrate how misleading the reflexive assumption "natural is good; synthetic is bad" can be. We hope to enrich your understanding of a scientific discipline, toxicology, and in doing so, to provide you with factual information for making informed choices about certain food safety issues. Before we begin, however, we offer an observation from our teaching experience. Even though the information we share is scientific data and our intention is purely educational, often when we present this information, some students become very angry. As scientists, we do not understand why sharing data has this effect, but it has happened so often that we know that the emotions triggered by the data are not unique to a few people. Even though we are uncertain why this happens, we find it very interesting. If nothing else, it makes clear the emotional connection many people have with food and reinforces the basic premise that emotions affect risk perceptions.

Chemical toxins and crop plants

When you read that heading, the first word that popped into your head was probably "pesticide," which is accurate and appropriate. However, like most people, you probably think of pesticides as synthetic chemicals marketed by companies and sold to farmers, which is only partially accurate. Long before companies began manufacturing pesticidal chemicals synthetically, plants were making them naturally. Most plants naturally contain chemicals that are toxic when tested on laboratory animals; some plant toxins naturally occur at levels high enough to seriously harm grazing livestock (Figure 18.2).

Plants did not evolve to serve as food for humans, livestock, or any other organism. For millions of years, they have done their very best *not* to be eaten by making hundreds of different chemicals, known as **secondary plant compounds**, to ward off would-be predators. Using breeding and mutagenesis, plant scientists have decreased the amount of natural toxins in most food crops. Without their efforts, many of the plants Mother Nature cooked up would not be fit for human consumption. However, despite years of work to strip food crops of these chemicals, many crop plants still contain small amounts of chemical toxins (Table 18.1). According to Bruce Ames, the biochemist who developed the most widely used toxicology test for measuring the **carcinogenic** (cancer-causing) potential of chemicals, in any given meal,

Figure 18.2 Natural plant toxins. Beautiful wildflowers often contain chemicals to deter herbivory. In the western states, wild lupines (*Lupinus* spp.) **(A)** and larkspurs (*Delphinium* spp.) **(B)** inflict significant losses on free-ranging livestock. Both are highly palatable to livestock and sheep, even though they contain alkaloids, which are neuromuscular poisons that lead to death from respiratory failure. Grazing sheep can die from consuming as little as 4 ounces of lupine each day for 3 days. (Photographs courtesy of William Cielsa, Forest Health Management International [http://www.forestry images.org] [A], and by Jack Dykinga, courtesy of the Agricultural Research Service, USDA [B].)

an individual consumes about 100 to 150 natural carcinogens and 10,000 times more natural carcinogens than synthetic carcinogens.

Does this mean you should stop eating plants that contain natural toxins? Of course not. You can see from the crops listed in Table 18.1 that some of the healthiest foods also contain natural toxins. The benefits of eating grains, fruits, and vegetables far outweigh the negligible to nonexistent risks posed

Table 18.1 Natural toxins[a]

Chemical	Plant examples	Physiological effects
Cyanogenic glycosides	Almond, cherry, lima bean	Chewing plant material releases hydrogen cyanide, which causes death by blocking cell respiration.
Glucosinolates	Broccoli, cabbage, peanut, soybean, onion	Inhibit thyroid function, leading to an enlarged thyroid gland, or goiter. Low levels of thyroid hormone inhibit growth and reproduction.
Glycoalkaloids	Potato, tomato	Interfere with nervous system function, leading to nausea, vomiting, difficulty breathing, and death; also cause birth defects.
Lectins	Most cereals, beans, potato	Stimulate mitosis, red blood cell agglutination, decreased nutrient uptake or absorption.
Oxalic acid	Spinach, rhubarb, tomato	Reduces availability of essential minerals, such as calcium and iron.
Phenols	Most fruits and vegetables, cereals, soybean, potato, tea, coffee	Destroys thiamine; raises cholesterol; estrogen mimic.
Coumarins	Celery, parsley, parsnip, fig	Light-activated carcinogens; skin irritation.

[a]Chemical toxins and carcinogens occur at low levels in most food crops. To assess the physiological effects and determine the toxic and carcinogenic doses, toxicologists fed laboratory animals the chemicals in amounts that are much higher than natural levels in plants. This table contains only a few examples of the chemical families known to be toxic or carcinogenic.

Table 18.2 Carcinogenic substances[a]

Substance	Carcinogenic potential
Red wine	5.0
Beer	3.0
Edible mushrooms	0.1
Peanut butter	0.03
Chlorinated water	0.001
Polychlorinated biphenyls (PCBs)	0.0002

[a]The higher the number, the greater is the cancer-causing potential. The carcinogenic potential of peanut butter is due to the toxin aflatoxin, produced by a mold that commonly infects peanuts and other crops.

by the naturally occurring chemicals they contain. Most plant toxins occur in concentrations that are so low they have no effect on human health, unless you eat large amounts of a single fruit or vegetable every day. The issue is not the safety of fruits and vegetables but the public's concept of toxicity.

Toxicity is the capacity of a substance to do harm or injury of any kind. In reality, *everything* is toxic if given in large enough amounts. Adults have killed children by forcing them to drink too much water, but you would never think of water as a toxic substance. No one has said it better than Paracelsus in the 16th century: "All substances are poisons, there is none which is not a poison. The dose differentiates a poison and a remedy." This is a fundamental principle in modern toxicology: the dose makes the poison. While you probably understand this for certain things, such as medicines or even vitamins, thinking of water, oxygen, and glucose as poisons or toxins is difficult, isn't it?

Many people are frightened by the use of synthetic chemicals on food crops because they have heard that these chemicals are toxic and cancer causing. But are all synthetic chemicals more harmful than substances we readily ingest, like coffee and soft drinks, not to mention alcohol? No (Table 18.2). For example, in a study to assess the toxicities of various compounds, half the rats died when given 233 mg of caffeine per kg of body weight, but it took more than 10 times that amount of glyphosate—the active ingredient in the herbicide Roundup—to cause the same percentage of deaths (4,500 mg of glyphosate/kg of body weight).

Does that mean that all synthetic chemicals are so safe they can be ignored as health risks? No. Our point is simply to demonstrate that it is in your best interest to keep issues of chemical toxicology and food safety in perspective. Things are never as simple as people would like them to be.

RISK ANALYSIS

The factors driving your concept of risk—emotion or fact—may or may not seem particularly important to you, yet they are. The risks you are willing to assume and the experiences or products you avoid because of faulty assumptions and misinformation affect the quality of your life and the lives of those around you. So even though it may be tempting to let misperceptions and emotions shape your ideas about risky products and activities, there are risks in misperceiving risks.

Fact-based risk analysis provides perspective

Risk is the *probability* of loss or injury. The science of risk analysis was developed in the 1970s in response to a new set of congressional acts giving federal regulatory authorities increased oversight and responsibility for protecting consumers and the environment. These laws assign different responsibilities and authorities to the various federal and state agencies charged with evaluating the risks of products and processes. The regulatory agencies can dictate how certain products can and cannot be used after they are commercialized. They also have the power to withdraw or recall products if circumstances later show that the risks were greater than expected. Regulatory agencies are held accountable for the responsibilities that are assigned to them by state and federal legislation. On the other hand, these agencies cannot exceed the authority that the legislation grants to them.

A formal **risk assessment**, which is a component of risk analysis, attempts to maximize the contributions of factual information and scientific data to regulatory determinations of risk. Regulators rely on empirical knowledge and scientific data and use methodical, objective analysis to establish a level of risk associated with a specific activity or product. Risk assessments are often quantitative evaluations, but we will skip the mathematics and limit our discussion to the conceptual aspects. Even though science and math are involved, because a risk is a probability statement, there are no absolute answers to questions about risk levels. This uncertainty often frustrates nonscientists.

Risk = hazard × exposure

The first step in a risk assessment is identifying the **hazard,** which is any thing that could go wrong or might lead to injury or harm. It is important to mentally distinguish hazard from risk. A hazard is the potential to cause harm and is similar to the concept of toxic described above. Identifying a possible hazard is not equivalent to identifying the risk level any more than saying water can be toxic means you should not drink water. Just as the probability of harm depends on the dose of a toxin, risk depends on both the hazard and exposure to the hazard. The risk (probability of loss or injury) increases as the severity of the hazard and/or exposure increases.

Exposure is a statement of the probability that there will be contact between the hazard and the thing the hazard might harm or injure. The uncertainty of exposure is one of the reasons risks are discussed in terms of probabilities instead of an absolute assessment of the seriousness of the hazard. In other words, a hazard, which is the potential for something to cause harm, becomes a risk only if there is exposure. Swimming pools are inherently hazardous but are not a risk to someone who has never been near one.

Risk management

The level of risk varies with the severity of the consequences of contacting the hazard and the probability that contact will occur. However, if effective safeguards can be established, the level of risk can be decreased dramatically, because:

$$\text{Risk} = \frac{\text{hazard} \times \text{exposure}}{\text{safeguards}}$$

Risk management is the set of activities that can be undertaken to control a hazard, minimize exposure, or both.

These two conceptual formulas illustrate the first steps in using rational risk analysis to assess and manage risks. By no means are these steps perfect or devoid of subjectivity, but they provide a starting point for comparing the relative risks of various products and activities.

You may ask why regulators worry about all of this. Why don't they overestimate risk to make sure they are fulfilling their responsibilities to protect the public and the environment? As the example of decaffeinated coffee (Box 18.2) demonstrates, responding to risks that don't exist can be costly, literally and figuratively, for people. Misperceptions about risks can limit your options and increase the money you spend. They also encourage the creation of an unnecessarily prohibitive regulatory process that impedes the development of beneficial products and also raises the costs of goods that obtain regulatory approval. It is possible to be overly cautious. For a ranking of the risks of familiar activities and products, see Table 18.3.

BOX 18.2 *The case of decaffeinated coffee*

In the mid-1980s, much attention was given to the use of methylene chloride to decaffeinate coffee. Members of the media reported that methylene chloride was a carcinogen in rats. They neglected to mention that it is a carcinogen when inhaled but has a much lower carcinogenic potential when ingested. This distinction is very relevant, because in the case of decaffeinated coffee, the only variable coffee drinkers need be concerned with is the carcinogenic potential when ingested. Nonetheless, in response to consumer fears, some manufacturers began decaffeinating coffee by a different process, and the price increased.

Coffee itself has over 300 chemical compounds, many of which are more toxic and more carcinogenic than methylene chloride. To ingest enough methylene chloride to reach the value shown to cause cancer in rats, a person would have to drink 50,000 cups of coffee a day. In those 50,000 cups, coffee's natural toxins and carcinogens would occur in far greater amounts than methylene chloride. More important, a person would die from drinking 50,000 cups of water a day. Long before the 50,000th cup was consumed, a person's kidneys would shut down.

The message is not that synthetic is safe and natural is unsafe. The situation is not that simple. The point of the story is that everyone works from misconceptions derived from inaccurate or misleading information. Unfortunately, misinformation made worse by emotionally driven evaluations of risk doesn't yield smart choices very often.

Compare the risks and the benefits

Regulators can institute safeguards to get the risk level as low as possible, but it will never reach zero, because there is no such thing as risk free. The essential question to be answered is whether the level of risk is acceptable compared to the benefits.

Table 18.3 Relative risks[a]

Risk	Source[b]
0.2[c]	PCB in diet
0.3[c]	DDT in diet
1.0[c]	1 quart of city water/day
8.0[c]	Swimming 1 hour/day in a chlorinated pool
18	Chance of dying by electrocution in a year
30[c]	2 tablespoons of peanut butter/day
60[c]	12 ounces of diet cola/day
100[c]	3/4 teaspoon of basil/day
367	Accidents in the home
600[c]	Indoor air that contains formaldehyde vapors from furniture
667	Respiratory illness caused by air pollution in the eastern United States
800	Chance of dying in an auto accident in 1 year
2,800[c]	12 ounces of beer/day
12,000[c]	One pack of cigarettes/day
16,000[c]	One tablet of phenobarbital/day

[a]The risks are listed from lowest to highest. Drinking a quart of city water, which contains chloroform as a by-product of chlorination, serves as a reference point. For example, the risk of eating three-fourths of a teaspoon of basil a day is 100 times greater than the risk in drinking a quart of city water a day. (Source, B. Ames and R. Wilson, *Science* **236**:267–279, 1987.)
[b]PCB, polychlorinated biphenyls; DDT, dichlorodiphenyltrichloroethane, a pesticide.
[c]Lifetime risk of getting cancer.

Assessing benefits

Assessing risks is a simple task compared to assessing benefits. Benefits are often very difficult to identify in the abstract. People need to experience them. Beyond the obvious and intended benefits of electric lights over gaslights, societies would never have foretold the unintended, direct effects of electrification of homes: electric washing machines, hot-water heaters, air conditioners, refrigerators, electric guitars, personal computers, and DVDs. Nor could they have predicted the secondary social impacts of the unintended effects: homes in the desert and frozen north, the Internet, home shopping, more free time, the disappearance of household servants, more women in the workforce, and as a result, a need for day care centers.

Are these unintended effects of electrification, whether direct or indirect, benefits? It depends on whom you ask. Benefits (and risks) are not distributed evenly throughout society. In addition to being difficult to predict and inequitable, many benefits are very subjective. They are based on emotions, values, and ethical principles and therefore vary from person to person and circumstance to circumstance. Some of the factors people use to measure benefits are:

- saving lives
- improving health
- solving problems
- improving quality of life
- increasing emotional well-being
- saving money

Putting risks and benefits together

Defining and assessing the risk level and then comparing it with the perceived benefits are components of a thoughtful, methodical analysis for decision making. If you use this methodology in an informal way, rather than evaluating risks solely based on emotions, you will be in a much better position to make judicious choices on issues involving the risks you are willing to take. However, once you have conducted our own risk-benefit analysis and decided on the acceptability of a risk, the person next to you may have arrived at a different conclusion. You may think that if the decision involves an individual's acceptance or rejection of personal risk, differences of opinion on the acceptability of risks should not present a problem. In other words, if someone wants to accept the risks of riding a motorcycle without a helmet or driving drunk, it's their business. Once again, it's not that simple. When does someone's personal decision about a personal risk they are willing to take infringe upon the rights of others?

For example, people who drive drunk may feel comfortable with their risk-benefit analysis for this behavior, but drivers sharing the road may not. Given that example, most people would probably come to similar conclusions regarding individual rights versus the rights of other members of society. Driving on public roads is a public matter made possible with public money. An individual's right to accept the risk of driving drunk is secondary compared with the public's right to use roads without fear of drunken drivers.

What about private matters, like safe sex and AIDS? If people risk HIV infection by having unprotected sex, isn't that their business and only their business? How can that choice infringe on the rights of others? This may not be as simple and straightforward as it first appears. As the cost of providing care for a growing population of HIV-infected and AIDS patients increases, everyone's health insurance costs increase. In addition to the cost increase in health insurance that each policyholder experiences directly, society will pay the price in ways that seem completely unrelated to AIDS. For example, many companies pay for their employee's health insurance. As health insurance costs increase and employers pay more money to insure their workers, they must compensate for increased costs by decreasing their costs or increasing their income. Cutting jobs is one way to cut costs. To increase income, companies usually raise the prices they charge for goods and services. So, an increase in health insurance costs could mean fewer jobs and/or higher prices. Should everyone pay the price for those who have chosen to have unprotected sex and, as a result, have been infected with HIV?

Issues of individual rights are never as simple as we would like. As members of society, all people have both rights and responsibilities to themselves and each other. Finding the proper balance between the two can be very, very difficult.

A CASE STUDY: MONARCH BUTTERFLIES AND Bt CORN

In May 1999, a story hit the front pages of major newspapers and popular magazines and got the attention of nature lovers around the world. Researchers at Cornell University had shown that corn pollen from plants genetically engineered to contain a gene from the bacterium *Bacillus thuringiensis* (Bt) killed monarch butterflies (Figure 18.3). Attention-getting headlines, such as "Attack of the Killer Corn," "Genetic Engineering Leads to Toxic Pollen," and "Nature at Risk," triggered strong emotional responses from the public that, in turn, led to a variety of political reactions. Some activists called for a worldwide ban on agricultural biotechnology.

The media coverage left readers with the misimpression that the Cornell scientists had made a new discovery that had been overlooked by the companies that developed **Bt corn** or government regulators who assessed its ecological risks. The companies, agricultural scientists around the world, and regulatory officials were not the least bit surprised by the findings of the Cornell study, which only reconfirmed a discovery made in 1918 and applied commercially for decades. Bt-based insecticidal sprays, which are toxic to only certain insect pests, have been used to control caterpillar pests in crops and forests since 1938. In addition, the Bt gene, the protein it encodes, and Bt corn were thoroughly reviewed and tested before commercialization. The results confirmed that Bt corn and its pesticidal protein are not harmful to most insects or other animals but provide an effective means of controlling caterpillar pests that feed on corn.

Nonetheless, special interest groups pressured U.S. regulatory agencies to reverse their decision and remove Bt corn from the market. The European Union's regulatory commission, which decides which crops can be grown and imported, halted the regulatory approval process for Bt corn varieties under review. They were also reluctant to allow farmers to plant the Bt corn varieties they had already approved. In response to media attention and political pressure, U.S. government agencies and the companies that marketed Bt corn provided university scientists with funding for additional research. Their results confirmed the regulators' original decision that Bt corn would pose minimal, if any, risks to monarch butterflies. Even though these studies did not receive much media coverage, especially compared to the Cornell study, the uproar eventually died down.

Although the controversy has blown over, we know of no better story for illuminating a number of points about emotions and risk assessment. If regulators had made a decision to ban Bt corn in response to political pressure and not real risks to monarch butterflies, who or what would have benefited:

Figure 18.3 Monarch butterfly. (Photograph courtesy of Jennifer E. Dacey, University of Rhode Island [http://www.forestryimages.org].)

monarch butterflies, other insects, farmers, the environment, the public, environmental organizations, or other special interest groups? Did the regulators make the correct decision when they approved Bt corn for commercial use? An unemotional, fact-based risk-benefit analysis can help provide the answers. The monarch butterfly-Bt corn story also provides another opportunity to practice the process for analyzing issues described in the previous chapter.

Gather the relevant facts

In order to analyze the risks to monarch butterflies associated with planting Bt corn and to evaluate the validity of the regulators' decision to approve large-scale planting of Bt corn, you need to gather basic background information on a variety of topics: the natural history of monarch butterflies, use of Bt-based pesticidal sprays, Bt corn, conventional corn, agricultural practices that U.S. corn growers use, and the regulatory approval process.

Facts about monarch butterflies

Monarch butterflies are members of the insect order Lepidoptera, which includes butterflies and moths. Like all lepidopterans, monarchs undergo **complete metamorphosis**, in which there are four distinct stages: egg (embryo), larva (caterpillar or "worm"), pupa (chrysalis or cocoon), and reproductive adult (butterfly). The development from egg to adult takes about 1 month in monarchs (Figure 18.4). In North America each year, there may be as many

Figure 18.4 Monarch life cycle. **(A)** Larvae emerge from eggs and immediately begin to feed on milkweed plant tissues. Like all lepidopteran larvae, monarch larvae molt a number of times as they grow. **(B)** When the larva reaches a certain size, hormonal changes trigger the formation of the pupa. **(C)** During the pupal stage, the adult butterfly forms. **(D)** Adult butterflies feed on nectar through highly specialized mouthparts. (Photographs by Peggy Greb, courtesy of the Agricultural Research Service, USDA [A]; Herbert A. "Joe" Pase III, Texas Forest Service [http://www.forestryimages.org] [B]; Peter Wirtz [http://www.forestryimages.org] [C]; and John Mosesso, courtesy of the National Biological Information Infrastructure of the U.S. Geological Survey [D].)

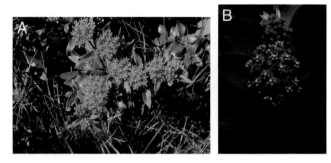

Figure 18.5 Milkweed. There are approximately 100 species of milkweed plants in North America, and most prefer to grow in open habitats, such as along roadsides. **(A)** Butterfly milkweed (*Asclepias tuberosum*) is common in the southern United States. **(B)** Common milkweed (*Asclepias syriaca*) is the monarch's preferred host plant in the area where almost all Bt corn is grown. (Photographs courtesy of David Stephens [http://www.forestryimages.org] [A] and Arnold Drooz, U.S. Fish and Wildlife Service [http://www.forestryimages.org] [B].)

as four to six generations of monarchs in the south and one to three generations in the northern states and southern Canada. Monarch butterflies also live throughout Central America and most of South America, in Australia, and in Hawaii and other Pacific islands.

Like some other lepidopterans, monarchs are **specialist feeders**, because the larvae will eat only leaves and flowers from a small set of plants, the milkweeds (Figure 18.5). The plant toxins in milkweeds make monarch larvae and adults unpalatable to many predators. Only the larval stage can feed on plant tissues, because butterflies have specialized mouthparts for sucking nectar from flowers. The adults prefer to drink nectar from milkweed flowers, but if they emerge from cocoons before milkweeds bloom, they will take nectar from a variety of flowers.

In North America, the adult (butterfly) stage of most generations lasts only 2 to 6 weeks (Figure 18.6). The butterfly is the only stage capable of reproduction. After mating, most females lay eggs, but only on milkweed plants, and then die. However, the last generation of adults migrates south for the winter, where they aggregate in trees. Millions of butterflies overwinter in huge aggregations for 6 to 9 months before leaving the roost in late February and beginning their migration north (Figure 18.7). As they migrate, they lay eggs on the newly emerging milkweed plants and then die. The monarchs migrating northward from Mexico usually arrive in the middle portion of Texas

Figure 18.6 Monarch life history. Monarchs in most generations in North America live approximately 2 months. The first month consists of immature stages that are not reproductive. The last generation of a summer overwinters and gives rise to the next year's first generation.

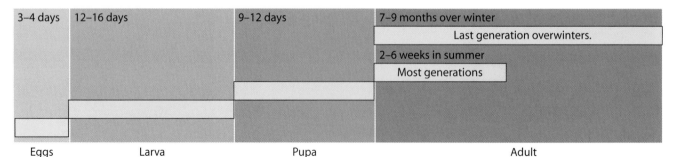

3–4 days	12–16 days	9–12 days	7–9 months over winter
			Last generation overwinters.
			2–6 weeks in summer
			Most generations
Eggs	Larva	Pupa	Adult

Figure 18.7 Monarch roost. In North America, the last generation of adults migrates to a warm location, where they overwinter in large aggregations. The weight of the butterflies is significant enough to bend the branches of mature fir trees. (Photographs courtesy of Harry O. Yates, U.S. Forest Service [http://www.forestry images.org].)

and the Gulf states in mid- to late March, where they lay eggs that produce the first new generation of adults by the end of April. Essentially tracking the emerging milkweed populations, these new adults migrate north and east, laying eggs along the way and establishing a large second generation.

There are two distinct populations of monarchs in North America: an eastern population that overwinters in Mexico and lives and breeds east of the Rocky Mountains and a western population that overwinters in California and lives and breeds west of the Rocky Mountains (Figure 18.8). The two populations overlap slightly, but interbreeding between the two populations is much rarer than within a population. Under such conditions, scientists would expect to see genetic differences in the two populations. Surprisingly, mitochondrial DNA studies reveal virtually no genetic variation, either between the eastern and western populations or within either population.

Figure 18.8 Monarch migration routes. At the end of the summer, monarch butterflies from North America migrate to overwintering sites in California and Mexico.

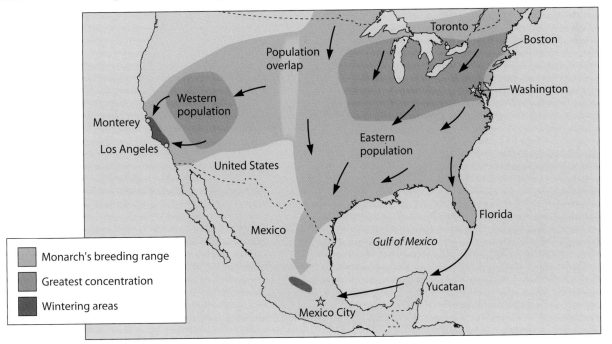

Facts about Bt

Bt is a naturally occurring bacterium found in soils in forests, savannahs, deserts, and cropland all over the world. It also occurs in enclosed insect-rich environments, such as grain storage facilities, and Bt has been found on tree leaves in some environments.

First commercialized as a microbial pesticide in France in 1938, Bt was approved by U.S. regulators for sale and use in 1961. Bt produces large amounts of a protein that, when ingested by certain insects, slows development, and it ultimately kills insects that consume enough of it (Figure 18.9). Unlike most pesticides used in commercial agriculture, the protein must be ingested to be toxic. Bt is harmful to a limited number of insect species but nontoxic to other insects, birds, mammals, and other organisms. Because of its selective toxicity, Bt does not harm beneficial insects, such as pollinators and insects that prey on caterpillars. Valued for this targeted pesticidal activity and environmental compatibility, Bt insecticidal sprays have been used by organic growers and home gardeners for decades, and foresters use them to try to control outbreaks of gypsy moths and other lepidopteran pests that feed on tree foliage.

Figure 18.9 Bt protein. The toxins in Bt are proteins that form crystals under certain conditions. **(A)** The Bt toxin genes vary slightly within each strain of the bacterium, as do the shapes of the crystallized proteins (Cry1A) they encode. **(B)** The mechanism of action of Bt protein. (Step 1) An inactive form of the crystallized Bt protein enters the insect's gut. (Step 2) An enzyme produced by the insect binds to the Bt protein crystal and activates it. (Step 3) The activated form binds to receptors on the membranes of the cells in the intestinal lining. (Step 4) When the activated Bt protein binds to the receptor, the receptor's shape changes and the ionic contents of the cells lining the insect's gut spill into the intestinal lumen. The shapes of the Bt crystal, enzyme active site, and receptor-binding site are responsible for the specificity of the Bt toxins.

A

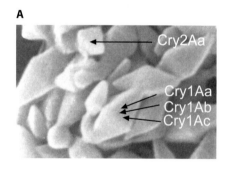

B

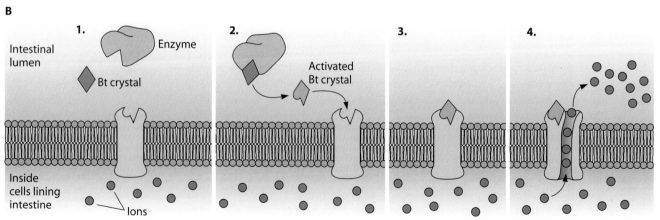

Table 18.4 Specificities of Bt subspecies^a

Bt subspecies	Insect order	Target pests
B. thuringiensis subsp. *kurstaki*	Lepidoptera (butterflies and moths)	Gypsy moths
		ECB
B. thuringiensis subsp. *tenebrionis*	Coleoptera (beetles)	Colorado potato beetle
B. thuringiensis subsp. *israelensis*	Diptera (flies and mosquitoes)	Black fly
		Yellow mosquito

^{a}The bacterium Bt exhibits targeted insecticidal effects. Different subspecies of Bt are selectively toxic to certain insect orders and are nontoxic to insects in all other orders. This selectivity has made Bt sprays valuable biopesticides for controlling certain targeted pests for more than 50 years. Genetic variation exists within each subspecies. In the case of Bt corn, the target organisms are lepidopteran caterpillars that are corn pests. Therefore, scientists isolated various Bt toxin genes from strains of *B. thuringiensis* subsp. *kurstaki* and used recombinant DNA techniques to produce Bt corn varieties.

Until the 1970s, scientists thought Bt was toxic only to lepidopteran caterpillars, but they have now identified different strains of Bt that exhibit equally selective toxicity to beetles, flies, and mosquitoes (Table 18.4). These strains are also used as biopesticidal sprays to control insect crop pests and insect vectors that carry certain tropical diseases. In the 1980s, Congress required the U.S. Environmental Protection Agency (EPA) to reassess the safety of all pesticides approved before 1984. Companies selling Bt microbial sprays provided the EPA with extensive data on Bt toxicity, environmental fate, and effects on **nontarget organisms**, which is any organism that is not the target of the pest control activity or product. After reviewing the data, the EPA issued a finding of no significant adverse effects and approved 180 Bt sprays for use in the United States.

In spite of their many benefits, Bt-based insecticidal sprays have a number of limitations. They break down soon after being applied to the crop, so repeated applications are necessary. Because they do not enter the plant tissue, they cannot control stem borers or root-dwelling insects. Finally, Bt sprays have caused allergic reactions in some people who spray them.

Facts about Bt corn

Bt corn is a generic name for a number of transgenic corn varieties that contain the gene encoding the insecticidal protein produced by the Bt strain *B. thuringiensis* subsp. *kurstaki*. Agricultural biotechnology companies developed Bt corn to protect field corn from a specific lepidopteran pest, the European corn borer (ECB). In the United States, field corn, which is used primarily for animal feed and secondarily for processed food, such as cornflakes and tortilla chips, is grown almost exclusively in the Midwest.

The ECB causes an estimated $1 billion worth of damage in lost field corn yields every year. Illinois corn growers alone lose $50 million a year from decreased yields resulting from ECB damage. The ECB is difficult to control, because it burrows into the corn stem and is inaccessible to insecticides (Figure 18.10). Little of the insecticide used on corn is for ECB control (approximately 2 to 3%, or 85,000 tons), but the Bt gene can also provide some protection against other lepidopteran pests that farmers try to control with insecticides, such as the corn earworm, fall armyworm, and southern corn borer. Together, these pests account for approximately 20 to 25% of the insecticides used on corn in the United States (870,000 to 1.2 million tons/year).

Currently, 11 varieties of Bt corn developed by five companies are available to U.S. farmers. In 2003, they planted a Bt variety on approximately

Figure 18.10 ECB damage. **(A)** ECB larva inside corn stalk. **(B)** These corn stalks were taken from plots of Bt and non-Bt corn grown adjacent to each other. ECB damage is clearly visible in the non-Bt variety on the right. (Photographs courtesy of Syngenta Agricultural Biotechnology Research Unit, Research Triangle Park, N.C.)

one-third of the corn acres in the United States. Globally that year, 30 million acres of corn containing the Bt gene were grown commercially in eight countries, five of which are developing countries. We will provide more information on Bt corn in chapter 23.

Facts about the regulatory approval process

We will discuss the U.S. regulatory framework for agricultural biotechnology products in detail later in this and other chapters. At this point, we briefly describe the regulatory process as it relates to assessing the impact of Bt corn on nontarget organisms, such as monarch butterflies.

In the United States, three regulatory agencies required crop developers to provide them with sufficient amounts of test data and information from the scientific literature so their scientists could determine the potential risks of Bt corn to the environment, agriculture, and human health. Two agencies were responsible for assessing environmental impacts on nontarget organisms.

- The U.S. Department of Agriculture (USDA) has regulatory authority over greenhouses, where the first studies were conducted, and approves or rejects applications for small-scale field tests and interstate transport of transgenic crops. It also must provide final regulatory clearance for large-scale commercial planting.
- The EPA requires approval for field trials larger than 10 acres if the transgenic crop contains a pesticidal compound, such as the Bt protein. Along with the USDA, the EPA also must issue a final permit before Bt crops can be sold or distributed.

Assessing risks and benefits to nontarget organisms. Before they permit field tests and commercial sale, the USDA and EPA require information on the possible negative effects of transgenic crops on insects and other animals. Companies conducted laboratory tests to establish the toxicity of the Bt protein and Bt corn plant material to a number of nontarget organisms, including the honeybee, earthworm, collembola (a soil insect), daphnia (an aquatic invertebrate), green lacewing, ladybird beetle, parasitic wasp, catfish, quail, and mouse. The results confirmed Bt's lack of toxicity to nonlepidopteran insects and other animals.

When the monarch story broke, the federal agencies were chided for not requiring the companies to submit data from research specifically focused on Bt corn and monarch butterflies. The agencies did not require studies of Bt corn and monarch butterflies because they already knew many facts about Bt pesticides, monarch butterflies, and corn growing that are relevant to assessing risks.

Because Bt pesticidal sprays have been tested extensively and used for many decades, they knew these facts.

- Any part of a plant expressing the Bt protein would harm lepidopteran larvae that eat it in sufficient amounts, including monarch larvae.
- Bt sprays are used widely to control gypsy moth outbreaks in forests and also by organic farmers and home gardeners. The impact of Bt sprays on monarchs is minor compared to the other factors threatening monarch populations, such as destruction of their overwintering habitat and mowing along roadsides, a common milkweed habitat.

They also knew these critical facts related to potential risks.

- Seventy-five percent of the North American monarch population does not live where farmers grow Bt corn.
- To harm monarchs, the larvae must consume the Bt protein. Monarch larvae eat only milkweed plants, so the only part of the Bt corn plant that could pose a risk is pollen that landed on milkweed plants.
- Throughout the breeding range of monarchs in the United States, most milkweeds do not occur near cornfields. Therefore, across the United States, comparatively few monarch eggs would be laid on milkweeds near Bt cornfields.
- Monarch females lay eggs before corn releases pollen throughout most of the United States.
- Lepidopteran larvae are capable of determining the palatability of food and, if given a choice, avoid harmful and distasteful substances.

The regulatory scientists looked at these facts and determined that monarchs were much more likely to be exposed to Bt from large-scale spraying to control outbreaks of gypsy moths and other forest lepidopteran pests than from Bt corn pollen (Figure 18.11).

Figure 18.11 Forest pests and defoliation. **(A)** Gypsy moth larvae feeding on an oak tree. **(B)** The nun moth, a close relative of the gypsy moth, completely defoliated evergreen trees in this European forest. (Photographs courtesy of Terry McGuire, Animal and Plant Health Inspection Service, USDA [A], and Beat Forster, Swiss Federal Institute for Forestry [http://www.forestryimages.org] [B].)

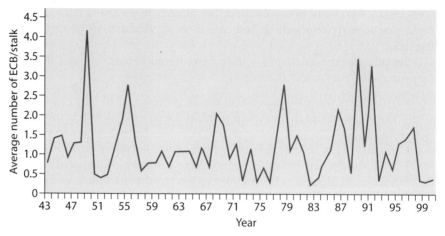

Figure 18.12 ECB densities in Illinois, 1943 through 1999. (Source, Gianessi and Carpenter, National Council for Food and Agricultural Policy, 1999.)

In conducting risk assessments, regulators must compare potential risks to potential benefits. They evaluated the benefits to farmers, agricultural ecosystems, the environment, and human health. They projected annual benefits to Midwestern corn growers of anywhere between $38 million and $200 million, in spite of the higher seed costs. The range is so wide because predicting when and where ECB infestations will occur is difficult, and their impact on yields varies greatly from year to year (Figure 18.12). Because most farmers do not use insecticides to control the ECB, the agencies did not expect to see a large decrease in the total amount of insecticides used per acre. Even so, the scientists at the EPA concluded that because of the large number of corn acres in the United States, even a small percentage decrease in insecticide use could have important positive effects on beneficial insects and farm workers.

In summary, USDA and EPA regulators who reviewed requests for testing and commercializing Bt corn recognized that widespread planting of Bt corn could potentially harm nontarget lepidopteran insects, including monarchs. However, they also felt that the benefits to lepidopterans and all other insects provided by fewer insecticide applications far outweighed the risks.

Facts provided by the Cornell study

In controlled laboratory feeding experiments, researchers at Cornell University clearly demonstrated that 3-day-old monarch larvae that ate milkweed leaves dusted with pollen from a Bt corn variety (176) ate less, grew more slowly, and suffered a higher mortality rate than larvae that were fed milkweed coated with non-Bt corn pollen or milkweed with no corn pollen. Only 56% of the larvae that were fed leaves with Bt corn pollen survived after 4 days compared with a survival rate of 100% for the other two treatment groups. The larvae feeding on pollen-free leaves ate significantly more leaf material than larvae that were fed leaves with either non-Bt corn pollen (1.61 versus 1.12) or Bt corn pollen (1.12 versus 0.57). After 4 days, the average weight of surviving larvae fed Bt corn pollen was less than half the average weight of the larvae fed leaves with no pollen (0.16 versus 0.38 g).

Because the harmful effects of Bt on lepidopterans have been well understood for 60 years, these results did not surprise the scientific community or regulatory agencies. The most important question is not whether monarch larvae that are forced to eat Bt pollen will be harmed but whether growing Bt corn will have negative impacts on monarch butterfly populations under natural conditions.

Delineate the risks of Bt corn to monarch butterfly populations

In deciding whether to grant approvals, the regulatory agencies try to consider a new product's potential risks and benefits carefully rather than reacting emotionally, overestimating risk, and reflexively blocking new products because they cannot prove there is zero risk. They work carefully and methodically, reviewing all of the data relevant to the question at hand. Their ultimate goal is to use scientific facts to make rational decisions in approving or denying requests.

We suspect that as you read this section, you will be struck by the tediousness of a science-based risk assessment, especially compared to those sexy headlines about toxic pollen and killer corn. This is precisely the point we hope to make. Although we are not interested in boring you, the truth is that risk assessments, done well, are *exceptionally* tedious. Please don't worry about learning the nuances of monarch butterfly biology, milkweed plant distribution, and growing corn in Iowa. We want you to see the types of questions that must be asked and answered before scientists are able to make definitive comments about the risks of Bt corn to natural populations of monarch butterflies.

Recall that risk depends on both the hazard and the probability of exposure to the hazard. To assess the risks of Bt corn pollen to monarchs, regulators must determine:

1. if Bt corn pollen is hazardous to monarch larvae (this step is unnecessary in this case)
2. the dose required for harm or injury
3. the likelihood that an individual larva will be exposed to that dose

Finally, in order to have an adverse effect on the monarch population (as opposed to an individual larva), a significant proportion of the larvae must be harmed.

Determining the hazardous dose

Bt corn contains the gene from *B. thuringiensis* subsp. *kurstaki* that encodes a protein specifically toxic to lepidopteran insects, including monarch larvae, that ingest any part of the plant expressing the Bt gene. You learned earlier that the concept of toxicity is meaningful only in reference to an amount. The Cornell study does not provide any information on the number of Bt pollen grains that must be consumed in order to observe the lethal or developmental effects of Bt pollen on 3-day-old larvae, nor does it provide other important information, such as the toxicity of Bt pollen to older, larger larvae, which are known to be less susceptible to Bt sprays. Subsequent research has provided specific information on (1) the Bt pollen dose that causes significant harm and (2) the relationship of the toxic dose to larval age (Table 18.5).

In addition, the level of hazard posed by Bt pollen varies with the transgenic corn variety, because some are much more toxic than others.

In summary, Bt corn pollen can be hazardous to monarch larvae, and the hazard level varies with the dose, Bt corn variety, and larval age.

Table 18.5 Bt toxicity[a]

Corn pollen density (grains/cm)	Corn variety	% Mortality for larvae aged (hours):	
		Less than 12	12–36
1,300	Bt 176	100	69
	Bt 11	60	56
	Non-Bt	60	12
135	Bt 176	70	37
	Bt 11	60	25
	Non-Bt	0	0
14		NA[b]	NA

[a]Scientists established the toxicity levels of two varieties of Bt corn (Bt 176 and Bt 11) by forcing monarch larvae to eat milkweed leaves coated with different amounts of corn pollen. A non-Bt corn variety served as the control. The percentages refer to percent mortality measured at a specific time (85 hours after feeding began). Note that newly hatched larvae are much more susceptible to Bt toxins than larvae that are 1 to 3 days old. Bt 176 is significantly more toxic than Bt 11, and non-Bt corn pollen can be toxic in high enough amounts. Finally, toxicity depends on the dose. At 14 grains of pollen/cm, Bt pollen is not toxic to larvae of any age.

[b]NA, not applicable; no differences in survival among the corn varieties or the ages of the larvae.

Assessing exposure

Not surprisingly, Bt corn pollen can be hazardous to monarch larvae, but what is its risk (probability of harm) to monarch larvae? To be harmed by Bt pollen, a larva must be in the wrong place (on a milkweed plant near a cornfield planted in certain Bt varieties, but not others) at the wrong time (when it is less than 3 days old). The probability that a young larva might be exposed to a sufficient dose of pollen depends on many variables, including corn pollen movement, milkweed abundance and distribution, female egg laying preferences in different habitats, temporal coincidence of pollen shedding with larval hatching and feeding, and larval feeding behavior. Therefore, to determine the risk, data on all of the variables that determine the degree of exposure must be collected.

Then, to have a population level effect, a significant proportion of larvae must be in the wrong place at the wrong time. The amount of geographic overlap between Bt corn-growing areas and monarch breeding populations defines the proportion of the population that might be at risk.

Do monarch larvae occur in areas where Bt corn is grown? Yes. Figure 18.13 shows where most corn (over 90%) is grown in the United States, which also coincides with the regions with the worst infestations of ECB. Farmers in areas with regular outbreaks of ECB are much more likely to plant Bt field corn than farmers in parts of the country where ECB infestations do not occur. Therefore, it appears that only certain monarchs in the eastern population would coexist with the Bt corn varieties currently marketed. Approximately 25% of the U.S. monarch population is concentrated in the Midwest, the primary corn-growing region of the country.

For purposes of our discussion, we will not differentiate among the different varieties of Bt corn, even though that information is essential for accurately estimating the level of risk to monarchs. As you can see in Table 18.5, Bt 176 is significantly more hazardous than all other Bt varieties. When the monarch-Bt corn story broke, Bt 176 was planted on less than 2% of the field corn acres. The Swiss company that marketed Bt 176, Syngenta, has since commercialized other Bt varieties with significantly lower toxicities, and Bt 176 is no longer on the market.

A. Corn for grain, harvested acres: 2002

B. Monarch breeding

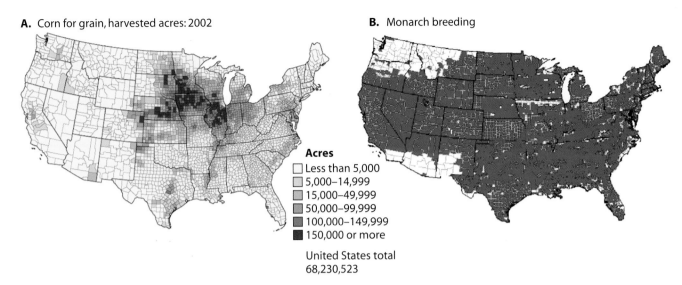

Acres

☐ Less than 5,000
▫ 5,000–14,999
▫ 15,000–49,999
▫ 50,000–99,999
▪ 100,000–149,999
■ 150,000 or more

United States total
68,230,523

Figure 18.13 Geographic distribution of corn-growing areas **(A)** and monarch breeding range **(B)**. (Maps redrawn from those of the USDA National Agricultural Statistics Service and the National Biology Information Infrastructure of the U.S. Geological Survey.)

What is the likelihood that milkweed plants will occur near Bt cornfields?
Monarch larvae eat only milkweed plants, so the degree of exposure of monarch larvae to Bt corn is related to the abundance, geographic distribution, and location of milkweed plants and the percentage of corn acreage planted with a Bt variety. The most common milkweed species are usually found in open areas: along roadsides, in ditches, at the edges of forests, and in open fields, such as pastures (Figure 18.14). A number of studies have looked at the relative probability that a female monarch butterfly will encounter a milkweed plant near a cornfield compared to other areas.

In Iowa, a female monarch is more likely to encounter roadside milkweeds than milkweeds next to cornfields, because the area around cornfields contains 85% fewer milkweed plants (22 milkweed patches/roadside acre

Figure 18.14 Milkweed habitat and distribution. Roadsides are a preferred habitat for milkweed plants (arrow). USDA scientists in Iowa found common milkweed (*Asclepias syriaca*) growing in 71% of the roadside plots they surveyed. In Iowa and other Midwestern states, many roadsides occur adjacent to cornfields, and on average, 30% of the corn grown in Iowa is a Bt variety. However, even during corn pollen shedding, the leaves of this milkweed plant would not contain enough Bt corn pollen to harm young monarch larvae, because very little corn pollen travels outside of the cornfield. (Photograph by Peggy Greb, courtesy of the Agricultural Research Service, USDA.)

compared to 3 milkweed patches/acre near a cornfield). However, the probability of encountering milkweeds near corn is greater than the 15% implied above, because in Iowa, 19% of the roadside area is next to a cornfield. Of those cornfields, approximately 17 to 20% were planted in Bt corn in 1999.

Other studies have compared milkweed densities, instead of numbers of patches, in agricultural and nonagricultural areas. In all study sites (central Minnesota, Wisconsin, Iowa, Ontario, and Maryland), milkweed densities are significantly higher in nonagricultural rural areas (prairies, pastures, and old fields) than agricultural areas, but the magnitudes vary from 4 to 7 times greater in the Midwest to over 100 times greater in Ontario. However, when the relative amount of agricultural and nonagricultural land is included in the calculations, female monarchs in Minnesota, Wisconsin, and Iowa are more likely to encounter milkweed in agricultural than nonagricultural areas; those in Ontario and Maryland are not.

(Reminder: don't worry about the details. Focus on the types of questions and the complexities of the answers. All of these details are fundamental to assessing risks, but they also extend well beyond the reasonable bounds of media coverage.)

What is the likelihood that Bt corn pollen will occur on milkweed plants growing near cornfields? Government regulatory scientists were able to get a general handle on this question by looking at many years of data on corn pollen movement. Because corn pollen is relatively large and heavy, it does not move very far outside the borders of the cornfield. Long ago, agricultural scientists had shown that 70% of the pollen released by corn plants stays within the confines of the cornfield and only 10% of the pollen travels farther than 3 m from the edge of the field (Figure 18.15).

Researchers measured the actual density of corn pollen on milkweed plants in and around cornfields and found that the average pollen density on milkweeds planted next to cornfields shedding pollen was significantly less than the dose that is hazardous to young monarch larvae.

In addition, rainfall significantly decreases the amount of corn pollen on milkweed plants, especially the upper leaves, which are the preferred feeding sites of young larvae. A single rain removed 54 to 86% of the pollen from milkweed plants, and 55% of the young larvae on milkweed plants are found on the upper leaves.

Figure 18.15 Corn pollen dispersal. Many studies to measure corn pollen dispersal from cornfields to milkweeds surrounding the field have consistently demonstrated that pollen levels decrease rapidly as the distance from the cornfield edge increases.

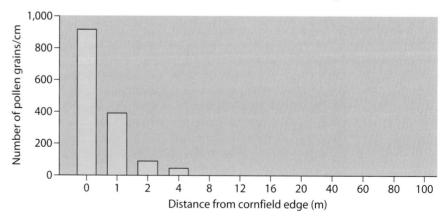

What is the likelihood that a female monarch will lay eggs on a milkweed plant with Bt corn pollen? Using only the data on corn pollen movement and milkweed abundance and distribution, insect ecologists assumed that most female monarchs would lay eggs on milkweed plants too far away from Bt cornfields to have much Bt corn pollen on the leaves. They also assumed that females would rarely lay their eggs on milkweeds within cornfields, because milkweeds, which are much shorter than corn plants, are presumably more difficult for females to locate.

Much to their surprise, female egg-laying behavior did not mirror milkweed distribution. Egg densities were higher on milkweeds located in cornfields in Iowa, Wisconsin, and Minnesota but were the same in nonagricultural areas and cornfields in Ontario and Maryland. Interestingly, the survival rate of larvae in cornfields, including Bt cornfields, was *higher* than for those not in cornfields, presumably due to less bird predation. Apparently, female monarchs have an easier time seeing milkweeds in cornfields than birds do.

What is the likelihood that Bt corn pollen will occur on milkweeds when larvae are feeding? The answer depends on the relative timing of monarch egg laying and corn pollen shedding. A given field of corn sheds pollen for 7 to 10 days sometime between June and August, the corn-growing season in North America. The approximate date of pollen shedding within that 2-month period varies with the latitude at which the corn grows. To be harmed by Bt corn pollen, young larvae must be feeding on milkweeds next to or in the cornfield during the period when pollen is shed.

Because of the migration rates of monarchs that overwinter in Mexico, no overlap occurs between monarch egg laying and corn flowering in the southern parts of the Bt corn-growing region. Some overlap between larval feeding and pollen shedding occurs in central Iowa (15%), and it reaches a maximum in Ontario (60%).

Drawing conclusions about risk

Finally, the potential risk of growing Bt corn to the monarch butterfly population has been clarified. The portion of the monarch population that may be harmed by Bt corn pollen is limited to those in the Bt corn-growing area that is north of central Iowa and east of the Rocky Mountains (Figure 18.16). In those areas, only very young larva feeding on milkweed plants within or immediately adjacent to cornfields planted in certain Bt varieties are at risk of consuming harmful amounts of Bt corn pollen.

What is the probability that a larva will consume enough Bt corn pollen to be harmed? Lepidopteran larvae exhibit clear feeding preferences. Data from the Cornell study show that monarch larvae do not readily eat milkweed leaves coated with pollen, whether or not that pollen contains the Bt gene. More than likely, if a feeding larva encounters pollen, the larva will avoid it and feed elsewhere. USDA scientists gave monarch larvae a choice between milkweeds with and without Bt corn pollen, and the majority chose the pollen-free leaves. In addition, a laboratory study gave a different lepidopteran species a choice between insect diet with and without Bt protein, and the larvae consistently avoided the diet with Bt.

Because of the low probability of exposure of susceptible monarch larvae to hazardous doses of Bt corn pollen, in 2001 the EPA stated, "While there is a small chance that one in 100,000 caterpillars could be affected by Bt corn pollen, research suggests even those larvae will mature into healthy butterflies."

Figure 18.16 Corn pollen shedding and larval feeding. The timing of corn pollen shedding and monarch egg laying affects the probability that young larvae will be exposed to Bt corn pollen. The red lines indicate the areas where overlap occurs between pollen shedding and larval feeding.

Establishing safeguards to decrease risks

The risk level depends on exposure to a hazardous dose and the safeguards that can be put into place to decrease the risk. What safeguards might decrease the risk of Bt corn pollen to monarchs?

- Because pollen from Bt corn varieties differs in toxicity level, in areas where corn pollen shedding and larval feeding overlap, farmers could plant varieties with the lowest toxicity to monarch larvae.
- It is now possible to genetically engineer plants to limit gene expression to specific tissues. Varieties expressing the Bt gene only in the cornstalk would offer protection against the ECB without harming monarchs, because monarchs do not feed on corn plants.
- The EPA requires farmers growing Bt crops, including corn, to plant non-Bt varieties to slow the evolution of resistance to the Bt protein. Planting a border of non-Bt corn around a Bt cornfield would further decrease the probability that Bt pollen would drift to milkweed plants near the field.
- Stop mowing roadsides in monarch-rich areas or plant milkweeds at sites distant from cornfields to increase the probability that an egg-laying female would encounter milkweeds uncontaminated with Bt corn pollen.
- Remove milkweeds from Bt cornfields so that larvae will not be exposed to high pollen levels. But is this wise, since it seems that monarch larvae survive better on milkweeds in cornfields, including Bt cornfields?

Assess the benefits of Bt corn

For a fair risk assessment of Bt corn, regulators must determine the potential benefits of Bt corn and contrast these benefits with the risks. Another way of framing the question is to consider the risks of not growing Bt corn, because Bt corn provides advantages to farmers, farm laborers, consumers, and the environment compared with current corn-growing practices.

- A number of studies have shown that Bt corn has significantly smaller amounts of mycotoxins than non-Bt corn has. These mycotoxins, which are fatal to livestock and harmful to humans, are produced by fungi that invade the corn plant, especially at sites of insect damage (Figure 18.17). Therefore, decreasing corn borers leads to a secondary benefit of improving food safety.
- Many feeding studies, including those submitted to U.S. regulatory agencies, have shown that the Bt protein, in contrast to chemical pesticides currently used, does not harm beneficial insects, such as ladybird beetles and honeybees, and is safe for freshwater invertebrates, earthworms, other soil organisms, birds, and mammals.
- Various studies have demonstrated that farmers using Bt corn have slightly decreased the amount of insecticides they apply to corn. This decrease in insecticide use is beneficial to all insects, including monarch butterflies, because spray drifts of pesticides can extend to large areas owing to the small size of droplets. For chemical insecticides, the typical impact on nontarget insects in or adjacent to treated fields approaches 100%.
- Fewer insecticide applications and a reduction in insecticide drift also benefit farm workers.
- Bt corn offers almost 100% protection against the ECB, providing U.S. farmers with an additional $1 billion of income per year.

Figure 18.17 Mycotoxins. Some molds and fungi that infect plants produce harmful substances know as mycotoxins. The fungus shown produces aflatoxin. A number of studies have shown that Bt corn has significantly lower levels of mycotoxins than non-Bt corn. (Photograph courtesy of Gary Payne, North Carolina State University.)

BIOTECHNOLOGY REGULATION IN THE UNITED STATES

Modern biotechnology was born under unique social and political circumstances. In 1973, soon after Herb Boyer and Stanley Cohen described their successful work recombining DNAs from very different organisms, a group of scientists who were responsible for seminal breakthroughs in molecular biology sent a letter to the prestigious National Academy of Sciences (NAS) and the widely read journal *Science*. They asked the scientific community to agree to a self-imposed moratorium on certain scientific experiments using recombinant DNA technology. Even though they had a clear view of its extraordinary potential for good and no evidence of any harm, they were uncertain about the risks some types of experiments posed.

In February 1975, 150 scientists from 13 countries, along with attorneys, government officials, and 16 members of the press, met to discuss recombinant DNA research. The conference attendees replaced the self-imposed moratorium with a complicated set of rules for conducting certain kinds of laboratory work with recombinant DNA but disallowed other experiments until more was known. (Although it may sound as if the scientists were in general agreement, the debate was quite contentious. For those interested in the politics and sociology of science, the interplay of scientific research and government policy, and the public's role in influencing technology development, the events and discussions about recombinant DNA research between 1972 and the mid-1980s are beautifully captured in two books, *The DNA Story*, by J. Watson and B. Tooze, and *The Recombinant DNA Controversy: a Memoir*, by D. S. Fredrickson.) At no other time has the international scientific community voluntarily ceased the pursuit of knowledge before any problems occurred, imposed regulations on itself, or been so open with the public. They adopted strict guidelines, and as the understanding of recombinant DNA grew and confidence in its safety became more certain, the guidelines were gradually relaxed. This approach, supported by private and public researchers, ensured the thoughtful, responsible, and very public inspection of this new technology in its earliest stages of development.

A substantial body of knowledge about laboratory research with recombinant DNA, accumulated over a 15-year period, led the NAS to issue a report that set the stage for moving research from basic laboratory work to commercial products. The academy came to the following conclusions.

- There is no evidence that unique hazards exist in the use of recombinant DNA (r-DNA) organisms or in the transfer of genes between unrelated organisms.
- The risks associated with r-DNA organisms are the same in kind as those associated with organisms modified by other genetic techniques.
- Assessments of the risks of using r-DNA outside of the laboratory should be based on the nature of the organism, not on the method used to produce it.

During the early 1980s, as biotechnology evolved from basic research into product development, federal regulatory agencies with responsibility for protecting human health and the environment stepped in and based their overall approach to biotechnology regulation on the NAS findings.

In June 1986, the federal government issued the *Coordinated Framework for the Regulation of Biotechnology* (Box 18.3). The U.S. Congress did not pass any new regulatory laws specifically granting authority over biotechnology development, because the regulatory agencies determined that they already had sufficient statutory authority to require product developers to seek regulatory approval before testing and commercializing biotechnology

B O X   1 8 . 3   *U.S. approach to biotechnology regulation*

In 1986, the *Coordinated Framework for the Regulation of Biotechnology* established the guiding principles that the regulatory agencies said they would use in regulating biotechnology products.

Product not process

The risk of a biotechnology product depends on the nature of the product and not on the process used to create it. In other words, it doesn't matter whether a disease-resistant crop was created by genetic engineering, mutagenesis, or plant breeding, the risks should be of essentially the same type. This premise provides the justification for using existing statutes and not enacting new laws that specifically focus on biotechnology. The agencies claiming regulatory oversight over a biotechnology product have jurisdiction because the product falls within the purview of statutes enacted prior to biotechnology development.

Risks of biotechnology products are not fundamentally different from those of similar products used in the same way

Because risk is based on the nature of the product and not on how it was produced, past experience and knowledge of similar products are quite pertinent and can be used to determine data requirements for new product approvals.

Degree of oversight proportional to degree of risk

The data requirements for assessing risks and the rigor and stringency of regulatory review should be commensurate with the risk level of the product. Having expedited review for biotechnology products that pose little or no risk allows regulatory agencies to focus their limited resources appropriately. Obviously, this premise assumes that the regulatory authorities have some idea about the degree of risk of a new product to begin with, which they do if it is similar to other products used in similar ways. If regulators have no experience with the product type, unfamiliarity triggers product review.

Science-based risk assessment

The criteria used to determine risks to health, the environment, and agriculture must be amenable to scientific investigation through either empirical testing or review of the scientific literature. Consistent with this philosophy, regulatory review is done on a case-by-case basis, since risk depends on the nature of the product and how it is used.

Flexible and transparent

The regulatory process must be transparent so that all interested parties—product developers, the public, government officials, and state regulators—can know the regulatory requirements the product must meet. What are the questions the regulators will ask, and what data will they need to be able to answer them? The regulatory process must be flexible so that, as the regulators learn more about certain products, data requirements can be adjusted accordingly.

products. The three federal regulatory agencies whose mandates captured almost all of the biotechnology products for the foreseeable future at that time were the U.S. Food and Drug Administration (FDA), the USDA, and the EPA (Table 18.6). They drew their authority not only from specific laws that gave them the authority and responsibility to protect human health, the environment, and agriculture, but also from the **National Environmental Policy Act**. This law requires all federal agencies to ensure that any decision they make does not have negative environmental impacts. Therefore, at a minimum, all regulatory agencies must conduct an environmental assessment (EA) for products they approve.

Follow a product through regulatory review

The regulatory process for the approval of biotechnology products is complicated and multifaceted. Specific details vary according to the product in question and how it will be used, so it can be difficult to wrap your mind around the entire regulatory apparatus. Even so, it is important to understand how these products and activities are regulated in the United States, because the regulatory process is a key factor determining which products, from the universe of possible products, make it to the market. Some products may not be approved because their risk-benefit profile makes their value to society questionable, but sometimes the regulatory process impedes the de-

Table 18.6 Biotechnology product regulation[a]

Regulatory agency	Responsibility	Product approval authority
FDA	To protect consumers by ensuring food and feed safety and quality; ensuring drug safety, quality, and efficacy; and requiring appropriate labels related to health and safety on food, feed, and drugs	Food, food additives (including microbes), animal feed, drugs, vaccines, diagnostics, medical devices
USDA	To protect consumers, agriculture, and the environment by assessing the impacts of a product on agriculture and the environment; ensuring food safety of meat, milk, and eggs; and ensuring the safety and efficacy of animal vaccines and assessing their potential impacts on the environment and human health	Plants, plant parts, animals, some microbes, animal vaccines
EPA	To protect consumers and the environment by assessing potential impacts of a product on the environment, human health, and food and feed safety	Pesticides, including insect-resistant and disease-resistant plants and some microbes

[a]In the United States, three regulatory agencies have the primary responsibility for assessing the safety and efficacy of biotechnology products and processes. The responsibilities and authorities of the different agencies were established by acts of Congress that were passed many years before the development of recombinant DNA techniques. The table lists only the transgenic products over which they have approval authority. The congressional acts grant regulatory agencies many more kinds of authority than product approval of transgenic organisms.

velopment of safe, efficacious, and beneficial products because the costs of amassing data to meet regulatory requirements are prohibitively high for small companies, public institutions, and nonprofit research institutes.

Walking a biotechnology product through the regulatory process allows you to get a handle on the complexity of the process, the roles of the agencies, data requirements, repeated opportunities for public participation, and other details of the extensive and iterative process. We will use the example of a Bt corn variety. While the product developer might be a university, a public research institution, or private industry, companies developed the Bt corn varieties on the market. We begin at the point when scientists isolated the Bt gene encoding the pesticidal protein and realized that a crop plant with built-in protection against caterpillars would be an excellent product, if they could get the gene to function in a plant (Figure 18.18). The business development staff did the math and determined that the costs of conducting laboratory and field research, performing tests required for regulatory approval,

Figure 18.18 Field testing. Before inserting the new gene into the crop that will be commercialized, researchers often use other plants as hosts to see if the gene will produce sufficient amounts of the new protein under field conditions. Tobacco is often the host crop that is used in early stages of transgenic-crop development, because regenerating tobacco plants from plant callus is much easier than it is for other plants, such as corn. This 1989 field test of Bt tobacco (left), which the company had no intention of commercializing, measured the efficacy of a transgene that was eventually used in a transgenic Bt corn variety. As you can see, the Bt gene did an excellent job of deterring lepidopteran pests of tobacco compared to tobacco without the Bt gene (right).

and marketing the product could be recouped, but only if the company chose to use the Bt gene in a large-volume commodity crop. Two commodity crops, field corn and cotton, suffer significant damage from lepidopteran larvae; our hypothetical company decides to move forward with Bt field corn.

Their first question: who has regulatory jurisdiction over the product and the development process? In the case of Bt corn, three agencies had to sign off on this product before it could enter commerce, the USDA, EPA, and FDA. The combined concerns of these agencies as they relate to Bt corn are:

- its potential impacts on the environment and agriculture
- food safety and nutrition for both humans and animals (74% of the field corn grown in the United States is used in animal feed)
- accurate, meaningful labels

Regulatory review stages

Stage 1: research concept
At the "idea" stage, the company's institutional biosafety committee, which includes representatives of the public, considered the possible environmental and health impacts of the potential product; determined if the potential risk level was acceptable, especially compared to the benefits to farmers; and notified the USDA, which had to inspect and approve the greenhouse facilities where early research was conducted.

Stage 2: small-scale field trials
Before the company could move their research from greenhouse to small-scale field tests, it needed permission from the USDA, which required detailed information on the host crop (corn), the new gene construct (Bt protein, promoter, and selectable markers), the new gene products (Bt protein), the origin of the new gene (*B. thuringiensis*), the purpose of the field, and the specific precautions the developer would use to prevent the escape of pollen, plants, or plant parts. The company's request also contained a large amount of additional technical information on ecological attributes so that USDA scientists could conduct the EA required by the National Environmental Policy Act, such as the potential for movement of genetic material to plants of the same or related species (outcrossing) and the potential to harm animals near and around the test site. We provide data requirements for this and other stages in product development in chapter 23.

If the USDA's decision was a "finding of no significant impact" (FONSI), it would issue a field test permit to the company and require that certain risk management procedures be followed, such as maintaining specific isolation distances from other crops, planting borders of nontransgenic corn, and monitoring a number of ecological factors. In addition, federal law requires the USDA to:

- provide an opportunity for public input during the EA process
- make both the EA and FONSI statements available to interested or affected parties
- delay action if a member of the public challenges the sufficiency of the EA or FONSI

Stage 3: large-scale field tests
Before the company could move to large-scale field tests (more than 10 acres), it needed permission from the EPA. Before granting the permit, the

EPA required information that allowed it to assess impacts on the environment and human health, including:

- product characteristics, such as the biochemistry and bioactivity of the pesticidal substance, the biology of both the recipient plant and the source of the introduced genetic components, and the expression levels of the pesticidal protein
- environmental fate of the pesticidal gene product, such as the amount found in soils when plants die and the length of time the protein persists in soil, water, and air
- human health effects
- ecological effects, such as the potential for gene flow to wild relatives, increased or decreased disease and herbivore resistance, and effects on nontarget organisms

Federal law requires the EPA to inform the public when it receives a permit request, to allow for public comment, and to publish its decision.

Stage 4: food safety and nutrition tests

The company then moved to the food safety and nutrition phase of product development. The large-scale field trials provided enough plant material for addressing the requisite food safety and nutrition questions required by the FDA and EPA. Around the time the company approached the EPA about a permit for large-scale field tests, it also met with FDA officials to discuss the proposed product so that the FDA could decide the required tests for meeting its statutory mandate of ensuring that food introduced to the market is safe and nutritious. Why isn't there a standard battery of tests for all transgenic crops? Because tests for ensuring food safety and nutrition vary with the host crop, the donor organism, the genetic construct, and the new protein encoded by the gene. We will discuss the food safety and nutrition tests required by the FDA and EPA in detail in chapter 21.

Stage 5: large-scale production

After the FDA and EPA assessed Bt corn's nutrition and safety for human and animal consumption, the company was ready to start large-scale production and sale of Bt corn seeds. It petitioned the USDA for "nonregulated status" for this Bt corn variety so that it and the farmers did not have to get permits from the USDA every time they planted it. The company had to provide the USDA with data from its EPA-approved large-scale tests to prove that the variety performed as predicted, exhibited the same expected biological and agronomic properties as the same corn variety without the Bt gene, and did not harm agriculture or the environment, especially as the environment relates to agriculture (e.g., impacts on beneficial insects).

When the USDA receives a request for nonregulated status, they must inform the public, make copies available to those who request them, and give the public 60 days to comment on the request. The USDA must also conduct a final EA, using the more extensive data acquired in large-scale field tests, and inform the public when the EA is available for public review and comment. The USDA scientists grant nonregulated status if they determine that the new variety is the same as its nontransgenic parental variety, except for the defined difference, which in this case is the Bt gene and the protein it encodes.

Once the USDA issued a formal determination of nonregulated status, that Bt variety could be produced, marketed, distributed, grown, and used in breeding without getting approval from the USDA. If, however, any problems ever arise, the USDA has the authority to remove it from the market.

Stage 6: final stage

Before the Bt variety could be grown and used commercially, additional reviews by the EPA and FDA were necessary. Before approving a pesticidal product, including Bt crops, for commercial use, the EPA is legally required to demonstrate that it presents "no unreasonable adverse effects on environment or human health when used according to label directions." The EPA looked at data acquired from many large-scale field trials at a number of locations before it made this determination for the Bt corn variety. Once the EPA issued its final approval, the Bt corn variety could be sold, but the EPA also has statutory authority to remove products it has approved from the market if problems arise.

The FDA reviewed the food safety and nutrition data to determine if there were any unresolved issues or unanswered questions. If so, they would require additional testing; if not, they would issue a letter to the product developer describing their findings and allowing the product to be used in food for humans and animals. In some cases, the FDA will require that the product and foods derived from the product be labeled, but this is not necessary with the Bt corn variety. The FDA has the authority to remove products from the market, as well.

For the earliest varieties of Bt corn approved for sale in the United States, the step-by-step regulatory approval process, from submitting a request to the USDA for a permit for a small-scale field test of the Bt gene to inserting the gene into the target crop (corn), and finally, to introducing Bt corn seeds to the commercial market, took approximately 7 to 8 years.

SUMMARY POINTS

The public perception of risk often bears little resemblance to a science-based assessment of risk, because many factors other than science affect a person's view of risk, such as whether the risk is voluntary or involuntary, can be controlled by the person, or is familiar or unfamiliar.

In general, people incorrectly associate natural with safe and synthetic with increased risk, especially with regard to food safety issues. However, the natural chemicals in plants often have higher toxicity levels than synthetic chemicals.

Toxicity is the capacity of a substance to do harm or cause injury of any kind. Therefore, everything, even water, can be toxic. The degree of harm or injury caused by toxic substances depends on the dose, because the amount of exposure determines the degree of risk.

A formal risk assessment is a science-based approach to determining the degree of risk. The first step is hazard identification. Like toxicity, a hazard is anything that could potentially be harmful. The risk posed by a hazard depends on exposure. The level of risk defined by the hazard and exposure can be decreased or managed by instituting safeguards that decrease either or both of these factors.

Because no activity or product is completely risk free, achieving a level of zero risk is impossible even when safeguards are used. Therefore, a risk assessment should also include an assessment of benefits of the activity or product, which is a

risk assessment of *not* using the product or engaging in the activity.

The much-publicized laboratory research on monarch butterflies and Bt corn led many members of the public to believe that a new, serious risk had been identified. However, for many decades the scientific community had known that monarch larvae forced to eat Bt would die. Before approving the commercial use of Bt corn, government regulators had conducted a risk assessment of the effect of Bt corn on nontarget organisms and decided the benefits exceeded the risks.

In response to media attention and political pressure, government agencies and private companies funded additional research specifically targeted to the effect of Bt corn on monarch butterfly populations. This research provided scientific data to estimate risk mathematically by assessing the hazard (toxic dose) and the probability of exposure of monarch larvae to Bt corn pollen. The data reaffirmed the decision of the regulatory agencies that the benefits of Bt corn outweighed the risks.

In 1986, the U.S. government established a framework for regulating the products and processes of biotechnology. Three agencies have the primary responsibility and authority for ensuring the safety and efficacy of biotechnology products: the FDA, USDA, and EPA. With regard to agricultural biotechnology products, the regulatory agencies review the product at various stages in development to assess its effects on the environment, agriculture, and human health.

KEY TERMS

Absolute risk

Bacillus thuringiensis

Bt corn

Carcinogen

Complete metamorphosis

Coordinated Framework for the Regulation of Biotechnology

Exposure

Hazard

National Environmental Policy Act

Nontarget organism

Relative risk

Risk

Risk assessment

Risk management

Secondary plant compounds

Specialist feeder

Toxicity

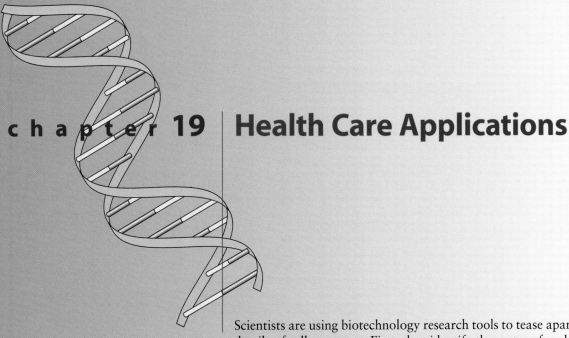

chapter 19 | Health Care Applications

Scientists are using biotechnology research tools to tease apart the nitty-gritty details of cell processes. First, they identify the roster of molecules that carry out a particular process. Then, they determine how the molecular players interact and identify the molecular referees that govern those interactions. Elucidating the molecular details of normal cell processes sheds light on why and how those processes fail in diseased cells, and studying the molecular basis of disease helps scientists understand normal processes. The more information that scientists can gather about cell physiology and pathology, the more skilled physicians will become in diagnosing, treating, and preventing disease.

Even though physicians now have access to much more information for answering their questions, the essential questions of clinical medicine remain the same.

1. What is the problem?
2. What is causing it?
3. What is the best way to treat, cure, or prevent it?

Answers to the first two questions provide information for answering the third.

In this chapter, we discuss how biotechnology-based products are helping physicians answer those questions more thoroughly than ever before. We focus on new tools and strategies for:

- diagnosing diseases and disorders
- treating, curing, and preventing diseases and disorders

The tools of biotechnology are altering many other aspects of health care provision, for example, by providing new methods for drug discovery and pharmaceutical manufacturing. Although we do not focus on these topics, we want to bring them to your attention because we allude to them periodically. Finally, throughout this chapter there are recurring themes, and these leitmotifs can provide additional frameworks for organizing the information we provide. One of the most important relates to comparing new medical technologies to those that are being used today. For new technologies to be worth developing, they should permit society either to do *new* useful things or to do the same things *better* (Figure 19.1). Deciding whether a medical technology gives clinicians a new capability is easy. For example, new biotechnology-based diagnostics can screen donated blood for viruses that cannot be detected by standard microbiological techniques. Decisions about whether a new technology allows clinicians to do the same thing better can be more ambiguous.

Figure 19.1 Improvements in karyotyping techniques. Technological innovations can provide more detailed answers to an old question. Physicians have used karyotyping to identify genetic problems for many decades. **(A)** Using low-resolution micrographs, they could identify major chromosomal disorders, such as extra chromosomes, as in trisomy 21, which is associated with Down's syndrome. **(B)** Better microscopy techniques allowed them to observe banding patterns, which helped them identify certain chromosomal disorders, such as translocations or deletions, as long as a significant length of the chromosome had been altered. **(C)** Spectral karyotype paints (SKY), a new technique that utilizes DNA probes labeled with fluorescent dyes, permits the identification of much smaller genetic changes. **(D)** Computer-assigned colors, based on the fluorescent dyes, are easier to analyze. **(E)** A side-by-side comparison of a spectral karyotype and a standard black and white karyotype. (Photographs courtesy of the Pathology Department, University of Washington [A], and H. Padilla-Nash and T. Ried, Genetics Laboratory, National Cancer Institute, National Institutes of Health [B to E].)

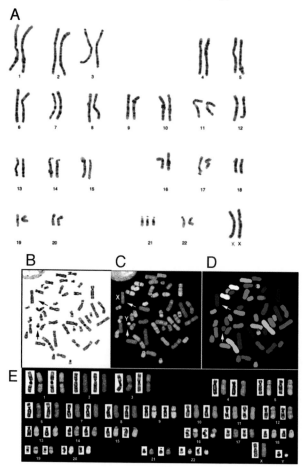

What are some general characteristics that translate to "better" in clinical medicine? A better technique:

- provides significantly more information for improving health
- provides the same information much sooner and/or more cheaply
- identifies the root cause, not a symptom
- allows intervention early in the disease process
- shifts the clinical strategy from disease management to cure and from disease treatment to disease prevention
- has fewer adverse side effects
- is more efficacious and/or more affordable

As you read about the new biotechnology-based products, try to assess whether they contribute to improved health care by meeting one or more of these criteria for doing the same thing better or by giving clinicians new capabilities.

MOLECULAR DIAGNOSTICS

A physician's success in managing or curing a disease depends on diagnosing it accurately. The diagnosis guides decisions about the appropriate treatment, and treating the wrong problem sometimes exacerbates the actual problem.

For centuries, physicians based their diagnoses on subjective information their patients provided and on symptoms they could see, feel, or hear (Figure 19.2). This diagnostic strategy often provides inaccurate and ambiguous answers. First, different diseases with very different causes often present the same symptoms. Second, by the time symptoms can be perceived by the physician's or patient's senses, the disease has often progressed to its later stages, leaving no hope of a cure or effective treatment. The physician's only option is to manage the symptoms.

Throughout the 20th century, advances in imaging technology provided diagnostic tools that made invisible symptoms visible and helped physicians catch some diseases earlier. For example, new visualization techniques could reveal cancers before clinical symptoms appeared and sometimes before they had spread to other organs (Figure 19.3). Other tests analyzed blood and

Figure 19.2 Disease diagnosis. **(A)** For centuries, the only available clues for diagnosing diseases were late-stage clinical symptoms that physicians could observe unaided. In this painting, the physician is attempting to diagnose asthma by using only his ear to hear the patient breathe. **(B)** In the early 1800s, physicians realized they could hear sounds in the lungs better by using a rolled up piece of paper. This led to the invention of a tube-like stethoscope by René Laënnec in 1830. **(C)** By 1880, the tube stethoscope had evolved into one that resembles the stethoscope physicians use today. The photograph in panel A, a reproduction of a painting by Theobald Chartan, was published in *Laënnec, à l'Hopital Necker, Ausculte un Physique* in 1853. (Photographs courtesy of the National Library of Medicine, National Institutes of Health [A], the National Museum of Health and Medicine, U.S. Armed Forces [B], and the National Museum of American History, Smithsonian Institution [C].)

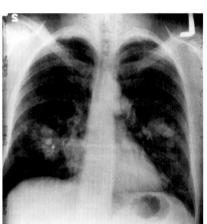
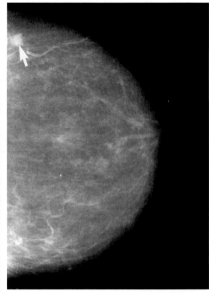

Figure 19.3 Imaging cancer. Lung X rays and mammograms allow physicians to detect cancer before clinical symptoms appear. **(Left)** The bright spots show that the patient has cancer in the lung on the right side of the X ray. **(Right)** The arrow points to a small tumor that may be breast cancer. (Photographs courtesy of National Cancer Institute, National Institutes of Health.)

urine for chemical imbalances and blood disorders that are symptomatic of certain diseases (Table 19.1). Not only could the chemical tests help to lessen disease damage by identifying some diseases early, in a few instances, such as sickle-cell disease, they also pinpointed the cause of the disease. For most clinical laboratory tests, however, different diseases can give the same test results. So, even though biochemical and cellular assays are closer to the root cause than visible symptoms, some ambiguity about the cause of the problem, and therefore the appropriate treatment, remains when physicians review laboratory test results.

In general, the new molecular diagnostics provide more information about the precise cause of a disease, catch diseases earlier in the disease process, and allow physicians to diagnose diseases that could not be diag-

Table 19.1 Clinical chemistry[a]

Class	Examples of substances and measurements
Ions	Sodium, potassium, chloride, calcium, carbonate
Organic metabolites	Glucose, lactate, citrate, bilirubin
Nitrogen-containing molecules	Ammonium, urea, creatine, uric acid, glutamine, phenylalanine
Lipids	Triglycerides, cholesterol, total fatty acids
Proteins	Total amount, albumin, globulins, lipoprotein
Enzymes	Amylase, acid and alkaline phophatases, creatine kinase, lipase, lactate dehydrogenase, galactosidase
Blood-clotting factors	Fibrinogen, plasminogen, prothrombin
Blood cells	Hemoglobin, red blood cell count, packed cell volume, proportion of different white blood cell types, platelets

[a]Hospital laboratories conduct a standard series of tests to measure the amounts of substances normally found in blood and urine samples. Deviations from normal values give physicians broad clues about possible diseases and disorders.

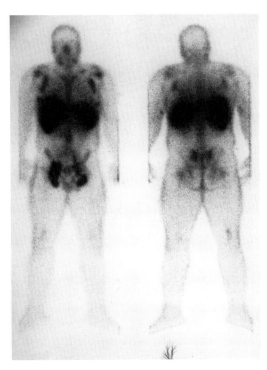

Figure 19.4 Monoclonal antibodies in cancer detection. Rather than surgically removing cells to determine whether a patient has cancer, physicians can use monoclonal antibodies that bind specifically to cancer cell surface antigens. In this imaging technique, researchers attach a radioactive isotope to monoclonal antibodies that bind to cancer cells. Radiolabeled antibodies confirm that the patient's cancer has spread to the lymph nodes, because radioactivity, indicated by dark areas, is detected in the lymph nodes of the armpits, neck, and groin. Not only does this technique remove the need for surgical biopsies, it also helps physicians identify the disease stage. In this image, the patient's liver and spleen are also darkened because all antibodies normally collect in those organs. **(Left)** Frontal view. **(Right)** Rear view. (Photograph courtesy of Jorge Carrasquillo, National Cancer Institute, National Institutes of Health.)

nosed previously. In addition, because the foundation of the diagnostic test involves very specific molecular interactions, such as antibodies binding to antigens or hybridization of homologous DNA sequences to each other, molecular diagnostics are typically more accurate than comparable diagnostics. Some of the new diagnostics are cheap and portable, which improves health care access; others are quite expensive. Finally, because molecular tests are looking for minute molecular differences, they can be conducted on urine, blood, or saliva, and therefore, they are often less invasive (Figure 19.4).

Early disease diagnosis improves prognosis

Whether the disease is infectious or degenerative, such as cancer, arteriosclerosis, or Alzheimer's, diagnostics that catch a disease in the earliest stage allow physicians to intervene in the disease process sooner rather than later. Early intervention lessens the harmful impacts of the disease and may even provide an opportunity to cure the disease rather than simply manage it.

Early detection of contagious diseases

You have already learned about new monoclonal antibody-based diagnostics for infectious diseases that can be passed from one person to another. These tests accurately identify pathogens, such as the bacteria that cause strep

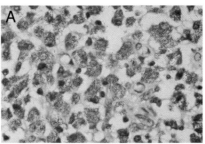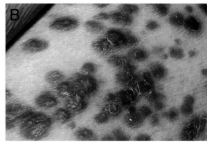

Figure 19.5 AIDS diagnosis. HIV-infected individuals display clinical symptoms many years after being infected. The symptoms include the appearance of rare infectious diseases and uncommon cancers that are AIDS-defining illnesses. **(A)** A micrograph shows lymph nodes containing a tremendous number of bacteria (dark-pink rods) that usually infect birds, not humans. **(B)** Kaposi's sarcoma. (Photographs courtesy of Edwin Ewing, Centers for Disease Control and Prevention [A], and National Cancer Institute, National Institutes of Health [B].)

throat, chlamydia, and gonorrhea, in minutes instead of the minimum 2 to 3 days typically needed to culture bacteria for identification. In addition, basing disease diagnosis on a unique molecular component rather than the whole organism allows physicians to detect pathogens that cannot be cultured. When antibiotic treatments are started as early as possible, the patient is much less likely to suffer from secondary infections or other medical complications.

Early detection and treatment of infectious diseases have important public health implications, as well. Biotechnology-based improvements in diagnosing human immunodeficiency virus (HIV) infections provide a striking example of the public health benefits of early detection. In the early 1980s, the only available HIV diagnostic was AIDS, the clinical manifestation of an HIV infection. A number of conditions that rarely occur in people are **AIDS-defining illnesses** that physicians use to clinically diagnose AIDS and, therefore, an HIV infection (Figure 19.5). For most HIV-infected individuals, signs of an AIDS-defining illness do not appear until 9 to 10 years after infection. During that time, infected individuals unknowingly transmit HIV to their sexual partners. In addition, some HIV-infected individuals never develop an AIDS-defining illness, which makes public health efforts to contain its spread even more problematic (see Box 19.1).

When AIDS first appeared in 1981, physicians did not know what caused it. They knew only that the causative agent must lead to a serious defect in the immune system, because all of the AIDS-defining illnesses are symptomatic of a collapsed immune system. When medical researchers identified the link between HIV infection and AIDS in 1983, they began developing tests for detecting HIV infections in presymptomatic individuals. If HIV had struck in previous decades, developing diagnostic tests for such individuals would have been very difficult, if not impossible, because culturing viruses is so difficult. The first HIV diagnostic tests, which were based on antibody-antigen interactions, could identify infected individuals 6 to 12 months after they contracted the virus (Figure 19.6). The diagnosis relied on finding HIV antibodies in a person and not on the patient having an AIDS-defining illness. The first HIV test shaved a remarkable 8 to 9 years from the amount of time it took to recognize an HIV-infected person. In the absence of this advancement, one can only imagine how much more horrific the current AIDS crisis would be.

BOX 19.1 *The stages of an HIV infection*

In 1980 and 1981, a sudden increase in the incidence of a very rare cancer, Kaposi's sarcoma, and uncommon infectious diseases alerted physicians to the emergence of a new disease. By 1983, researchers had identified the under-lying cause: infection by a virus the researchers named human immunodeficiency virus (HIV). HIV infects a specific type of immune system cell, the CD4$^+$ T cell. Because CD4$^+$ cells orchestrate many interrelated and interdependent immune system activities, loss of a significant number of CD4$^+$ cells eventually leads to an immune system melt-down (see the figure).

An HIV infection progresses through a series of stages. (A) During the acute stage, immediately after entering a person's bloodstream, HIV begins to locate, infect, and kill CD4$^+$ cells. The CD4$^+$ cell in the scanning electron micrograph has a cluster of HIV on its surface. This causes an abrupt decrease in the number of CD4$^+$ cells. The person's immune system, recognizing that the body is under attack, rallies. The increase in CD4$^+$ cells leads to a momentary decrease in the number of viruses. Very soon, however, the viral load begins to increase as they continue to infect and kill CD4$^+$ cells. (B and C) During the clinical latency stage, the CD4$^+$-cell count continues to decrease as viral numbers rise. Note the cluster of viruses that has emerged from the infected CD4$^+$ cell at the top of the image (enlarged in panel C). (D and E) Micrographs show an HIV beginning to emerge from a CD4$^+$ cell and budding from the cell membrane to become a free-living virus that will infect another CD4$^+$ cell. (F) During the clinical latency stage, because the number of CD4$^+$ cells is decreasing, infected individuals become more susceptible to common infections, such as strep throat or influenza. However, because these illnesses are not unusual, the person still does not realize that he or she is HIV infected. In the AIDS stage, the CD4$^+$ cell numbers continue to fall, and the patient begins displaying signs of very rare, AIDS-defining diseases. In this example, the patient's lungs are filled with cysts of the fungus *Pneumocystis carinii*. (Photographs courtesy of Cecil Fox, National Cancer Institute, National Institutes of Health [A], Matt Gonda, National Cancer Institute, National Institutes of Health [B to E], and Edwin Ewing, Centers for Disease Control and Prevention [F].)

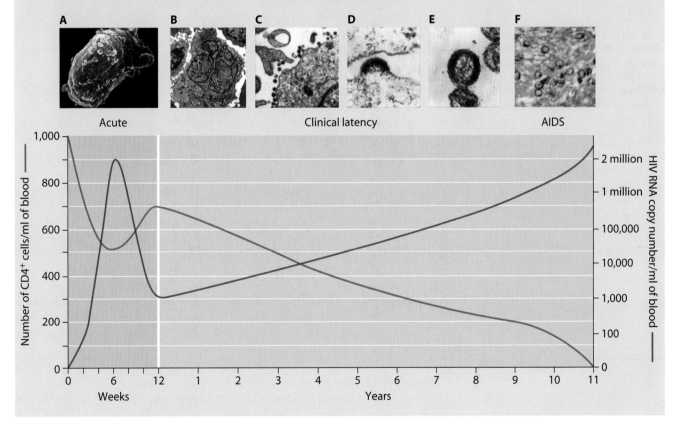

Unknowingly spreading the infection for 6 months still constitutes a major public health problem, but it took that long for an HIV-infected individual to generate a sufficient number of antibodies to be detected. Improvements in detection technologies heightened test sensitivity to fewer and fewer antibodies, decreasing the time from infection to detection to 3 months. As

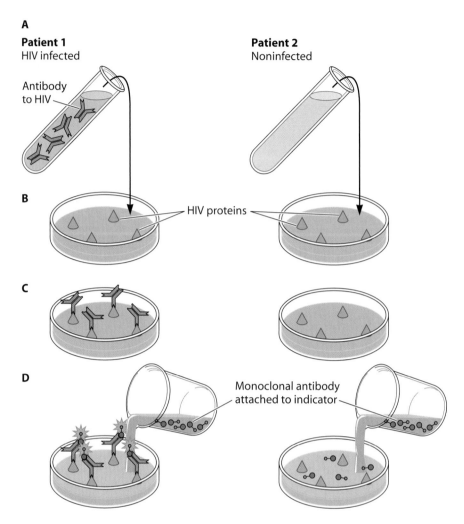

Figure 19.6 An antibody-based HIV diagnostic test. **(A)** Blood is drawn from two people. Patient 1 is infected, so the blood contains antibodies to HIV. **(B)** The patient's antibodies specifically bind to the purified HIV proteins in the diagnostic test. **(C)** The blood is washed off, but HIV antibodies stay bound to the HIV antigens. **(D)** In the final step, monoclonal antibodies with a fluorescent tag are added. These bind specifically to HIV antibodies. When the monoclonal antibody binds to an HIV antibody, the indicator tag fluoresces. The lack of fluorescence indicates a lack of HIV antibodies. According to this test, patient 1 is HIV positive. When the physician obtains a positive result, a different type of test is conducted to verify that the person is HIV infected.

scientists identified HIV gene sequences, a new diagnostic, based on HIV genetic material, improved the speed and sensitivity of HIV detection by many orders of magnitude. No longer was it necessary to give an HIV-infected individual a number of months to synthesize a detectable number of antibodies. Instead, a PCR machine took over the synthesis task by producing a detectable amount of an HIV-specific gene sequence (see Figure 15.13). As a result, the PCR-based diagnostic for HIV can detect as few as 20 viral particles per milliliter of blood, dropping the time for detecting HIV infection from months to days.

Other diseases and disorders

Cardiovascular disease, cancer, and other degenerative diseases are by far the major cause of death for people living in the industrialized world. Most of these diseases leave unique molecular footprints, or **biomarkers**, as they progress from one stage to the next. When medical researchers identify and characterize these telltale molecular markers, physicians will be able to use the information to diagnose a disease at the earliest stages, sometimes before they can detect diagnostic changes in cell morphology (Figure 19.7). In addition, these biomolecular diagnostics will help physicians make informed de-

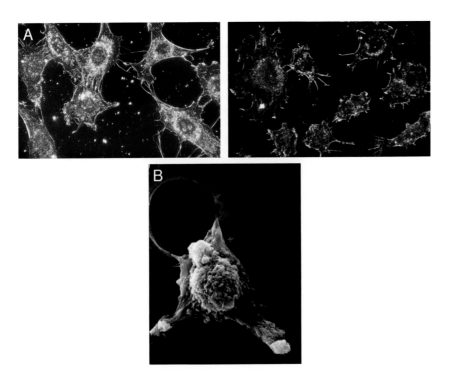

Figure 19.7 Cancer and normal cells. **(A)** Human connective tissue. The large, variably shaped nuclei; small cytoplasmic volume compared to the nucleus; and loss of normal cell specialization features are all characteristic of cancer cells (right) compared to normal cells (left). (Photograph courtesy of Cecil Fox, National Cancer Institute, National Institutes of Health.) **(B)** Metastasis. Locomotion is integral to the metastasis process, and metastatic cancer cells develop long arms, or pseudopodia, for that purpose. Scientists have recently identified the protein that causes cancer cells to grow arms. (Scanning electron micrograph by Susan Arnold, courtesy of Raouf Guirgus, National Cancer Institute, National Institutes of Health.)

cisions in selecting treatments because medicines and procedures that work at one disease stage may be ineffectual at others. We will use colon cancer to illustrate.

You learned in chapter 10 that for a cell to become cancerous, a number of mutations have to accumulate in somatic cells. Some are accelerator mutations that continuously encourage cell division, and others inhibit the cell division braking mechanisms. Others are not malfunctions in the mechanisms regulating cell division but instead are mistakes in the DNA repair machinery.

These mutations would not be detectable in microscopic examinations of cell morphology during the time frame in which the cells are accumulating mutations but are still precancerous (Figure 19.8). Nonetheless, those mutations lead to changes in DNA sequences, messenger RNAs (mRNAs), and proteins encoded by the mutated gene. All of these would not only serve as biomarkers of a precancerous condition, they would also help physicians know whether the root problem in need of fixing was a mutation of the accelerator, brake, or DNA repair genes. Knowing the specifics allows physicians to target treatment to the cause of the cancer. Under certain circumstances, identifying biomarkers and using them to predict the eventual onset of disease would shift the medical intervention from disease treatment to disease prevention.

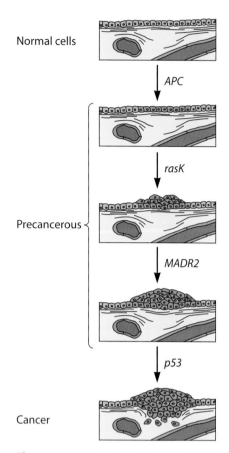

Normal cells

APC

Precancerous

rasK

MADR2

p53

Cancer

Figure 19.8 Biomarkers. As normal cells accumulate mutations, changes in their molecular products such as mRNA and protein also occur. Identifying the biochemical changes can help physicians diagnose cancers much earlier. If they are able to identify precancerous cells, they may be able to prevent the cancer from developing.

Genetic information is increasingly important in disease diagnosis

Although the relationship between genes and health is receiving more public attention now than ever before, using genetics as a diagnostic component of health care is not new. For centuries, physicians have used a person's hereditary history to shed light on current and future health, because many medical conditions and diseases are **familial**, which means they "tend to" run in families.

Simple and complex genetic diseases

Long before they knew genes existed, medical scientists studied family trees, or pedigrees, and recognized some disorders as hereditary. Their ability to classify a disorder as an inherited condition depended on the tightness of the link between an invisible genetic mutation and an obvious disease phenotype that was easy to track through many generations. As Mendel, Morgan, and their successors uncovered the basic laws of heredity, scientists realized that mutations in a single gene caused these **simple genetic diseases**. By the middle of the 20th century, scientists had identified hundreds of genetic diseases and disorders because they were monogenic traits, inherited in typical Mendelian fashion (Table 19.2).

The link between having disease genes and displaying disease symptoms is greatest when the disorder results from a single dominant allele that is relatively impervious to environmental influences and has 100% penetrance. But as you learned in chapter 12, even with this seemingly absolute degree of genetic predetermination, the severity of a genetic disorder can vary considerably among people having precisely the same genetic mutation (variable expressivity). The link between genes and a certain disorder becomes increasingly vague and ambiguous when either many genes contribute to the disorder (**multigenic disorder**) or genes and environmental factors interact and lead to the disorder (**multifactorial disorder**). As ambiguity in the link increases, the predictive power of hereditary history on medical diagnoses lessens. Unfortunately, the overwhelming majority of diseases and disorders that are responsible for mortality in the industrialized world are both multigenic and multifactorial.

Uncoupling genetic disease phenotypes from disease diagnosis

Until very recently, diagnosing the most straightforward inherited disorder depended upon observing certain phenotypic traits. Even though medical researchers did not need to understand the molecular basis of a disorder in order to either diagnose the disorder or identify disease gene carriers, they had to have a measurable or observable phenotype that was consistently associated with the genetic defect. For certain disorders, the phenotypic correlate might be a biochemical change that occurs prior to clinical manifestations, so the disease can be diagnosed early. Under the very best of circumstances, early detection of the biochemical phenotype permits early intervention, which can sometimes prevent the development of clinical symptoms. For example, a biochemical test for phenylketonuria (PKU), conducted soon after birth, identifies babies with the PKU mutation. The disease can be prevented if the amount of phenylalanine in the diet is kept to a minimum.

In most cases, however, the phenotypic trait that allows physicians to diagnose even simple genetic disorders or identify disease gene carriers is a clinical symptom of the disease. A mother and father, for example, suddenly realize they both have an allele for a mutant form of the cystic fibrosis transmembrane protein only after one of their children begins displaying clinical symptoms.

Table 19.2 Simple genetic diseases[a]

Disease (occurrence)	Protein function	Clinical symptoms
Autosomal dominant disorders		
Huntington's disease (1/10,000)	Cytoskeleton; vesicle transport	Degenerative nervous system function; dementia
Myotonic dystrophy (1/8,000)	Protein kinase enzyme	Muscle loss; defective heart contractions
Hypercholesterolemia (1/500)	Low-density lipoprotein receptor	Arterial blockage; severe coronary heart disease
Autosomal recessive disorders		
Cystic fibrosis (1/2,500)	Chloride channel protein	Respiratory disease; digestive disorders
Gaucher's disease type I (1/50,000)	Glucocerebrosidase enzyme	Enlarged spleen and liver; fragile bones
Beta-thalassemia (1/20,000)	Hemoglobin oxygen transport	Anemia; enlarged spleen; bone deformities
Alpha-antitrypsin deficiency (1/3,500)	Protective enzyme in lungs	Emphysema
PKU (1/10,000)	Phenylalanine degradation enzyme	Mental retardation
Sex-linked disorders		
Hemophilia A (1/5,000 males)	Blood clotting factor VIII	Excessive bleeding from small injuries
Duchenne muscular dystrophy (1/3,300 males)	Muscle contraction protein	Muscle degeneration
Color blindness (8/100 males)	Visual pigment molecule	Inability to detect red and green

[a]These are examples of disorders caused by mutations in a single gene. For decades, physicians have known that these diseases are inherited because they exhibit clear, predictable inheritance patterns and have obvious, definitive clinical symptoms. Using biotechnology-based research tools, scientists have now mapped, isolated, and sequenced mutant alleles responsible for these diseases and identified the defective proteins they encode. The proteins include enzymes, receptors, transporters, and structural proteins.

The wealth of genomics information made available by the Human Genome Project is providing diagnostic tests for early detection of many hereditary diseases. With only partial sequences of disease genes or nearby markers, physicians can diagnose genetic diseases before symptoms appear. At first, using sequence information for predictive diagnosis of genetic defects will be limited to disorders caused by one or a very few genes, such as cystic fibrosis or early-onset Alzheimer's disease. Over time, as researchers use genomics to identify genes involved in multigenic disorders, genetic tests will help people learn their disease tendencies. Diagnostic tests will also identify people who have genetic propensities for diseases with strong environmental components. This information will provide them with an opportunity to make appropriate lifestyle changes, such as avoiding tobacco or changing their diets, to lessen the probability they will contract the disease.

In summary, not only will genetic information push disease diagnosis to the earliest possible stage (predisposition), it also has the potential to shift the focus of clinical medicine from disease treatment to prevention (Figure 19.9).

Figure 19.9 Progress in disease diagnosis. A disease process begins with molecular and cellular changes, moves through a series of stages, and, in the later and more severe stages, becomes manifested as visible clinical symptoms. Technological advances in the 20th century made it possible to diagnose diseases earlier, before the patient showed clinical signs of having the disease. Technological advances in the coming century will make it possible to identify diseases at increasingly earlier stages. In certain cases, the diagnosis may occur before the disease process has begun, which increases the prospect of disease prevention.

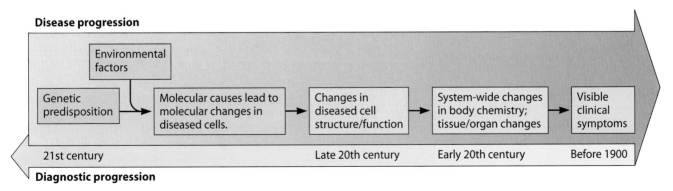

Identifying molecular variation can maximize therapeutic efficacy

One of the key messages we have tried to drive home in this book is that visible phenotypic similarity often obscures underlying molecular variability. Often molecular variability has important implications for human health. The new molecular diagnostics help physicians identify hidden variation that can affect their decisions about disease management. This allows them to target treatment to the patient's genetic makeup, the molecular nature of the disease, and in the case of infectious diseases, the genetic makeup of the pathogen.

Variation among patients

As you learned in chapter 16, a medicine's safety often varies from one person to the next. Much of the variation can be traced to small sequence differences in genes encoding enzymes that metabolize medicines. Physicians can use DNA-based diagnostic tests to help them determine the safest medicine to prescribe. Small genetic differences among patients also affect drug efficacy. Closely matching the specific treatment to specific patient subpopulations maximizes its effectiveness. The capacity to tailor treatments to a patient's genetic makeup—pharmacogenomics—depends upon physicians having diagnostics that are powerful enough to reveal those minute differences (Figure 19.10).

For example, even though a certain type of cancer may be caused by a mutation in one gene, not all mutations alter the shape of the encoded protein in the same way. Slight differences in shape can affect the binding of a drug to its target. The lung cancer drug Iressa, which targets the epidermal

Figure 19.10 Pharmacogenomics. Just as a prism subdivides visible light into separate wavelengths that are invisible to the human eye without the prism, the new molecular diagnostics reveal underlying, invisible genetic variation in groups of patients who appear uniform.

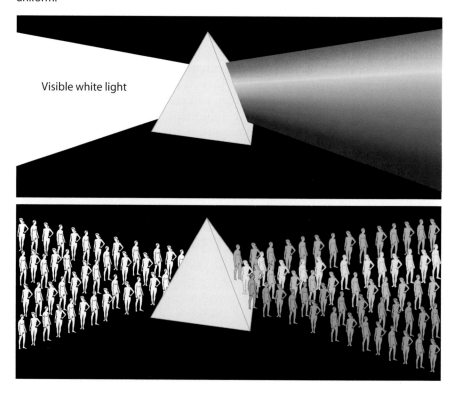

growth factor receptor, shrinks tumors by 50%, but only in 10% of the thousands of people who have non-small-cell lung cancer. Those who respond to Iressa carry similar mutations in a certain area of the gene encoding the epidermal growth factor receptor protein; nonresponders do not. Because researchers have identified the specific genetic mutation that explains the drug's variable efficacy, physicians can identify the subpopulations of non-small-cell lung cancer patients that should and should not receive Iressa.

Variation in response can also be explained (and identified) by genetic variability in genes other than the gene encoding the drug target. Most women with estrogen-dependent breast cancer receive tamoxifen, which halts tumor growth by binding to the estrogen receptor without stimulating cell division. Approximately 50% not only are resistant to the drug, but tamoxifen encourages tumor growth. Recently, scientists discovered that these women produce large amounts of a protein kinase enzyme that alters the shape of the estrogen receptor. When tamoxifen binds to the receptor, like estrogen, it promotes cell division and tumor growth. Mutations in two different genes can be used to predict tamoxifen resistance. Diagnostic tests for these mutations are being developed. They will allow physicians to identify patients with estrogen-induced breast cancer who should not be given tamoxifen.

Variation in the molecular basis of a disease

In addition to slight genetic variations among patients who have the same disease with the same molecular basis, such as breast cancer that is estrogen dependent, diseases that appear the same clinically can have strikingly different causes at the molecular and cellular levels. Whenever possible, physicians try to address the cause, not simply manage the symptoms. Different molecular causes of diseases that are clinically similar require fundamentally different therapeutic approaches. Diagnostic tests of cellular and molecular differences among the "same" diseases help physicians select the most appropriate course of treatment.

Because you now understand the molecular bases of a number of clinically similar diseases that have different causes, the importance of targeting the treatment to the molecular problem should be obvious to you. For example, as you learned in chapter 8, people with diabetes cannot regulate blood glucose levels. As their blood glucose increases, insulin-driven absorption of glucose from blood into muscle and fat cells does not occur. Physicians can use the results of clinical chemistry tests that measure glucose to identify a patient with diabetes, but which type of diabetes? In type I diabetes, the pancreas does not make insulin, while in type II diabetes, the pancreas makes plenty of insulin, but cells lose the ability to respond to insulin's signal (see Figures 8.7 and 8.9). The insulin injections that help type I diabetics manage their blood glucose levels would only exacerbate the problem in type II diabetes. For decades, physicians have used a clinical symptom, age of onset, to help them distinguish type I (juvenile onset) and type II (adult onset) diabetes. Age of onset is no longer as informative as it once was, however. More and more children are developing type II diabetes because they are overweight.

Variation among pathogens

Pathogens that are identical under a microscope can be so distinct genetically that they belong to different strains, and different strains often require different treatments. For example, herpes simplex virus (HSV) comes in a variety of types. HSV type 1 (HSV-1) causes cold sores, while HSV-2 causes genital herpes, and antiviral drugs that work on one strain have little effect on

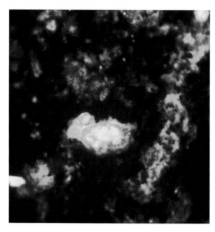

Figure 19.11 Pathogen identification. Monoclonal antibodies with fluorescent tags quickly confirm the presence of HSV in cultured cells. In this case, the monoclonal anitibodies are specific for HSV-2 infections and do not react with HSV-1-infected cells. (Photograph courtesy of Craig Lyeria, Centers for Disease Control and Prevention.)

the other. Even though the two types differ in their preferred infection locations, the difference is not absolute, because 15 to 20% of genital herpes infections are caused by HSV-1. Therefore, in treating genital herpes, it is essential that physicians have a test to distinguish the two types, and the specificity of monoclonal antibodies provides the level of distinction that is necessary (Figure 19.11).

Small genetic differences in pathogens are also responsible for variation within a bacterial species in susceptibility or resistance to a given antibiotic. Knowing whether a patient is infected with a resistant strain is essential to effective treatment. Even though laboratory technicians can identify resistant strains of some bacteria by culturing them on selective media, the technique is time-consuming, and some species of bacteria are difficult to culture (Figure 19.12). New molecular techniques based on immunofluorescence, hybridization, or polymerase chain reaction (PCR) can identify resistant strains in a matter of hours instead of the 2 to 3 days required to culture bacteria.

BIOTECHNOLOGY THERAPEUTICS

Advances in cell and molecular biology are providing society with improved versions of existing therapies, such as human insulin in place of pig insulin, as well as therapeutics that would not be possible without biotechnology. The U.S. Food and Drug Administration has approved many biotechnology-based therapeutics, but medical biotechnology is still in the earliest stages of development. Some of the therapeutics described below have been commercialized, but most are still in the research and development phase of product development. As we discuss some of the new biotechnology-based therapeutics, you will notice a few recurrent themes.

Specificity. In general, understanding the specificities of molecular interactions leads to therapeutic compounds that are more targeted. An example of the value of specificity comes from some of the newer, biology-based cancer treatments. As you know, chemotherapy can severely disable cancer patients because the toxic chemicals intended for the cancer cells do not distinguish cancer cells from normal cells. Researchers have discovered some of molecular differences between cancer cells and normal cells and are exploiting those differences to target the chemotoxins to cancer cells. One approach uses monoclonal antibodies that bind very selectively to cancer cells. Physicians link the antibody to a chemotherapeutic toxin, and the antibody delivers it quite specifically to cancer cells. In addition, the monoclonal antibodies can locate and kill small colonies of cancer cells that have metastasized (Figure 19.13). In other examples, the cancer therapeutic is not a toxin but a molecule, such as Iressa or Gleevec, that interferes with growth signals binding to receptors.

Biological therapeutics. Many of the new therapeutics are natural products synthesized by plants, microbes, insects, and other animals. Others capitalize on the human body's innate ability to heal itself. For example, the body synthesizes growth factors that stimulate the growth of new blood vessels and other molecules that suppress this growth. The first molecule might be used to treat patients with blocked coronary arteries. The second could be a cancer therapeutic, because blocking blood vessel growth in tumor cells would prevent them from receiving nutrients or ridding themselves of waste products.

Figure 19.12 Assessing resistance to an antibiotic. These bacteria are growing on media that contains the antibiotic tetracycline. All of the strains, except the one that is circled, are resistant to tetracycline. (Photograph courtesy of Linda Bartlett, National Cancer Institute, National Institutes of Health.)

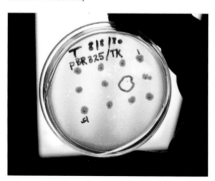

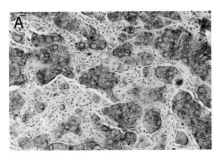
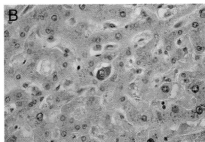

Figure 19.13 Targeted therapy with monoclonal antibodies. Monoclonal antibodies can deliver chemotherapeutic toxins specifically to cancer cells. **(A)** The cytoplasm of tumor cells in breast tissue is stained brown with a monoclonal antibody that recognizes an antigen that occurs in cancer cells but is rare in normal, differentiated cells. **(B)** The same monoclonal antibody is able to locate a single breast cancer cell that has metastasized to the patient's liver. (Photographs courtesy of Jeffrey Schlom, National Cancer Institute, National Institutes of Health.)

New production methods. In the absence of new manufacturing methods, many of the new therapeutics described below would never have become commercial realities. They occur in minute amounts in tissues or are derived from organisms that cannot be kept in laboratory facilities. As a result, large-scale, economically feasible production of some natural compounds with therapeutic potential would be impossible. Recombinant DNA technology, cell culture, and other biomanufacturing technologies provide new ways to manufacture natural molecules. For example, plant cell culture can produce taxol, a drug for treating breast and ovarian cancers (Figure 19.14).

Many of nature's molecules have medicinal properties

Living organisms produce compounds that coincidentally have therapeutic value for people. Microorganisms are the sources of most antibiotics, and a number of medicines that have been prescribed for many years, such as digitalis, are plant products.

Having new ways to manufacture sufficient amounts of rare molecules encourages scientists to investigate many plants and animal as sources of new medicines. Once researchers identify the genes required for synthesizing therapeutic compounds, they can use recombinant techniques to engineer that pathway into microbes for large-scale manufacture. Ticks could provide anticoagulants, and poison arrow frogs might be a source of new painkillers.

Researchers have only recently turned their attention to the extraordinarily diverse ecosystems found in the sea (Figure 19.15). They have discovered compounds that heal wounds, destroy tumors, prevent inflammation, relieve pain, and kill microorganisms. As exciting as these developments are, their therapeutic potential could not become reality without the new biomanufacturing processes. For example, researchers would have to collect 2,400 kg of a sponge in order to obtain 1 mg of a molecule that has anticancer potential. In the absence of cell culture or recombinant DNA technology, using these sponges as sources of a pharmaceutical would not be feasible for both economic and ecological reasons.

The immune system needs a boost at times

Like the armed forces that defend countries, the immune system is made up of different branches, each containing different types of "soldiers" that

Figure 19.14 Taxol production. The anticancer drug taxol occurs in the bark of the Pacific yew tree. **(A)** The slow-growing yew tree reaches 5 feet in a number of decades. **(B)** It takes 30,000 pounds of bark to produce 1 kg of taxol. Between 2,000 and 4,000 trees had to be cut down to obtain that much bark. **(C)** Taxol is extracted from ground bark and purified. Plant cell culture provides another method for taxol production. (Photographs by Mike Trumball, Hauser Northwest, courtesy of National Institutes of Health.)

Figure 19.15 Medicines from the ocean. Marine organisms, such as this sponge, are rich sources of potential therapeutic molecules. Like plants, many invertebrate organisms on coral reefs are sedentary and defend themselves with chemicals. (Photograph courtesy of the National Underwater Research Program, National Oceanic and Atmospheric Administration.)

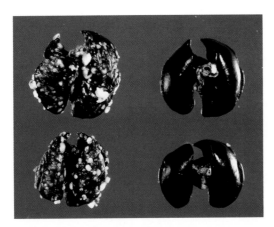

Figure 19.16 Treating cancer with immune system molecules. In this experiment, investigators injected mice with cancer cells. Within a few weeks, more than 250 tumors were evident in their lungs (left). The lungs on the right are from mice treated with interleukin-2, a protein normally secreted by the immune system, and a type of T cell. On average, the treated mice had fewer than 12 tumors. (Photograph courtesy of Steven Rosenberg, National Cancer Institute, National Institutes of Health.)

interact with each other in complex, multifaceted ways. Researchers are discovering ways to enlist the help of the immune system in fighting a variety of diseases.

For example, when the body is under attack, a variety of chemical soldiers, proteins known as **cytokines**, stimulate the cellular soldiers of the immune system and coordinate the complex sequence of events in an immune system attack. Cytokines typically occur fleetingly and in minute amounts, but these proteins can now be produced in amounts that are sufficient for therapeutic use. Physicians have begun using them to treat a variety of diseases (Figure 19.16). Small doses of the cytokine interleukin-2 have been effective in treating various cancers and AIDS, while interleukin-12 has shown promise against some infectious diseases.

Although we typically think of the immune system as a defense against infectious organisms, its protective functions are much broader. Antigens from tumor cells sometimes stimulate a weak attack by the immune system. **Cancer vaccines** help the immune system find and kill the tumor by intensifying the reactions between the immune system and the tumor antigen (Figure 19.17). Unlike vaccines against infectious organisms, cancer vaccines are given *after* the patient has contracted the disease, so they are not preventative.

Some diseases are caused by missing or malfunctioning proteins

You have learned about a number of very serious diseases caused by having insufficient amounts of a single protein. In the past, once medical researchers identified the missing or defective protein, physicians could treat a few of these diseases by giving patients the protein from other mammalian sources, if they could obtain large amounts of it. They gave diabetics insulin extracted from animal pancreatic tissue, collected at slaughterhouses; hemophiliacs relied on human blood transfusions to obtain the protein they lacked.

These life-saving tactics have some downsides, however. The amino acid sequence of an animal protein is usually not identical to that of its human counterpart, so injecting these proteins triggers an immune response in some people. In addition, acquiring the missing protein from extraneous sources carries with it the risk of contamination with pathogens or other harmful substances. Today, recombinant microbes and mammalian cells manufacture

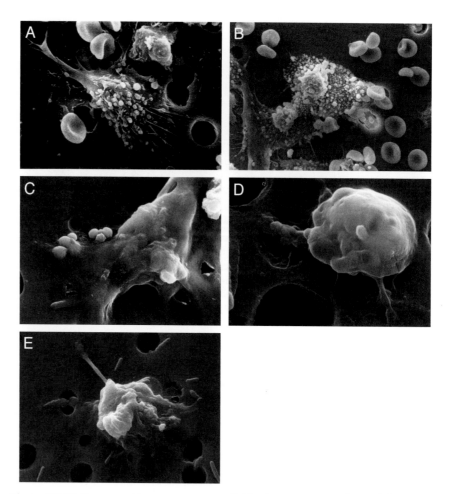

Figure 19.17 Cancer and immune system cells. The immune system has a difficult time distinguishing cancer cells from normal cells because they are both "self." Cancer vaccines teach the immune system to recognize the tumor as foreign. **(A)** A metastatic cancer cell (note the pseudopods). **(B)** Macrophages recognize the cancer cell and begin to stick to it. **(C)** Macrophages inject toxins into the cancer cell, which begins losing its pseudopods. **(D)** The macrophages fuse with the cancer cell, giving it a lumpy appearance. **(E)** The cancer cell shrinks up and dies. (Scanning electron micrographs courtesy of Raouf Guirgus and Susan Arnold, National Cancer Institute, National Institutes of Health.)

human forms of the missing proteins under carefully controlled conditions. In addition to using recombinant proteins to treat diseases described in previous chapters, such as insulin for type I diabetes and glucocerebrosidase for type I Gaucher's disease (Figure 19.18), physicians are replacing missing proteins to treat a number of disorders. Two examples are hemophilia and some cases of emphysema.

Hemophiliacs lack certain proteins in the cascade of molecular events that terminates in the formation of a blood clot. The missing protein in hemophilia A, factor VIII, is now synthesized by recombinant mammalian cells, while the missing protein in hemophilia B, factor IX, is synthesized by a recombinant microbe. Using pure forms of the missing proteins to treat hemophilia obviates the need for blood transfusions that can unknowingly transmit viruses to the recipient.

Some people who have never smoked tobacco begin to show signs of emphysema in their early 30s because they lack a functional form of a protein,

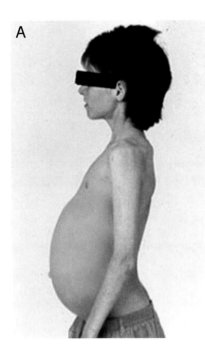

Figure 19.18 Enzyme replacement therapy for Gaucher's disease. In 1991, Roscoe Brady of the National Institutes of Health isolated enough glucocerebrosidase from human placentas to give injections of the missing enzyme to 12 patients. Even though the injections had dramatic effects, isolating the enzyme from placental tissue was not cost-effective. Using recombinant DNA techniques, the gene that encoded the enzyme was engineered into yeast cells, making glucocerebrosidase replacement therapy a viable option. **(A)** Before. **(B)** After. (Photographs courtesy of Roscoe Brady, National Institute of Neurological Disorders and Stroke, National Institutes of Health.)

alpha 1-antitrypsin. This protein protects the lining of the lungs from a destructive enzyme produced by white blood cells. A 1-amino-acid substitution creates a form of alpha 1-antitrypsin that cannot be secreted from the cells that synthesize it, so it never reaches the lungs.

Researchers are testing gene therapy for hereditary and nonhereditary disorders

The ability to isolate and clone specific genes opens up the possibility of doing more than simply replacing missing or dysfunctional proteins. **Gene therapy** would use genes, or related molecules, such as RNA, to treat diseases. Rather than giving daily injections of missing or malfunctioning proteins, medical researchers dream of supplying patients with accurate instruction manuals—nondefective genes. This solution, known as **replacement gene therapy**, would address the cause of the genetic defect. In addition, the patient's body, not a manufacturing facility, would make the protein.

Only some hereditary diseases are amenable to correction via replacement gene therapy. Primary candidates are diseases caused by the lack of a protein, such as hemophilia and severe combined immunodeficiency disease (SCID), commonly known as the "bubble boy disease" (Figure 19.19). Other options are simple genetic disorders that are caused by production of a defective protein, such as Huntington's disease. Small pieces of RNA would interfere with the mRNA transcribed from the mutant gene and block production of the defective protein.

Figure 19.19 Gene therapy. **(A)** Children with SCID, which can have a number of causes, must spend their lives in germ-free environments. **(B)** In 1991, two children with SCID received gene therapy to correct the genetic defect. (Photographs courtesy of National Institute of Allergy and Infectious Diseases, National Institutes of Health [A], and Michael Blaese, National Institutes of Health [B].)

Table 19.3 Gene therapy trials for nongenetic diseases

Brain tumors

Liver cancer

Prostate cancer

Colon cancer

Melanoma

Head and neck cancer

Breast and ovarian cancer

Non-small-cell lung cancer

Hemophilia

Hemoglobin disorders

Hyperlipidemia

Metabolic storage disorders

AIDS

Asthma

Muscle atrophy

Graft vs. host disease (organ or marrow transplants)

Cardiovascular disease

Hypercholestemia

Infectious diseases

Neurodegenerative diseases

Gene therapy has proved to be much simpler in theory than it is in practice. The initial enthusiasm for the extraordinary promise of replacement gene therapy has been tempered with a large dose of reality. Although researchers have tried to use replacement gene therapy to treat various monogenic disorders, few of the clinical trials have demonstrated improvements significant enough to qualify as gene therapy and not simply gene transfer. Significant technical barriers must be overcome before replacement gene therapy can live up to its potential to cure inherited genetic disorders. The barriers include getting replacement genes into the appropriate cells, inserting them into the proper site within the genomes of those cells, and getting them to function and respond to normal physiological signals (Figure 19.20). Also, as is always true of any medical intervention, gene transfer has risks. The clearest demonstration of the benefits of gene therapy to date—gene transfer for a sex-linked type of SCID—also led to serious health problems. European researchers gave correct copies of the mutant gene to 16 children, and 15 were leading normal lives as long as 4 years after receiving the gene. However, two developed leukemia because the gene inserted itself into a tumor suppression gene.

Even though technical impediments may have tempered initial optimism about using replacement gene therapy for inherited genetic disorders, enthusiasm for the potential of more **transient gene therapy** as a tool for treating other diseases has grown. Researchers are investigating the use of briefly introduced genes as therapeutics for a variety of diseases that are not typically thought of as hereditary disorders (Table 19.3).

Some diseases can be treated with donated organs and cells

In the United States, more than 60,000 people are on organ recipient lists. Another 100,000 need organs but are not on lists. As a result, in the United States, approximately 12 people die each day waiting for organs to become available for transplantation. To lessen this problem, medical researchers are investigating alternative organ sources. They are also using human cell culture to increase the number of patients who might benefit from a single organ donor or to extend survival until an organ is donated. For example, implanting liver cells grown in cell culture into patients has kept them alive until a donated liver became available.

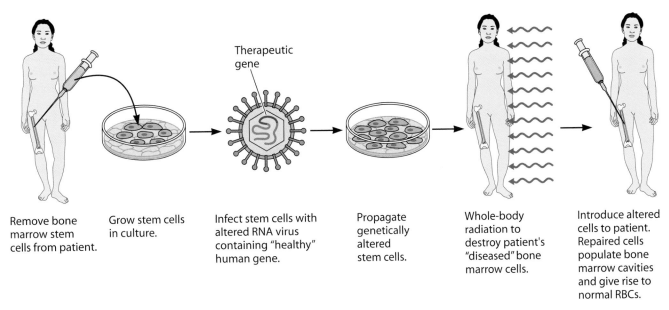

Therapeutic gene

Remove bone marrow stem cells from patient.

Grow stem cells in culture.

Infect stem cells with altered RNA virus containing "healthy" human gene.

Propagate genetically altered stem cells.

Whole-body radiation to destroy patient's "diseased" bone marrow cells.

Introduce altered cells to patient. Repaired cells populate bone marrow cavities and give rise to normal RBCs.

Figure 19.20 Gene replacement therapy. For genetic defects that affect cells derived from bone marrow stem cells, such as SCID, bone marrow is removed from the patient, and the stem cells are multiplied in cell culture. A correct copy of the gene is inserted into a viral vector, using recombinant DNA techniques. The virus is cultured with the bone marrow cells. It infects the cells and inserts the replacement gene into some of them. Radiation destroys the patient's defective bone marrow cells, and the physician injects cultured cells, now containing the correct gene, into empty bone marrow cavities. To date, only a few of the gene replacement trials have been therapeutic.

Xenotransplantation

Heart surgeons have used pig heart valves to repair faulty human heart valves for many decades. Organs from pigs or other mammals could also be a potential source of donor organs and therapeutic cells, but there are significant obstacles to **xenotransplantation** (xeno, from the Greek word meaning foreign). The first is the body's self-protective response. When nonhuman tissue is introduced, the body cuts off blood flow to the donated organ. The most promising method for overcoming this rejection may be genetic modification of the donor animals. One approach deletes the pig gene for the protein that triggers the rejection; another adds a few genes for human membrane proteins to disguise the pig cells as human cells.

A more significant obstacle may be the risk of infectious viruses or retroviruses that have been permanently incorporated in pig DNA. Even though a 1999 study of 160 people who had received pig cells as part of their treatment showed no signs of ill health related to having received the cells, the spread of disease from mammals to humans through xenotransplantation is a cause for concern that must be addressed before xenotransplantation becomes a viable alternative to organ donation.

Cell transplant therapy

The most familiar form of cell transplant therapy is the 20+-year practice of transplanting bone marrow cells into cancer patients. In some cases, the patient's own bone marrow cells are removed, grown in culture, and reimplanted after chemotherapy to save the bone marrow cells from the cell toxin. However, if the patient suffers from leukemia or another blood cell cancer, the transplanted bone marrow must come from a healthy donor who is genetically similar to the patient.

For certain diseases, cell therapy comes closer to curing a disease than do repeated injections of missing proteins. For example, to treat type 1 diabetes, researchers have implanted insulin-producing cells from organ donors into patients. Eighty percent of the patients required no insulin injections 1 year after receiving the pancreatic cells; after 2 years, 71% had no need for insulin injections.

As is true of whole-organ transplants, donated cells stimulate the patient's immune system responses, mediated by T cells, that recognize and reject foreign cells. Researchers have used their understanding of the molecular details of the immune system to develop monoclonal antibodies that suppress this response by binding to various receptors on T cells. In addition, they have developed molecular capsules to encase the implanted cells. **Cell encapsulation** allows cells to secrete hormones or provide a specific metabolic function without being recognized by the immune system.

REGENERATIVE MEDICINE

The human body has a remarkable capacity to repair and maintain itself. If necessary, your liver can regenerate up to 50% of itself, and by the time you finish reading this paragraph, you will have manufactured 200 million brand new red blood cells. Your body's toolbox for self-repair and maintenance contains many growth factors, signaling molecules, and various types of stem cells. Medical scientists are excited about the prospect of using the body's natural healing processes not simply to treat debilitating diseases but perhaps to cure them. Two scientific breakthroughs in the 1990s that are generating this optimism are the development of methods for maintaining human embryonic stem (hES) cells in culture and somatic cell nuclear transfer, which we discussed in chapter 17.

Endogenous proteins promote cell division and differentiation

The human body produces small amounts of a wide array of growth factors that promote cell growth, stimulate cell division, and, in some cases, guide cell differentiation. As proteins, they are prime candidates for large-scale production by recombinant cells. Injections of these endogenous proteins can heal wounds and regenerate injured tissue.

- Epidermal growth factor stimulates skin cell division and could be used to encourage wound healing in burn victims.
- Fibroblast growth factor, which stimulates cell growth and division, has been effective in healing ulcers and broken bones and growing new blood vessels in patients with blocked coronary arteries.
- Transforming growth factor beta helps cells differentiate into different tissue types.
- Nerve growth factor encourages nerve cells to grow and could repair damage resulting from head and spinal cord injuries or degenerative neural diseases, such as Alzheimer's.

Stem cells can repair and regenerate tissue

As you learned in chapter 10, during development from a fertilized egg to an organism, cells differentiate and organize themselves into tissues and organs. However, some tissues and organs always maintain a population of adult stem (AS) cells to continually replenish the supply of certain cells with a high turnover rate and to replace cells that have died or been injured. Other tissues have no resident stem cell populations.

AS cells are partially differentiated **progenitor cells**, which are cells that give rise to other cells. When AS cells receive a cue to become fully differentiated, they first divide in two. One daughter cell differentiates, and the other remains undifferentiated, ensuring a continual supply of stem cells for regenerating that tissue type (Figure 19.21). All AS cells are multipotent, but different types of AS cells display various degrees of plasticity. Bone marrow contains AS cells capable of differentiating into all cell types found in blood, as well as bone. Liver AS cells can become any of the specialized cells of the liver—bile-secreting cells, glycogen storage cells, or cells that line the bile duct. Under normal conditions, however, AS cells in the liver cannot differentiate into white blood cells, and bone marrow stem cells do not become liver cells.

In November 1998, two groups of researchers finally succeeded in establishing pure cultures of hES cells. One group isolated and cultured certain cells from a frozen blastocyst, which you recall is the 5-day-old ball of 100 to 150 cells that eventually develops into an embryo (Figure 19.22). The other group isolated and cultured ES cells from progenitor germ cells (gametes) isolated from fetuses that had been aborted. Unlike AS cells, ES cells are pluripotent; they can differentiate into any kind of cell in the body. In addition to their total developmental plasticity, researchers believe hES cells in culture may be able to reproduce without limit.

Therapeutic potential of stem cells

The combined characteristics of developmental versatility and unlimited capacity for self-renewal make stem cells excellent therapeutic tools. Now that researchers know how to maintain different types of AS and ES cell lines in culture and are also learning how to direct their development into specific cell types, their enormous therapeutic potential may become reality. Although significant technical impediments exist and some very basic scientific questions remain unanswered, in the future, physicians may be able to use AS or ES cells to replace damaged or dead cells; reestablish function in stroke victims; cure diabetes; regenerate damaged heart muscle, spinal cord, or brain tissue; and treat diseases associated with aging, such as Alzheimer's.

For example, medical researchers in Germany have used cultured AS cells to repair tissue severely damaged after heart attacks. They injected the

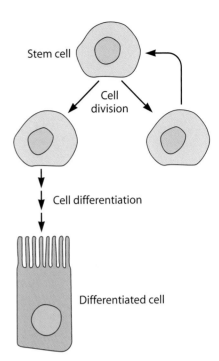

Figure 19.21 Stem cell proliferation. To maintain a constant supply of stem cells while continuing to provide differentiated cells for renewing tissue, a single stem cell divides into two daughter cells. One daughter differentiates, and one daughter remains a stem cell.

Figure 19.22 ES cell culture. To generate a culture of ES cells, researchers remove the inner cell mass from a blastocyst. Inner cell mass cells and their derivative ES cells have the potential to become any cell type. If placed into the uterus, however, neither will develop into a complete organism, because inner cell mass and ES cells cannot implant into the uterine wall. Trophoblast cells are required for implantation.

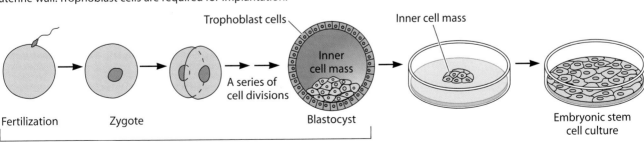

patients' own heart stem cells into their coronary arteries and saw improvements within a few weeks. On average, the area of damaged heart tissue decreased by 36% and heart function improved 10%.

The therapeutic potential of ES cells surpasses that of AS cells because ES cells can become any type of cell. In addition, ES cells are the only available stem cells for repairing tissues that do not retain a group of AS cells that can be mobilized for treating injuries and diseases. Encouraging results using ES cells to cure some nervous system disorders have been observed in laboratory animals. A number of researchers successfully treated an animal form of Parkinson's disease by injecting the animals with ES cells that differentiated into neurons that could synthesize and release dopamine, the neurotransmitter Parkinson's patients lack. Other laboratories used ES cells to restore certain motor functions in animals that had had strokes. However, in spite of the encouraging results from animal research, using ES cells to treat human diseases is proceeding slowly.

Research on differentiation

To maximize the therapeutic potential of AS or ES cells, medical researchers must tease apart the intricate (sometimes almost quirky) combination of environmental factors, molecular signals, and internal genetic programming that interact in deciding a cell's fate. In a fairly straightforward example, Israeli scientists successfully orchestrated ES cell differentiation into a number of cell types by providing cocktails of different growth factors and chemical differentiation signals at different times. Other scientists, working with a type of AS cell, mesenchymal stem cells, gave them the proper mix of nutrients and growth factors, and the cells differentiated into three terminally differentiated cell types: fat, bone, and cartilage (Figure 19.23). However, the mesenchymal stem cells must be touching each other to become fat cells, or if the cell density is too high, they will not differentiate into bone cells even when provided with the appropriate nutrients and chemical signals. Neural stem cells require a shot of vitamin A to nudge them along certain developmental paths, but not others, while endothelial cells in blood vessels release chemical signals that regulate the multiplication and differentiation of the neural stem cells in the vicinity.

Another line of research assesses the developmental plasticity of a certain type of AS cell. For example, scientists are determining where in the differentiation process the identity of an AS cell type becomes fixed. Under certain conditions, can a mesenchymal stem cell be coaxed into becoming a cardiac stem cell? Is a broader degree of flexibility possible? For example, can a mesenchymal stem cell become a type of stem cell that derives from an ectodermal stem cell, such as a neural stem cell? Converting one AS cell type into another is known as **transdifferentiation**.

Research on dedifferentiation

Another approach to developing stem cell therapies takes a different tack. Rather than determining the molecular events that occur during differentiation and turn a stem cell into a specific cell type, scientists are studying the **dedifferentiation** process. In other words, what are the factors that wipe out the identity of a fully differentiated somatic cell and return it to the embryonic state of complete plasticity? As we discussed in chapter 17, before Dolly's birth, scientists did not know they could ask that question, much less attempt to answer it. Biologists had long assumed that a fully differentiated

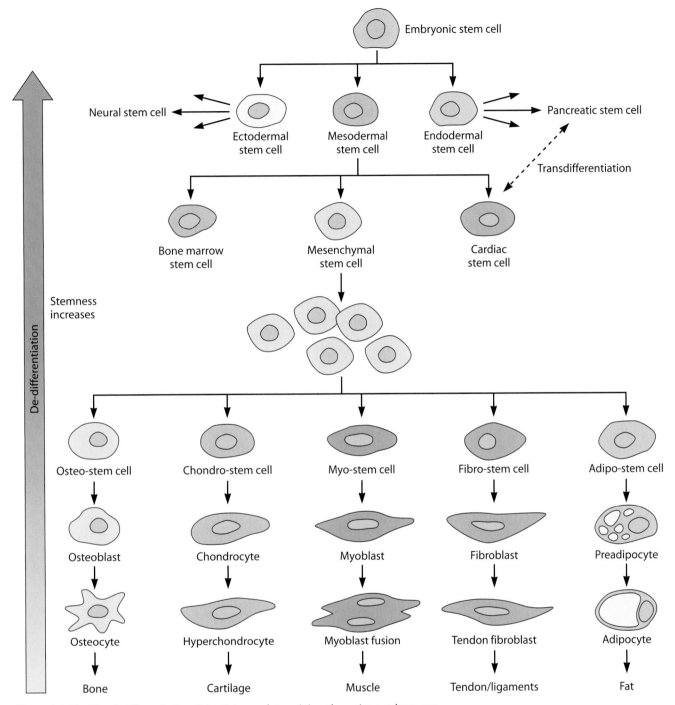

Figure 19.23 AS cell differentiation. Scientists are determining the unique culture conditions, necessary growth factors, and nutritional requirements that turn a specific type of adult progenitor stem cell into terminally differentiated cells. Note that cells have different levels of stemness in the differentiation process. Scientists are also studying the factors responsible for dedifferentiation and transdifferentiation.

somatic cell could not revert to the unspecialized status it had relinquished in the first few days after fertilization. Even though Dolly was a single, hard-won success in decades of experimental attempts to create a clone of an adult animal, her existence generated great excitement in the scientific community by proving that the genetic programming of a specialized somatic cell could be erased.

Discovering the factors that reverse the differentiation process would solve two problems—one political, the other biological—that are impeding the development of stem cell therapies: using embryos and immune system rejection of foreign cells. If medical researchers learned how to take fully differentiated somatic cells from patients, return them to an undifferentiated state, and then implant them, as either pluripotent or multipotent cells, the therapeutic cells would be genetically identical to the patient. In addition to generating a sufficient number of cells, this method would not depend on using an embryo.

However, to identify the factors responsible for dedifferentiation, researchers will need to identify the genes that turn on and off as a somatic cell returns to its undifferentiated state. Unfortunately, the best way to return a differentiated somatic cell to its undifferentiated state is to fuse it with an unfertilized egg that has no nucleus or to remove the somatic cell nucleus and place it in an unfertilized, enucleated egg. In other words, to learn how to create therapeutic ES cells without using embryos, researchers will need to conduct research on ES cells isolated from embryos. We discuss this paradox in more detail in the next chapter.

Some researchers have found that certain terminally differentiated blood cells and neurons revert to a state of "stemness" if starved. However, as you can see in Figure 19.23, there are various levels of stemness. Unless the degree of stemness that is achieved by starving cells is embryonic and not simply some level of adult stemness, this research will do little to help scientists understand that most critical stage in dedifferentiation: from multipotency back to pluripotency. This is the stage that scientists must tease apart in order to avoid using embryos in the future.

Researchers are using cell culture to grow new tissues in the laboratory

Tissue engineering, a marriage of cell biology and recent advances in materials science, allows scientists to create semisynthetic tissues and organs consisting of biocompatible scaffolding materials and living cells grown in culture (Figure 19.24). It will be at least 10 to 20 years before tissue-engineering technologies produce routine clinical procedures.

The most basic form of tissue engineering uses fully differentiated cells, taken from the patient and multiplied using cell culture techniques, and either synthetic polymers or natural biological materials, such as collagen, for scaffolding. The tissue is grown in a dish in the laboratory and then implanted. For example, physicians now use sheets of skin grown in the laboratory to heal wounds and burns. The two-layered skin is produced by first infiltrating a collagen gel with a patient's connective tissue cells and then adding a layer of tougher protective epidermal cells to create the outer skin. In another example, tubes of a synthetic, flexible polymer provided the structural guide for creating blood vessels, again using differentiated cells from the patient. After implantation, adjacent cells invade the new, laboratory-grown tissue, integrating it into the surrounding tissue, and the scaffolding is degraded and absorbed by the body.

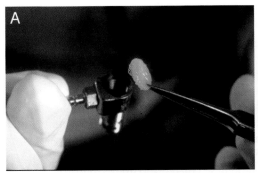

Figure 19.24 Tissue engineering. **(A)** The small tissue construct uses cartilage stem cells, grown in culture, embedded in a biocompatible scaffolding material. When implanted into damaged bone, the construct stimulates the growth and differentiation of bone stem cells, which regenerate healthy bone. **(B)** Scientists are culturing stem cells to develop an implantable artificial pancreas that will regulate insulin for more than a year before needing to be replaced by a minor surgical procedure in hope of developing healthy pancreatic tissues for diabetics. (Photographs by Gary Meeks, courtesy of the Bioengineering Department, Georgia Institute of Technology [A], and courtesy of the Georgia Institute of Technology-Emory Center for Engineering Living Tissues [B].)

Simple tissues and organs, such as skin, cartilage, and urinary bladders, which have simple three-dimensional structures and consist of one or two cell types, were the first tissues to be engineered successfully. Ultimately, the goal is to use stem cells, treated with the appropriate growth factors and chemical signals, to create whole organs consisting of a number of tissue types and intricate internal structures. Scaffolding, shaped like the diseased or injured organ, containing appropriate growth factors, and spiked with stem cells grown in the laboratory prior to implantation, is placed in the body where needed. Once implanted, the laboratory-grown stem cells multiply and differentiate. For certain tissues, the patient's AS cells also populate the scaffolding with cells by migrating to the implant, invading the structure, reproducing, and differentiating into the appropriate tissue type. The fully differentiated cells near the implant make the researcher's job easier. They "teach" laboratory-grown and endogenous stem cells what type of cell to become, because the cellular environment always influences stem cell differentiation.

Although this may sound like science fiction, results from research on laboratory animals support this hope for the future. Researchers at the Massachusetts Institute of Technology seeded three-dimensional scaffolds with hES cells and treated different sections of the scaffold with the appropriate growth factors to induce the cells to differentiate into different tissue types. After 2 weeks, the scaffold contained three different layers of tissue, representing each of the three germ layers: nervous tissue (ectoderm), cartilage (mesoderm), and liver (endoderm) with blood vessels embedded in it. After they implanted the scaffold under the skin of mice, the stem cells continued to differentiate and the mouse's blood vessels made connections with the tissue in the scaffold.

In another experiment, researchers removed a piece of spinal cord from a number of rats and replaced it with scaffolding containing neural stem cells. Interestingly, the stem cells did not differentiate further. Instead, they released chemical signals that stimulated the adjacent neurons to reconnect the spinal cord. These same researchers implanted a scaffold containing neural stem cells but no growth factors into the brains of mice that had suffered strokes. The stem cells differentiated into two types of nervous system cells, while neurons and blood vessels from healthy brain tissue nearby infiltrated the scaffold.

The experimental results of tissue engineering in animals are encouraging, but what about using human stem cells and growth factors to create complex organs in people? Medical researchers have achieved remarkable results using a biohybrid kidney to maintain patients with accident-induced kidney failure until their injured kidneys repaired themselves. The "kidney" was made of hollow tubes seeded with kidney stem cells that proliferated until they lined the tube's inner wall. These cells differentiated into the type of kidney filtration cell that releases the hormone aldosterone. In addition to carrying out the expected metabolic functions of filtration, urine transportation, and hormone release, the cells in the hybrid kidney also responded to hormonal signals produced by the patient's healthy organs and tissues. A group of patients with only a 10 to 20% probability of survival due to acute renal failure regained normal kidney function and left the hospital in good health because the hybrid kidney prevented the fatal events that typically follow kidney failure: infection, sepsis, and multiorgan failure.

Somatic cell nuclear transfer can circumvent the problem of tissue rejection

The biohybrid kidney just described was not implanted in the patient's body. Instead, the physical setup resembled the relationship between a patient and a kidney dialysis machine. This arrangement prevents the patient's immune system from rejecting kidney stem cells harvested from a genetically dissimilar organ donor. Even though we alluded to this earlier, we want to make it clear that the potential therapeutic value of stem cell therapy and tissue engineering will be maximized only if the therapeutic stem cells and any tissues derived from them are genetically identical to the recipient. This criterion can be met easily if the damaged cells, tissues, or organs have a healthy population of AS cells that physicians can harvest from the patient. If not, then what are the options for generating immune-compatible stem cells to repair or replace damaged and diseased tissue?

Recently, some parents, planning for their newborn babies' future medical needs, have begun asking physicians to freeze umbilical cord blood. Those blood cells have greater plasticity than AS cells and may be pluripotent, like ES cells. Another option for generating immune-compatible stem cells will open if scientists successfully uncover the series of environmental factors that return somatic cells to their embryonic state. Until that time, generating immune-compatible ES cells for therapeutic applications will rely on somatic cell nuclear transfer—the laboratory procedure that led to Dolly. Recall that in reproductive cloning, researchers remove a fully differentiated somatic cell (or only its nucleus) and fuse it with an enucleated egg cell. Under carefully controlled conditions, this erases the identity of the somatic cell. The egg begins to divide as if it had been fertilized. After 4 to 5 days of cell divisions, the resulting 100- to 150-cell blastocyst is implanted into an adult female. In Dolly's case, the somatic cell was a sheep udder cell, and Dolly was genetically identical to the female sheep that provided it (except for the mitochondrial DNA of the egg donor) (see Figure 17.7).

To create ES cells that are genetically identical to a patient's, medical researchers follow the first few steps just described. They remove a somatic cell from the patient and insert its nucleus into an egg cell. The egg divides until it becomes a blastocyst, but the blastocyst is not implanted (Figure 19.25).

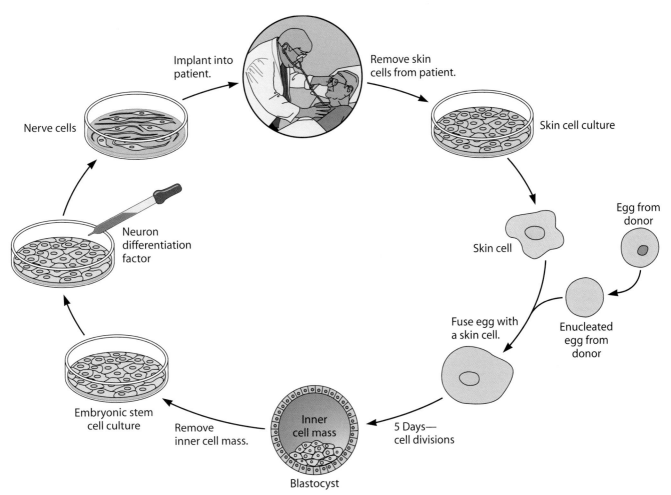

Figure 19.25 Immune-compatible stem cells. To maximize the therapeutic potential of stem cells, the cells must not be seen as foreign by the immune system. This is best achieved by using the patient's own cells. In this example, the patient has a disease or injury that is repairable by neural stem cells, such as Parkinson's disease or a spinal cord injury. Nuclei removed from the patient's skin cells are implanted in a donated, enucleated egg. After the cells are cultured for approximately 4 to 5 days, a blastocyst is produced. The inner cell mass of the blastocyst is removed and cultured, creating a line of hES cells that are genetically identical (except for the mitochondrial DNA from the egg donor) to the patient. Treatment with appropriate differentiation factors converts the ES cells into neural stem cells and then into fully differentiated nerve cells. These are implanted in the patient in hope of replacing the diseased or injured cells with healthy nerve cells.

Instead, the researchers remove the inner cell mass and culture those cells to create a line of hES cells that is genetically identical to the patient (except for mitochondrial DNA—unless female patients also provide the egg cells). These stem cells would serve as the source of cells for treating the patient. The process just described is commonly called **therapeutic cloning** to distinguish it from reproductive cloning, in which the blastocyst is implanted.

Even though ES cells derived from genetically dissimilar donors have therapeutic value, they are foreign to the patient's immune system, which will reject them. To maximize the therapeutic potential of ES cells, they must be genetically identical to the patient. Therefore, even though it is possible to discuss the ethical implications of hES cell research and cloning research separately, as you can see, often these topics become inextricably intertwined.

VACCINES

There is much truth to the adage "an ounce of prevention is worth a pound of cure." The best way to battle diseases is not to develop new therapies for treating or curing them but to prevent them. Researchers are developing better preventative agents by improving on a practice that has been used for 200 years: vaccination. Vaccines, which help the body recognize and fight infectious diseases, are analogous to the threat of war that incites countries to build up a supply of weapons.

Biotechnology is improving existing vaccines and creating new ones

The vaccines you received as a child, such as those for polio, diphtheria, and measles, contain either killed or live, but weakened, viruses. When vaccinated with these nonvirulent viruses, your body produces antibodies to those organisms, but you don't get the disease. If you are exposed to the virus again, your body has a supply of antibodies for defending itself. For the most part, vaccines cause no serious problems, but they do have side effects: allergic reactions, aches and pains, and fever. On rare occasions, the vaccine causes the disease it is intended to prevent.

A second problem with this method of vaccination is the method of production. Growing large amounts of some human-pathogenic viruses outside of the human body is not easy, because viruses need the biochemical machinery of a living cell to reproduce (Figure 19.26). Finally, developing any pharmaceutical, including vaccines, requires human testing, which always carries a certain amount of risk.

Using recombinant DNA technology, scientists are able to overcome some of these barriers to vaccine development. Usually, a few proteins on the surface of the pathogen are responsible for triggering antibody production. By isolating the gene for the pathogen's cell surface protein(s) and inserting it into a yeast or a bacterium, such as *Escherichia coli*, pharmaceutical companies can use biomanufacturing to produce large quantities of this protein to serve as the vaccine. When the protein is injected, the immune system responds and produces antibodies that can recognize the pathogen. However, because the whole virus is not injected, the adverse side effects of vaccination are minimized. Using this new strategy, companies have developed vaccines against life-threatening diseases, such as hepatitis B and meningitis. Because these vaccines do not rely on whole organisms, developing and testing vaccines for the deadliest viral diseases have become more feasible.

In addition to creating vaccines against pathogens that cause familiar infectious diseases, researchers are also developing vaccines against viruses that cause cancer. For example, the great majority of cervical cancer cases are caused by a virus that is transmitted from males to females during sexual in-

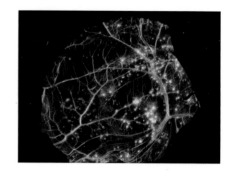

Figure 19.26 Vaccine production. To produce vaccines for viral diseases, the virus must be grown in living tissue. Typically, companies that manufacture vaccines use the embryos in chicken eggs. In this photograph, pockmarks in chick embryonic tissue indicate colonies of the smallpox virus. (Photograph courtesy of John Noble, Centers for Disease Control and Prevention.)

tercourse. Recent tests have shown a significant decrease in the incidence of cervical cancer in females who received a vaccine against this virus. Vaccinating young girls before they become sexually active could decrease the incidence of cervical cancer by approximately 70%. Finally, researchers are also broadening the vaccine concept beyond protection against infectious organisms. Various studies of vaccines against diseases such as diabetes, chronic inflammatory disease, Alzheimer's disease, and a number of cancers are being conducted.

DNA can also trigger an immune response

Much to the surprise of many scientists, injecting naked DNA into muscles or skin cells also elicits an immune response. In early gene therapy trials, the immune response against the therapeutic protein was too strong for the gene therapy to be effective. While this result was disappointing to the gene therapy researchers, other scientists saw the positive side. The exciting and unexpected discovery of **DNA-based vaccines** could lead to more advances in vaccine production, improved vaccines with fewer side effects, and more organisms for which we can develop effective vaccines.

How do DNA vaccines work? Researchers insert genes for one or more of the pathogen's proteins into a small, circular, noninfectious piece of DNA called a plasmid. After introduction into the host, the host cells synthesize the pathogen protein(s) encoded by the injected DNA (Figure 19.27). Recognizing the protein as foreign, the immune system produces both antibodies and T cells specific for that antigen. DNA vaccines against AIDS, malaria, herpes, hepatitis B, and influenza are currently in clinical trials.

Figure 19.27 DNA vaccines and the immune response. Plasmids altered to carry a gene for a protein (antigen) produced by a pathogen are injected into muscle cells. The gene encoding the antigen is transcribed into mRNA, which moves to the ribosome, where it is translated into the antigen. The cell secretes some copies of the antigen and chops others into small pieces. Proteins that identify every cell in the body as self carry the antigen pieces to the cell surface. In response, the immune system synthesizes T cells that will recognize the pathogen's antigen, while the secreted antigens trigger the production of antibodies by the B cells of the immune system.

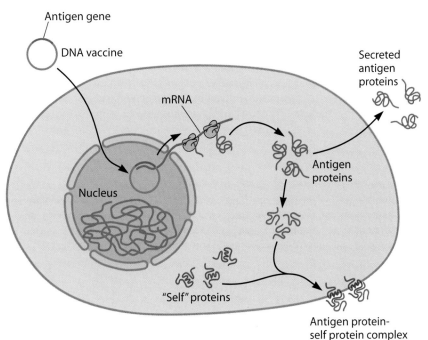

Researchers recently developed food-based vaccines

Whether the vaccine is a live virus, a coat protein, or a piece of DNA, vaccine production requires elaborate and costly facilities and procedures—and then there's the issue of painful injections. How could society use the tools of biotechnology to lessen these problems, making vaccines more available to all people?

Industrial and academic researchers are circumventing both of these problems by using recombinant DNA technology to create edible vaccines, manufactured by plants and animals. Earlier, we mentioned the prospect of using genetically engineered plants and animals for the production of therapeutic proteins. Transgenic organisms could also produce antigens for vaccines made of a pathogen's cell surface proteins. Company scientists have genetically engineered goats to produce a malaria antigen in milk. University researchers have obtained positive results using human volunteers who consumed hepatitis vaccines in bananas and *E. coli* and cholera vaccines in potatoes. If edible vaccines become a reality, being vaccinated by drinking a glass of milk and eating a banana is much more appealing than a shot in the arm, isn't it? In addition, because these vaccines are genetically incorporated into food plants and need no refrigeration, sterilization equipment, or needles, they may be especially important for developing countries.

S U M M A R Y   P O I N T S

Biotechnologies provide medical researchers with new tools for investigating the molecular causes and symptoms of disease at the cellular level and create new products that clinicians can use in diagnosing, treating, and preventing health problems.

For centuries, disease diagnosis depended on visible clinical symptoms. During the 20th century, technological innovations helped physicians identify diseases earlier in the disease process. The new biotechnology-based diagnostic tests allow them to diagnose diseases earlier than ever before, and the earlier a disease is identified, the greater the probability of successful treatment. In addition, biotechnology-based diagnostics rely on molecular interactions, which increases accuracy.

Genetic information is becoming increasingly important in disease diagnosis because the Human Genome Project has provided researchers with a much broader informational base for linking diseases with specific mutations. Genetic information can help physicians diagnose certain diseases before any clinical or measurable symptoms appear. The first available genetic tests will be for simple genetic disorders that are inherited in a Mendelian fashion. As researchers discover more about the genetic and environmental bases of more complex diseases, genetic tests will help physicians identify patients with genetic predispositions to certain diseases.

The new molecular diagnostics are quite sensitive to minute changes in a molecule's structure and function. This level of detail makes it possible to target treatment more specifically to the genetic makeup of the person, the molecular basis of the disease, or the molecular biology of the specific pathogens. All of these factors increase the probability that the treatment will be effective and have fewer adverse effects on the patient.

Advances in biotechnology are also improving existing therapeutics and providing physicians with many new therapeutic options. Many of the newer therapeutics are more targeted to the specific problem or disease cause rather than to secondary and tertiary symptoms, which minimizes detrimental effects and maximizes efficacy.

New manufacturing methods, based on biotechnology, make it possible to use many molecules that would not have been available in large enough amounts to warrant consideration as potential therapeutics. For example, many endogenous proteins produced by the immune system can now be used therapeutically, and diseases caused by a malfunctioning or absent protein can be treated by providing patients with the proteins they lack.

The medical research community's initial enthusiasm for the prospect of curing or treating genetic defects by providing patients with correct versions of genes has been dampened by the lack of clinical successes. To date, gene transfer to correct a form of SCID provides the only example of a gene being therapeutic, and a small percentage of those who benefited from the gene transfer developed leukemia caused by the gene transfer. However, researchers have successfully treated other "nongenetic" diseases with transient gene therapy.

S U M M A R Y   P O I N T S *(continued)*

The goal of regenerative medicine is to mobilize the body's innate capacity to heal itself. Endogenous growth factors, signaling molecules, and various types of stem cells can be used to regenerate or repair diseased or injured tissues. In addition, researchers are using biomaterials and cells grown in cell culture to create replacement organs. To maximize the therapeutic value of regenerative medicine, a patient's cells should serve as the source of genetic material. At present, this requires insertion of a somatic cell nucleus into a donated egg to generate ES cells with the same nuclear genetic material as the patient.

If scientists can discover the factors that cause a somatic cell to dedifferentiate, then it will not be necessary to create ES cells.

Biotechnology provides new options for developing antigen-based vaccines rather than vaccines that use whole organisms. This advance should make it possible to develop more and safer vaccines. In addition, scientists have recently discovered that injected DNA can sometimes trigger an immune response. This discovery further broadens the list of possible vaccines.

K E Y   T E R M S

AIDS-defining illness	Dedifferentiation	Progenitor cell	Transient gene therapy
Alpha 1-antitrypsin	DNA-based vaccines	Replacement gene therapy	Xenotransplantation
Biomarkers	Familial	Simple genetic disease	
Cancer vaccines	Gene therapy	Therapeutic cloning	
Cell encapsulation	Multifactorial disorder	Tissue engineering	
Cytokines	Multigenic disorder	Transdifferentiation	

chapter 20 | Medical Biotechnology in Society

Nowhere is the interplay of science, technology, and society described in chapter 1 more apparent than in the application of molecular biology and biotechnology to human health care. In chapters 17 and 18, we introduced methodologies and strategies for rationally discussing the societal impacts of advances in biotechnology, such as issues related to product safety and risks, as well as societal benefits and costs. In subsequent chapters, we will amplify our discussion of the risks and safety of biotechnology applications. While there are many health and safety issues related to medical biotechnology applications, in this chapter we limit our discussion to some of the ethical issues and political ramifications. As always, our goal is to help you learn *how* to think about these issues by using concrete examples. We focus on two types of medical biotechnology applications that have received a substantial amount of attention from politicians and ethicists, genetic diagnostic tests and human embryonic cell therapies.

As you learned in chapters 17 and 18, when it comes to assessing a product's safety, regulatory authorities and product developers depend on science-based answers to questions about risk. While scientific information can contribute factual substance to discussions of ethical issues, factors other than science play a much more significant role in each person's assessment of the ethical implications of various medical practices. In this chapter, before describing some of the societal issues surrounding genetic testing and human embryonic stem (hES) cell therapies, we will provide relevant scientific and historical information to ground the discussion. Typically, the media cannot convey this level of detail to audiences, no matter how essential it is to a full understanding of the debate, because of space and time constraints. In describing just a few of the legal, ethical, and political issues of medical biotechnology, we raise certain questions, describe the issues, and point out problems, but we do our best not to offer answers or opinions.

Table 20.1 Disease screening in asymptomatic populations

Blood pressure measurements for hypertension

Blood cholesterol for cardiovascular health

Mammography for breast cancer

Papanicoulaou smears for cervical cancer

Fecal occult blood testing or sigmoidoscopy for colorectal cancer

RISKS AND BENEFITS OF DISEASE SCREENING

In the last chapter, you learned that early detection of a disease or disorder allows physicians to intervene sooner rather than later, which can limit damage and sometimes save lives. Knowing the benefits of early disease detection, it may seem logical to conclude that physicians should screen patients for as many diseases as possible, whether or not the patient has any disease symptoms. But out of hundreds of available tests that physicians could use to diagnose diseases and disorders in presymptomatic patients, they routinely screen for only a handful (Table 20.1).

Even though it may be tempting to regard their inaction as negligent, a better descriptor is responsible. All actions carry some level of risk, even those that seem innocuous or beneficial, such as trying to identify a disease before any clinical symptoms appear. The risks of invasive diagnostic tests that require physicians to remove tissue samples are obvious, as are those based on X rays and injected dyes, radioactive molecules, or other imaging agents. Having blood drawn is much less risky than being subjected to a surgical procedure that removes tissue, but the risk of having a blood sample collected is not zero.

These risks are easy to grasp because they are associated with doing bodily harm. Other risks associated with diagnostic testing are much less obvious because they are rooted in statistics. The results of diagnostics tests can be inaccurate. A **false positive** indicates that a person has the disease when he or she does not; a **false negative** tells a person who has the disease that he or she does not (Table 20.2). The risks associated with these two types of inaccurate diagnoses can harm patients in different ways. False negatives might lull a patient into a state of complacency in which they ignore disease symptoms, dismiss the need for future tests, and neglect to make lifestyle changes that could slow disease progress. People who receive a false-positive diagnosis can become unnecessarily stressed, which causes health problems. Their anxiety may cause them to change their lifestyle, which in turn might detract from their health. Even more serious, physicians must respond to a positive diagnostic test by conducting more tests. Typically these tests are more invasive and more expensive. All in all, the risks of receiving a false-positive diagnosis are generally more serious than those associated with a false negative.

A number of factors affect clinical risks and benefits of a diagnostic test

The medical community screens asymptomatic people for the diseases listed in Table 20.1 because epidemiological evidence has proven that the benefits of conducting these tests outweigh the direct risks of bodily harm plus the indirect risks to health resulting from inaccurate diagnoses. Physicians use different parameters to assess the risk/benefit ratio of routinely using a diagnostic test on any patient, whether or not they notice outward signs of a disease.

Nature of the disease. Is the disease treatable or preventable? Are the effects of early treatment significantly better than those of later treatment? Will late detection harm public health? What are the risks of not conducting the test?

Disease prevalence. How common is this disease in the population? Does it have a significant impact on population mortality or morbidity?

Patient attributes. How old is the patient? What is his or her current health status? Does the patient belong to a demographic group with increased risks?

Table 20.2 Possible outcomes of a diagnostic test

Test result	Disease condition	
	Present	Absent
Positive	True positive	False positive
Negative	False negative	True negative

Test costs. Is the monetary cost justified by the potential health benefits? Will incurring the high cost now save money in the long run?

Test accuracy. How sensitive is the test? How specific is the test?

The risk of receiving an incorrect diagnosis increases as both test sensitivity and test specificity decrease. Tests with high sensitivity give few false negatives, because few true positives escape detection. Tests with high specificity give few false-positive results because few of the true negatives are read as positive (Figure 20.1).

Figure 20.1 Diagnostic sensitivity and specificity and disease prevalence. The probability of receiving an inaccurate diagnosis is affected by the specificity and sensitivity of the diagnostic test and the prevalence of the disease being diagnosed. **(A)** Tests with higher specificity give fewer false positives, and tests with higher sensitivity give fewer false negatives. In both cases shown, 20% of the population had the disease. **(B)** The probability of inaccurately diagnosing a disease is related to the prevalence of the disease in the population being screened. Note the significant decrease in the number of false positives as disease prevalence increases, even though the sensitivity and specificity of the test do not change. This explains why disease prevalence affects a physician's decision to use diagnostic tests.

A. Diagnostic specificity and sensitivity

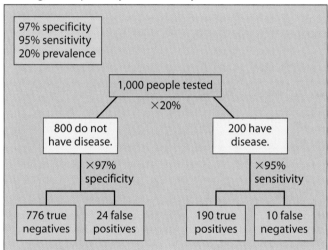

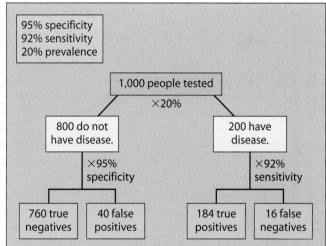

B. Disease prevalence

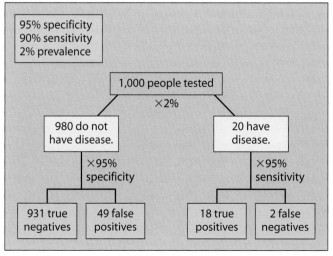

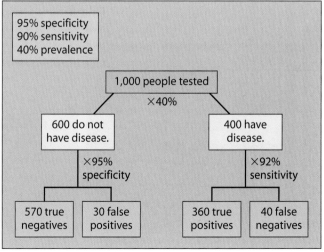

BIOTECHNOLOGY DIAGNOSTICS

New biotechnology-based diagnostic tests can both alleviate and exacerbate the health risks and financial costs of asymptomatic disease screening. First and most obviously, physicians will have access to many more diagnostic tests. Patients may request these tests if they are not conscious of the downside of asymptomatic screening: the risks associated with receiving an inaccurate diagnosis. If the new tests are inaccurate (the results are not predictive) or unreliable (the results are not reproducible), then they could do more harm than good.

Because diagnostic tests based on cells and molecules are typically more specific and often more sensitive, they should be more accurate than similar tests that are currently used to diagnose the same disease. In addition, molecular tests that allow physicians to identify incipient diseases before clinical symptoms appear could be a blessing, provided there is a way to prevent the disease, retard its development, or lessen its impacts.

As we mentioned earlier, physicians have used genetic information to assess a patient's current and future health for a very long time, but new gene identification technologies will greatly expand the importance of genetic information in health care (Figure 20.2). Greater availability of gene-based diagnostic tests, made possible by the Human Genome Project (HGP), carries with it the same problems shared by all diagnostic tests (Box 20.1). However, because of the nature of genetic information, these tests also introduce unique concerns, as well as unique benefits.

Results of genetic tests can be ambiguous

Researchers are in the earliest stages of linking disease genotypes with disease phenotypes. In chapter 12, you learned that the path from genotype to phenotype is an indirect one, filled with opportunities for myriad factors other than a single gene to exert influences (Figure 20.3). The final expression of a

Figure 20.2 Genetic diagnostics. Physicians have used karyotyping to identify major chromosomal alterations for many decades. The sequence information provided by the HGP greatly amplifies the diagnostic power of karyotyping, because DNA probes, tagged with fluorescent tags, can be used to identify changes in single genes. In this case, some chromosomal changes that cannot be visualized with standard staining techniques are revealed. These cancerous cells have many numerical and structural chromosome abnormalities, including aneuploidy (changes in chromosome number) and deletions and translocations (multicolored chromosomes). Note that the chromosomal changes differ in the two types of cancer. **(A)** Normal. **(B)** Bladder cancer. **(C)** Brain cancer. (Micrographs courtesy of H. Padilla-Nash and T. Ried, National Cancer Institute, NIH.)

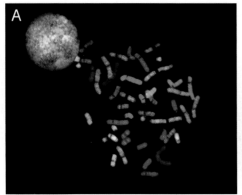
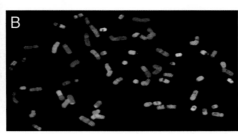
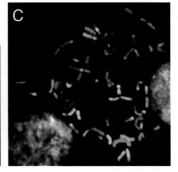

gene is affected by many variables: the allelic forms of the gene in question, other genes, and environmental influences. In addition, having a genetic defect that has been consistently linked to a disease does not guarantee that a person will actually develop the disease.

Ambiguity exists even when the relationship between a gene and a disease seems very straightforward. For example, if the disease is caused by a single gene with clear dominance relationships between heterozygous alleles and little, if any, environmental contribution, individuals inheriting the same number of mutant alleles can exhibit an array of symptoms. Recall the examples of cystic fibrosis and sickle-cell anemia. Symptoms may range from none to severe, and they may first appear in infants or older adults. Variability in disease severity occurs even when two people have identical mutations in the same gene.

If some degree of phenotypic variability, and therefore uncertainty about the clinical significance of a genetic defect, exists in the simplest, most straightforward examples, then how informative will genetic tests for multigenic or multifactorial diseases be? Researchers have made remarkable progress in linking certain mutant genes with certain diseases, but the DNA sequence, encoded protein, or phenotypic trait associated with thousands of other genes will remain unknown for decades. Any of those unidentified genes might interact with identified genes in ways that significantly alter the expected phenotypic trait expressed by the identified gene. Without knowing how identified disease genes interact with other, unidentified genes or with the environment, the results of a diagnostic test for the disease gene could lead to unnecessary heartache or unwarranted relief. On the other hand, if the research community waits until all genes have been identified and their interactions have been elucidated before it develops a diagnostic test for a genetic disease, thousands of opportunities to save lives will be missed.

For example, let's say researchers have linked a nucleotide substitution mutation in gene A with disease A, so they develop a test capable of identifying that gene A mutation. If you received a test result indicating you had that mutation, you might well think you would eventually develop that disease. But think back on the genetics of Labrador retriever coat color, in which one gene prevents the expression of genes at another locus. What if a certain form of gene B interacts with gene A epistatically and *prevents* disease A but researchers have not identified gene B, much less determined its function? If you have the form of gene B that blocks phenotypic expression of gene A, you would not get the disease (Figure 20.4). On the other hand, is it unethical not to develop a diagnostic test for the A mutation until the entire human genome is decoded?

Becoming well-informed and inquisitive patients can help resolve some questions about the appropriate time for developing a genetic test for a certain disease. Before jumping to conclusions about your risk of developing a disease that has been associated with a certain gene, always remember to ask about penetrance and expressivity. How many people who have the gene A mutation do *not* get disease A? Even though medical researchers or physicians may not know *why* the gene A mutation is not 100% penetrant (that is, they don't know that gene B is epistatic to gene A), they will know that it isn't.

How about the flip side of the coin? Does discovering you do not have the nucleotide substitution mutation in gene A mean that you will not get disease A? No. Remember the discussion of genetic heterogeneity. Other unidentified mutations in that gene or mutations in different genes, neither of which the test would reveal, could cause that disease. The relevant question to ask

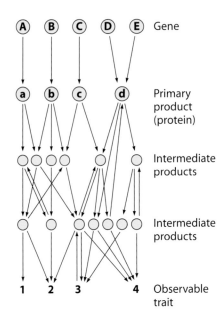

Figure 20.3 Genes and observable traits. A single gene affects many traits, and any one trait results from the actions of many genes.

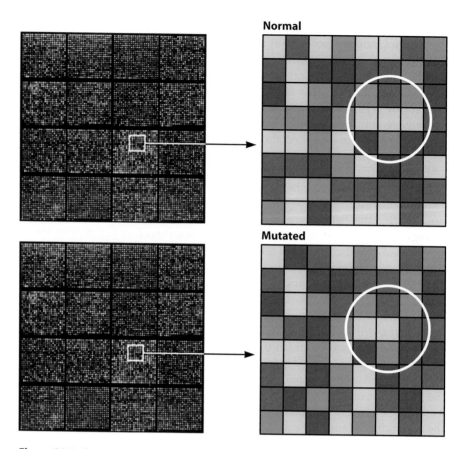

Figure 20.4 Genetic testing. The microarray test shown focused on a gene known to be associated with cancer. As you can see, there are other differences between the two genomes, but researchers know very little about how other genetic differences might affect the propensity to develop cancer. (Photograph courtesy of Jason Lang, National Cancer Institute, NIH.)

when you discover you do *not* have the mutation is what percentage of the people who get disease A do not have the nucleotide substitution mutation in gene A. The genetics of breast cancer described in various chapters provides an excellent real-world example. If a woman is tested for mutations in the "breast cancer gene" (*BRCA1* and *BRCA2*) and the test results reveal she has neither mutation, does that mean she will not get breast cancer? No, because only 5 to 10% of breast cancer cases are associated with the *BRCA1* and *BRCA2* mutations.

Genetic test results can be predictive

As you see, using genetic information to identify the causes of a disease can be as ambiguous as using the results of clinical tests to determine disease causes. However, with clinical tests, physicians are measuring the physiological, biochemical, and morphological correlates of health and disease, or in other words, the phenotypic expression of genetic information. If a patient has begun to develop cardiovascular disease, the physician will be able to discern the problem through blood tests, electrocardiograms, and blood pressure measurements. The same is true of a lung disease, such as emphysema. Measurable factors, such as lung capacity and force of exhalation, indicate that the patient is developing or has developed the disease.

Genetic information differs from clinical measures because sometimes genes shed light on future health problems. The results of genetic tests might allow physicians to predict the possible appearance of cardiovascular or lung disease in the absence of any clinical symptoms. This capacity provides physicians with new tools for improving their patients' health, especially in those diseases with a significant environmental component that the patient can control. On the other hand, shedding more light on a person's future health concerns also poses new problems, which we discuss below.

Genetic testing raises both old and new questions

Physicians, scientists, ethicists, attorneys, and the general public have voiced a number of complex, interrelated concerns associated with new genetic testing capabilities. Some of the societal issues raised by advances in genetic testing are not unique to new developments in biotechnology but have existed for a number of decades. Others are brand new, and some are not easy to resolve in ways that are equally acceptable to all viewpoints. Because the concerns are often inextricably interrelated, not to mention technically complex and often emotionally charged, analyzing them rationally can be difficult. To help, we have divided large issues of genetic testing into a few smaller subissues.

Some people have expressed concern that the ever-growing capacity to identify genetic defects will cause more harm than good by placing individuals in psychological, emotional, and social quandaries without easy answers (Table 20.3). They also believe that the widespread availability of genetic screening presents new ethical dilemmas for individuals, societies, and governments. As described in chapter 17, one of the most productive techniques in analyzing societal issues raised by new technology is to place the issue in a historical context. We use this approach in addressing each concern.

Some fear that providing people with information about personal genetic disorders could harm them psychologically if the disease cannot be cured at the time the diagnosis is given. In addition, they feel that telling parents they have conceived a child who suffers from an incurable genetic disease would force parents into a psychologically and emotionally difficult decision about terminating the pregnancy.

What about this issue is new or unique to the new genetic testing capability that biotechnology provides? First, let's consider what is not new, because people and their physicians have grappled with many of the same issues for many decades.

If a routine physical examination reveals that you have an incurable cancer, do you expect your physician to share that information with you? Probably so. But just a few decades ago, physicians did not tell their patients they had cancer because they thought not knowing was best for the patient's emotional well-being. The tension between withholding and disclosing information is not new to medicine and varies with the disease, the physician, the patient, and the prevailing practice. It is not specific to new genetic testing capabilities.

The same can be said about **prenatal testing** for inherited disorders. In 1952, a medical researcher in the United Kingdom developed amniocentesis, a method for detecting certain genetic defects in embryos (Figure 20.5). Another prenatal genetic diagnostic technique, chorionic villus sampling, was developed a few decades later. Tests on fetal cells collected in these procedures can detect more than 200 diseases, congenital disorders, and chromosomal abnormalities. Physicians can treat or prevent a few, such as phenylketonuria,

Table 20.3 Genetic tests[a]

Disease	Symptom(s)
Alpha 1-antitrypsin deficiency	Emphysema and liver disease
Amyotrophic lateral sclerosis (Lou Gehrig's disease)	Progressive loss of motor function; paralysis; death
Alzheimer's disease	Late-onset senile dementia
Ataxia telangiectasia	Progressive brain disorder; loss of muscle control; death
Gaucher's disease	Enlarged liver and spleen
Inherited breast and ovarian cancer	Early-onset tumors of breast and ovaries
Hereditary nonpolyposis colon cancer	Early-onset tumors of colon and other organs
Charcot-Marie-Tooth	Loss of feeling in extremities
Congenital adrenal hyperplasia	Hormonal deficiency; ambiguous genitalia; excessive salt loss; death
Cystic fibrosis	Diseases of lung and pancreas; chronic infections; poor digestion
Duchenne muscular dystrophy	Muscle wasting; deterioration; weakness
Dystonia	Muscle rigidity
Fanconi's anemia	Leukemia; anemia; skeletal deformities
Factor V-Leiden	Blood clotting disorder
Fragile X syndrome	Mental retardation
Hemophilia A and B	Bleeding disorders
Hemochromatosis	Excess iron storage disorder
Huntington's disease	Progressive, lethal, degenerative neurological disease
Myotonic dystrophy	Progressive muscle weakness
Neurofibromatosis	Multiple, benign, disfiguring nervous system tumors
Phenylketonuria	Progressive mental retardation
Polycystic kidney disease	Kidney failure and liver disease
Prader-Willi syndrome	Decreased motor skills; cognitive impairment; early death
Sickle-cell disease	Blood cell disorder; chronic pain and infections
Spinocerebellar ataxia	Reflex disorder; involuntary muscle movement; explosive speech
Spinal muscular atrophy	Severe, lethal muscle-wasting disorder in children
Thalassemias	Decreased blood cell levels; anemia
Tay-Sachs disease	Fatal neurological disease; seizures; paralysis

[a]Some of the 900 DNA-based tests made possible by the HGP.

once the baby is born. However, most of the disorders detected by these tests are incurable. Their progress cannot be subverted by any known treatments. For many decades prior to the development of the new genetic testing methodologies, hundreds of thousands of parents have voluntarily had these tests performed and have been faced with difficult decisions about pregnancy termination in the second trimester.

One way parents who know they are carriers of serious genetic diseases can avoid having to make decisions about aborting fetuses is through **preimplantation genetic testing** for diseases. Every year, thousands of couples, unable to conceive a child, seek help from an in vitro fertilization (IVF) clinic. Embryos produced by mixing sperm and eggs in a petri dish are screened for genetic defects prior to implantation to improve the probability that the fetus will be carried to term. However, this "solution" to the problem of prenatal genetic testing ending in abortion provides no solace for couples who believe life begins when egg and sperm fuse, because embryos with genetic defects are discarded.

You may be asking why these issues are potential ethical problems that society as a whole must contemplate. Can't people avoid the potential psychological trauma posed by these difficult decisions by simply *not* subjecting themselves or their unborn offspring to genetic testing? After all, no one re-

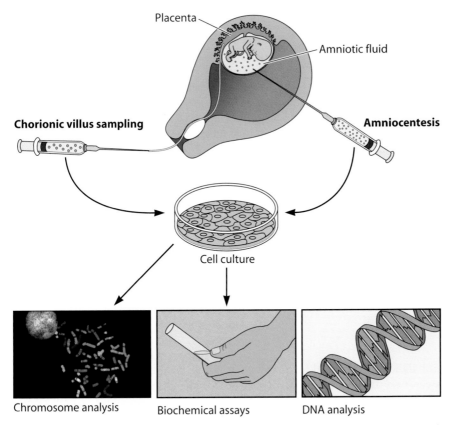

Placenta

Amniotic fluid

Chorionic villus sampling

Amniocentesis

Cell culture

Chromosome analysis

Biochemical assays

DNA analysis

Figure 20.5 Prenatal testing. Amniocentesis and chorionic villus sampling are prenatal tests for genetic and biochemical defects. In both procedures, fetal cells are withdrawn from the mother's uterus, cultured, and tested. Amniocentesis uses fetal cells that are floating in the amniotic fluid. Chorionic villus sampling removes a small piece of tissue from the placenta.

quires them to take the tests, do they? Unfortunately, history shows us that technologies that are supposed to increase a person's options sometimes increase their obligations. A choice at one time may become a requirement later on. For example, parents are sometimes surprised to learn that the hospital staff tested their newborn babies for a number of inherited disorders immediately after delivery because states legally require these **neonatal tests** (Table 20.4). Even though it seems inconceivable that prenatal genetic testing would ever be required in this country, it is important to go through the mental exercise of analyzing the potential uses and abuses of a technology to ensure that the technology develops in the most responsible way.

Methodologies to identify disease genes can create ethical dilemmas

Before knowing the exact locations of genes that cause disorders, scientists first develop markers associated with defective genes by widespread screening of families in which the disorder is prevalent. As you know, the process of developing markers depends on acquiring blood samples from many members of a family and analyzing their DNA. As a part of this gene discovery process, people discover information about their genetic makeup. What if someone discovers they have a disease gene but they do not want anyone to know? This poses an ethical dilemma between a person's right to genetic privacy and a responsibility to family members. Where would you draw the line

Table 20.4 Neonatal screening for genetic disorders[a]

Disorder	U.S. incidence	Description	Potential effects	Treatment
Phenylketonuria	1/13,947	Deficiency of phenylalanine breakdown enzyme	Mental retardation; seizures	Low-phenylalanine diet
Hypothyroidism	1/3,004	Deficiency of thyroid hormone	Mental retardation; stunted growth	Thyroid hormone
Galactosemia	1/53,261	Deficiency of galactose breakdown enzyme	Brain and liver damage; cataracts; death	Galactose-free diet
Sickle-cell disease	1/3,721	Hemoglobin abnormalities	Organ damage; stroke; delayed growth	Penicillin; vaccinations
Adrenal hyperplasia	1/18,987	Deficiencies of cortisol and aldosterone	Death due to salt loss; growth difficulties	Hormone replacement; salt replacement
Biotinidase deficiency	1/61,319	Biotin-recycling deficiency	Mental retardation; seizures; hearing loss	Biotin supplements
Maple syrup urine disease	1/230,028	Deficiencies of leucine, valine, isoleucine breakdown enzyme	Mental retardation; coma; seizures; death	Dietary management and supplements
Homocystinuria	1/343,650	Deficiency of homocysteine breakdown enzyme	Mental retardation; skeletal abnormalities; stroke	Dietary management and vitamin supplements

[a]These are the genetic disorders for which testing is most commonly mandated by state law. All 50 states require neonatal testing for the first three disorders. Approximately 35 to 40 states mandate testing for the remaining five. Eight states require that hospitals test newborns for 30 genetic disorders.

in balancing your need for privacy with their need to know? Would your answer vary depending upon the nature of the disease, its genetic basis, or whether it can be treated? If you feel you have a responsibility to tell family members, how broadly would you define family? Siblings, parents, cousins, second cousins?

Once a disease gene is identified and screening of asymptomatic populations becomes available, some people will want to know if they have that disease gene, and others will not. If people share their test results with family members, people who do not want to know their probability of having disease genes may well end up knowing. How does someone balance their right to know with the right of others not to know? At first, the answer may seem obvious. People being tested should keep the information to themselves. But what if the genetic screening reveals that a person has a disease gene, but getting the disease depends on environmental factors related to individual choices, such as diet or smoking? Is it unethical *not* to tell a family member that they or their child may have a gene for a disease that could be avoided through behavioral changes?

What are ethically acceptable ways to use test results?

Some issues raised by the widespread availability of genetic testing are not novel issues, spawned by new technological advances, but are more related to slight changes in degree. However, at some point in gradually changing the degree, new and difficult ethical issues arise.

As we said above, physicians have been able to conduct prenatal and preimplantation tests for genetic disorders for many years. Most of those tests are for incurable diseases or serious medical problems that leave little or no room for human intervention. In the future, physicians will be able to detect many more defective genes than those that cause incurable illnesses. Some genetic defects will cause serious problems that can be treated with appropriate therapies (hemophilia or diabetes). Other defective genes will simply indicate a propensity to develop a certain disease (emphysema). Still others will cause defects that are in no way life threatening, but they do make

BOX 20.1 *The Human Genome Project*

With the birth of recombinant DNA technology in the 1970s, scientists began to dream of identifying and locating all of the genes in the human genome. As recently as the mid-1980s, however, such an undertaking seemed impossible. Using the available technologies, it would have taken 1,000 researchers 200 years and cost $15 billion to $20 billion. Even so, a few scientists began devising strategic approaches that might turn their dream into reality. The project's initial goals were to develop detailed maps of the human genome and determine its complete nucleotide sequence by 2005 for $3 billion. In 1990, the U.S. Congress formally initiated the HGP by providing funding for 5 years at less than $200 million per year.

New technologies and new techniques

Without technological improvements, the scientists would never achieve the project's goals, so developing better technologies and techniques was their first priority. They knew that the HGP would generate massive amounts of data, and they needed consistent methods for storing, managing, accessing, and integrating the data. Therefore, before they began collecting significant amounts of sequence data, they turned to software engineers to create programs to make it easy to manage, share, and integrate the data electronically. The HGP scientists also needed new technologies for large-scale sequencing based on the chain termination method of DNA sequencing (see chapter 15), because state-of-the-art laboratories could sequence only 500 bases each day. Automated sequencing replaced manual sequencing, and laboratory robots performed some steps, such as preparing DNA templates for sequencing. With these technological advances, 15,000 bases could be sequenced each day. Finally, other scientists focused on improving mapping techniques and developing better methods for cloning large amounts of DNA that could be assembled into overlapping fragments.

Because of technological innovations, the development of new laboratory techniques, increasing experience, and collaboration with scientists around the world, the HGP was completed early (in 2003) and under budget ($2.7 billion). Although the U.S. Congress may have taken the first step in 1990, sequencing and mapping the human genome was not strictly a government or U.S. initiative. Private companies had research groups dedicated to genome sequencing, and 2,000 scientists at 20 institutions in six countries made significant contributions.

Mapping and sequencing

Mapping a chromosome requires two types of map, a physical map and a genetic linkage map, both of which provide useful, but different, information. Physical maps give researchers a more detailed view of a chromosome segment, because they show the actual physical distance between two points on a chromosome. Scientists use the size of a restriction fragment to deduce the distance between points, and they determine a fragment's size by comparing its migration rate in gel electrophoresis to those of fragments with known lengths. Genetic linkage maps are not derived from direct physical measurements but by analyzing how often two genes are inherited together. The closer two genes are on a chromosome, the less likely they will be separated by crossing over and the more

DNA sequencing. Today's DNA-sequencing machines can sequence over 420,000 bases in 1 day. (A) At the Department of Energy's 57,000-square-foot Production Genomics Facility in Walnut Creek, Calif., row after row of DNA-sequencing machines churn out sequences. (B) In the chain termination method of DNA sequencing, modified versions of each of the four nucleotides are labeled with uniquely colored fluorescent dyes to create the DNA sequence. (C) A computer reads the sequence information provided by the fluorescent dyes and converts the fluorescent signals into a sequence of nucleotides. (Photographs courtesy of the U.S. Department of Energy Human Genome Program [http://www.ornl.gov/hgmis] [A and C] and the Virginia Tech Sequencing Facility [B].)

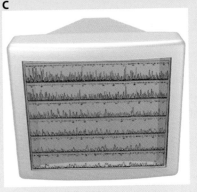

ATTAGAGGC TCACC GATTCATGTCGG AGATGG TCAGAAAA AC
CGTT TCAGA AGCAACCTTGGGC TTAGTCC CACC CTTTT TAGG C
GTGCCTAGAA AGATGACAACTCAAGCACC GACGTT TACGCAG
GTACTGG AGGG TAGTACC GCAACCTT TGAGGC TCACATTAGT
GTGAGC TGG TT TAGGG ATGG CCAGG TGATT TCCACTTCCACTC
TCCTT TAGCGATGG CC GCGC TAA ACTGACGATCCCC GCC GTAGA
CGATATTCC CTGAA AGCCACCAATGG ATCTGG ACAAGCGACT
GTGAA AGC TGAGACAGCACCACC CAACTTCGR RCAACGACTG
CAAGGA AGCCAAGTGAGACTCCAAGTGAGAGTGACTGGA ATC
TTCTACC GGG ATGG AGCC GAA ATCCAGAGC TCC CTTGATT TCC
GACCTCTACAGC TTACTGATTGCAGA AGCATACC CTGAGG ACT
AATGCCACCAATAGCGTTGGA AGAGC TACTTCGACTGC TGA AT
GA AGA AGTACCTGC TAAAA AGACAA AGACAATTGTT TCGACT
AGACAA ACCC GA ATTGAAA AGA AGATTGA AGCC CACTT TGAT
GTTGAGATGG TCATAGATGG TGCC GC TGGG CAACAGC TGCCAC
ATTCCTCCGATCATAGA

(continued)

B O X 2 0 . 1 *The Human Genome Project* (continued)

likely they will be inherited together. Linkage maps are important because they allow researchers to locate genes associated with particular diseases or other observable physical characteristics. Ultimately, the maps are examined in conjunction with each other to obtain the detailed nucleotide-by-nucleotide sequence and to locate specific genes on the chromosome.

The final step in making a physical map is determining the actual DNA sequence between two points. To read a sentence, you begin at the beginning and read one word after another until you reach the end. DNA sequencing doesn't work that way. Because DNA molecules are so large, scientists must cut DNA molecules into manageable stretches of sequence by using restriction enzymes. Unfortunately, this disrupts the sequence order. To reorder them, they treat many, many copies of the same DNA molecule with different restriction enzymes and identify overlapping nucleotide sequences, because a bit of sequence on one end of a restriction fragment is the same as a bit on the other end of another.

After more and more fragments cut with different restriction enzymes are sequenced and the overlaps are matched up, the pieces can be placed in order.

Only the beginning

You may hear people compare the completion of the HGP to landing on the moon, but this analogy is misleading. Both were expensive, huge, interdisciplinary, publicly funded science and technology projects that attracted a great deal of media attention, but the impacts of these landmark achievements differ.

Landing on the moon was an end point. Sequencing the human genome is just the beginning, more akin to the Wright Brothers' first flight than to a moon landing.

If the only output of the HGP consisted of huge stretches of nucleotide sequences, the information would not be very meaningful. Even though sequencing and mapping may have been the explicit goals when the project began, the implicit, understood, and widely accepted goals then and now are to identify and fully understand our genes—where they are, what they do, how they are regulated, and what happens when they malfunction. The best way to discover what a gene does and how it is regulated is to study proteins, not genes. Scientists are now trying to identify all of the proteins in a cell type, determine their function, and map their interactions. The success of this field will depend on developing technologies to rapidly identify the types, functions, and amounts of thousands of proteins in a cell.

Societal issues

The HGP is surely unique among big government science projects in that a portion of the funding is earmarked for the ethical, legal, and social implications (ELSI) that will arise as a result of newfound knowledge of genetics. Some of the issues addressed by ELSI are:

- fairness in the use of genetic information by insurers, employers, courts, schools, adoption agencies, the military, and other government institutions

- privacy and confidentiality of genetic information

- psychological impacts and stigmatization due to an individual's genetic differences

- genetic testing of an individual for a specific condition due to family history (prenatal, carrier, and presymptomatic testing) and widespread population screening for inherited disorders (newborn, premarital, and occupational)

- reproductive issues, including informed consent for procedures, use of genetic information in decision making, and reproductive rights

- effective use of genetic information in clinical settings, including education of health service providers, patients, and the general public and implementation of standards and quality control measures in testing procedures

- commercialization of products, including property rights (patents, copyrights, and trade secrets) and accessibility of data and materials

- conceptual and philosophical implications regarding human responsibility, free will versus genetic determinism, and the concepts of health and disease

Overlapping fragments

Fragment 1:   CCGTATTGCTTGATTGCGCCTTCGAAATTGGGCT

Fragment 2:   AATGCCGTAGCTGGGTACCGTATTGCTTG

life more difficult (color blindness or albinism). Will each couple be free to decide the fate of offspring carrying a "defective" gene, or will the government or insurance agencies develop policies that encourage certain choices?

Where will society draw the line when defining a genetic defect and determining how to handle it? When is a defect truly a defect? At what point will eliminating genetic defects inch toward selecting desired characteristics, such as great height or attractive physical appearance? Many couples have used prenatal and preimplantation genetic testing to select the sex of their children, unrelated to any consideration of sex-linked diseases, such as sex-linked severe combined immunodeficiency disease. This practice has skewed the sex ratio toward males so greatly in certain countries that governments have passed laws prohibiting sex selection.

Should information on genetic predisposition affect employment?

Many companies screen potential employees for health-related factors, especially for certain positions. At least 50% of the employers in the United States require job applicants to have medical examinations. In those examinations, many of the tests conducted are indirect measures of the applicant's genetic makeup, because the tests measure phenotypic manifestations of genetically based traits.

Companies are interested in the health of workers for reasons related to decreasing costs.

- Time lost due to sickness or disability costs money.
- Hiring healthy people allows corporations to hold down health care costs.
- Identifying workers susceptible to certain occupational hazards helps companies protect themselves against lawsuits.

During the earliest stages of the HGP, many people expressed concern that companies would use the results of genetic tests to unfairly deny people employment. Some groups pressured Congress to pass laws that would make it unlawful for companies to use genetic testing results in hiring decisions. On the surface, such a law would seem to be beneficial to the public, but once again, the issue deserves further consideration.

The type of genetic information companies might be interested in would vary with the position and company. For example, earlier we discussed how genetic variation in the ability to metabolize certain compounds helps explain why only certain people have adverse reactions to medicines. If a job involves working with hazardous chemicals, a company might be interested in hiring people with alleles encoding certain versions of drug-metabolizing enzymes because they assume those enzymes might also do a better job of breaking down hazardous chemicals. If there is no relationship between having certain genes for drug metabolism and the ability to break down hazardous chemicals, some people might be denied employment based on a misconception about gene function. This type of abuse of genetic information inspires advocacy groups to push for laws prohibiting the use of genetic test results in hiring decisions.

Would having laws that prohibit companies from using genetic testing information *always* be in the best interest of a potential employee? If the test just described would inform you and your employer that you were genetically unable to break down hazardous chemicals and therefore your body might be concentrating them in certain tissues, wouldn't you want to know that? Would it be unethical for a company *not* to test a potential employee in that situation, if such a test were available?

HUMAN EMBRYONIC CELL THERAPIES

The availability of reliable sources of undifferentiated human cells opens up new avenues for:

- treating, curing, and even preventing human diseases and injuries for which neither effective therapies nor cures exist
- conducting research on cell differentiation and early embryonic development

- discovering new medicines
- assessing drug safety more accurately during drug development to identify adverse effects before the drug is marketed

However, some people object to laboratory research or therapeutic strategies that use fertilized human eggs or blastocysts, both of which can develop into a viable fetus if implanted in a human female.

Using factual information and specific language in discussions of complex technical issues is always important; both become even more essential if the topic elicits strong emotional feelings about moral issues. The methods, framework, and emotional detachment described in chapter 17 are especially helpful in this regard. Therefore, before we begin to broach questions about the ethical and political issues surrounding the use of human embryonic cells, we need to provide you with additional scientific background and clear, specific definitions.

Gather relevant background information

In November 1998, two different research groups simultaneously reported they had successfully maintained human embryonic cells in culture. One group used the inner cell masses (ICM) of blastocysts to establish hES cell lines. The other group derived embryonic cell lines from primordial germ cells that had been taken from aborted fetuses. Embryonic cells derived from primordial germ cells, which are the cells that develop into eggs and sperm in adults, are correctly referred to as **embryonic germ (EG) cells**. The distinction between EG cells and ES cells is not trivial, because one of the ethical issues some find troubling is the original source of human embryonic cells. Much more research has been conducted on hES cells than human EG (hEG) cells, so hES cells derived from blastocysts will be our primary focus.

Blastocysts consist of an outer layer of cells, the trophoblast, and the ICM. The blastocyst implants in the uterine wall 5 or 6 days after fertilization as the trophoblast cells begin to infiltrate the uterine lining. At day 5, the ICM consists of around 30 to 35 cells. After the trophoblast cells become integrated into the uterus, the ICM cells divide and differentiate into the three distinctive germ layers—ectoderm, mesoderm, and endoderm—found in the gastrula (Figure 20.6). Ultimately, the ICM cells develop into the fetus and the trophoblast cells become the placenta. Even though ES cells or ICM cells

Figure 20.6 Early embryonic development. The schematic representation depicts human embryonic development from blastocyst to gastrulation. Five days after fertilization, the blastocyst begins to implant into the wall of the uterus. By day 14, gastrulation has occurred, implantation is complete, and the pluripotent cells of the ICM have already differentiated into ectoderm, mesoderm, and endoderm. Therefore, they are no longer pluripotent. The trophoblast cells differentiate into placental tissues.

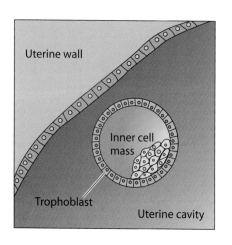

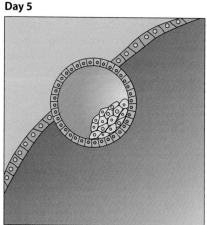
Day 5

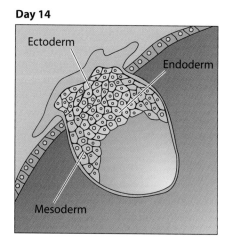
Day 14

can differentiate into any cell type found in the body, neither cell type can develop further if inserted into a female because they lack the cells required for implantation into the uterine wall. Therefore, neither ES cells nor cells from the ICM are totipotent.

Drawing on decades of stem cell research on other animals, scientists assume that hES and hEG cells maintained in laboratory cultures are pluripotent and can exist almost indefinitely under appropriate laboratory conditions. When implanted in mice, hES cells and hEG cells have differentiated into tissues resembling neural epithelium, bone, cartilage, gut, striated muscle, and kidney. By administering chemical signals known to be released by embryonic cells during development, scientists have created the three tissues that give rise to all cell types, as well as many distinct cell types.

Although we do not discuss safety issues in this chapter, it is important to note that if human embryonic cells are implanted in a person for therapeutic purposes, they often become cancerous. When you recall what you have learned about the molecular basis of cancer, you will see that in many ways cancer cells resemble stem cells. To decrease their risk of becoming cancerous, ES cells need to be treated with differentiation factors before they are implanted.

Place the topic in a relevant context

To place issues related to research on human embryonic cells in perspective, it is important to gather information on IVF and other unnatural methods of human reproduction, some of which were discussed in chapter 17. Scientists fertilized human eggs under laboratory conditions for the first time in 1968, and the first human baby created through IVF was born in England in 1978.

When couples hoping to conceive a child through IVF seek assistance from clinics that specialize in reproductive technologies, physicians begin by administering hormones to the female. The hormones stimulate the development and ovulation of approximately 10 to 20 eggs, which are removed from the female surgically and fertilized in petri dishes (Figure 20.7). After fertilization, the eggs are maintained under cell culture conditions for 3 to 5 days so that clinicians can observe the earliest stages of development. Fertilized eggs that develop abnormally are discarded, and the others are saved. At day 5, the blastocyst must be implanted or frozen or it will not survive. Excess blastocysts are frozen for future implantation, if needed, to spare the mother another round of hormone injection and egg removal. After a number of years, IVF clinics request permission from the parents to discard excess frozen embryos. Therefore, a number of embryos are created in vitro and discarded, either before or after being frozen, so that a couple can have a single child.

The history of IVF is relevant to discussions concerning human embryonic cell research because virtually all hES cell lines created to date were derived from excess blastocysts about to be discarded from IVF clinics.

Different cell sources give rise to human embryonic cell lines

Certain people object to research and therapeutic uses of human embryonic cells because some are derived from blastocysts. Therefore, it is important to delineate the issue of source quite thoroughly when evaluating ethical uses of human embryonic cells. As we mentioned, some embryonic cells (ES cells) are derived from blastocysts and others (EG cells) are not. Each source raises unique ethical questions that are more or less problematic to different groups of people.

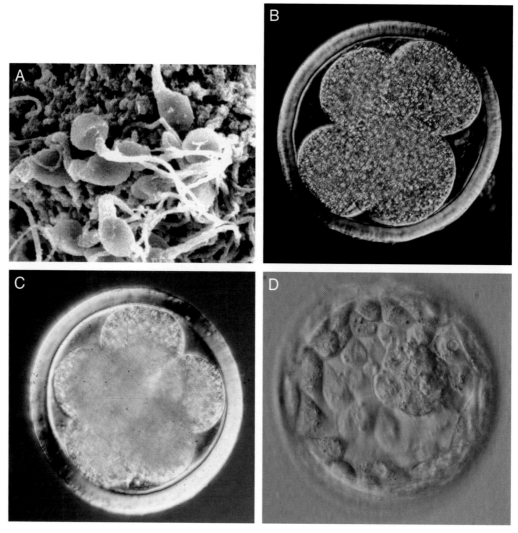

Figure 20.7 IVF. After fertilizing a number of eggs in petri dishes, the clinicians nurture the fertilized egg through the first 5 days until it becomes a blastocyst. At that point, the embryo must be either implanted or frozen to survive. **(A)** Human sperm on the surface of the egg. **(B)** Day 2—four cells. **(C)** Day 3—eight cells. **(D)** Day 5—blastocyst. (Photographs courtesy of D. Waddell, Center for Microscopy and Microanalysis, University of Queensland, Brisbane, Australia [A], and Michael Vernon, West Virginia University Center of Reproductive Medicine, Morgantown, W.Va. [B to D].)

Ethical issues associated with sources of ES cells

Contrary to a pervasive misconception, no hES cell line is derived from aborted fetuses. All are derived from blastocysts. To date, two methods for deriving hES cells from blastocysts have been successful. In all cases except one, the source of ES cells was frozen blastocysts from IVF clinics. In a single instance, a fully differentiated somatic cell nucleus was implanted in an enucleated egg, which was cultured to the blastocyst stage.

Blastocysts from fertilized eggs.   At any point, the IVF clinics in the United States collectively have around 400,000 frozen embryos in storage. Researchers obtained frozen embryos that were about to be discarded from IVF clinics after the parents had given written informed (and unpaid) consent, al-

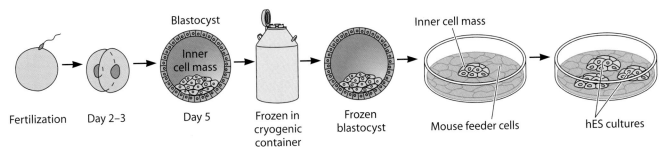

Fertilization   Day 2–3   Day 5   Frozen in cryogenic container   Frozen blastocyst   Mouse feeder cells   hES cultures

Figure 20.8 Creating hES cell lines. Virtually all hES cell lines were derived from frozen blastocysts from IVF clinics that were being discarded with the parents' permission. The first successful attempts to establish hES cell lines from ICM (in 1998 to 2001) relied on a layer of mouse cells to support the human cells. As a result, the early lines are unsuitable for therapeutic uses because of possible contamination with mammalian viruses.

lowing the blastocysts to be used for research. Researchers isolated ICMs from frozen blastocysts and, to encourage growth and maintenance of hES cell lines, they grew the cells on a layer of mouse cells (Figure 20.8).

As we mentioned above, many people object to using the ICM from blastocysts to create ES cell lines. Because creating an hES cell line from a blastocyst necessitates destroying the blastocyst, they believe it is unethical and disrespectful of potential human life. This concern has elicited a variety of political initiatives by governments all over the world. Governmental bodies in different countries have responded in different ways. While the U.S. government has placed significant restrictions on hES cell research, the governments in Japan, South Korea, China, Singapore, France, Switzerland, and the United Kingdom have been supportive, though not without restrictions. We discuss government policies later in the chapter.

Moral arguments and policy decisions rooted in ethical concerns about hES cell research raise other ethical issues. Recall that ES cells have the potential to treat, cure, and even prevent a wide variety of diseases that, when combined, affect over 100 million Americans. Some of these are fatal diseases for which there are no effective treatments. ES cell research would also provide information on developmental problems that cause birth defects, which might help to prevent them. Although these applications are at least a decade away, the scientific community agrees that the potential is there. What are the ethical implications of denying living human beings these lifesaving cures while focusing on the rights of potential human beings? Is it ethical to value potential human life more than existing human life? Would your answer change if you, your child, or your parent suffered from a disease that might be cured by hES research?

In addition, hES cell lines are derived from frozen blastocysts that are about to be destroyed, with the parents' approval. So, even though you may hear politicians say that scientists are "creating a life to destroy it," the embryos that are the source of virtually all hES cell lines are not being created and destroyed for research purposes. The embryo is about to be destroyed by its parents, who no longer need it for the purposes for which it was created. Which of these fates—discarded because it is no longer needed by the parents or salvaged to be used in research—is more respectful of the potential human life? Are both equally disrespectful? If both are unethical, then to be ethically consistent, shouldn't governments with policies that discourage hES cell research also be developing laws or policies that accomplish one or all of the following?

- Close down IVF clinics.
- Ban the practice of discarding frozen embryos from IVF clinics.
- Punish parents who have allowed their frozen embryos to be discarded.

- Require implantation of all frozen embryos currently stored in IVF clinics.
- Restrict IVF of eggs to the number a female is willing to have implanted.

A 2001 public perception survey conducted by Virginia Commonwealth University revealed quite clearly how important the issue of source is to a person's ethical view of hES cell research. When asked if it was acceptable to conduct medical research on stem cells derived from 5-day-old human embryos, 48% of the respondents found it acceptable and 43% did not. However, when asked if it is acceptable to conduct medical research on stem cells from frozen 5-day-old embryos that were about to be discarded from IVF clinics, the number in favor of such research increased to 75% and the number opposed decreased to 19%.

Blastocysts from somatic cell nuclear transfer.   Research on many different mammals, initiated by the birth of Dolly the sheep, has repeatedly demonstrated that nuclei from fully differentiated somatic cells will dedifferentiate if placed in unfertilized, enucleated eggs and subjected to certain laboratory conditions (Figure 20.9). The egg receiving the somatic cell nucleus acts as if it has been fertilized, and it begins the embryonic development process, creating a blastocyst containing an ICM capable of creating an ES cell line.

Scientists have only recently (February 2004) proven that somatic cell nuclear transfer (SCNT) can be used to create an hES cell culture. Beginning with 242 eggs voluntarily donated by 19 females, Korean scientists successfully created a single hES cell line by removing and culturing ICMs from the 16 blastocysts that developed from the 242 eggs. This poor success rate is similar to research on animal cloning through SCNT, so much work needs to be done before the technique could be used to create hES cell lines for use in therapeutic cloning.

People who may not object to using frozen blastocysts that are about to be discarded from IVF clinics to create ES cell lines may object to SCNT for at least three reasons.

- Some people fear that allowing the use of SCNT to create ES cells for research or therapeutic purposes, which they find ethically acceptable, opens the door to reproductive cloning, which they find ethically unacceptable. You will often hear these sorts of ethical concerns characterized as "slippery slope" issues.

Figure 20.9 SCNT. The nucleus is removed from the egg on the left. A nucleus from a somatic cell is injected on the right. Note the small tear in the egg membrane and the tiny nucleus between the membrane and the cytoplasm. A nucleus is also visible in the pipette. (Photographs courtesy of Roslin Institute, Edinburgh, United Kingdom.)

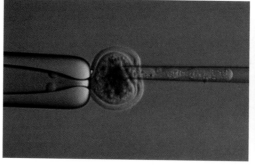 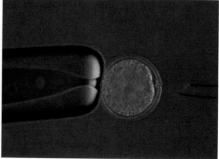

- Those who object to females donating their eggs to IVF clinics for reproductive purposes may also object to this method for creating ES cell lines, since it relies on donated eggs.
- In contrast to using IVF blastocysts, in this case researchers would be creating a blastocyst (and therefore a potential life) in order to destroy it for research or therapeutic purposes.

Ethical issues associated with other sources of embryonic cells

Scientists have successfully derived pluripotent embryonic cells from two other sources.

Primordial germ cells.   The third source of human embryonic cells does not rely on using blastocysts, and therefore, no embryos are destroyed. Instead, primordial germ cells are extracted from an aborted fetus that is approximately 6 to 8 weeks old. hEG cell lines are then derived from these primordial germ cells. People who do not object to using blastocysts because the embryo is so young may feel very differently about using cells from aborted fetuses. Those who object to using 5-day-old embryos would surely object to this method if the fetuses were aborted intentionally. However, people who disapprove of using 5-day-old embryos may approve of using fetal tissue, even though the embryo is much older, if the fetus had been miscarried naturally.

Unfertilized eggs?   The right chemical concoction can fool an unfertilized egg into beginning the cell division and differentiation process as if it had been fertilized. The development of an organism from an unfertilized egg is known as parthenogenesis, and the developing organism is not an embryo but a **parthenote**. In mammals, a parthenote cannot develop into a fetus if placed in a uterus, but researchers have successfully derived pluripotent ES cells from the ICM of animal parthenotes.

As this book goes to press, no researcher has derived hES cells from an unfertilized egg; therefore, we do not know whether it is possible. Using unfertilized eggs as the source of hES cells, if it is possible, could circumvent the problem of using embryos, since parthenotes cannot develop into a fetus if implanted, as well as any additional ethical concerns related to SCNT. However, those whose objections hinge on the fear of "opening doors" to future problems would most certainly be upset with the idea of research that devises schemes to trigger parthenogenetic development in a human egg that has not been fertilized by a sperm!

SCIENCE, LAW, AND POLITICS

In addition to ensuring that medical products on the market are safe and efficacious, governments are responsible for determining how the products of medical biotechnology can and cannot be used. We discuss some of the policy initiatives related to genetic testing in Box 20.2.

What should be the role of politicians in determining which paths of scientific inquiry are pursued and therefore which medical breakthroughs are most likely to occur? This question is not unique to biotechnology research. Below, we focus primarily on ES cell research to describe some of the ways that government bodies influence which scientific questions are pursued and, therefore, which potential technological applications become commercial realities.

BOX 20.2 *Genetic discrimination policies*

In 1996, the Joint Working Group on Ethical, Legal and Social Implications of Human Genome Research (ELSI Working Group) and the National Action Plan on Breast Cancer (NAPBC) published a set of recommendations for policies that would protect against genetic discrimination in insurance coverage and employment practices. In general, the recommendations prohibit:

* insurance providers from requiring genetic tests, disclosing genetic information, or using genetic information to affect insurance premiums or deny or limit coverage
* employers from requiring genetic tests as a condition of employment, disclosing genetic information, or using genetic information in hiring or promotion decisions

To date, the U.S. Congress has not enacted federal legislation related specifically to genetic discrimination in the workplace or insurance coverage. Over the past decade, several bills were introduced. Some were new bills based on the ELSI Working Group-NAPBC recommendations. Others attempted to amend existing civil rights statutes and labor laws, because some attorneys believe existing antidiscrimination laws could be interpreted to cover genetic discrimination. These laws include:

* Americans with Disabilities Act (1990)
* Health Insurance Portability and Accountability Act (1996 and 2002)
* Title VII of the 1964 Civil Rights Act

The only federal policy explicitly prohibiting genetic discrimination is an executive order signed by President Clinton in 2000. However, the order protects only current or potential employees of the federal government by prohibiting federal employers from:

* requiring or requesting genetic tests as a condition of being hired or receiving benefits
* using protected genetic information to deprive employees of advancement opportunities

Various state legislatures have enacted laws to prevent genetic discrimination, but none are comprehensive with regard to coverage, protections afforded, or enforcement authority.

Government policies influence the direction of scientific research

Even though researchers are sometimes frustrated when politics spills over into science, policymakers have a responsibility to guide the direction of research in addition to deciding how its results and products should or should not be applied. Governments use a variety of tools, both carrots and sticks, to encourage some areas of research while discouraging others. The most obvious is the carrot of money.

Providing government funding

The U.S. government, or more accurately, the U.S. taxpayer, is far and away the primary funding source for basic research and a major contributor to applied research, which is one of the reasons policymakers have a right and a responsibility to determine research priorities. Both the executive and legislative branches have mechanisms for altering, relatively quickly, allocations of research monies to respond to political pressure, unforeseen problems (bioterrorism or emerging infectious diseases), and shifts in philosophies and priorities that often accompany changes in the makeup of Congress or new administrations.

Funding agencies, such as the National Institutes of Health (NIH), are overseen by cabinet-level agencies headed by presidential appointees who

BOX 20.3 *Federal funding of medical research*

The federal agencies that are the primary sources of funding for research on human health are the NIH and the National Science Foundation (NSF).

Founded in 1887, the NIH is the federal focal point for medical research. The NIH, comprising 27 separate institutes and centers, is a component of the U.S. Department of Health and Human Services. The NIH mission is to uncover new knowledge that will lead to better health for everyone by preventing, diagnosing, and treating diseases and disorders. The NIH works toward that mission by conducting research in its own laboratories; supporting the research of non-Federal scientists in universities, medical schools, hospitals, and research institutions throughout the country and abroad; helping in the training of research investigators; and fostering communication of medical and health sciences information.

The NSF is an independent agency of the U.S. government. The NSF consists of the National Science Board of 24 part-time members and a director (who also serves as ex officio National Science Board member), each appointed by the president with the advice and consent of the Senate. Other senior officials include a deputy director, who is appointed by the president with the advice and consent of the Senate.

The National Science Foundation Act of 1950 established the NSF and set forth its mission, which is to promote the progress of science; advance the national health, prosperity, and welfare; and secure the national defense. In addition, the act granted very specific authorities to NSF, including, but not limited to, the authority to:

- initiate and support, through grants and contracts, scientific and engineering research and programs
- determine the total amount of federal money received by universities and appropriate organizations for the conduct of scientific and engineering research
- initiate and support specific scientific and engineering activities in connection with matters relating to international cooperation, national security, and the effects of scientific and technological applications upon society
- strengthen research and education innovation in the sciences and engineering, including independent research by individuals, throughout the United States
- at the direction of the president, support applied research by organizations

must carry out the administration's policies (Box 20.3). In addition, every year, when Congress passes its appropriations bills that divvy up grant monies among the various federal funding agencies, funds are earmarked for research areas favored by certain members and their constituents. Appropriations bills can also use money as a stick and explicitly prohibit funding for some research areas that are perfectly legal but out of favor. Most research scientists rely solely on public funds to finance their research, so they often alter their research programs to varying degrees in response to government allocation of funds and any policy shifts being executed by the various funding agencies.

Regulating research with laws, policies, and guidelines

Governments can also influence the course of scientific discovery by establishing regulatory requirements that govern certain types of research but not others. In chapter 23, we discuss how the financial costs of obeying regulatory requirements can deter some areas of academic research, often unintentionally. However, sometimes bans on certain areas of research or strict regulatory guidelines can advance research by creating clarity about what is and is not permissible. A well-established and fair regulatory regime, no matter how strict, is preferred by both academic and industrial researchers, because it creates a known, predictable environment within which they can function. For example, the United Kingdom has strict government guidelines regarding ES cell research: the ES cells must be derived from frozen blastocysts about to be discarded, no research can be conducted on human embryos older than 14 days, and so forth. ES cell research is flourishing in the United Kingdom, which is viewed internationally as a leader in the field.

Controlling research through funding bans

Soon after the 1973 *Roe v. Wade* Supreme Court decision legalizing abortion, the executive and legislative branches of the U.S. government began issuing a series of policies, moratoria, and funding bans that discouraged or prohibited research on embryos, IVF, and fetal tissue. In 1993, President Clinton issued an executive order to remove the funding ban on fetal tissue research. Later that year, the U.S. Congress enacted the NIH Revitalization Act, which permitted funding for research on both fetal tissue transplants and IVF.

However, in 1995, Congress attached a ban to the NIH appropriation bill (and each successive NIH appropriation bill) that prohibits the use of federal funds for "the creation of a human embryo or embryos for research purposes; or research in which a human embryo or embryos are destroyed, discarded, or knowingly subjected to risk of injury or death greater than that allowed for research on fetuses *in utero*." As a result, since 1974, private companies and research foundations have financed almost all research on IVF and fetal tissue in the United States, as well the hES and hEG cell work described above.

In response to the 1998 breakthroughs that allowed researchers to create and maintain hES cell lines, the NIH director, recognizing the potential medical benefits of research on hES cells, asked for a legal opinion regarding research on hES cell lines. He wanted to know if the congressional ban on research that destroyed human embryos applied to hES cell lines, because no embryos were destroyed once the cell lines were established. In 1999, the Office of the General Counsel of the Department of Health and Human Services, which oversees the NIH, ruled that hES cell lines should be excluded from the funding ban because they lack the natural "capacity to develop into human beings" (i.e., lacking trophoblast cells, they would not be able to implant) and therefore "are not embryos."

Following the determination that the NIH could legally fund some types of hES cell research, NIH began drafting guidelines to specify research-funding requirements. In August 2000, the NIH released guidelines that placed the following stipulations on hES research that it could fund.

1. The *derivation* of hES cell lines was *not* fundable because derivation of ES cells from blastocysts destroys embryos.
2. Research on hES cells derived and established by privately funded entities was acceptable, but only if the following conditions were met.
   - The private entity derived the hES cell lines from frozen blastocysts that were about to be discarded.
   - The frozen blastocysts were donated with the informed, written consent of the parents.
   - The parents were not compensated for donating the blastocysts.
   - The private entity did not profit from the sale of blastocysts used for stem cell derivation.

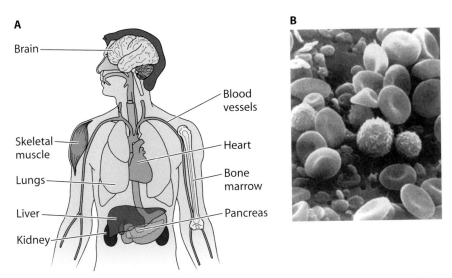

Figure 20.10 Adult stem cells. **(A)** Some organs maintain a population of stem cells to replenish the cell types found in that organ. **(B)** For example, bone marrow stem cells can give rise to any of the cell types found in blood: red blood cells, various types of white blood cells, and platelets. Not all organs have a resident population of stem cells. Adult stem cells are multipotent but not pluripotent. Recent research indicates that adult stem cells may have more developmental flexibility than previously thought. (Photograph courtesy of NIH.)

Soon after the NIH established these guidelines, George W. Bush was elected president. A few months after he took office in 2001, his administration withdrew the guidelines by publishing a *Notice of Withdrawal of the NIH Guidelines for Research Using Pluripotent Stem Cells*. On 9 August 2001, President Bush announced a more restrictive funding policy than the one established by the previous administration. Federal funding is now allowed only for the hES cell lines that had been derived prior to 9 August 2001. No federal research funds can be spent on any hES cell line derived after his announcement. In his announcement, President Bush also promised to increase federal funding for adult stem cell research (Figure 20.10).

Controlling research through federal legislation

The past three sessions of Congress have witnessed a flurry of legislative activity related to advances in hES cell research and the closely related topic of SCNT. Various members of Congress in both houses have introduced dueling bills. Some attempt to expand the ban established in the Bush administration's 9 August 2001 policy statement, while others try to counter the policy with a bill expressly supporting hES cell research.

The thorny scientific, technical, and ethical issues surrounding hES cell research become even more intractable when they are linked to those associated with SCNT. Even though everyone in Congress is opposed to SCNT (commonly referred to as cloning) for reproductive purposes, they have been unable to pass a law prohibiting it. The bills that attempt to ban reproductive cloning also include one of the following:

1. a ban on therapeutic cloning
2. support for both therapeutic cloning and research on hES cells derived from IVF blastocysts destined for destruction

Therefore, because members of Congress are divided on the issues of therapeutic cloning and the use of hES cells, they have not banned reproductive cloning, as have other governments throughout the world, even though every member of Congress is opposed to it. The standoff in Washington leaves the United States open to a practice that virtually everyone on Earth opposes, because it has no law that bans reproductive cloning or specifically gives a federal agency authority over activities in private research laboratories, IVF clinics, or any commercial operation dealing with human embryos.

At least 10 bills related to hES research and cloning were introduced and referred to various congressional committees between 1999 and 2004, but to date, only one has passed. In 2001, and again in 2003, the House of Representatives passed a bill criminalizing SCNT for research or therapeutic purposes. Any scientist engaged in such research would be imprisoned for up to 10 years and fined a minimum of $1 million. A companion bill has been introduced to the Senate every year since 2001, but to date, none has passed. In addition to the criminal and civil penalties mandated by the House bill, the Senate bill also criminalizes the act of receiving therapies that use SCNT in countries that permit it, such as Japan and the United Kingdom. According to some interpretations of the bill's language, not only would U.S. citizens who sought SCNT-based treatments for diseases or spinal cord injuries be subject to imprisonment and fined at least $1 million, but they would also not be able to reenter the United States.

Many members of Congress who are antiabortion do not support funding restrictions or bans on hES cell research and therapeutic cloning. In May 2004, over 200 members of the House, including members of President Bush's party and more than 30 abortion opponents, sent a letter urging him to reverse his August 2001 decision and reinstate federal funding for research that adheres to the guidelines the NIH issued in August 2000.

Government policies have impeded ES cell research and training

At first, the scientific community breathed a collective sigh of relief on the evening of 9 August 2001, because they feared President Bush would ban federal funds for *any* research using ES cells. Stem cell researchers would rather have as many hES cell lines as possible to study, because the reliability of research results always improves as the number of samples increases. But many thought 70 hES cell lines seemed like a sufficient number, especially in the initial stages of research, when so many basic questions need answering. Others were more skeptical, and unfortunately, the skeptics were right.

After thoroughly investigating the quality and availability of the existing hES cell lines, the NIH issued a statement that only 15 lines are both suitable and available for research purposes. Some of those are becoming difficult to maintain because cell division has tapered off, and some have developed severe genetic abnormalities. In addition, by now you know enough about genetic variation to question the applicability of research findings obtained from studying only 15 cell lines, especially if the cell lines become increasingly abnormal as mutations begin to add up.

Finally, none of the 15 lines will ever be useful for therapeutic purposes because they were grown on mouse feeder cells and may be contaminated with viruses. Since Bush's announcement, privately funded researchers have improved the techniques for deriving and maintaining hES cells, and the newer cell lines are not grown on mouse cells. Unfortunately, these cell lines are not available to the great majority of academic researchers in the United States. Using them would jeopardize the researchers' access to federal funds on which they depend.

Because the current political environment (at the federal level) is not supportive and so many uncertainties exist about the direction of future policies and laws, stem cell researchers in the United States are accepting offers from academic institutions or biotechnology companies in countries with known requirements and government support for their research. Not only is the United States losing this generation of stem cell researchers, but the government's policies are also jeopardizing the future. With the prospect of 10 years in prison hanging over their heads, graduate students training for careers in cell and molecular biology are opting out of stem cell research. Because a great deal of graduate level training in cell and molecular biology is devoted to learning hands-on skills, the United States will lack a cadre of scientists, trained in the art of stem cell culture, to do the research or to train the next generation of cell biologists.

State governments and private funds are filling the void

Irrespective of the activities in Washington, many people in the United States are supportive of both hES cell research and therapeutic cloning. As we mentioned above, 75% of people surveyed supported hES cell research and therapies when they understood that the embryos providing the ES cells had been frozen and were about to be discarded. In addition, another survey of more than 1,000 Americans showed that 67% were in favor of SCNT for therapeutic purposes.

State governments have responded to the stagnation at the federal level, the substantial amount of public support, and the considerable potential benefits of hES cell research and therapies. Two states, California and New Jersey, have passed laws specifically supporting research on hES cells and therapeutic cloning while banning reproductive cloning in their states. Another seven states are considering similar legislation. Much of the legislative activity at the state level is spawned and sponsored by patient groups that would benefit from stem cell research. In addition, to compensate for the lack of federal funds, state governments with supportive laws are also providing funding. New Jersey appropriated $50 million to hES cell and therapeutic-cloning research, and in November 2004, California voters passed a bond referendum that commits $3 billion over a 10-year period to research on hES cell and therapeutic cloning.

Private money is also plugging the funding gap.

- The Michael J. Fox Foundation for Parkinson's Research has given more than $5 million to researchers.
- Harvard researchers created 17 new hES cell lines with monies provided by the Juvenile Diabetes Foundation and the Howard Hughes Medical Institute.
- The founder of Intel donated $5 million to University of California stem cell researchers.
- An anonymous donor gave Stanford $12 million to start a new institute devoted to hES cell research.

Research on hES cells could obviate the need to use embryos

Somewhat paradoxically, if researchers were allowed to study a sufficient number of hES cell lines and create new lines with SCNT, their research would probably solve the ethical dilemma of using embryos to create hES cell lines for research or therapeutic purposes.

For example, if scientists can tease apart the molecular steps that convert an ES cell into an adult stem cell, then they should be able to undo those steps and turn an adult stem cell into an ES cell. Using microarray technology, one group of researchers observed 532 genes turn off and 140 genes turn on as ES cells differentiated into neural stem cells. In addition, when a somatic cell nucleus is placed in an egg, under certain conditions, the genetic program of the somatic cell is erased. If scientists discover the steps in the dedifferentiation process, they will not need to use SCNT to create hES cells for therapeutic purposes. Instead, they could erase the genetic programming of the somatic cell not by putting it into an egg but by replicating the steps of the dedifferentiation process. Once the deprogramming had taken the somatic cell back to an embryonic stage, the cell could be reprogrammed to specific cell types that are genetically compatible with the cell's original source.

But to discover how ES cells become adult stem cells or the molecular steps in dedifferentiation—in order to avoid using embryonic cells in the long run—scientists must use embryos in the short run. If U.S. scientists are allowed to use only a handful of aging ES cell lines, how much can they discover? Will their research results be generally applicable to most cells and indicative of activities in healthy cells or biased because the cells are old and the samples are not large enough to be genetically representative? If the hES cell research that occurs in the United States is conducted primarily at private companies and not publicly funded research institutions, will stem cell therapies be so expensive that only wealthy people will be able to afford them?

SUMMARY POINTS

Although it may seem to be a good idea to be screened for diseases whether or not symptoms are present, the risks of widespread screening of asymptomatic populations are greater than the benefits for all but a few diseases. The risks are greatest with invasive diagnostic procedures, but the probability of receiving inaccurate results also carries risk, especially for false-positive results.

The HGP, an international project conducted from 1990 to 2003, produced a complete sequence of the human genome. It has also led to the identification and mapping of many disease genes. However, the HGP is only the first step in elucidating a more complete picture of the complex relationship between genes and disease.

Many of the new biotechnology-based diagnostics are genetic tests for disorders and diseases that previously could not be diagnosed until symptoms appeared. Predictive diagnostic tests are especially beneficial if early diagnosis can prevent, delay, or lessen the disease. However, predictive tests may also lead to unnecessary concerns, because the results of genetic tests can be ambiguous. Patients may misinterpret genetic predisposition as genetic predetermination.

The new methodologies that scientists use to identify and map genes associated with diseases can create ethical dilemmas because they can set up a conflict between equally important rights: the right to privacy and the right to know.

There are also questions about appropriate and ethical uses of information provided by genetic tests, such as the abortion of fetuses for traits unrelated to health and the unfair use of information by employers and insurance agencies.

Human embryonic cells have the potential to treat a variety of serious diseases, but some people find the derivation of embryonic cell lines to be ethically troubling. Embryonic cells can be derived from a variety of cell sources, each of which raises its own set of ethical concerns. Decisions to stop or restrict hES cell research for ethical reasons create a new set of ethical concerns.

Government policies influence the direction of scientific research through funding, laws, and policies that encourage some types of research and restrict others. By directing the course of scientific inquiry, governments determine which medical advances occur.

KEY TERMS

Embryonic germ cells

False negative

False positive

Neonatal testing

Parthenote

Preimplantation genetic testing

Prenatal genetic testing

chapter 21 | Biotechnology in the Food Industry

You have probably heard the expression "you are what you eat." But if you eat fruit and salads for 2 days and then exist on pizza, potato chips, and Twinkies for the next 2 weeks, you continue to look and feel like yourself. So how can you be what you eat? Your body's metabolic machinery breaks food down into its basic atomic and molecular units and then reassembles those units into you. By the time Twinkies and tomatoes are broken down and absorbed into your bloodstream, they look the same to your cells. Almost.

A healthy life depends on a healthy diet. Getting enough calories to build cells and carry out life's processes is not a problem for most people, but having access to the right kind of food is. According to the World Health Organization, more than 3 billion people—over 50% of the current world population—are malnourished because their diets lack essential macronutrients or micronutrients (Table 21.1). In industrialized countries, where balanced diets are available and cheap, people are malnourished because they make poor nutritional choices. In developing countries, the idea of dietary choice is a luxury unavailable to many people.

Scientists are using the new biotechnologies to improve the food supply in many ways. Rather than enumerating the many research avenues they are pursuing, we limit our discussion to three of their strategic objectives:

- enhancing the nutritional value of food and animal feed
- making foods healthier
- improving the safety of the food supply

After describing only a few of the tactical approaches for achieving each objective, we discuss:

- the science underlying food safety assessments for crops modified with recombinant DNA gene technology
- the U.S. system for regulating transgenic food crops

Table 21.1 Micronutrients[a]

Nutrient	Function(s)	Sources
Fat-soluble vitamins		
A	Vision, immune system, bone, skin	Yellow, orange, and red fruits and vegetables; green leafy vegetables; liver; eggs
D	Bone, calcium absorption	Egg yolk, fortified milk, fish liver oil
E	Cell membrane	Green vegetables, vegetable oils, whole grains
K	Blood clot formation; ATP[b] synthesis	Bacteria in colon, cabbage
Water-soluble vitamins		
Thiamin (B$_1$)	Connective tissue, Krebs cycle coenzyme	Whole grains, green leafy vegetables, legumes, organ meats
Riboflavin (B$_2$)	Krebs cycle coenzyme	Whole grains, eggs, milk
Niacin	Krebs cycle coenzyme	Peanuts, meat, fish, green leafy vegetables
B$_6$	Amino acid and fatty acid metabolism	Whole grains, spinach, meat
Pantothenic acid	Glucose, lipid metabolism	Meat, eggs
Folic acid	Nucleic acid synthesis, red blood cells	Green vegetables, whole grains, meat
B$_{12}$	Red blood cells, amino acid metabolism	Poultry, fish, organ meat, dairy
Biotin	Starch, fat, amino acid metabolism	Legumes, eggs
Ascorbic acid (C)	Collagen, bone, cartilage, teeth, immune system	Fruits and vegetables, especially berries, citrus, cantaloupe, broccoli, cabbage
Minerals		
Calcium	Bone, teeth, blood clotting, nerves, muscle	Dairy, green vegetables, legumes
Chloride	Acid in stomach, nerves	Table salt
Copper	Enzyme cofactor, ATP synthesis	Nuts, legumes, seafood
Iodine	Thyroid hormone	Marine fish, shellfish, iodized salt
Iron	Hemoglobin, ATP synthesis	Whole grains, green leafy vegetables, legumes, nuts, eggs, meat, shellfish
Magnesium	Enzyme cofactor, muscle, nerves	Whole grains, legumes, nuts, dairy
Phosphorus	Bone, nucleic acids, ATP, phospholipids	Whole grains, poultry, red meat
Potassium	Muscle, nerves, protein synthesis	Many dietary sources
Sodium	Muscle, nerves	Table salt
Sulfur	Some amino acids	Dietary proteins
Zinc	Enzyme cofactor, wound healing	Whole grains, legumes, nuts, meat, seafood

[a]People must consume sufficient amounts of macronutrients (proteins, carbohydrates, and lipids) and micronutrients (vitamins and minerals), or they develop diet-associated diseases. The most common macronutrient deficiencies are the essential amino acids and essential lipids the human body cannot synthesize. This table lists the sources of essential micronutrients and the tissues and functions that rely on micronutrients.
[b]ATP, adenosine triphosphate.

ENHANCING FOOD AND FEED NUTRITIONAL VALUE

The best solution to malnutrition is a diverse diet. This option is not realistic for many people in developing countries, who subsist on a few staple cereal crops, such as rice or corn. Broadening their diets to include fruits, vegetables, and grains that are rich in essential macro- and micronutrients is the best solution to malnutrition, but in the short term, scientists are taking a more pragmatic approach. They are using molecular genetics to improve the nutritional content of the staple foods that constitute the bulk of the dietary intake of malnourished people.

Diets must include the eight essential amino acids

Because all cell functions rely on proteins, each cell needs a ready supply of all 20 amino acids. The human body synthesizes 12 amino acids from scratch and depends on dietary sources for essential amino acids (Table 21.2). Insufficient protein is the most pervasive macronutrient deficiency, but often the problem is not the total amount of protein consumed, but one or two amino acids. Virtually all plant foods, including the major staples, are poor sources of at least one essential amino acid (Table 21.3). Animal products, such as meat, milk, and eggs, provide a ready source of complete protein, but only if the animal's diet contained the full array of essential amino acids. Vegetarians use complementary protein sources to make sure they've got their bases covered when it comes to essential amino acids, but many people whose diets consist of a single staple do not have this option. In addition, their farm animals often consume the same staple crop, lacking the same essential amino acids. To serve as a complete protein source, animal feed must be supplemented with amino acids. Livestock and poultry farmers in industrialized countries can afford this expense; subsistence farmers cannot.

Scientists are using new genetic modification techniques, with mixed success, to improve the amino acid balance in staple crops by:

- giving the crop plant a new gene encoding a protein rich in the amino acid(s) it lacks
- reengineering the crop plant's amino acid biosynthesis pathways

Adding new proteins to staple crops

Researchers in India successfully increased the amino acid content of the potato by providing it with a gene from amaranth, a high-protein grain used in South American and Asian cultures for centuries (Figure 21.1). Potatoes are the most important noncereal staple crop in the world, because they are cheap and easy to grow and store and children eat them readily. The researchers' goal was *not* to turn the potato into a primary protein source, which would be difficult because potatoes contain very little protein. Instead, they viewed the potato as a vehicle for delivering essential amino acids. The scientists isolated an amaranth gene that encoded a protein that is high in some essential amino acids. When they inserted it into potatoes, it increased lysine, methionine, and tyrosine concentrations 2.5 to 4 times while simultaneously tripling potato yields.

In a less successful but equally important venture, industry scientists attempted to improve the methionine content of poultry feed by incorporating a gene encoding a protein from the Brazil nut into soybeans. During the requisite testing for possible allergenicity, the company discovered that, unfortunately, the high-methionine protein happened to trigger allergic reactions in some people with food allergies to Brazil nuts. Even though the company was developing methionine-rich soybeans for poultry feed, they knew some

Table 21.2 Essential amino acids

Isoleucine
Leucine
Lysine
Methionine
Phenylalanine
Threonine
Tryptophan
Valine

Figure 21.1 Amaranth. Amaranth was a dietary staple of the Aztec and Inca Indians in pre-Columbian times. Each plant produces 40,000 to 60,000 seeds per growing season, and the leaves are also edible. The protein content of the seeds is significantly greater than that of other cereal grains, and the proteins contain large amounts of lysine, an essential amino acid. (Photograph courtesy of Cooperative Extension Service, University of California—Davis.)

Table 21.3 Incomplete proteins in plants[a]

Amino acid	Deficient foods
Lysine	Legumes: peanuts
	Cereals: corn, oats, rice, wheat, rye
	Nuts/seeds: almonds, cashews, walnuts, pecans, sunflower seeds
Methionine and threonine	Legumes: dried beans, black-eyed peas, chickpeas, lentils, lima beans, peanuts
	Vegetables: asparagus, broccoli, green beans, potatoes, soybeans
Tryptophan	Legumes: dried beans, chickpeas, lima beans, peanuts
	Cereals: corn
	Nuts: almonds, walnuts
	Vegetables: green peas

[a]Most food crops are deficient in one or more of the essential amino acids.

of those soybeans would inadvertently end up in shipments of soy intended for human consumption, and so they terminated the research project. Although the company had no saleable product to show for their work, their discovery was quite valuable scientifically, because it identified at least one of the Brazil nut's allergenic proteins. This information directed scientists to one of the genes that must be silenced to create nonallergenic Brazil nuts. Later in the chapter, we discuss allergenicity testing and research on decreasing food allergenicity in more detail.

Metabolic engineering of amino acid synthesis

A reasonable strategy for increasing essential amino acid concentrations might be to provide a crop plant with many copies of the genes encoding enzymes that synthesize the amino acids it lacks. Why not load up a staple crop with genes encoding biosynthesis enzymes from the pathway shared by a number of essential amino acids (Figure 21.2)? In theory, this might work, but not in practice, because amino acid synthesis is tightly regulated through end product feedback inhibition. To increase the essential amino acid end products shown in Figure 21.2, researchers must release the pathways from regulatory control. Using genetic modification to shift existing enzymatic pathways so that certain products are favored over others is known as metabolic engineering.

Figure 21.2 Metabolic pathway engineering. To increase the amount of lysine produced by crop plants, molecular biologists inserted genes encoding enzymes (AK and DHDPS) that are resistant to lysine feedback inhibition control.

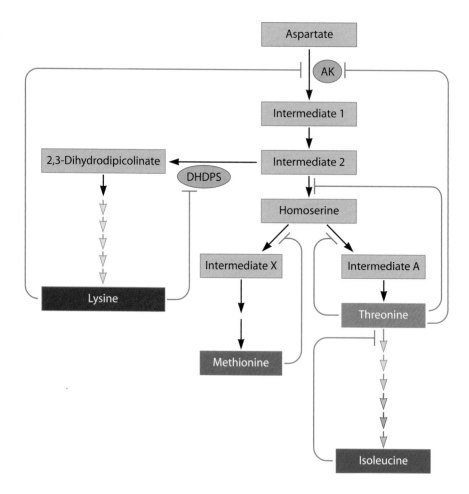

The enzyme catalyzing the first step in the pathway, aspartate kinase (AK), is a feedback inhibition target for two essential amino acids, threonine and lysine. Scientists inserted a bacterial AK gene that is much less sensitive to feedback inhibition into canola and, as expected, significantly increased the amounts of four essential amino acids derived from aspartate. But typically a crop plant is not deficient in all of these amino acids (Table 21.3), and releasing AK from feedback inhibition does not allow the plant to fine-tune its control of amino acid synthesis. Interfering with end product feedback inhibition after the branch point in the pathway provides a much finer level of control. Some bacteria also have a lysine-insensitive form of the enzyme dihydropicolinate synthase (DHDPS) that catalyzes the first step after the branch point. Scientists inserted genes encoding both feedback-resistant enzymes into canola and, by liberating lysine synthesis from feedback inhibition at two points, increased lysine 100% without significantly changing threonine or methionine levels. Inserting the same genes into soybeans increased lysine levels fivefold.

Staple crops are deficient in micronutrients

Micronutrients are food components, such as vitamins and minerals, needed in small amounts because they play key roles in specific cellular processes, such as oxygen transport (iron), hormone function (iodine), enzyme catalysis (zinc), or vision (vitamin A). Researchers from around the world are using recombinant DNA technology and plant breeding to increase the amounts of micronutrients in staple grains. This is known as **biofortification**.

Enhancing the amount and availability of iron

Two billion people worldwide suffer from iron deficiency, which impairs mental and physical development, increases susceptibility to infections, and is responsible for approximately 20% of all maternal deaths after childbirth. Unlike most malnutrition, iron deficiency is not necessarily associated with poverty, because 20% of U.S. female college students are iron deficient. Iron deficiency in industrialized countries results from poor food choices; in developing countries, the staple foods do not provide enough iron. Scientists cannot simply add a gene to a crop to increase iron synthesis because plants don't make iron but absorb it from the soil. Once iron is absorbed, plants store it in specialized storage proteins or attach it to organic molecules to protect cells from iron's destructive effects. These protective mechanisms benefit the plant but make it difficult for the human digestive system to extract iron from plant material.

Rice, the primary food source of many malnourished people, contains almost no usable iron. Using recombinant DNA technology, scientists have improved the amount and availability of iron in rice. To increase the amount of iron in each grain, they inserted a gene for an iron storage protein, **ferritin**. Each ferretin molecule can store 4,500 iron atoms (Figure 21.3). They also inserted two additional genes to improve iron's **bioavailability**, which is a measure of the amount of iron that is digested and absorbed from the small intestine in a form cells can utilize. Rice plants attach as much as 95% of their iron to phytate, an organic molecule humans cannot digest, so phytate-bound iron passes through the digestive tract and is excreted. Scientists gave rice a gene for a fungal enzyme that degrades phytate, which increases the iron available for absorption. The final gene encoded the protein **metallo-thionein**, which aids in iron absorption and transport in the bloodstream.

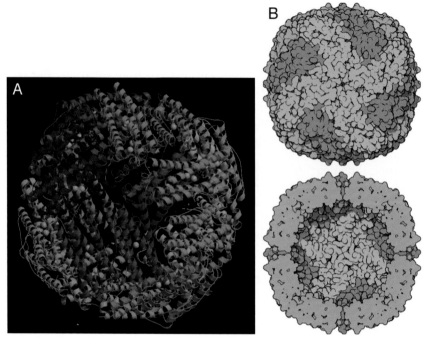

Figure 21.3 Ferritin. Iron is an essential micronutrient found in hemoglobin and a number of enzymes responsible for ATP synthesis. When exposed to water and oxygen, it forms rust-like molecules that are insoluble in water, so iron atoms must be stored and transported within watertight proteins. Ferritin is an iron storage molecule consisting of 24 identical subunits surrounding a hollow core that stores thousands of iron atoms. **(A)** A ribbon model shows how the tightly packed atoms protect the iron atoms from water. **(B)** A cross section of the molecule reveals the hollow core and tiny pores on the top, bottom, and sides that allow iron atoms to enter and leave when needed. (Images courtesy of the Protein Data Bank [http://www.pdb.org]. Image A is PDB entry 1FHA, submitted by D. M. Lawson. Image B is by David Goodsell, The Scripps Institute.)

Figure 21.4 Provitamin A synthesis. Provitamin A, also known as beta-carotene, is a lipid molecule. Rice naturally contains one of its precursors, geranylgeranyl-diphosphate. Scientists created transgene constructs that contained three enzymes that can convert geranyl-geranyl-diphosphate to provitamin A: phytoene synthase, phytoene desaturase, and lycopene β-cyclase. They used *Agrobacterium*-mediated transformation to insert the transgene into rice tissue cultures.

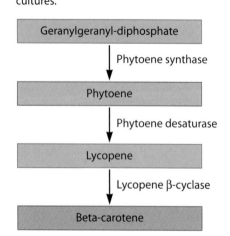

Increasing vitamin A in rice

Every year, approximately 1 to 2 million children die from severe vitamin A deficiency, and another 5 million suffer permanent eye damage, including total blindness. Another 125 million children with moderate vitamin A deficiency suffer from learning disabilities and are much more vulnerable to infections, such as malaria and measles. Rice provides the primary food for people in countries with the most significant vitamin A deficiency problems. To add more vitamin A to diets that are heavily dependent on rice, an international group of scientists used recombinant DNA techniques to create rice that is high in **beta-carotene**. The human body easily converts beta-carotene, which is also known as **provitamin A**, to vitamin A. Rice has the necessary enzymes to produce one of the intermediates in the provitamin A pathway, geranylgeranyl-diphosphate (Figure 21.4). Scientists used geranylgeranyl-diphosphate as the precursor and inserted new genes encoding the remaining three enzymes in the pathway into rice. They then crossed the provitamin A rice variety with the high-iron variety described above, creating a new variety known as **Golden Rice** (Figure 21.5).

Companies that own patents on various genes in Golden Rice gave the patent rights to a public research institution that is the world's leading center for rice research and breeding, the International Rice Research Institute (IRRI) in the Philippines. Using funds from the Rockefeller Foun-

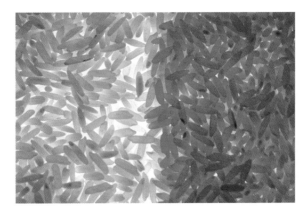

Figure 21.5 Golden Rice. High levels of beta-carotene, a pigment molecule found in many fruits and vegetables but not in rice grains, are responsible for the yellow rice grains on the right. Conventional rice is on the left. (Photograph courtesy of Louisiana State University *AgCenter News*.)

dation, IRRI scientists are increasing provitamin A amounts further, incorporating the Golden Rice genes into local varieties, and conducting field tests. The IRRI plans to distribute golden forms of local varieties to farmers for free.

Golden Rice does not address the underlying causes of vitamin A deficiency in developing countries: poverty, lack of access to a diverse diet, and poor or nonexistent infrastructure. As a result, many activists oppose the IRRI's plan to give poor farmers seeds for growing Golden Rice. Instead, Greenpeace favors government programs to distribute vitamin A pills and promote red palm oil, a crop rich in vitamin A grown in developing countries. The Union of Concerned Scientists believes the dietary problems Golden Rice is meant to fix would be solved better by distributing iron tablets, encouraging people to grow vitamin-rich gourds in their yards, and building roads to improve access to rural communities that typically suffer the most malnutrition.

MAKING FOODS HEALTHIER

While many people in developing countries have difficulty getting access to enough of the right kinds of food, the dietary problems for most people in the industrialized world stem from making poor choices. The two most important measures of your future health, tobacco intake and weight, are both within your control. The choice regarding tobacco intake and health is unambiguous: don't use it at all. Making healthy choices about food is much more difficult because, unlike using tobacco, people have to eat. However, the relationship between health and food is just as clear as that between tobacco and health. Being overweight affects your chances of having a wide variety of health problems (Table 21.4). Below, we talk about the health benefits of antioxidants and nutraceuticals, but their impacts are negligible to nonexistent if a person is overweight.

In an effort to lose weight, consumers often go from one fad diet to the next, hoping to find the quickest and easiest solution to being overweight. No matter what their hook is, all fad diets share a common trait: if a diet leads to weight loss, it does so simply because the dieters are burning more calories than they are consuming. The calories being consumed are contained

Table 21.4 Obesity and health: conditions related to excess weight[a]

Coronary heart disease

Stroke

Hypertension

Non-insulin-dependent (type 2) diabetes

Osteoarthritis

Sleep apnea and other breathing problems

Gall bladder disease

Stress incontinence or urine leakage

Depression

Cancer of the uterus, cervix, ovary, breast, colon, rectum, and prostate

[a]In the United States, 280,000 adult deaths/year are attributed to obesity. Overweight individuals have an increased risk of developing one or more of these conditions.

in three classes of biological molecules: proteins, fats, and carbohydrates. Because medical researchers have established a link between dietary lipids and cardiovascular disease, we limit our discussion to a few biotechnology applications that affect food lipid content.

The relationship between dietary lipids and heart disease is complex

True or false: fats are bad for you. If you pay minimal attention to public discussions of diet and health, you probably picked "true." Medical organizations began publicizing the relationship between a high-fat diet and obesity, heart disease, high blood pressure, diabetes, and other health problems at least 30 years ago. Americans heard the message loud and clear and significantly reduced the *percentage* of calories in their diet that came from fats. But this decrease coincided with a 100% *increase* in the number of Americans classified as obese. To confuse matters further, a number of scientific studies comparing various European countries have shown that populations with the lowest rate of heart disease have the highest fat intake.

Clearly, the relationship between dietary fats and health is more nuanced than "fats are bad for you." All fats are not created equal. Some decrease heart disease risks, while others increase them. The same can be said of cholesterol, which is a lipid, but not a fat. People associate cholesterol with health problems, because high cholesterol levels can alert physicians to patients at risk of heart disease. But all cholesterol is not created equal, either. The most useful health measure is not the absolute level of blood cholesterol but the ratio of "good cholesterol" (HDL, or high-density lipoprotein) to "bad cholesterol" (LDL, or low-density lipoprotein), because HDL actively protects against heart disease by removing excess cholesterol from the blood. The probabilities of heart disease in two patients with cholesterol levels of 200 mg/dl differ vastly if the 200 mg are distributed as 170 LDL-30 HDL in one patient and 110 LDL-90 HDL in the other.

The take-home message: some lipids are good for you, because they actively protect you from heart disease; others are definitely linked to heart disease and other health problems.

A healthy diet must include some lipids, because they are key components of cell membranes, enclose nerve cells in a protective sheath, and carry messages within and between cells. In addition, just like dietary amino acids, the body cannot synthesize all of the lipids it needs, so diets must include these "essential lipids." Making smart nutritional choices about lipids boils down to opting for a diet that decreases the levels of LDL and fats, or triglycerides, in the blood and increases HDL levels. In general, this translates into avoiding saturated fats and partially hydrogenated oils and using monounsaturated and polyunsaturated vegetable oils (Tables 21.5 and 21.6). Anyone who has dieted knows that changing eating habits can be extremely difficult because it requires a great deal of will power. Researchers are trying to help people eat healthier without having to change ingrained dietary preferences through metabolic engineering of fatty acid synthesis pathways in the crop plants that provide oils.

Metabolic engineering can make cooking oils more healthful

To obtain the oils used in cooking and food processing, the seeds of certain crops, such as soybeans, canola, and sunflowers, are pressed. As you can see from Table 21.7, the oilseed crops have varying amounts of the different fatty acids. Most of the fatty acids in oilseed crops are chains of 16 and 18

Table 21.5 Dietary lipids

Triglycerides

Three fatty acids bonded to a molecule of glycerol.

Most dietary lipids are in this form.

High blood levels are associated with heart disease.

The fatty acids in triglycerides are chains of carbon atoms bonded to hydrogen atoms.

The length of the chain and the number of hydrogen atoms bonded to each carbon differ from one fatty acid to the next.

Cholesterol

Obtained from animals; plants do not synthesize cholesterol.

Key component of cell membranes.

The precursor of testosterone, estrogen, and corticosterone.

High levels of LDL in blood are associated with heart disease.

HDL removes cholesterol from the bloodstream.

Other essential lipids

The fat-soluble vitamins, A, D, and E, or their precursors must be included in the diet.

Vitamin K, another fat-soluble vitamin, is synthesized by intestinal microbes.

---

carbons, but the relative proportions of those that are saturated, monounsaturated, and polyunsaturated vary from one crop to the next. Using molecular genetics, scientists can change the fatty acid profiles of oilseed crops to minimize the intake of unhealthy fatty acids. To understand their tactics, you need some basic information on the fatty acid synthesis pathways they are changing.

Fatty acid synthesis

In virtually all organisms, fatty acid synthesis consists essentially of hooking together a number of acetyl coenzyme A molecules, the two-carbon molecule you met in chapter 6 that plays a central role in metabolism. Fatty acid synthesis enzymes, which we will call FAsyns, add each two-carbon unit to the

Table 21.6 Fatty acids[a]

Fatty acid type	Examples of dietary sources	Effect on lipid level in blood
Saturated All carbons have maximum number of hydrogens Solid at room temperature	Butter, beef fat, cocoa butter	Increase triglycerides Increase LDL and HDL
Monounsaturated 2 carbons do not have maximum number of hydrogens Liquid at room temperature	Vegetable oils	Decrease LDL Increase HDL
Polyunsaturated—essential fatty acids ≥ 4 carbons do not have maximum number of hydrogens Liquid at room temperature	Vegetable oils	Decrease LDL Increase HDL
trans A special type of fatty acid created primarily by food processors by bubbling hydrogen through polyunsaturated vegetable oils to improve their stability Solid at room temperature	Partially hydrogenated oils	Increase triglycerides and LDL

[a]Biochemists categorize fatty acids based on the number of hydrogens bonded to each carbon. The _trans_-fatty acid category is not related to the numbers of hydrogen atoms per carbon atom but to their positions.

Table 21.7 Proportions of fatty acid types

Lipid source	% Saturated	% Monounsaturated	% Polyunsaturated	% *trans*
Plant oils				
Canola	7	62	27	0
Olive	12	73	10	0
Soybean	16	23	55	0
Sunflower	11	21	64	0
Palm	52	35	9	0
Coconut	85	10	3	0
Animal fats				
Beef tallow	40	47	3	9
Butter	59	25	6	4
Chicken fat	29	40	29	0
Hydrogenated plant oils				
Vegetable shortening	24	28	25	19
Margarine, stick	25	14	28	26
Margarine, tub	20	25	42	10

growing carbon chain (Figure 21.6). Therefore, synthesis of an eight-carbon fatty acid requires three different FAsyn enzymes: FAsyn1, FAsyn2, and FAsyn3.

$$C—C + C—C \xrightarrow{FAsyn1} C—C—C—C + C—C \xrightarrow{FAsyn2}$$

$$C—C—C—C—C—C + C—C \xrightarrow{FAsyn3} C—C—C—C—C—C—C—C$$

To maximize the efficiency of fatty acid synthesis, cells organize FAsyn enzymes into a cluster, so that each FAsyn enzyme can hand off its product to the next one. When the chain consists of 16 fully hydrogenated carbon atoms, which is the saturated fatty acid palmitic acid, a number of things can occur (Figure 21.7).

- If palmitic acid is the end product, it is released from the FAsyn enzyme complex by an enzyme that we will call, for simplicity's sake, palmitic-releasing enzyme.
- The enzyme palmitic desaturase removes two hydrogens from palmitic acid, converting it into a 16-carbon monounsaturated fatty acid.
- Another two-carbon unit can be added to palmitic acid. The resulting 18-carbon saturated fatty acid, stearic acid, can be used as is or converted to unsaturated forms by its own desaturase enzymes.

Figure 21.6 Synthesis of a fatty acid. Fatty acid synthesis begins as two molecules of acetyl coenzyme A (acetyl-CoA), shaded blue and orange, are joined by FAsyn1 to create a four-carbon molecule. Three more FAsyn enzymes each add an acetyl-CoA unit to the growing chain, creating a 10-carbon fatty acid.

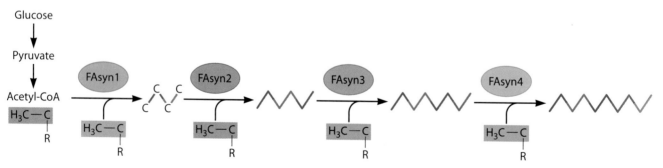

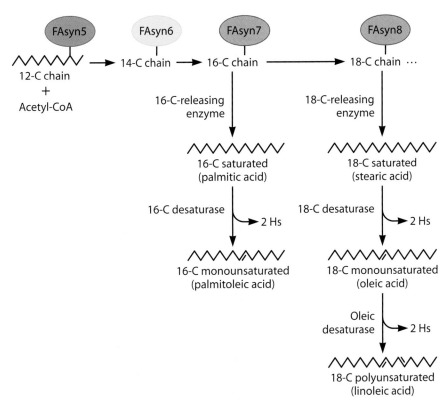

Figure 21.7 Synthesizing saturated and unsaturated fatty acids. In a continuation of the process illustrated in Figure 21.6, a 16-carbon saturated fatty acid (palmitic acid) is synthesized by a number of FAsyn enzymes that are organized into a complex. If palmitic acid is released from FAsyn7 by the 16-carbon releasing enzyme, it can be used as is or converted to a monounsaturated fatty acid by the 16-carbon desaturase enzyme, which removes two hydrogen atoms. If it is not released, FAsyn8 converts the 16-carbon fatty acid to stearic acid, an 18-carbon fatty acid. Stearic acid can be converted to a 20-carbon fatty acid by FAsyn9 or released from FAsyn8 by the 18-carbon releasing enzyme. Desaturase enzymes convert stearic acid to oleic acid, linoleic acid, and ultimately linolenic acid (not shown), which has three double bonds.

Decreasing the proportion of saturated fatty acids in oils

To change the ratio of saturated to unsaturated fats in salad oils, scientists use various molecular techniques to block certain enzymes and rev up others. Using gene-silencing technology, they block the palmitic-releasing enzyme and the stearic-releasing enzyme. This forces the 18-carbon saturated fatty acid, stearic acid, to take the desaturase route. Stearic acid desaturase removes two hydrogens, creating the 18-carbon monounsaturated acid oleic acid, associated with olive oil's health benefits. Blocking both releasing enzymes decreases the percentage of saturated fats in soybean oil by 50 to 60%. Other scientists add palmitic desaturase genes from yeast and bacteria, which have higher activity levels. These genes preferentially route palmitic acid to the pathway that converts it to a monounsaturated 16-carbon fatty acid.

Eliminating *trans*-fatty acids

When the medical community established the link between saturated fatty acids and heart disease, food processors began replacing the saturated fats they had been using, such as beef tallow and lard, with vegetable oils. Rich in polyunsaturated fatty acids, vegetable oils are much more healthful than saturated fats, but the attribute that makes them more healthful also causes

Saturated fatty acid

***cis* double bond (common)**

***trans* double bond (rare)**

Figure 21.8 *cis* and *trans* double bonds. The terms *cis* and *trans* refer to the placement of the hydrogens on the double-bonded carbons. In nature, most double bonds are *cis*, so enzymes have specific shapes that bind to *cis* bonds to break them or to place hydrogens on the same sides of the double bond. Hydrogenating double bonds by bubbling hydrogen through a container of fatty acids creates many *trans*-fatty acids, because hydrogen placement is random.

problems for food processors. The high proportion of polyunsaturated fatty acids makes vegetable oils unstable at the high temperatures required for frying and causes them to break down and become rancid when exposed to oxygen. In addition, vegetable oils are liquid at room temperatures, and some food products that food processors developed to replace saturated fatty acids, such as margarine for butter, had to be solid.

To solve these problems while continuing to replace saturated fats with vegetable oils, food processors began bubbling hydrogen through oils to convert some of the polyunsaturated fatty acids to monounsaturated fatty acids—hence the term "partially hydrogenated" oil that you may have noticed on food labels. They did not know that a minuscule difference between naturally occurring monounsaturated oils and the monounsaturated oils they created by chemical hydrogenation would lead to health problems, caused by the presence of ***trans*-fatty acids** (Figure 21.8).

Today, scientists are changing polyunsaturated to monounsaturated fats genetically rather than through chemical hydrogenation. These monounsaturated oils have the same desirable properties as partially hydrogenated oils—they are solid at room temperature and stable when exposed to oxygen or heated during frying and baking, and they have a long shelf life—but the natural fatty acids are not *trans*-fatty acids because highly specific plant enzymes, not chemical hydrogenation, are responsible for their synthesis.

Plant breeders at the U.S. Department of Agriculture (USDA) altered the sunflower's fatty acid profile through random mutagenesis and selective breeding. One new sunflower line has a fatty acid profile that has been shifted from 20% monounsaturated oils to 80% by decreasing the proportion of polyunsaturated fatty acids. However, these techniques provide no information about which enzymes in the sunflower's fatty acid synthesis pathway changed. The breeders focus solely on the final result: higher percentages of monounsaturated fatty acids (Figure 21.9).

Using recombinant DNA techniques, scientists produced soybean varieties with high percentages of monounsaturated oils. First, they used gene silencing to block the desaturase enzyme that converts the monounsaturated oil, oleic acid, to polyunsaturated oils. This shifted the fatty acid profile from the 20 to 25% monounsaturated oil typical of soybeans to 75 to 80% while simultaneously decreasing the percentage of polyunsaturated oil from 55 to

Figure 21.9 Altering fatty acid profiles by breeding. Scientists have used recombinant DNA technology, mutagenesis, and breeding to shift the proportions of saturated, monounsaturated, and polyunsaturated oils in the most common oilseed crops. In this photograph, USDA geneticists are hand pollinating sunflowers to increase the amount of oleic acid, the fatty acid found in olive oil. (Photograph by Russ Hanson, courtesy of Agricultural Research Service, USDA.)

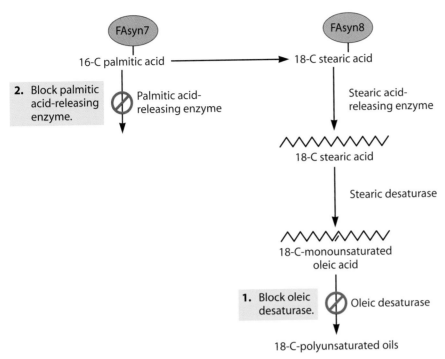

Figure 21.10 Altering fatty acid profiles with molecular techniques. Recombinant DNA techniques for blocking gene expression increased the percentage of monounsaturated fatty acids in soybean oil from approximately 23 to 90%.

3 to 4%. To increase the percentage of monounsaturated fats even further, they shifted their attention to the saturated oils. They blocked the palmitic-releasing enzyme so that virtually all of the 16-carbon, fully saturated palmitic acid was converted to the 18-carbon stearic acid. Then, approximately one-half to two-thirds of the stearic acid was converted to its monounsaturated form (oleic acid) by stearic desaturase, raising the total percentage of monounsaturated fatty acids to approximately 90% (Figure 21.10).

Molecular techniques can increase amounts and availability of nutraceuticals

Nutraceuticals, or **functional foods**, are foods or parts of foods that provide a medical or health benefit beyond the basic human need for sufficient amounts of macronutrients and micronutrients to prevent medical problems caused by their deficiencies. Health claims linked to nutraceuticals include prevention of infectious diseases and chronic degenerative diseases associated with aging. Nutraceuticals include the following.

- Micronutrients, such as vitamins A, C, and E, in larger doses than the recommended daily allowance (RDA). The RDA values are the minimal amounts needed to protect against nutritional disorders, such as scurvy and rickets. Consuming more than the RDA of certain vitamins and minerals seems to reduce the risk of contracting some diseases.
- **Phytochemicals**, which are molecules that occur in plants in small amounts (Table 21.8). In small amounts, they have health benefits, but in large amounts, phytochemicals can cause health problems,

Table 21.8 Phytochemicals

Phytochemical class	Potential benefit	Dietary source(s)
Carotenoids		
Beta-carotene	Neutralizes free radicals	Carrots, cantaloupe
Lycopene	Reduces risk of cancer, heart disease	Tomato, watermelon
Lutein	Maintains healthy vision	Leafy green vegetables
Xanthophylls	Maintains healthy vision	Yellow fruits and vegetables
Flavonoids		
Anthocyanidins	Improves circulatory function	Blueberries, red grapes
Flavonones	May reduce cancers	Citrus fruits
Catechins	May reduce cancers	Green tea
Proanthocyanidins	Prevents infections	Cranberries
Glucosinolates		
Sulforanes	Stimulates anticancer enzymes	Broccoli, cabbage
Glucoraphanin	Protects against inflammation and infection	Cabbage, kale, mustard
Isoflavonoids		
Genistein	Protects against osteoporosis	Soybeans
Daidzein	Eases menopausal symptoms	Soybeans
Phytosterols		
Stanol	Reduces risk of heart disease, lowers LDL	Corn and soy oils
Saponins		
Sapogenin	Immune booster, lowers LDL	Alfalfa sprouts, spinach

proving once again that the dose makes the poison. For example, broccoli contains glucosinolates, which neutralize free radicals and may protect cells from becoming cancerous. On the other hand, eating too much broccoli or its relatives can lead to enlargement of the thyroid gland, or goiters.

• **Probiotics**, which are beneficial, symbiotic microbial species, such as the intestinal bacteria that aid digestion and synthesize vitamin K (Figure 21.11). They also attack invading pathogens and keep other symbiotic microbes that can cause health problems in check. **Prebiotics** are foods, such as leeks, onions, and fruits, that selectively encourage the growth of beneficial symbiotic microbes because they contain certain sugars that humans cannot digest but microbes can.

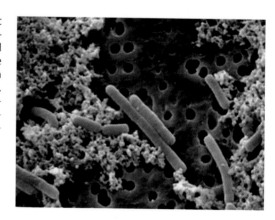

Figure 21.11 Probiotics. Lactic acid bacteria, such as *Lactobacillus casei*, pictured here, are helpful residents of the human digestive system. Yogurt and sour cream contain live *Lactobacillus* cultures. (Photograph courtesy of Department of Education Human Genome Project and Utah State University.)

Vitamin E and beta-carotene

You have probably heard about **antioxidants**, the scavenger molecules that protect cells from destructive free radicals and in doing so help prevent the onset of various cancers and slow the aging process. Vitamins A, C, and E all appear to have antioxidant properties, so scientists are using molecular techniques to enhance the levels naturally found in foods and are developing methods to manufacture novel antioxidants not found in food crops.

Vitamin E occurs in four forms, one of which, alpha, is most important to human health, but most oilseed crops have little alpha-vitamin E. For example, only 7% of the vitamin E in soybean oil is alpha, and most is in the gamma form, a precursor to alpha. Scientists have isolated the gene encoding the enzyme that converts gamma to alpha. They created a recombinant molecule that linked the gene encoding the enzyme to the carrot plant's promoter of the gene. Transgenic plants containing this gene construct overexpressed the enzyme, which increased the amount of alpha more than 80-fold without increasing the total amount of vitamin E.

The **carotenoids** are fat-soluble pigment molecules; they include beta-carotene, which makes carrots and cantaloupes orange, and lycopene, which gives tomatoes and watermelon their red color. Both have health benefits in amounts that are higher than the RDA, and beta-carotene is also the precursor for vitamin A. Some microbes and plants make novel carotenoids, some with very high antioxidant activities. Even though these novel carotenoids are not found in crop plants, their biosynthetic pathways could be engineered into food crops. Scientists have isolated genes encoding the biosynthetic enzymes and successfully transferred them to both microbes and plants. One transgenic microbe, which is a well-known workhorse in industrial fermentation, produced 12 different carotenoids simultaneously. Microbial fermentation would provide a way to manufacture these carotenoids so they could be sold as supplements.

IMPROVING FOOD SAFETY

In the United States, people expect the food they buy to be safe, because in general, that is what their experience tells them they should expect. The U.S. food supply is safer than it has ever been, but even so, two significant food safety problems remain: food-borne illness caused by microbial contamination and food allergies. How might the new biotechnologies lessen these problems?

Microbes cause food-borne illnesses

Foods contain nutrients that are just as attractive to microbes as they are to people. Unfortunately, some microbes that chow down on your food are pathogens capable of causing illness and sometimes even death. Food-borne illnesses caused by pathogenic microbes are by far the most significant food safety issue (Table 21.9). The Centers for Disease Control and Prevention

Table 21.9 Food-borne illness statistics for 2000

Pathogen	No. of cases	No. of hospitalizations	No. of deaths	Cost ($ billion)
Campylobacter spp.	1,963,141	10,539	99	1.2
Salmonella spp.	1,341,873	15,608	553	2.4
Escherichia coli O157	62,458	1,843	52	0.7
E. coli non-O157	31,229	921	26	0.3
Listeria monocytogenes	2,493	2,298	499	2.3
Total	3,401,194	31,209	1,229	6.9

estimate that microbial contamination of food causes 75 million cases of food-associated diarrhea, 325,000 hospitalizations, and 5,000 deaths in the United States annually. The cause of most of these illnesses is undercooked meat or raw fruits and vegetables that have been fertilized with animal manure.

Stories about food-borne illnesses seem to be appearing more often, and after reading the chapter on assessing risks, you might wonder if the increase can be explained by greater publicity. No, not completely. Food-borne illness has increased due to a number of reasons.

- Increased consumption of raw foods, especially seafood.
- Globalization of the food supply. Thanks to refrigeration and airplanes, farmers all over the world provide a remarkable variety of fresh (and therefore raw) fruits and vegetables year round. Often, however, farmers in other countries use production practices that are not as safe as those that U.S. growers follow.
- More meals eaten out. More people eat meals that they have not prepared, so they are depending on the food-handling and hand-washing habits of the staff preparing and serving the food.
- New food-borne pathogens. Microbes can adapt to almost any control measure. Those whose numbers had been kept in check can escape control and emerge as a new problem even though they have always been present in low numbers.

Improvements in animal and plant health

To reduce the incidence of microbial contamination of food, microbes must be controlled at every step along the food production and distribution chain, whether the food product is plant or animal.

Decreasing microbe levels in meat, milk, and eggs begins on the farm by controlling animal disease. Improvements in animal health care, comparable to those for humans, are finding their way into veterinary clinics and animal production facilities. Biotechnology-based diagnostics for infectious disease allow farmers to identify sick animals (on site) early and contain them before the pathogen spreads. Endogenous therapeutics, similar to those developed for people, boost the animal's immune system and fight infectious diseases. Veterinary scientists are also developing new animal vaccines, and crop developers are using recombinant techniques to incorporate vaccines and therapeutics into animal feed crops.

Plants are also susceptible to infections. Plant pathogens cannot infect people, but they produce compounds, such as mycotoxins, that harm people and kill livestock and poultry. Microbes, including fungi, invade the plant by entering holes created by insect feeding (Figure 21.12), and the crop plant, its grains, and food and feed products may all become contaminated. Mycotoxins are a significant problem for people in developing countries but not in the United States, because U.S. farmers have more options for preventing fungal infections and controlling insects that feed on crops (Figure 21.13). Biotechnology now provides more options for creating disease- and insect-resistant crops. As we mentioned earlier, Bt (for *Bacillus thuringiensis*) corn, which is resistant to certain insects, has significantly lower levels of mycotoxin contamination than non-Bt corn.

Detection before entering the marketplace

For decades, food processors have used visual inspection, analytical chemistry, biochemistry, and microbiology to test for food contamination, and

Figure 21.12 Fungal pathogen. Fungi are responsible for the majority of plant diseases. This scanning electron micrograph shows fungal hyphae invading barley. (Micrograph courtesy of the Center for Microscopy and Microanalysis, University of Queensland.)

biotechnology now provides them with a host of new methods to detect harmful microbes (Figure 21.14). The diagnostic kits based on biotechnology have increased detection sensitivity by orders of magnitude and decreased detection time from 4 to 7 days to 22 to 48 hours. In addition to providing rapid, accurate results, the kits are portable, cheaper, and less labor-intensive. Many are based on a simple color change, obviating the need for expensive laboratory equipment, which makes them especially useful in developing

Figure 21.13 Mycotoxins and food safety. Regular consumption of grains infected with mycotoxin-producing fungi can lead to chronic diseases. Corn grown in the Guatemalan highlands is harvested and graded as clean, spoiled, or rotten (right). Cornmeal for making tortillas consists of equal parts of each grade. Not surprisingly, this meal contains levels of mycotoxins (26 parts per million [ppm]) that are 100 times greater than cornmeal from corn grown in the United States (0.2 ppm). The toxicity problem is amplified by dosage. The average consumption of tortillas each day (14 to 16 ounces) is over 100 times greater than U.S. consumption (0.12 ounce). (Photographs courtesy of Ronald Riley, USDA Mycotoxins Laboratory.)

Federal Inspections of Animal Products in 1999	
Cattle	43,891,921
Swine	105,755,405
Other livestock	5,420,077
Poultry	8,365,372,345
Egg products	3,400,000,000

Figure 21.14 Meat inspection. The USDA Food Safety Inspection Service (FSIS) is responsible for ensuring that meat, poultry, fish, milk, and eggs are not contaminated by microbes when the products leave the processing facility. More than 7,500 FSIS inspectors inspect 6,000 privately owned processing plants annually. In addition, approximately 3 billion pounds of meat and poultry from 32 countries passes inspection for entry into the United States annually. (Photograph courtesy of FSIS, USDA.)

countries. Until recently, most kits allowed the detection of one or a few microbial species, but due to the development of microarray technology, companies are creating gene chips to test for many food-borne pathogens simultaneously (Figure 21.15).

Food allergies can be life threatening

We must first clarify the concept of food allergy, which must be widely misunderstood, because approximately 30% of American adults think they are allergic to certain foods while actually only 1 to 2% have food allergies. (This percentage applies to adults only. Approximately 5 to 6% of children under the age of 3 years have a food allergy. Many children outgrow certain food allergies, especially those involving milk or eggs.) This small percentage should not mislead you into dismissing the significance of food allergies as a public health problem, because their effects can be life threatening.

Physicians distinguish food allergies from other adverse reactions to foods. A food allergy is an abnormal response of the immune system to naturally occurring food components that most people can consume without having an adverse reaction. A protein(s) in the food, the allergen, triggers the production of allergen-specific antibodies, much like the immune response to other antigens. The initial exposure to the food allergen does not cause a reaction, but subsequent exposures trigger the release of histamine within minutes or, at most, a few hours. Food intolerances, like lactose intolerance, or food sensitivities, such as those to amines in some wine and fermented foods, are not food allergies, because they do not involve antibody-producing cells of the immune system.

Any food is capable of triggering an allergic response in someone, but only eight foods, the major allergens, are responsible for 90% of all food allergies worldwide: peanuts, soybeans, eggs, milk, tree nuts (Brazil nuts, cashews, etc.), fish, shellfish, and wheat. The remaining 10% are due to one of the over 170 minor allergens known to have elicited an immune response from a few people, sometimes only a single person. The allergenicity of a

Figure 21.15 Food-borne illness. **(A)** A *Salmonella* species. The Centers for Disease Control and Prevention estimates that *Salmonella* bacteria cause from 1.5 to 4 million cases of food-borne illness annually in the United States. As few as 15 to 20 bacteria constitute an infectious dose. **(B)** *Listeria monocytogenes*, the bacterium that often triggers food recalls, causes an average of 1,600 infections and 415 deaths/year in the United States. The infectious dose is much higher than that for *Salmonella* (900 to 1,100 bacteria), but the symptoms of listeriosis are more severe. *L. monocytogenes* is an intracellular parasite. The bacterial cells (stained red with fluorescent dye) move throughout the infected cell using tail-like structures, which look like comet tails in this micrograph. (Photographs courtesy of National Institute of Allergy and Infectious Diseases, National Institutes of Health [A], and Julie Theriot, Whitehead Institute [B].)

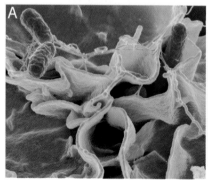
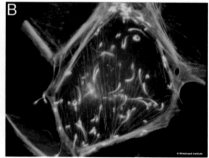

food may vary among societies with different diets: rice allergies are common in Japan, but not in North America. Approximately 200 of hundreds of thousands of dietary proteins are allergenic, a remarkably small number. People with food allergies are usually allergic to a few specific proteins within one or two foods. Within a group sharing a food allergy, the amount of allergen required to trigger the reaction can vary considerably, as does the severity of the response. In addition, the allergenic protein may differ among people allergic to the same food.

Detecting food allergens

People with food allergies must learn to avoid the offending food. Whole foods, like peanuts and eggs, are easy to identify and avoid, but in processed foods, where individual components are not always identifiable, consumers with food allergies must read the product ingredient labels. A much more serious problem associated with identifying allergens in processed foods is related to the *inadvertent* presence of trace amounts of allergens when different food products are made at a single facility. You may have noticed that many food processors in the United States use the ingredients label to warn consumers if the food product was manufactured with equipment that might have trace amounts of one of the eight major allergens. A biotechnology diagnostics company recently developed a kit that allows food processors to detect trace amounts (at the level of parts per million) of the eight major allergens so that food companies can catch the inadvertent presence of an allergenic food before the product enters the marketplace.

Decreasing allergenicity of problem foods

A better solution would be to lessen the severity of the allergic reaction or, better yet, remove the allergenic proteins from the food. Molecular techniques can identity the specific protein in an allergenic food that triggers the immune response. Once the allergen is known, its gene can be identified by back translating from the amino acid sequence to the DNA sequence. Knowing the sequence provides a number of routes for decreasing the food's allergenicity. The most obvious uses gene-silencing techniques to prevent gene expression in the part of the plant that is consumed. This approach has been used to virtually eliminate one of the soy allergens from its seeds. Another involves changing the shape of the allergen so it can no longer bind to its antibodies. Any one of three proteins in peanuts, Ara h1, h2, and h3, might trigger an allergic response in people with a peanut allergy. Changing a single amino acid alters their ability to bind to their specific antibodies, and unbound antibodies do not trigger histamine release. The altered Ara h1, h2, and h3 proteins reduced or completely eliminated the allergic response in mice with peanut allergies.

Rather than trying to correct the problem by changing the food, scientists are also using another angle of attack: the person with the allergy. Recently they have developed a therapeutic monoclonal antibody that binds to the type of antibody released in allergic responses. This impedes binding between the allergen and antibody and consequently prevents histamine release on exposure to the allergen. This strategy has significantly improved tolerance to peanuts and is useful for lessening all allergies, not just food allergies. Another approach uses a DNA vaccine encoding the allergenic protein to habituate the person to the allergen, preventing it from triggering an immune response. This is essentially the strategy underlying the practice of giving small doses of the allergen (allergy shots) to people with allergies to things they cannot avoid, such as pollen, mold, or animal hair.

Figure 21.16 Mutagenesis breeding. Plant breeders subject crop seeds to mutagens, such as chemicals and X rays; plant the seeds; and determine if any of the plants that result have the desired trait.

ISSUES ASSOCIATED WITH FOOD BIOTECHNOLOGY

To date, food biotechnology has captured significant attention from the media and general public, but only one biotechnology application has received the lion's share of the attention: genetically engineered, or transgenic, crops. The two concerns the public voices most often are the following.

- It's not natural.
- How do we know food made from a transgenic crop is safe and nutritious?

As is true of all applications of modern biotechnology, these concerns should be analyzed within the context of past experience. This is especially true for transgenic crops, because people had a long history of genetically altering crops before recombinant DNA technology was developed. If the new recombinant DNA techniques disappeared tomorrow, would concerns about naturalness and food safety disappear? You need more background on crop genetic modification prior to recombinant DNA technology before you can answer that question.

In chapter 13, you learned that nature constantly recombines genetic material from different sources without any assistance from humans. Sexual reproduction recombines genetic material through crossing over in meiosis and cell fusion in fertilization; conjugation, transduction, and transformation are bacterial recombination tactics. Nature has invented many methods for recombination, because combining genetic material from different sources increases genetic diversity, which is the raw material for evolutionary change. Crop scientists use existing genetic diversity when they genetically modify crops by selective breeding or genetic engineering, both of which recombine genes in new combinations. When nature's supply of genetic material does not suit their needs, they generate genetic diversity by creating novel genes with mutagens (Figure 21.16).

Selective breeding relies on natural genetic diversity

Plant genetic diversity, created naturally through sexual reproduction, is a valuable natural resource that humans have exploited for centuries. Individual plants had traits people valued. They saved the seeds of those plants, grew them into parent plants for the next generation, selected the best of their offspring, and discarded the rest. By selecting certain genotypes from a population and excluding others, humans orchestrated the recombining of genetic material just as natural selection does.

Scientific progress in the past 100 years provided plant breeders with better tools for purposeful recombination. Armed with knowledge of plant reproduction and genetics, they directed genetic change with greater predictability by controlling pollination, and they broadened access to genes in other plant species. They expanded the size of the gene pool within which they could fish for useful traits to improve crops by developing laboratory techniques for crossbreeding plants incapable of hybridizing without human matchmakers (Table 21.10). At first, they hybridized crops to plants in different species but the same genus. Increasingly sophisticated techniques allowed them to extend those boundaries and cross crops with plants in different genera.

When breeders incorporate new useful traits into an established crop variety, they cross it with a plant having the desirable trait. The plant donating the gene could be a different variety of the same crop, a different variety treated with a mutagen, a wild relative, or a wild relative treated with a mu-

Table 21.10 Laboratory techniques for crossbreeding plants[a]

Barrier	Techniques for overcoming barrier
Prefertilization barriers	
Failure of pollen germination	Remove pistil, then pollinate exposed end
	Use recognition mentor pollen
Slow pollen tube growth	Chemical treatment with organic solvents or growth regulators
Pollen tube growth stops	In vitro fertilization
	Use of plant growth hormones and chemicals, like chloramphenicol and acriflavin
Failure to obtain sexual hybrids	Protoplast fusion
Different no. of chromosomes	Chemically induce chromosome doubling
Postfertilization barriers	
Embryo abortion (immediate)	In vivo/in vitro embryo rescue/implantation
Embryo abortion (early stages of development)	Culture ovaries in petri dishes
Lethality of F_1 hybrids	Use cell culture to regenerate plants
Chromosome elimination	Alter genomic ratios of species
	Induce chromosomal exchanges with tissue culture or irradiation
Hybrid sterility	Chemically induce chromosome doubling

[a]During the second half of the 20th century, plant breeders developed increasingly sophisticated laboratory techniques for crossbreeding plants in different species and different genera. The table provides a few examples of natural physiological barriers to wide crosses and the techniques plant breeders have used to overcome the barriers.

tagen (Figure 21.17). Breeders focus on the phenotypic trait and, prior to breakthroughs in plant genomics in the past decade, have been less concerned with identifying the genes or proteins responsible for the trait. Before a crop variety with the new trait can be commercialized, breeders must rid it of undesirable traits produced by the thousands of genes with unknown functions added during crossbreeding, assuming the traits can be identified. To accomplish this, they typically spend 10 to 12 years backcrossing the new crop variety with its well-established parental variety.

Mutagenesis breeding creates new genes through mutation

Certain desirable traits cannot be found in the gene pools available to plant breeders even when they implement the unnatural crossbreeding tricks of the trade mentioned above. In the 1940s, crop developers began to create new genes in crops and related plants by using mutagenic agents, such as X rays and mutagenic chemicals. To date, they have used mutagenesis and subsequent breeding to give new, desirable traits to more than 1,500 crop varieties, including certain varieties of all major crop plants. Because mutagens cause genetic changes randomly, plant breeders do not know which or how many genes mutated to create the new trait. They also do not know how many other genetic changes, perhaps involving scores of genes with unidentified functions, might have occurred in response to the chemicals or radiation.

Recombinant DNA technology relies on natural genetic diversity

In addition to directing recombination of genetic material by intentionally uniting certain pollen and eggs, crop scientists now recombine genetic material at the molecular level. Genetic engineering combines pieces of DNA from

Figure 21.17 Genetic diversity and plant breeding. Plant breeders incorporate naturally occurring desirable traits into existing crop varieties. In this instance, a mutant variety of cauliflower, with levels of beta-carotene more than 100 times greater than those of normal cauliflower, was discovered in a cauliflower field in Canada. (Photograph courtesy of David Garvin, USDA.)

different sources by using some enzymes to cut DNA in specific places and others to reassemble the pieces, creating a transgene. Vectors, such as plasmids, then ferry the transgene into plant cells, where it recombines with the crop's genome. The plant cell is cultured and regenerated into a whole plant.

Because reproductive barriers do not exist at the molecular level, the capacity to improve crops is no longer restricted to the relatively small gene pool within which crop scientists can manipulate breeding (see Box 21.1). A desirable gene in any organism may theoretically be placed into a crop plant, no matter how distantly related the gene donor and crop plant are. The ability to move genes from any organism provides unprecedented flexibility and expands the prospects for improving crops, because now crop developers have access to all of nature's astounding genetic diversity.

The final steps in developing new crop varieties are identical

The techniques for generating genetic diversity differ, but plant breeding, mutagenesis, and genetic engineering share the same objective: creating a crop variety with a new trait.

During the first stages of new-variety development, only some plants have the desirable trait, irrespective of the genetic modification method that is used. (1) In breeding, the crop genome recombines with the genome of the plant containing the new trait when egg and pollen fuse in fertilization. Because chromosomes segregate independently during gamete formation, in subsequent crosses, some of the offspring will have the trait and others will not. (2) In mutagenesis breeding, crop seeds are treated with chemicals or radiation, causing random mutations in the DNA molecules. Some mutations may lead to beneficial traits, but most will not. (3) In genetic engineering, new genes for the beneficial trait are inserted into the genome at the cellular level using molecular techniques. Some cells will take up the gene; others will not. Those that do are cultured and grown into plants capable of reproduction.

No matter how they created the subset of plants with the new, desirable trait, the breeders' job is to incorporate the new trait into a high-performance, or **elite**, crop variety that has a known history with respect to food safety, nutrition, and field performance. All of the plant populations created through breeding, mutagens, or genetic engineering contain both superior and inferior plants. Crop developers discard inferior plants and cross the superior plants that also possess the new trait with the elite variety of that crop. Some offspring will have the new trait; others won't. Some with the new trait will be superior, others inferior. The breeder retains only the superior plants with the new trait and crosses them with the elite variety. This cycle of crossing, discarding, and recrossing occurs until the breeder is certain that the new trait is stably inherited and the new variety performs as well in the field as its parental, elite variety does.

Concern: it's unnatural to move genes between different species

People have expressed concern that genetic engineering allows crop developers to cross the normal reproductive and genetic barriers that exist in nature, and they fear that in doing so, humans violate one of nature's rules. Is the capacity to move genes between different species unique to biotechnology? No, in two regards; one you can guess, but one might surprise you.

You already know that many food crops were created by using unnatural processes long before genetic engineering appeared on the scene. Using laboratory techniques, plant breeders forced cross-pollinations between plant species incapable of interbreeding under natural conditions (Figure 21.18).

BOX 21.1 *Golden Rice*

Because of the unity of life at the molecular level, all organisms are able to read the genetic design manuals of other organisms. Recombinant DNA technology provides crop developers with a method for introducing instructions for certain design features into a crop plant. This gives crop developers the capacity to improve the nutritional value of staple crops in ways that they could only dream about until recently. However, the remarkable flexibility that recombinant DNA technology grants to crop developers is precisely the same power that concerns people who are uncomfortable with transgenic crops because they seem unnatural to them.

To create Golden Rice, scientists incorporated genes from five different species into a commonly used rice variety. To increase the amount and availability of iron, they used genes from the green bean, *Phaseolus vulgaris* (A); basmati rice (B); and a fungus, *Aspergillus fumigatus* (C). These genes encode the proteins ferritin, metallothionein, and phytase, respectively. Genes from two species, the daffodil, *Narcissus pseudonarcissus* (D), and a bacterium, *Erwinia uredovora* (E), provided genes for enzymes in the beta-carotene synthesis pathway.

For the moment, let's put aside discussions of nature's unnatural tendency to move genes among species, as well as considerations about the nutritional benefits that Golden Rice could bring to millions of people, and return to a question we have repeatedly asked you to consider in this book. Where would you draw the line, and why would you draw it there, when it comes to moving genes between two species?

Most everyone would automatically accept the addition of the gene from basmati rice to the host rice variety because both plants are rice. However, if someone objects to moving genes between two species that would not interbreed in nature, it's important to note that often two crop plants that seem—to people—to be closely related will not interbreed without help from plant breeders. The reverse is true as well. Some plants that taxonomists have placed in different species, and even different genera, will interbreed naturally, in spite of the fact that they seem to people to be very different.

Other people would not be concerned about the gene from the green bean because both genes are from plants, even though they would never interbreed naturally, and both plants have been in the food supply for thousands of years. But what about daffodils? Daffodils are more closely related to rice genetically than green beans are, but daffodils have never been in the food supply. In fact, all home gardeners know that small mammals that love to eat bulbs completely avoid daffodils, so daffodils probably contain secondary chemicals that make them distasteful and perhaps toxic. Finally, most people would probably balk at the idea of having genes from a fungus and a bacterium added to their food. However, because *Aspergillus* and *Erwinia* are ubiquitous pests of many wild and domesticated plants, people all over the world have been unintentionally eating these pathogens longer than they have been eating either rice or green beans.

Images courtesy of Horticulture Department, Texas A&M University (A); Agricultural Research Service, USDA (B); International Union of Microbiological Societies (C); Animal and Plant Health Inspection Service, USDA (D); and Cooperative Extension Service, Oregon State University (E).

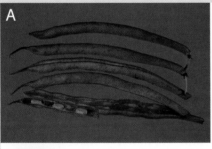

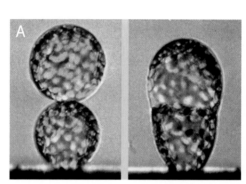

Figure 21.18 Laboratory techniques in plant breeding. **(A)** In protoplast fusion, scientists remove the cell walls from plant cells and force them together to combine their genomes. **(B)** Modern plant breeding relies heavily on tissue culture to coax plant cells and tissues into becoming whole plants capable of reproduction. (Photographs courtesy of the International Seeds Consortium [A] and Agricultural Research Service, USDA [B].)

Table 21.11 Overcoming the species barrier in plant breeding[a]

Crop species	Trait
Corn	Fungal disease resistance
Canola	Altered fatty acid ratios
Oats	Increase yield 25–30%
Beets	Nematode resistance
Tomato	Virus resistance
	Harvesting traits
	Nematode resistance
	Fungal disease resistance
Rice	Virus resistance
Potato	Fungal disease resistance
	Virus resistance
	Insect resistance
	Nematode resistance
Wheat	Fungal disease resistance
	Increase protein
	Insect resistance
	Drought tolerance
	Winter hardiness

[a]In the early 1900s, plant breeders began producing crop varieties by crossbreeding plants in different species and also different genera, the next level of genetic difference. This table provides selected examples of crops developed using wide crosses and the traits transferred into the crop species through the crosses.

They then rescued embryos that resulted from the forced matings so that the plant would not reject them. Thus, breeders crossed genetic barriers that would not be breached in nature long ago. As a result, at least some varieties of all major crop plants contain genes from different species that cannot hybridize naturally (Table 21.11).

The second reason this concern is not unique involves the "unnaturalness" of nature. While some people may object to moving genes across species because it perturbs their sense of propriety, nature apparently disagrees with their notion of what is and is not appropriate. She regularly violates species boundaries.

First of all, she lets different species interbreed under natural conditions.

- Field ecologists regularly describe naturally occurring hybrids between different species.
- Interspecific hybridization was crucial to the evolution of higher plants.
- For thousands of years, genes have moved from crops to wild plants and, more often, from wild plants to crops.

Second, almost all of the molecular techniques in the genetic engineer's toolbox come from nature, the most sophisticated genetic engineer of all, having had billions of years to perfect her interspecific gene transfer technologies. **Horizontal gene transfer**, the natural exchange of genetic material between different species through a mechanism other than sexual reproduction, is most common in the prokaryotes. Gene movement among bacterial species through conjugation, transduction, and transformation played an important role in their evolution.

Small pieces of genetic material also move between species that are very different genetically. Many viruses cause diseases by inserting part of their genetic material into their hosts' genomes. New DNA sequence information provided by the Human Genome Project shows that various bacteria left some of their own genes in the human genome when they infected people in past millennia. Plant scientists see evidence of similar gene transfers from microbes to higher plants. *Agrobacterium*, a naturally occurring plant pathogen commonly used in plant genetic engineering, integrates a portion of its genome into plants it infects. As a result, some plant genes actually originated in *Agrobacterium*. Finally, microbes are not the only organisms that move genes among species. DNA sequence data show that a parasitic mite transferred genes between two fly species and kept a few fly genes for itself.

Of course, significant obstacles impede the transfer of genetic information across species boundaries, so blocking interspecific gene flow is more prevalent than permitting it. Even so, the issue of natural versus unnatural gene transfer is not as clear-cut as you might think.

Concern: are new foods from transgenic crops safe and nutritious?

Because issues of food safety and good nutrition can trigger deep-seated feelings, before beginning the discussion, we want to remind you that scientists can *never* prove something is 100% safe. Often, nonscientists become unnecessarily fearful when scientists will not commit to a statement that a food crop is risk free. The public interprets this reluctance as an indicator of relative risk or safety when it only indicates the nature of science. After reading the above section on food allergies, you might see why scientists always seem to be hedging their bets. Over 170 minor allergens exist, but each one causes a reaction in only a small number of people. Sometimes, only one person has that allergy. Think of the millions of people who have eaten that food and not had an allergic response. Such an astronomically large number might tempt you into describing the food as 100% nonallergenic—but you'd be wrong. Because biologists are so conscious of genetic variability in a population in response to all environmental factors, including diet (Box 21.2), they assume at least one person has a unique allele that would disprove a conclusion of risk free.

B O X   2 1 . 2   *Nutrigenomics*

A healthy life depends on a healthy diet, but the constituents of a healthy diet vary from one person to the next. A familiar example of this type of variation is lactose intolerance, discussed in chapter 7. For Caucasians of northern European descent, the parental plea for teenagers to drink milk instead of soda makes perfect sense, because the calcium and vitamin D in milk are essential to healthy bone development. However, for most of the world population consuming raw dairy products after the first few years of life does more harm than good, because 95 and 90% of Asians and Africans, respectively, are lactose intolerant.

At least part of the variation among people regarding dietary requirements is due to genetic variation. Genetic dif-

ferences lead not only to differences in breaking down nutrients, but also to differences in other protein-mediated cell processes that affect nutrition, such as absorption from the intestine, assimilation of nutrients into tissues, and nutrient storage capacity. The study of the interaction between someone's genetic makeup and nutrition is known as **nutrigenomics**. Like pharmacogenomics, nutrigenomics recognizes that genetic variation can affect the safety and efficacy of the nutrients people consume. Scientists who study nutrigenomics use the same biotechnology research tools as pharmacogenomic researchers, such as single-nucleotide polymorphism identification and gene expression assays.

It is easy to see the parallel between nutrigenomics and pharmacogenomics in the lactose intolerance example. Just as some medicines are harmful to the set of people who have a malfunctioning version of an enzyme that breaks down the medicines, lactose is harmful to people who cannot metabolize it, because it leads to diarrhea. In many ways, however, nutrigenomics is much more complicated than pharmacogenomics. Most people take medicines intermittently, in pure form and exact doses, to cure and prevent diseases. Nutrition not only prevents disease, it can also cause certain diseases. The effect a specific nutrient has on someone's health varies with both the genetic makeup of the person and the amount of nutrient

(continued)

consumed. Very few of the gene variants associated with the variable health effects of diet have been identified. In addition, nutrients are consumed in very complex mixtures, so linking a single nutrient to a human health effect is very difficult.

In spite of the very small number of diet- and health-associated genes that have been identified and the complex relationship between nutrition and genetics, a number of companies have begun marketing individualized nutritional profiles directly to consumers. Similar in concept to designer drugs tailored specifically to a patient's genotype and the genetic basis of the disease, designer diets are developed according to the consumer's genetic makeup, determined from a swab of cheek cells mailed to the company. Most of the designer diets that we have read advise the client to eat more fruit and green leafy vegetables; minimize intake of alcohol, saturated fats, and *trans*-fatty acids; avoid tobacco; and maintain a healthy body weight through exercise and controlling calorie intake. Interestingly, even though this is precisely the same dietary advice physicians have been dispensing for decades, surveys show that a higher percentage of people follow this advice when it is couched as a personalized, gene-based diet.

Returning to the issue at hand, is this concern unique to biotechnology? If scientists eliminated recombinant DNA technology from their list of genetic modification techniques, would questions about the safety and nutritional value of new foods evaporate? No. Every year, new foods enter the food supply. Most have been derived from the scores of new crop varieties, genetically modified using breeding and mutagenesis, that seed companies commercialize each year. How do you know these new varieties are safe and nutritious? A handful of new foods are new imports that are dietary components of other cultures. Even though kiwis, cilantro, and arugula may seem like old hat to you, they are relatively new to the American diet. Some people developed food allergies to kiwis very soon after they entered the U.S. marketplace. They could not have known beforehand that they would be allergic to kiwis, never having eaten one.

Scientists agree that the risks intrinsic to food derived from the transgenic crops developed to date are the same in kind as those associated with conventional food crops (Table 21.12). All food crops, genetically modified or not, may contain compounds that affect food safety, such as:

- proteins that are allergens for some people
- secondary plant compounds that can be toxic if the food is eaten in excessive amounts or prepared incorrectly
- antinutrients, which are molecules that inhibit food digestion or absorption

For example, using conventional plant breeding practices, crop scientists developed a new potato variety for food processors and realized that the genetic enhancement of certain processing traits had inadvertently increased the natural potato toxin, solanine. Although modern recombinant DNA techniques may broaden the number and types of genes and traits that crop developers can access, that extension does not, in and of itself, change the type of risk or its magnitude. Remember the *product not process* principle: risk depends on the nature of the product, not on how it was produced.

A better way to frame the concern about the safety of transgenic food crops is to ask, "What are the food safety and nutritional value issues of genetically modifying a crop plant, irrespective of the method, and what is their likelihood of occurring?" You probably recognize this as the first two steps in a basic risk assessment: identifying risks and determining their probability.

Table 21.12 Comparing the safety and precision of plant breeding and biotechnology

At least 14 professional scientific societies and expert advisory panels have assessed the relative risks of crops genetically modified through recombinant DNA technology or plant breeding and mutagenesis, and all have come to the same conclusion.

Modern biotechnology broadens the scope of the genetic changes that can be made in food organisms, and broadens the scope of possible sources of foods. This does not inherently lead to foods that are less safe than those developed by conventional techniques. Therefore, evaluation of food and food components obtained from organisms developed by the application of the newer techniques does not necessitate a fundamental change in established principles, nor does it require a different standard of safety. (Organisation for Economic Co-operation and Development [OECD], *Safety Evaluation of Foods Derived through Modern Biotechnology: Concepts and Principles,* Paris, France, 1993.)

All methods of plant breeding can induce unexpected or unintended changes in plants, including pleiotropic effects. It is not possible to design a test to identify these effects. Instead, the FDA has encouraged developers to examine whether important nutrients, toxicants, and other components are present in the new plant variety at levels that are within the range expected for commercial varieties. (D. A. Kessler, M. R. Taylor, J. H. Maryanski, E. L. Flamm, and L. S. Kahl, Policy statement on food derived from biotechnology, *Science* **256:**1747, 1992.)

Traditional plant breeding methods include wide crosses with closely related wild species and may involve a long process of crossing back to the commercial parent to remove undesirable genes. A feature of GM [genetic modification] technology is that it involves the introduction of one or, at most, a few well-defined genes—rather than the introduction of whole genomes or parts of chromosomes as in traditional plant breeding. This makes toxicity testing for transgenic plants more straightforward than it is for conventionally produced plants with new traits, because it is much clearer what the new features are in the modified plant. On the other hand, GM technology can introduce genes from diverse organisms, some of which have little history in the food supply. (National Academies of Sciences from seven countries, *Transgenic Plants and World Agriculture,* U.S. National Academy of Science, 2000.)

While r-DNA [recombinant DNA] techniques may result in the production of organisms expressing a combination of traits that are not observed in nature, genetic changes from r-DNA techniques will often have inherently greater predictability compared to traditional techniques because of the greater precision that the r-DNA technique affords; [and] it is expected that any risks associated with the applications of r-DNA organisms may be assessed in generally the same way as those associated with non–r-DNA organisms. (OECD, *Recombinant DNA Safety Considerations,* National Experts on Biotechnology, Paris, France, 1986.)

Recombinant DNA methodology makes it possible to introduce a piece of DNA consisting of either single or multiple genes, that can be defined in function and even in nucleotide sequence. With classical techniques of gene transfer, a variable number of genes can be transferred . . . ; but predicting the precise number or the traits that have been transferred is difficult, and we cannot always predict the phenotypic expression that will result. With organisms modified by molecular methods, we are in a better, if not perfect, position to predict phenotypic expression. (National Research Council, *Field Testing Genetically Modified Organisms: Framework for Decisions,* National Academy Press, Washington, D.C., 1989.)

Step 1: identify the possible risks

In chapter 17, we emphasized the importance of taking complex, multifaceted issues and breaking them into manageable-size chunks so that you could focus your analysis on one facet at a time. In analyzing the possible safety and nutrition risks of genetically modifying crop plants, it is helpful to:

- differentiate the genetic modification techniques from one another
- separate the risks of the *intended* genetic change from those presented by *unintended* changes in the crop
- distinguish the issues associated with currently marketed crops from those of genetically modified crops in the future

What are the intended genetic changes in current crops?

Crops genetically modified by recombinant DNA techniques

The intended changes in the transgenic crops currently on the market are very similar to those incorporated into crops for decades using breeding and mutagenesis: disease resistance, herbicide tolerance, insect resistance, enhanced nutritional quality, and improved processing characteristics (Table 21.11). Those descriptors refer to the intended *trait* change. Food safety and nutrition considerations need to get down to the biochemical basis of the change—the DNA and proteins underlying the trait.

Before a transgenic crop is brought to market, the developer subjects the new protein(s) encoded by the transgene to a series of tests to determine if it is toxic or a possible allergen when ingested (Box 21.3). The regulatory agencies

BOX 21.3 *Regulation of transgenic food crops*

The U.S. FDA has the primary regulatory responsibility for ensuring that transgenic food crops are safe and nutritious. The U.S. Environmental Protection Agency also requires tests to assess the human health effects of those transgenic crops that are resistant to insects, pathogens, or other pests. Various tests can establish the crop's nutritional value, but no amount of testing can prove it is 100% safe. Instead, the FDA determines if the transgenic crop is as safe and nutritious as similar varieties that have been in the food supply for many years and that have a history of safe use by using a series of tests designed to answer two questions.

- Is the transgenic crop materially different from its nontransgenic parent (also known as the host crop), which has a well-established record of safe use?

- Did the genetic change raise any safety concerns?

Is the transgenic crop materially different from the nontransgenic host?

Two types of tests, crop biochemical analysis and animal performance studies, address the question of material difference. Biochemists assess and compare the nutritional value of both the host crop and the transgenic crop by measuring the amounts and bioavailability of the micronutrients and macronutrients, such as carbohydrates, including measurements of both starch and sugars; fats, including specified fatty acids; proteins, including the amounts of individual amino acids; percent fiber; total calories; and percent moisture. If the transgenic crop differs nutritionally from its host crop, the FDA mandates that the transgenic crop and all food derived from it be labeled to describe the nutritional difference, whether or not the change is intentional. The second set of biochemical tests addresses food safety.

If scientists know that the host crop produces compounds that affect food safety, such as solanine in potatoes or the enzyme inhibitors or allergens in soy, these compounds must be measured to ensure that their levels are equal in the transgenic and the nontransgenic crops. Because the nontransgenic host crop has a history of safe use, the FDA assumes that if the levels are equal, the transgenic crop is as safe as its nontransgenic parent. In animal health and performance studies, one group of animals is fed the transgenic crop, while a control group is fed the nontransgenic host. Various indicators of animal health and performance, such as weight gain and litter size, are measured, and the groups are compared. To date, 42 studies comparing various farm animals fed host crops and transgenic crops have demonstrated that all measures of health and performance were equal in the two groups.

If no differences between the transgenic crop and the nontransgenic host are found in the biochemical tests or animal performance studies, the transgenic crop and the nontransgenic host are substantially equivalent. The testing then shifts from the whole crop to the safety of the transgene and the proteins it encodes.

Does the genetic change introduce new concerns?

With transgenic crops, the crop developer has a second responsibility after establishing substantial equivalency: to ensure that the new additions to the host crop—the transgene(s) and transgenic protein(s)—do not introduce any new safety concerns. As mentioned above, this requirement is imposed on transgenic crops but not on crops genetically modified by other techniques simply because it *can* be. The crop developers have the gene and its encoded protein in hand and therefore know what to test. In the premarket regulatory

approval process, they first determine if the gene donor has a history of safe consumption. If not, the protein products of the inserted genetic material must be tested to ensure that they are nontoxic and are not potential allergens. If the transgene has a history of safe use, the FDA does not require further safety testing of the protein. The new genes in most transgenic crops currently on the market have a history of safe consumption, so the crop developers were not required to test the transgenic protein to obtain approval to sell the crop. However, because some consumers have expressed concern about the safety of transgenic crops, the developers opted to test the new proteins in all transgenic crops approved to date to provide additional assurance to consumers.

You might be questioning how the gene donors of the transgenic crops you are familiar with, such as *B. thuringiensis* (Bt corn) and the papaya ring spot virus (PRSV) (disease-resistant papaya), could possibly have a history of safe consumption. Even though the gene donor may not be considered part of the food supply, people have been consuming the gene and its encoded protein unintentionally for millennia. For example, one of the first transgenic products approved by the FDA was a virus-resistant squash that, like virus-resistant papaya, was resistant because it had been given a coat protein gene from the squash virus. In determining whether the coat protein gene has a history of safe consumption, the developer discovered small amounts of the gene donor, the squash virus, in one-third of the squash in grocery stores. Thus, it is in the food supply and has a history of safe, albeit unintentional, consumption.

Testing transgenic proteins for toxicity and allergenicity

To test the toxicity of transgenic proteins, mice or other laboratory animals receive very large doses of the transgenic protein for 7 to 14 days. Scientists observe the animals daily to monitor acute adverse effects, and at the end of the study, they sacrifice the animals and microscopically examine their organs and tissues. To date, none of the transgenic proteins has had any observable adverse effects at doses as high as 2,000 to 1 billion times the expected maximum daily intake for humans.

Determining if a protein is an allergen is much trickier than toxicity testing. As explained above, scientists can *never* prove conclusively that a protein is *not* an allergen to someone, somewhere. It may surprise you to learn they also cannot prove a protein *is* an allergen, unless it comes from a food known to be allergenic. If you think that sounds circular, you're right. The definition of an allergen is a substance that triggers an allergic reaction, and scientists cannot know whether a protein will cause an allergic reaction before the fact. The allergic response defines the protein as an allergen. The only time a scientist can prove a protein *is* a food allergen before it enters the marketplace is if it comes from a food to which a substantial number of people have already reacted. We'll use an example to decrease the confusion level.

Let's assume that a crop scientist discovers a wild peanut relative that is resistant to the fungus that produces the mycotoxin aflatoxin, which causes cancer. Luckily, a single gene encodes the fungus resistance trait; scientists isolate it and identify its encoded protein. They hope to use recombinant DNA techniques to incorporate the gene into the many crops the fungus infects, but first they must make sure that the wild-peanut gene does not encode a protein that contributes to the peanut's well-established allergenicity. People with allergies have donated their sera for scientific research purposes, and serum

samples from at least 14 different people with food allergies to peanuts are used to assess the allergenicity of the wild-peanut protein. If antibodies in a serum sample bind to the wild-peanut protein, then unfortunately, the protein that provides fungus resistance is also a peanut allergen, so the project will not go forward. This occurred with the Brazil nut protein that researchers hoped would enhance soy's essential amino acids. If the serum does not contain antibodies to this protein, can the scientists be sure the wild-peanut protein is not an allergen? No. Even though it is nonallergenic to those who donated serum samples, that does not prove it is not an allergen to someone.

The situation becomes more complicated when the desirable gene comes from a source with no history of allergenicity. If the gene source has been in the food supply for a long time with no reports of allergic reactions, then it is not allergenic to the great majority of the population, perhaps everyone. For example, earlier you read about PRSV resistance and concerns that the viral coat protein might trigger allergic reactions. For as long as people have been consuming papayas, they have been consuming PRSV, and therefore PRSV coat proteins, without allergic reactions.

How do transgenic crop developers determine the potential allergenicity of a protein new to the food supply? They always begin by assuming that the transgenic protein may be a food allergen, and they run it through a series of tests. After each test, if the results are encouraging, the developers have increased the level of certainty that the transgenic protein is probably not an allergen. First, they compare the amino acid sequences of the transgenic protein to those of *all* known allergens, not just food allergens. If it shares any sequence of eight amino acids with a known allergen, such as ragweed

pollen, the transgenic protein is tested on sera from people with that allergy. If this test is negative, or if no sequence of eight amino acids corresponds to an allergen's, the developer then tests the stability of the transgenic protein to heat, because known food allergens tend to be heat stable. If the transgenic protein breaks down at high temperatures, the developer then tests its stability to digestive fluids, because if a protein is digested by enzymes and stomach acids, it is not available to the immune system for antibody formation. Stability to heat or digestive fluids does not prove that a transgenic protein *is* an allergen, nor does susceptibility to these conditions prove that a transgenic protein is *not* an allergen. Transgenic-crop developers will never be able to claim with absolute certainty that the transgenic protein is not an allergen to someone, somewhere.

You may not find it particularly comforting that the best they can do is establish that the transgenic protein has a low probability of being an allergen. But don't forget, if crop developers did away with transgenic crops tomorrow, the risk of introducing new allergens into the food supply would not disappear. Being in a better position to assess the risk, simply because more is known, does not mean the risk is greater. Instead, the increased knowledge has created a paradox. Scientists know much more about the genetic changes that occur with recombinant DNA technology than with selective breeding and mutagenesis. Because they understand the genetic basis of the intended changes, not only do they know the identity of the new protein, but they are better able to predict possible unintended changes. As understanding increases, the risk probability decreases, but the amount of premarket testing, and therefore the cost of regulatory approval, increases.

that assess the safety of transgenic crops decided there was no need to require safety tests for the new DNA. DNA is found in virtually all of our food, is readily digested and absorbed, and is not associated with food allergies or other adverse reactions to food. In addition, the amount of DNA in food is minuscule compared to other food molecules (approximately 0.005 to 0.02%); of that amount of DNA, only 0.00042% would be transgenic if the transgenic food is eaten raw, and significantly less if the food is processed.

Crops genetically modified through selective breeding or mutagenesis

With crops genetically modified through breeding and mutagenesis, developers cannot test the safety of the proteins responsible for the new trait because they do not even know the number of proteins involved, much less their identities. All they know is that the crop now has the desirable trait and that the trait must have a genetic basis because it passes from one generation to the next. Could the biochemistry underlying the new phenotypic trait increase the crop's allergenic, anti-nutritive, or toxic potential? Yes, it could.

For example, perhaps mutagenesis led to the disease resistance trait because it created a gene that encodes a novel protein that turns out to be allergenic to some people. Perhaps crossing a crop with its wild relative created an insect-resistant variety because the wild relative had genes for synthesizing secondary plant compounds that kill insects. Those same compounds may also be toxic to people in large amounts. Even the safest plant-breeding strategy—crossing two crops with histories of safe use—could potentially cause problems. For example, a cross to enhance the levels of essential amino acids may be successful because it increases the endogenous level of an allergenic protein. Unfortunately, the amount of exposure to an allergen can affect the development and severity of a food allergy.

What are the intended changes for future transgenic crops?

As described above, the intended change may actually be enhanced safety and improved nutritional value (Figure 21.19). All genetic modification techniques can be directed to these dual objectives. For example, plant breeders recently announced the release of a high-protein corn variety they had spent over 20 years developing. As you will learn in chapter 23, crop scientists developed canola to provide a healthier source of vegetable oil. Canola would not exist without genetic modification through breeding and mutagenesis that not only changed the oil profile but also decreased the levels of natural toxins.

Plant-made pharmaceuticals (PMP)

Recombinant DNA technology gives crop developers unique abilities not provided with other genetic modification techniques, and as a result, some of the intended changes of future transgenic crops raise unique concerns. For example, the new techniques make it possible to use crop plants as "manufacturing plants" to produce new medicines. Even though wild plants have provided humans with natural therapeutic compounds for centuries, they were not food plants (Figure 21.20). The pharmaceutical plants of the future will include food crops that are capable of synthesizing a medicine. Because they are not intended for food consumption, they will be grown by a few specially trained and tightly supervised growers and harvested, transported, and processed under strict confinement, with harsh penalties imposed for any breaches of the segregation processes and procedures.

Figure 21.19 Genetic modification through breeding. Using modern plant-breeding techniques, USDA scientists have produced carrot varieties in a range of colors, reflecting varying amounts of the different carotenoid pigments. The National Plant Genome Initiative has helped the breeders determine that 20 different genes are responsible for the rainbow of colors. However, they have not identified the genes or the proteins encoded by the genes. (Photograph by Stephen Ausmus, courtesy of the Agricultural Research Service, USDA.)

Figure 21.20 Medicinal plants. **(A)** Foxglove (*Digitalis purpurea*). **(B)** Wormwood (*Artemesia annua*). A number of pharmaceuticals are derived from plants. The foxglove is the source of digitalis, a heart medicine, and compounds extracted from wormwood have antimalarial properties. (Photographs courtesy of Thomas Barnes, University of Kentucky [A], and by Scott Bauer, courtesy of the Agricultural Research Service, USDA [B].)

No matter how many safeguards are established to ensure that manufacturing plants are isolated from food crops, the system will not be foolproof. Prudence requires the regulatory agencies to assume that small quantities might enter the food supply, so the regulators must consider the safety implications of consuming PMP before they grant approval to field test the crop. Some PMP will be protein molecules, such as insulin, that cannot survive stomach acid and digestive enzymes. On the other hand, some PMP will be designed to ensure that they make it through the digestive system unscathed and enter the body in an active form. Because these molecules are pharmaceuticals, they must undergo the U.S. Food and Drug Administration (FDA) premarket safety testing. When the extensive safety testing is combined with the assumption that small amounts of the PMP will be consumed because its presence in the food supply is inadvertent, the risks are negligible. If, however, major breaches of containment processes and procedures occur, the risk to some people could be substantial. Recognizing this, the U.S. regulatory agencies responsible for approving field tests and commercialization of transgenic crops developed an approval process and postapproval monitoring requirements that are much more stringent than those applied to transgenic crops intended as food.

Allergenic crops

A significant concern due to the unique capabilities provided by recombinant DNA technology is related to food allergenicity. People allergic to certain foods become aware of their allergies early on and avoid those foods. Now crop developers can move genes from allergenic foods (e.g., Brazil nuts) to foods not typically seen as allergenic (e.g., corn). This could pose major health risks to people with allergies to the gene donor (e.g., Brazil nuts) unless the crop developer can prove *conclusively* that the transferred gene does not encode the donor's allergenic protein. We discuss allergenicity tests required for regulatory approval of transgenic crops in Box 21.3.

What are the unintended changes due to genetic modification?

Identifying the risks associated with unintended changes due to genetic modification is much more difficult. Plants have thousands of chemical compounds whose amounts might be altered incidentally by the intended genetic modification, but because many have not been identified, their amounts cannot be measured. In addition, the huge expense of measuring a plant's total chemical makeup before and after genetic modification would make developing new varieties cost prohibitive. Rather than concern themselves with measuring every chemical compound in a plant, crop developers focus on those compounds known to affect food safety and nutritional value. They use the parental crop (the elite variety), which has been in the food supply for many years, as a benchmark and determine if the genetic modification has inadvertently changed the levels of the compounds associated with food safety or nutritional value.

Step 2: assess the probability of unintended effects

To get a handle on the *likelihood* that unintended effects may occur, scientists must understand the biochemical mechanics that could lead to an intended genetic change having unintended effects that increase risk. A number of possible explanations for unintentional changes come to mind.

- The gene(s) encoding the protein(s) responsible for the new, intended trait has pleiotropic effects (a single gene affects a number of phenotypic traits) because it plays a role in a number of biochemical pathways and cell processes that are linked or interdependent.
- The gene(s) for the intended change interacts with other genes in the parental crop. In chapter 12, you learned that genes interact, often producing unexpected and novel traits. One example of gene interaction is epistasis, in which one gene masks another gene's expression.
- The genetic modification may disrupt a gene's DNA sequence. For example, in genetic engineering, when the new gene recombines with the plant genome, it inserts itself randomly.

Are any of these genetic phenomena unique to a specific genetic modification technique? No. Even though public discussion of the third—physically disrupting a DNA sequence—seems to surface only in relation to recombinant DNA technology, the phenomenon also occurs during breeding and mutagenesis. After all, a mutation is an alteration in the sequence of nucleotides, and that alteration may involve a single nucleotide or a large section of the chromosome. In breeding, during gamete formation, when chromosomes exchange DNA sequences by breaking and rejoining (crossing over), sometimes genes get disrupted and other times entire genes are lost or duplicated. In addition, the natural processes of chromosomal translocation and inversion disrupt DNA sequences; and don't forget transposable elements, or jumping genes.

Which of the three genetic modification techniques is most likely to lead to unintended effects through pleiotropy, unexpected genetic interactions, or gene disruption? Pleiotropy and unexpected genetic interactions are *less* likely to occur with recombinant DNA techniques than with the other two genetic modification methods, because scientists move single genes with known functions into a parental variety that has a long history of use. In selective breeding and mutagenesis, they do not know the number or identities of the genes or proteins responsible for the trait newly incorporated into the parental line. In addition, in plant breeding, the gene(s) for the desirable trait drags hundreds of genes, many with unknown functions, with it because they are all located on the same chromosome. With mutagenesis, scientists do not know how many other genes may have mutated. By increasing the precision (single genes) and certainty (known functions) of the genetic modification, the risk of producing organisms with unexpected traits decreases. This led the International Food Biotechnology Council to the following conclusion.

> When recombinant DNA methods are used the genetic material has been precisely identified, the amount of genetic material introduced is controlled and the result can be fully characterized. Thus if the level of knowledge concerning the genetic material permits, recombinant DNA should be the method of choice, especially if the source of the new genetic material is not a component of the food supply.
>
> *Regulatory Toxicology and Pharmacology* **12**:3, 1990

Even though selective breeding and mutagenesis are more likely to have unexpected effects due to pleiotropy and gene interactions than recombinant DNA techniques, plant breeders have an excellent safety record. Of the thousands of new varieties they have introduced to the market in the past century, only four crop varieties had an unintended change that led to an adverse health effect: one variety each of potato, squash, spinach, and celery.

What about the third cause of unintended effects, disrupting the integrity of a gene's sequence of nucleotides? Not long ago and without hesitation, we would have declared that gene disruption was much less likely to occur in selective breeding than in mutagenesis or genetic engineering, because biologists thought of the genome as a fairly static entity physically. They knew that natural processes occur during selective breeding that could disrupt the DNA sequence of a gene or alter the positions of genes relative to each other, such as chromosomal translocations, inversions, and uneven exchanges in crossing over, but they also knew that the events occur rarely.

In the past 25 years, a number of discoveries have forced biologists to reevaluate the idea of a steadfast gene, firmly seated at its assigned locus on the chromosome, composed of a DNA sequence that is rarely altered. One was the revelation that genomes contain transposable elements that excise themselves from a certain location in the genome and randomly insert themselves elsewhere, sometimes disrupting gene function. Then, scientists discovered that antibody production by the immune system results from the genome physically rearranging itself.

The latter discovery, while unrelated to plant genetics, calls into question the view of a static eukaryotic genome. When new discoveries challenge existing mental models that biologists assume are universally applicable, such as the physical constancy of the genome, the biologists step back from the model and begin to poke at it experimentally to see how many other exceptions they can discover. Ultimately, they replace the old model with a more accurate one. On the issue of natural processes that rearrange the genome, the biological community has stepped back from their earlier model and begun the process of reassessing how static the genome is. Thus, we do not know how often natural forms of genome rearrangement go awry in plants and lead to gene disruption, causing unintended changes, and we are reluctant to guess, since biologists are at the earliest stages of model reevaluation. Surely it happens less often in selective breeding than it does when plant breeders subject crop seeds to heavy doses of known mutagens to induce changes in the DNA sequence significant enough to produce new genes encoding new traits. What about selective breeding and genetic engineering? With an understanding of the intricacies of recombinant DNA technology, and comparing that process to current knowledge of natural genome rearrangements, if we had to guess, we would say that gene disruption probably happens more often in genetic engineering. However, that guess may be more reflective of the tendency people have to hold on to their old mental models than it is of biological reality.

SUMMARY POINTS

Scientists are using the new biotechnologies, especially genetic engineering, to improve the food supply in many ways.

Using recombinant DNA technology, they are increasing the macronutrients and micronutrients in staple cereal grains that most of the world population relies on for food. They add novel genes encoding proteins rich in essential amino acids to crops. By blocking feedback inhibition in certain amino acid synthesis pathways, they increase essential amino acids that the crop lacks.

They are also increasing the vitamin and mineral content of staple crops. For example, an international collaboration created a rice variety that will increase the amount of iron and vitamin A in the diets of people who subsist primarily on rice.

In industrialized countries, obesity is a more significant problem than malnutrition. Scientists are altering the fatty acid profiles of the oilseed crops to increase the proportion of unsaturated fats in salad oils and decrease *trans*-fatty acids in processed foods. Metabolic engineering also allows scientists to increase the nutraceutical levels in food crops.

Biotechnology provides new tools and strategies for improving the safety of the food supply. New detection technologies that are based on polymerase chain reaction, DNA probes, and monoclonal antibodies improve food safety by identifying sick animals before they enter the food supply and assessing the microbial contamination of plant and animal food products during processing.

Scientists are trying to decrease the allergenicity of the eight major allergens that are responsible for 90% of food allergies. They are blocking expression of genes that encode allergens, using genetic engineering to change the shape of allergenic proteins, and developing monoclonal antibodies to block the binding of allergens, which triggers the cascade of events that culminates in the allergic reaction.

Public concerns about food biotechnology are focused primarily on naturalness, changes in nutritional content, and food safety. Crop developers began using unnatural methods to improve crop varieties more than a century ago. Genetic engineering can lead to crops that raise unique concerns about nutrition and safety; however, most issues associated with transgenic crops are also relevant to crops genetically modified through the conventional means of breeding and mutagenesis.

The U.S. FDA has the primary responsibility for ensuring that new crop varieties are as safe and nutritious as foods currently in the food supply. Before a transgenic food crop can be marketed, crop developers conduct tests of nutritional content, toxicity, and allergenicity.

KEY TERMS

Antioxidant	Elite crop variety	Metallothionein	Probiotic
Beta-carotene	Ferritin	Nutraceutical	Provitamin A
Bioavailability	Functional foods	Nutrigenomics	*trans*-fatty acid
Biofortification	Golden Rice	Phytochemical	
Carotenoids	Horizontal gene transfer	Prebiotic	

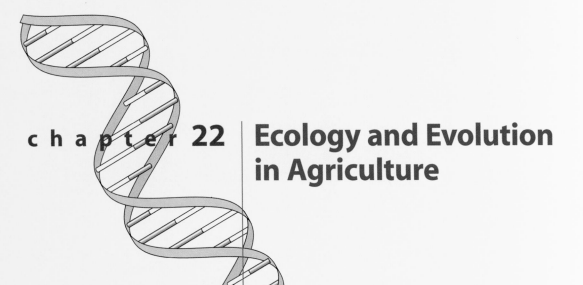

Ecology and Evolution in Agriculture

A recent study of knowledge and attitudes about food derived from transgenic crops revealed that only 30% (range, 22 to 41%) of the consumers in the seven largest European Union (EU) countries know that all crop plants have genes. On average, 32% (range, 21 to 44%) are certain that the only crops that contain genes are those developed by biotechnology companies. In other words, they believe that scientists in those companies gave genes to crop plants that were geneless before the scientists set to work on them. The remaining 38% are undecided. Paradoxically, students in these same countries repeatedly score exceptionally well on tests assessing their knowledge of biology.

How can these conflicting results be reconciled? Do EU citizens not know that the crop plants that provide food are biological organisms, or have they simply forgotten that all living organisms have genes? Whatever the explanation, one conclusion is inescapable. EU citizens, and virtually all citizens of industrialized countries, are divorced from the realities of the agricultural production systems that sustain human societies. When agricultural technologies replaced manual labor and fewer hands were needed on the farm, people left the farm and joined the industrial workforce, usually in urban areas far from farms. They and their descendents became increasingly divorced from the realities of agriculture as time passed. In the United States, only 1% of the population is responsible for growing the crops that feed the other 99% (Figure 22.1).

Historical context contributes essential information to productive thought and discussion about societal issues raised by technological developments. Because modern life is so separate from agriculture, most people have no context in which to anchor information and viewpoints on the appropriate uses of technology in today's agriculture. The lack of familiarity with agriculture might not be a problem in and of itself if it were not accompanied by deep-seated emotional responses to food. These feelings may be

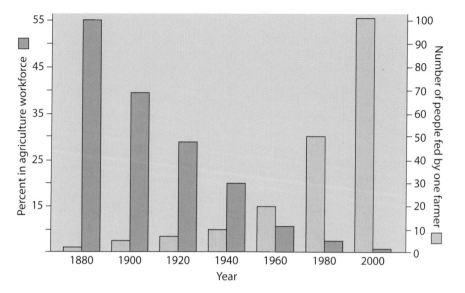

Figure 22.1 Changes in the U.S. agricultural workforce. As modern technologies replaced manual laborers, people left farming and joined the nonagricultural workforce. Technological improvements allowed one farmer to feed more and more people. The blue bars in the histogram represent the percentage of people directly involved in agriculture, and the yellow bars represent the number of people in the United States who are fed by one American farmer.

innate or, as biologists say, "hardwired" because the need for food trumps all others. This combination of deep, innate feelings and insufficient knowledge can make analysis and discussion of agricultural technologies especially difficult.

In this chapter, we discuss plant agriculture through the lenses of ecology and evolution. This provides us with an opportunity to teach you more about ecological interactions and evolutionary adaptations using row crop agriculture as the hook. We also want to lessen the information gap that exists between farmers and citizens in the industrialized world.

To meet both goals, we provide information on the:

- evolution of crop plants from wild relatives
- conflicts between plant adaptations for increased fitness and human requirements for food crops
- developments in agriculture that have allowed crop productivity to keep pace with the exponential growth of the human population
- ecological problems caused by agricultural activities

THE BEGINNINGS OF AGRICULTURE

A farmer's field is an artificial ecosystem that is subject to nature's laws governing species interactions, population dynamics, energy economics, and materials accounting. Crop plants are the primary producers, and like all autotrophs, they involuntarily nourish all of the other species in the ecosystem, including humans. Farmers compete with herbivores and plant parasites that try to use crop plants as food, but the farmer, as manager of these artificial ecosystems, has the upper hand in this fight—most of the time. In addition, farmers must also battle abiotic forces that harm plants, such as droughts, floods, and poor soil nutrient levels. As a result, even though some people who are unfamiliar with farming tend to romanticize that lifestyle, farming is incredibly difficult work, made more difficult by nature's unpredictability.

Humans can thank their early ancestors for creating many of the problems that make farming so difficult. They opted to domesticate certain wild plants from thousands of available plants, and then the early agriculturists genetically modified the plants to suit human needs. Even though the choice of plants and genetic changes made perfect sense then (and now), the choices and changes also created the problems that farmers have continually battled for the past 10,000 years.

Crop plants and human societies coevolved

The ecological relationship binding one species to another can sometimes be so close that when one species changes genetically, the other responds with a genetic adaptation of its own. The forces of natural selection driving coevolution have created many natural relationships in which the survival and reproductive success of two species are so tightly and intricately woven together that one species cannot exist without the other (see chapter 13). The relationship between humans and domesticated crop plants is very much like that.

Around 8,000 BC, Stone Age hunter-gatherers began cultivating some of the wild grasses they had been gathering. Humans specifically chose to domesticate certain wild grass species for a number of good reasons (Figure 22.2). They are **annual plants**, which means they grow, reproduce, set seed, and die. Because they invest most of their energy in reproduction rather than in living long lives and becoming large plants, annual grasses are very prolific. Grasses are to oak trees as house mice are to elephants; both are opportunistic species.

Grasses produce many seeds, and seeds are a perfect food source for people and many other organisms. Because a seed must provide nutritional sustenance to the plant embryo it contains, seeds are high in protein and oils. They pack a lot of nutrients and energy into a small package, especially when compared to the parent plants, which consist primarily of cellulose and other biological molecules that humans cannot digest. The seeds of wild grasses are biochemically designed to enter a period of dormancy if they end up in an environment that is not conducive to survival. They lose 95% of their water content and wait until conditions improve, even if it takes many, many years. Consequently, seeds can be dried easily without losing nutritional value, and these concentrated pellets are then easy to store and transport.

Figure 22.2 Wild grasses. All major cereal staples, such as corn, wheat, rice, oats, and barley, belong to the plant family commonly referred to as the grass family (Poaceae). Approximately 40 of the 200 weeds listed in the U.S. Department of Agriculture (USDA) *Handbook of Weeds* are in this plant family. The traits that make many grasses weeds also made them good crops for the first agriculturists. Wild grasses, such as this relative of wheat, *Aegilops cylindrica*, are prodigious seed producers that rapidly infest disturbed areas. (Photograph by Phil Westra, Colorado State University, courtesy of http://www.forestry.com.)

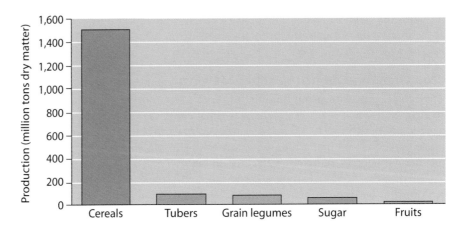

Figure 22.3 Global crop production. The cereal crops, such as corn, rice, and wheat, are by far the most important food source for most of the human population.

As soon as humans began cultivating wild grasses, they began to change them genetically. The word "domesticate," which is derived from the Latin word "domus" (house) and means to "bring into the house," is inseparable from the concept of genetic modification. By selecting certain plants with useful traits from a population of genetically variable plants, they directed the evolution of wild grasses into the domesticated cereal crops, such as wheat, rice, and corn, that constitute the primary source of food for most of the world's population today (Figure 22.3).

Domesticated plants, which eventually included legumes (beans), fruits, and vegetables, responded genetically to the selective pressure humans imposed, but the human coevolutionary response to crop evolution was **cultural evolution**, not biological evolution (genetic change). As described in chapter 14, with the birth of agriculture, people became more sedentary, established large settlements, and created a multisectored economy with division of labor. At some point, the gradual shift from a hunter-gatherer economy based on food procurement to an agricultural economy based on food production became irreversible. Humans became dependent on crop plants for survival, and crop plants became dependent on humans.

The evolution of gathered plants from their wild, ancestral forms to domesticated crops was irreversible. In the process of domesticating plants, people changed their genetic makeup so significantly that crop plants could no longer survive "outside of the house." In terms biologists use, they were no longer "fit." Recall that the fittest organisms are those that produce the most offspring that survive to reproductive maturity. Survival matters, but only in the service of reproduction. Crop plants lost the capacity to make it in the real world that their wild ancestors had inhabited because the reproductive and survival strategies of the wild plants conflicted with the interests of the people who domesticated them. The more people shaped wild plants for human consumption, the less able the plants were to survive and reproduce. The selective pressures humans imposed reflected the needs of the cultivator and not the evolutionary needs of the plant. Table 22.1 provides many examples of the ways in which domestication decreased plant fitness, and we discuss a few in detail below.

PLANT REPRODUCTIVE STRATEGIES

Although it may be difficult for you to visualize, plants have sex—that is, they reproduce sexually. Some also reproduce asexually under certain conditions, but whenever possible, plants engage in sexual reproduction because it

Table 22.1 Domestication changes versus plant needs

Generic changes in almost all crop plants

Reduced tendency to outcross decreases genetic diversity.

Nonshattering seed capsule decreases seed dispersal.

Lower levels of endogenous toxins increase susceptibility to herbivores and pathogens.

Changing fruit ripening and seed germination from sequential to synchronized increases probability all offspring will succumb to environmental adversities.

Eliminating seed hairs and spines increases seed loss to predators.

Crop-specific changes

Seed crops, such as legumes and cereals, increased seed size at expense of seed number.

Fruits decreased number of seeds and increased pulp.

Root crops, such as carrots, condensed the finely branched root system, which provides more stability, into a single structure.

Tubers, such as potatoes, condensed dispersed, thin tubers into a single structure; moved deeply embedded, well-protected eyes to the tuber surface for easier peeling.

Leafy vegetables, such as spinach, decreased materials that provide plant with structural support, such as cellulose and lignin, because they are indigestible.

serves the highly desirable function of generating genetically variable offspring. Asexual reproduction, while quicker and easier, produces offspring that are genetically identical to each other and to the parent (clones).

Virtually all crops belong to the flowering plant group of the plant kingdom, the **angiosperms**. Flowers are the reproductive organs of the angiosperms. In the process of pollination, pollen, which is the functional analog of sperm, fertilizes the flower's eggs, or ovules. Fertilization triggers biochemical changes that lead to fruit and seed formation (Figure 22.4).

Cross-pollination creates genetically diverse offspring

Flowering plants are almost always hermaphroditic. More often than not, a single flower contains both male and female organs. In other plants, such as corn and squash, the male and female organs are housed in different flowers on the same plant. Although many hermaphroditic plants are capable of fertilizing themselves, plants favor outcrossing because it maximizes the probability of future evolutionary success by creating genetically variable offspring. Consequently, nature has provided plants with a variety of mechanisms to facilitate cross-fertilization and other mechanisms to impede self-fertilization. In many plants, the pollen is capable of fertilization before the female organs in that flower become receptive. By the time the female organs become receptive, pollen from that plant is no longer viable (Figure 22.5). In other cases, the male and female structures are positioned too far away from each other for the pollinator to place the flower's pollen on its stigma

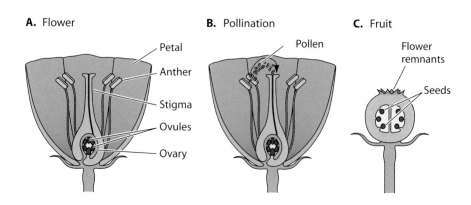

A. Flower
- Petal
- Anther
- Stigma
- Ovules
- Ovary

B. Pollination
- Pollen

C. Fruit
- Flower remnants
- Seeds

Figure 22.4 Plant reproduction. In the flowering plants (angiosperms), flowers and fruits are reproductive organs.

Figure 22.5 Hermaphroditic plants. **(A)** Most flowers have both male and female reproductive organs. In this flower, *Centropogon talamancensis*, the white female stigma is surrounded by male anthers, which appear blue. Even though this flower is self-compatible, self-fertilization rarely occurs because the female stigma becomes receptive and pushes through the anther tube only after the pollen is no longer fertile. **(B)** Squash plants are also hermaphroditic, but male (right) and female (left) reproductive structures occur in different flowers on the same plant. (Photograph in panel B courtesy of Purdue University Cooperative Extension Service.)

(Figure 22.6). A few flowering plants, such as papaya, achieve the pinnacle of forced outcrossing by housing male and female flowers on completely different plants, which ensures that genetically diverse offspring will be produced.

Seed dispersal lessens competition

It is in a plant's best interest not only to produce large numbers of genetically diverse seeds by outcrossing but also to have mechanisms for dispersing them over the greatest area possible. Widespread seed dissemination increases the reproductive success of the parent plant by minimizing competition among young seedlings and also between the parent plant and its offspring. Broad seed dispersal also makes it less likely that herbivores will destroy all of a plant's offspring. Plants have evolved a number of different strategies for disseminating their offspring far and wide. Some rely on wind power, and their seeds have structural adaptations to keep them airborne as long as possible (Figure 22.7). Other plants, such as the impatiens plants that homeowners grow, have fruits that seem spring-loaded when ripe, bursting open with remarkable force.

Animals help plants reproduce

When it comes to finding sexual partners (other than themselves) and putting distance between themselves and their offspring, plants are at a distinct dis-

Figure 22.6 Avoiding self-pollination. When a passion flower first opens, the male anthers are positioned well below the female stigma. The flower's nectar is in a donut-shaped trough at the base of the reproductive organs. Pollinators, such as this bee, run around the trough, and pollen is deposited on their backs **(A)**. A few hours later, the female stigma bends down and is at a perfect level to receive pollen from bees that have visited other flowers **(B)**.

advantage. For all practical purposes, plants are immobile. Some use the wind to spread their pollen, but this is very inefficient use of the plant's resources. Many rely on animals for uniting the pollen of one flower with the eggs of another and for moving their offspring away from home to avoid parent-offspring competition.

Plants produce brightly colored flowers and fruits to grab the attention of pollinators and seed disseminators and lure them in. Once there, the plants reward the pollinators with nectar and protein-rich pollen and the seed disseminators with sweet, fleshy fruits (Figure 22.8). The seeds of some plants that rely on animals for transportation services are capable of germinating only after a trip through the digestive tract of a bird or mammal. Plants like beggar's lice and sand spurs have Velcro-like structures or barb-like appendages for attaching their seeds to fur and feathers so they can hitch a ride to a new habitat. Still other plants make sure their fruits are positioned on the plant in a way that makes it as easy as possible for the animal to take the fruit and disperse the seeds (Figure 22.9).

Figure 22.8 Rewarding pollinators. Nectar, which is sugar water, and high-protein pollen are rewards that pollinators receive for helping plants reproduce. There is a downside to rewarding pollinators with pollen, because the plant's primary objective, reproduction, depends on the pollinator transferring pollen and not eating it. This plant, *Couroupita guainesis*, solves that conflict by producing infertile (false) pollen on the brightly colored structures on one part of the flower and fertile pollen on the small, unobtrusive structures surrounding the female stigma.

Figure 22.7 Seed dispersal. Wind-dispersed seeds, such as those from dandelions **(A)** and maple trees **(B)**, have structural adaptations that keep them airborne for a long time. This decreases competition among offspring and between the parent plant and its offspring. (Photographs by Kenneth Gale [A] and Dave Powell [B], courtesy of http://www. forestryimages.org.)

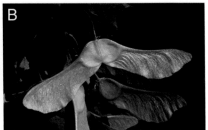

Figure 22.9 Seed dispersal. In tropical forests, bats play important roles in both pollination and seed dispersal. This plant (*Piper* sp.) relies on bats for seed dispersal. Ripened fruits are so easy to remove from the plant that bats fly through the rainforest and grab mouthfuls of piper fruits without stopping. This ensures that the piper plant's offspring are carried to a different part of the forest.

Figure 22.10 Domestication and seed size. Seeds from domestic crops (inner circle) are usually larger, lighter in color, and more uniform than their wild relatives. Clockwise from top: peanuts, corn, rice, coffee, soybean, hops, pistachio, and sorghum. (Photograph by Stephen Ausmus, courtesy of Agricultural Research Service, USDA.)

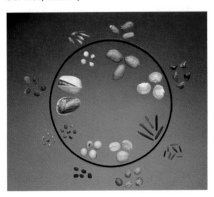

So, while people may egocentrically view flowers as decorations and fruits as food, for a plant they represent strategic means of achieving their most important goal: reproducing offspring that survive and reproduce.

HUMAN INTERFERENCE IN PLANT EVOLUTION

Many of the plant's strategies for maximizing reproductive output and offspring survival conflicted with human needs for certain crop traits, so early agriculturists imposed selective pressures on plants to shape them to their liking. They decreased crop genetic variability and impeded the dispersal of offspring (seeds). In the case of fruits and vegetables, artificial selection decreased the number of offspring (seeds) because people were interested in eating the pulp, not the seeds. In seed crops, such as cereals and legumes, artificial selection often increased seed size at the expense of seed number (Figure 22.10). In addition, many of the plant's survival mechanisms made it difficult for both herbivores and people to eat them. Early farmers eliminated many of these survival mechanisms and set the stage for the never-ending competition between people and crop pests.

Early farmers decreased plant reproductive success and genetic variability

When you compare domesticated fruits to their wild relatives, the most consistent, observable change is larger size. Selection for increased fruit size could have led to increased fitness for the plant, but it did not because people selected fruits with more pulp but fewer seeds. Creating varieties with fewer seeds directly conflicts with the evolutionary needs of a plant.

Decreased seed number

For example, the fruits of wild tomatoes consist of a thin layer of skin packed full of seeds embedded in the familiar gelatinous material we associate with tomato seeds, and it is clear the fruit is a mechanism for making a new generation of tomato plants. Virtually 100% of the internal volume is devoted to seeds; there is almost no pulp. The primitive cultivated forms of wild tomatoes resemble cherry tomatoes, and they are about twice as large as ancestral tomatoes and have four times more pulp and half as many seeds. A turn-of-the-century tomato variety is approximately 20 times larger than the ancestral wild tomato, and 60% of its internal volume is pulp. Finally, today's beefsteak tomato is 90 times larger than the ancestral wild tomato, and 87% of its internal volume is pulp (Figure 22.11). People have created a tomato plant that devotes almost all of its energy to making fruit pulp for feeding humans and none to making seeds to ensure its own perpetuation.

Humans have taken the trend to fewer seeds to an extreme by creating sterile varieties. Certain plants spontaneously mutated, creating cultivars able to produce fruit in the absence of fertilization. Navel oranges, bananas, pineapples, and certain varieties of grapes, figs, and cucumbers produce rudimentary seeds or none at all. In the wild these variants would have no future, because their genes would immediately be removed from the population by natural selection. However, people intervened in nature and propagated these varieties using vegetative reproduction.

Decreased genetic diversity

Cross-fertilization may be in the plant's best interest, but humans are more concerned with higher yields than genetically variable plant offspring. Self-

 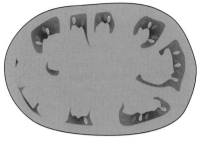

Wild
tomato

Cherry
tomato

Ordinary
cultivated
tomato

"Beefsteak" tomato

Figure 22.11 Tomato domestication. Genetic modification through artificial selection and plant breeding has made the tomato an excellent food for people, but it has made the domesticated tomato completely dependent on humans for its survival. In addition to increasing the amount of pulp at the expense of seeds (offspring), the genetic changes significantly increased the amounts of vitamin A and vitamin C in tomatoes and decreased the amounts of naturally occurring tomato toxins.

pollination does not depend on pollen transport by wind, insects, or other animals, so seed production tends to be higher and fruit set more certain in plants that can "self." Therefore, whenever possible, plant breeders have removed barriers to self-fertilization and discouraged cross-fertilization in many crops, such as wheat, rice, and soybeans.

Early farmers prevented seed dispersal

Ripened ears of the wild relatives of cereals shatter into individual seeds that are scattered widely by wind, landing far from the parent plant and each other. Widely scattered seeds may increase the plant's reproductive success, but harvesting them must have been extraordinarily time-consuming. Is it any wonder that "no seed dispersal" was one of the first genetic modifications Stone Age farmers insisted on when they transformed wild grasses into domesticated cereal crops? Cultivated cereals, such as wheat and corn, have rigid spikes of seeds that remain intact when ripe, but their ancestors have seeds that are easily scattered. In addition, the ancestors of oats, barley, and wheat have long, appendage-like structures that enclose the kernels and aid in seed dispersal. Domesticated varieties lack these because people selected seeds with the least amount of armature so that they could get to the seed more easily (Figure 22.12).

Early farmers minimized crop defense mechanisms

As we mentioned above, plants are immobile. When attacked, they cannot run and hide, so they have no choice but to fight back. The arsenal of weapons plants use to defend themselves and their offspring against a variety of organisms—herbivores, parasites, and pathogens—includes both physical and chemical weapons.

Some plants armor themselves against herbivory with thorns or small hairs that secrete gummy substances that act like fly paper (Figure 22.13). Wild grasses also have structures that protect the grain from seed-eating animals—and also make it very hard to remove kernels from the plant. As a result, manual threshing is an arduous task that entails repeatedly beating stalks of grain to free the seeds. Any spontaneous mutation that lessened the need to thresh would have been highly valued by early agriculturists. Archaeological records show very early selection for free-threshing varieties—mutants that readily release the kernel from its protective covering.

Plants also defend themselves against pathogens, parasites, and hungry insects with a variety of chemical tactics. When infected with a fungal disease, many plants release enzymes that specifically destroy the cell wall of fungal cells. Other secondary plant compounds work by inhibiting protein-digesting

Figure 22.12 Domestication and seed dispersal. **(A)** Wild grasses, such as this wild relative of barley, have structural adaptations for seed dispersal. **(B)** Those same structures made it difficult for early farmers to get to the seed, so they genetically modified wild barley by artificial selection to create domesticated barley, which lacks these structures. (Photographs by Dave Powell, courtesy of http://www.forestry.org [A], and Robert Sonerg, courtesy of the Smithsonian Institution [B].)

Figure 22.13 Mechanical defense mechanisms. Many plants protect themselves with thorns, spines, and sticky glandular secretions, which have been removed or minimized in crop plants.

enzymes in the insect's stomach. Many crop ancestors contained bitter, harmful chemicals in high concentrations (Figure 22.14). Early agriculturists selected against these protective chemicals and created crop varieties with low levels of distasteful and unhealthy chemicals. Lettuce, potatoes, cucumbers, spinach, beans, and cabbage contain only small amounts of the extremely bitter chemicals found in their ancestors.

In selecting for crops with low levels of bitter or toxic chemicals, people stripped plants of their ability to defend themselves against herbivores, parasites, and pathogens. Improving the taste of our crop plants also made them tastier to insect pests, which partially explains why farmers must constantly do battle with these pests.

Figure 22.14 Chemical defense. In Tanzanian grasslands in the dry season, there is little, if any, vegetation for primary consumers. Even so, the fruits of this tomato relative remain untouched. The tomato family (Solanaceae), which is commonly known as the nightshade family, is infamous for containing high levels of many toxins.

AGRICULTURAL ECOSYSTEMS

Removing a crop plant's physical and chemical defense mechanisms was only one of the choices early farmers made in the hope of making their lives better through crop improvement, but in solving some problems, they created others.

In some ways the continual conflict between agriculture and nature began with the early agriculturists' initial domestication choice—changing wild grasses into cereal crops. As we mentioned above, they selected wild grasses as the first crop to domesticate for excellent reasons, but that choice had important ecological implications that grew increasingly problematic as human population numbers swelled.

The ancestors of cereal crops, the wild grasses, belong to the category of species we analogized to sprinters in chapter 14. They are opportunistic species that invade disturbed habitats. Wild grasses cannot compete with other plants in well-established ecosystems. To create conditions that will maximize productivity, farmers must give cereal crops the environment in which their relatives do best, which is a disturbed habitat. Disturbing agricultural ecosystems every year ultimately leads to a number of inevitable changes in the physical and biological environments. Understanding these changes will help you see why farmers, whether ancient or modern, continually resemble the Red Queen in *Alice in Wonderland*, who ran as fast as she could just to stay in the same place.

Disturbing the habitat causes soil erosion and nutrient depletion

To create an environment that fosters crop growth, farmers clear the land of other plants that compete with crops for nutrients, water, and light and loosen the soil to make life easier for the embryo in the newly planted seed. Young seedlings meet no resistance as they emerge through the loosened soil and send out fragile new roots to anchor and nourish the plant (Figure 22.15). Loosening the soil for planting every year causes the disturbed habitat to lose topsoil, which is an agricultural necessity that provides a plant with much more than something to hold onto while it grows. Topsoil is rich in the organic matter that gives soil a porous structure, which determines the soil's capacity to retain water and allows roots to penetrate and air to circulate. Organic matter also provides the nutritive and energetic foundation of the soil ecosystem and the crop ecosystem above ground. The soil populations of detritivores and decomposers rely on decaying organic matter to provide not only the materials and energy source they need, but also nitrogen in a form they can use—ammonia and ammonia-like substances. They, in turn, release material building blocks and provide usable nitrogen to the crop

Figure 22.15 Seedling emerging. (Photograph courtesy of the National Park Service.)

plants. Soil erosion washes away not only the organic matter, but also the nitrification and nitrogen-fixing bacteria.

In summary, not only does topsoil erosion deplete the ecosystem of nutrients and organic matter, which creates an environment inhospitable to crop growth, but as the soil erodes, it also takes with it the biotic elements—decomposers, nitrification bacteria, and nitrogen fixers—that could potentially restore the physical environment to one that could sustain plant life (Figure 22.16).

The problem of nitrogen availability becomes even more critical because farmers must harvest annual crops every year. During the growing season, the crop plant accumulates soil nutrients, and these nutrients leave the ecosystem when the farmer harvests the crop. Unlike natural ecosystems, in which materials cycle between the abiotic and biotic environments, some proportion of the nutrients in an agroecosystem is siphoned off every time a crop is harvested. As the soil becomes depleted of nutrients, crop yields decrease from year to year. Modern farmers replenish their soil with fertilizers, but early farmers responded by clearing another area near their settlement and beginning the process anew, which increased the land area that was subject to erosion.

The chronic loss of nutrients, especially nitrogen, from the soil through crop harvesting and soil erosion creates an unsustainable system. One of the reasons that American settlers continued to move west was that the depleted soils of the East Coast could no longer support the increasing population numbers. Eventually farmers learned to replenish soil by incorporating organic matter. They applied animal manure to their fields or let fields lie fallow and plowed the standing vegetation back into the soil to rot. This "green manure" functions like animal manure, replenishing the soil's nutrient supply, restoring its biotic community, and maintaining its structure.

Agriculture transformed organisms into pests

Not only did agriculture change aspects of the abiotic environment and the soil's biotic community, it also created a field of dreams for organisms that rely on plants as food, such as herbivores, pathogens, and plant parasites.

With the birth of agriculture, humans gave herbivores a dense resource around which they could congregate. Not only was there no need to forage

Figure 22.16 Soil erosion. Runoff from a heavy rain carries topsoil from an unprotected, highly erodable field in Iowa. (Photograph by Lynn Betts, courtesy of Natural Resources Conservation Service, USDA.)

widely in search of food, but herbivores were also presented with large blocks of a single species, a **monoculture**. In natural ecosystems, herbivores cause the most damage to solid stands of a single species, and plants that grow intermingled with other plants show less damage from diseases and herbivory. In addition, a consistent effect of habitat disturbance in nature is a reduction in the number of species present. This simplified community structure typically involves loss of species at higher trophic levels, such as predators that prey on insects. Finally, to make matters just about perfect for herbivores and pathogens, people store the crops after harvesting. The plant pests are not subjected to a period of resource scarcity that would help to keep their numbers in check.

In summary, agriculture gives plant pest populations the best of all worlds: large expanses of a single plant species, stripped of its self-defense mechanisms; a year-round food supply; and a simplified community structure that tends to reduce the number of predators and competitors. All of these factors make it easy for populations of herbivores and plant pathogens to reach huge population sizes that are not typically found in natural ecosystems (Figure 22.17).

Is it any wonder that the earliest records of human history describe pest outbreaks, such as the swarms of locusts described in the Old Testament, as well as methods for trying to bring them under control? The Sumerians, Greeks, and Romans concocted various mixtures of sulfur and aromatic plants, such as garlic, cedar, and bay. The Ancient Chinese mixed herbs with soot and were the first to use inorganic metals, such as mercury and arsenic, to control insect pests. Early American settlers spread ground-up tobacco on stored grains because nicotine has pesticidal properties.

Perhaps the various tactics for decreasing plant herbivory, controlling plant pathogens, and restoring exhausted soils with manure would have been sufficient to maintain adequate levels of crop production had the human population size remained at pre-1800 levels. When the population began to expand exponentially, people had to devise better strategies for improving agricultural productivity. Scientific discovery and technological innovation provided farmers with the necessary boost for maximizing the productivity of their artificial ecosystems.

AGRICULTURE AND THE INDUSTRIAL REVOLUTION

The first agricultural revolution, the birth of agriculture, led to an increase in the human population size that was so dramatic it irreversibly changed the structure of societies. These same variables—large population size and a new social organization—were not the effects but the drivers of the second

Figure 22.17 Crop infestations. Large blocks of genetically identical organisms encourage pest outbreaks. In this example, a healthy tobacco crop (left) became infested with a viral pathogen, the yellow dwarf gemini virus (right). (Photograph by Gary Baxter, Department of Primary Industries, courtesy of http://www.forestryimages.org.)

agricultural revolution, which is often referred to as the **Green Revolution**. The explosive population growth that began during the last part of the 18th century was accompanied by an expanding industrial economy that needed to be fed with workers who could leave the farm. Both factors necessitated increased agricultural productivity, and farmers in the industrialized world and some developing countries met the challenge. The rise in productivity came about:

- by converting more land into cropland
- by increasing the productivity of existing croplands by applying fertilizers; constructing irrigation systems; using chemicals to control herbivores, pathogens, and weeds; and genetically improving crop plants

Most of these strategies were made possible by technological developments in a completely different industry—the petroleum industry (Box 22.1).

In a sense, the second agricultural revolution had three scientific and technological threads: mechanization, chemicals, and crop genetics. We discuss them separately below, but operationally they are intertwined.

Machines and chemicals improve crop productivity

For many centuries, agriculture was practiced in essentially the same way: people performed all tasks by hand or with small tools. When animal labor replaced human labor, the farm tools became larger and heavier. Then, technological innovation born during the industrial revolution led to revolutionary changes in agriculture, beginning with the extensive mechanization of tasks that had been performed by people or animals. Discovering how to use new power sources—first steam and then gasoline—led to the development of engines that could pull farm implements that were much larger than those pulled by animals. Because of these mechanical changes, farmers were able to greatly expand agricultural acreage in order to increase productivity (Figure 22.18). This expansion led to more food, but also to more problems. Pushing the land to yield more and more crops can lead to soil erosion, nutrient depletion, water management problems, and pest outbreaks.

Larger, heavier tools for turning up soil during seed planting and weed cultivating led to tremendous losses of topsoil, which irrigation exacerbates. On-farm sources of nitrogen—animal and green manure—were insufficient for restoring soil on so many acres. Germany had developed an electricity-driven chemical process to fix nitrogen for explosives that they used in World War I, but it was very expensive. Large-scale manufacture of nitrogen fertilizers was not possible until an energy source significantly cheaper than electricity was available, because the chemical reaction that produces ammonia from nitrogen must be conducted at very high pressures and temperature (600°C), even with the aid of a catalyst. Developing a manufacturing process based on fossil fuels made economical production of fertilizer possible, and cheap fertilizer caused yields to soar in countries whose farmers applied it to their crops. Without these fertilizers, farmers would have been forced to increase productivity by clearing more land, including land that was marginal and therefore more prone to erosion.

Insect herbivory and plant disease also worsened as crop acreage expanded. Rather than relying on concoctions of herbs, plant extracts, and drying agents, like dust and ashes, farmers began to develop cheaper pesticides because they needed large quantities to use on the ever-expanding crop acreages. The first two, developed for grape pest problems by French wine-

BOX 22.1 *Fossil fuels in agriculture*

Most of the technology-based strategies that farmers adopted in the 20th century owe their success to the petroleum industry. Oil, the liquid, preserved remains of organisms that lived long ago, is a complex mixture of hundreds of hydrocarbons with very few uses in its crude form. In 1850, a British scientist developed a process that used heat to separate the set of hydrocarbons with low boiling points from all other compounds in the mixture. This fraction contained lamp oil and kerosene, which could be used as fuel to light and heat homes. These compounds, which replaced the increasingly scarce whale oil, provided the foundations of the future petroleum industry. The heating process also produced useful by-products with high boiling points, such as candle wax and lubricants, as well as a highly flammable by-product that was nothing more than a dangerous nuisance, gasoline. All of the other hydrocarbons in crude oil had lower boiling points and did not separate from the mixture. These fractions were burned as trash or dumped into rivers with the equally worthless gasoline fraction.

The industry's view of the usefulness of crude oil's various fractions was turned on its head in the late 1800s, when electric lighting supplanted kerosene lamps and engineers developed the internal combustion engine. The instantaneous popularity of automobiles made gasoline the most valuable fraction. The petroleum industry now viewed the lamp oil fraction as worthless, in addition to the fractions with lower boiling points.

This point of view did not last long either. Engineers developed a process, oil refining, that uses a series of decreasing temperatures to sequentially collect crude oil's many fractions (see the figure). Each fraction now gives rise to an extraordinary array of products, including the organic chemicals farmers use to control insects, pathogens, and weeds. According to some estimates, crude oil's various fractions have given rise to over 5 million industrial products.

However, as you will recall from the discussion of ecosystems in chapter 14, society's excessive reliance on nonrenewable fossil fuels not only leads to environmental problems but is also a fundamental reason why so many human activities are nonsustainable. This is especially true in modern agriculture, where fossil fuel energy has replaced manpower.

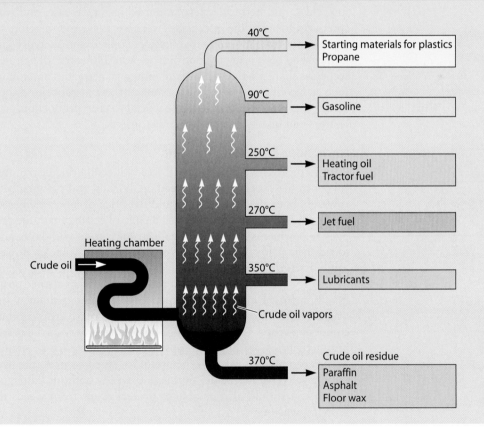

Figure 22.18 Steam plow. In the 1800s, farmers replaced animals with mechanical power, such as this steam-powered plow. (Photograph courtesy of USDA.)

makers, were copper sulfate, which controls fungal diseases, and a copper-arsenate mixture for protection against insects. A lead-arsenate compound for controlling soil pests soon followed. All of the pesticides based on inorganic elements may have been affordable, but they were also highly toxic to all animals, including mammals and humans, and often harmful to the plant they were meant to protect.

Interest in maximizing the number of useful products derived from crude oil's "waste" fractions led to the development of petroleum-based agrochemicals for controlling pests (see Box 22.1). Organic (as in carbon-based) insecticides, fungicides, and herbicides revolutionized insect, disease, and weed control, respectively. Some of the insecticides were less toxic and therefore represented improvements over the inorganic arsenic, copper, and mercury compounds that farmers had been using to control crop pests. However, all petroleum-derived insecticides share a handicap with all inorganic pesticides and natural insecticides, such as nicotine and pyrethrum. None of these insecticides distinguish between insect pests and insects that are beneficial to crop production, such as crop pollinators and insects that prey on other insects.

Herbicides help conserve soil

Prior to the discovery of chemical weed killers (herbicides) in the 1930s, farmers controlled weeds by hand weeding, hoeing, and plowing between the crop rows. As both the crop acreage and the price of labor increased, using weeding and hoeing as the primary weed control strategy became prohibitively expensive. Farmers turned to mechanical cultivation with large farm implements to rid their fields of weeds. This strategy has drawbacks, however, because it can worsen the soil erosion problems that plague farmers (Figure 22.19).

To try to circumvent these problems, scientists turned to chemistry. The development of chemicals to kill weeds presented them with unique challenges. Weeds are pests that significantly decrease crop productivity by outcompeting crops. Unlike the biochemical machinery of the pests that decrease productivity by feeding on crops, weed biochemistry is essentially the same as the crop biochemistry. To be useful for weed control, the herbicide needed to be a **selective herbicide**, that is, kill weeds without damaging the crop. If possible, it should also be able to protect crops from weeds throughout the growing season.

In the early 1950s, when chemists discovered an herbicide with a very short lifespan that killed all plants, including crops, they thought it had no commercial potential. However, this herbicide, and others with similar properties, revolutionized agriculture because it eliminated the need to plow weeds under prior to planting the crop. Farmers use **preplant herbicides** to

Percent of time in severe and extreme drought

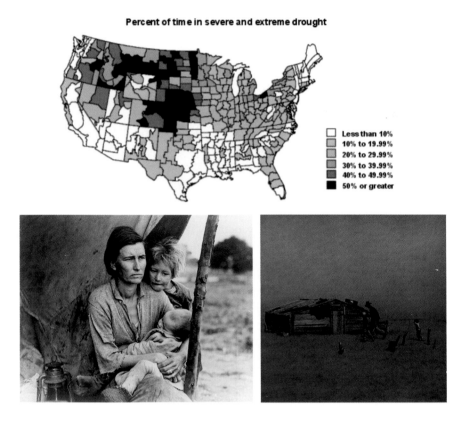

Less than 10%
10% to 19.99%
20% to 29.99%
30% to 39.99%
40% to 49.99%
50% or greater

Figure 22.19 Wind erosion. Poor land management practices combined with four severe droughts in 10 years (1930 to 1940) created the Midwest Dust Bowl that led to massive migration and the creation of migrant camps from Missouri to California. (Map courtesy of U.S. Geological Survey Drought Mitigation Service, 1930–1940. Photographs by Arthur Rothstein [right] and Dorthea Lange [left], courtesy of Library of Congress.)

rid their fields of weeds without fear of harming their crops because the herbicides become inactivated very soon after application. By allowing no-till or minimum-till agriculture, the preplant herbicides prevent soil erosion caused by plowing and conserve soil moisture, lessening the need for irrigation (Figure 22.20).

Genetically improved plant varieties significantly increase productivity

The third and final thread of the second agricultural revolution was the creation of improved crop varieties. Rather than using fossil fuels to manufacture, transport, and apply chemicals in order to improve yields and protect crops, this approach to increasing productivity relies on genetic information within the crop plant.

As we explained in chapters 1 and 21, genetic variation within and among plant populations has been a natural resource that humans have exploited for centuries. The Stone Age farmers who created domesticated crops with desirable traits did so by shaping existing genetic diversity that is inherent in all wild plant populations. Farmers and plant breeders then began to increase genetic diversity in crop plant populations by cross-pollinating plants having different desirable characteristics. From the genetically diverse population generated by crossing plants, they selected certain plants with the most desirable suite of traits. Through these techniques of controlled pollination and selection, they were able to create crop cultivars that were specifically adapted to local conditions.

At first, farmers and breeders restricted their cross-pollinations to plants that could interbreed easily, but often the gene pool within which crosses could easily occur did not provide enough variation. Plant breeders wanted

Figure 22.20 No-till agriculture. Preplant herbicides allow farmers to plant crop seeds without tilling the soil to plow weeds under. Not only does no-till agriculture reduce soil erosion, but the dead plants conserve moisture and nourish the crop. (Photograph by Linda Betts, courtesy of Natural Resources Conservation Service, USDA.)

to extend the gene pool boundaries to encompass plants closely related to the crop, but they often ran into difficulties because nature has many strategies, known as **reproductive isolating mechanisms**, for inhibiting reproduction between two different species.

A critical step in **speciation**, which is the evolution of a new species, is the development of mechanisms that reproductively insulate the incipient species from the influx of genes from other species. As you recall, the concept of species is actually based on reproductive isolation. A species is a population of organisms that interbreed and therefore share a gene pool; the gene pool they share is separated from other gene pools. In the most common form of speciation, a geographic barrier divides a population in two, and over time, the two populations accrue enough differences that if the barrier is removed they can no longer interbreed.

Plant breeders that wanted to incorporate traits from different species into a crop needed to develop techniques for vaulting over these natural reproductive barriers. A wide variety of laboratory techniques have allowed breeders to crossbreed plants, not only in different species, but also in different genera (see Tables 21.10 and 21.11). But, even with these new techniques, breeders could push the breeding envelope only so far. As a result, they were not able to incorporate certain desirable traits into crops because genes for the trait were not present within the new, expanded gene pool. Not easily thwarted, plant breeders took a new route: they created new genes in crops through mutagenesis.

Armed with this battery of sophisticated techniques and a knowledge of genetics, during the last century, plant breeders have done a remarkable job creating crop varieties that are resistant to diseases, insects, nematodes, and environmental stresses, such as drought and cold. What is particularly remarkable is the degree to which a specific cultivar is adapted to local conditions. Why should a plant breeder spend time, energy, and money trying to incorporate genes for resistance to diseases caused by pathogens that don't even occur where a certain cultivar is grown?

In addition to increasing agricultural productivity by developing better-protected crops, plant breeders also developed **high-yielding varieties** (HYV) that led to remarkable increases in yields (Table 22.2 and Figure 22.21); but often, in order to get the maximum possible yield, farmers need to apply large amounts of fertilizer and water to HYV. As a result, many poor farmers in developing countries could not benefit from the new HYV, which required large capital investments. However, even though only some of the farmers in developing countries reaped the economic benefits these varieties provided, the global increase in agricultural productivity managed to stave off the mass starvation that many people had predicted (Figure 22.22).

Although genetic improvement through breeding and mutagenesis increased crop productivity in ways that seem to be more sustainable because

Table 22.2 The Green Revolution and wheat production

Country	Production (metric tons)		
	1950	**1970**	**1985**
Mexico	270,000 million	2.4 million	5.2 million
India	9 million	20 million	37 million
Pakistan	4.6 million	8.5 million	17 million
United States	29 million	45 million	57 million

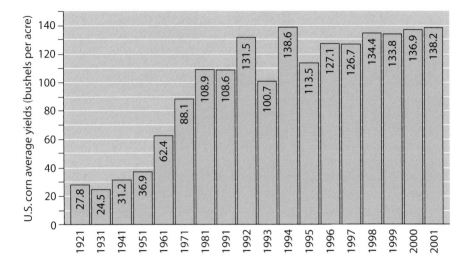

Figure 22.21 U.S. corn production.

neither technique relies on fossil fuels, in some ways these varieties also fall short of the mark. As mentioned above, some of the improved varieties continued to require irrigation and high nitrogen inputs. In addition, the striking productivity improvements provided by genes for superior yields and pest protection encouraged farmers to plant **monoclonal cultures**. Not only did agriculture provide herbivores and pathogens with very dense stands of a single crop (monoculture), farmers began growing genetically uniform versions of that crop (at that location).

You will recall from the chapter on evolution that genetic diversity is key to a population surviving and adapting to environmental changes. In nature, species constantly hedge their bets by using sexual reproduction to generate variable offspring. Monoclonal agriculture is particularly susceptible to massive crop losses if all farmers over a large area grow the same genetic variety. If one plant is susceptible to a pathogen or insect pest, all of the surrounding plants are as well. U.S. corn growers experienced widespread crop losses in 1970 and 1971 when a new genetic strain of the fungus that causes southern corn blight suddenly appeared in the late 1960s. The variety of corn most farmers in the Midwest were growing, which produced very high yields under most conditions, was uniformly susceptible to the new strain. The southern corn blight epidemic reduced their yields by 40%, which cost those corn growers hundreds of millions of dollars.

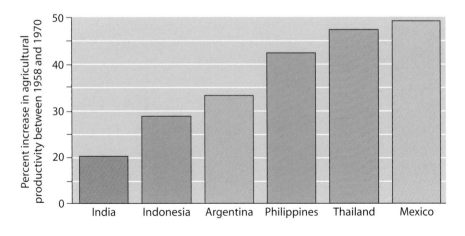

Figure 22.22 Productivity gains in developing countries. Between the late 1950s and the early 1970s, agricultural production increased in many developing countries. Unfortunately, the population size of the developing world increased even more rapidly.

Farmers understand better than anyone else the risks they assume when they present pests with a field planted in a single, genetically uniform variety. So why do they take the risk? Productivity. Remember that the goal of the technological innovations discussed in this chapter is increasing productivity to meet the needs of an ever-expanding population. Farmers opt for genetic uniformity within their fields (but not across the country or globe) because of the unrelenting need for more food.

The value of genetic diversity is related to the long-term adaptability of a species, not necessarily its immediate survival (productivity). In the 1930s and 1940s, recognizing the interdependency of the long-term survival of the crop provided by genetic diversity and the immediate increase in productivity provided by the genetically uniform HYV, a number of organizations developed programs and procedures for preserving the genetic diversity of crop plants and their wild ancestors (Table 22.3). Those organizations could never have imagined the extraordinary impact their foresight would have on crop improvement. They established mechanisms, invented techniques, and built facilities to preserve and share (internationally) genetic diversity decades before the development of recombinant DNA technology (Figure 22.23). Genes they began preserving in the 1940s can now be incorporated into any crop by using molecular techniques not dreamed of at the time (Figure 22.24).

Agriculture is full of trade-offs and compromises

The duality between short-term productivity through genetic uniformity and long-term productivity through genetic diversity exemplifies the types of trade-offs and imperfect solutions that permeate modern agriculture. Because the problems plaguing agriculture are so complex, often a benefit gained by using a certain technology or strategy is diminished and sometimes even nullified by costs. Some examples are listed below.

- Using chemically synthesized fertilizers prevents deforestation by increasing the productivity of land that has already been cleared. Therefore, using chemical fertilizers preserves genetic diversity. On

Table 22.3 Number of collections worldwide and approximate number of germ plasm samples (accessions) for major crops

Crop	No. of collections	No. of accessions
Barley	65	256,700
Beans (*Phaseolus*)	70	159,000
Brassicas[a]	45	40,500
Cucurbits[b]	59	46,100
Maize (corn)	61	152,100
Oats	38	109,200
Peanut	40	66,100
Pea	33	52,100
Peppers	39	36,500
Potato	90	62,100
Rice	45	343,400
Soybean	70	137,000
Sorghum	34	137,800
Tomato	39	57,800
Wheat	115	509,000

[a]The brassicas include crops such as canola, cabbage, broccoli, cauliflower, mustard, and kale.
[b]The cucurbits include crops such as melons, cucumber, squashes, and pumpkins.

Figure 22.23 Preserving genetic diversity. **(A)** The USDA's National Seed Storage Laboratory preserves more than 1 million samples of plant germplasm in an 18°C storage vault. **(B)** Containers of seeds, cryopreserved in vats of liquid nitrogen, can remain viable for thousands of years. Each vat contains approximately 10,000 seeds. (Photographs by Scott Bauer, courtesy of Agricultural Research Service, USDA.)

Figure 22.24 Crop genetic diversity. When combined, the genetic diversity of different cultivars of the same species, such as *Phaseolus vigna*, is significant, even though the gene pool has been depleted by centuries of breeding. Conserving wild relatives of crops greatly expands the available gene pool. (Photograph by Keith Weller, courtesy of Agricultural Research Service, USDA.)

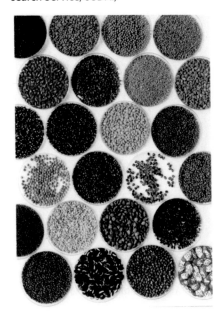

the other hand, because they are so expensive energetically, societies use much more energy manufacturing, transporting, and applying fertilizers than they acquire from the crops grown in the fertilized field.
- Insecticides kill pests that destroy agricultural crops, but they also kill the beneficial insects that agriculture depends on, such as pollinators.
- Herbicides conserve soil by permitting no-till agriculture, but they are made from petroleum derivatives, which are nonrenewable.

At the present size of the human population and its rate of increase, it is difficult to see how current agricultural practices (in either industrialized or developing countries) can be sustainable over the long term, especially in areas with fragile soils and limited water availability.

Could the tools, techniques, and products of biotechnology help make agriculture more sustainable?

The mere mention of technology as a solution to the problem of nonsustainable practices often elicits an immediate negative response, because many people blame technology as the source of the problem. If you feel a need to assign blame to someone or something for the problems besetting the natural environment because of agriculture, blame the people who changed humans from a hunter-gatherer society to one based on agriculture. However, many anthropologists interested in the origin of agriculture believe our Stone Age ancestors were forced into agriculture due to environmental pressures that were caused by massive climatic changes (sound familiar?). It doesn't seem fair to blame people who had no choice in the matter, does it? It makes more sense to blame the many scientists who discovered that microbes cause infectious diseases; recognized the relationship among high infant mortality, sewage treatment, and public hygiene; and developed antibiotics and vaccines to combat infectious diseases. But blaming them doesn't seem appropriate either, even though they are the group that is most responsible for the human population's exponential growth.

The crux of the problem is that the size of the human population is large and getting larger every minute. At its current rate of increase, approximately 180 people are added to the population every minute, 10,800 every hour, 260,000 every day, and 1.8 million every week. What option does society have other than to turn to technology in the hope of making agriculture more sustainable?

In the next chapter, we discuss some of the applications of biotechnology that could help make agriculture more sustainable and two of the ecological issues that are often discussed in relation to modern biotechnology: the evolution of resistance to pesticides and gene flow.

SUMMARY POINTS

The first crops that humans domesticated were wild grasses. With significant modification through selective breeding, Stone Age farmers converted wild grasses into the cereal crops that are the primary source of nutrition for most people. The seeds were the plant part they were interested in, because seeds are very nutritious and can be dried for storage.

The process of domestication modified plants so significantly that crop plants are no longer able to survive and reproduce successfully in competition with wild plants.

Flowers and fruits are tactics for meeting a plant's primary objective: reproduction. To encourage cross-pollination in order to create genetic diversity, many plants have evolved mechanisms to prevent or minimize self-fertilization. Many flowering plants depend on animals for pollination and seed dispersal and reward them for their services.

Human domestication of plants is a form of genetic modification. Using artificial selection to shape crop plants to human needs not only decreased the plants' capacity to survive and reproduce in the real world, it also created a number of problems that farmers have struggled with since agriculture's beginnings.

Basing agriculture on annual plants necessitates annual habitat disturbance. This leads to soil erosion, agriculture's most significant negative impact on both the environment and the long-term sustainability of agricultural ecosystems.

Growing large blocks of the same crop encourages pest outbreaks. This problem is amplified when the crop consists of genetically identical individuals.

The exponential growth of the human population in the past two centuries placed significant pressure on agricultural ecosystems. Increasing their productivity required modern technological innovations in mechanization, chemical discovery and manufacture, and plant breeding. These advances gave rise to the Green Revolution, which prevented massive starvation.

These innovations, which are requisite in the short term, make many modern agricultural practices unsustainable over the long term. Continuing to feed the ever-growing human population while attempting to improve the sustainability of agricultural ecosystems is a monumental problem that societies must address as soon as possible.

KEY TERMS

Angiosperm
Annual plants
Cultural evolution
Green Revolution

High-yielding varieties
Monoclonal culture
Monoculture
Preplant herbicides

Reproductive isolating
  mechanism
Selective herbicide
Speciation

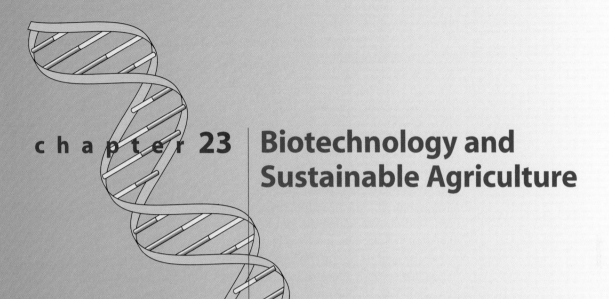

chapter 23 | Biotechnology and Sustainable Agriculture

If farmers could design the perfect crop plant, what features would they give it? First, they would make sure it could survive biotic and abiotic stresses without the farmer's help. Then, because the plant products of reproduction, seeds and fruits, are usually the farmer's commercial product, farmers would maximize the energy and materials the crop devoted to reproduction. In other words, the farmer's essential design features conform to the fundamental goals of all living organisms: survive and generate many offspring that also survive and reproduce.

In the quest for long-term reproductive success, all organisms share common objectives—get enough nutrients, avoid being eaten, find mates, and maximize numbers of offspring—but their strategies and tactics vary. Evolution crafted a wonderfully diverse set of tactics, because each species gave natural selection different genetic material to shape, and the specific problems each species faced, that is, the selective pressures, varied as well. The conjunction of different selective pressures acting on different genetic material for millions of years provides a cornucopia of adaptations, which are genetic solutions to the challenges organisms face.

We ended the last chapter asking if the new biotechnologies could make agriculture more sustainable. Using biotechnology techniques in research could improve agricultural management practices, because having a clear and detailed view of the biological processes that farmers want to maximize (plant growth), control (pathogenesis and herbivory), or exploit (nitrogen fixation) should lead to efficacious agricultural products that improve yields and decrease costs. But will the products that are developed necessarily encourage farmers to adopt sustainable management practices? As we have mentioned many times, science and technology may define the universe of possible products, but other variables determine which possible products become real products that farmers can buy. Economics, business strategies, government policies, and consumer preferences are analogous to evolutionary forces.

They select for some product ideas from the pool of possible products and select against others. These factors provide direction to technology development much as natural selection leads to some adaptations and not others.

In this chapter, we describe how these forces could be applied to produce products to support the goal of sustainable agriculture.

- We describe management practices that decrease the sustainability of modern agriculture.
- We discuss how an understanding of basic plant biology could lead to better agricultural products.
- We describe the ways that the current group of transgenic crops encourages or discourages sustainable practices.
- We discuss two of the potential environmental impacts of transgenic crops.

BIOTECHNOLOGY AND SUSTAINABLE AGRICULTURE

Sustainability, in the broadest sense, means meeting the population's current needs without sacrificing the ability of future generations to meet theirs. Human activities are not sustainable if they:

- degrade the quality of essential resources
- deplete essential resources because they rely on nonrenewable resources, such as fossil fuels, or use renewable resources at a rate higher than the renewal speed

The concept of sustainable agriculture has a number of different facets (Box 23.1), but we will focus on only one: the ecological sustainability of agroecosystems.

B O X   2 3 . 1 *What is sustainable agriculture?*

The concept of sustainable agriculture means different things to different people. Specific details may vary, but all share a common theme: a longer-term view than the one dictated by short-term economic considerations and a tripartite set of interdependent dimensions for measuring sustainability— ecological, economic, and social. Each one of the three dimensions has many facets or levels of impact.

Ecological sustainability is not only sustainability within the agroecosystem; it extends outward from the farmer's field to encompass adjacent ecosystems affected by the agroecosystem. Measures of ecological sustainability include the quantity and quality of resources essential to agriculture and environmental conservation of surrounding ecosystems.

Economic sustainability is related first to agricultural productivity, and its primary measures are the yield per acre and the economic value of output compared to costs of inputs (farm profit). Other aspects of economic sustainability include consumer food prices, the economic vitality of agribusiness suppliers in rural communities, and even variables such as government subsidies, international markets, and trade deficits.

Social sustainability encompasses those variables that affect the quality of life in rural communities, such as the number of family farms, emigration from rural to urban areas, and demographic variables, such as the age distribution and educational level of the rural population. The second-order variables extend to society as a whole and include components such as high-quality food with the least risk to human health and the environment.

As is true of all systems with many interacting and interdependent parts, the goal of agricultural sustainability is best served by recognizing the dynamic tension existing among the three dimensions. Often practices that foster sustainability along one dimension detract from sustainability along the other two. An ecologically sustainable practice may not be profitable and therefore not economically sustainable. A practice that is economically sustainable may encourage the consolidation of a number of family farms into one large farm. Choices that try to maximize sustainability along a single dimension can cause the system to collapse. Consequently, trade-offs between dimensions are inescapable. When the sustainability triangle is overlaid on a venture as complex, variable, and unpredictable as agriculture, it becomes clear that sustainability is best achieved by adopting a mindset that relies on systems thinking, honors the dynamic nature of the whole, allows for continual readjustment over time, and acknowledges the validity of making different choices from one situation to the next.

Can biotechnology improve agricultural sustainability?

Before we try to answer the question of whether biotechnology can improve agricultural sustainability, we need to acknowledge that different groups would take issue with our asking about biotechnology's potential role in sustainable agriculture. Some people disagree with our fundamental assumption that agriculture today is not sustainable. Other people believe that biotechnology, especially transgenic crops, has no place in a discussion of sustainable agriculture.

Reporters, science writers, and textbook authors typically handle these differences of opinion by stating them, as we just did: one group thinks this; another group thinks that. We agree with the necessity of presenting different viewpoints. We are also tempted to take the easy way out and avoid controversy by following their lead and simply describing the different viewpoints. However, we believe that merely describing different positions offers little useful information to people who are struggling to understand the reasoning behind the positions.

So that we could add a little meat to a bare-bones restatement of the two opinions regarding sustainable agriculture and biotechnology, we spent time analyzing the positions as rigorously and objectively as possible. After gathering the necessary information for supporting or refuting both opinions, we have to confess that we don't understand either point of view.

How can the intensive agriculture currently practiced in industrialized countries today be sustainable?

- It depends heavily on nonrenewable resources, both directly and indirectly (Figure 23.1). Farmers need fuel to run their machinery and irrigation systems, and fossil fuels are the feedstock chemicals in fertilizer and pesticide manufacturing.
- U.S. farmers lose 2 billion tons of topsoil annually to water and wind erosion. On average, each acre of cropland in the United States loses almost 5 tons of topsoil every year (Table 23.1).

Figure 23.1 Energy use in agriculture. Agricultural practices use fossil fuels to power farm equipment and other vehicles, dry crops, and control the temperature in animal production facilities. Fossil fuels are also used in the production of chemical inputs, such as fertilizers and pesticides.

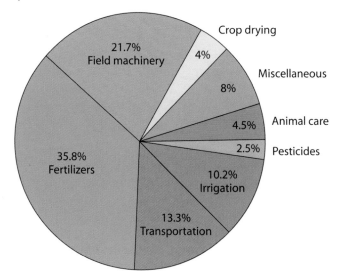

Table 23.1 Wind- and water-driven soil erosion from U.S. cropland[a]

Estimate and type	1982	1987	1992	1997	2001
Total (billion tons/yr)					
Wind	1.39	1.31	0.92	0.81	0.79
Water	1.68	1.49	1.18	1.05	1.0
Per acre (tons/yr)					
Wind	3.3	3.2	2.4	2.2	2.1
Water	4.0	3.7	3.1	2.8	2.7

[a]These estimates apply only to cropland and do not include erosion from pastureland and urban areas, even though those sources make significant contributions to the total amount of erosion. Erosion on cropland is of particular interest because of its relationship to agricultural productivity and long-term cropland sustainability. These data were collected by the USDA's Natural Resources Conservation Service from the same 800,000 sample sites every 5 years.

- Crop irrigation accounts for approximately 80% of the water consumed in the United States. Even though water is a renewable resource, society's demands on the water supply are outpacing the rate at which it is renewed. In the United States, approximately 90% of the total water used is drawn from renewable surface and groundwater, but the remaining 10% is an overdraft from stored, nonrenewable groundwater.

- Not only do current agricultural practices deplete the supply of essential resources, they degrade resource quality. For example, water runoff from agricultural fields that erodes topsoil also causes **sedimentation**, which is the deposition of material suspended in water and the number one water pollution problem (Figure 23.2). In addition, irrigation not only depletes stored groundwater, it leads to soil **salination**, which is mineral salt buildup in poorly drained soils. Salination decreases yields, and if salt concentrations are high enough, crops die.

Figure 23.2 Water pollution. The primary water pollution problem in the United States is sedimentation. The major contributor to sedimentation is water runoff from cropland and overgrazed pastures. (Photograph by Linda Betts, courtesy of Natural Resources Conservation Service, USDA.)

On the other hand, why do environmental activists and consumer advocates oppose the use of transgenic crops on the grounds that they are unnatural and may be unsafe in the long term?

- Certified organic growers have banned the use of crops genetically modified with recombinant DNA, but at least 75% of them grow insect- and disease-resistant crops that were genetically modified through random mutagenesis and breeding techniques that force hybridization between two plants that will not interbreed naturally.
- Some transgenic crops can decrease the amount of synthetic chemicals that farmers apply to their crops. Decreases in chemical inputs have been particularly striking in developing countries, where farmers can least afford to buy chemicals.
- Transgenic crops with the Bt (*Bacillus thuringiensis*) gene have significantly lower levels of fungal pathogens and the toxins they secrete (Figure 23.3). Scientists have demonstrated that mycotoxins harm humans by increasing the rate of neural tube defects in developing embryos and the probability of developing some types of cancer. They also are fatal to livestock.

Scientifically, neither point of view makes sense to us, primarily because they are both rigid and absolute. Here's our opinion. The tools of biotechnology have much to offer sustainable agriculture. However, we are not certain how much of that potential will be realized because so many variables affect which products make it to market and are commercially successful.

Biotechnology is not the only way to improve sustainability. In fact, some people argue that agriculture could become completely sustainable, while maintaining current levels of agricultural production globally, without

Figure 23.3 Fungal pathogens, Bt corn, and mycotoxins. Because Bt corn provides better protection against insect damage, there are fewer holes in the plant that can be invaded by fungal pathogens, such as *Fusarium*. In the study shown, researchers at Iowa State University measured both the incidence of infection and the amount of mycotoxins secreted by *Fusarium*. Comparable results showing decreased levels of mycotoxins in Bt corn have been documented by scientists in other U.S. universities and in Argentina, France, Italy, Spain, and Turkey.

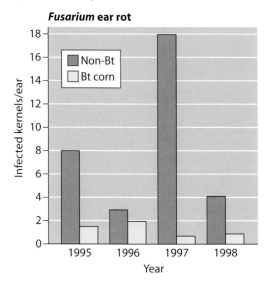

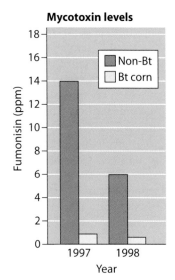

Figure 23.4 Soil conservation. Low-tech forms of conservation that decrease erosion and increase the efficiency of water use include contour farming, strip farming, and terracing (upper right corner). The contours redirect water laterally, thus eliminating runoff. Interspersing thin strips of alfalfa (green) between large expanses of highly erodible cornfields (tan) creates a buffer, especially against wind erosion. Terrace farming helps prevent erosion and water runoff when the field is located on a slope. (Photograph by Tim McCabe, courtesy of Natural Resources Conservation Service, USDA.)

using biotechnology. They may be right. Agricultural scientists have improved crop yields with simple, affordable, low-tech practices that do not deplete or degrade essential resources (Figure 23.4). The more intractable problem, however, is determining how agriculture will meet the food demands of the 8 to 9 billion people forecast for 2030 without biotechnology.

Modeling agroecosystems after natural ecosystems would improve sustainability

What are the essential resources of agroecosystems that humans should not deplete or degrade in order to achieve sustainability? As in all ecosystems, energy and materials are requisite resources. The materials required are both abiotic resources, such as water, carbon dioxide, and nutrients, and biotic resources, such as decomposers and nitrogen-fixing bacteria. Biotechnology will contribute to sustainable agriculture if it helps farmers maintain the quality and quantity of the biotic and abiotic resources they depend upon or decreases agriculture's consumption of nonrenewable resources.

A smart strategy for discovering and developing products for sustainable systems is to study nature's solutions to the problems of life on earth. The problems that decrease agricultural production are essentially the same problems organisms have battled for millennia. Because the farmer's goals are consistent with the crop plant's goals—survival and reproduction—why not use nature's strategic approach as a framework for new product development, since natural systems tend to be much more sustainable than those humans design? Better yet, in attempting to make agriculture more sustainable, why not just co-opt the molecules and molecular processes underlying nature's solutions and turn them into products? Millions of years of engineering have gone into designing, testing, revising, and retesting nature's solutions, or adaptations. In addition, they would be more sustainable because they are based on a renewable resource—the information in a molecule that renews itself, DNA.

The remarkable versatility provided by biotechnology provides many new approaches for improving sustainability. All of the biotechnologies described in chapter 1 can make important contributions to sustainable agriculture. For example, with new diagnostic tests based on DNA probes or monoclonal antibodies, farmers can identify diseased plants much earlier. Early diagnosis minimizes disease spread, and this in turn decreases fungicide application. Similar detection techniques can help farmers determine if insect pests are susceptible or resistant to the insecticides they are planning to use, decreasing the application of ineffective chemicals. Metabolic engineering can improve yields, without applying more fertilizer, by using RNA interference to block certain genes and shunt most of the plant's resources to the part of the plant that is the crop.

Even though many biotechnologies could be used to improve sustainability, we will limit our discussion to recombinant DNA technology.

CROP PLANT BIOLOGY

As we mentioned in the introduction to this chapter, biotechnology provides researchers with many tools and techniques for elucidating the molecules and processes that are essential to plant survival and reproductive success. A detailed understanding can lead to a number of different tactics for solving a single problem. In this section, we discuss how greater knowledge of basic plant biology could lead to tactics and products that lessen three barriers to sustainability: water usage, fertilizer application, and chemical control of insect pests.

Drought-resistant plants could decrease water usage

Most scientists view water scarcity as the most serious impediment to feeding the growing world population. Forty percent of the world's current food supply is produced on irrigated land, and irrigation accounts for approximately 70% of the water consumed globally (Figure 23.5). The United States uses 100 billion gallons of water per day to irrigate crops, and as we mentioned

Figure 23.5 Irrigated cropland. Approximately 40% of the world's food supply grows on irrigated croplands. The number of cropland acres that depend on irrigation has increased by approximately 70% in the past 25 years. According to the United Nations Food and Agriculture Organization, only four countries account for 50% of the irrigated cropland: India, China, Pakistan, and the United States.

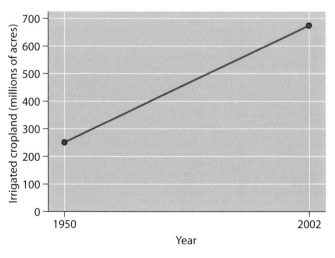

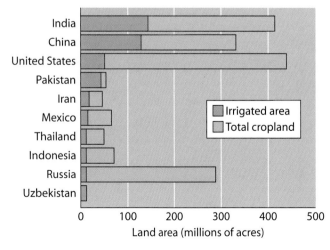

above, 10% of the water withdrawals in the United States are depleting non-renewable water resources. Transgenic crops that allow farmers to decrease water usage would improve sustainability by:

- slowing the current depletion of groundwater supplies
- decreasing erosion caused by irrigation
- lessening the use of nonrenewable fossil fuels to power irrigation systems
- decreasing soil salination caused by irrigation

Water conservation in plants

When organisms moved from water to land millions of years ago, coping with water loss became the primary problem in need of evolutionary solution. Terrestrial organisms, including plants, stingily guard their internal water supplies. More than any other factor, water availability limits the abundance and distribution of plants, and it is also the primary constraint on global agricultural productivity. The importance of water to terrestrial plants is reflected in the amount of resources they devote to root production. The roots of a 4-month-old rye plant have a combined surface area of approximately 650 square meters. Plants absorb water from the soil through root hairs, which are threadlike, tubular extensions of root epidermal cells that greatly increase the amount of surface area for absorbing water and nutrients (Figure 23.6). The rye plant described above has approximately 14 billion root hairs that, if laid end to end, would stretch more than 5,000 miles.

Because the solute concentration of root cell cytoplasm greatly exceeds that of soil, water enters the root hairs easily, but even under the best of circumstances, over 90% of the water absorbed by roots is lost to the atmosphere as water vapor. This water loss, known as **transpiration**, is an unfortunate and unavoidable consequence of the mechanism that plants use to acquire carbon dioxide (Figure 23.7).

Figure 23.6 Plant root hairs. **(A)** Plants absorb water through microscopic root hairs that cover their roots. Like the microvilli in your intestine, root hairs greatly increase the surface area available for absorbing nutrients. **(B)** Scanning electron micrograph of a single root hair showing clearly that it is an extension of root epidermal cells (red). The fluorescence stain the electron microscopist used reveals actin filaments (green) that are identical to those in animal cells. (Micrograph A courtesy of Nina Allen, North Carolina State University; micrograph B by Elison Blancaflor of the Nobel Organization, courtesy of the Samuel Roberts Noble Foundation, Ardmore, Okla.)

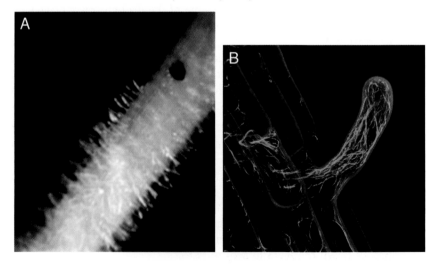

Because of water's fundamental importance to survival, all plants have evolved many different mechanisms to conserve it, improve its uptake, and cope with its loss. In spite of these many adaptations, plant breeders have had much less success breeding plants that are drought resistant than they have had developing pest-resistant varieties, primarily because so many genes contribute to drought tolerance. Using microarray technology to monitor gene expression, scientists have observed at least 20 genes switch on when the plant is water stressed. The proteins encoded by these genes are responsible for getting the plant through water shortages.

All plants have at least some of this genetic infrastructure for tolerating water shortages, but only some plants survive droughts. Those that survive are more responsive to external changes in water levels. Perhaps their genetic regulatory mechanisms become activated by smaller changes in water availability, or maybe they have gene duplications for the drought resistance genes. Whatever the mechanism, many drought-resistant plants survive tough times not because they have unique genes but because they respond earlier and more strongly to decreasing water availability.

Scientists have successfully transformed drought-susceptible plants into drought-tolerant plants by overexpressing the regulatory genes that switch on the plant's constellation of drought-responsive genes. However, under field conditions, crop yields are low because the plant routes too much energy into the continual production of 20 or more drought-responsive proteins rather than into greater yields of fruits and seeds. Scientists are characterizing genes encoding the drought-responsive proteins so that they can manipulate the expression of one or two rather than keeping all of them turned on all of the time. They have found that the drought-responsive genes reflect a number of different strategies for surviving water shortages, including the following.

Figure 23.7 An open plant stomate. Plants acquire carbon dioxide for photosynthesis and release oxygen through highly specialized structures, the stomata. Leaves contain thousands of stomata per square inch. Unfortunately, when stomates are opened for gas exchange, water that has been absorbed by the roots is lost. (Photograph copyright Dennis Kunkel Microscopy, Inc.)

Water conservation. One gene that is activated in response to water stress encodes a protein that increases production of the waxy cuticle that lessens water loss from leaves.

Water access. Another water stress gene encodes a structural protein that causes root cells to elongate. This causes the roots to grow deeper rather than laterally.

Osmotic protection. Water loss causes the concentration of ions inside the cell to increase, and high ionic concentration disrupts cell membranes and denatures proteins. Some of the drought-responsive genes encode proteins that prevent the destructive effects of high ionic concentration by increasing the production of transport proteins that pump ions out of the cytoplasm into the plant cell vacuole (Figure 23.8). Other genes encode proteins responsible for synthesizing small molecules (osmoprotectants) that shield protein molecules and stabilize cell membranes.

Oxidative stress protection. Under stress, including water stress, cells increase production of **oxygen radicals**, which are also known as reactive oxygen species (ROS) or by the more familiar term free radicals. Oxygen radicals are small, destructive molecules that contain oxygen atoms. They are out of balance electronically and try to restore balance by stealing electrons from other molecules, which explains their destructive effects. One of the genes turned on by the drought-induced transcription factors is the one encoding **catalase**, an enzyme that breaks down the ROS hydrogen peroxide.

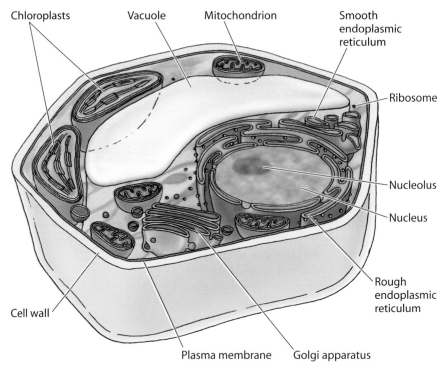

Figure 23. 8 Plant cell.

Improving nutrient uptake could decrease fertilizer use

Even though photosynthesis allows plants to create their own food from carbon dioxide and water, plants are often stressed nutritionally because they lack certain minerals in specific ionic forms, especially nitrogen as nitrate (NO_3) or ammonium (NH_4) and phosphorus as phosphate (PO_4). To maintain crop yields, U.S. farmers must add millions of metric tons of chemically manufactured fertilizer to the soil annually (Figure 23.9). Manufacturing chemical fertilizers is so expensive energetically that fertilizers account for over 35% of the energy budget in U.S. agriculture (Figure 23.1). Current global food production is equally dependent on added nutrients. Globally, farmers use 90 million metric tons of chemically synthesized nitrogen fertilizer, 50 million metric tons of nitrogen fixed by soil bacteria, and 20 million metric tons of phosphate fertilizer. Transgenic crops that decrease the amount of fertilizer that farmers must apply would improve the sustainability of agriculture by decreasing the use of nonrenewable fossil fuels and decreasing water pollution caused by fertilizer runoff into lakes and rivers.

Improving uptake and availability of essential nutrients

Interestingly, nitrogen and phosphorus deficiencies often occur in the midst of plenty because of poor nutrient uptake mechanisms. Cereal crops absorb only 50% of the available nitrate and ammonium, on average, and phosphate absorption in some crops is a dismal 20%. Therefore, one strategy for decreasing the use of commercial fertilizers is to increase the efficiency of uptake. Because the ionic concentration inside the root cells exceeds the soil concentration, plants must actively transport nitrogen and phosphorus by using membrane transporter proteins as channels. As a testament to the importance of nitrogen and phosphorus to plant survival, the *Arabidopsis*

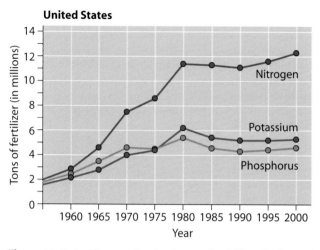

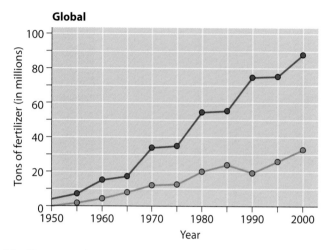

Figure 23.9 Fertilizer application to cropland. Chemically synthesized fertilizers contain three essential plant nutrients: nitrogen, phosphorus, and potassium. The two most important elements that limit agricultural productivity globally are nitrogen and phosphorus.

genome encodes at least 16 different PO_4, 16 different NO_3, and 4 different NH_4 transporter proteins. Increasing their expression levels leads to greater plant growth even when soil PO_4, NO_3, and NH_4 concentrations are low.

Phosphorus is often inaccessible to plant roots because it is tightly attached to soil particles or bound up in organic matter. Scientists have created transgenic crops with roots that secrete citric acid and malic acid, two intermediates of glucose breakdown that cause soil particles to release their hold on phosphate ions. Other scientists have increased phosphate availability 40% by genetically modifying crops to produce enzymes, such as phytase, that release phosphorus from organic molecules.

Nitrogen fixation

Rather than using biotechnology to simply decrease the amount of fertilizer farmers apply, why not think big and use biotechnology to exploit nature's solution to nitrogen shortages: bacteria that fix nitrogen using the enzyme nitrogenase? You will recall from chapter 14 that a handful of bacterial species are able to fix nitrogen by converting nitrogen in the air to ammonia, which requires breaking the triple bond found in nitrogen gas (N_2). Some nitrogen-fixing bacteria are free-living soil bacteria, and others form symbiotic relationships with plants, most notably legumes, such as alfalfa, clover, and soybean (Table 23.2).

Farmers have long understood the benefit of growing legumes that harbor nitrogen-fixing bacteria in root nodules. These crops thrive in nitrogen-poor soil, and during the growing season, they enrich the soil by releasing fixed nitrogen. For centuries farmers have rotated legumes with nonleguminous crops. Some farmers improve productivity by inoculating legume seeds with commercially available nitrogen fixers, while others rely on indigenous soil bacteria to locate crop roots and initiate the nodulation process.

Over the past century agricultural scientists have enhanced legume nitrogen-fixing capacity by genetically modifying both members of the symbiotic pair, the crop and the nitrogen-fixing bacteria. Using plant breeding and mutagenesis, they have improved legume nodulation ability and enhanced

Table 23.2 Examples of nitrogen-fixing bacteria are found in each of the major prokaryotic evolutionary branches

Cyanobacteria	Fix nitrogen in specialized structures (heterocysts); some are free living, and some associate with higher plants, such as cycads and the water fern
Azobacteria	Free-living nitrogen fixers
Rhizobial species	Symbiotic bacteria that fix nitrogen in legume root nodules; many display very specific associations in which one legume is associated with only one *Rhizobium* species
Purple-green bacteria	Photosynthetic bacteria that can also fix nitrogen
Frankia	An antinomycete (a fungus-like bacterium) that forms symbiotic relationships with trees, such as alders, and other woody plants
Azobacterium	Symbiotic nitrogen fixer that lives in loose association with roots of grasses, such as sorghum
Clostridium	An obligate anaerobe
Archaebacteria	Distant relatives of all other bacterial species; live in extreme environments, and some use methane, rather than glucose, as their primary energy molecule

the secretion of nourishing carbohydrates from plant roots to resident bacteria. Microbiologists have used mutation and selection to improve the nitrogen-fixing abilities of the commercial inoculants, but often these inoculants cannot compete well with indigenous nitrogen fixers.

In modern biotechnology's early years, scientists were almost giddy with the prospects provided by recombinant DNA technology. They dreamed of converting all crop plants, not just legumes, to nitrogen fixers by giving them the gene for nitrogenase. They knew that nitrogenase was a group of molecules and not a simple protein encoded by a single gene, but they still felt it might be feasible if they could identify the genes involved in nitrogenase synthesis and regulation. Over the years, this early enthusiasm has been tempered by the discovery that at least 20 genes encode essential elements of the nitrogenase complex and its regulators. As of now, scientists are not as hopeful as they once were about giving crops the capacity to fix their own nitrogen using nitrogenase.

If biotechnology provides so many strategies for attacking problems, then why haven't scientists thought about genetically modifying crops so they all could form symbiotic relationships with nitrogen-fixing bacteria rather than giving them 20 or more genes required for nitrogenase synthesis? Scientists have been reluctant to pursue this research avenue because the task of characterizing all of the plant genes involved in establishing and maintaining the symbiosis is much more difficult than isolating nitrogenase genes. Not only are plant genes harder to identify, isolate, and characterize than are bacterial genes, but the problems of engineering them into crop plants and getting them to work in a coordinated fashion seemed insurmountable.

Recent findings provide encouragement that this line of research may very well be the one that gets scientists closer to the dream of crop plants that fertilize themselves. Certain fungal species also form a symbiotic relationship, known as **mycorrhizae**, with plants (Figure 23.10). Many scientists believe that mycorrhizal fungi were the key to plants colonizing the land millions of years ago, because they provide plants with phosphorus and improve water uptake. In return, the plants provide the fungi with carbohydrates, much as they do for nitrogen-fixing bacteria. However, unlike the symbiosis between nitrogen-fixing bacteria and plants, 85 to 90% of the plant families form relationships with mycorrhizal fungi.

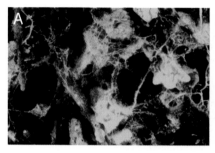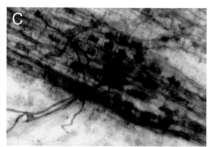

Figure 23.10 Mycorrhizae. **(A and B)** The fungal threads, or hyphae, of the fungus species in mycorrhizal associations form a dense network around the plant roots. **(C)** Some hyphae invade the root cells, which is similar to the first stage of nodulation exhibited by nitrogen-fixing bacteria that develop symbiotic relationships with plants. The dark-blue dots in panel B are nutrient storage vesicles of the fungus. (Photographs courtesy of Randy Molina, U.S. Forest Service, USDA [A], and Agricultural Research Service Eastern Regional Research Center, USDA [B and C].)

Recently, scientists who are isolating plant and fungal genes responsible for mycorrhizal relationships realized that some of the plant genes involved in establishing these associations are the same as those that establish the symbiosis between legumes and nitrogen fixers. In other words, it seems that 85 to 90% of the plant kingdom, including virtually all of our crops, already has at least some of the genetic machinery they need in order to establish relationships with nitrogen-fixing bacteria. The goal of making all crops capable of forming symbiotic relationships with nitrogen fixers may not be as crazy as was once thought.

Natural pest control methods could improve species diversity

Developing better methods of pest control would improve sustainability by preserving biological resources and not by lessening fossil fuel usage because manufacturing pesticides accounts for little of agriculture's energy expenditures (2%). The problem with most chemical pesticides is that they are not very selective. Most insecticides and fungicides kill both good and bad insects and fungi. Insecticides and fungicides that selectively target crop pests would improve sustainability by preserving beneficial organisms. Interestingly, the opposite is true of the third class of pesticides, herbicides. As long as the herbicide does not harm the crop, then the broader the spectrum the better.

Biological control, or **biocontrol**, is the suppression of pests through biological agents. The biological agent might be a whole organism or one of its molecules. For example, some organisms, including predators, parasitoids, and microbial pathogens, kill insect pests. Plant pathogens can be deterred by microbes that kill or outcompete them. Other insects and microbes can control weed populations (Figure 23.11). In addition, a form of biocontrol can be provided by the crop plant if it produces molecules for defending itself against herbivory and infections.

The various biotechnologies provide a number of options for broadening the use of biocontrol in agriculture. For example, one of the drawbacks of using microbial pesticides is that sunlight breaks down their pesticidal molecules, giving them a very short half-life in the field. Using recombinant DNA technology, scientists can give the microbe's pesticidal genes to a different bacterial species that forms spores resistant to ultraviolet light. A different biotechnology, insect cell culture, permits the production of insect viruses, such as baculoviruses, which specifically infect single insect species, without

Figure 23.11 Biological control. **(A)** Predatory insects. Ladybird beetles are beneficial insects in agroecosystems because they prey on aphids, many of which are crop pests that transmit plant pathogens. **(B)** Parasitoid wasps. A number of parasitoid wasp species lay eggs in lepidopteran larva, such as this tobacco hornworm. The wasp larvae feed on the internal tissues of the lepidopteran larvae and then emerge as pupae. **(C)** The USDA scientist shown is examining species of fungi that may be useful in weed control. (Photographs A and C by Scott Bauer, courtesy of Agricultural Research Service, USDA.)

having to keep a colony of insects to manufacture the viruses. This is analogous to using mammalian cells for commercial-scale production of human vaccines as discussed in chapter 19.

The greatest capacity for broadening the use of biocontrol in agriculture comes from using crop genes as the biological agents that suppress pests. For over a century plant breeders have purposefully developed disease- and insect-resistant crop varieties by crossing cultivars that are locally adapted to their unique environments with other plants that have the desirable pest resistance genes. The plants that donated the resistance gene might be other commercial cultivars; primitive cultivars, which are known as **landraces**; the wild ancestor of the crop plant; or a close relative of the wild ancestor. Laboratory techniques that were developed during the past century (see Table 21.10) widened the crop's potential gene pool and gave breeders access to a larger, but finite, number of resistance genes. In the case of wheat, breeders found useful pest resistance genes in 20 different species, both wild and domesticated, within five genera. Creating new genes by mutagenesis also increased the number of available resistance genes. The new recombinant DNA technology expands the boundaries of the accessible gene pool to encompass resistance genes found in almost any species. Just as important, because of the universality of life's genetic code, pest-resistant genes developed for major crops, such as corn and wheat, can be moved into crops that have not received much research attention, such as subsistence crops of the developing world.

The natural defense systems of plants

In chapter 19, we described various ways the new molecular techniques allow physicians to use the human body's natural defense mechanisms. An analogous approach for protecting crop plants relies on self-defense mechanisms in plants. Plants have two endogenous defense systems, and each is equipped with a diverse arsenal of molecular weapons for protection against disease and herbivory. One branch, the **hypersensitive response** (HR), is a localized response to infection. Plants respond to molecules from the infecting pathogen through ROS, which in turn trigger at least three biochemical responses.

1. The ROS cause cross-linking of molecules in the plant cell wall, which reinforces the plant's physical protection against pathogens.
2. The programmed cell death cascade (described in chapter 9 for animal cells) is initiated in the infected plant cell. When the infected cell commits suicide, the pathogen is unable to survive and reproduce, thwarting the pathogen's spread to other cells.
3. ROS act as warning signals to cells surrounding the dying cell, which stimulates the activation of a second defense system, the **systemic acquired response** (SAR). In the SAR, uninfected cells begin to biochemically equip themselves for the coming onslaught.

Herbivory also triggers the SAR, independent of the HR, because herbivores have molecules in their saliva that activate the SAR pathway. Irrespective of the triggering event, the SAR leads to the production of a variety of molecules, such as:

- secondary plant compounds that deter herbivory and kill pathogens
- proteins that inhibit the activity of an insect's digestive enzymes
- enzymes, such as chitinase, that degrade fungal cell walls
- proteins that inhibit protein translation in the pest
- various antiviral compounds that degrade viral RNA, block cell entry, or stop replication

All plants seem to have the genetic infrastructure for both the HR and SAR defense mechanisms, but like drought tolerance, the ability to respond to smaller pathogen loads or less herbivory makes some plants less susceptible to pest attacks. Scientists isolated the gene that is the SAR master switch, and not surprisingly, its overexpression enhances disease resistance. However, as is true of the water stress response, continuous expression of the entire constellation of molecular defense weapons leads to yield reduction. Scientists are attempting to improve endogenous pest resistance by giving crops extra copies of only one or two genes encoding a SAR defense molecule.

The SAR and HR are inducible by triggering molecules generated by the attacking organism, which opens a new avenue for crop protection. Spraying the triggering molecules, in the absence of pathogens or herbivores, activates the plant's own defense systems (Figure 23.12). In addition to the triggering molecules secreted by pathogens and herbivores, volatile compounds made

Figure 23.12 Triggering molecules for self-defense. Grapefruit leaves that were sprayed with naturally occurring molecules that trigger the plant's self-defense systems defended themselves against fungal diseases. Those that were not sprayed were infected. (Photograph by Keith Weller, courtesy of Agricultural Research Service, USDA.)

by virtually all plants also activate the SAR. These volatile compounds provide additional alternatives for triggering molecules but also raise an interesting question. If pathogens and herbivores do not secrete these compounds, then why do long-distance signaling molecules, secreted by other plants, trigger the plant's defense system?

Plants under attack release these chemicals, which travel by air and bind to cells of other plants, including those in different species. The binding of the signaling molecules to the plant cells triggers the SAR. In other words, plants alert other plants in the neighborhood to the presence of a predator or pathogen, and plants receiving the signal muster their arsenal of defensive molecules before they are attacked. How cool is that? The story gets even better. Some insects that feed on other insects are sensitive to these chemicals, and they also respond to the plants' distress calls. The volatile signaling molecules alert predators and parasitoids to insect herbivores nearby and guide them to their next meal (Figure 23.13). Even more amazing, the plant's volatile signals are often species specific. Parasitoid wasps can be very selective about the insect species they parasitize. The exact mixture of chemicals that the plant secretes changes according to the insect pest species that is eating it so that the *correct* wasp species receives the plant's SOS.

Figure 23.13 Recruiting parasitoids. When plants are attacked by herbivores, they secrete volatile compounds that trigger the self-defense systems of other plants and also alert parasitoid wasps and insect predators. This parasitoid wasp, which is only 1/4 inch long, is laying an egg in an insect pest. (Photograph by Scott Bauer, courtesy of Agricultural Research Service, USDA.)

TRANSGENIC CROPS AND SUSTAINABILITY

Have the transgenic crops that are currently on the market improved sustainability? The scorecard on sustainability for the first generation of transgenic crops is mixed. In some ways, they are conducive to sustainable agriculture, and in other ways they are not.

The first transgenic crops commercialized were genetically modified for better pest management and included varieties with new genes for resistance to insects, herbicides, or diseases. In every country that has granted commercial approval for a transgenic variety, the farmers have rapidly adopted this new technology. In 1996, which was the first year transgenic crops were grown on a commercial scale, six countries grew 4.6 million acres (1.7 million hectares) of transgenic crops, with the United States and China accounting for 90% of that acreage (Figure 23.14). By 2003, 7 million farmers in 18 countries grew 167 million acres (67.7 million hectares) of transgenic crops (Table 23.3). Most of that acreage was in only six countries—two industrialized countries and four developing countries—and the developing countries accounted for almost one-third of the 2003 transgenic-crop acreage (20.4 million hectares). From 2002 to 2003, the growth in the number of transgenic crop acres was essentially identical in industrialized and develop-

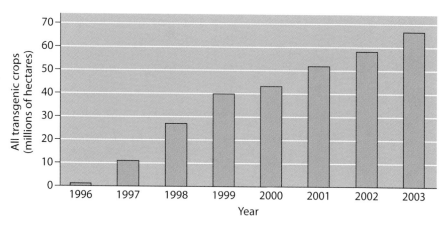

Table 23.3 Countries that grew transgenic crops commercially in 2003

Country	No. of hectares (millions)
United States	42.8
Argentina	13.9
Canada	4.4
Brazil	3.0
China	2.8
South Africa	0.4
Australia	0.1
India	0.1
Romania	<0.1
Uruguay	<0.1
Spain	<0.1
Mexico	<0.1
Philippines	<0.1
Colombia	<0.1
Bulgaria	<0.1
Honduras	<0.1
Germany	<0.1
Indonesia	<0.1

Figure 23.14 Global area of transgenic crops. The amount of cropland planted in transgenic varieties has increased every year since the first transgenic crop became available for large-scale commercial use in 1996. One hectare is approximately 2.47 acres. Therefore, the total number of acres planted in transgenic crops in 2003 was 167.2 million acres.

ing countries (Table 23.4), but the percent increase was significantly higher in developing countries (28 versus 11%). The global market value of transgenic crops in 2003 was approximately $4.5 billion.

Since their introduction in 1996, the number of acres devoted to transgenic crops has increased 40-fold. This rate of adoption of a new agricultural technology is unprecedented, because typically farmers are conservative about changing crop production practices. Faced with so many risks outside of their control, they try to reduce the uncertainty of farming by sticking with practices they know work well. The high adoption rate says one thing for certain: transgenic crops must be more profitable than comparable conventional varieties even though the seeds cost significantly more. If they weren't, farmers would not continue to buy them. Either the yields of transgenic crops surpass those of their conventional counterparts or they require fewer inputs, such as water and chemicals, or both. A number of independent analyses of U.S. farmers' yields and expenditures demonstrated an average economic gain of $16/acre for the three transgenic crops with the greatest acreage, which are the commodity crops soybean, field corn, and cotton (Figure 23.15). This amount may not seem impressive until you consider that the sizes of more than 70% of the U.S. farms growing commodity crops exceed 2,000 acres. In 2001 and 2002, the annual economic gains of transgenic varieties for U.S. farmers were $1.5 billion and $1.9 billion, respectively.

Have all farmers who have grown transgenic crops benefited economically every year? No. In some years, at some locations, yield differences or

Table 23.4 Number of hectares of transgenic crops grown in industrialized and developing countries from 1997 to 2003[a]

Countries	Global area in transgenic crops (millions of hectares)					
	1997	1998	1999	2000	2001	2002
Industrialized	9.5	23.4	32.8	33.5	39.1	42.7
Developing	1.5	4.4	7.1	10.7	13.5	16.0
Total	11.0	27.8	39.9	44.2	52.6	58.7

[a]Data provided by the International Service for the Acquisition of Agri-Biotech Applications (http://www.ISAAA.org).

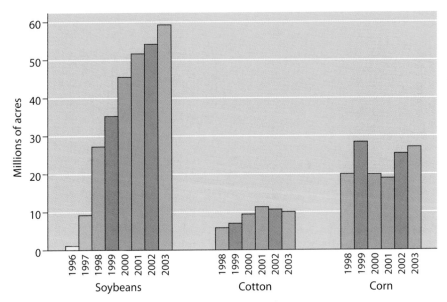

Figure 23.15 Acres of transgenic crops grown by U.S. farmers.

the cost of other inputs did not compensate for the higher seed costs. Farmers expect this sort of year-to-year variation, especially for some insect-resistant crops, because as you recall from the story of monarch butterflies and Bt corn, insect outbreaks can be much more difficult for farmers to predict accurately.

For the first generation of transgenic crops, we believe a good case might be made for improved sustainability if a transgenic crop variety had the same or greater yields than its conventional counterpart with fewer inputs of resources, such as water, fertilizers, or pesticides. This was the case for transgenic varieties of corn, soybean, canola, and cotton (Table 23.5). Even so, these crops have come under fire from many quarters for a variety of reasons.

Criticism: herbicide-tolerant crops keep farmers on the "chemical treadmill"

This is true. Herbicide-tolerant crops allow farmers to use herbicides to control weeds, but this is not a new development brought about by biotechnology. U.S. farmers have used herbicides on 95% of the acres planted in soy, corn, and other commodity crops since the 1970s. Most fruit and vegetable crops are also treated with herbicides. Is this necessarily bad?

Table 23.5 Annual impacts of transgenic crop varieties on yields and pesticide use in the United States for 2001[a]

Transgenic crop	Reduction of pesticide use[b]	Yield difference[c]
Corn	8,408,456	+3,540,992
Cotton	8,039,100	+185,373
Soybean	28,703,001	No difference
Canola	531,000	No difference

[a]Data provided by the National Council for Food and Agriculture Policy (http://www.ncfap.org).
[b]Pounds of active ingredient per year.
[c]Between transgenic and nontransgenic varieties (pounds).

Farmers must control weeds, and they have a number of available options. Herbicides are very expensive, and if farmers can avoid that expense and choose a cheaper weed management option, most will. Unfortunately, one of the most effective alternatives, soil tillage, causes significant soil erosion, while herbicides permit no tillage, which decreases erosion. Minimizing erosion is a goal of sustainable agriculture. This is a perfect example of the trade-offs that must sometimes be made in managing systems for sustainability. In addition, not all herbicides are created equal. Some have very low toxicities, do not persist in the soil or leach into groundwater, and are biodegradable. Therefore, in addition to improving sustainability by decreasing tillage and soil erosion, if an herbicide-tolerant transgenic crop has allowed farmers to switch to herbicides with lower toxicities, greater environmental compatibility, or acceptable performance in smaller amounts, then those crops have improved sustainability by minimizing negative environmental impacts and decreasing the use of nonrenewable fuels and feedstocks.

Criticism: society does not need crops with higher productivity because there is already enough food for everyone on the planet

Let's assume this argument is correct and current crop yields are sufficient to feed everyone. Increased productivity does not necessarily mean more product. Productivity is a measure of the ratio of output to input, or yield/input. If society already has enough of an agricultural product (yield) and has no desire for more, genetically improving productivity lets farmers decrease inputs needed in its production. In other words, if the productivity of soybeans is improved, farmers can get the same amount of soybeans from fewer acres or from smaller inputs of water and chemicals.

As you can see in Figure 23.16, when the Green Revolution's genetically improved, high-yielding varieties became available to farmers in the middle of the 20th century, the amount of cropland stayed essentially constant even though the amount of cereal grains the world produced more than doubled.

Therefore, by allowing farmers to produce more on less land, transgenic crops with higher productivity can improve sustainability by decreasing deforestation and returning cropland to its natural state, both of which offer protection from soil erosion.

Figure 23.16 Cropland acreage changes from 1700 to 2000. (Data courtesy of the United Nations Food and Agriculture Organization.)

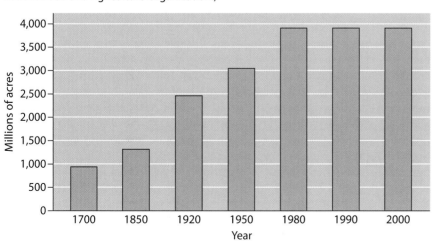

Criticism: these crops encourage farmers to continue farming in the same way, with large acreages devoted to very few crops

This criticism is right on target. The first wave of transgenic crops did nothing to broaden the diversity of crops that U.S. farmers grow. Crop diversification has been shown repeatedly to improve agricultural sustainability, and the transgenic crops currently on the market did not help U.S. farmers break out of the "big commodity" box. What is so discouraging about this failing is that biotechnology's versatility provides remarkable potential to increase the number of crop options available to farmers. Why haven't crop developers broadened the options? The answer is complex, but it brings home the point that many factors other than science and technology influence which transgenic crops get developed from the universe of potential candidates.

For example, markets affect which crops, transgenic or not, get developed. Companies need products they can sell, especially if they spend millions of dollars researching and developing the product. The large seed companies that produced the first wave of transgenic crops concentrated their research and development activities on the highest-volume crop U.S. farmers grow. With or without modern biotechnology, the major agribusiness companies that conduct research and new product development consistently choose this business strategy. Unlike the pharmaceutical industry, which can charge very high prices for its products, agriculture's profit margins are minuscule. Companies can recoup the costs of research and development only by selling large volumes of seeds, which means commodity crops.

Thus, our question about the frustratingly narrow, unimaginative direction of transgenic-crop development can be reframed as a question about why U.S. farmers are wedded to growing large acreages of a few commodity crops. The reasons are complex and varied, but for the purposes of our discussion, one reason is sufficient: for over 50 years the federal government has rewarded farmers for growing only one or a few commodity crops. In general, the various farm subsidy programs were structured so that a farmer's reward was directly related to yields, so farmers tried to grow as much as they could of the crop they were being paid to grow. Large yields meant large acreages. Large acreages required large equipment and massive irrigation systems, which required large capital investments, financed by borrowing. In order to make loan payments, farmers needed to keep those large yields and government subsidies flowing. Thus, many farmers became boxed in, and the federal government's agricultural policies regarding subsidies, exports, and lending had built the box.

If companies with research and new product development divisions are under financial pressure to limit their focus to large-volume commodity crops grown by most farmers (not only in the United States but worldwide), then who is responsible for research that improves other, noncommodity crops? Traditionally, genetic improvement of "small-acreage crops," such as fruits and vegetables, has been the responsibility of public research institutions (e.g., land grant universities and the U.S. Department of Agriculture [USDA]) and other nonprofit organizations (e.g., the Rockefeller Foundation). Over the past century, crop scientists at these institutions have used traditional plant breeding and mutagenesis to improve the yields and pest resistance of small acreage crops. Recently, they began incorporating the new tools of biotechnology into their crop improvement research. They have made remarkable progress in isolating and characterizing useful genes and incorporating these genes into small-acreage crops by using recombinant DNA technology.

Unfortunately, however, public research institutions and nonprofit organizations cannot afford the cost of obtaining regulatory approval for transgenic crops, nor can the small seed companies that might be interested in transferring new improved transgenic-crop varieties from research to the marketplace. Current estimates of the regulatory costs associated with the risk assessments, product approval process, and postmarket monitoring of a new transgenic crop variety in the United States are approximately $3 million to $10 million, depending on the crop. Note that this estimate covers only the regulatory costs and not the additional costs of research, development, seed production, and marketing. Therefore, in spite of the remarkable potential of recombinant DNA technology for improving all crop varieties, the only organizations able to afford the regulatory costs for testing transgenic crops are large multinational corporations, and as we explained, the only crops they can afford to be interested in are large-scale commodity crops.

This is a perfect example of the hidden costs of regulation and the risk of technology versus the risk of no technology discussed in chapter 18. In terms an evolutionary biologist would use, our society has created a regulatory environment that selects *for* large seed companies with deep pockets and large-scale farmers who grow huge volumes of a few commodity crops and *against* small seed companies, small-scale farmers, and crop diversification on large farms.

Is the idea that biotechnology could be used to improve small-acreage crops a valid one? Often the nature of a technology limits its usefulness to a certain group. For example, the agricultural technologies of the past century definitely favored the owners of the largest farms, because they could borrow money (using their land as collateral) to make large capital investments in the technologies, such as irrigation systems and expensive farm machinery.

The technology behind transgenic crops is in the seed. The seeds cost more, but no huge capital expenditure is required to be able to access the technological benefits. A number of studies have shown that the economic benefit of growing transgenic crops does not accrue only to farmers with very large acreages in industrialized countries. These same studies have shown that often the farmers enjoying the greatest per capita increase in income are small-scale farmers in developing countries who have fewer than 10 acres and an annual income under $1,000.

Transgenic crops may not widen the gap between large- and small-scale farmers like earlier technologies, but they also will not automatically reverse the trend toward increasingly large farms. If a society was committed to helping small-scale farmers stay in business, could biotechnology help achieve that goal?

The universe of possible transgenic crops includes minor crops that would improve the profits of small-scale farmers and also improve sustainability by giving farmers with large acreages viable options for diversifying crop production. However, as explained above, the high cost of getting regulatory approval to market those crops precludes their development. In chapters 18 and 21, we described some of the scientific and technical information transgenic crop developers must submit to regulatory authorities for assessing environmental and health risks. Conducting the laboratory and field tests to acquire this information leads to the high cost of regulatory approval. The degree of regulatory oversight (and therefore, regulatory costs) should be proportional to the degree of risk. Therefore, if a society wants its small farmers to succeed and all farmers to improve sustainability through crop diversification, then it must ask if the cost of no technology—no transgenic small-acreage crops—is warranted by the potential risks of the technology.

Below, we discuss in detail only two of those potential risks, gene flow and the evolution of resistance. In chapter 18, you learned that the first step in a risk analysis is a risk assessment, which begins by identifying the risks. Government regulatory agencies must dissect a large, ill-defined risk, such as gene flow, into its component parts so they can focus their resources on those instances where the risk is real. They approve a product if they determine it is safe and the benefits outweigh the costs. However, as you also learned, science can never prove a product is 100% safe, and even the safest products are not risk free. Regulatory authorities often approve a product but place restrictions on its sale or use in an attempt to minimize any risks associated with it. Therefore, the second step in a risk analysis is developing risk management strategies.

To help you understand the thought processes that regulators use in assessing risk, we explain the types of scientific questions about gene flow that regulatory authorities must be able to answer satisfactorily before they grant approval for commercial production of a transgenic crop. This list of questions also dictates the data requirements and consequently the testing costs that must be incurred by the crop developer. To explain the role that regulators play in managing known risks, we use the example of the evolution of resistance. Risk management requirements may also increase the cost of a product if the product developer must monitor how the product is being used to ensure their consumers are complying with the conditions of use. Often, the regulatory agencies are responsible for the monitoring requirement, and in those cases, the cost is borne by taxpayers.

As you read the next section, you will learn more about plant reproduction, population genetics, ecology, and evolution, which is our primary intention. While you are learning more biology, we also want you to be practicing the method for analyzing issues outlined in chapter 17. In addition, think about the various questions and the data that are necessary for answering them. The regulatory approval processes discussed below are mechanisms for protecting the environment and agriculture, but the regulatory testing requirements can also impede the development of crops that could help small farmers and improve agricultural sustainability through crop diversification. Looking at them through two different filters—impeding product development and protecting the environment and agriculture—should illuminate the contrast between the risk of a technology and the cost of no technology described in chapter 18.

GENE FLOW FROM CROPS TO OTHER PLANTS

The use of transgenic crops has triggered questions about the possibility that the transgene will move to other, nontransgenic plants, both wild plants and crop plants, via pollination. Gene flow between different plants can occur with any crop. Indeed, there is abundant evidence that it occurred with other crops long before modern biotechnology appeared on the scene. Therefore, the answer to the requisite question of whether this issue is unique to modern biotechnology is no.

Define the issue as specifically as possible

The moniker "the gene flow problem" encompasses two very different problems: gene flow to wild plants and gene flow from a transgenic crop variety to a comparable nontransgenic variety. These two issues differ in all relevant dimensions that characterize a risk analysis: the probability that gene flow will occur, the nature of the adverse effect if it does, and the methods to lessen the probability or mitigate the adverse effects. The principles guiding the risk assessment for gene flow are the same, however, because the formula for risk level (risk = hazard × exposure) applies to both.

The first question a risk assessment must address is the probability that the gene will move from a transgenic crop to unintended recipients (exposure); the second is deciding whether it matters if it does (hazard, or adverse effect). Gene flow can occur without having adverse effects. If the answer to the first question is zero, then there is no need to ask the second. If the answer to the second is no, the hazard is essentially zero and gathering data to estimate the risk level is unnecessary. If, however, the answers to both questions are positive, then there is a hazard and a possibility of exposure and therefore some level of risk. The job of the regulator is to define the nature, probability, or severity of the risk as specifically as possible by asking the following questions.

- Who will be adversely affected—the farmer growing the transgenic crop, neighboring farmers, food processors, or consumers?
- What is the nature of the adverse impact—agricultural, environmental, or economic?
- Is the adverse impact easily manageable?
- Does the potential adverse effect override the potential benefits?

The answer to each of the above questions varies with the crop, the recipient plant, the environmental factors, and the nature of the trait encoded by the transgene. Therefore, to thoroughly assess the issue of gene flow, regulators must evaluate each situation systematically, on a case-by-case basis.

Gene flow depends on cross-pollination

Whether the unintended recipient of the gene is a wild plant or nontransgenic crop, the probability of gene flow from a transgenic crop depends first and foremost on the potential for cross-pollination between the transgenic crop (pollen donor) and the other plant. Pollination includes transfer of pollen and subsequent fertilization of the recipient plant's eggs. If transgenic pollen simply lands on the stigma of the recipient plant but fertilization does not occur, there can be no gene flow. For cross-pollination to occur, viable, mature pollen must reach the recipient's viable, mature eggs before other pollen fertilizes the eggs. Many factors affect the probability of successful cross-pollination (Table 23.6).

At this point in the risk assessment, the recipient plant—wild plant or crop—begins to play a significant role, which affects the questions asked and judgments about risk levels. To keep the discussion to a manageable size, we will limit our focus to gene flow to wild plants.

Assess exposure: the probability of gene flow from crops to wild plants

Pollen transfer and successful fertilization are necessary, but not sufficient, for gene flow to occur because there are many postfertilization barriers to reproduction if the two plants do not have a sufficient amount of genetic similarity. Gene flow between crops and wild relatives can occur only if the postfertilization barriers are overcome and the resulting offspring are not only viable but also fertile. Cross-pollination between two different plants that produces fertile, viable offspring is known as **hybridization**. To hybridize with each other (in the absence of human intervention), plants must be very closely

Table 23.6 Factors that affect the probability of cross-pollination

Physical proximity of the plants to each other

Tendency to self-fertilize

Pollen longevity, which can be minutes for some crops and hours for others

Degree of synchronicity between pollen shedding and recipient's receptivity to fertilization

Relative amounts of pollen produced by the two plants

Pollination vector of the two plants (wind, insects, or both)

Table 23.7 Probable centers of origin for some important food crops

Central America	North America	Europe
Maize	Sunflower	Oats
Common bean	Blueberry	Cabbage
Papaya		Sugar beet
South America	**Central Asia**	**Southern Asia**
Potato	Soybean	Asian rice
Peanut	Onion	Banana
Sweet potato	Alfalfa	Sugarcane
Peppers	Peach	Citrus
Tomato	Apple	
Africa	**Middle East**	
Sorghum	Wheat	
Pearl millet	Barley	
African rice	Rye	
Cowpea	Pea	
Coffee	Lentil	
	Plum	

related to each other. Therefore, gene flow between crops and wild plants is not possible if no wild relatives of the crop occur in the near vicinity. **Centers of diversity** are specific geographic locations that have many wild relatives of a crop. Each crop has its own center(s) of diversity, which is usually, but not always, the **center of origin** of that crop (Table 23.7). The center of origin is the geographic area where the crop was domesticated from wild relatives.

In addition, reproductive compatibility is often asymmetric. Gene flow from the wild relative to the crop plant typically occurs much more often than the reverse. Demonstrating that cross-pollination and hybridization are possible if the wild relative provides the pollen does not guarantee that the reverse is true.

In summary, in assessing the risk of gene flow from a crop to wild plants, the first question regulators must answer is whether the crop will be planted near wild relatives with which it could hybridize successfully. If the answer is no, then exposure is zero and the gene flow risk assessment is finished—for that crop in that location. If the answer is yes, then the regulators must determine the probability of gene flow occurring by systematically evaluating the factors that affect the potential for the three steps required for gene flow: cross-pollination, fertilization, and production of viable seed. To illustrate the necessity of looking at each crop on a case-by-case basis to assess the probability of gene flow, it helps to use specific crop examples. We have chosen corn, soybean, and canola. As in the step-by-step analysis of Bt corn risks to monarch butterflies in chapter 18, we include the details only so that you can see the complexities, nuances, and qualifications contained in the broad issue of gene flow.

Corn and the teosintes

The center of origin and center of diversity of corn are the same: Mexico and Central America. Gene flow from corn to wild plants is possible only in Mexico and Central America, because corn's ancestor and close relatives, the teosinte species, are restricted to those areas. Assessing the risk of gene flow from corn is unnecessary outside of Mexico and Central America.

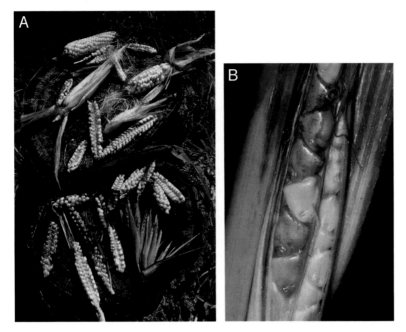

Figure 23.17 Corn and teosinte. In Mexico, the center of origin for corn (or maize), corn **(A, upper left)** readily hybridizes with its wild relatives, the teosintes **(A, lower right, and B)**. (See Figure 1.6 for the relative sizes of modern corn and teosinte.) (Photographs copyright Klaus Ammann.)

Gene flow from corn to some of the teosintes can occur. In fact, botanists have been studying it for a century (Figure 23.17). A corn plant can either self-pollinate or cross-pollinate, and pollen dispersal is driven by wind speed and direction. Because corn pollen does not stay airborne for very long, the probability of cross-pollination is directly related to the distance between the crop and cross-compatible teosintes and the number of teosinte plants near the cornfield. Cross-pollination can lead to successful hybridization, but the success rate varies according to the direction of the hybridization. A number of studies have shown that genes are more likely to move from teosinte to corn than in the opposite direction.

Soybean and wild soy relatives

The center of origin for soybeans is northeastern China, and its wild relatives are found only in central and northeast China, Taiwan, and certain regions of Siberia, Korea, and Japan. Therefore, gene flow from soybeans to wild plants is possible only in these areas. How probable is gene flow between soybean and its wild relatives? Very low. Soybean and its relatives are self-fertile and self-pollinated. Pollination typically occurs before the flower even opens, in which case foreign pollen cannot reach the plant before it fertilizes itself. Cross-pollination, which depends on insects, and hybridization are possible, but not probable. Scientists trying to maximize the probability of hybridization between soybean and its relatives by using controlled laboratory conditions have a difficult time creating hybrids.

Canola and its wild relatives

Gene flow from canola is much more complicated for a number of reasons. First of all, there is no such thing as a canola plant. Canola is a special type

Table 23.8 Using pre-recombinant DNA genetic modification techniques to remove the noxious, long-chain fatty acids eicosenoic (20:1)[a] and erucic (22:1) acids from *Brassica* species and to increase amounts of the healthy fatty acid oleic acid (18:1)

Genetic modification technique	Fatty acid composition (%)							
	16:0	18:0	18:1	18:2	18:3	20:1	22:1	24:0
Original *Brassica*	4	2	17	13	9	14	41	1
Conventional plant breeding	4	2	62	20	9	2	<1	<1
Mutagenesis breeding	5	2	64	23	<3	<1	1	<1
Ovule mutagenesis and culture[b]	5	2	85	3	3	1	<1	<1

[a]This is the conventional form that biochemists use to represent fatty acids. The first number refers to the number of carbons in the chain, and the second number indicates the number of double bonds. Therefore, 16:0 is the 16-carbon saturated palmitic acid and 18:2 is the 18-carbon polyunsaturated linoleic acid.

[b]Ovule culture is one of a number of laboratory techniques that use plant reproductive structures as the starting material. Plant breeders use tissue culture to convert the reproductive structure into a callus, which eventually differentiates into a whole plant. Because pollination does not occur, plants derived from ovules, anthers, and pollen are haploid, as are the seeds they produce.

of oil that is produced by any of three *Brassica* species, *Brassica napus*, *Brassica rapa*, and *Brassica juncea*. Interestingly, none of these species produced canola oil before plant breeders set to work on them in the 1970s, so there was no such thing as a canola crop until then. In nature, these three species have large amounts (25 to 40%) of undesirable long-chain fatty acids, as well as high levels of sulfur-containing compounds that cause goiters and other health problems. As a result, none of the species was fit for human consumption. Using a host of techniques, plant breeders developed varieties of these three plants that have high percentages of healthy fatty acids, such as oleic oil; less than 2% of the nutritionally undesirable oils; and minuscule amounts of the sulfur-containing compounds (Table 23.8).

Because the question of gene flow now involves three crop species (collectively known as the oilseed *Brassica* crops) instead of one, as you might expect, the number of locations with wild relatives is much greater. Various studies describe wild *Brassica* relatives in Canada, Australia, the United States, a number of countries in Africa and Central and South America, and the Mediterranean countries, and there may well be more locations where wild relatives of *Brassica* occur. Out of prudence, regulators must assume that at any location where canola is grown, a sexually compatible wild relative might be present.

The oilseed *Brassica* species are both wind and insect pollinated. *B. rapa* is self-incompatible, so it must be cross-pollinated. *B. napus* and *B. juncea* are normally self-fertile, and on average, 80% of the seed arises from self-fertilization. *Brassica* pollen is smaller than corn pollen, but like that of corn, a large proportion of the pollen falls to the ground within a few feet of the field. Nonetheless, a small percentage (5 to 10%) becomes airborne and under ideal conditions may drift 2 to 3 miles away. Insects can transport *Brassica* pollen long distances as well. All of these factors combine to make the possibility of pollen flow between an oilseed *Brassica* and a wild relative quite high.

Pollen movement does not constitute gene flow, however. Numerous scientists have studied the potential for hybridization between the three oilseed *Brassica* species and various wild relatives. Most studies were conducted in the laboratory or under field conditions that maximized cross-pollination by blocking the self-fertilization that usually occurs in the self-compatible *Bras-*

Table 23.9 Gene flow between the oilseed *Bassica* species and some wild relatives[a]

Laboratory—hand pollination with wild relatives *Sinapsis arvensis* and *Brassica nigra*

Pollen donor	Pollen recipient	Results
B. rapa	S. arvensis	No seeds
B. napus	S. arvensis	No seeds
B. juncea	S. arvensis	No seeds
S. arvensis	B. rapa	No seeds
S. arvensis	B. napus	No seeds
S. arvensis	B. juncea	2.5 seeds/100 pollinations (all sterile)
B. rapa	B. nigra	No seeds
B. napus	B. nigra	0.1 seed/100 pollinations
B. juncea	B. nigra	0.5 seed/100 pollinations
B. nigra	B. rapa	No seeds
B. nigra	B. napus	0.9 seed/100 pollinations
B. nigra	B. juncea	3 seeds/100 pollinations

Field experiments—cocultivation

3 yr within the same field

All oilseed *Brassica* spp. and *S. arvensis*—no hybrids

All oilseed *Brassica* spp. and *B. nigra*—no hybrids

Between two varieties of one species in fields that were 46, 137, and 366 m apart

	Outcrossing rate (%)		
	46 m	**137 m**	**366 m**
B. napus (self-compatible)	2.1	1.1	0.6
B. rapa (self-incompatible)	8.5	5.8	3.7

[a]Gene flow can occur with hand pollination or natural pollination but is rare under both laboratory and field conditions.

sica species. Therefore, these studies maximized gene flow estimates because they were conducted to determine if gene flow from *Brassica* to a wild relative is possible. As you can see from experiments on only two wild relatives (Table 23.9), no simple generalizations can be made about the possibility of gene flow from "canola" to its many wild relatives.

Few studies representative of natural field conditions have been conducted to assess the *probability*, not the *possibility*, of gene flow from the oilseed *Brassica* species to wild relatives. All show that hybridization can occur with some wild relatives, but it seems to be rare and the direction is asymmetric. For example, French researchers planted a wild relative (wild radish) throughout a *Brassica* field and along its borders, which maximized opportunities for cross-pollination. They found that 1 of 189,084 (5.3^{-6}) wild radish seeds contained a marker gene from the crop, while the genomes of 5 of 73,847 (6.7^{-5}) *Brassica* seeds revealed evidence of gene flow from wild radish to the crop. These experimental results coincide with 50 years of observations by Canada's commercial and pedigree seed producers, who report seeing frequent crosses between two of the oilseed *Brassica* crops (*B. napus* and *B. rapa*) but no evidence of gene flow from wild relatives that commonly grow around *Brassica* production fields.

Assess the hazard: the nature of adverse impacts

Gene flow (successful pollination plus fertile hybrids) does not necessarily lead to adverse environmental impacts. In order to pose an environmental problem, the crop genes must become established in the wild plant population and enhance the hybrid's ability to outcompete wild plants lacking the crop gene and to leave more offspring. Reproductive output is evolution's bottom line and the essential determinant of the potential for adverse impacts. As we noted above, gene flow from crops to wild relatives has been occurring for centuries. At times, this has made the farmer's life more difficult if a weedy wild relative has become a worse weed. We could not find reports of crop genes transforming a wild relative into a "superweed" that has either caused the extinction of a wild plant or given an herbivore unlimited food resources, leading to an explosion of the herbivore population. Perhaps this has occurred, but if so, plant scientists have not studied it or reported it in the literature.

If crop genes have been transferred to wild relatives for centuries, then why hasn't gene flow led to serious ecological problems that are significant enough to be identified and investigated? After all, for a century plant breeders have intentionally incorporated genes that improve survival and reproductive success—abiotic stress resistance, pest resistance, and higher yields—into crops. These genes, if transferred to a wild plant, could increase its fitness. The following are a few possible explanations for the apparent lack of serious ecological impacts caused by gene flow from crops to wild relatives.

Because gene flow from crops to their wild relatives is a relatively rare event, crop genes may be lost from the wild plant population due to genetic drift. This is especially true if:

- the crop genes offer no selective advantage to the hybrid
- the hybrid can no longer interbreed with its wild parent
- a crop gene offering an advantage is linked to other crop genes that handicap the hybrid

For example, you recall that one of the genetic characteristics early farmers insisted on was no seed dispersal. Every time corn hybridizes with teosinte, the hybrid offspring loses its ability to disperse seeds, because "no seed dispersal" is a one-gene dominant trait. This places the hybrid at a distinct disadvantage, because populations of the wild parent plant disperse their seeds widely.

On the other hand, if the hybrid interbreeds freely with its wild parent and produces fertile offspring, the crop gene can become established at a low frequency in the wild plant population even if it offers no selective advantage at that point in time. This **introgression** of crop genes into a wild plant population does not necessarily constitute an adverse ecological impact. Low levels of introgression have surely occurred between some crops and wild relatives for centuries. The key determinant of adverse effects is not simply the *presence* of a crop gene in a wild plant population.

Adverse environmental impacts become much more likely if the hybrid plants survive better and reproduce more than wild plants lacking the crop gene(s). The potential to increase fitness depends upon the nature of the transferred crop genes and certain traits of the recipient plant, as well as the population characteristics of the wild plant population that now has new genes. The potential of gene flow to cause adverse impacts is circumscribed by the ecological factors limiting the wild plant population size at that time.

For example, at first glance, a gene for insect resistance might appear to offer a competitive advantage (less herbivory) that would automatically lead to a population explosion of the wild plants with the gene. But if the factor limiting the size of the wild plant population is water availability or a plant pathogen, the population will continue to be held in check by those factors irrespective of this new advantageous gene for insect resistance. If insect herbivory decreased the fitness of that wild plant, then acquiring a gene for insect resistance could improve survival and lead to higher reproductive output of certain individuals. The gene frequencies within the population would change without the population size increasing. A change in gene frequencies, however, can have impacts down the road if it alters the population's ability to respond adaptively to new selective pressures.

Herbicide tolerance and superweeds

The public seems to have focused most of its concern about gene flow from transgenic crops to wild relatives on the new herbicide-tolerant varieties. People fear that a wild plant will acquire herbicide tolerance genes and become an uncontrollable superweed that is resistant to that herbicide. For 50 years, U.S. farmers have been growing herbicide-tolerant crop varieties, which allowed them to decrease soil erosion by using herbicides rather than tillage to kill weeds. Many weed species in crop fields have naturally evolved resistance to one or more herbicides. In becoming resistant to one herbicide, a weed does not become resistant to all herbicides, so farmers have responded to resistant weeds by selecting a different herbicide from the 100 or so that the Environmental Protection Agency (EPA) has approved. Interestingly, in virtually all cases where the biochemical mechanism underlying the resistance trait is understood, the ever-adaptable weeds have evolved the resistance trait on their own and not by acquiring genes from the herbicide-tolerant crops with which they have coexisted for decades.

Irrespective of the way the weeds became resistant—gene flow from crops to weeds or selective pressure on genes in the weed population—the essence of the problem remains the same. If gene flow from a transgenic herbicide-tolerant crop to a wild plant occurs, the solution to the problem is the same: change to a different herbicide.

Even though the public's imagination has been caught up in the specter of superweeds, we are less concerned with herbicide tolerance than we are with the impacts of plants acquiring genes for other traits. Society has abundant experience controlling plants that become resistant to a few herbicides; controlling plants that acquire a gene for drought tolerance when their population has been limited by water availability would be much more difficult. In recent experiments, scientists gave the Bt gene to wild sunflowers, and those plants produced significantly more seeds under natural conditions than those lacking the Bt gene. If insect herbivory limits wild sunflower populations, then acquiring the Bt gene could lead to an ecological release of that population. If this occurred, would an increase in the population size of wild sunflowers by itself constitute an adverse environmental effect? Maybe the increase in sunflower seeds would increase the number of songbirds whose primary food is seeds. Is that an adverse impact? Why or why not? Maybe the increase in sunflowers causes the extinction of a rare prairie plant, known to only a few botanists. Is that an adverse effect? Why or why not? We pose these questions not because they have a correct answer, but to help you see the complexity of the issue. Science can help us assess the probability of gene flow, but it offers society no help in assessing the value of songbirds and rare plants.

To date, the transgenes in crop plants have encoded traits similar to those incorporated by plant breeders—pest resistance, drought tolerance, and improved processing traits. Recombinant DNA technology provides the capacity to give crops genes for novel traits with which farmers, and the plants, have no experience, such as the ability to manufacture vaccines and pharmaceuticals. Assessing the risks of gene flow from these crops involves questions regulators have never tackled, and the assessment's primary focal point must shift from wild plants to nontransgenic varieties that might inadvertently be pollinated by these pharmaceutical plants. Rather than ecological impacts, the primary concern becomes human health effects. With this change comes a major shift in the risk/benefit ratio, not only for farmers, but also for consumers, especially those in developing countries who would benefit most from having crops that produce vaccines. We discuss these crops in more detail in the next chapter.

EVOLUTION OF RESISTANCE TO PEST CONTROL TACTICS

Farmers struggle continuously to protect their crops from pests—weeds, insects, and pathogens. While the crop protection battle is a constant one, specific pest control tactics must change frequently, because pests can evolve resistance to almost any control measure farmers use (Figure 23.18). Even though public discussions of pest resistance usually limit their focus to synthetic chemical pesticides, and more recently transgenic crops, pests can evolve resistance no matter how farmers try to kill them.

- The height of dandelions in a lawn decreases in response to mowing.
- Weed seeds that evolve to look like crop seeds survive the winnowing process.
- Young weed seedlings mimic young crop seedlings, and in doing so, escape hoeing.
- Insects have developed self-defense mechanisms against biocontrol agents, such as parasitoid wasps.

Figure 23.18 Many crop pests have become resistant to synthetic chemical pesticides. Over 500 insect pests are resistant to at least one insecticide, and over 100 weed species have evolved resistance to at least one herbicide.

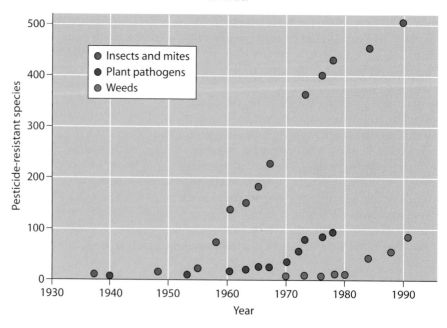

- Insects and pathogens have overcome gene-based pest resistance bred into crops by plant breeders.
- Insects, weeds, and pathogens that evolve resistance to a chemical pesticide, whether natural or synthetic, often develop **cross-resistance** to chemically similar pesticides.

For farmers to stay a step ahead of the pest, they need to have as many control options as possible. But, as you will see, even then there are no guarantees.

The corn rootworm exhibits a number of resistance strategies

The fascinating story of one pest, the corn rootworm, captures the remarkable adaptability of pests. Even though mature rootworms are only 1/2 inch long, they have earned the nickname "billion dollar bug" for the damage they inflict on U.S. corn growers ($800 million in lost yields and $200 million for insecticide treatments). The corn rootworm's name is misleading, because it is actually a beetle (Figure 23.19). The adult beetles feed on corn leaves, silks, and pollen; in late summer, they mate, and females lay their eggs below the soil surface. The eggs overwinter in the soil in a state of suspended animation called **diapause**. After hatching in early summer, the larvae begin feeding on the tender roots of new corn plants (Figure 23.20). When they reach maturity, the larvae pupate in the soil and emerge into the cornfield as adults around midsummer.

Figure 23.19 An adult corn rootworm feeding on corn silk. (Photograph by Tom Hvalty, courtesy of Agricultural Research Service, USDA.)

Historical development of the corn rootworm problem

The corn rootworm probably originated in Central America, alongside its chosen food, corn. Central American farmers, both then and now, grow corn in small plots, along with beans, squash, and other vegetables (Figure 23.21). Crop diversity, small acreages, and regular alternation of crops grown in particular fields probably kept the corn rootworm population from exploding and warranting pest status. After the Spanish conquered Central America and began using the European system of agriculture—large acreages of a single crop and **continuous cropping**, which is planting the same crop in the same field every year—things began to change. The rootworm population increased, and they began to extend their range north and east. They first appeared in Colorado in 1910 and were causing serious economic damage in all of the corn belt states by 1970. Currently, 30 million corn-producing acres in the United States suffer from rootworm infestation. In the early 1990s, the corn rootworm was discovered for the first time in Europe, in a small cornfield next to the Belgrade airport. Eight European Union countries now have significant rootworm problems.

Figure 23.20 A corn rootworm larva. (Photograph courtesy of Agricultural Research Service, USDA.)

Figure 23.21 Corn rootworm control. For centuries, the agricultural management practices of Central American farmers prevented widespread outbreaks of the corn rootworm. Today's farmers in the Lake Atitlan area of Guatemala continue the practice of growing small plots of diverse, regularly rotated crops.

When the rootworm began to inflict significant damage in the 1950s and 1960s, farmers who were locked into continuous corn cultivation (a legacy of price support programs) had no choice but to use persistent soil insecticides to control the larvae. Unfortunately, they also sprayed the same pesticide if silk herbivory by adults was decreasing yields. Over 90% of the continuous corn growers in the Midwest used (and still use) soil insecticides in the spring as an insurance policy, because the larvae are capable of causing such great damage. It's little wonder that within 10 to 15 years, rootworms had become resistant to these pesticides (chlorinated hydrocarbons), causing farmers to switch to different types (organophosphates and carbamates) in the early 1970s. Perhaps because these pesticides are not as persistent, resistance was delayed (25 to 30 years), but inevitably farmers noticed that rootworm populations in continuous cornfields had become resistant to the new classes of pesticides.

Other farmers, free to plant different commodity crops on acres covered by the price support programs, tried to control rootworms by alternating the crops grown in a field from one year to the next, which is known as **crop rotation**. Most corn growers chose a corn-soybean rotation so they could benefit from soy's nitrogen-fixing capability. Female beetles, munching happily away on corn, descended to the ground and laid their eggs, but when their offspring emerged from winter diapause, they were greeted, unhappily, by soybean roots. The larvae cannot develop on crops other than corn, so they died soon after hatching.

Evading crop rotation control

The corn-soybean rotation provided growers with excellent rootworm control for a number of decades, but in the 1980s and 1990s, they began to observe rootworm damage on corn plants. Different rootworm populations had evolved two different strategies to sidestep the grower's crop rotation scheme. The larvae in Iowa and southern Minnesota skirted the control by lengthening their diapause from 1 year to 2. On hatching, they happened to find just the food they needed—corn roots. The solution for rootworms in eastern Illinois was much more inventive. After feeding on corn plants, adult females would fly over to the nearest soybean field to lay their eggs. Their offspring emerged to a field of corn roots the next spring. This phenomenon was first discovered in a few Illinois counties in 1992, and by 2000, rootworm populations on 5 million acres in Illinois, Indiana, Michigan, and Ohio contained females with this new egg-laying behavior.

Our word choice in describing the rootworm's adaptations may unintentionally give you the impression that rootworms made these choices with a goal in mind. Evolution is not goal oriented but opportunistic. For millennia, there may well have been a small number of females who were not rigidly programmed to lay their eggs in corn and only in corn. At first, some of those females left cornfields and chose other egg-laying sites, but they had no offspring that survived and reproduced. They were selected against until the selection pressure changed because of crop rotation. At that point, females with genes for inflexible egg-laying behavior who were wedded to cornfields were selected against, because when their offspring hatched, there was no corn in sight, only soybeans.

While evolutionary ecologists are fascinated by the rootworm's adaptability, corn growers who had been able to control this pest with crop rotation were not amused. By the late 1990s in some areas of the Midwest, they needed to apply soil insecticides to 90% of the acres of rotated corn, just like the growers who used continuous corn cultivation. According to the EPA, the rootworm is responsible for the single largest use of insecticides in the United States, because to cope with the insecticide resistance problem, many corn growers must spray three different insecticides in one growing season. To control corn rootworm, Midwest farmers apply more than 50 million pounds of formulated insecticides (approximately 25 million pounds of active ingredient) every year at a cost of $155 million. On average, approximately 50% of the total amount of insecticides applied to all row crops every year in the United States is for corn rootworm control.

Bt corn for rootworm control

The farmers' plight may be improving, however. Recently, industry scientists developed a transgenic crop variety that expresses a gene from a *B. thuringiensis* strain specifically toxic to beetles. This Bt protein differs from the one that controls lepidopteran larvae, such as the European corn borer. Field tested by the company, USDA, and university researchers for at least 5 years prior to its commercial release in 2003, this corn variety provided superior rootworm control compared to the conventional insecticide program that farmers use. In 2001, the National Council for Food and Agricultural Policy predicted the rootworm-resistant Bt corn variety would decrease insecticide use by 14 million pounds of active ingredient per year. In a few years, farmers will know how close their estimate is to the actual decrease.

Will rootworms become resistant to this control tactic as well? Given their past history, we would be foolish to say no. Humans are powerless to stop evolution or change the mechanism through which it acts—natural selection. Repeated and continuous use of any control tactic leads inevitably to the development of resistance in certain pest populations. The only way to prevent the evolution of resistance to control methods is to stop trying to control the pests, which is not a viable option. The best that farmers can do is attempt to slow evolution's course by decreasing the selective pressure they place on populations. Some control measures are easier to evolve resistance to than others. Even so, a prudent course of action means always assuming that at least a few individuals in the population have genes, randomly generated through mutation, that provide resistance to the control tactic. The rate at which those genes increase in the population depends on a number of factors:

- the number of individuals with the resistance gene when a farmer begins using the tactic
- the number of genes that encode the resistance trait
- the amount of mating that occurs between resistant and susceptible individuals
- the degree to which farmers use other control tactics
- the degree of exposure of the pest population to the control tactic

Recent studies of other Bt crop varieties provide encouraging news about the probability of pests becoming resistant to Bt. These crop varieties contain a gene from the Bt strain specifically toxic to larvae of a different insect family, the lepidopterans. After 7 years of widespread use in the United States, Australia, and China—a cumulative total of at least 130 million acres—scientists who had predicted a slight increase in Bt resistance, at the very least, were surprised to find none. Farmers cannot be certain this pattern will hold as Bt crop acreages and the number of years of exposure increase, because the scientists are not entirely sure why the predicted increase in resistance has not materialized. At least a portion of the credit no doubt goes to the **insect resistance management** plans that companies must develop and implement in order to obtain EPA regulatory approval for Bt crops.

No single resistance management plan works for all insect pests or pathogens, because each must be tailored to the control measure in question, the pest species, the cropping system and acreage, and the genetic basis of resistance to the control tactic. Of the many strategies for deterring resistance that were proposed to the EPA, the agency adopted a plan that requires farmers to grow Bt crops containing a very high dose of Bt near a non-Bt variety of the same crop. The Bt crops must produce a sufficient amount of Bt to kill all but the most resistant insects. The few resistant individuals then mate with susceptible or partially resistant individuals that live in the "refuge" provided by the non-Bt variety, effectively diluting the genes for resistance within the population. This high-dose–structured refuge strategy requires farmers to grow Bt crops in very specific ways.

- A farmer growing Bt corn must plant 20% of the corn acreage to a non-Bt variety in areas where cotton is not grown and 50% in cotton-growing regions. Why the different percentages? Because a few corn pests also feed on cotton; a smaller percentage of Bt crops lessens their possible exposure and therefore the selection pressure for Bt resistance (Figure 23.22).
- The refuge must be planted within 1/2 mile of the Bt corn.

As a condition of product approval, the EPA also requires farmers and companies to monitor populations of the target pests of Bt corn so they can identify the first signs of Bt resistance and respond in a way that maintains the long-term effectiveness of Bt products. A comparable insect resistance

Figure 23.22 The corn earworm also feeds on cotton bolls. (Photograph by Keith Weller, courtesy of Agricultural Research Service, USDA.)

management plan accompanied the approval of Bt for corn rootworms. Corn growers who plant the Bt variety must set aside 20% of their corn acres as a refuge by growing a non-Bt variety.

Have these new transgenic varieties finally solved the problem of pest resistance to control tactics? No. Farmers will always battle pests, and evolutionary biologists will continue to be fascinated by the remarkable adaptability of the pests. The transgenic varieties are useful not because they magically solve the problem of pest resistance but because they give farmers more tools to use in the fight.

SUMMARY POINTS

Sustainability means meeting the human population's current needs without sacrificing the ability of future generations to meet theirs. Depleting nonrenewable resources and degrading the quality of essential resources are not sustainable.

Modern agriculture is heavily reliant on nonrenewable fossil fuels, leads to soil erosion, requires chemical inputs, and depends to a great extent on irrigation, none of which are sustainable activities.

The increased understanding of plant biology provided by biotechnology research applications may lead to products that promote sustainable agricultural practices, such as drought-resistant crops, crops that require less fertilizer, and biological methods of pest control.

The first generation of transgenic crops has a mixed record on improving agricultural sustainability. Most allow farmers to decrease chemical inputs while maintaining or, in some cases, increasing yields. Decreasing chemical inputs and increasing yields per acre improve sustainability. However, one way to make agroecosystems more sustainable is through crop diversification, and the currently marketed transgenic crop varieties do nothing to increase crop diversity.

Two ecological impacts of transgenic crops that have received a great deal of media attention are gene flow from crops to wild relatives and the evolution of resistance to pest control mechanisms. Neither is unique to biotechnology. Gene flow between crops and wild relatives has been occurring for centuries. As soon as humans began to try to control crop pests, the pests began evolving resistance to the control measures.

Gene flow depends on cross-pollination, production of fertile offspring, and introgression of the gene into the wild plant population. The probability of gene flow varies with the crop, the location where it is grown, and the wild relatives that are present. Gene flow in and of itself is not necessarily an adverse ecological effect. Many variables affect whether the gene flow from a crop to a wild relative is harmful.

Pests will evolve resistance to any control measure humans use. The best strategy for staying one step ahead of crop pests is to provide farmers with many pest control options. The speed with which a pest population evolves resistance depends on a number of factors, including the number of genes that control the resistance trait and the degree of exposure of the pest population to the control tactic.

KEY TERMS

Biocontrol
Catalase
Centers of diversity
Centers of origin
Continuous cropping
Crop rotation

Cross-resistance
Diapause
Hybridization
Hypersensitive response
Insect resistance
    management

Introgression
Landrace
Mycorrhizae
Oxygen radicals
Salination
Sedimentation

Sustainability
Systemic acquired
    response
Transpiration

chapter 24 | Environmental Sustainability and Biotechnology

In the minds of many, technology is inextricably linked to environmental degradation and is the primary contributor to unsustainable activities. The technological innovation that spawned the Industrial Revolution of the past two centuries led to environmental problems, such as depletion of nonrenewable natural resources, generation of large volumes of nondegradable waste products, and pollution of air, soil, and water. In light of these facts, is it any wonder that the public's view of the relationship between the environment and technology is often simple and linear: technology ruins the environment?

It is true that technology has a profound effect on our relationship with the natural environment. That, after all, is its intent. People use technology as insulation from the often harsh realities of nature, and some of those technologies have harmful effects on the natural environment. For example, burning fossil fuels to generate energy for heating and cooling homes pollutes the air supply and produces acid rain. But are good environmental quality and technology development inherently incompatible? Not necessarily. If technologies are engineering solutions to the problems of life on earth, then technology exists in nature. Nature's technological solutions to life's challenges are known as evolutionary adaptations.

In this chapter, we continue the train of thought established in the last chapter. The environmental problems caused by human activities are interrelated and interdependent, so one problem often amplifies another. Therefore, the only viable solutions to most environmental problems must encompass the whole system and be sustainable in the long term. Natural systems and processes provide models of sustainability that are worth emulating. How can society use its ever-expanding knowledge of the natural world to develop products and process that are more sustainable? To address this problem, we will:

- describe attributes of nature's sustainable systems and processes
- discuss past and future industrial uses of biological organisms and their molecules

627

- provide examples of human exploitation of life's biosynthetic and catabolic pathways
- describe environmental benefits of incorporating biotechnology into manufacturing

USING NATURE'S PRINCIPLES, PRODUCTS, AND PROCESSES

Other species have evolved technologies that are much more harmonious with nature than ours. An obvious example is the conversion of solar radiation into chemical energy (glucose) by plants. This is a solution to a problem all living organisms must face: obtaining energy. Getting water to the top of a 150-foot tree is no small feat of engineering, either. Tree frogs and other animals defy gravitational forces and hang upside down with the aid of suction cup-like structures (Figure 24.1); and don't forget the water strider, described in chapter 5, that can stand and walk on water.

Nature's design process is the evolutionary process. Genetic recombination and mutation submit new designs, all of which are minor variations on previously successful engineering solutions. Natural selection, the head of Mother Nature's engineering department, picks the winning designs. Like most engineers, natural selection is impressed by elegant designs that solve technical problems effectively and efficiently. Unlike most department directors, however, natural selection is remarkably indifferent to power games. Because natural selection cares nothing about politics, economics, or public opinion, these factors have neither driven nor shaped the technological solutions devised by nature. So maybe it's not technology per se but the technological solutions, out of all possible technologies, that humans have elected to develop that have had detrimental environmental effects.

A few design principles underlie all natural technologies

The engineering solutions produced by billions of years of natural selection share certain design principles and features. Interestingly, the design principles are utilized and reinforced from nature's smallest to its largest operational scales, that is, from molecules to ecosystems:

> Raw materials. The building materials, biological molecules, are biodegradable and recyclable. Therefore, the flow of materials is cyclical, not linear. Building materials are used and reused again and again, and the concept of waste is meaningless.

Figure 24.1 Nature's technologies. The frog (*Hyla ebraccata*) **(A)** and bat (*Thyroptera tricolor*) **(B)** have evolved similar adaptations for maintaining a fixed location. All tree frogs differ from most other frogs by having dilated toe tips that act as suction cups. Most bats hang by their sharp claws when they roost in caves or trees, but the disk-winged bat roosts inside a large tropical plant leaf as it unfurls. These bats use suction cups on their wrists and ankles to protect the fragile leaves.

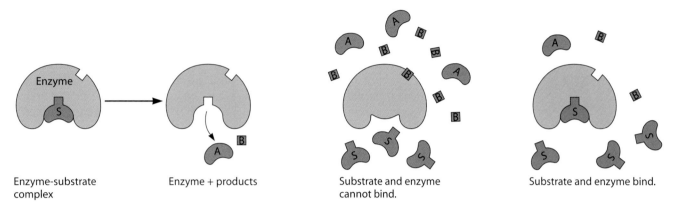

Enzyme-substrate
complex

Enzyme + products

Substrate and enzyme
cannot bind.

Substrate and enzyme bind.

Figure 24.2 Controlling production with feedback inhibition. Cells can control the amount of products (A and B) they make because one product (B) inhibits the binding of enzyme (E) to its substrate (S), which is the precursor of B.

Processes. The processes are exceedingly efficient. Materials and energy are not squandered. Processes are tightly controlled because a series of internal feedback loops provides real-time monitoring, instantaneous feedback, and immediate adjustments, if needed (Figure 24.2). In addition, the processes also use relevant external environmental cues to maximize efficiency (Figure 24.3).

By using nature's design principles as a guide in product and process development, engineers would:

- conserve natural resources through efficient use of materials and energy
- minimize environmental degradation by decreasing the amount of waste and eliminating the use and production of hazardous or toxic substances

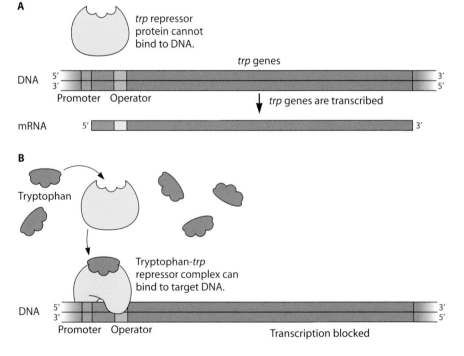

Figure 24.3 Controlling production with gene regulation. **(A)** Bacteria manufacture the amino acid tryptophan when none is available from the environment. **(B)** When environmental concentrations of tryptophan are high, bacteria do not need to synthesize it. Under those conditions, the tryptophan-repressor protein complex binds to DNA and prevents the transcription of genes involved in tryptophan biosynthesis.

In other words, human activities would become more sustainable environmentally and economically if they were modeled after nature.

Rather than using nature's design principles and features as inspiration for new technological products and processes, why not do the simpler thing—steal them: co-opt nature by using biological molecules as raw material and cellular metabolic machinery to carry out certain processes. Does this sound familiar? It should, because it is essentially the definition of modern biotechnology.

Biomolecules provide people with a great deal more than food

As plants and animals go about their business of breaking down and synthesizing biological molecules for their own purposes, they coincidentally provide people with molecules for meeting food, shelter, clothing, and energy needs. Less known or appreciated are the subtle ways in which they have added to the quality of human life beyond basic sustenance. Living organisms synthesize many molecules that serve a variety of functions, such as self-defense and communication. As it turns out, many of those secondary metabolites serve some human purposes as well. Table 24.1 provides a few examples of plant molecules used by various industries.

The list of useful plant molecules is minuscule in size and importance compared to that of microbial molecular products. Of all the organisms on the planet, microbes are the least appreciated, both for their roles in nature and for their usefulness to people (Box 24.1). In chapter 14, you learned that microbes are the unsung heroes of the planet because they keep it chugging along. Aquatic microbes fix 25% of the total amount of carbon that is fixed through photosynthesis, and others play essential roles in the nitrogen cycle. In addition to these life-sustaining functions, microbes naturally produce an extraordinary array of molecular products that humans rely on. Table 24.2 provides a very abbreviated list of the useful molecular products synthesized commercially by microbes.

Table 24.1 Molecules produced by plants and used by people

Industry	Molecular product	Source	Purpose
Pharmaceuticals	Digoxin	Foxglove	Heart disorders
	Codeine	Poppy	Sedative, pain
	Vinblastine	Periwinkle	Leukemia
	Atropine	Nightshade	Anti-spasmodic
Food	Vanilla	Orchid fruit	Flavor
	Cloves	Myrtle tree flower buds	Spice
	Cinnamon	Laurel tree bark	Spice
	Alginate	Brown seaweed	Emulsifier
	Carrageenan	Red seaweed	Texturizer
	Pectin	Citrus peel	Jelly, gelled candy
	Bromelin	Pineapple	Enzyme
Agriculture	Pyrethrins	Chrysanthemum	Insecticide
	Nicotine	Tobacco	Insecticide
Textiles	Indigo	Legume	Dye
	Starch	Corn	Sizing
Personal care	Papain	Papaya	Contact lens cleaner
	Shikonin	*Lithnospermin* root	Cosmetic color
	Jasmine	Tea olive tree	Perfume

BOX 24.1 *What are microbes?*

In biology, group names usually refer to organisms that are related by virtue of genetic and morphological similarity. The term vertebrate, for example, indicates organisms that share the characteristic of having a backbone. Vertebrates also form a genetically similar group. The term microbe, in contrast, is a nonscientific term used to designate organisms that share only the characteristic of being very small—you need a microscope to see them. Organisms that have been called microbes represent enormously diverse biological groups. For comparison, you could make up a similar term to refer to organisms you could see with the naked eye, perhaps "macrobes." Worms, flies, mushrooms, trees, grasses, fish, and people all would be macrobes, to give you an idea of the range of organisms that could fall under that label. Microbes are, if anything, even more diverse.

Organisms commonly referred to as microbes represent six major groups.

Bacteria, perhaps the best known of the microbes, are prokaryotes. They are unicellular and reproduce asexually by dividing. Some bacteria can get energy from the sun, as plants do, but others require a source of food, like animals. Many play essential roles in ecosystems. A few bacteria infect people and other animals, causing disease, but the vast majority are not pathogenic. Bacteria have very diverse metabolic capabilities and are therefore extremely useful in commercial and research applications of biotechnology. They were the first organisms to be genetically engineered with recombinant techniques, and they are still important hosts for cloning DNA and producing foreign proteins.

Like bacteria, archaea are prokaryotic and unicellular, and they reproduce by fission. However, archaea are a distinct group. Genetically, they are as far removed from bacteria as they are from human beings. Archaea are often found in extreme environments where no other type of organism could live, such as hot springs or deep-ocean hydrothermal vents. Not surprisingly, they have unique enzymes and biochemistries that biotechnologists want to understand and use.

Viruses are not cells and are not really alive. They use their protein coat to bind to specific host cells. Once inside a host, their nucleic acid is translated by the host metabolic machinery, leading to the production of new virus particles. The process of virus reproduction is often harmful to the host cell, so viruses cause disease in the organisms they infect, from bacteria to plants to animals (including humans). Scientists use viruses as nucleic acid delivery vehicles.

They remove the native viral nucleic acid and replace it with genes they want delivered into the virus's host cell.

Yeasts and molds are members of the group of eukaryotic organisms called fungi. Yeasts are unicellular; molds are multicellular, consisting of long branched filaments. Their cells are surrounded by walls composed of the polysaccharide chitin instead of the cellulose found in plant cell walls. Fungi require an external food source. Like bacteria, many are decomposers. Fungi have many uses in biotechnology. Yeasts cells are used as hosts to produce recombinant proteins. Yeasts and molds are also used in manufacturing many pharmaceuticals, such as antibiotics and birth control pills, and in food fermentation. A few fungi can infect humans and cause disease, but the vast majority do not.

Protozoa are unicellular eukaryotic organisms that typically live in water or moist environments. Their cells have membranes but no cell walls. A few cause disease, notably malaria, sleeping sickness, and amebic dysentery, but most are free living and not pathogenic.

Algae include both unicellular and multicellular forms. Their cells are eukaryotic, photosynthetic, and surrounded by walls. They contain a variety of different pigment molecules that capture sunlight. They use solar energy to produce carbohydrates from carbon dioxide gas. Algae form the base of the food chain in the oceans. No known algae infect humans, but some are toxic to animals that ingest large amounts.

As you can see, the term microbe is very imprecise and can refer to vastly different organisms. A few microbes cause disease in humans (earning them the additional imprecise designation of "germ"), but most do not. Their diverse biochemistries mean that their genomes offer a myriad of different enzymes that could potentially be useful to humans, and their small size and fast reproduction make them easy to use in the laboratory and in manufacturing.

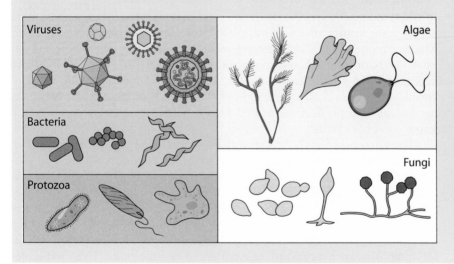

Table 24.2 Examples of the types and categories of microbial molecules that people and industries use

Whole organism	**Food additive**
Vaccines	Glycerol
Probiotics	Dextran
Baker's yeast	Xanthan gum
Microbial pesticides	Acidulants
Nitrogen-fixing inoculants	Beta carotene
Mycorrhizal inoculants	Monosodium glutamate
Composting decomposers	**Flavor**
Bulk chemical	Almond
Acetone	Licorice
Citric acid	Wintergreen
Acetic acid	Coconut
Propylene glycol	Butter
	Honey
Fine chemical	Peach
Enzymes	
Plant hormones	**Aroma**
Perfume fragrances	Rose
Polysaccharides	Hyacinth
	Gardenia
Pharmaceutical	Jasmine
Antibiotics	Lavender
Birth control pills	Citronella
Enzyme inhibitors	
Anti-inflammatory compounds	**Energy**
Immunomodulators	Ethanol
Anticoagulants	Methane
Nutritional supplement	
Vitamins	
Amino acids	

Table 24.3 Food fermentation and products[a]

Cheeses	Wine
Yogurt	Beer
Cottage cheese	Distilled liquors
Kefir	Cider
Buttermilk	Cocoa
Sour cream	Coffee
Soy sauce	Tea
Tamari	Vinegar
Tofu	Bologna
Miso	Salami
Tempeh	Sauerkraut
Bread/baked goods	Pickles
Pastrami	Olives

[a]The end product of food fermentation, which is a $30 billion industry in the United States, varies with both the starting material and the microbes used in the fermentation.

Humans also rely on microbial metabolic processes

In addition to using specific molecules extracted from microbes, humans have a very long history of using microbial processes to perform certain tasks. Much as our early ancestors began genetically modifying wild plants and animals without the slightest understanding of heredity and reproductive processes, they also began using microbes without understanding the biochemical processes they were exploiting. In fact, when humans first began using microbes to produce fermented food 8,000 years ago, they didn't even know microbes existed.

The microbial metabolic pathway being exploited in food fermentation, then and now, is the glucose breakdown pathway. Yeasts break down the glucose in wheat flour to carbon dioxide, which causes bread to rise. If the starting material is fruit sugar, the product of glucose breakdown in yeast is wine. Many different microbial species in addition to yeast produce the list of fermented foods in Table 24.3. Irrespective of the microbe responsible for the fermentation, microbial degradation of glucose always leads to small biological molecules, such as ethanol or a variety of organic acids. These molecules inhibit the growth of other microbes, so fermentation preserves raw agricultural products, such as wheat, fruit, and milk, by preventing food spoilage.

Louis Pasteur's discovery that microbial biochemical processes were changing raw foods to their fermented forms broadened exploitation of the glucose breakdown pathway found in different microbes (see Table 6.1). Mirroring the story line of plant and animal agriculture, the understanding provided by Pasteur's discovery allowed people to exercise more control over the processes they had been exploiting. It also broadened the range of microbial metabolic pathways being exploited well beyond glucose catabolism.

As with plant and animal agriculture, even though humans have always used microbial processes, the new biotechnologies, especially the bioprocess technologies and recombinant DNA technology, greatly expand the ways in which microbial processes contribute to human society.

BIOPROCESS TECHNOLOGIES

Bioprocess technologies are a set of technologies that use **biocatalysts**, which are living cells or their enzymes, to:

- produce a molecular product
- change one molecule into another
- degrade a molecule

In other words, they use cellular biochemical machinery to do what cells always do: break down large molecules to obtain energy and to generate a supply of chemical building blocks, which the cell then uses to synthesize new molecules. The living cells most frequently used in bioprocesses are microbes, although the use of cells from multicellular organisms has increased considerably in the last decade as scientists have become more adept at keeping cultured cells happy and healthy.

Bioprocesses have some advantages over chemical processes

Bioprocesses based on biocatalysts offer a number of advantages for making, changing, or breaking down molecules, especially compared to their chemical process counterparts.

- Biocatalysts are the ultimate in sustainability. Microbes and cells reproduce, so they continually renew themselves and their molecular components. Enzymes are masters of the reuse-and-recycle concept, because they catalyze chemical changes without being changed or used up in the process.
- Unlike many chemical processes, bioprocesses must occur under conditions that are compatible with life. Biocatalysts are water soluble and function best at comparatively low temperatures, normal atmospheric pressure, and neutral pH. Consequently, bioprocesses typically consume less energy than chemical processes. Bioprocess waste products are usually less destructive to the environment than those of chemical processes, because organic solvents, strong acids and bases, and highly concentrated salt solutions are not used in bioprocesses.
- Because enzyme-catalyzed reactions are very specific, they generate fewer undesirable by-products than chemical processes. Enzymes are also highly selective. Because they bind to only one or a few molecules in a mixture of molecules, often the raw materials fed into the process do not have to be as pure as those in chemical processes. Not having to purify raw materials decreases energy consumption.

- Bioprocesses can be continually improved, because biocatalytic organisms can be genetically modified to optimize various factors that affect process efficiency and yields, such as enzyme activity levels, tolerance of environmental conditions, and reproductive rate.

Genetic optimization through selection and mutagenesis

Scientists began genetically modifying the microbes used in bioprocesses as soon as they learned that microbes were responsible for the chemical reactions they had been exploiting. Microbial populations, like all groups of living organisms, are composed of genetically variable individuals. At first, like early agriculturists improving crops and livestock, industries that relied on microbes used selection of the very best strains to genetically modify microbes and improve yields and efficiency. When they learned that mutations could be induced with radiation and chemicals, they began creating strains with novel genes and selected mutant strains that were superior performers. In the past century, techniques for genetically modifying microbes became more sophisticated technically and increasingly unnatural, paralleling the evolution of genetic modification in plants and animals. For example, scientists began to use cell fusion to forcefully join two useful microbial species to create a new microbe that contained the production characteristics of both parental species.

Optimization with molecular techniques

A host of new molecular techniques broaden the capacity to continually improve and optimize a certain bioprocess. Industrial scientists have used recombinant DNA technology to increase the copy number of the gene encoding the commercial product in the production organism. Another genetic modification technique for ramping up production takes a different tack. RNA interference blocks the expression of specific genes and improves yields by rerouting metabolic pathways to favor synthesis of the commercial product. Other molecular techniques, such as protein engineering, enhance the activity of the enzymes that catalyze the bioprocess being maximized. We discuss the critical role recombinant DNA technology has played in transforming bioprocesses in more detail below.

Biocatalysts also have some disadvantages

Many of the advantages that biocatalysts provide become limitations if the processes must be carried out under extreme conditions. Industrial processes, such as paper making and textile processing, sometimes require organic solvents, high or low temperatures, or extreme chemical conditions. Unfortunately, most biocatalysts cannot function well at low temperatures, and they fall apart at temperatures above 110°F, or approximately 40°C. In addition, extreme chemical conditions, such as high or low pH, high salt concentration, or organic solvents, disrupt the hydrogen bonds or hydrophobic-hydrophilic interactions that keep proteins folded properly. Once high temperatures or extreme chemical conditions ruin the protein's shape, they ruin the biocatalyst's function.

Because biocatalysts can offer so many environmental and economic advantages, scientists are trying to develop strategies to circumvent these limitations so that biocatalysts can be used under unfavorable conditions. As you learned in chapter 5, if the manufacturing process is carried out under conditions that are not too extreme, the solution might be as simple as adding a few cysteine molecules to the enzyme to increase the number of disulfide

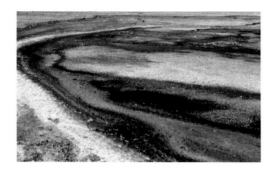

Figure 24.4 Extremophiles. The red stream has high concentrations of sulfuric acid and iron, but the green photosynthetic microbes can tolerate these extreme conditions. (Photograph by D. E. White, courtesy of U.S. Geological Survey.)

bonds. In this type of protein engineering, scientists must know a great deal about the enzyme, especially its shape. They mutate, quite specifically, the gene encoding the enzyme in a process known as **directed mutagenesis** or **site-specific mutagenesis**. In this case, the goal of directed mutagenesis is to stabilize the enzyme's structure by increasing the number of disulfide bonds without affecting the enzyme's active site. Therefore, they insert a few codons for the amino acid cysteine into the gene encoding the enzyme.

Often, however, the conditions are extreme enough to require solutions other than site-directed mutagenesis. Nature's wealth of genetic diversity provides one solution; another comes from a different type of protein engineering.

Discovering novel biocatalysts

Scientists have recently begun discovering microbial species that live under conditions formerly thought to be incompatible with life (Figure 24.4). As a result, these microbes have forced scientists to redefine the environmental boundaries within which life is possible. Over the billions of years they have been on earth, microbes have adapted to every imaginable environment; no matter how harsh, some microbe has found a way to make a living there. Scientists have found microbes in hot springs that are the source of the geysers in Yellowstone, thousands of feet underwater in hydrothermal vents, in the Dead Sea, and living off inorganic materials, such as copper sulfide in copper mines. Microbes capable of living in such extreme environments are called **extremophiles**, and their enzymes are called **extremozymes** (Table 24.4).

Life in unusual habitats makes for unique biocatalysts, and the great majority of that biochemical potential remains untapped. Less than 1% of the planet's microbes have been cultured and characterized. Looking for microbes with desirable metabolic capabilities is known as **bioprospecting**.

The best place to begin a search for a biocatalyst capable of withstanding an extreme process condition is in a natural environment that mimics that condition. Large-scale studies of microbial genomes, similar to the Human

Table 24.4 Microbes have been discovered living in extreme environments[a]

Bare basalt rock 1 mile beneath the Columbia Plateau

Acid mine drainage at pH <1 and 75°C

Sewage sludge at pH 10 and 56°C

Six feet under permanent ice-covered lake in Antarctica at −45°F

South African gold mines, 2 miles deep at 120°F

Hydrothermal vents on ocean floor, 200°F and two times atmospheric pressure

[a]In the metric system, water boils and freezes at 100 and 0°C, respectively.

Genome Project, have provided the complete genomic sequences for over 100 microbes. As a result, researchers may already know part of the sequence of the gene encoding the enzyme they are hoping to find. This makes the search much less difficult, because they can use DNA probes to fish for that sequence in the extreme environment. Perhaps they are looking for microbes with a very specific capacity, such as the ability to break down a pesticide in polluted water at high temperatures, but they know nothing about the gene(s) encoding an enzyme capable of this feat. They would collect many soil samples from fields where that pesticide had been used and culture the microbes on media containing the pesticide as the primary food source. The microbes that survived would be capable of breaking down that pesticide. The scientists would then try to grow those cultures in aqueous solution at high temperatures and save any microbes that survived. Using these techniques, scientists have discovered novel biocatalysts that function optimally at the extreme levels of acidity, salinity, temperature, and pressure that are found in some industrial processes (Figure 24.5).

As you might expect, many extremophiles cannot be cultured in the laboratory, nor do scientists know the identities of the genes encoding the enzymes they hope to find. Before the days of recombinant DNA, any useful biochemical potential housed in the genomes of these microbes would have remained inaccessible to scientists. Now, however, using techniques you learned about in chapter 15, scientists can collect microbes that cannot be cultured, determine if any have genes for the desirable trait, isolate and identify those genes and their products, and give the useful genes to microbes that are easy to culture.

A few scientists have recently begun using a novel approach to bioprospecting, known colloquially as **total community genomics**. They collect samples from any natural environment, and then they use enzymes to break up the DNAs of *all* of the organisms in the sample. They make no attempt to separate the organisms in the sample or identify them, much less culture them. They are not interested in the organism that is the source of the gene,

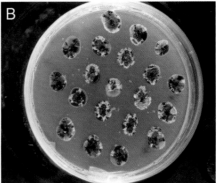

Figure 24.5 Discovering extremophiles. **(A)** A researcher collects samples from a hot spring in the hope of finding microbes with genes that can degrade cellulose at very high temperatures. **(B)** Back in the laboratory, the microbes are cultured on media containing cellulose as the sole source of carbohydrates. Colonies that grow can degrade cellulose. (Photographs by Mike Himmel [A] and Warren Gretz [B], courtesy of the National Renewable Energy Laboratory.)

only the gene itself. The DNA fragments from the entire community of organisms are recombined with plasmid DNA and inserted into a bacterium that is easy to culture, such as *Escherichia coli*. The host, *E. coli*, will express unique genes its new plasmid might have, and the proteins encoded by the novel genes are screened for the desired activity. Recombinant colonies with genes for the desirable enzyme are then isolated so that the gene can be identified and characterized.

Improving biocatalysts

Above, we described a technique for making slight changes in a protein's amino acid sequence to enhance the stability of its tertiary structure under conditions often found in industrial manufacturing. Another method for improving natural biocatalysts is modeled after the evolutionary process, but on a molecular level. A new extremozyme, or any other enzyme, that scientists may discover did not evolve to catalyze human industrial processes but to maximize its owner's survival and reproductive success. Therefore, once a new enzyme is discovered, scientists may need to tweak its three-dimensional structure a bit to optimize its catalytic activity for industrial purposes.

Because the enzyme is newly discovered, its tertiary structure is unknown, as are the precise changes in the amino acid sequence that would generate the tertiary structure that is most effective and efficient. Therefore, scientists cannot use directed mutagenesis to enhance the enzyme. In this situation, scientists use a number of different techniques to create a large number of genetic variants of the gene encoding the enzyme. They then screen the enzymes encoded by these genetic variants to identify slight variations of the enzyme that exhibit the greatest activity levels. This technique, known as **directed protein evolution**, optimizes the effectiveness and efficiency of the newly discovered enzyme. Once the enzyme with the best properties is identified, they determine the nucleotide sequence that will create that amino acid sequence. Directed protein evolution has been used to modify the specificity of enzymes, improve catalytic properties, and broaden the conditions under which enzymes can function. It has also been used to improve the functionality of other proteins, such as monoclonal antibodies or therapeutic proteins.

LARGE-SCALE BIOMANUFACTURING

Biomanufacturing exploits the anabolism (synthesis) branch of metabolic machinery to produce substances with commercial value. Throughout this book, we have mentioned the essential role that large-scale biomanufacturing plays in making some biotechnology products commercially viable. In the absence of bioprocess technologies that allow companies to manufacture a sufficient amount of product at an affordable price, the impact of recombinant DNA technology would be limited primarily to the research laboratory. Biomanufacturing technologies allow greater access to the rich biochemical diversity that nature provides.

The biomanufacturing workforce is composed of cells, typically microbial cells or cells from multicellular organisms. Not only does this cellular workforce make a high-quality product quite efficiently and with no job training, it also creates additional workers. The two primary methodologies used in commercial-scale biomanufacturing are **microbial fermentation** and **mammalian cell culture**. In the scientific community, the word fermentation refers to the anaerobic steps in glucose breakdown. In the industrial manufacturing community, fermentation means using microbes, usually bacteria and fungi, to manufacture a commercial product.

Whether the production cell is a bacterium, a yeast, or a mammalian cell, its job description is the same: use biosynthetic pathways to convert raw materials into a specific product. Just like employees everywhere, cells require certain resources and a safe supportive working environment to do their job well and be as productive as possible. Creating these working conditions for a beaker full of cells is quite easy, but keeping billions and billions of cells happy is a challenge. Consequently, large-scale biomanufacturing is a triumph of both chemical engineering and biology (Figure 24.6).

Cells require resources and certain environmental conditions to be productive

Special containers known as **bioreactors** provide the safe environment that cells need to do their job. Bioreactors, which are also known as **fermentors**, exclude microbes that might kill the production organism or compete with it for resources. In addition to the cells, the bioreactor contains a nutrient soup that bathes the cells. Most of the resources that the cells need in order to complete their assigned tasks are contained in the nutrient soup. Bioreactors are designed so that all cells in the bioreactor are subjected to essentially the same conditions. In addition to their need for resources and a safe environment, cells need a supportive environment. To a cell, the factors that combine to create a supportive environment are temperature, pH, and oxygen levels (for most cells).

Temperature. Cell productivity is maximized within a fairly narrow range. Higher temperatures kill cells and can denature protein products. If the temperature is too low, cells won't die, but they also won't reproduce or make much product.

Oxygen. The great majority of cells used in biomanufacturing must have oxygen to survive. Oxygen is poorly soluble in water and diffuses slowly in a liquid. In large-scale biomanufacturing, oxygen availability is usually the primary factor limiting production. One of the engineering challenges is constantly providing oxygen to every cell in the tank, by mixing the contents or bubbling air into the tanks, without harming the cells.

Figure 24.6 Large-scale biomanufacturing. A large-scale biomanufacturing facility contains many bioreactors. (Image courtesy of the Evans U.S. Army Community Hospital.)

pH. The production cell or organism will grow best at a certain pH, but it will also have a narrow range of pH values over which it can survive. Outside of that range, the cell will die. To maximize the yield of the product, the manufacturer wants the cells to be very active metabolically. However, cells produce acids as a result of their metabolic activity, much like the lactic acid people produce when they exercise. If cells secrete the acid into the surrounding solution, they create an environment that kills other cells. To minimize the harmful effects of high metabolic rates, **buffers** are added to the tank. Buffers are chemicals that neutralize acids and keep the pH fairly constant.

This sheds light on one of the many challenges bioprocess engineers face. By doing their job well, cells are constantly making the manufacturing environment less supportive.

Biomanufacturers need to maximize yields and minimize costs

The goal of large-scale biomanufacturing is to maximize product yields, minimize waste products, and do both as cheaply as possible. Biomanufacturers tinker with each of the components in Figure 24.7 as they try to achieve this goal.

The inputs in a biomanufacturing process are the cells, the nutrient soup that will provide the raw materials for making the commercial product and the resources to support cell growth and reproduction, oxygen (in most cases), and energy for running the manufacturing process.

The outputs include not only the product but also a large amount of water, waste products produced by the cells, and any leftover inputs that were not utilized in the conversion. In some operations, the cells are also a component of the output stream. To separate the product from everything else, the output must also be processed. It sounds simple, but in many biomanufacturing facilities, recovering the commercial product from the output stream and purifying it account for 90% of the manufacturing costs.

The heart of the biomanufacturing process is the middle component in Figure 24.7. In the conversion step, the biochemical machinery of the production cell converts raw materials into product. Choosing the right cell type for the conversion step is one of the most critical decisions that will determine whether the production goal is met. Companies want low-maintenance,

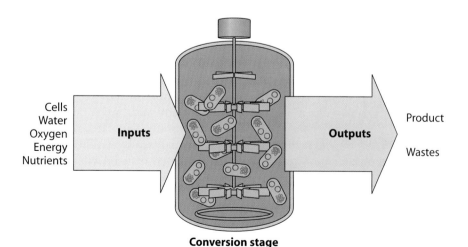

Conversion stage

Cells
Water
Oxygen
Energy
Nutrients

Inputs

Outputs

Product

Wastes

Figure 24.7 Biomanufacturing. The nutrients contain the raw materials that the production cells will convert into products. The conversion of inputs into products occurs in a bioreactor, which comes equipped with sensors for monitoring the key environmental factors—temperature, oxygen level, and pH—so that the process can be adjusted to maintain optimal conditions for product manufacture. The outputs contain not only the commercial product, but also a waste stream that must be separated from the product stream.

adaptable cells that are cheap to house and feed and that make a useful product in large amounts. Of the millions of different microbial species on Earth, biomanufacturers have recruited only a few species into their production workforce. Biomanufacturers use different cell types because, just like human employees, different cell types have different capabilities.

Before the development of recombinant DNA techniques, microbes joined the workforce if they made a useful product with commercial value and could also be cultured. If many different microbes made the useful product, such as the enzyme amylase, biomanufacturers chose those that either made a better-quality product or converted raw materials into product more efficiently.

With the advent of recombinant DNA technology, the ability to manufacture a useful product has been uncoupled from other attributes that make certain species good job candidates. As a result, some microbes that have never been part of biomanufacturing because they made no useful product are being recruited if they have other important attributes. Once hired, they can be given genes for making useful products. For example, **baculoviruses**, which are viruses that specifically infect one or a few insect species, have recently been recruited because they devote a large percentage of their resources to protein production, including those encoded by transgenes inserted by recombinant techniques. On the other hand, high-maintenance microbes that have been retained in the microbial workforce because of some unique biochemical ability have had their manufacturing jobs outsourced to microbes that are less finicky.

Scientists have also used recombinant DNA technology to broaden the job descriptions of some of the cheap, easy-to-grow microbes that have been used in bioprocesses, such as food fermentation, for centuries. These microbial workhorses now have the genetic machinery to make molecular products that are naturally synthesized by microbes that are too expensive to maintain or impossible to culture. In addition, these microbes are also used to produce human proteins, such as insulin (Figure 24.8).

Plants and animals can also function as bioreactors

Although many of the commercial products synthesized by microbes (Table 24.2) can be manufactured by using chemical processes rather than bioprocesses, synthesizing proteins from scratch is very expensive and often impossible. In the absence of recombinant DNA technology, few therapeutic proteins would ever have become commercial realities. Even so, building a biomanufacturing facility takes at least 5 years and $500 million. The annual costs of manufacturing the product and maintaining the facility are also in the millions.

The flexibility provided by recombinant DNA technology opens the door to using any organism, including animals and plants, as the production organism. Using plants or animals as bioreactors to produce therapeutic proteins or other high-value molecules presents several clear advantages. First, the costs of building the large-scale manufacturing plant are eliminated. Costs of production and facility maintenance would also be significantly less. Scaling up the production process to meet increased demand would not be handicapped by the many issues that impede scale-up in cell-based manufacturing.

Because of these many benefits, scientists have inserted genes encoding therapeutic proteins into livestock and a variety of commonly grown crop plants, such as tobacco, corn, rice, and soybeans, as well as in noncrop

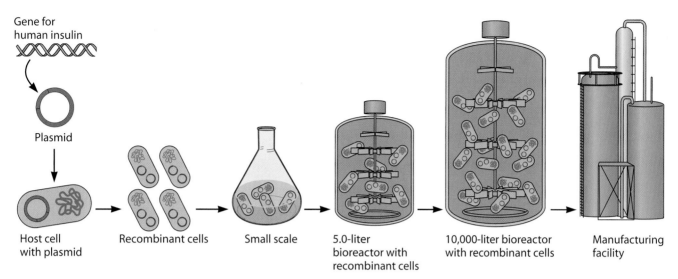

Gene for
human insulin

Plasmid

Host cell
with plasmid

Recombinant cells

Small scale

5.0-liter
bioreactor with
recombinant cells

10,000-liter bioreactor
with recombinant cells

Manufacturing
facility

Figure 24.8 Biomanufacturing of proteins. In the example shown, the company is using a bacterial host cell to manufacture human insulin. The gene encoding insulin must first be isolated and inserted into a plasmid. The plasmid carries the insulin gene into a bacterial host cell. The host cell is cloned to increase cell and plasmid numbers. The recombinant host cells grow and reproduce in a 1-liter flask. Eventually, the microbial population is so large it must be transferred to a 5-liter bioreactor. The next stage can be quite tricky as the manufacturing process is scaled up to thousands of liters. The insulin-manufacturing facility (not drawn to scale) contains many 10,000-liter bioreactors.

plants, such as duckweed and mushrooms. Therapeutic proteins that have been produced by transgenic plants include antibodies, antigens, growth factors, hormones, enzymes, blood proteins, and collagen. The proteins maintain their stability for more than 2 years in dried seeds of the transgenic plants. The target diseases of these proteins include heart disease, diabetes, human immunodeficiency virus infection, multiple sclerosis, obesity, and arthritis. Several plant-produced therapeutics are in clinical trials.

In addition, scientists have made excellent progress in using plants as both vaccine-manufacturing facilities and delivery systems. They have used tomatoes, potatoes, and bananas to produce vaccines against microbes causing infectious diseases, including cholera; a number of microbes that cause food poisoning and diarrhea; hepatitis B virus; and the bacterium that causes dental cavities.

One of the companies developing plant-produced antibodies estimates that this production method is 25 to 100 times less expensive than existing cell culture or microbial fermentation methods, which decreases production costs from the current $200 to $1,000 per gram to $10 to $100. In addition, this method increases the volume of antibodies that can be produced 100 times. Because plant manufacturing of pharmaceuticals or other useful biological molecules, such as enzymes and industrial oils, would require relatively little capital investment, and the costs of production and maintenance are minimal, they may provide the only economically viable option for independent production of therapeutic proteins in developing countries. In addition, if plant-produced vaccines become a reality, there will be no need for sterilized needles and refrigeration, making these vaccines more readily available to people in developing countries at a fraction of the present cost.

In 2003, the U.S. Department of Agriculture, which is the federal regulatory agency responsible for approving field tests of transgenic plants, granted 20 permits for growing plant-made pharmaceuticals at 34 field test sites in 14 different states.

USING BIODEGRADATION PATHWAYS

All living cells break down large organic molecules to access the energy and small building block molecules they contain. Humans have exploited these catabolic pathways for centuries. They began by using the microbial machinery for breaking down glucose to preserve raw foods. Microbial catabolic machinery has also lessened the impact human activities have had on the environment. Communities have long relied on complex populations of microbes for sewage treatment. As with food fermentation, at first people had no idea that they should be thanking microbes for breaking down human waste. Once they realized microbes were involved, engineers and biologists began developing various technologies for making waste degradation more efficient and effective; one avenue for improvement was genetic modification.

Increased understanding of molecular biology and the tools it provides greatly expands the opportunities for using the molecular degradation machinery of microbes. Below, we discuss only two areas of application: breaking down biomolecules for energy and molecular building blocks and using biodegradation pathways to clean up environmental problems.

Biomass can be a source of energy and building block molecules

The economies of the industrialized world are based on fossil fuels, such as oil, coal, and natural gas. Fossil fuels provide the energy that drives the engines of economic growth and keeps people in the comfortable lifestyle to which they have become accustomed. They also provide the building block molecules, which are also called **feedstock chemicals**, for thousands of consumer products. Society's dependence on fossil fuels leads to environmental, political, and economic problems.

- Fossil fuels are nonrenewable resources, as well as the major contributor, by far, to greenhouse gas emissions. Processes based on fossil fuels generate unwanted by-products, some of which are hazardous substances that pollute the air, soil, and water.
- Because most of the oil that U.S. citizens and industries rely on is located outside of our borders, the U.S. economy and our quality of life depend on a handful of oil-producing countries. America's heavy reliance on resources found in other countries represents a threat to national security and also contributes to the balance-of-trade deficit.
- The stability of the governments in countries with considerable amounts of petroleum affects our economic well-being. No matter how stable the U.S. government is, political turmoil in the oil-producing countries raises the price of petroleum and threatens access to that essential resource.
- As developing countries become industrialized, worldwide demand, and therefore competition, for this limited and nonrenewable resource will increase quite significantly.

All of these forces place considerable pressure on society to develop new, reliable, and affordable sources of energy and raw materials.

Life, like fossil fuels, is also carbon based. Fossil fuels are nothing more than living organisms from millions of years in the past. The fossilized carbon that industrialized societies depend on today is the carbon that was contained in the biological molecules of organisms when they died. The source of that carbon was the same then as it is today: carbon dioxide. Billions of years ago, photosynthetic organisms mastered the engineering feat of capturing solar energy and converting it into a usable form by hooking together the carbons found in carbon dioxide into glucose and the other biological molecules. Green plants and other photosynthetic organisms continue to fix carbon in the very same way as those ancient organisms. The biological molecules they synthesize contain energy and molecular building blocks, just like the fossil fuels. Unlike fossil fuels, however, the energy and building block molecules are renewable resources because photosynthetic organisms reproduce themselves. Increasing the use of renewable resources is a key pillar of ecological sustainability.

Using living biomass as a source of energy and feedstock chemicals is definitely not a new or sophisticated concept. Many people in developing countries still rely on gathered wood for cooking fuel. In the early 1900s, chemists developed biological and chemical processes to convert living biomass into feedstock molecules, such as acetone, glycerol, and alcohols, used by many industries. In the 20th century, technological improvements in oil refining decreased the cost of using petroleum as the starting material for producing chemical feedstocks and maximized the amounts of energy and feedstock chemicals that could be extracted from petroleum. Consequently, most processes that converted living biomass into feedstock molecules, either chemically or biologically, could not compete with comparable processes based on petroleum. This still holds true today. Thus, even though it has always made sense environmentally to shift from fossilized biomass to living biomass, economically that has not been the case.

Now, however, recognizing both the economic and political imperatives to lessen society's reliance on fossil fuels, the government has begun to establish policies and commit more money to basic and applied research into alternative sources of energy and feedstock chemicals. Living biomass could serve as the raw material for both resources, and living organisms could provide the metabolic machinery to convert biomass into fuels and feedstocks. Many technical barriers will need to be overcome before biomass can make a significant contribution to the country's demand for energy and feedstock molecules. To solve some of these problems, government laboratories and industrial scientists have devoted significant resources to research projects that use biotechnologies, such as recombinant DNA technology, protein engineering, molecular evolution, and bioprocess engineering, to make biomass-derived energy and feedstocks economically competitive.

Biofuels

Biomass has always provided societies with energy directly, but biomass can also be used to create **biofuels**. With biofuels, the energy in the biomass is not released immediately through burning but is stored in other organic molecules that will not rot. Fuel ethanol, or bioethanol, biogas, and biodiesel, are examples of biofuels (Figure 24.9).

Current U.S. production of fuel ethanol, which is added to gasoline to create gasohol, is approximately 2 billion gallons per year. The amount of ethanol in the gasoline-ethanol mixture is only 10%, but, even so, that small percentage has environmental benefits because it decreases air pollution by lowering ground level ozone. Using a higher percentage of bioethanol would offer significantly more direct environmental benefit (Table 24.5). The U.S.

Table 24.5 Environmental benefits of gasohol

Decreased need for tetraethyl lead to increase octane ratio

No need to add MTBE, a water pollutant, to oxygenate the fuel

57% less CO

64% less hydrogen gas

13% less nitric oxide

Figure 24.9 Biofuels. The bus uses diesel fuel derived from soybean oil. (Photograph courtesy of the National Renewable Energy Laboratory and the Nebraska Soybean Board.)

government committed significant resources to bioethanol production technologies during the energy crisis of the 1970s, but when the energy crisis subsided, so did the government's interest in biofuels.

Brazil has used ethanol, which is half the price of gasoline, as their primary motor fuel for a number of decades. The Brazilian government committed the country to fuel ethanol and established the National Fuel Alcohol Program, which built hundreds of plants that converted biomass into ethanol. The factor driving this total commitment was not environmental concern but economics. Brazil, like other developing countries, has little money to buy oil. They do, however, have a cheap and readily available raw material for ethanol production: sugar from sugarcane.

In the United States, the starch molecules in corn provide the starting material for the great majority of bioethanol produced (Figure 24.10). Starch must be converted into glucose before it can be converted into ethanol. The cost of converting cornstarch into sugar is one barrier to increasing the use of bioethanol. Genetically engineered microbes that decrease the cost of this step have been developed, but even so, in the absence of the government subsidies that are now provided for bioethanol, the cost of ethanol derived from cornstarch cannot compete with that of petroleum.

Figure 24.10 From cornstarch to ethanol. The man is unloading corn at an ethanol plant. The starch in the corn will be broken down into corn sugar, which will be converted into ethanol. (Photograph by Warren Gertz, courtesy of the National Renewable Energy Laboratory.)

Biofeedstocks

Most chemical feedstocks are also derived from petroleum. Many petrochemicals, such as the molecular feedstocks for polyethylene and plastic, are small organic (carbon-containing) molecules that are hooked together to create polymers. Glucose, rather than petroleum, can be used as the started material for producing these building block molecules (monomers), and biocatalysts, rather than chemical synthesis, can be responsible for creating the monomers from glucose (Figure 24.11).

Almost all large chemical companies are building partnerships with biotechnology companies to develop enzymes that can break down plant sugars into useful building block molecules that they can use in polymer synthesis. The first biomaterial came to market in 2004. Produced by a joint venture between Cargill, a grain commodity company, and Dow Chemical Company, this biomaterial is a polymer of lactic acid. The Cargill-Dow process converts cornstarch into sugar and sugar into lactic acid. The lactic acid molecules are hooked together to create the polymer polylactic acid, which is marketed under the name NatureWorks. The manufacturing process uses 20 to 50% less fossil fuels than manufacturing a comparable petrochemical polymer. NatureWorks is a biodegradable polymer that can be used as a starting material for manufacturing packaging materials and fibers for clothing, pillows, and comforters. The bedding company Pacific Coast Feather plans to replace 80% of its polyester products with polylactic acid over the next 5 years.

The source of biomass

Three possible sources of biomass could provide the resource base of a biobased economy built on biofuels and biochemicals:

- harvesting naturally occurring vegetation
- growing agricultural crops and trees specifically for biomass utilization
- using biological waste products generated by agriculture, municipalities, and industries

Figure 24.11 Biodegradable plastics. A process developed at the Pacific Northwest National Laboratory converts the carbohydrates in potatoes into building block monomers that are used as feedstock chemicals in manufacturing plastics, adhesives, and textiles. (Photograph courtesy of the Pacific Northwest National Laboratory.)

All have some environmental benefits simply because the starting material is living biomass and not fossil fuels. Not only does that change decrease the use of nonrenewable resources, it also balances carbon dioxide output with input. As you learned in chapter 14, the amount of carbon dioxide released into the atmosphere through human activities greatly exceeds the amount of carbon dioxide that can be fixed by photosynthetic organisms. This imbalance is responsible, in part, for the greenhouse effect.

Of the three options, the first—harvesting naturally occurring vegetation—is obviously the least attractive economically and environmentally. Biomass derived from natural vegetation would be impossible to replenish quickly enough to keep pace with society's needs. We will limit our discussion to the other two options.

The second option has both political and economic advantages. U.S. farmers would benefit because there would be a large new market for agriculture commodity crops. This could help revitalize rural economies, which are suffering; in turn, this revitalization would abate the loss of people from rural areas and their immigration to urban centers. It relies on U.S. strengths: agricultural production, a large landmass, and an existing forestry industry. However, if this option entails clearing additional land to grow crops to create the biomass that would be converted to biofuels and biochemicals, then its environmental attractiveness is diminished. In addition, all of the environmental problems associated with row crop agriculture must be added to the equation. Finally, as the world population continues to expand exponentially, would row crop agriculture for biomass production be competing with row crop agriculture for food production?

We need to emphasize that in option 2, the source of the starting material is not agricultural refuse, such as corn stalks. Crops with high sugar content, such as sugarcane and sugar beets, are the most attractive options because sugar, not starch, is the starting material for manufacturing either bioethanol or biobased feedstock chemicals. If plant starch, such as the starch in grain and root crops, is the starting material, then the economic costs of option 2 must include the cost of converting starch into sugar. Converting cornstarch into corn sugar is easily accomplished either with biocatalysts or by chemical hydrolysis. Currently, the chemical process is used most often, but genetic improvement of biocatalysts will eventually shift the conversion to a bioprocess.

Figure 24.12 Cellulosic biomass. Biofine Corporation in Waltham, Massachusetts, has developed a process that can convert cellulose into a small organic molecule that can serve as a monomer for synthesizing many chemicals. (Photograph courtesy of the Pacific Northwest National Laboratory.)

The third option, using waste products, is clearly the most attractive environmentally and politically. It solves an environmental problem—getting rid of large amounts of wastes. The cost of the material is negligible compared to the cost of grain and root crops. However, other costs are very significant, as are the technical barriers that must be overcome.

Most of the carbohydrate molecules in agricultural residues and forestry by-products are not starch, but cellulose. Woody materials also contain a large amount of another molecule, lignin. As a result, most of the molecules that would serve as the starting material for the biocatalysis to glucose are biologically unavailable (Figure 24.12). Very few organisms can digest cellulose, the primary molecule in plant cell walls, or lignin. Before recombinant DNA technology, using agricultural or forestry refuse as the starting material for a bioprocess would have been virtually impossible. Now, however, scientists can move the gene for **cellulase**, a cellulose-digesting enzyme, into microbes commonly used in bioprocesses and use other genetic modification techniques to increase cellulase yields. Protein engineers have also improved the catalytic power of naturally occurring cellulases. To date, the various biotechnologies have decreased the costs of converting cellulosic biomass to sugar, which is then used to produce ethanol, from $5 per gallon of ethanol to 30 cents per gallon.

Catabolic pathways can remediate environmental problems

Communities have depended on complex populations of naturally occurring microbes for sewage treatment for many years. Microbes help purify water by breaking down solid organic wastes before the water is recycled (Figure 24.13). Solid organic wastes are not the only type of pollutants that need to be removed from our water supplies. Human activities have polluted the water, soil, and air, and microbes are helping to clean up those pollutants as well. The use of microbes to remove pollutants, known as bioremediation,

Figure 24.13 A sewage treatment plant in Hawaii. (Photograph courtesy of the National Renewable Energy Laboratory.)

is most often used to remediate oil spills, toxic waste sites, and leakage from underground storage tanks (Figure 24.14). Plants are also being drafted into the cleanup service, which is known as **phytoremediation**. Bioremediation and phytoremediation clean up many hazardous wastes more efficiently than conventional methods. In addition, they greatly reduce the use of unsatisfying waste cleanup methods, such as incineration or hazardous waste dump sites.

Environmental engineers use two basic bioremediation methodologies. In **bioaugmentation**, they provide nutrients to stimulate the activity of bacteria already present in the soil or water. In the other method, new microbes, specifically chosen because of their ability to break down the pollutant, are added to the site. In some cases, the by-products of the pollution-fighting microorganisms are themselves useful. Methane, for example, can be derived from a form of bacteria that degrades sulfur liquor, a waste product of paper manufacturing. After the microbes consume the waste materials, they die off or return to their normal population levels in the environment.

The biodiversity of microbes described in "Discovering novel biocatalysts" above is an exceptionally valuable resource in bioremediation activities. Indigenous microbes that contribute to the natural cycling of metals, such as mercury, offer prospects for removing heavy metals from water. Other naturally occurring microbes that live on toxic waste dumps can degrade wastes, such as polychlorobiphenyls (PCBs) to harmless compounds.

The vast majority of bioremediation applications use naturally occurring microorganisms either to identify and filter manufacturing waste before it is introduced into the environment or to clean up existing pollution problems. Using only naturally occurring microbes can be somewhat limiting, however. For example, scientists have discovered a number of species of aquatic microbes that degrade some of the 200 naturally occurring halogenated hydrocarbons found in the ocean. Some major soil pollutants belong to this class of chemical compounds, but marine microbes cannot survive in soil. Recombinant DNA technology provides the best of both worlds. Scientists can isolate the genes encoding the enzymes that degrade halogenated hydrocarbons and insert them into soil microorganisms. These genetically engineered microorganisms might then be able to clean up hazardous waste sites.

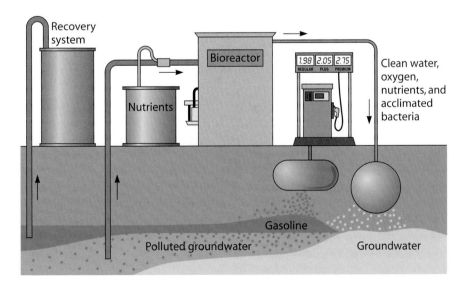

Figure 24.14 Bioremediation of a gasoline spill. Gasoline from an underground storage tank seeps through the soil to the water table. After the leak is stopped, the free-floating gasoline is pumped out to a recovery tank. Polluted groundwater is pumped into a bioreactor tank with oxygen, nutrients, and hungry microbes. After the microbes eat the gasoline, the mixture of clean water, nutrients, and microbes is pumped back into the ground so that more of the pollutant can be degraded.

BOX 24.2 *Environmental monitoring*

The techniques of biotechnology are providing novel methods for diagnosing environmental problems and assessing normal environmental conditions. Companies have developed methods for detecting harmful organic pollutants in the soil by using monoclonal antibodies and the polymerase chain reaction, while scientists in government laboratories have produced antibody-based biosensors that detect explosives at old munitions sites. Not only are these methods cheaper and faster than the present laboratory methods, which require large and expensive instruments, they are portable. Rather than gathering soil samples and sending them to a laboratory for analysis, scientists can measure the level of contamination on site and know the results immediately.

Even more impressive, scientists have used recombinant techniques to create novel metabolic pathways in bioprocess microbes. One of the best examples of this metabolic engineering involves a catabolic pathway. In chapter 6, you learned about oil-eating bacteria that environmental management companies are increasingly using to clean up oil spills, and you also know that oil contains many different compounds. Some bacteria have a penchant for one molecule or another, but a single strain is incapable of breaking down all hazardous compounds in oil. Using various biotechnologies, scientists have combined a number of catabolic pathways into one microbe.

Not all pollution problems that threaten human health result from human activities. During the hot summer months in Australia, aquatic cyanobacterial populations increase rapidly and secrete carcinogenic chemical compounds into the water supply. To counter these population blooms, copper sulfate, which kills the cyanobacteria but doesn't break down the toxins, is dumped into the water. As you might expect, microbes that coexist with the cyanobacteria have evolved a better solution to the problem. Recently, scientists discovered a bacterium that produces three enzymes that sequentially break down the cyanobacterial toxins until they are harmless. Conveniently for scientists, the three genes that code for the enzymes are clustered together in a single functional unit, an operon.

The remarkable ability of microbes to break down chemicals is proving useful, not only in pollution remediation but also in pollutant detection. A group of scientists at Los Alamos National Laboratory work with bacteria that degrade a class of organic chemicals called phenols. When the bacteria ingest phenolic compounds, the phenols attach to a receptor. The phenol-receptor complex then binds to DNA, activating the genes involved in degrading phenol. The Los Alamos scientists added a reporter gene that, when triggered by a phenol-receptor complex, produces an easily detectable protein, thus indicating the presence of phenolic compounds in the environment (Box 24.2).

INDUSTRIAL SUSTAINABILITY

Industrial sustainability includes processes and products that meet the present consumer demand for products without compromising the resources and energy supply of future generations. How can industrial manufacturing achieve sustainability? The key words are "clean" and "efficient." Any change in production processes, practices, or products that makes production cleaner and more efficient per unit of production or consumption is a move toward sustainability. In practical terms, industrial sustainability means utilizing technologies and know-how to:

- reduce material and energy inputs while maximizing renewable resources as inputs
- minimize the generation of harmful pollutants or waste during product manufacture and use
- produce recyclable or biodegradable products

Many people believe that biotechnology will reduce the environmental impact of industrial manufacturing. Their optimism is rooted in the belief that the biotechnologies will increase the use of renewable biological resources in place of nonrenewable chemical ones. Also, due to the greater specificity, precision, and predictability of biologically based technologies, fewer adverse environmental impacts should occur.

According to the Organization for Economic Cooperation and Development, industrial sustainability requires the continuous innovation, improvement, and use of clean technology to reduce pollution levels and consumption of resources. Modern biotechnology provides a number of avenues for achieving these goals. The chemical, textiles, pharmaceutical, pulp and paper, food and feed, metal and minerals, and energy industries have all experienced improved efficiency made possible by incorporating biotechnology into their production processes. Almost always, these improvements are due to biocatalysts. Improved efficiency translates into less use of fossil fuels to drive the manufacturing process. In addition, manufacturing processes that use biodegradable molecules as biocatalysts, solvents, or surfactants are less polluting.

SUMMARY POINTS

A few design principles underlie all natural processes. Raw materials are reduced, reused, and recycled, and the processes are circular and exceptionally efficient.

In developing technologies, society should use nature as both a model and a source of molecules and metabolic machinery.

Human societies have relied on microbial products and processes for many millennia. The bioprocess technologies and other biotechnologies will broaden our use of microbes.

Biological processes have some advantages over chemical manufacturing processes because they are carried out under conditions that are compatible with life. In certain situations, these same attributes can be detrimental. Protein engineering techniques help to decrease the limitations of biological processes.

Microbes that live in extreme environments have unique biochemical capacities that are proving very useful in biotechnology research and commercial applications.

Microbial catabolic pathways make it possible to use biomass as a raw material for generating biofuels and feedstock chemicals. The optimal source of biomass would be agricultural wastes and forestry by-products, but the technical challenges of using cellulose as a starting material are significant. Nonetheless, progress has been made.

Both microbes and plants are being used to remediate polluted environments and prevent pollution.

By using the many biotechnologies discussed in this chapter, industrial manufacturing should become more sustainable.

KEY TERMS

Baculovirus	Bioprospecting	Extremophiles	Phytoremediation
Bioaugmentation	Bioreactor	Extremozymes	Site-specific mutagenesis
Biocatalyst	Buffer	Feedstock chemical	Total community genomics
Biofuel	Cellulase	Fermentor	
Biomanufacturing	Directed mutagenesis	Mammalian cell culture	
Bioprocess technologies	Directed protein evolution	Microbial fermentation	

Index